# Studies in Systems, Decision and Control

Volume 556

The series "Studies in Systems, Decision and Control" (SSDC) covers both new developments and advances, as well as the state of the art, in the various areas of broadly perceived systems, decision making and control–quickly, up to date and with a high quality. The intent is to cover the theory, applications, and perspectives on the state of the art and future developments relevant to systems, decision making, control, complex processes and related areas, as embedded in the fields of engineering, computer science, physics, economics, social and life sciences, as well as the paradigms and methodologies behind them. The series contains monographs, textbooks, lecture notes and edited volumes in systems, decision making and control spanning the areas of Cyber-Physical Systems, Autonomous Systems, Sensor Networks, Control Systems, Energy Systems, Automotive Systems, Biological Systems, Vehicular Networking and Connected Vehicles, Aerospace Systems, Automation, Manufacturing, Smart Grids, Nonlinear Systems, Power Systems, Robotics, Social Systems, Economic Systems and other. Of particular value to both the contributors and the readership are the short publication timeframe and the world-wide distribution and exposure which enable both a wide and rapid dissemination of research output.

Indexed by SCOPUS, DBLP, WTI Frankfurt eG, zbMATH, SCImago.

All books published in the series are submitted for consideration in Web of Science.

Vladik Kreinovich · Woraphon Yamaka ·
Supanika Leurcharusmee
Editors

# Applications of Optimal Transport to Economics and Related Topics

Springer

*Editors*
Vladik Kreinovich
Department of Computer Science
University of Texas at El Paso
El Paso, TX, USA

Woraphon Yamaka
Faculty of Economics
Center of Excellence in Econometrics
Chiang Mai University
Chiang Mai, Thailand

Supanika Leurcharusmee
School of Public Policy
Chiang Mai University
Chiang Mai, Thailand

ISSN 2198-4182 ISSN 2198-4190 (electronic)
Studies in Systems, Decision and Control
ISBN 978-3-031-67769-4 ISBN 978-3-031-67770-0 (eBook)
https://doi.org/10.1007/978-3-031-67770-0



This Springer imprint is published by the registered company Springer Nature Switzerland AG
The registered company address is: Gewerbestrasse 11, 6330 Cham, Switzerland

# Preface

Often, when a new successful data processing technique appears in one of the application areas, it then proves to be useful in many other areas. This was the case of optimal transportation techniques: these techniques were first developed for transportation problems, but now they have been shown to be successful in many statistical applications, including applications to economics.

These techniques are the main focus of this book, but this book also contains papers that use other techniques, ranging from more traditional statistical approaches to more recent ones such as stochastic frontier methods, multivariable quantiles, random forest, and deep learning. Applications include all aspects of economics, from production (including agricultural) to trade (including international) and finances, with relation to issues of crime (including computer crime and cyberbullying), demographics, economic freedom, environment, health, and tourism.

We hope that this volume will

- help practitioners to become better knowledgeable of the state-of-the-art econometric techniques, especially optimal transport techniques and
- help researchers to further develop these important research directions.

We want to thank all the authors for their contributions and all anonymous referees for their thorough analysis and helpful comments.

The publication of this volume was partly supported by the Faculty of Economics of the Chiang Mai University, Thailand. Our thanks to the leadership and staff of the Chiang Mai University for providing crucial support. Our special thanks to Prof. Hung T. Nguyen for his valuable advice and constant support.

We would also like to thank Prof. Janusz Kacprzyk (Series Editor) and Dr. Thomas Ditzinger (Senior Editor, Engineering/Applied Sciences) for their support and cooperation in this publication.

El Paso, USA — Vladik Kreinovich
Chiang Mai, Thailand — Woraphon Yamaka
Chiang Mai, Thailand — Supanika Leurcharusmee
December 2023

# Contents

# Theoretical Results

# Structural Model for US Gun Violence

**Arnab Bhattacharjee and Swagatam Sen**

**Abstract** The US has alarmingly high, and rising, gun violence and fatality rates. While it is generally believed that supply side policy is critical in addressing the problem, legislation has been relatively ineffective, for example the Gun Violence Prevention and Community Safety Act of 2020. Previous literature has identified hotspots and proximate determinants, particularly urban locations. Effective supply side policy would require application of optimal transport methods. We apply novel and current generation structural models to identify the spatial network for supply of guns and propagation of violence risks. This provides clean quantification of spatial supply chains and associated costs of transport between nodes, which are necessary for mitigation using optimal transport and logistics network management methods.

## 1 Introduction

The US has had a huge problem with gun violence, not least the highest gun ownership per capita of any nation in the world and a gun violence rate at almost 20 times higher than other comparable developed countries. The fatality per day is about 100 lives and civilian mass shootings are increasing annually [17]. Urban areas with high inequality have the highest incidence rates [17], and prevalence of gun culture is also closely related [14, 23]. The Gun Violence Prevention and Community Safety Act of 2020 was an important legislative development in advancing supply side policies, but its ability to effectively control gun violence in the US appears limited [15].

Supply side policies are almost universally accepted to be most important for mitigation, with availability and supply of guns highly prominent in public and academic debate [5, 10, 15, 20]. Understanding the supply chain of guns is therefore

A. Bhattacharjee (✉)
Heriot-Watt University and National Institute of Economic and Social Research, London, United Kingdom
e-mail: a.bhattacharjee@hw.ac.uk

S. Sen
Santander Investment Bank, London, United Kingdom

 
V. Kreinovich et al. (eds.), *Applications of Optimal Transport to Economics and Related Topics*, Studies in Systems, Decision and Control 556,
https://doi.org/10.1007/978-3-031-67770-0_1

key to effective mitigation policy [5, 11, 15, 20]. Importance of cross border purchase of guns in the context of gun violence and gun related crimes in US cannot be overstated. It has been increasingly clear that such migration of guns play a key role in undermining the effects of firearm regulations even in the states where they are relatively stringent.

One of the persistent debates in this area has been that of the influence of policies and regulations aimed at curbing the level of violence that has plagued the US [15, 22]. At the core of such debates is the fact that crime can be conceived as much a result of demographic drivers as that of the availability of means. There has been considerable work around the preventive side of curbing gun violence which focuses on developmental and demographic issues at individual and community levels that works as risk factors for conflicts that eventually uses guns as a means of violent resolution [2]. Any analysis solely focusing on gun sales and accessibility without coming to grips with the innate risk factors for the predicate crimes is not expected to produce the full picture.

At the same time, there is acute awareness that gun related violence and crime show significant signs of spatial concentration [4, 9, 19]. However, such 'hot spots' are not static over time. There is ample evidence in favour of a spatio-temporal diffusion of gun violence as opposed to purely spatial clustering [18, 21]. This is also similar to studies on narcotics where spatial supply chains are critical in effective control [12, 13].

In [21], the relationship between rate of change in gun related crime rate in a state, and firearm regulations in the state and neighbouring ones, was investigated as an evolutionary diffusion process. A key assumption in the model was that crimes diffuse across state borders as the firearms do. However, movement or change in crime may not necessarily follow the trail of guns even if the guns remain one of the prime drivers. Particularly it is important to consider demographic factors while studying crime patterns to compliment the effects of firearms [17]. As in the case of narcotics [13], this is critical in isolating demand side factors from the supply side ones.

In addition to that, the evolutionary model assumed that effects of any changes in the regulation in one state propagate to neighbouring states over a duration of time. Given data constraints, this diffusion was assumed to happen over a year's time. But, in reality any such changes are much likely to impact overall gun related crime landscape within the year the changes take place. In this context, a structural model that allows simultaneous interactions across the factors present in the states is much more well-suited. This is the domain of spatial econometrics, where a branch of the literature has developed methods to identify and estimate the spatial (supply chain) network structure. Specifically, the network structure is only partially identified from the data, and hence discovery of the network requires structural assumptions [6–8].

Finally, the original paper investigated potential paths of crime movement across a string of neighbouring states. The findings are insightful for trafficking of guns, and hence crime, across borders by small spatial units. However, such assumptions of physical proximity is relatively low in relevance when it comes to large supply chain of firearms across the country. At the same time, robust identification of the supply chain network is critical in isolating the optimal transport (supply) of guns across the spatial domain, and thereby also mitigation against such spatial diffusion of crime [13].

In the present study, our goal is two-fold.

First of all, we aim to look at gun related crime as a composite factor of guns available in the market and demographic indicators for crime propensity. Here we bring together the demand and supply side of crime in a single coherent framework. By moving away from a singular focus on guns to explain gun crimes, we develop a more balanced and realistic understanding of where, why and how crime evolves.

Furthermore, we wish to particularly understand how firearms availability in a market is affected by both 'in-state sales' (guns purchased within the state) as well as 'out-of-state traffic' (guns purchased outside the state). The hypothesis we want to test is whether looking separately at in-state sales, we can observe a significant and conclusive evidence of the impact of regulations.

Finally, for the out-of-state traffic we intend to analyse the 'local' (traffic across physical borders of neighbouring states) and 'global' (traffic across states that are not physically close but connected through infrastructure e.g. Interstate freeways, bus routes, flights etc) structure of how guns are traded. Particularly we are interested in tracing out global structures. There has been significant findings by the law enforcement that indicate that smuggling rings tend to operate along interstate freeways.

## 2 Methodology

In the present study we have taken the same data as the previous paper [21] which looks into rates of gun crimes across all the US states over the period 2000-2017. The central subject of interest is the change in rates of firearm related crimes over these 18 years. Henceforth, we denote gun crime rate as $C_{it}$ where $i$ denotes the state and $t$ denotes the year.

We model $C_{it}$ as

$$C_{it} = N_{it} P_{it}$$

where $N_{it}$ is the net influx of guns into the $i$-th state in year $t$, and $P_{it}$ is the likelihood that a new gun made available in the market would be used for criminal purposes. Here, at the very first step, we are improving on the assumptions of the original paper which took any change in gun crime to be proportional to accessibility of guns in the market. Any regulation would influence the market availability of guns. However, actual crime rates involving those guns would simultaneously depend on intrinsic structures prevalent in individual states.

Here we have assumed $P_{it}$ to be shaped by a number of intrinsic factors:

- Poverty rate
- Population density

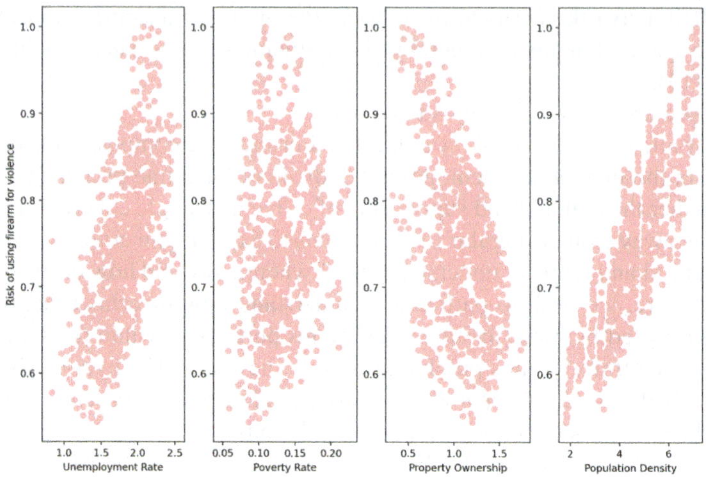


**Fig. 1** Risk of individual guns made available in the market being used for violence was estimated as the dominant effect across Unemployment, Poverty, Property ownership and Population density. Each of the individual factors are depicted in log-scale

- Unemployment rate
- Home ownership
- Percentage of population above age 65.

All of the mentioned factors were adjusted for skewness by applying logarithmic transformation. Then, we estimate $P_{it}$ by applying Principal Component Analysis (PCA) to the transformed data (Fig. 1).

This allowed us in turn to estimate the net influx of guns in the states in a given year.

$$N_{it} = C_{it}/P_{it}$$

Our goal from here on is to understand how much of this influx of guns is due to guns being purchased in other states and brought across the borders. As discussed earlier, since we are considering large scale supply chains of such trafficking, we are not restricting ourselves merely to geographical proximity.

We capture this network of influence using a structural equation model.

$$N_{it} = \sum_{j \neq i} w_{ij} N_{jt} + \alpha R_{it} + \beta F_{it}$$

where $R_{it}$ is the number of gun regulations in $i$-th state in the $t$-th year, and $F_{it}$ is the number of licensed firearm dealers in the state. The matrix $W = (w_{ij})$ is the structural dependence (spatial weights) matrix with zero diagonal elements, and this describes the familiar spatial autoregressive (or spatial lag) model .

From this equation, we proceed to estimate the free parameters $W, \alpha$ and $\beta$. For this purpose, we employ an iterative algorithm with a regression step followed by optimization for $W$. As the first step we set $W = W^0$ using a normalized haversine distance between states based on their centroid latitude/longitude. This initial estimate focusing on physical proximity allowed us to build on the foundations of previous works as well as to empirically determine eventual deviation from this starting point; this approach is related to [24].

We used $W^0$ to then calculate change in guns sold exclusively within the state as $n = (I - W^0)N$. In the regression step, $n_{it}$ is regressed on number of regulations in the state ($R_{it}$) and number of licensed firearm dealers in the state ($F_{it}$). The rationale behind this model is to understand individual effects of regulations and size of firearm market in the state on the in-state gun sales unfettered by the effects of cross border movement of guns.

The regression step would thus produce estimates of $\alpha$ and $\beta$ as $\tilde{\alpha}$ and $\tilde{\beta}$ respectively. Using these estimates it is in turn possible to further optimise $W$ to minimise the residual error

$$\mathcal{E} = \|N_{it} - \sum_{i \neq j} w_{ij} N_{jt} - \tilde{\alpha} R_{it} - \tilde{\beta} F_{it}\|$$

**Algorithm 1:** Iterative scheme to estimate structural dependence

**Data**: $N_{it}, R_{it}, F_{it}, W^0$
**Result**: $W, \alpha, \beta$
**while** $\mathcal{E} > 0.0001$ **do**
  **Estimate** $n_{it} = N_{it} - \sum_{i \neq j} w^0_{ij} N_{jt}$, where $n_{it}$ is the new guns sold exclusively within the state. ;
  **Regress** $n_{it} = \alpha R_{it} + \beta F_{it}$ to find effects of regulations and firearm licenses ($\tilde{\alpha}$ and $\tilde{\beta}$ respectively). ;
  **Optimize** weights $w_{ij}$ such that $\mathcal{E} = \|N_{it} - \sum_{i \neq j} w_{ij} N_{jt} - \tilde{\alpha} R_{it} - \tilde{\beta} F_{it}\|$ is minimized. Set $W^0 = W$.
**end**

As we repeat the regression and optimisation steps, we observed that each step in this evolutionary process de-noises the structural dependence matrix and focuses on fewer, stronger relationships across states. Some of these relationships are symmetrical while the others are directional. Note that the spatial network (weights) matrix $W$ is only partially identified [7]. Unlike [7], we do not assume symmetry as the identifying assumption, but rather achieve identification through a nested sparse network structure [6, 8].

At the same time, we found a cleaner statistically significant reciprocal relationship between in-state gun-sales and regulations. Importantly, we find that the effect of existing licensed forearm dealers is not statistically significant (p-value $= 0.101$) (refer Table 2). As evident from Table 1, the final regression step in the process produced a significant goodness-of-fit with the data (Fig. 3).

**Table 1** Model Summary for final regression step on in-state gun sales

| Adjusted RSS | Goodness of Fit (p-value) | AIC | Likelihood |
|---|---|---|---|
| 0.018 | 0.000 | 2438 | –1217 |

**Table 2** Variable effects for final regression step on in-state gun sales

| Variable | Effect | Standard error | Significance (p-value) |
|---|---|---|---|
| Regulations | –0.1317 | 0.034 | 0.000 |
| Firearm licenses | –0.0553 | 0.030 | 0.101 |

## 3 Discussion of Results

One of the stated goals of our study was to discover the structure of major interstate gun trafficking (supply chain) network. Our algorithm was able to refine that structure into a relatively sparse dependency matrix (see Fig. 2).

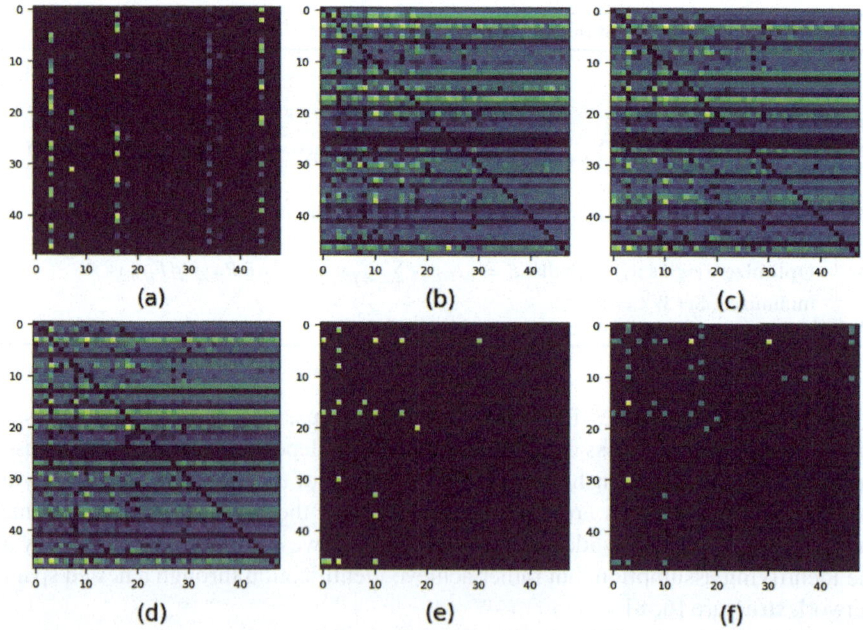


**Fig. 2** **a** Initial estimate based on normalized haversine distance between states. **b–d** Progressive refinement of the dependence structure. At each step localized signals are amplified and diffused lines are toned down. **e** Final dependence structure after a soft threshold of 80 **f** Symmetrised dependence structure

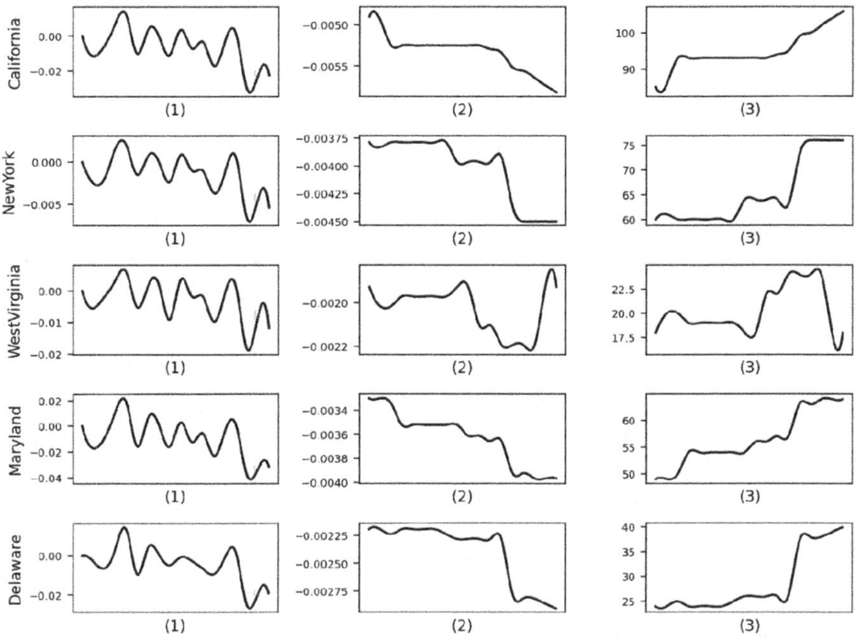


**Fig. 3** (1) Rate of change in Gun related crime rate in the state over last 18 years. (2) Final estimate of new guns sold exclusively within state over the timeframe. This is an unscaled indicator of net changes in guns sold within border and not the absolute number of guns sold. This estimate was then used to investigate effects of firearm regulations within state. (3) Number of firearm regulations implemented within state.

One natural way to decode and understand this structure is to look for the top importer and exporter states with reference to firearms trafficking. As Fig. 4 shows, the states that stand out as importers of guns (Michigan, Maryland, New Jersey, and California) have very little outflow into other states. This clearly indicates that the flow is far from isotropic and there is a systemic pattern in the movement of guns. With the exception of Michigan, the top importers have all adopted relatively strong regulatory regime to curb gun crimes. However, the significant inflow of guns across state borders tend to undermine that effort.

Interestingly, we found moderately significant evidence of negative correlation between firearm regulations in a state and how much guns are being sold and moved outside of the state. For states like Montana, Alabama or Idaho where gun regulations have been relatively less stringent, we found greater movement of guns out of these states. On the other hand, states with traditionally stricter firearm purchase laws (e.g. California, New York, Massachusetts, New Jersey or Maryland) have very little contribution to overall guns flow across their borders and on to other states.

However, that does not stop them from being subject to incoming gun traffic. States like Maryland, Michigan and California are some of the top importers of

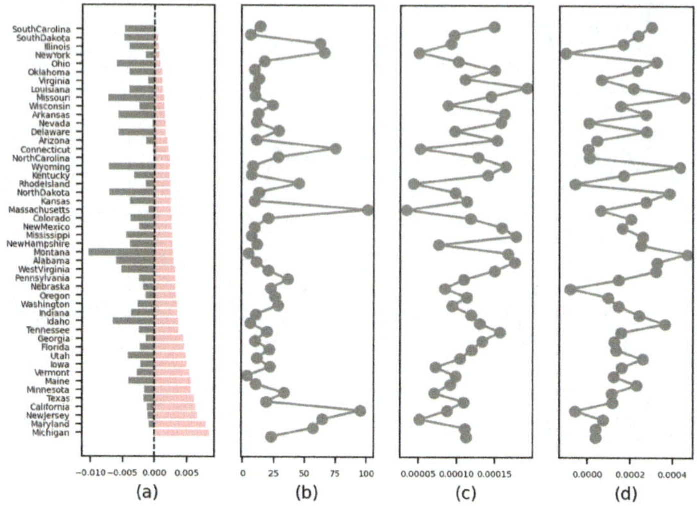


**Fig. 4** **a** Overall inflow and outflow of guns across border from individual states. Top importers are usually not top exporters. **b** Average number of regulations over the 18 years. Regulations have a significant negative correlation (coefficient = −0.48, p-value = 0.0005) with outflux of guns. **c** Average gun related crime rate over 18 years. There is a significant positive correlation (coefficient = 0.43, p-value = 0.001) between crime rate and outflux of guns. **d** Average year-on-year changes in crime rate over 18 years. There is a very high correlation (coefficient = 0.92, p-value = 0.00) between changes in crime rate and gun traffic flowing out of the state

guns from outside of state. Here the regulations have a limited impact which is not unexpected. Correlation between regulations and inflow of guns is not significant (coefficient = −0.2, p-value = 0.21). This is best exemplified by California and Michigan. While California has one of the strictest regulatory regimes it also suffers from high inflow of guns from across its borders. On the other hand, Michigan with considerably fewer and more relaxed gun control evidenced a similar level of out-of-state gun purchases.

We also looked into how the flow of guns eventually relate to gun crimes in these states. We observe low correlation between inflow of guns and crime rate. This is not entirely unexpected since our methodological premise is that gun related crimes may not be explained singularly by the presence of guns made available through out-of-state traffic. There are states like Louisiana where a lax regulation allows guns to be sold within state quite easily driving its gun related crime rate up to the highest among all the states.

However, there was a very strong correlation (coefficient = 0.92, p-value = 0.00) between average year-on-year changes in gun crime rates and outflow of guns from state. This observation presents a possibility of endogeneity between changes in crime rate and gun sales out-of-state. It requires an extension of the present model

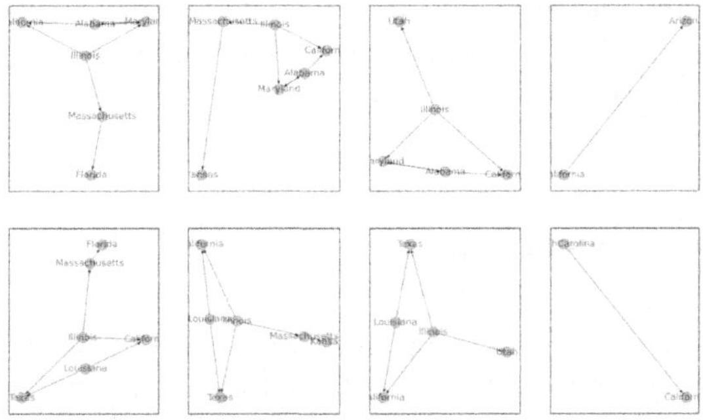

**Fig. 5** Gun traffic network around California shows persistent traces of a ring of operations across the state groups Alabama, Texas and Louisiana—all located along the Interstate 10

to investigate any causal effect of gun crime rate changes on outflow of guns. Such analysis is retained outside the scope of the present study.

To better understand the nature of gun traffic beyond the importer/exporter roles of individual states we were able to identify possible paths of gun traffic supply chain. The results allow us to focus on individual states and investigate the supply chain through that state. Depending on the overarching role that the state plays in the overall migration of guns with criminal intent, these sub-structures could take different topological forms.

For example, in case of California, one of the states with highest cross-border gun migration, we were able to identify various different modes of relationships with other states (see Fig. 5). Most noteworthy among these patterns is a quadrilateral network between California, Illinois, Alabama and Maryland spanning across the country from coast to coast with Illinois and Alabama playing the role of source of guns. Another similar cluster is observed between California, Illinois, Louisiana and Texas. Even though these states are not geographically neighbouring they are nevertheless connected by major interstate freeways.

In a similar manner when we look into the case of Michigan, another top importer, curiously we find a very different local topology. In this case, we observed an explicit hub-and-spoke structure where Michigan is attracting guns from a number of states

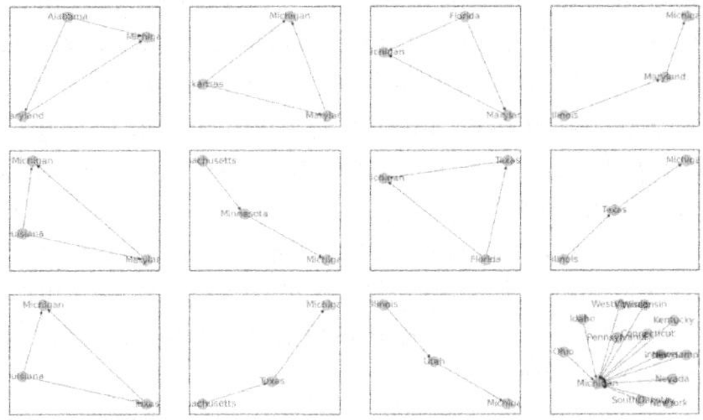

**Fig. 6** Local network structure around Michigan predominantly shows a hub-and-spoke topology reflecting a steady inflow of guns from a multitude of different states

with relaxed gun regulations including Idaho, Ohio, Wisconsin, West Virginia and Kentucky. The structure suggests that these are geographically local trafficking of guns (Fig. 6).

## 4 Conclusion

Aligned with the objectives of this study, we have estimated the effects of cross border gun sales in a state and found significant and systematic patterns of such trafficking across different routes. Adjusting for these migratory effects, we have estimated the 'in-state' gun sales over 18 years for each state and studied how these sales are influenced by firearm regulations and presence of licensed firearm dealers within state.

Our analysis shows that such influences are indeed significant for firearm regulations while licensed firearm dealers have very marginal impact on in-state gun sales. This result is significant as the importance of gun regulations is often masked when looked purely through the lens of gun related crime rate as a vast proportion of these crimes are committed using guns procured out-of-state.

An absence of a significant correlation between in-state gun sales and number of licensed firearm dealers is also important as it is suggestive of a shadow market present disproportionately across states that operate outside the control of regulations. Although currently we did not include a further study of the mechanism of such a shadow market, it does present interesting scope for a future study along these lines.

In terms of out-of-state trafficking of guns, our analysis showed a non-uniform pattern across states with some of the states like California and Michigan emerging as top importers of such traffic. We have also observed that the mechanism of such traffic is widely different for these states. While traffic into California is not correlated with geographical proximity and evidences guns coming from states that are connected by interstate freeways, Michigan faces its share of imports from local operations across its physical borders with neighbouring states.

Finally we were able to establish a direct correlation between regulations and out-of-state gun traffic. We found that not only do the states with more relaxed firearm regulations have more in-state gun purchases, but that they are also likely to be source of guns trafficking out of state. However, we find no evidence of the regulations having any correlations with inflow of guns traffic in a state. This observation reinforces the evidence that even though regulations play a significant role in controlling in-state gun purchases as well as outflow of guns, it has very limited control on inflow of such gun traffic.

Analysis of results also revealed a high degree of correlation between Year-on-Year changes in gun crime rate in a state and how much gun traffic flows out from that state. Causal structure of this relationship is interesting and will be investigated in depth in later work.

In summary, the present study reinforces a causal structure that connects gun crime with gun sales (both in-state and out-of-state), regulations and exogenous demographic factors in a state. It establishes that presence of guns in a state is an important driver for crime even though it is not the only one. It also illustrates that in understanding the effect of regulations it is critical to study the effect of both in-state and out-of-state guns; while regulations are critical they are not adequate to stem the traffic of guns coming from outside the states.

Finally, while it is fairly simple to set up an optimal transport problem here [1, 3, 16], effective mitigation and enforcement would require robust quantification of transport (trade) supply chain network of out-of-state purchase. This is where our work makes novel contributions that are new to the literature. This offers exciting insights on the limits of state level regulations in gun control.

**Acknowledgements** The lecture notes of Hung T. Nguyen were very helpful in our conceptualisation of the spatial dynamics of gun violence problem in terms of optimal transport. Discussions with Siddharth Chandra on conceptualizing spatial supply chains for narcotics was also very helpful. The authors thank both for their generous help. Both authors also thank the organisers and editors for their generous welcome and help. The views expressed in the paper owe solely to the authors and do not relate to their institutions. The usual disclaimer applies.

## References

1. Ambrosio, L., Gigli, N., Savaré, G.: Gradient Flows in Metric Spaces and in the Space of Probability Measures. In: Lectures in Mathematics ETH Zürich. Birkhäuser Verlag, Basel (2005)
2. American Psychological Association: Gun violence: Prediction, prevention, and policy. http://www.apa.org/pubs/info/reports/gun-violence-prevention.aspx (2013)
3. Angenent, S., Haker, S., Tannenbaum, A.: Minimizing flows for the Monge-Kantorovich problem. SIAM J. Math. Anal. **35**(1), 61–97 (2003)
4. Barboza, G.: A secondary spatial analysis of gun violence near Boston schools: a public health approach. J. Urban Health **95**(3), 344–360 (2018)
5. Bartley, W.A., Williams, G.F.: The role of gun supply in 1980s and 1990s youth violence. Contemp. Econ. Policy **40**(2), 323–348 (2022)
6. Bhattacharjee, A., Holly, S.: Understanding interactions in social networks and committees. Spat. Econ. Anal. **8**(1), 23–53 (2013)
7. Bhattacharjee, A., Jensen-Butler, C.: Estimation of the spatial weights matrix under structural constraints. Reg. Sci. Urban Econ. **43**(4), 617–634 (2013)
8. Bhattacharjee, A., Castro, E., Maiti, T., Marques, J.: Endogenous spatial regression and delineation of submarkets: a new framework with application to housing markets. J. Appl. Economet **31**(1), 32–57 (2016)
9. Braga, A.A., Papachristos, A.V., Hureau, D.M.: The concentration and stability of gun violence at micro places in Boston, 1980–2008. J. Quant. Criminol. **26**, 33–53 (2010)
10. Braga, A.A., Wintemute, G.J., Pierce, G.L., Cook, P.J., Ridgeway, G.: Interpreting the empirical evidence on illegal gun market dynamics. J. Urban Health **89**, 779–793 (2012)
11. Butt, N., Lambrick, F., Menton, M., Renwick, A.: The supply chain of violence. Nat. Sustain. **2**(8), 742–747 (2019)
12. Chandra, S., Barkell, M.: What the price data tell us about heroin flows across Europe. Int. J. Comp. Appl. Crim. Justice **37**(1), 1–13 (2013)
13. Chandra, S., Peters, S., Zimmer, N.: How powdered cocaine flows across the United States: Evidence from open-source price data. J. Drug Issues **44**(4), 344–361 (2014)

14. Cukier, W., Eagen, S.A., Decat, G.: Gun violence. Aggression and violence. In: Bushman, B.J. (ed.) Aggression and Violence: A Social Psychological Perspective, pp. 179–193. Routledge (2016)
15. Eisenhauer, J.G.: An economic approach to gun control: analysis of the Gun Violence Prevention and Community Safety Act of 2020. Appl. Econ. Lett. (2023) (in press)
16. Galichon, A.: Optimal Transport Methods in Economics. Princeton University Press (2016)
17. Johnson, B.T., Sisti, A., Bernstein, M., Chen, K., Hennessy, E.A., Acabchuk, R.L., Matos, M.: Community-level factors and incidence of gun violence in the United States, 2014–2017. Soc. Sci. & Med. **280**, 113969 (2021)
18. Loeffler, C., Flaxman, S.: Is gun violence contagious? A spatiotemporal test. J. Quant. Criminol. **34**, 999–1017 (2018)
19. Mazeika, D.: The effect of unreported gun-related violent crime on crime hot spots. Secur. J. **36**(1), 101–117 (2023)
20. Semenza, D.C., Stansfield, R., Link, N.W.: The dynamics of race, place, and homicide context in the relationship between firearm dealers and gun violence. Justice Q. **39**(1), 134–151 (2022)
21. Sen, S., Sen, A., Liu, Y., Sen, B.: On The Dynamics of Interstate Diffusion of Firearm Violence and Impact of Firearm Regulations. https://www.medrxiv.org/content/early/2021/11/16/2021.11.15.21266358 (2021)
22. Webster, D.W., Vernick, J.S.: Reducing Gun Violence in America: Informing Policy with Evidence and Analysis. The Johns Hopkins University Press (2013)
23. Yamane, D.: The sociology of US gun culture. Sociol. Compass **11**(7), e12497 (2017)
24. Zhang, Z., Bhattacharjee, A., Marques, J., Maiti, T.: Spatio-temporal patterns in Portuguese regional fertility rates: a Bayesian approach for spatial clustering of curves. J. Off. Stat. **37**(3), 611–653 (2021)

# Extending Jaynes: How Scientific and Economic Claims are Judged

**William M. Briggs**

**Abstract** ET Jaynes work on probability theory, which he subtitled the logic of science, should be much better known. We cast some light on a small segment of his work, demonstrating how logical probability can be used to analyze people judge scientific and economic pronouncements and claims of cause. This has important consequences in economics, and any area in which people are asked to believe a cause has been discovered or verified in science, especially when they are asked to act on the discovery.

**Keywords** Cause · Decision analysis · Explanation · Probability · Scientific evidence

## 1 A Logical Beginning

E.T. Jaynes's book *Probability Theory: The Logic of Science*, [1], has been deeply influential in resurrecting logical probability, an idea began by Laplace [2], and which found some initial success. But an idea which by the late Twentieth Century had fallen by the wayside. This is odd, especially in economics, because one of the most important economists, John Maynard Keynes, also wrote a seminal work in logical probability, yet his book scarcely known today, [3].

This neglect was corrected to some extent first in Australia by David Stove, [4, 5], and later by James Franklin and others, [6, 7]. Others soon took the idea up, e.g. [8].

Physicists seeking empirical improvements to probability models were the first to embrace the return of logical probability, mostly through the techniques of maximum entropy, e.g. [9–11]. These techniques, of course, provide certain kinds of limited-information priors on parameters in probability models, priors which can often be justified on physical and logical reasoning.

Logical probability has also been recognized in economics and econometrics, albeit to a more limited extent, see e.g. [12, 13].

W. M. Briggs (✉)
New York, USA
e-mail: matt@wmbriggs.com



V. Kreinovich et al. (eds.), *Applications of Optimal Transport to Economics and Related Topics*, Studies in Systems, Decision and Control 556,
https://doi.org/10.1007/978-3-031-67770-0_2

Discussions of maximum entropy techniques would take us too far afield. These are anyway not of central interest to us here. Our interest is in extending Jaynes's more fundamental ideas of evidence and how probability helps us understand how judgments in science are made and understood. In particular, we want to understand how people judge, and then accept or reject, pronouncements in science, including econometrics. To do this, we expand some curious results in Chap. 5 in Jaynes's book, "Queer Uses of Probability". This is a companion paper to [14], which examines how disagreements in the face of new evidence can arise.

## 2 Which to Believe?

The public, bureaucrats and rulers usually do not participate in scientific research, but they are often presented the results from this research and asked to accept or act on it. Some research is challenging to believe: that which goes against common sense or experience. Which is not to say it is wrong, but that it can be hard to believe. How do people think about this kind of research? Logical probability can help us analyze this.

Jaynes's original example revolved around the question whether a certain person really did exhibit extrasensory perception (ESP, or "mind reading"). Most people thinking about this question start with prior evidence, however vague, and form some kind of probability of the proposition "This person has ESP". This prior probability does not need to be quantified with any precision. Most questions about uncertainty are not quantified; probability is left as vague as the prior information.

It is only researchers who insist on probability being quantified. Usually they do this by calling to theories of subjective probability, e.g. [15]. But we reject that interpretation, at least for the intolerable fluidity and unfixed nature of guesses (see [8]), and at least for the this paper insist probability is a matter of logic. Like with all logical statements, once the premises are fixed, the conclusion from *just those premises*, and none other, should be sought and logically follows.

Back to the ESP example. We start with some vague prior belief. Evidence about an experiment subsequently arises, which formally can be married to prior information using Bayes's theorem, and a posterior probability of the proposition is formed. So much is usual practice.

But with non-researchers, or rather with those not offering the new evidence, the way the new evidence is married to the old is not the same as it is with researchers and those offering the new evidence. Researchers take a strict mathematical view, bound to a formal model. They usually take that formal model as the last word, rejecting all prior evidence, and supposing the model exactly, or near enough, represents the real world. To researchers, the model is reality.

However, non-researchers are free to question the entire research process, including the formal models used. This means that non-researchers do not always agree with the conclusions of researchers. Jaynes points out that evidence offered by some quarters can have a depressing effect; which is to say, that instead of bolstering belief in the research proposition in question, the belief is quashed. This happens when the sources presenting the evidence is distrusted, or when other causes than the ones the researchers proposed are considered by non-researchers.

As an example of this, we examine the claims of Zhu and others, [16], in their paper "Including intangible costs into the cost-of-illness approach: a method refinement illustrated based on the $PM_{2.5}$ economic burden in China". Their claims are simple: dust of the certain size, i.e. "particulate matter with aerodynamic diameters less than 2.5 μm", causes in unspecified ways certain health maladies, like acute death. Which is to say rapid, unexpected deaths. These deaths and other health problems, said to be caused by $PM_{2.5}$, also have large economic costs, which can be quantified in various ways.

The question before us is whether $PM_{2.5}$ really does *cause*, in some cases, acute deaths. Is the probability of that proposition high enough to act on politically? Any politician will start out with some prior probability of the proposition, again, likely not quantified. He will then hear evidence, almost surely from interested parties, on the question. And he will hear the evidence of the researchers. All this will be used to form a posterior probability, which will be used to make decisions. Non-politicians will follow the same path, though their intermediate evidence will likely consist of press and friend reports.

Before we begin the exercise, the reader should pause and consider his own prior probability, conditional on whatever background information he may have, on the proposition "$PM_{2.5}$ causes acute deaths."

We now turn to the researcher's claim. It's useful to examine Zhu's model of incidence rate of maladies said to be caused by $PM_{2.5}$.

$$I_0 = \frac{I}{\exp\left(\beta \times (C - C_0)\right)}, \tag{1}$$

where $I_0$ is the background incidence rate of the disease (one incidence rate per disease) at ambient levels of $PM_{2.5}$, $\beta$ is the exposure-response coefficient, $C$ is the measured $PM_{2.5}$ concentration, $C_0$ is the ambient background $PM_{2.5}$ concentration, and $I$ the incidence rate at (higher) levels of $C$. From this we can calculate

$$I = I_0 e^{(\beta(C-C_0))} \tag{2}$$

If our interest is, for instance, in $PM_{2.5}$ levels $\alpha$ times the background concentration, then we can simplify this to

$$I = I_0 e^{(\beta(\alpha-1)C_0)}. \tag{3}$$

We now need only the background incidence rate and $PM_{2.5}$ concentration. This model becomes economically important because, as is well known, incidence rates of disease translates directly to medical costs. The authors therefore provide a way to tie incidence rates to costs, but the specific details are not relevant here.

It turns out, according to the authors, $C_0 = 35\mu g/m^3$, which is the "second grade standard" in the China Ambient Air Quality Standard. They also publish incidence rates for several maladies, but arguably the most interesting one is acute mortality. Is dust killing people outright, and not just exacerbating chronic conditions (a separate question)? The$\beta$ for acute mortality was given at 0.00019. The authors publish a rate for acute mortality, but it's not clear if this of hospitalizations, or acute deaths divided by illness rate, or an overall population rate. The latter seems unlikely because the published rates are too high for the population as a whole.

We can calculate an intelligible rate from their data, however, by noting that in 2013 there were in the area of China they studied some 24,820 acute mortality cases (fatalities in their Table 2). The author's published incidence rate was 0.0716 (Table 1), for a (calculated) outpatient population of 339,106. The precise numbers are not of special concern, as will become clear. Even if these are wrong, the appropriateness of the example holds.

Assume the 0.0716 rate holds for the higher values of $PM_{2.5}$ above the background value. Their Fig. 2 lists observed rates, depending on the city and year, of $PM_{2.5}$ of anywhere from the high 30s to about $160\mu g/m^3$, with the average from 60 to $70\mu g/m^3$, which are rates from about 2 to 4.5 times higher than the background. Let's use the worst case scenario of $PM_{2.5}$ being 4.5 times higher than background. Using their formula gives for a baseline or background $PM_{2.5}$ $I_0 = 0.007$, an absolute increase of $\sim 0.0016$, or a relative increase in incidence rate of about 2.4%.

If $PM_{2.5}$ had *no* effect on acute mortality (AM), then all 24,820 deaths would be due to other causes. But if $PM_{2.5}$ did cause acute mortality, then something like $0.007 \times 339,106 = 23,737$ deaths would be due to other causes, and $24,280 - 23,737 = 543$ outpatient deaths would be caused by $PM_{2.5}$ for the worst case scenario. All on average, of course.

Here is where Jaynes's result finally comes in. Should we believe that $PM_{2.5}$ causes acute mortality, at about these rates?

If the true background incidence rate was 0.007, then the probability of more than 24,280 deaths out of 339,106 is roughly 0.00013. And if the rate of 0.0717 is correct, which assumes $PM_{2.5}$ *is* a cause of acute death, then there is a chance of 0.588 for seeing the same result. A difference of 4,367 times higher.

This then seems like solid evidence in favor of the hypothesis that $PM_{2.5}$ is causing acute mortality. Call that hypothesis $H_1$, and the hypothesis that $PM_{2.5}$ is not a cause of acute mortality $H_0$.

As discussed above, anybody considering these hypotheses would have some background evidence they use to judge the likelihood either is true. Call the background evidence one has $E$.

Now it is perfectly clear that this paper has not proved, in a logical or mathematical sense, that $PM_{2.5}$ is *certainly* a cause of AM. That is, it not true that the following equal is exactly equal to 1:

$$\Pr(H_1|DE) = \frac{\Pr(D|H_1E)}{\Pr(D|E)}\Pr(H_1|E), \tag{4}$$

where $\Pr(H_1|E)$ is the prior probability brought to the hypothesis that $PM_{2.5}$ causes AM, $\Pr(D|H_1E)$ the probability of the observed data assuming $H_1$ is true and given E (here equal to 0.588), and $\Pr(D|E)$ the probability of the data assuming only the background information E.

The denominator can, of course, be expanded:

$$\Pr(D|E) = \Pr(D|H_1E)\Pr(H_1|E) + \Pr(D|H_0E)(1 - \Pr(H_1|E)).$$

In order for (4) to approach one, regardless of the prior, the cause $H_1$ must make the data $D$ vastly more likely than the background cause(s) $H_0$, all conditional on E.

Let's see this by assuming two levels of prior belief, $\Pr(H_1|E_a) = 0.1$ and $\Pr(H_1|E_b) = 0.9$, a skeptic ($a$) and believer ($b$). Then, after the data, and assuming Zhu's model, we have $\Pr(D|E_a) = 0.0589$ and $\Pr(D|E_b) = 0.529$.

Therefore $\Pr(H_1|DE_a) = 0.998$ and $\Pr(H_1|DE_b) \sim 1$. So both the skeptic and the believer move toward assessing the hypothesis that $PM_{2.5}$ is a cause of AM with high probability.

It would thus seem future skepticism is now entirely unwarranted, and that just about anybody, save those who have previously ruled the proposition impossible must conclude $PM_{2.5}$ is at least very likely a cause of AM.

But this formal procedure is, of course, not how evidence is considered. We next examined how it is done in real life.

## 3 The Other Side of the Story

It's very curious that nearly every published paper using statistics asks us to conclude something very like belief, as provided in the formal analysis above, or at least they ask us to form probabilities very near 1. Researchers show us results, which are in line with the hypothesis they set out to prove, and nothing short of agreement on the part of the reader is warranted. We are rarely, or even never, asked to continue the analysis as we now precede to do. And when we do, this analysis is not welcomed.

It is obvious there are more causes of AM than $PM_{2.5}$. One could be a change (increase) in population ($H_2$), which could lead to increased hospitalization rates. Another is the changes in the population themselves, ($H_3$). Since this data was collected over various cities, each with different demographics, some populations could be older, or suffer other illness that cause AM. And then $PM_{2.5}$ could be mismeasured, ($H_4$), or the exposure levels do not correspond to actual dose levels ($H_5$). The covid panic could have changed people's behavior and whether they sought hospitalizations or not ($H_6$), which is important here because the incidence rates appear to be tied to hospitalizations.

One of the biggest, or at least most usual, alternatives is we might suspect the authors of "cherry picking" their data ($H_7$), making the case more in their favor than is warranted in real life. After all, they might have been chasing after wee p-values, ($H_8$, "p-hacking"). Or they may have adjusted their model after seeing initial results, so that it fit better ($H_9$).

These are just a few of many possible alternative hypothesis, all of which explain the data, even better than the formal model is. That is, $\Pr(D|H_j E) \approx 1$ for $j = 2, 3, \ldots$.

Let's rewrite (4) with the alternate explanations in mind:

$$\Pr(H_1|DE) = \frac{\Pr(D|H_1 E)\Pr(H_1|E)}{\Pr(D|H_1 E)\Pr(H_1|E) + \Pr(D|H_0 E)\Pr(H_0|E) + \sum_i \Pr(D|H_i E)\Pr(H_i|E)}. \quad (5)$$

It's now much more difficult for $\Pr(H_1|DE)$ to be anywhere close to 1. As Jaynes points out, for us to give $H_1$ high probability, and therefore act as if it were true, it must be that

$$\Pr(D|H_1 E)\Pr(H_1|E) \gg \Pr(D|H_0 E)\Pr(H_0|E) + \sum_i \Pr(D|H_i E)\Pr(H_i|E). \quad (6)$$

Suppose, as is plausible when we are first asked to believe some new or "shocking" hypothesis, such as ours, that except for the "null" each hypothesis we entertain more or less equally likely supports the observed data, then in order to have high confidence in $\Pr(H_1|DE)$

$$\Pr(H_1|E) \gg \sum_i \Pr(H_i|E). \quad (7)$$

The null drops out in research that makes it through the publishing process, its probability of $D$ being too low. In our case, as we saw, $\Pr(D|H_j E) \approx 1$ for $j = 2, 3, \ldots$, so there is no way the model can "beat" the alternatives.

It is thus apparent that in the face of plausible competing hypotheses, it is very difficult to reach a high posterior probability of $H_1$. Each one has to be ruled out for there to be any hope in believing $H_1$. And that takes a lot of work.

On the other hand, it is easy to get a high posterior of $H_1$ as long as those competing hypotheses are rejected or assigned minuscule probabilities. And that it just what supporters of research results do, usually by failing to consider them at all, or only to present some as "limitations" while also tacitly dismissing them.

## 4 Be Cause

The conclusion is that cause matters most. Research does not count for nothing, and could count for everything. But research presented by itself alone is rarely convincing, especially since it has to be presented by researchers, and nobody on earth can be trusted completely. Logical probability analysis shows why.

We expect that researchers will tell their own stories well, and present their case in the best way, according to their lights. But their word cannot be the final word. One reason for this is the sheer number of causal claims made.

Reports vary, but one estimate is that the number of scientific papers published every year is about 8 million—and rising fast, see inter alia [17]. It depends on how "scientific papers" are designated, of course, but these kinds of details do not change the mighty flow of science. Nearly all of these papers ask us, implicitly or explicitly, tacitly or by direct claim, to believe a cause has been discovered or confirmed. Based on the number alone, it cannot be plausible that all such claims are correct, or even most. Again, see [17], as this relates to the replication crisis in economics, caused in great part by researchers misunderstanding cause.

Since non-researchers usually do not have the background to judge intricacies of the models used in science, they have to rely on pronouncements by scientists. Scientist overwhelming rely on null hypotheses significance testing, and these so-called "nulls" are almost always presented as extremely implausible. They are the straw men of science.

Non-researchers see how often claims in science fail, and many recognize the flimsiness of these nulls, and so it is natural for them to question all new claims. We see here how easy distrust develops.

## References

1. Jaynes, E.T.: Probability Theory: The Logic of Science. Cambridge University Press, Cambridge (2003)
2. de Laplace, M.: A Philosophical Essay on Probabilities. Dover, Mineola, NY (1996)
3. Keynes, J.M.: A Treatise on Probability. In: Dover Phoenix Editions, Mineola, NY (2004)
4. Stove, D.: Probability and Hume's Inductive Scepticism. Clarendon, Oxford (1973)
5. Stove, D.: The Rationality of Induction. Clarendon, Oxford (1986)
6. Franklin, J.: Erkenntnis **55**, 277 (2001)
7. Campbell, S., Franklin, J.: Synthese **138**, 79 (2004)
8. Briggs, W.M.: Uncertainty: The Soul of Probability, Modeling & Statistics. Springer, New York (2016)
9. Bretthorst, G.L.: Bayesian Spectrum Analysis and Parameter Estimation. Springer, New York (1988)
10. Briggs, W.M.: Bayesian Inference and Maximum Entropy Methods. In: Knuth, K.H., Caticha, A., Center, J.L., Giffin, A., Rodriguez, C.C. (eds.) Science and Engineering, pp. 101–108. American Institute of Physics, Melville, New York (2007)
11. Jaynes, E.T.: Bayesian Inference and Maximum Entropy Methods. In: Knuth, K.H., Caticha, A., Center, J.L., Giffin, A., Rodriguez, C.C. (eds.) Science and Engineering, pp. 101–108. American Institute of Physics, Melville, New York (2007)

12. Zellner, A.: Bayesian Methods and Entropy in Economics and Econometrics, pp. 17–31. Springer Netherlands, Dordrecht (1991). https://doi.org/10.1007/978-94-011-3460-6_2
13. Shen, E.Z., Perloff, J.M.: Journal of Econometrics **104**(2), 289 (2001)
14. Briggs, W.M.: Partial Identification in Econometrics and Related Topics. Springer (2024), p. forthcoming
15. Cohen, J.: Scientific American, pp. 128–139 (1957)
16. Zhu, B., Pang, R., Chevallier, J., Wei, Y.M., Vo, D.T.: Eur. J. Health Econ. **20**(4), 501 (2019). https://doi.org/10.1007/s10198-018-1012-0.
17. Briggs, W.M.: Asian J. Econ. Bank. **7**(2), 180 (2024)

# The A Priori Procedure for Estimating the Location Parameter Under Elliptical Settings

**Xiangfei Chen, Tonghui Wang, S. T. Boris Choy, David Trafimow, and Tingting Tong**

**Abstract** The a priori procedure (APP) provides minimum sample sizes for estimating parameters of the population distribution that ensure precision and confidence in sample statistics. In this paper, we extend the APP to include the family of elliptical distributions, which is a member of a broad family of probability distributions and is an extension of the family of normal distributions. Properties of the elliptical distribution are discussed. Under the uncorrelated assumption and with a given precision and a confidence level, the desired sample size for estimating the location parameter is obtained for generalized elliptically symmetric logistic, symmetric Kotz Type, and *t* distributions, respectively. The confidence interval for the location parameter is constructed based on the desired sample size using elliptical distribution. Three real data examples are given to illustrate our main results. In addition, the Shinyapp program links for some of the elliptical distributions are provided for researchers and practitioners to use. In the program, the desired sample size is calculated.

**Keywords** Elliptical distribution · A priori procedure · Estimation of location · Required sample size

X. Chen · T. Wang (✉) · D. Trafimow · T. Tong
New Mexico State University, Las Cruces, NM, USA
e-mail: twang@nmsu.edu

X. Chen
e-mail: xiangfei@nmsu.edu

D. Trafimow
e-mail: dtrafimo@nmsu.edu

T. Tong
e-mail: tongtt@nmsu.edu

S. T. B. Choy
The University of Sydney, Camperdown, NSW, Australia
e-mail: boris.choy@sydney.edu.au



V. Kreinovich et al. (eds.), *Applications of Optimal Transport to Economics and Related Topics*, Studies in Systems, Decision and Control 556,
https://doi.org/10.1007/978-3-031-67770-0_3

# 1 Introduction

The A priori procedure (APP) is designed for researchers to determine the required sample sizes before collecting data. When using the APP, two issues that should be addressed:

- How close do I want my sample statistics to be to their corresponding population parameters?
- What probability do I want to have of meeting the precision criterion?

For example, if a researcher is interested in having a 95% probability of obtaining a sample mean within a tenth of a standard deviation of the population effect size under normality, Trafimow [22] suggests it is necessary to obtain a sample size of 385 by using the APP. Not only does the APP run well for estimating the single means under normality, but it provides solutions for estimating location, scale, shape, correlation coefficient, and difference of locations under skew normality, see [23, 25–28]. Moreover, the APP contributes to the estimation of the Cohen's effect size [2], one-way analysis of variance paradigms [29], the mean in both log-normal and gamma distribution [1], and estimation of regression coefficients in linear models [21]. The foregoing citations indicate a substantial and rapidly expanding APP literature.

The goal of this paper is to expand the APP to the family of elliptical distributions. As a generalization of normal distributions, elliptical distributions possess many of the tractable and nice statistical properties of normal distribution. It is introduced by Kelker [11] and systematically studied in [5, 6, 12, 17]. The elliptical family contains some heavy-tail classes, like $t$ distributions and Laplace distributions; it also possesses some light-tail classes, like exponential power distributions when the power parameter is greater than 2. The abundant flexibility of elliptical distributions is applied to financial risk management and many other fields, see [13–16].

This paper is organized as follows. In Sect. 2, the definition of the elliptical distributions, and some sub-classes of this family of distributions are reviewed. The general APP equation for an elliptical distribution is obtained in Sect. 3 and the required sample sizes under different cases are specified in Sect. 4. In addition, plots of density curves are provided, which include the $t$ distribution, elliptically symmetric logistic distribution, and a special Kotz-type distribution. Finally, three real data examples are given for illustration of our main results in Sect. 5 and the conclusion is given in Sect. 6.

## 2 Family of Elliptical Distributions and Some of Its Sub-classes

In this section, we introduce the definition and properties of the family of elliptical distributions. In addition, some sub-classes of elliptical distribution are exhibited.

**Definition 1** (*Spherical distribution*) The $p$ dimensional random vector $\mathbf{X}$ is said to be *spherically distributed* if for every $p \times p$ orthogonal matrix $\Gamma$, $\Gamma\mathbf{X}$ and $\mathbf{X}$ have the same distribution, denoted by $\Gamma\mathbf{X} \stackrel{d}{=} \mathbf{X}$. Moreover, let

$$\Phi_p = \{\phi(\cdot) : \phi(\mathbf{t}'\mathbf{t}) \text{ is a p-dimensional characteristic function}\},$$

then there exists a function $\psi(\mathbf{t}) \in \Phi_p$ such that $\phi(\mathbf{t}'\mathbf{t}) = \psi(\mathbf{t})$, where $\psi(\mathbf{t})$ is the characteristic function of $\mathbf{X}$ and $\mathbf{t}'$ is the transpose of $matht$. We denote $\mathbf{X} \sim S_p(\phi)$, and $\phi(\cdot)$ is called the *characteristic generator* of the spherical distribution. Then $\mathbf{X}$ has a stochastic representation

$$\mathbf{X} \stackrel{d}{=} R\,\mathbf{U}^{(p)},$$

where $\mathbf{U}^{(p)}$ is uniformly distributed on the unit sphere surface in $\Re^p$, $R$ and $\mathbf{U}^{(p)}$ are independent where $R \geq 0$ is the random variable with $R \sim F$ in $[0, \infty)$ called the generating variate and $F$ is called the generating distribution function.

**Definition 2** (*Elliptical Contoured Distribution*) A $p$ dimensional random vector $\mathbf{X}$ has a *multivariate elliptical contoured distribution (elliptical distribution)*, denoted by $\mathbf{X} \sim ELL_p(\boldsymbol{\mu}, \Sigma, \phi)$ if the stochastic representation of $\mathbf{X}$ is of the form

$$\mathbf{X} \stackrel{d}{=} \boldsymbol{\mu} + R\,\mathbf{A}'\mathbf{U}^{(k)},$$

where $\boldsymbol{\mu} \in \Re^p$, $\mathbf{U}^{(k)}$ is uniformly distributed on the unit sphere surface in $\Re^k$, $R$ is a generating variate, and $\mathbf{A}$ is a $k \times p$ matrix such that positive-definite matrix $\Sigma = \mathbf{A}'\mathbf{A}$.

Two remarks on this definition are given below.

**Remark 1** It is not always true that the density of $\mathbf{X}$ exists if $\mathbf{X} \sim ELL_p(\boldsymbol{\mu}, \Sigma, \phi)$; however, if it exists, the density of $\mathbf{X}$ is of the form

$$f_{\mathbf{X}}(\mathbf{x}) = \frac{c_p}{\sqrt{|\Sigma|}} g_p\left[(\mathbf{x}-\boldsymbol{\mu})'\Sigma^{-1}(\mathbf{x}-\boldsymbol{\mu})\right], \ \mathbf{x} \in \Re^p, \tag{1}$$

where function $g_p$ is called the density generator; meanwhile, $g_p$ satisfies

$$\int_0^\infty u^{\frac{p}{2}-1} g_p(u)du < \infty, \tag{2}$$

and $c_p$ is called the normalizing constant which can be computed by

$$c_p = \frac{\Gamma(\frac{p}{2})}{\pi^{p/2}} \left[ \int_0^\infty u^{\frac{p}{2}-1} g_p(u) du \right]^{-1}.$$

Now, if the condition in (2) holds, then the density of elliptical distribution with the density generator $g_p$ can be written by $\mathbf{X} \sim ELL_p(\boldsymbol{\mu}, \Sigma, g_p)$. In addition, condition in (2) holding does not imply that the mean and covariance of $\mathbf{X}$ exist; for example, the Cauchy distribution with density generator $g_p(u) = (1+u)^{-(p+1)/2}$. It can be proved that if

$$\int_0^\infty g_1(u) du < \infty, \tag{3}$$

then the mean of $\mathbf{X}$ exists with $E(\mathbf{X}) = \boldsymbol{\mu}$. Moreover, if

$$\int_0^\infty \sqrt{u} g_1(u) du < \infty, \tag{4}$$

the covariance matrix exists with $Cov(\mathbf{X}) = -2\phi'(0)\Sigma$.

It is easy to prove the following result.

**Proposition 1** (Linear Transformation) *Let* $\mathbf{X} \sim ELL_p(\boldsymbol{\mu}, \Sigma, \phi)$, *$B$ is the $m \times p$ matrix of rank $m \leq p$, and $\mathbf{v} \in \Re^m$, then*

$$B\mathbf{X} + \mathbf{v} \sim ELL_m(B\boldsymbol{\mu} + \mathbf{v}, B\Sigma B', \phi). \tag{5}$$

*It is to say that the linear combination of elliptical distribution is still elliptical distribution with the same characteristic generator $\phi$; however, note that the new distribution may not have the same density generator as* $\mathbf{X}$.

**Remark 2** If $\mathbf{X} \sim ELL_p(\boldsymbol{\mu}, \Sigma, g_p)$, then all the marginal density exist. For $B$ and $\mathbf{v}$ defined in Proposition 1,

$$B\mathbf{X} + \mathbf{v} \sim ELL_m(B\boldsymbol{\mu} + \mathbf{v}, B\Sigma B', g_m), \tag{6}$$

where

$$g_m(u) = \int_u^\infty (\omega - u)^{\frac{p-m}{2}-1} g_p(\omega) d\omega = \int_u^\infty g_{m+2}(\omega) d\omega, \; u \geq 0. \tag{7}$$

**Proposition 2** *Let* $\mathbf{X} = (X_1, X_2, \ldots, X_p)'$ *and* $\mathbf{1}_p = (1, 1, \ldots, 1)' \in \Re^p$ *where* $X_i \sim ELL_1(\mu_i, \sigma_i^2, g_1)$, *$i = 1, 2, \ldots, p$, a column vector of ones with dimension $p$, then*

$$\bar{X} = \frac{1}{p}\mathbf{1}'_p\mathbf{x} \sim ELL_1(\bar{\mu},\ \sigma^2,\ g_1), \tag{8}$$

*where*

$$\bar{\mu} = \sum_{i=1}^{p}\mu_i/p,\ \ \sigma^2 = \frac{1}{p^2}\sum_{i=1}^{p}\sum_{j=1}^{p}\sigma_{ij},\ \ \sigma_{ii} = \sigma_i^2,\ \ \textit{and}\ \ -2\phi'(0)\sigma_{ij} = cov(X_i, X_j)$$

*with* $i \neq j,\ i,\ j = 1, 2, \ldots, p$.

Now we will discuss some sub-classes of the elliptical distributions.

## 2.1 Classical Student t Distributions

The multivariate $t$ distribution is a special symmetric multivariate Pearson Type VII distribution [5]. The $n$ dimensional random vector $\mathbf{X}$ has a symmetric multivariate Pearson Type VII distribution as $\mathbf{X} \sim MPVII_n(\boldsymbol{\mu}, \Sigma, g_{N,\nu})$, where the density generator is given by

$$g_{N,\nu}(u) = (1 + u/\nu)^{-N},\ u > 0,$$

with $c_n = (\pi\nu)^{-n/2}\Gamma(N)/\Gamma(N - n/2)$.

If $N = \frac{1}{2}(n + \nu)$, $\mathbf{x}$ is called the multivariate t distribution, denoted by $\mathbf{x} \sim Mt_n(\nu, \boldsymbol{\mu}, \Sigma)$. Here, the density of $\mathbf{x}$ is given by

$$f_{\mathbf{x}}(\mathbf{x}) = \frac{c_n}{\sqrt{|\Sigma|}}\left(1 + \frac{(\mathbf{x} - \boldsymbol{\mu})'\Sigma^{-1}(\mathbf{x} - \boldsymbol{\mu})}{\nu}\right)^{-(n+\nu)/2},\quad \mathbf{x} \in \Re^n, \tag{9}$$

where $c_n = \dfrac{\Gamma((n + \nu)/2)}{(\pi\nu)^{n/2}\Gamma(\nu/2)}$. Moreover, to make conditions (2)–(4) hold, we ask $\nu > 2$ and so

$$Cov(\mathbf{x}) = -2\phi'(0)\Sigma = \frac{\nu}{\nu - 2}\Sigma. \tag{10}$$

## 2.2 Generalized Elliptically Symmetric Logistic Distributions

Wang and Yin [24] define a new class of multivariate elliptical distributions, called the family of generalized elliptically symmetric logistic distributions (GESL). An elliptical distributed random vector $\mathbf{X}$ follows GESL distribution with location parameter $\boldsymbol{\mu}$ and scale matrix $\Sigma$, denoted by $GL_n(\boldsymbol{\mu}, \Sigma, g_n)$ if its pdf and density generator have the forms

$$f_{GL}(\mathbf{x}) = \frac{c_n}{\sqrt{|\Sigma|}} g_n\left[(\mathbf{x}-\boldsymbol{\mu})'\Sigma^{-1}(\mathbf{x}-\boldsymbol{\mu})\right], \qquad \mathbf{x} \in \Re^n, \tag{11}$$

$$g_n(u) = \frac{e^{-au}}{(1+e^{-bu})^s}, \qquad u > 0, \tag{12}$$

respectively, where $a$, $b$, and $s > 0$ are constants. The normalizing constant is computed by

$$c_n = \left(\frac{b}{\pi}\right)^{\frac{n}{2}} \left[\Psi_s^*\left(-1, \frac{n}{2}, \frac{a}{b}\right)\right]^{-1},$$

where $\Psi_s^*$ is the generalized Hurwitz-Lerch zeta function. The elliptically symmetric logistic distribution, denoted by $LO_n(\boldsymbol{\mu}, \Sigma)$, is one of the special cases of generalized elliptical logistic distribution if $a = b = 1$ and $s = 2$.

Landsman and Valdez [13] provide the expression of the normalizing constant of $LO_n$, which is given by

$$c_n = \frac{\Gamma(\frac{n}{2})}{\pi^{n/2}} \left[\int_0^\infty u^{\frac{n}{2}-1} g_n(u) du\right]^{-1} = \pi^{-n/2} \left[\sum_{j=1}^{\infty} (-1)^{j-1} j^{1-n/2}\right]^{-1}. \tag{13}$$

## 2.3 Symmetric Kotz-Type Distributions

If $\mathbf{X} \sim ELL_n(\boldsymbol{\mu}, \Sigma, g_n)$ and the density generator $g_n$ is of the form

$$g_n(u) = u^{\nu-1} exp(-ru^s), \quad r, s > 0, \ 2\nu + n > 2. \tag{14}$$

Thus the density of $\mathbf{X}$ which is called the Kotz-type distribution, denoted by $Kt_n(\boldsymbol{\mu}, \Sigma, r, s, \nu)$, is given by

$$f_{\mathbf{X}}(\mathbf{x}) = c_n |\Sigma|^{-1/2} g_n\left((\mathbf{x}-\boldsymbol{\mu})'\Sigma^{-1}(\mathbf{x}-\boldsymbol{\mu})\right), \quad \mathbf{x} \in \Re^n, \tag{15}$$

where the normalizing constant $c_n = \dfrac{s\,\Gamma(n/2)}{\pi^{n/2}\Gamma((2\nu+n-2)/2s)} r^{(2\nu+n-2)/2s}$.

Several special cases of this family of distributions are listed below.

- If $\nu = 1$, $s = 1$ and $r = \frac{1}{2}$, the distribution possesses a multivariate normal distribution, denoted by $N_p(\boldsymbol{\mu}, \Sigma)$.
- If $s = 1$, the distribution is, studied by Kotz [10], called the original Kotz distribution, denoted by $Kotz_p(\boldsymbol{\mu}, \Sigma, r, \nu)$. In the following section, we will use this distribution to show the results of our main theorem and connect it to the real example.

- If $\nu = 1$, the distribution is said to have the exponential power distribution.
- If $\nu = 1$, $r = 1$, and $s = \frac{1}{2}$, Naik and Plungpongpun [18] use this special Kotz type distribution, the multivariate Laplace distribution, to study its multivariate statistical inference.

## 3 The APP for Estimating Parameter $\mu$

First of all, we denote $I_n$ as an $n \times n$ identity matrix and $\mathbf{J}_n = \mathbf{1}_n \mathbf{1}_n'$, the $n \times n$ matrix with all elements of 1.

**Theorem 1** *Let* $\mathbf{X} = (X_1, X_2, \ldots, X_n)' \sim ELL_n(\boldsymbol{\mu}, \Sigma, g_n)$, *where* $X_i \sim ELL_1(\mu, \sigma^2, g_1)$, $i = 1, 2, \ldots, n$, $\Sigma = \sigma^2 I_n$, *and* $g_1$ *satisfies conditions* (2)–(4). *Let* $c$ *be the confidence level and* $f$ *the precision defined as a fraction of the standard deviation which satisfies*

$$P(-f\sigma_{x_1} \leq \bar{X} - \mu \leq f\sigma_{x_1}) = c, \tag{16}$$

*where* $\sigma_{X_1}$ *is the standard deviation of* $X_1$ *and* $\bar{X}$ *is the sample mean. Then the required sample size* $n$ *can be obtained by*

$$n = \left\lceil \frac{z^2_{\frac{1}{2}(1-c)}}{(-2\phi'(0))f^2} \right\rceil, \tag{17}$$

*where* $z_{\alpha/2}$ *is the* $(1-\alpha/2)^{th}$ *percentile of the spherical distribution* $ELL_1(0, 1, g_1)$.

***Proof*** By Proposition 2, since $\bar{X} \sim ELL_1(\mu, \sigma^2/n, g_1)$ then $Z \equiv \dfrac{\bar{X} - \mu}{\sigma/\sqrt{n}} \sim ELL_1(0, 1, g_1)$, a spherical distribution $S_1(\phi)$. Note that the variance of $Z$ is $-2\phi'(0)$. The confidence interval with level $c$ is given by

$$\begin{aligned}
c &= P(-f\sigma_{x_1} \leq \bar{X} - \mu \leq f\sigma_{x_1}) \\
&= P\left(-f\sqrt{-2\phi'(0)}\sigma \leq \bar{X} - \mu \leq f\sqrt{-2\phi'(0)}\sigma\right) \\
&= P\left(-f\sqrt{-2\phi'(0)n} \leq \frac{\bar{X} - \mu}{\sigma/\sqrt{n}} \leq f\sqrt{-2\phi'(0)n}\right).
\end{aligned}$$

Thus it is easy to see that the desired sample size $n$ is given in (17) and the proof is completed.

**Remark 3** Note that the $c \times 100\%$ confidence interval for $\mu$ based on Theorem 1 is

$$\left[\bar{X} - f\sigma\sqrt{(-2\phi'(0))}, \bar{X} + f\sigma\sqrt{(-2\phi'(0))}\right]. \tag{18}$$

We provide freely available online calculators for some families of elliptical distributions listed below. In these programs, the desired sample size is given, together with the variance, $-2\phi'(0)$, of the corresponding spherical distribution.

Kotz-type distribution:
https://apptaildistribution.shinyapps.io/APPforKotz

GESL distribution:
https://apptaildistribution.shinyapps.io/APPforGESL/.

The $t$ distribution:
https://appelliptical.shinyapps.io/APPforStudent-t/.

## 4 Main Results

In this section, density curves of Normal distribution($N$), logistic distribution ($LO$), Kotz-type distribution with $\nu = 1$, $r = 0.5$, and $s = 5$ ($Kt_1(\cdot, \cdot, 0.5, 5, 1)$), and $t_3$, the $t$ distribution with degrees of freedom $\nu = 3$ are given in Fig. 1. Without loss of

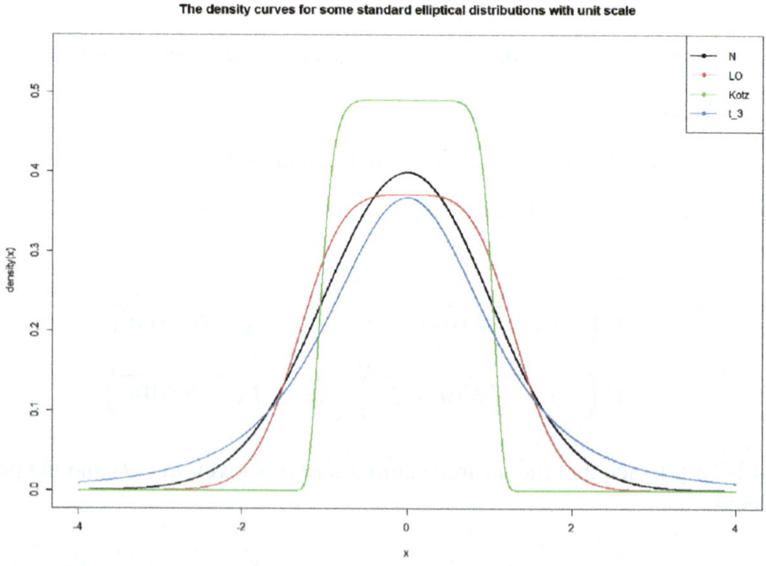


**Fig. 1** The density curves with location parameter 0 and scale parameter 1

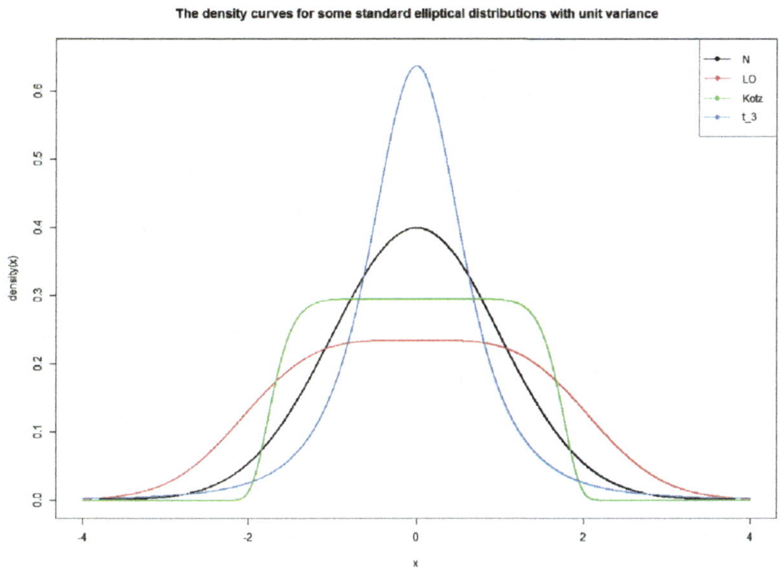


**Fig. 2** Same distributions with mean 0 and variance 1

generality, the location parameter $\mu$ is set to be 0 and the scale parameter to be 1. From Fig. 1, we can see that the standardized density curves are different. Meanwhile, in order to find the required sample sizes for estimating the location parameter $\mu$, density curves of these distributions with mean 0 and variance 1 are needed, which are given in Fig. 2. From Theorem 1, the required sample sizes $n$ for each of the distributions mentioned above for precision $f = 0.1, 0.15, 0.2, 0.25$, and 0.3 and confidence level $c = 0.90$ and 0.95, respectively, are calculated and given in Table 1.

**Table 1** The desired sample sizes $n$ for $f = 0.1, 0.15, 0.2, 0.25$, and 0.30, and $c = 0.95, 0.9$ under specific distribution settings

| c | f | $Kt_1(\cdot, \cdot, 0.5, 5, 1)$ | $LO$ | $t_3$ | $N$ |
|---|---|---|---|---|---|
| | | n | | | |
| 0.95 | 0.1 | 286 | 339 | 338 | 385 |
| | 0.15 | 127 | 151 | 151 | 171 |
| | 0.2 | 72 | 85 | 85 | 97 |
| | 0.25 | 46 | 55 | 55 | 62 |
| | 0.3 | 32 | 38 | 38 | 43 |
| 0.90 | 0.1 | 245 | 257 | 185 | 271 |
| | 0.15 | 109 | 115 | 83 | 121 |
| | 0.2 | 62 | 65 | 47 | 68 |
| | 0.25 | 40 | 42 | 30 | 44 |
| | 0.3 | 28 | 29 | 21 | 31 |

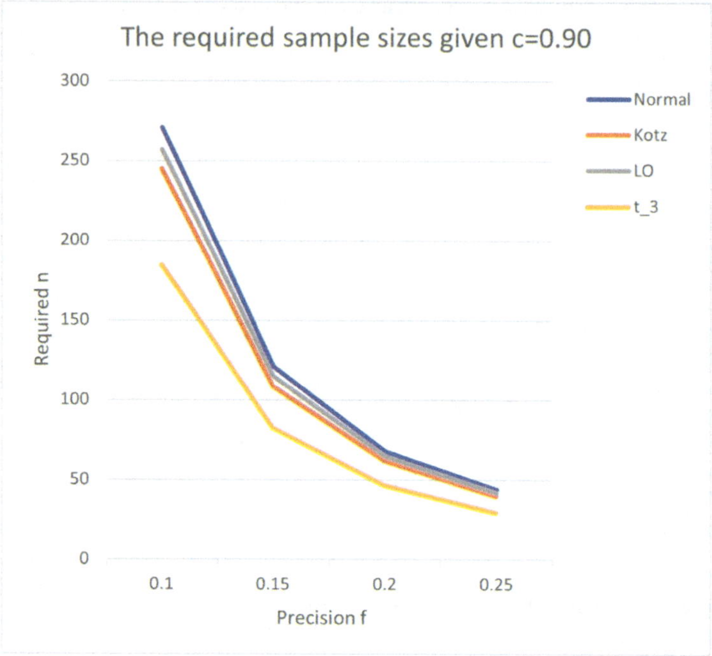


**Fig. 3** The relationship between the desired sample sizes $n$ and precision $f$ while $c = 0.90$

Also for the relationships between the required sample size $n$ and precision $f$, a line chart to shows that as $f$ increases, the required sample size $n$ decreases for all four cases of elliptical distributions, which is given in Fig. 3.

## 5 Real Data Examples

In this section, without additional annotation, the parameters $r$ and $s$ in Kotz-type distribution are set to be 0.5 and 1, respectively; moreover, the parameters $a$, $b$, and $s$ in the GESL are set to be 1, 1, 2, respectively.

For illustration of our main results, three real data sets are considered. The maximum likelihood estimation is considered to find estimates of parameters using the whole data set and is numerically computed in the $R$ environment. Besides, for fitting the $t$ distribution, we have used the specialized routine `t.scaled` in the R package `metRology` [19]; for fitting $GESL$ and $Kt_1$, we have used the local optimizer `optim` with method `L-BFGS-B` in the R packages `stats`. To assess the performance of our models, we use the Akaike's information criterion (AIC), namely $AIC = -2\log(\hat{L}) + 2k$, and the Bayesian information criterion (BIC), $BIC = k\log(n) - 2\log(\hat{L})$, where $\log(\hat{L})$ means the maximized loglikelihood, $k$ is the number of parameters estimated by the model.

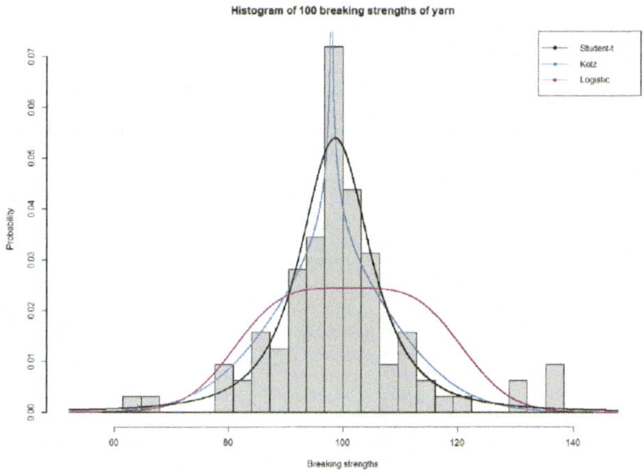


**Fig. 4** Histogram of 100 breaking strengths of yarn and its fitted density curves

**Table 2** Estimates of parameters, $log(\hat{L})$, AIC, and BIC for $t_\nu$, GESL, and $Kt_1$

| | $\hat{\mu}$ | $\hat{\sigma}$ | $\hat{\nu}$ | $log(\hat{L})$ | AIC | BIC |
|---|---|---|---|---|---|---|
| $t_{\hat{\nu}}$ | 98.6685 | 6.6085 | 2.1264 | −381.5328 | 769.0656 | 776.8811 |
| **$Kt_1$** | 97.9998 | 12.8209 | 0.8617 | **−374.4175** | **754.8350** | **762.6505** |
| GESL | 100.9510 | 15.2366 | | −406.3007 | 816.6014 | 821.8117 |

**Example 1** Duncan [3] presents a dataset containing 100 breaking strengths of yarn. This data was also analyzed by Korkmaz et al. [9] and Tahir et al. [20]. The mean of the dataset (size of 100) is $\mu^* = 99.430$, and the variance is $\sigma^{*2} = 153.685$. The histogram of the data and its fitted $t_\nu$, $Kt_1$, and GESL density curves are given in Fig. 4.

The MLEs of parameters, $\log\hat{L}$, AIC, and BIC for $t_\nu$, GESL, and $Kt_1$ are given in Table 2. It can be shown, from Table 2, that the Kotz-type distribution is the best one to fit the data among three candidates of distributions. Based on the values of the extra parameters $\nu = 0.8617$, $r = 0.5$, and $s = 1$ in the Kotz-type distribution, the required sample size for $f = 0.25$ and $c = 0.95$ is $n = 69$ using the Shinnyapp program provided. Next, we select a random sample of size $n = 69$, the sample mean $\bar{x} = 98.7826$ and the standard deviation of related Kotz-type distribution is 0.8505. The MLEs of parameters based on the sample of size $n = 69$, together with $\log\hat{L}$, AIC, and BIC for $t_\nu$, GESL, and $Kt_1$ are given in Table 3 and the conclusion is consistent with Table 2. Thus, the 95% confidence interval for the location parameter $\mu$ under precision $f = 0.25$ is [96.0565, 101.5087], which contains $\mu^* = 99.430$.

**Example 2** The Fama-French three-factor model is a statistical model designed in 1992 by Eugene Fama and Kenneth French to describe stock returns [4]. SMB is one of the three factors in the Fama-French model, which stands for small minus big, and it is calculated by subtracting the returns of a portfolio of large-cap stocks

**Table 3** Comparison among the $t_{\hat{\nu}}$, GESL, and $Kt_1$ for a sample of size $n = 69$ in Example 1

| | $\hat{\mu}$ | $\hat{\sigma}$ | $\hat{\nu}$ | $log(\hat{L})$ | AIC | BIC |
|---|---|---|---|---|---|---|
| $t_{\hat{\nu}}$ | 98.1534 | 7.3132 | 2.4275 | −265.7092 | 537.4183 | 544.1206 |
| **Kotz** | 98.0000 | 16.6218 | 0.7830 | **−257.1861** | **520.3722** | **527.0745** |
| GESL | 99.8553 | 15.2414 | | −280.2563 | 554.5126 | 568.9808 |

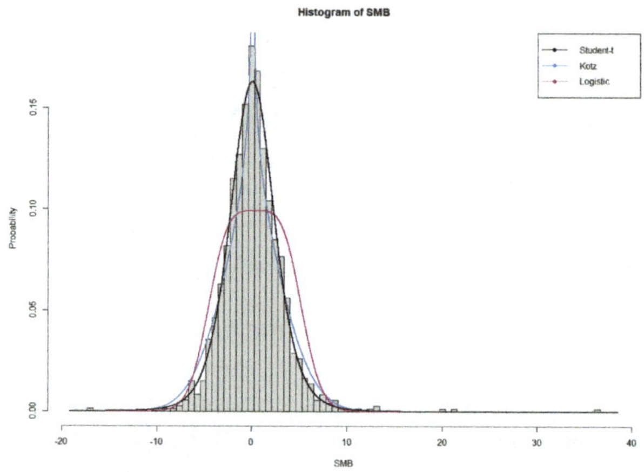


**Fig. 5** Histogram of US monthly SMB factors and its fitted density curves

**Table 4** Comparison among the $t_{\hat{\nu}}$, GESL, and $Kt_1$ in Example 2

| | $\hat{\mu}$ | $\hat{\sigma}$ | $\hat{\nu}$ | $log(\hat{L})$ | AIC | BIC |
|---|---|---|---|---|---|---|
| $t_{\hat{\nu}}$ | 0.0726 | 2.2550 | 4.5047 | **−2871.08** | **5748.161** | **5763.342** |
| $Kt_1$ | 0.0700 | 3.6609 | 0.8653 | −2946.668 | 5899.336 | 5914.517 |
| GESL | 0.3401 | 3.7421 | | −3121.36 | 6246.72 | 6256.841 |

from a portfolio of small-cap stocks. SMB measures the effect of size on portfolio returns, where smaller companies tend to outperform larger ones over the long term. French provided a data library [7] to construct US monthly Fama-French three-factor files. We use the dataset of US monthly SMB factors given from July 1926 through July 2023, which is of size 1165. The mean of the dataset is $\mu_{SMB} = 0.1898$ and the variance is 10.0471. The histogram of the data and its fitted $t_\nu$, $Kt_1$, and GESL density curves are given in Fig. 5.

The MLEs of parameters, $\log \hat{L}$, AIC, and BIC for $t_\nu$, GESL, and $Kt_1$ are given in Table 4. It can be shown, from Table 4, that the $t$ distribution with degrees of freedom $\nu = 4.5047$ is the best one to fit the data among three candidates of distributions.

The required sample size for $f = 0.10$ and $c = 0.98$ is $n = 692$ using the Shinnyapp program provided. Next, we select a random sample of size $n = 692$, the sample mean $\bar{x} = 0.2902$ and the standard deviation of $t$ distribution is 1.7985. The

**Table 5** Comparison among the $t_{\hat{\nu}}$, GESL, and $Kt_1$ for a sample of the data set in Example 2 with $f = 0.1$, $c = 0.98$, and $n = 692$

| | $\hat{\mu}$ | $\hat{\sigma}$ | $\hat{\nu}$ | $log(\hat{L})$ | AIC | BIC |
|---|---|---|---|---|---|---|
| $t_{\hat{\nu}}$ | 0.2150 | 2.2615 | 4.7471 | **−1698.907** | **3403.814** | **3417.433** |
| $Kt_1$ | 0.2200 | 3.4578 | 0.8794 | −1721.984 | 3449.968 | 3463.587 |
| GESL | 0.2000 | 3.5454 | | −1806.3 | 3616.6 | 3625.679 |

MLEs of parameters based on the sample of size $n = 692$, together with $\log\hat{L}$, AIC, and BIC for $t_\nu$, GESL, and $Kt_1$ are given in Table 4 and the conclusion is consistent with Table 5. Thus, the 98% confidence interval for the location parameter $\mu$ under precision $f = 0.10$ is $[-0.0123, 0.5926]$ so that $\mu_{SMB}$ is included in this confidence interval.

**Example 3** According to Yin et al. [24], GESL with constants $a = 0.5$, $b = 25$, and $s = 0.5$ can fit the single fibers data set of 10 mm in gauge lengths with size 63 concluded by Gupta and Kundu [8] better than Normal distribution and Logistic distribution. The mean of the set is $\mu_* = 3.0593$. We list the $log(\hat{L})$, AIC, and BIC for $t_{\hat{\nu}}$, GESL with defined $a$, $b$, and $s$, and Kotz-type distribution in the following table. The histogram of the data with size 63 and its fitted $t_\nu$, $Kt_1$, and GESL density curves are given in Fig. 6.

The MLEs of parameters, $\log\hat{L}$, AIC, and BIC for $t_\nu$, GESL, and $Kt_1$ are given in Table 6. It can be shown, from Table 6, that the GESL with constants a = 0.5, b = 25, and s = 0.5 can modal the data better than others based on AIC and BIC and their $\log(\hat{L})$'s are very close.

Now for $f = 0.25$ and $c = 0.90$, the required sample size $n = 43$ using the Shinnyapp program provided for GESL with constants $a = 0.5$, $b = 25$, and $s = 0.5$. Thus a random sample of size $n = 43$ is selected. The sample mean $\bar{x} = 3.0180$ and

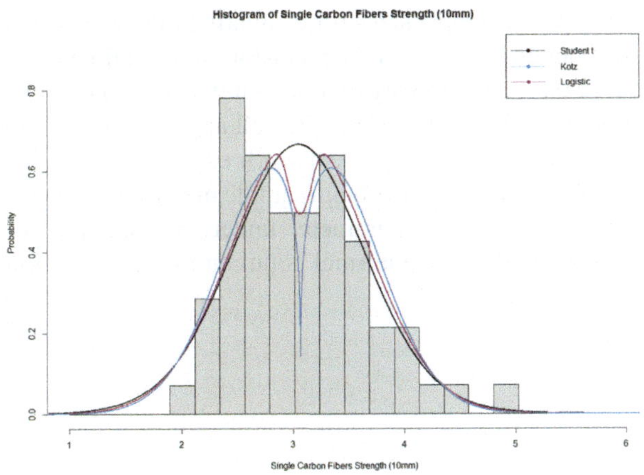


**Fig. 6** Histogram of Single Carbon Fibers Strength (10mm) and its fitted density curves

**Table 6** Comparison among the $t_{\hat{\nu}}$, GESL, and $Kt_1$ in Example 3

| | $\hat{\mu}$ | $\hat{\sigma}$ | $\hat{\nu}$ | $log(\hat{L})$ | AIC | BIC |
|---|---|---|---|---|---|---|
| GESL | 3.0616 | 0.5997 | | −58.36 | **120.72** | **129.9006** |
| $t_{\hat{\nu}}$ | 3.0474 | 0.5942 | 28.8739 | −58.7922 | 123.5845 | 130.0139 |
| $Kt_1$ | 3.0684 | 0.5527 | 1.1210 | **−57.8867** | 121.7734 | 135.5443 |

**Table 7** Comparison among the $t_{\hat{\nu}}$, GESL, and $Kt_1$ for a sample of the data set in Example 3 with $f = 0.25$, $c = 0.9$, and $n = 43$

| | $\hat{\mu}$ | $\hat{\sigma}$ | $\hat{\nu}$ | $log(\hat{L})$ | AIC | BIC |
|---|---|---|---|---|---|---|
| GESL | 3.0189 | 0.6110 | | −40.3041 | **84.6083** | **89.5942** |
| $t_{\hat{\nu}}$ | 2.9966 | 0.5808 | 15.2022 | −40.5225 | 87.0451 | 92.3287 |
| Kotz | 3.0577 | 0.5627 | 1.1176 | **−40.1910** | 86.3818 | 91.6654 |

the standard deviation is 0.5243. The MLEs of parameters based on the sample of size $n = 43$, together with $\log\hat{L}$, AIC, and BIC for $t_\nu$, GESL, and $Kt_1$ are given in Table 7 and the conclusion is consistent with the one given Table 6. Therefore, the 90% confidence interval for the location parameter $\mu_*$ under the precision 0.25 is [2.9094, 3.1266] so that $\mu_*$ is included in this confidence interval.

## 6 Discussion

We extended the APP for estimating parameters under normal, skew normal, gamma, and log-normal settings to another popular family of distributions, called elliptical (contoured) distributions. Properties of this family of distributions and its three sub-classes, such as multivariate $t$ distributions, generalized elliptically symmetric logistic distributions, and symmetric Kotz-type Distributions are discussed. Specifically, the required sample sizes for estimating the location parameter for these sub-classes are obtained for specified precision and confidence level. We provided three real data examples, showing that best-fit density curves may change from one data set to another data set, even if they are all approximately symmetric.

In conclusion, the present work expands the ability of researchers to determine sample sizes needed to obtain sample means that are good estimates of corresponding population location parameter under the family of elliptical distributions, which includes multivariate normal populations, $t$ distributions.

## References


1. Cao, L., Tong, T., Trafimow, D., Wang, T., Chen, X.: The A Priori procedure for estimating the mean in both log-normal and gamma populations and robustness for assumption violations. Methodology **18**(1), 24–43 (2022)
2. Chen, X., Trafimow, D., Wang, T., Tong, T., Wang, C.: The APP procedure for estimating the Cohen's effect size. Asian J. Econ. Bank. **5**(3), 289–306 (2021)
3. Duncan, A.J.: Graphical methods for process control. Quality Control and Industrial Statistics, p. 1019. Irwin, Homewood (1986)
4. Fama, E.F., French, K.R.: The cross-section of expected stock returns. J. Finance **47**(2), 427–465 (1992)
5. Fang, K.W.: Symmetric Multivariate and Related Distributions. CRC Press (2018)
6. Fang, K.T., Zhang, Y.T.: Generalized Multivariate Analysis (1990). (No Title)
7. French, K.R.: Data library. Tuck School of Business at Dartmouth (n.d.). Accessed 11 April 2023. https://mba.tuck.dartmouth.edu/pages/faculty/ken.french/data_library.html
8. Gupta, R.D., Kundu, D.: Generalized logistic distributions. J. Appl. Stat. Sci. **18**(1), 51 (2010)
9. Korkmaz, M.Ç., Altun, E., Yousof, H.M., Hamedani, G.G.: The odd power Lindley generator of probability distributions: properties, characterizations and regression modeling. Int. J. Stat. Probab. (2019)
10. Kotz, S.: Multivariate distributions at a cross road. In: A Modern Course on Statistical Distributions in Scientific Work: Volume 1-Models and Structures Proceedings of the NATO Advanced Study Institute held at the University of Calgagry, Calgary, Alberta, Canada July 29–August 10, 1974, pp. 247–270. Springer Netherlands, Dordrecht (1975)
11. Kelker, D.: Distribution theory of spherical distributions and a location-scale parameter generalization. Sankhyā: Indian J. Stat. Seri. A, 419–430 (1970)
12. Kotz, S., Ostrovskii, I.: Characteristic functions of a class of elliptic distributions. J. Multivariate Anal. **49**(1), 164–178 (1994)
13. Landsman, Z.M., Valdez, E.A.: Tail conditional expectations for elliptical distributions. North Amer. Actuarial J. **7**(4), 55–71 (2003)
14. Landsman, Z., Makov, U., Shushi, T.: Multivariate tail conditional expectation for elliptical distributions. Insur.: Math. Econ. **70**, 216–223 (2016)
15. Landsman, Z., Makov, U., Shushi, T.: Tail conditional moments for elliptical and log-elliptical distributions. Insur.: Math. Econ. **71**, 179–188 (2016)
16. Landsman, Z., Makov, U., Shushi, T.: A multivariate tail covariance measure for elliptical distributions. Insur.: Math. Econ. **81**, 27–35 (2018)
17. Liang, J.J., Bentler, P.M.: Characterizations of some subclasses of spherical distributions. Stat. & Probab. Lett. **40**(2), 155–164 (1998)
18. Naik, D.N., Plungpongpun, K.: A Kotz-type distribution for multivariate statistical inference. In: Advances in Distribution Theory, Order Statistics, and Inference, pp. 111–124. Birkhäuser Boston, Boston, MA (2006)
19. Stevenson, M., Brown, R.: metRology: Support for Metrological Applications (Version 0.9-38) [Software] (2021). https://cran.r-project.org/web/packages/metRology/index.html
20. Tahir, M.H., Zubair, M., Cordeiro, G.M., Alzaatreh, A., Mansoor, M.: The Weibull-Power Cauchy distribution: model, properties and applications. Hacettepe J. Math. Stat. **46**(4), 767–789 (2017)
21. Tong, T., Trafimow, D., Wang, T., Wang, C., Hu, L., Chen, X.: The A Priori Procedure (APP) for estimating regression coefficients in linear models. Methodology **18**(3), 203–220 (2022)
22. Trafimow, D.: Using the coefficient of confidence to make the philosophical switch from a posteriori to a priori inferential statistics. Educ. Psychol. Meas. **77**(5), 831–854 (2017)
23. Trafimow, D., Wang, T., Wang, C.: From a sampling precision perspective, skewness is a friend and not an enemy! Educ. Psychol. Meas. **79**(1), 129–150 (2019)
24. Wang, Y., Yin, C.: A new class of multivariate elliptically contoured distributions with inconsistency property. Methodol. Comput. Appl. Probab. **23**, 1377–1407 (2021)

25. Wang, C., Wang, T., Trafimow, D., Chen, J.: Extending a priori procedure to two independent samples under skew normal settings. Asian J. Econ. Bank. **3**(02), 29–40 (2019)
26. Wang, C., Wang, T., Trafimow, D., Myüz, H.A.: Necessary sample sizes for specified closeness and confidence of matched data under the skew normal setting. Commun. Stat.-Simul. Comput. **51**(5), 2083–2094 (2022)
27. Wang, C., Wang, T., Trafimow, D., Myüz, H.A.: Desired sample size for estimating the skewness under skew normal settings. In: Structural Changes and their Econometric Modeling, vol. 12, pp. 152–162. Springer International Publishing (2019)
28. Wang, C., Wang, T., Trafimow, D., Li, H., Hu, L., Rodriguez, A.: Extending the a priori procedure (APP) to address correlation coefficients. Data Sci. Finan. Economet. 141–149 (2021)
29. Wang, C., Wang, T., Trafimow, D., Talordphop, K.: Extending the a priori procedure to one-way analysis of variance model with skew normal random effects. Asian J. Econ. Bank. **4**(2), 77–90 (2020)

# Estimation of Under-Reported COVID19 Cases with Susceptible-Infected-Removed Epidemiological Model via Stochastic Frontier Analysis

**Nene Coulibaly, Zheng Wei, and Tonghui Wang**

**Abstract** As the COVID-19 pandemic has evolved, it has become increasingly evident that the actual number of cases has likely been underestimated. In this study, we review an econometric method to estimate the true scale of COVID-19 cases for 40 countries spanning from January 1, 2020, to November 3, 2020. The method centers around the 'structural' model, which is an expansion of the SIR epidemiological model, and is designed to incorporate the notion of underreporting. The findings of our analysis reveal substantial underreporting, aligning with prior research and expert opinions within the field of public health.

## 1 Introduction

Efforts to gauge the extent and public health implications of COVID-19, both in the United States and globally, have faced significant challenges. Official figures for COVID-19 cases are widely believed to fall short of the actual numbers. Researchers have pointed out the considerable issue of missing data in characterizing the pandemic accurately. To address this, we apply a "structural" model, an extension of the Susceptible-Infected-Removed (SIR) epidemiological model, and an econometric approach to account for measurement errors and missing data.

As of November 1st, 2020, nearly 46 million cases[1] have been reported globally by the World Health Organization (WHO) in which it acknowledged the number of

[1] https://iris.who.int/bitstream/handle/10665/336478/nCoV-weekly-sitrep01Nov20-eng.pdf.

N. Coulibaly · Z. Wei
Texas A&M University, Corpus Christi, USA
e-mail: nene.coulibaly@tamucc.edu

Z. Wei
e-mail: zheng.wei@tamucc.edu

T. Wang (✉)
New Mexico State University, Las Cruces, USA
e-mail: twang@nmsu.edu

 
V. Kreinovich et al. (eds.), *Applications of Optimal Transport to Economics and Related Topics*, Studies in Systems, Decision and Control 556,
https://doi.org/10.1007/978-3-031-67770-0_4

potential global cases are far exceeding its official count [1]. These estimates, while cautionary, underscore the challenges in obtaining accurate data amid a pandemic. Measurement challenges surrounding COVID-19 cases can be attributed to multiple factors. First, during the early stages of the pandemic, the lack of comprehensive testing, testing inaccuracies, and a scarcity of information may have led to numerous cases being either misdiagnosed or missed altogether [2]. This shortage of resources and capacity in many countries hindered accurate reporting. Second, a considerable number of individuals infected with the virus are asymptomatic, making them difficult to capture in official counts, especially when universal or randomized testing isn't in place [3]. Third, concerns about political corruption and institutional influences in underreporting are prevalent in various nations, including the United States [4]. Fourth, cases occurring outside of public health surveillance systems go uncounted. This can happen due to a lack of access, personal choice, or fear, particularly in rural areas [5]. Fifth, data collected at the local level needs to be aggregated and reported officially, which can introduce inconsistencies and errors in the reporting process [6]. Moreover, variations in the determination of who may have COVID-19 by health officials in the United States, for example, can lead to discrepancies in reporting. This discretion has occasionally resulted in ad hoc corrections to official counts, as seen in instances where COVID-19 cases were retroactively counted. These complexities in data collection and reporting have led to undercounting.

To address underreporting issues, researchers have employed various methods. In this paper, an extended SIR epidemiological model via Stochastic Frontier Analysis is utilized to assess undercounting. Stochastic frontier Model (SFM) is a suitable framework to determine the true scale of the pandemic. By estimating stochastic frontier models on weekly data from numerous countries, the findings indicate substantial underreporting. The estimates suggest that the actual number of cases may be approximately two to fifteen times higher than the officially reported figures, respectively. Furthermore, we establish a strong correlation between testing and the number of cases with increased testing associated with more cases.

This paper is organized as follows: a review of the epidemiology model is given in Sect. 2; a review of SFM is provided in Sect. 3; an estimation is given in Sect. 4. The data is discussed in Sect. 5, and the results are presented in Sect. 6.

## 2 An Epidemiological Model with Underreporting Based on SFM

An epidemiological model with underreporting, based on the Susceptible-Infected-Removed (SIR) model, can be extended to incorporate the concept of underreporting. This type of underreporting can occur when not all cases of a disease are detected and reported, leading to an incomplete picture of the epidemic. A common way to model underreporting is by introducing a factor that accounts for the proportion of cases that go unreported. This can be achieved by incorporating a Surveillance Factor Multiplier (MF) into the model.

The classical SIR is a popular model for many epidemiology of infectious disease for practical problems [7]. The disease is such that the population can be divided into three distinct classes: the susceptibles, S, who can catch the disease; the infectives, I, who have the disease and can transmit it; and the removed class, R, namely, those who have either had the disease, or are recovered, immune or isolated until recovered. The progress of individuals is schematically represented by:

$$S \rightarrow I \rightarrow R$$

The standard SIR model describes the dynamics of the epidemic without considering underreporting. The SIR model assumes that the population is closed, meaning that there are no new individuals entering or leaving the population during the time period of interest. The model also assumes that the disease is transmitted only by direct contact between susceptible and infected individuals. The SIR model can be used to make predictions about the spread of infectious diseases and to study the effects of interventions such as vaccination or social distancing on disease transmission.

The SIR model can be extended to find the under-reporting number of COVID-19 within different countries around the world. The SIR equation is given below:

$$S_{it} + I_{it} + R_{it} = N_{it} \tag{1}$$

where $S_{it}$, $I_{it}$, and $R_{it}$ are defined as the number of individuals for $i$th country of week $t$ in each class, and $N_{it}$ is the overall population of $i$th country.

The assumptions made about the transmission of the infection and incubation period are crucial. These are reflected in the terms in the equations and the parameters. With $S_{it}$, $I_{it}$, and $R_{it}$ as the number of individuals in each class, it is assumed that: (i) the gain in the infective class is at a rate proportional to the number of infectives and susceptibles, that is, $\beta S_{it} I_{it}$, where $r \geq 0$ is a constant parameter. The susceptibles are lost at the same rate; (ii) the rate of removal of infectives to the removed class is proportional to the number of infectives, that is, $\gamma I_{it}$ where $\gamma_{it} \geq 0$ is a constant; $1/\gamma_{it}$ is a measure of the time spent in the infectious state; and (iii) the incubation period is short enough to be negligible; that is, a susceptible who contracts the disease is infective right away. The model mechanism based on the above assumptions for the change in the number of susceptible individuals at time $t$ can be summarized in the following equation:

$$\frac{dS_{it}}{dt} = -\beta_{it} I_{it} S_{it} \tag{2}$$

where $\beta_{it} \geq 0$ is the transmission rate that occur in country $i$ in period $t$ between infected and susceptible individuals. The equation for the change in the number of infectives at time $t$,

$$\frac{dI_{it}}{dt} = \beta_{it} I_{it} S_{it} - \gamma_{it} I_{it}, \tag{3}$$

where $\gamma_{it} > 0$ is the removal rate of infectives. Removal happens in two ways: recovery or death. This removal happens when fatality rate is subtracted from the recovery rate. Finally, the change in the number of recovered and dead at time $t$,

$$\frac{dR_{it}}{dt} = \gamma_{it} I_{it}. \tag{4}$$

Millimet and Parameter [8] extended the SIR model to estimate the number of new cases at time $t$, denoted by $Y^*_{it}$, from Eq. (3) above,

$$Y^*_{it} = \beta_{it} I_{it} S_{it} \tag{5}$$

When there are underreported cases, we can model the relationship between the true number of new cases called, $Y^*$, and the reported number of cases, $Y$, as

$$Y^*_{it} = Y_{it} \exp(u_{it}) \tag{6}$$

and $Y^*_{it}/Y_{it} = \exp(u_{it})$ is the multiplication factor (MF) for new cases. Note that $MF > 1$ indicates that new cases at time $t$ are underreported. To model the unreported value, we substitute Eq. (5) into Eq. (6), and take the logarithms of $Y_{it}$. In sum, the new equation is given by,

$$\ln(Y_{it}) = \ln(\beta_{it}) + \ln(I_{it}) + \ln(S_{it}) - u_{it} \tag{7}$$

Assume that $S_{it}$, $I_{it}$ are unobserved, the equation becomes,

$$\begin{aligned}
\ln(N_{it}) &= \ln(S_{it} + I_{it} + R_{it}) \\
&= \ln\left[S_{it}\left(1 + \frac{I_{it} + R_{it}}{S_{it}}\right)\right] \\
&= \ln(S_{it}) + \ln\left(1 + \frac{I_{it} + R_{it}}{S_{it}}\right) + \ln(I_{it}) \ln(I_{it}) \\
&= \ln(S_{it}) + \ln(I_{it}) + \ln\left(\frac{N_{it}}{S_{it} I_{it}}\right),
\end{aligned}$$

which gives

$$\ln(N_{it}) = \ln(S_{it}) + \ln(I_{it}) + v^*_{it} \tag{8}$$

where $v^*_{it} \in \mathcal{R}$. Substituting Eq. (8) into Eq. (7) lead to

$$\ln(Y_{it}) = ln(\beta_{it}) + \ln(N_{it}) - u_{it} + v^Y_{it} \tag{9}$$

where $v_{it}^{Y} = -v_{it}^{*}$. The assumption here is that $\beta_{it}$ is modeled partially in a deterministic fashion. We can let

$$\ln(\beta_{it}) = X_{it}\beta_X + v_{it}^{\beta} \tag{10}$$

where $v_{it}^{\beta} \in R$. $X$ are the independent variables of country $i$ that includes variables that are measuring the extent of COVID-19 like population density, median age, diabetes prevalence, and gross domestic product (GDP) taken by the country prior to $t$. We also include region specific time trends in $\beta_{it}$ and use linear trend to allow the effectiveness of each to vary over time. This is a similar model developed by Arroyo-Marioli et al. [9]. The only difference is that they used $\beta_{it}$ as random walk and we are using it as deterministic function with a proportional error. Then, take the logarithms of the model to get an additive error by inserting Eq. (11) into (10), we get

$$\ln(Y_{it}) = X_{it}\beta_X + \beta_N \ln(N_{it}) - u_{it} + v_{it} \tag{11}$$

where $v_{it} = v_{it}^{Y} + v_{it}^{\beta}$, $v_{it}^{Y}$ and $v_{it}^{\beta}$ are symmetric distribution around 0 and if $u > 0$, then $Y_{it} < Y_{it}^{*}$, the $MF > 1$. So far we rewrite the SIR model with underreporting in terms of Stochastic Frontier Model (SFM) and we provide a review of SFM in the next section.

## 3 Review of Stochastic Frontier Model

A Stochastic Frontier Model is a statistical and econometric model used to analyze and measure the efficiency of production processes in various industries, such as agriculture, manufacturing, healthcare, and services, etc. The SFM introduces stochasticity to account for unobservable factors and measurement errors affecting production or cost efficiency. The basic idea behind SFM is that there are two sources of variation in the data: random noise and inefficiency. Random noise refers to the natural variation in inputs and outputs that cannot be controlled by the firm, while inefficiency refers to the extent to which the firm is not using its inputs as efficiently as it could be. The model assumes that observed outcomes (e.g., production levels or costs) result from a combination of deterministic factors (the production or cost frontier) and random, unobservable factors (efficiency and error terms).

The original formulation, proposed in [10], assumes the production frontier $h(\boldsymbol{x})$ describing the maximum output with the given input $\boldsymbol{x}$. The output function would lie below the frontier of a producer or a firm because the outputs $y$ are often less than it could feasibly produce in reality. Thus, the real production $Y$ is proportional to the production frontier,

$$Y = h(\boldsymbol{x}) \cdot TE.$$

Here $TE$ is called the *technical efficiency* and $0 \leq TE \leq 1$. The producer obtains its production frontier if and only if $TE = 1$, otherwise $TE < 1$. For firm $i$, TE can be written using exponential transformation of error $\varepsilon_i$ and we have:

$$Y_i = h(\boldsymbol{x}_i|\boldsymbol{\beta}) \exp(\mathcal{E}_i) \qquad i = 1, \ldots, n,$$

where $Y_i$ is the production of the $i$th firm, $\boldsymbol{x}_i$ is a $k$-vector of input quantities of the $i$th firm, $\boldsymbol{\beta}$ is a vector of unknown parameters. The error term can be decomposed as the difference of two independent error terms, $\mathcal{E}_i = V_i - U_i$, where $U_i$s are non-negative (one-sided) random variables for the possible inefficiency of firms and $V_i$ are the random variables representing the possible measurement errors, and $U_i$ and $V_i$ are assumed to be independent random variables [11]. The inefficiency score of the firm $i$ can be obtained as

$$\exp(-U_i) = \frac{Y_i}{h(\boldsymbol{x}_i|\boldsymbol{\beta}) \exp(V_i)}, \qquad i = 1, \ldots, n.$$

Let $V_i \sim N(0, \sigma_v^2)$, the normal distribution with mean 0 and variance $\sigma_v^2$, $U_i \sim HN(0, \sigma_u^2)$, the half-normal distribution with scale parameter $\sigma_u^2$, and $V_i$s and $U_i$s are independent. And if we further assume $h(\boldsymbol{x}|\boldsymbol{\beta})$ to be Cobb-Douglas production function, i.e., $h(\boldsymbol{x}|\boldsymbol{\beta}) = \exp(\beta_0) x_1^{\beta_1} x_2^{\beta_2} \ldots x_k^{\beta_k}$, where $\boldsymbol{\beta} = (\beta_1, \ldots, \beta_k)^T$. Then, taking log-transformation, the SFM with Cobb-Douglas production function can be written as [12],

$$\log Y_i = \beta_0 + \log(\boldsymbol{x}_i)^T \boldsymbol{\beta} + V_i - U_i. \tag{12}$$

The probability density function (PDF) of the composite error term $\mathcal{E}_i$, which will be useful in derivation of the maximum likelihood function for estimation, is

$$f_{\mathcal{E}}(\varepsilon) = \frac{2}{\sqrt{\sigma_v^2 + \sigma_u^2}} \phi\left(\frac{\varepsilon}{\sqrt{\sigma_v^2 + \sigma_u^2}}\right) \left[1 - \Phi\left(\frac{\frac{\sigma_u}{\sigma_v}\varepsilon}{\sqrt{\sigma_v^2 + \sigma_u^2}}\right)\right], \tag{13}$$

where $\phi(\cdot)$ and $\Phi(\cdot)$ are the PDF and CDF of the univariate standard normal distribution. There is a close connection between the density of $\mathcal{E}_i$ and the skew normal distribution [13].

Since the Covid-19 data is panel data with the weekly reported new cases information of each country, the SFM with independent error terms can be written as,

$$\begin{aligned} &\log Y_{it} = \beta_0 + \log(\boldsymbol{x}_{it})^T \boldsymbol{\beta} + V_{it} - U_{it}, \\ &V_{it} \sim_{i.i.d.} N(0, \sigma_v^2), \\ &U_{it} \sim_{i.i.d.} HN(0, \sigma_u^2), \\ &V_{it} \text{ and } U_{it} \text{ are independent.} \end{aligned} \tag{14}$$

Our paper applies SFM in Eq. (14) to SIR model in Eq. (11). The two equations provided represent different models used in distinct fields-economics (the Stochastic Frontier Model, SFM) and epidemiology (the Susceptible-Infectious-Recovered, SIR model with underreporting). In this context, it assumes that not all cases are being accurately reported, and the degree of underreporting is not related to the disease dynamics. Stochastic elements might be introduced to account for randomness in the transmission of the disease or the likelihood of reporting cases. In this stochastic framework, we use methods like random effects models to account for unobserved individual-specific or time-specific factors. The model considers the temporal dimension and can help assess the impact of time-varying variables on the outcomes of interest within a panel COVID-19 data. The log-linear form of the model is particularly useful for econometric analysis, as it helps make the relationships between variables more interpretable and allows for more straightforward estimation of parameters, such as elasticities. In the SIR model, $\log Y_{it}$ represents the natural logarithm of disease prevalence, while $X_{it}\beta$, $\beta_N \ln N_{it}$, and $u_{it} - v_{it}$ represent different components of the model while $\beta_X$ and $\beta_N$ are parameters to be estimated.

The model in Eq. (14) uses the Cobb-Douglas production function which is a widely used mathematical equation in economics to model production or output as a function of inputs. After the log transformation, the equation becomes linear in its parameters, making it easier to estimate and interpret. Therefore, in the SIR model equation, the natural logarithm is applied to $Y_{it}$ and $N_{it}$. The stochastic error terms $V_{it} - U_{it}$ in the SFM capture unexplained variations and under-reported cases based on the observed data.

## 4 Estimation

The objective of this section is to provide the estimation method for the multiplication factors of countries based on SIR model derived in Eq. (11). As one can see that SIR model with under-reported cases in Eq. (11) share a similar structure as SFM in Eq. (14). If we further assume the latent variable $u_{it}$ in Eq. (11) follows the positive one-sided distribution, SFM could be a valuable tool to estimate MF based on $u_{it}$ for each country considered [14].

SFMs can be estimated by using the maximum likelihood techniques and it requires the distribution of the composite error $\varepsilon_{it} = v_{it} - u_{it}$. According to [15, 16], inverse hyperbolic sine (IHS) transformation is frequently applied in econometric studies to transform right-skewed variables that include zero or negative values. The IHS transformation of a variable $z$ is defined as:

$$\operatorname{arcsinh}(z) = \log\left(z + \sqrt{z^2 + 1}\right).$$

To account for the zeros values in dependent variables, weekly reported new cases are transformed using the IHS transformation. We implemented MLE estimations in R [17].

**Likelihood Function of SFM in Eq.** (14)**:**
By assuming that the error $u_{it}$ follows i.i.d. half-normal distribution, HN(0, $\sigma_u^2$), $v_{it}$ follows i.i.d. normal distribution, N(0, $\sigma_v^2$), and $v_{it}$ and $u_{it}$ are independent, we can derive the maximum likelihood function for estimating SIR model based on SFM under cross-sectional setting. The density of the distribution of the composite error $\varepsilon_{it} = v_{it} - u_{it}$ is given in Eq. (13), and this give rise to the following log-likelihood function:

$$\ln L = \sum_{i=1}^{N}\sum_{t=1}^{T_i}\left\{-\ln\sigma + \ln\Phi\left[-\frac{\lambda\epsilon_{it}}{\sigma}\right] - \frac{\epsilon_{it}^2}{2\sigma^2}\right\}, \tag{15}$$

where $\sigma^2 = \sigma_v^2 + \sigma_u^2$, $\lambda = \sigma_u/\sigma_v$.
**Inefficiency and Multiplication Factor Estimates:**
Since $V_{it}$'s and $U_{it}$'s are assumed to be independent and $\mathcal{E}_{it} = V_{it} - U_{it}$ is composite residual for $i$th country at week $t$. The non-negative term $U_{it}$ is used to measure the multiplication factor of the $i$th country at week $t$:

$$MF_{it} = \frac{\log(f(\boldsymbol{x}_{it}|\boldsymbol{\beta})) + U_{it}}{\log(f(\boldsymbol{x}_{it}|\boldsymbol{\beta}))} = \exp(U_{it})\,. \tag{16}$$

Since $U_{it}$ are latent variables and not directly observed, multiplication factors are alternatively estimated by using conditional expectations on residual $\varepsilon_{it}$ based on estimation of Eq. (12) :

$$\hat{M}F_{it} = \exp\left(E[U_{it}|\mathcal{E} = \varepsilon_{it}]\right). \tag{17}$$

By using Bayes' theorem, the conditional density of $U_{it}$ given $\mathcal{E} = \varepsilon_{it}$ is given by

$$f_{U|\mathcal{E}}(u_{it}|\varepsilon_{it}) = \frac{\phi_V(\varepsilon_{it} + u_{it})\phi_U(u_{it})}{\phi_{\mathcal{E}}(\varepsilon_{it})}, \tag{18}$$

where $\phi_V$ denotes the PDF of normal distribution with mean zero and variance $\sigma_v^2$, $\phi_U$ is the PDF of the half-normal distribution with pre-truncation mean 0 and variance $\sigma_u^2$, and $\phi_\varepsilon$ is the PDF of $\varepsilon$ given in (13). And after some algebra, $E[U|\varepsilon]$ is given by,

$$E(U|\varepsilon) = \mu_* + \sigma_*\frac{\phi(\mu_*/\sigma_*)}{\Phi(\mu_*/\sigma_*)}, \tag{19}$$

with

$$\mu_* = -\varepsilon \frac{\sigma_u^2}{\sigma_u^2 + \sigma_v^2} \quad \text{and } \sigma_*^2 = \frac{\sigma_u^2 \sigma_v^2}{\sigma_u^2 + \sigma_v^2}.$$

Once we obtain the MLEs $(\hat{\boldsymbol{\beta}}, \hat{\sigma_v}, \hat{\sigma_u})$ of $(\boldsymbol{\beta}, \sigma_v, \sigma_u)$ in the SFM, their associated multiplication factors can be estimated by

$$\hat{MF}(\hat{\varepsilon}_{it}; \hat{\boldsymbol{\theta}}) = \int_0^\infty \exp(u) f_{U|\varepsilon}(u|\hat{\varepsilon}_{it}; \hat{\boldsymbol{\theta}}) du, \tag{20}$$

where $\hat{\varepsilon}_i = \log y_i - \boldsymbol{x}_i^T \hat{\boldsymbol{\beta}}$, and $f_{U|\varepsilon}(u|\varepsilon; \boldsymbol{\theta})$ is given in Eq. (18).

## 5 Data

We sourced our data on daily COVID-19 cases from Our World in Data,[2] covering the period from January 1, 2020, to November 3, 2020 (i.e., the first 44 weeks of 2020). At first, the data are from over 210 countries, but to ensure data consistency, we focused on countries that had data available for every week during this time-frame. Then, we further restrict the data to top 60 develop countries[3] based on the annual Human Development Index (HDI) score resulting in a sample of 40 countries listed in Table 1. The term "develop country" is described as having a sovereign state that possesses a mature economy, a technologically advanced infrastructure and good across a vast range of sectors like GDP per capita, literacy rate, life expectancy, political stability, and access to electricity. Then all of these things are put into a single score called HDI.

In our analysis, we further refined our sample based on two criteria. For the number of cases, we limited the sample to weeks following the first week with at least seven newly reported cases. This sample restriction, while not strictly necessary, was essential for facilitating the convergence of the stochastic frontier models. This approach also aligns with the common practice of discussing a country's COVID-19 trajectory after surpassing certain initial milestones. However, it's worth noting that our estimates of underreporting may be conservative, as they do not account for missed cases.

With our sample in hand, our dependent variables comprise new cases on a weekly basis. Independent variables encompass population size, population density, median age, the percentage of the population aged 65 and above, diabetes prevalence, and

[2] https://github.com/owid/covid-19-data/tree/master/public/data#the-complete-our-world-in-data-covid-19-dataset.

[3] https://worldpopulationreview.com/country-rankings/developed-countries.

**Table 1** Countries included in sample

| Country name | | |
|---|---|---|
| United Arab Emirates | Australia | Austria |
| Belgium | Bahrain | Belarus |
| Canada | Switzerland | Czech Republic |
| Germany | Denmark | Spain |
| Estonia | Finland | France |
| United Kingdom | Georgia | Greece |
| Croatia | Ireland | Iceland |
| Israel | Italy | Japan |
| South Korea | Kuwait | Lithuania |
| Luxembourg | Malaysia | Netherlands |
| Norway | New Zealand | Oman |
| Qatar | Romania | Russia |
| Singapore | Sweden | Thailand |
| United States | | |

per capita gross domestic product, all obtained from the same source as the COVID-19 data. Additionally, we included Transparency International's 2020 Corruption Perception Index as a country-level measure of corruption.[4]

Furthermore, the models incorporate fixed effects for various global regions, which encompass East Asia and the Pacific, Europe and Central Asia, Latin America and the Caribbean, the Middle East and Africa, North America, and South Asia. Additionally, the models include region-specific linear time trends. These time trends are measured in relation to the duration of data collection within the sample, specifically in terms of weeks following the initial week with reported cases surpassing a predefined threshold. Notably, these trends are not aligned with conventional calendar time. Furthermore, we include regions fixed effects (East Asia, and the Pacific, Europe and Central Asia, Latin America and the Caribbean, Middle East and Africa, North America, and South Asia) in our calculation.

Finally, we also include in our independent variables data on new tests which are measured in different units across countries (people tested, samples tested, tests conducted, or unclear). When modeling new cases we consider the average of the index computed over the third and fourth lags.

[4] https://www.transparency.org/en/cpi/2020.

**Table 2** Stochastic frontier results

| | Estimate | Std. Error |
|---|---|---|
| Log population | 0.490*** | 0.053 |
| Population density | 0.035 | 0.044 |
| Median Age | 25.385** | 7.843 |
| Percent of population Age 65 and Older | 0.060** | 0.020 |
| Percent of population with Diabetes | 0.027 | 0.023 |
| Log per capita GDP | 0.242 | 0.130 |
| Region | 1.086*** | 0.084 |
| Country Corruption Index | 0.054*** | 0.003 |
| New $\text{Tests}_{t-1}$ | −0.022 | 0.131 |
| $(\text{New Test}_{t-1})^2$ | −0.007 | 0.010 |
| $(\text{New Tests}_{t-1}) * (\text{Linear Time Trend})$ | 0.006 | 0.003 |
| $(\text{New Tests}_{t-1} * (\text{Test Units} = \text{People}))$ | −0.044 | 0.089 |
| $(\text{New Tests}_{t-1} * (\text{Test Units} = \text{People})) * (\text{Linear Time Trend})$ | −0.001 | 0.003 |
| $(\text{New Tests}_{t-1} * (\text{Test Units} = \text{Samples}))$ | −0.338 | 0.229 |
| $(\text{New Tests}_{t-1} * (\text{Test Units} = \text{Samples}))^2$ | 0.012 | 0.023 |
| $(\text{New Tests}_{t-1} * (\text{Test Units} = \text{Samples})) * (\text{Linear Time Trend})$ | 0.002 | 0.003 |
| $(\text{New Tests}_{t-1} * (\text{Test Units} = \text{Tests}))$ | −0.420*** | 0.127 |
| $(\text{New Tests}_{t-1} * (\text{Test Units} = \text{Tests}))^2$ | 0.034*** | 0.010 |
| $(\text{New Tests}_{t-1} * (\text{Test Units} = \text{Tests})) * (\text{Linear Time Trend})$ | −0.001 | 0.003 |
| sigmaSq | 5.207*** | 0.529 |
| gamma | 0.723*** | 0.070 |

## 6 Results

- Model: COVID-19 cases using SIR with Stochastic frontier models assuming independent underreporting.

Table 3 gives the multiplication factor for the many of the countries. For this model, we make distributional assumptions for $v_{it}^q$ and $u_{it}^q$, $q \in Y$, using the same approach as Askitas et al. [15] and estimate the maximum likelihood where the dependent variables are to be transformed using the inverse hyperbolic sine transformation (IHS) to take care of the many zeros we have in our data. This will make the coefficients estimates consistent regardless of the distributional assumptions Papadopoulos, and Parameter [18]. Our estimation is performed in R using linear trend and SFA. Table 2 gives the maximum likelihood estimates for the coefficient of the dependent variables. From this, we can see that the dependent variables population, population density, people with medium age and 65 or older, region specific, country corruption.

Overall, with the Half normal model specification, we can conclude that from January 1, 2020 to November 3, 2020, the total number of cases reported for the United States was underreported by 2.56 times. The model's estimations can also

**Table 3** Country-specific multiplication factors

| Country name | MF | Country name | MF |
|---|---|---|---|
| United Arab Emirates | 3.00 | Iceland | 3.63 |
| Australia | 3.87 | Israel | 2.97 |
| Austria | 3.30 | Italy | 3.37 |
| Belgium | 2.90 | Japan | 3.02 |
| Bahrain | 3.07 | South Korea | 2.95 |
| Belarus | 2.31 | Kuwait | 2.93 |
| Canada | 4.13 | Lithuania | 3.38 |
| Switzerland | 2.93 | Luxembourg | 3.08 |
| Czech Republic | 2.98 | Malaysia | 2.96 |
| Germany | 3.30 | Netherlands | 2.87 |
| Denmark | 3.46 | Norway | 3.11 |
| Spain | 2.77 | New Zealand | 4.55 |
| Estonia | 3.21 | Oman | 3.23 |
| Finland | 3.01 | Qatar | 2.92 |
| France | 3.11 | Romania | 3.04 |
| United Kingdom | 3.15 | Russia | 2.92 |
| Georgia | 4.29 | Singapore | 3.27 |
| Greece | 3.29 | Sweden | 2.45 |
| Croatia | 4.25 | Thailand | 5.01 |
| Ireland | 3.33 | United States | 3.40 |

serve as a tool for evaluating variations in pandemic experiences among different countries. To initiate this assessment, Table 2 illustrates the country-specific MFs for number of cases derived from each specification across the entire sample period. Several notable observations emerge. Initially, the rankings of data accuracy among the various models exhibit a high degree of correlation. The MFs generated assuming independence range from 2.31 for the country of Belarus to 5.1 for the country of Thailand.

In this model we assume that the underreporting is independent of the time. Table 2 gives the multiplication factor for the many of the countries. For this model, we make distributional assumptions for $v^q_{it}$ and $u^q_{it}$, $q \in Y$, using the same approach as Askitas et al. [15] and estimate the maximum likelihood where the dependent variables are to be transformed using the inverse hyperbolic sine transformation (IHS) to take care of the many zeros we have in our data. This will make the coefficients estimates consistent regardless of the distributional assumptions Papadopoulos, and Parameter [18]. Our estimation is performed in R using linear trend and SFA. Most of our multiplication factor (MF) ranges from 1.96 for Belarus to 2.56 for the United States. Overall, with the Half normal model specification, we can conclude that from

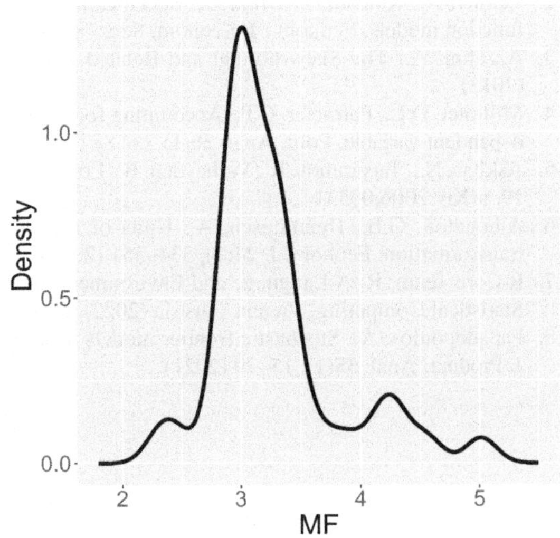


**Fig. 1** Kernel density of multiplication factors for countries

January 1, 2020 to November 3, 2020, the total number of cases reported for the United States was underreported by 2.56 times (Fig. 1).

## References

1. Organization, W.H. et al.: Covid-19 weekly epidemiological update. Accessed from 3 November 2020
2. Manski, C.F., Molinari, F.: Estimating the covid-19 infection rate: Anatomy of an inference problem. J. Econom. **220**(1), 181–192 (2021)
3. Reese, H., Iuliano, A.D., Patel, N.N., Garg, S., Kim, L., Silk, B.J., Hall, A.J., Fry, A., Reed, C.: Estimated incidence of coronavirus disease 2019 (covid-19) illness and hospitalization-united states, february-september 2020. Clin. Infect. Dis. **72**(12), e1010–e1017 (2021)
4. Greer, S.L., King, E.J., da Fonseca, E.M., Peralta-Santos, A.: The comparative politics of covid-19: the need to understand government responses. Global Public Health **15**(9), 1413–1416 (2020)
5. McCulloh, I., Kiernan, K., Kent, T.: Inferring true covid19 infection rates from deaths. Front. Big Data **3**, 565589 (2020)
6. Dubrow, J.K.: Local data and upstream reporting as sources of error in the administrative data undercount of covid 19. Int. J. Soc. Res. Methodol. **25**(4), 471–476 (2022)
7. Murray, J.D.: An Introduction. Springer (2002)
8. Millimet, D.L., Parmeter, C.F.: Covid-19 severity: a new approach to quantifying global cases and deaths. J. Roy. Stat. Soc. Ser. A: Stat. Soc. **185**(3), 1178–1215 (2022)
9. Arroyo-Marioli, F., Bullano, F., Kucinskas, S., Rondón-Moreno, C.: Tracking r of covid-19: A new real-time estimation using the kalman filter. PloS one **16**(1), e0244474 (2021)
10. Aigner, D., Lovell, C.K., Schmidt, P.: Formulation and estimation of stochastic frontier production function models. J. Econom. **6**(1), 21–37 (1977)
11. Coelli, T., Henningsen, A., Henningsen, M.A.: Package 'frontier'. Tech. rep, Technical Report (2020)

12. Zellner, A., Kmenta, J., Dreze, J.: Specification and estimation of cobb-douglas production function models. Econom.: J. Econom. Soc. 784–795 (1966)
13. Azzalini, A.: The Skew-normal and Related Families, vol. 3. Cambridge University Press (2013)
14. Millimet, D.L., Parmeter, C.F.: Accounting for skewed or one-sided measurement error in the dependent variable. Polit. Anal. **30**(1), 66–88 (2022)
15. Askitas, N., Tatsiramos, K., Verheyden, B.: Lockdown strategies, mobility patterns and covid-19. arXiv:2006.00531
16. Aihounton, G.B., Henningsen, A.: Units of measurement and the inverse hyperbolic sine transformation. Econom. J. **24**(2), 334–351 (2021)
17. R Core Team: R: A Language and Environment for Statistical Computing, R Foundation for Statistical Computing, Vienna, Austria (2022). https://www.R-project.org/
18. Papadopoulos, A.: Stochastic frontier models using the generalized exponential distribution. J. Product. Anal. **55**(1), 15–29 (2021)

# AI and Econometric Modeling: Deep Reinforcement Learning in Predictive Modeling

**Do Huu Hai and Pham Van Tuan**

**Abstract** Artificial intelligence (AI) has significantly impacted many different industries, including finance and economics. These technologies are increasingly being used to improve economic forecasting and analysis, providing more accurate predictions and better decision-making. The article delves into the expanding use of artificial intelligence (AI) in economic forecasting and analysis, including subjects such as big data, predictive analytics, and econometrics. It will also examine the problems of using AI in econometrics research, as well as the future of these technologies. Traditional forecasting approaches rely on econometric models for economic forecasting and analysis. These models evaluate correlations between economic indicators such as GDP, inflation, and unemployment using statistical approaches and mathematical equations. However, these classic models have limitations in that they may not be able to explain more complex interactions between economic variables. Furthermore, they require a lot of data as well as certain assumptions about the structure of the model, which can affect their ability to adapt to changing economic conditions. The use of AI and Machine Learning (ML) in economics can help overcome these problems and improve the accuracy of predictions about future trends.

**Keywords** AI · Big data · Econometric · Digital technologies · Econometric · Machine learning

---

D. H. Hai (✉)
Viet Hung Industrial University, Ha Noi City, Vietnam
e-mail: haidh1975@gmail.com

P. Van Tuan
National Economics University (NEU), Hanoi, Vietnam

 
V. Kreinovich et al. (eds.), *Applications of Optimal Transport to Economics and Related Topics*, Studies in Systems, Decision and Control 556,
https://doi.org/10.1007/978-3-031-67770-0_5

# 1 Econometrics, Big Data and Deep Reinforcement Machine Learning Algorithms

The term "big data" refers to enormous and complicated data collections that are challenging to process and analyze using traditional methods. Government statistics, financial records, and social media trends are all important data sources. Predictive analytics is a technique for identifying patterns in data and forecasting future consequences. There are numerous advantages to employing big data and predictive analytics for economic analysis. It offers more accurate and timely forecasting of economic trends, for example, by swiftly spotting correlations that may not be visible using standard methods. Furthermore, predictive analytics provides insight into possible economic risks and possibilities, enabling policymakers to make informed decisions. These techniques have been applied in many different aspects of econometrics, such as predicting stock prices, digital currencies, measuring consumer demand, and recognizing economic cycles. They have also proven useful during recessionary economic cycles, or more recently the COVID-19 pandemic, assisting policymakers in promoting practical relief efforts or stimulus packages.

Because of their tremendous ability to process and analyze enormous data sets, machine learning algorithms have grown in popularity in recent years. These algorithms are designed to find important patterns and relationships in data that typical econometric models that rely on a priori assumptions about associations between variables. Deep reinforcement learning models, in addition to their ability to handle nonlinear data correlations, can be utilized for a number of economic research and forecasting tasks, such as predicting stock prices, estimating growth growth, and estimating the inflation rate. Algorithms can also be used to better understand consumer behavior and economic trends by applying them to big data sets such as social media and search engine data. Although deep reinforcement machine learning algorithms have numerous advantages over classic econometric models, they do have certain drawbacks, including the requirement for high-quality training data and problems in interpreting models and algorithms. As a result, when employed in econometric analysis, they should be handled with caution.

# 2 Use of Artificial Intelligence and Deep Machine Learning in Economic Forecasting

The applications of AI and augmented deep learning in economic forecasting and analysis are numerous. Here are some examples of how these technologies can be put to use:

Predict economic indicators—Artificial intelligence and deep machine learning algorithms can identify trends in past data, allowing for more accurate predictions of GDP, inflation, and unemployment rates.

Financial Market Analysis—Augmented artificial intelligence and deep machine learning algorithms can sift through financial market data to forecast stock prices, currency rates, and other indicators.

Risk management—In financial institutions, algorithms can predict potential economic risks such as a recession or financial crisis, as well as credit and market hazards.

Understand consumer behavior—AI algorithms and augmented deep machine learning can provide insight into customer spending habits and preferences, which can help with pricing, marketing, and design decisions. Product design for businesses.

Artificial intelligence text processing and augmented deep learning can be used to discover trends in news stories or social media posts, assisting in gauging public opinion on economic concerns.

Supply Chain Optimization—These technologies enable supply chain processes to be optimized, lowering costs while increasing efficiency.

The application of artificial intelligence and augmented deep learning for economic forecasting has brought much-needed insights into complicated linkages while keeping potential limits in mind, such as data quality or capacity interpretation and analysis.

Here are some examples of how these technologies can be put to use:

Creating a model of the economy using AI involves a multidisciplinary approach, combining economic theory, data analysis, and machine learning techniques. Here's a general guide to help get started [3]: 1. Gather Data: Collect relevant economic data from reliable sources. This could include GDP, inflation rates, unemployment rates, interest rates, stock market data, and more. Ensure that the data is clean, consistent, and covers the time period of interest; 2. Clean and preprocess the data. Handle missing values, outliers, and ensure that the data is in a format suitable for analysis. This may involve transforming time series data, normalizing variables, etc.; 3. Identify relevant features that may impact the economy. This could include economic indicators, policy variables, global events, etc. Create new features if necessary; 4. Choose a Modeling Approach: Decide on the type of model want to use. Common approaches include traditional statistical models (regression, time series analysis) and machine learning models (neural networks, decision trees, etc.). The choice depends on the complexity of your problem and the nature of data; 5. Train the Model: Split your data into training and testing sets. Train your model on the training data, adjusting parameters as needed. Evaluate the model on the testing set to ensure it generalizes well to new data; 6. Interpret Results: Understand the insights gained from your model. Identify the key factors influencing economic outcomes and assess the model's performance; 7. Iterate and Refine: Based on the results, iterate and refine your model. This may involve tweaking parameters, adding new features, or exploring different modeling approaches; 8. Deploy and Monitor: If your model performs well, consider deploying it for ongoing analysis. Monitor its performance over time and update it as needed.

We have the data and use OLS and Python to process the data:

| GDP | Investment | Consumption | Unemployment |
|---|---|---|---|
| 5.2 | 2 | 3 | 8 |
| 6 | 2.5 | 3.5 | 7.5 |
| 7.5 | 3 | 4 | 7 |
| 8.2 | 3.5 | 4.2 | 6.5 |
| 9 | 4 | 4.5 | 6 |

```
# Import necessary libraries
import pandas as pd
import statsmodels.api as sm
# Create a hypothetical dataset
data = {
    'GDP': [5.2, 6.0, 7.5, 8.2, 9.0],
    'Investment': [2.0, 2.5, 3.0, 3.5, 4.0],
    'Consumption': [3.0, 3.5, 4.0, 4.2, 4.5],
    'Unemployment': [8.0, 7.5, 7.0, 6.5, 6.0]
}
df = pd.DataFrame(data)
# Add a constant term for the intercept
X = sm.add_constant(df[['Investment', 'Consumption', 'Unemployment']])
y = df['GDP']
# Fit the linear regression model
model = sm.OLS(y, X).fit()
# Print the model summary
print(model.summary())
```

Get results:

```
OLS Regression Results
==============================================================================
Dep. Variable:                    GDP   R-squared:                       0.991
Model:                            OLS   Adj. R-squared:                  0.982
Method:                 Least Squares   F-statistic:                     112.9
Date:                Fri, 17 Nov 2023   Prob (F-statistic):            0.00878
Time:                        19:44:48   Log-Likelihood:                  3.069
No. Observations:                   5  AIC:                            -0.1384
Df Residuals:                       2  BIC:                             -1.310
Df Model:                           2
Covariance Type:            nonrobust
==============================================================================
                 coef    std err          t      P>|t|      [0.025      0.975]
------------------------------------------------------------------------------
const          0.0769      0.104      0.738      0.538      -0.372       0.526
Investment     0.8657      0.895      0.967      0.436      -2.985       4.717
Consumption    1.3488      0.999      1.351      0.309      -2.948       5.646
Unemployment-0.0962      0.157     -0.614      0.602      -0.771       0.578
==============================================================================
Omnibus:                          nan   Durbin-Watson:                   2.973
Prob(Omnibus):                    nan   Jarque-Bera (JB):                0.876
Skew:                          -0.991   Prob(JB):                        0.645
Kurtosis:                       2.477   Cond. No.                     1.67e+17
==============================================================================
```

In this example, we have a simple dataset with GDP as the dependent variable and Investment, Consumption, and Unemployment as independent variables. The code uses the Ordinary Least Squares (OLS) method to fit a linear regression model.

In case the number of observations is increased to n = 20, the results will be given:

```
# Import necessary libraries
import pandas as pd
import statsmodels.api as sm
import numpy as np
# Generate a larger hypothetical dataset
np.random.seed(42)  # for reproducibility
n_obs = 20
data = {
    'GDP': np.random.uniform(5, 10, n_obs),
    'Investment': np.random.uniform(2, 5, n_obs),
    'Consumption': np.random.uniform(3, 6, n_obs),
    'Unemployment': np.random.uniform(5, 10, n_obs)
```

```
}
df = pd.DataFrame(data)
# Add a constant term for the intercept
X = sm.add_constant(df[['Investment', 'Consumption', 'Unemployment']])
y = df['GDP']
# Fit the linear regression model
model = sm.OLS(y, X).fit()
# Print the model summary
print(model.summary())
```

Get results:

```
OLS Regression Results
==============================================================================
Dep. Variable:                    GDP   R-squared:                       0.370
Model:                            OLS   Adj. R-squared:                  0.252
Method:                 Least Squares   F-statistic:                     3.133
Date:                Fri, 17 Nov 2023   Prob (F-statistic):             0.0548
Time:                        19:45:48   Log-Likelihood:                -31.860
No. Observations:                  20   AIC:                             71.72
Df Residuals:                      16   BIC:                             75.70
Df Model:                           3
Covariance Type:            nonrobust
==============================================================================
                 coef    std err          t      P>|t|      [0.025      0.975]
------------------------------------------------------------------------------
const          9.6625      2.022      4.778      0.000       5.375      13.950
Investment    -1.1530      0.392     -2.941      0.010      -1.984      -0.322
Consumption    0.2302      0.322      0.715      0.485      -0.452       0.913
Unemployment0.0652      0.209      0.311      0.760      -0.379       0.509
==============================================================================
Omnibus:                        0.676   Durbin-Watson:                   2.080
Prob(Omnibus):                  0.713   Jarque-Bera (JB):                0.532
Skew:                          -0.363   Prob(JB):                        0.766
Kurtosis:                       2.666   Cond. No.                         64.1
==============================================================================
```

We can adjust the size of dataset according to needs. Larger datasets generally provide more reliable estimates and better represent the underlying population, but keep in mind that the quality of the data is crucial for meaningful results. If we have access to real-world economic data, we can replace the generated dataset with actual data for more meaningful analysis.

## 3 Difficulties and Constraints

While the application of AI and augmented deep learning in forecasting and econometric analysis can be advantageous, there are a number of obstacles and constraints to consider.

The accuracy of AI and augmented deep learning models is strongly dependent on the quality of the data used to train them. If the data is missing, incorrect, or biased, the results may not adequately reflect the underlying factors between variables.

In contrast to econometric models, which are built on known economic links, AI and advanced deep learning algorithms can detect patterns that are not immediately evident. This makes understanding what drives predictions challenging, as well as detecting potential flaws or biases in the model. Detail as (1) Overfitting: If an algorithm is trained with a data set that is too small or specific, it will likely incorrectly target the sample rather than actual samples, leading to inaccurate predictions and limits its general applicability to other data sets; (2) Box models: Certain AI and deep reinforcement learning models offer limited explanations for how they achieve their results; this limits their use as interpretability can be very important in specific applications; (3) Human bias: If the data used to train AI and deep machine learning algorithms contains human bias (due to the decisions of individuals who created the data), these algorithms may perpetuate that bias, leading to non-significant results; (4) Technological constraints: The use of AI and deep machine learning necessitates large computational resources and knowledge, which can limit access for organizations lacking these resources.

When applying these tools for economic forecasting and analysis, various problems and constraints must be considered. Consideration of these factors can assist guarantee that the models generated are dependable, free of bias, and suitable for decision-making.

## 4 Conclusion

AI and augmented deep learning have a promising future in forecasting and econometric analysis, with numerous chances for innovation and improvement. Here are several implications and developments to keep an eye on: (1) Real-time data will become more common: With the availability of real-time data, augmented AI and deep machine learning algorithms can be employed more successfully in economic forecasting and analysis, allowing for more accurate forecasts and faster response to changes in economic situations; (2) Improved interpretability: Research is being conducted on approaches to make augmented AI and deep machine learning algorithms easier to interpret; this can boost confidence in these models and make them more effective in decision making; (3) Natural language processing advances: As natural language processing improves, it will be possible to swiftly evaluate text data and discover economic trends; (4) Other technologies integration: By merging AI and augmented deep machine learning, blockchain technology, and IoT (Internet of Things), new economic research and forecasting prospects may emerge. Collecting real-time data from IoT devices, for example, can lead to better predictions of future financial situations; (5) Improved collaboration between academia and business: Academics will collaborate closely with businesses and policymakers to develop and deploy new types of artificial intelligence and deep reinforcement learning approaches to economic forecasting and analysis.

The future of AI & augmented deep machine learning in econometrics is likely to be characterized by increased data availability, improved interpretability, and better collaboration between academia and business, which will provide greater accuracy when predicting economic trends and present new possibilities—both business-wise and policy-wise—when it comes time to make informed decisions.

Because the obstacles that must be overcome, the future of AI and augmented deep machine learning in forecasting and econometric analysis appears bright. We expect economic predictions to become more accurate and valuable when new methodologies are developed and linked with other fields of research. Businesses and policymakers should actively follow these developments, assessing the possible benefits and risks when making future decisions.

## References

1. $H_2O$ AutoML: Automatic Machine Learning. https://docs.h2o.ai/h2o/latest-stable/h2o-docs/automl.html; https://maseconomics.com/
2. Hai, D.H.: Corporate Culture. National Economics University Press (2019)
3. Hai, D.H.: Statistical Probability and Quantitative Methods in Economics. Hanoi Labor Publishing House (2018)

# Multivariate Quantiles: Geometric and Measure-Transportation-Based Contours

**Marc Hallin and Dimitri Konen**

**Abstract** Quantiles are a fundamental concept in probability and theoretical statistics and a daily tool in their applications. While the univariate concept of quantiles is quite clear and well understood, its multivariate extension is more problematic. After half a century of continued efforts and many proposals, two concepts, essentially, are emerging: the so-called *(relabeled) geometric quantiles*, extending the characterization of univariate quantiles as minimizers of an $L_1$ loss function involving the *check functions*, and the more recent *center-outward quantiles* based on measure transportation ideas. These two concepts yield distinct families of *quantile regions* and *quantile contours*. Our objective here is to present a comparison of their main theoretical properties and a numerical investigation of their differences.

## 1 Introduction

Quantiles are a fundamental concept in probability and theoretical statistics and a familiar tool in their applications, ranging from descriptive statistics and data analysis to statistical inference. The univariate concept of a quantile function (the inverse of a distribution function) is well understood and well studied in dimension $d = 1$. Despite half a century of efforts, however, the multivariate extension (dimension $d \geq 2$) of this essential notion remains quite problematic—the obvious difficulty being the absence of a canonical ordering of the real space for dimension two and higher.

M. Hallin (✉)
Université libre de Bruxelles, Brussels, Belgium
e-mail: mhallin@ulb.be

Czech Academy of Sciences, Prague, Czech Republic

D. Konen
University of Cambridge, Coventry, UK
e-mail: dk738@cam.ac.uk

 
V. Kreinovich et al. (eds.), *Applications of Optimal Transport to Economics and Related Topics*, Studies in Systems, Decision and Control 556,
https://doi.org/10.1007/978-3-031-67770-0_6

Various proposals have been made in the literature, none of which is fully agreed upon. Two concepts, essentially, are emerging from these many attempts: the so-called *(relabeled) geometric quantiles* [2], extending the characterization of univariate quantiles as minimizers of an $L_1$ loss function involving the *check functions*, and the more recent *center-outward quantiles* [7] based on measure transportation ideas.

The objective of this note is a brief comparison of some of the properties of the quantile regions and quantile contours associated with these two concepts.

## 2 Quantiles: From Univariate to Multivariate

Throughout, the inner product of two vectors $x, y \in \mathbb{R}^d$ is denoted as $x'y$; $\mathbb{B}_d$ and $\mathbb{S}^{d-1}$ stand for the open unit ball and the $(d-1)$-dimensional unit sphere, respectively, in $\mathbb{R}^d$.

### 2.1 *Definitions and Basic Properties*

Let $X$ denote a Lebesgue-absolutely continuous real-valued random variable with distribution P over $\mathbb{R}$ and distribution function

$$F_{\mathrm{P}} : x \mapsto F_{\mathrm{P}}(x) := \mathrm{P}(X \leq x).$$

To simplify the exposition, assume that $X$ admits a nonvanishing density $f := \mathrm{dP}/\mathrm{d}\lambda$ with respect to the Lebesgue measure $\lambda$ so that, in particular, $F_{\mathrm{P}}$ is continuous and strictly monotone increasing, hence invertible. The quantile function of P (equivalently, the quantile function of $X \sim \mathrm{P}$) then is defined as the inverse $Q_{\mathrm{P}} := F_{\mathrm{P}}^{-1}$ of the distribution function $F_{\mathrm{P}}$ and the quantile of order $\alpha \in (0, 1)$ of P (of $X \sim \mathrm{P}$) is the value $Q_{\mathrm{P}}(\alpha)$ of $Q_{\mathrm{P}}$ at $\alpha \in (0, 1)$.

Along with quantiles of given order $\alpha \in (0, 1)$, it is often convenient to consider central quantile regions and contours. In dimension $d = 1$, call *quantile region of order* $\tau \in [0, 1)$ *of* P (of $X \sim \mathrm{P}$) the closed interval

$$\mathscr{R}_{\mathrm{P}}(\tau) := \left[Q_{\mathrm{P}}\Big(\frac{1-\tau}{2}\Big), Q_{\mathrm{P}}\Big(\frac{1+\tau}{2}\Big)\right]$$

(an interquantile interval) and *quantile contour of order* $\tau$ *of* P (of $X \sim \mathrm{P}$) the boundary

$$\mathscr{C}_{\mathrm{P}}(\tau) := \left\{Q_{\mathrm{P}}\Big(\frac{1-\tau}{2}\Big), Q_{\mathrm{P}}\Big(\frac{1+\tau}{2}\Big)\right\}$$

of $\mathscr{R}_{\mathrm{P}}(\tau)$.

The quantile region of order 0 is the degenerate interval $\mathscr{R}_{\mathrm{P}}(0) = \{Q_{\mathrm{P}}(1/2)\}$ and has Lebesgue measure zero: call it the *median region*. Similarly, the quantile region of order 1/2 is the traditional interquartile interval: $\mathscr{R}_{\mathrm{P}}(1/2) = [Q_{\mathrm{P}}(1/4), Q_{\mathrm{P}}(3/4)]$. Contrary to the quantile halflines $(-\infty, Q_{\mathrm{P}}(\alpha)]$, which very much depend on the orientation of the real line (an orientation that is no longer meaningful in $\mathbb{R}^d$ for $d \geq 2$), the concepts of quantile regions and contours of order $\tau$ are *"center-outward"* concepts (the "center" being the median $Q_{\mathrm{P}}(1/2)$) and are invariant under a change of the orientation of $\mathbb{R}$ (an orthogonal transformation). The collection $\{\mathscr{R}_{\mathrm{P}}(\tau) : \tau \in [0, 1)\}$ of quantile regions is stricly nested (in the sense of set inclusion) as $\tau$ ranges from zero to one, with $\mathscr{R}_{\mathrm{P}}(0) = \bigcap_{\tau \in [0,1)} \mathscr{R}_{\mathrm{P}}(\tau)$. It is easily seen that $\mathscr{C}_{\mathrm{P}}(|2F_{\mathrm{P}}(x) - 1|)$ is the unique quantile contour running through $x \in \mathbb{R}$; actually,

$$\mathscr{R}_{\mathrm{P}}(\tau) = \{x : |2F_{\mathrm{P}}(x) - 1| \leq \tau\} \text{ and } \mathscr{C}_{\mathrm{P}}(\tau) = \{x : |2F_{\mathrm{P}}(x) - 1| = \tau\} \quad \tau \in [0, 1). \quad (1)$$

An essential property of a quantile function is that the P-probability content of a quantile region $\mathscr{R}_{\mathrm{P}}(\tau)$ is $\tau$ irrespective of P: namely, $\mathrm{P}(\mathscr{R}_{\mathrm{P}}(\tau)) = \tau$ for all P. What would be the relevance, indeed, of an interquartile region $\mathscr{R}_{\mathrm{P}}(1/2)$ with probability content 0.5 under $\mathrm{P} = \mathrm{P}_1$ (as should be) but probability content 0.6 under $\mathrm{P} = \mathrm{P}_2$, and probability content 0.7 under $\mathrm{P} = \mathrm{P}_3$? This property, in dimension $d = 1$, is equivalent to the property of $F_{\mathrm{P}}(X)$ being uniform over [0, 1) for $X \sim \mathrm{P}$.

When the center-outward quantile regions $\mathscr{R}_{\mathrm{P}}(\tau)$ and contours $\mathscr{C}_{\mathrm{P}}(\tau)$ are the main points of interest, it is natural to replace $F_{\mathrm{P}}$ and $Q_{\mathrm{P}}$ by center-outward counterparts: call $F^{\pm}_{\mathrm{P}} := 2F_{\mathrm{P}} - 1$ and $Q^{\pm}_{\mathrm{P}} := (F^{\pm}_{\mathrm{P}})^{-1}$ the *center-outward distribution* and *quantile functions* of P (or $X \sim \mathrm{P}$), respectively. Just like $F_{\mathrm{P}}$ and $Q_{\mathrm{P}}$, the maps $F^{\pm}_{\mathrm{P}}$ and $Q^{\pm}_{\mathrm{P}}$ are monotone increasing and carry the same information about P, which they fully characterize. Instead of being uniformly distributed over [0, 1), $F^{\pm}_{\mathrm{P}}(X)$, when $X \sim \mathrm{P}$, is uniform over $(-1, 1)$, which is the open unit ball $\mathbb{B}_1$ in $\mathbb{R}$. In fact, denoting by $\mathrm{U}_1$ the uniform distribution over $\mathbb{B}_1$, we have $F^{\pm}_{\mathrm{P}}(X) \sim U_1$ when $X \sim \mathrm{P}$, hence $Q^{\pm}_{\mathrm{P}}(U) \sim \mathrm{P}$ if $U \sim \mathrm{U}_1$. In the terminology of measure transportation, we say that the *transport* $F^{\pm}_{\mathrm{P}}$ is *pushing* P *forward to* $\mathrm{U}_1$ and that its inverse $Q^{\pm}_{\mathrm{P}}$ is *pushing* $\mathrm{U}_1$ *forward to* P, which we conveniently denote as

$$F^{\pm}_{\mathrm{P}}\#\mathrm{P} = \mathrm{U}_1 \quad \text{and} \quad Q^{\pm}_{\mathrm{P}}\#\mathrm{U}_1 = \mathrm{P},$$

respectively. The quantile region $\mathscr{R}_{\mathrm{P}}(\tau)$ and quantile contour $\mathscr{C}_{\mathrm{P}}(\tau)$ then take the simple forms (equivalent to (1))

$$\mathscr{R}_{\mathrm{P}}(\tau) = Q^{\pm}_{\mathrm{P}}(\tau\mathbb{B}_1) \quad \text{and} \quad \mathscr{C}_{\mathrm{P}}(\tau) = Q^{\pm}_{\mathrm{P}}(\tau\mathbb{S}^0), \quad \tau \in [0, 1), \quad (2)$$

respectively, where $\mathbb{S}^0 := \{-1, 1\}$ stands for the (0-dimensional) unit sphere in $\mathbb{R}$.

### 2.2 Characterization of Univariate Quantiles as Minimizers of Expected Check Functions

When P is a probability measure over $\mathbb{R}$ and $X \sim \mathrm{P}$, and given a level $\alpha \in [0, 1)$, a quantile of order $\alpha$ of $X$, by definition, is any $x \in \mathbb{R}$ satisfying

$$\mathrm{P}(X < x) \leq \alpha \leq \mathrm{P}(X \leq x).$$

It turns out that such $x$'s are exactly the minimizers (with respect to $\theta \in \mathbb{R}$) of the objective function

$$\theta \mapsto M_\alpha^{\mathrm{P}}(\theta) := \mathrm{E}_{\mathrm{P}}[\rho_\alpha(\theta - X) - \rho_\alpha(X)] \tag{3}$$

where, denoting by $z^+$ and $z^-$ the positive and negative parts, respectively, of $z \in \mathbb{R}$,

$$\rho_\alpha : z \mapsto \rho_\alpha(z) := |z| + (2\alpha - 1)z = 2\big\{\alpha z^+ + (1-\alpha)z^-\big\}, \quad z \in \mathbb{R},\ \alpha \in [0, 1)$$

stands for the so-called *check function*. Noting that $M_\alpha^{\mathrm{P}}(\theta)$ rewrites as

$$M_\alpha^{\mathrm{P}}(\theta) = \mathrm{E}_{\mathrm{P}}\big[|\theta - X| - |X|\big] - (2\alpha - 1)\theta \tag{4}$$

for all $\theta \in \mathbb{R}$, it follows from the triangular inequality that $\big||X - \theta| - |X|\big| \leq |\theta|$, which is integrable. The objective function $M_\alpha^{\mathrm{P}}$, hence the correspondence between quantiles and the minimizing procedure in (3), thus remains well defined for an *arbitrary* probability measure P, without any moment assumption. In this general setting, we have, even in the absence of a finite moment of order one,

$$Q_{\mathrm{P}}(\alpha) = \mathrm{argmin}_{\theta \in \mathbb{R}} M_\alpha^{\mathrm{P}}(\theta), \qquad \alpha \in [0, 1).$$

For $\alpha = 0.5$, $M_\alpha^{\mathrm{P}}(\theta) = \mathrm{E}_{\mathrm{P}}[|\theta - X| - |X|]$; in case $\mathrm{E}_{\mathrm{P}}(X) < \infty$, thus, we recover the familiar characterization of the median $Q_{\mathrm{P}}(1/2)$ as an $\mathrm{L}_1$ location parameter, the minimizer of the expected absolute deviation $\mathrm{E}_{\mathrm{P}}[|\theta - X|]$.

Recalling that $F_{\mathrm{P}}^{\pm} := 2F_{\mathrm{P}} - 1$, we see that the natural labeling for center-outward quantiles is provided by $\beta := 2\alpha - 1 \in (-1, 1)$. Therefore, the center-outward version of the characterization of quantiles as minimizers is obtained by relabeling $2\alpha - 1$ as $\beta$ in (3). This leads to the characterization of center-outward quantiles $Q_{\mathrm{P}}^{\pm}(\beta)$ as minimizers with respect to $\theta \in \mathbb{R}$ of the objective function

$$\theta \mapsto O_\beta^{\mathrm{P}}(\theta) := \mathrm{E}_{\mathrm{P}}\big[|\theta - X| - |X|\big] - \beta\theta, \qquad \beta \in (-1, 1). \tag{5}$$

We then have (still, without any moment assumption)

$$Q_{\mathrm{P}}^{\pm}(\beta) = \mathrm{argmin}_{\theta \in \mathbb{R}} O_\beta^{\mathrm{P}}(\theta), \qquad \beta \in (-1, 1).$$

The value $\beta = 0$ yields the median $Q_{\rm P}^{\pm}(0) = Q_{\rm P}(1/2)$ of P, while positive (resp. negative) values of $\beta$ are associated to quantiles sitting to the right-hand (resp. left-hand) side of the median.

### 2.3 *Characterization of Univariate Quantile Functions as Monotone Transports*

Consider a probability measure P with non-vanishing density $f$ with respect to the Lebesgue measure on $\mathbb{R}$. An alternative to the characterization, in dimension $d = 1$, of $Q_{\rm P}^{\pm}$ (hence of $Q_{\rm P}$ and $F_{\rm P}^{\pm}$) as minimizers of expected check functions is obtained by noting that $Q_{\rm P}^{\pm}$ is monotone increasing and pushes $\rm U_1$ forward to P. This fully characterizes $Q_{\rm P}^{\pm}$ up to a set of values with Lebesgue measure zero.[1] The center-outward quantile function $Q_{\rm P}^{\pm}$, thus, can be characterized as the almost everywhere unique monotone map from the open unit ball $\mathbb{B}_1 = (-1, 1)$ to $\mathbb{R}$ such that $Q_{\rm P}^{\pm}\#{\rm U}_1 = {\rm P}$.

## 3 Multivariate Quantile Functions, Regions, and Contours

Extending to dimension $d > 1$ the characterizations of quantile functions developed in Sect. 2.2 leads to the concept of *geometric quantiles* first proposed by [2] while the measure-transportation-based characterization developed in Sect. 2.3 leads to the concept of multivariate *center-outward quantiles* proposed in [7].

### 3.1 *Geometric Quantiles and Geometric Quantile Contours*

Consider a probability measure P supported on $\mathbb{R}^d$, with $d \geq 2$. Extending to a $d$-dimensional framework the characterization of univariate quantiles as minimizers of expected check functions is obtained by replacing, in (5), the absolute values with Euclidean norms and $\beta \in (-1, 1) = \mathbb{B}_1$ with $\mathbf{v} \in \mathbb{B}_d$. Letting $\mathbf{v} =: \tau\mathbf{u} \in \mathbb{B}_d$ with $\tau = \|\mathbf{v}\| \in (0, 1)$ and $\mathbf{u} = \mathbf{v}/\|\mathbf{v}\| \in \mathbb{S}^{d-1}$, define the *geometric quantile* of order $\tau > 0$ in direction $\mathbf{u} = \mathbf{v}/\|\mathbf{v}\|$ of P (of $\mathbf{X} \sim {\rm P}$) as an arbitrary minimizer, over $\mathbf{z} \in \mathbb{R}^d$, of the objective function

$$\mathbf{z} \mapsto O^{\rm P}_{\tau,\mathbf{u}}(\mathbf{z}) := {\rm E}_{\rm P}\big[\|\mathbf{z} - \mathbf{X}\| - \|\mathbf{X}\|\big] - \tau\mathbf{u}'\mathbf{z} \qquad (6)$$

[1] As we shall see, the existence and (almost everywhere) uniqueness, in dimension $d = 1$, of a monotone non-decreasing mapping (that is, the derivative, hence the gradient, of a convex function) pushing $\rm U_1$ forward to P is a particular case of a famous and more general measure transportation result by [13].

where $\mathbf{X} \sim \mathrm{P}$; throughout, boldface is used to stress that the variables take values in $\mathbb{R}^d$. The objective function $O^{\mathrm{P}}_{\tau,\mathbf{u}}$ being convex, it can be shown that the geometric quantiles of P are unique as soon as P is not supported on a single line of $\mathbb{R}^d$—a situation which essentially boils down to a univariate setting: see Theorem 1 in [15] for a discussion. For $\tau = 0$, define the geometric median as the minimizer of $\mathrm{E}_{\mathrm{P}}\big[\|\mathbf{z}-\mathbf{X}\| - \|\mathbf{X}\|\big]$ which, still in view of the triangular inequality, exists without any moment assumption and is better known as the *Fréchet median.*

When uniqueness holds (P not supported on a line), we denote by $\mathbf{Q}^{\mathrm{g}}_{\mathrm{P}}(\tau\mathbf{u})$ the geometric quantile of order $\tau \in [0, 1)$ in direction $\mathbf{u}$ of P (of $\mathbf{X} \sim \mathrm{P}$) and by $\mathbf{Q}^{\mathrm{g}}_{\mathrm{P}}(\mathbf{0})$ the geometric or Fréchet median, which can be interpreted as the geometric quantile of order $\tau = 0$ in any direction $\mathbf{u}$. Here again, boldface is used to stress that $\mathbf{Q}^{\pm}_{\mathrm{P}}$ takes values in $\mathbb{R}^d$.

Strongly related to the gradient of $O^{\mathrm{P}}_{\tau,\mathbf{u}}$, we define the *geometric distribution function* $\mathbf{F}^{\mathrm{g}}_{\mathrm{P}}$ of P as the mapping

$$\mathbf{z} \mapsto \mathbf{F}^{\mathrm{g}}_{\mathrm{P}}(\mathbf{z}) := \mathrm{E}_{\mathrm{P}}\Big[\frac{\mathbf{z}-\mathbf{X}}{\|\mathbf{z}-\mathbf{X}\|}\mathbb{1}[\mathbf{X} \neq \mathbf{z}]\Big], \qquad \mathbf{z} \in \mathbb{R}^d, \tag{7}$$

where $\mathbb{1}[A]$ stands for the indicator function of $A$. While the definition of $\mathbf{Q}^{\mathrm{g}}_{\mathrm{P}}$ is motivated by the characterization of univariate center-outward quantiles as minimizers of an objective function, let us stress that the geometric distribution function in (7) is a natural analogue of the univariate center-outward distribution as well. Indeed, recalling that, for $d = 1$, $F^{\pm}_{\mathrm{P}} := 2F_{\mathrm{P}} - 1$ and observing that

$$2\mathbb{1}[X \leq z] - 1 = \mathrm{sign}(z - X),$$

one obtains

$$F^{\pm}_{\mathrm{P}}(z) = \mathrm{E}_{\mathrm{P}}[\mathrm{sign}(z - X)], \quad z \in \mathbb{R}, \tag{8}$$

to which $\mathbf{F}^{\mathrm{g}}_{\mathrm{P}}$ in (7) reduces when $d = 1$. The direction $(\mathbf{z}-\mathbf{X})/\|\mathbf{z}-\mathbf{X}\|$, thus, can be interpreted as a multivariate sign for $\mathbf{z} - \mathbf{X} \neq \mathbf{0}$.

Provided that P is non-atomic and is not supported on a single line of $\mathbb{R}^d$, Theorem 6.2 in [12] entails that $\mathbf{F}^{\mathrm{g}}_{\mathrm{P}}$ is a homeomorphism between $\mathbb{R}^d$ and the open unit ball $\mathbb{B}_d$ of $\mathbb{R}^d$, with inverse $\mathbf{Q}^{\mathrm{g}}_{\mathrm{P}}$. Consequently, one also can define $\mathbf{Q}^{\mathrm{g}}_{\mathrm{P}}$ as $\big(\mathbf{F}^{\mathrm{g}}_{\mathrm{P}}\big)^{-1}$ with $\mathbf{F}^{\mathrm{g}}_{\mathrm{P}}$ defined in (7), or define $\mathbf{F}^{\mathrm{g}}_{\mathrm{P}}$ as $\big(\mathbf{Q}^{\mathrm{g}}_{\mathrm{P}}\big)^{-1}$ with $\mathbf{Q}^{\mathrm{g}}_{\mathrm{P}}$ defined as the minimizer of $O^{\mathrm{P}}_{\tau,\mathbf{u}}$ in (6). Further results about the regularity of $\mathbf{F}^{\mathrm{g}}_{\mathrm{P}}$ and $\mathbf{Q}^{\mathrm{g}}_{\mathrm{P}}$ are provided in [11].

With this, one can define *geometric quantile regions* and *contours*, which are multivariate extensions of the univariate center-outward quantile regions and contours, by letting

$$\mathscr{R}^{\mathrm{g}}_{\mathrm{P}}(\tau) := \mathbf{Q}^{\mathrm{g}}_{\mathrm{P}}(\tau\mathbb{B}_d) \quad \text{and} \quad \mathscr{C}^{\mathrm{g}}_{\mathrm{P}}(\tau) := \mathbf{Q}^{\mathrm{g}}_{\mathrm{P}}(\tau\mathbb{S}^{d-1}), \quad \tau \in [0, 1). \tag{9}$$

For $\tau = 0$, $\mathscr{R}^{\mathrm{g}}_{\mathrm{P}}(0)$ and $\mathscr{C}^{\mathrm{g}}_{\mathrm{P}}(0)$ reduce to the singlepoint set consisting of the *geometric* or *Fréchet median*; as in the univariate case, the geometric median is an $\mathrm{L}_1$ location parameter that minimizes, when finite first-order moments exist, the expected absolute deviation (*absolute* deviations, here, are to be understood as *Euclidean norms* of deviations).

### 3.2 Measure-Transportation-Based Quantiles and Quantile Contours

Denote by $\mathrm{U}_d$ the spherical uniform over the open unit ball $\mathbb{B}_d$, that is, the product of a uniform distribution over the unit sphere $\mathbb{S}^{d-1}$ and a (univariate) uniform over the distances from the origin; in other words, $\mathrm{U}_d$ is the probability measure with density $f_{\mathrm{U}_d}(\mathbf{x})$ proportional to $1/\|\mathbf{x}\|^{d-1}\mathbb{1}[\mathbf{0} < \|\mathbf{x}\| \leq 1]$ for $0 < \|\mathbf{x}\|$ and $f_{\mathrm{U}_d}(\mathbf{0}) = \infty$.

A celebrated result by [13] implies that, for any distribution P over $\mathbb{R}^d$, there exists an almost everywhere unique gradient of a convex function pushing P forward to the spherical uniform $\mathrm{U}_d$ over the (open) unit ball $\mathbb{B}_d$ of $\mathbb{R}^d$. In dimension $d = 1$, a gradient of convex function is a monotone non-decreasing function and the unique monotone non-decreasing function pushing P forward to $\mathrm{U}_1$ is the center-outward distribution function $F_{\mathrm{P}}^{\pm} := 2F_{\mathrm{P}} - 1$. For arbitrary dimension $d \geq 1$, this leads [7] to define the *center-outward distribution function* of P (or $\mathbf{X} \sim \mathrm{P}$) as the almost everywhere unique gradient of a convex function $\mathbf{F}_{\mathrm{P}}^{\pm}$ pushing P to the spherical uniform $\mathrm{U}_d$.

Assuming that P belongs to the class $\mathcal{P}_d^+$ of Lebesgue-absolutely continuous distributions with density bounded away from zero and infinity on any compact[2] subset of $\mathbb{R}^d$, [5] showed that $\mathbf{F}_{\mathrm{P}}^{\pm}$ is a homeomorphism between $\mathbb{R}^d \setminus (\mathbf{F}_{\mathrm{P}}^{\pm})^{-1}(\mathbf{0})$ and the punctured unit ball $\mathbb{B}_d \setminus \{\mathbf{0}\}$. The restriction of $\mathbf{F}_{\mathrm{P}}^{\pm}$ to $\mathbb{R}^d \setminus (\mathbf{F}_{\mathrm{P}}^{\pm})^{-1}(\mathbf{0})$ is thus continuously invertible. We call its (continuous) inverse $\mathbf{Q}_{\mathrm{P}}^{\pm} := (\mathbf{F}_{\mathrm{P}}^{\pm})^{-1}$ the *center-outward quantile function* of P (or $\mathbf{X} \sim \mathrm{P}$). This yields *center-outward quantile regions* and *contours*, of order $\tau$ which are multivariate extensions of the univariate center-outward quantile regions and contours defined in (2), of the form

$$\mathscr{R}_{\mathrm{P}}^{\pm}(\tau) := \mathbf{Q}_{\mathrm{P}}^{\pm}(\tau\mathbb{B}_d) \quad \text{and} \quad \mathscr{C}_{\mathrm{P}}^{\pm}(\tau) := \mathbf{Q}_{\mathrm{P}}^{\pm}(\tau\mathbb{S}^{d-1}), \qquad \tau \in (0, 1), \tag{10}$$

respectively. As for $\tau = 0$, define $\mathscr{R}_{\mathrm{P}}^{\pm}(0) := \bigcap_{0<\tau<1} \mathscr{R}_{\mathrm{P}}^{\pm}(\tau)$ as the *center-outward median region*, which is shown to be convex and compact, with Hausdorff dimension at most $d - 1$ [5]. Note the parallel between (10) and (9).

---

[2] Formally, for any compact subset $K \subset \mathbb{R}^d$, there exist constants $0 < c_K^- \leq c_K^+ < \infty$ such that $c_K^- \leq \mathrm{dP}/\mathrm{d}\lambda(\mathbf{x}) \leq c_K^+$ for all $\mathbf{x} \in K$. This assumption is maintained here for clarity of exposition, but it can be relaxed: see [3, 4].

## 4 Main Properties

### *4.1 Geometric Quantile Regions and Contours*

Consider a probability measure P which is non-atomic and is not supported on a single line of $\mathbb{R}^d$; in particular, $\mathbf{F}^{\mathrm{g}}_{\mathrm{P}}$ and $\mathbf{Q}^{\mathrm{g}}_{\mathrm{P}}$ are homeomorphisms with $\mathbf{Q}^{\mathrm{g}}_{\mathrm{P}} = (\mathbf{F}^{\mathrm{g}}_{\mathrm{P}})^{-1}$. Then, geometric quantile regions and contours have the following topological properties:

(i) the regions $\mathscr{R}^{\mathrm{g}}_{\mathrm{P}}(\tau)$ are compact and arc-connected for all $\tau \in [0, 1)$;
(ii) the regions $\mathscr{R}^{\mathrm{g}}_{\mathrm{P}}(\tau)$ are stricly nested as $\tau$ increases: for any $0 \leq \tau_1 < \tau_2 < 1$, we have $\mathscr{R}^{\mathrm{g}}_{\mathrm{P}}(\tau_1) \subsetneq \mathscr{R}^{\mathrm{g}}_{\mathrm{P}}(\tau_2)$;
(iii) if P has a bounded density $f$, then the contour $\mathscr{C}^{\mathrm{g}}_{\mathrm{P}}(\tau)$ is a $(d-1)$-dimensional manifold of class $\mathcal{C}^{d-1}$ for all $\tau \in (0, 1)$; if, in addition, $f$ is of class $\mathcal{C}^{k,\alpha}$ for some $k \in \mathbb{N}$ and $\alpha \in (0, 1)$, then $\mathscr{C}^{\mathrm{g}}_{\mathrm{P}}(\tau)$ is of class $\mathcal{C}^{d+k}$ for all $\tau \in (0, 1)$;
(iv) for $d = 1$, $\mathbf{F}^{\mathrm{g}}_{\mathrm{P}}$, $\mathbf{Q}^{\mathrm{g}}_{\mathrm{P}}$, $\mathscr{R}^{\mathrm{g}}_{\mathrm{P}}$, and $\mathscr{C}^{\mathrm{g}}_{\mathrm{P}}$ reduce to $F^{\pm}_{\mathrm{P}}$, $Q^{\pm}_{\mathrm{P}}$, $\mathscr{R}^{\pm}_{\mathrm{P}}$, and $\mathscr{C}^{\pm}_{\mathrm{P}}$.

While properties (i)–(ii) are straightforward consequences of the continuity of $\mathbf{Q}^{\mathrm{g}}_{\mathrm{P}}$, property (iii) requires more advanced mathematical tools. That property was established in [11], where it is shown that geometric distribution functions satisfy a linear partial differential equation, which was the key to the derivation of the regularity properties of geometric contours. Among other things, this PDE implies that geometric distribution functions fully characterize *arbitrary* probability measures (no moment or non-atomicity assumptions required), namely,

(v) if $\mathrm{P}_1$ and $\mathrm{P}_2$ are *arbitrary* probability measures such that $\mathbf{F}^{\mathrm{g}}_{\mathrm{P}_1} = \mathbf{F}^{\mathrm{g}}_{\mathrm{P}_2}$, then $\mathrm{P}_1 = \mathrm{P}_2$.

This characterization result has been long known, though: the first proof was provided in Theorem 2.5 of [10] and was refined recently in [11].

Geometric quantile and distribution functions further enjoy the following natural equivariance properties; see [6, 12].

(vi) Let $\boldsymbol{\theta} \in \mathbb{R}^d$ and denote by $\mathbf{O}$ an arbitrary $d \times d$ orthogonal matrix. For $\mathbf{X} \sim \mathrm{P}$, write $\mathbf{F}^{\mathrm{g}}_{\mathbf{X}}$, $\mathbf{F}^{\mathrm{g}}_{\mathbf{X}}$, etc. instead of $\mathbf{F}^{\mathrm{g}}_{\mathrm{P}}$, $\mathbf{F}^{\mathrm{g}}_{\mathrm{P}}$, etc. We have

$$
\begin{aligned}
\mathbf{F}^{\mathrm{g}}_{\boldsymbol{\theta}+\mathbf{OX}}(\boldsymbol{\theta} + \mathbf{Ox}) &= \mathbf{OF}^{\mathrm{g}}_{\mathbf{X}}(\mathbf{x}) && \mathbf{x} \in \mathbb{R}^d, \\
\mathbf{Q}^{\mathrm{g}}_{\boldsymbol{\theta}+\mathbf{OX}}(\mathbf{Ou}) &= \boldsymbol{\theta} + \mathbf{OQ}^{\mathrm{g}}_{\mathbf{X}}(\mathbf{u}) && \mathbf{u} \in \mathbb{B}_d, \\
\mathscr{R}^{\mathrm{g}}_{\boldsymbol{\theta}+\mathbf{OX}}(\tau) &= \boldsymbol{\theta} + \mathbf{O}\mathscr{R}^{\mathrm{g}}_{\mathbf{X}}(\tau) && \tau \in [0, 1), \quad \text{and} \\
\mathscr{C}^{\mathrm{g}}_{\boldsymbol{\theta}+\mathbf{OX}}(\tau) &= \boldsymbol{\theta} + \mathbf{O}\mathscr{C}^{\mathrm{g}}_{\mathbf{X}}(\tau) && \tau \in [0, 1).
\end{aligned}
$$

In particular, if P enjoys symmetry properties (rotational, central, or with respect to some hyperplanes), $\mathbf{F}^{\mathrm{g}}_{\mathrm{P}}$, $\mathbf{Q}^{\mathrm{g}}_{\mathrm{P}}$, $\mathscr{R}^{\mathrm{g}}_{\mathrm{P}}(\tau)$, and $\mathscr{C}^{\mathrm{g}}_{\mathrm{P}}(\tau)$ enjoy *the same* symmetry properties as P for all $\tau \in [0, 1)$.

Recall that an essential property of distribution functions in dimension $d = 1$ is their non-decreasing monotonicity. In higher dimensions, the concept of monotonicity still makes sense; we distinguish two notions of monotonicity, which both reduce to the usual concept of (non-decreasing) monotonicity when $d = 1$: *monotonicity* and *cyclical monotonicity*. We say that a map $\mathbf{G} : \mathbb{R}^d \to \mathbb{R}^d$ is *monotone* if

$$(\mathbf{G}(\mathbf{y}) - \mathbf{G}(\mathbf{x}))'(\mathbf{y} - \mathbf{x}) \geq 0$$

for all $\mathbf{x}, \mathbf{y} \in \mathbb{R}^d$. We say that $G$ is *cyclically monotone* if for any $m \geq 1$ we have

$$\sum_{k=1}^{m} \big(\mathbf{G}(\mathbf{x}_{k+1}) - \mathbf{G}(\mathbf{x}_k)\big)' \mathbf{x}_{k+1} \geq 0$$

for every cycle $x_1, x_2, \ldots, x_{m+1} := x_1$. For $m = 2$, cyclical monotonicity yields

$$\big(\mathbf{G}(\mathbf{x}_2) - \mathbf{G}(\mathbf{x}_1)\big)' \mathbf{x}_2 + \big(\mathbf{G}(\mathbf{x}_1) - \mathbf{G}(\mathbf{x}_2)\big)' \mathbf{x}_1 \geq 0,$$

and cyclical monotonicity thus implies monotonicity. Theorem 24.8 in [16] essentially entails that $\mathbf{G}$ is cyclically monotone precisely when it writes as the gradient $\mathbf{G} = \nabla\psi$ of some convex (potential) function $\psi : \mathbb{R}^d \to \mathbb{R}$.

The following then holds.

(vii) The maps $\mathbf{F}^{\mathrm{g}}_{\mathrm{P}}$ and $\mathbf{Q}^{\mathrm{g}}_{\mathrm{P}}$ are cyclically monotone, hence also monotone, over $\mathbb{R}^d$.

Indeed, similarly to the center-outward distribution functions based on measure transportation introduced in Sect. 2.1, geometric distribution functions are gradients of convex functions. This is clear from (7), which straight forwardly yields $\mathbf{F}^{\mathrm{g}}_{\mathrm{P}} = \nabla\phi^{\mathrm{g}}_{\mathrm{P}}$, where

$$\mathbf{z} \mapsto \phi^{\mathrm{g}}_{\mathrm{P}}(\mathbf{z}) := \mathrm{E}\big[\|\mathbf{z} - \mathbf{X}\| - \|\mathbf{X}\|\big], \quad \mathbf{z} \in \mathbb{R}^d$$

is a convex function and $\mathbf{X} \sim \mathrm{P}$. It follows that $\mathbf{F}^{\mathrm{g}}_{\mathrm{P}}$ is cyclically monotone, which implies that its inverse $\mathbf{Q}^{\mathrm{g}}_{\mathrm{P}}$ is cyclically monotone too (hence the gradient of a convex function).

However, the geometric distribution function $\mathbf{F}^{\mathrm{g}}_{\mathrm{P}}$, in general, does not push P forward to the spherical uniform $\mathrm{U}_d$ over $\mathbb{B}_d$. Would this be case, $\mathbf{F}^{\mathrm{g}}_{\mathrm{P}}$ and $\mathbf{F}^{\pm}_{\mathrm{P}}$ both being gradients of convex functions, the a.s. uniqueness result of McCann (1955) would imply $\mathbf{F}^{\mathrm{g}}_{\mathrm{P}} = \mathbf{F}^{\pm}_{\mathrm{P}}$ a.s. for all $\mathrm{P} \in \mathcal{P}^{+}_{d}$ (see Sect. 2.2), which cannot hold in general. Indeed, $\mathbf{F}^{\mathrm{g}}_{\mathrm{P}}$ can be defined as the solution of the previously mentioned *linear* PDE (see [11]), while $\mathbf{F}^{\pm}_{\mathrm{P}}$ can be defined as the solution of a PDE of the Monge-Ampère type which is *nonlinear* (see [7]). Therefore, the solutions $\mathbf{F}^{\mathrm{g}}_{\mathrm{P}}$ and $\mathbf{F}^{\pm}_{\mathrm{P}}$, for general P, do not coincide and $\mathbf{F}^{\mathrm{g}}_{\mathrm{P}}$, in general, does not push P forward to $\mathrm{U}_d$.

Nevertheless, recalling that $\mathbf{F}^{\mathrm{g}}_{\mathrm{P}}$ is a homeomorphism between $\mathbb{R}^d$ and $\mathbb{B}_d$, we see that each point $\mathbf{x} \in \mathbb{R}^d$ belongs to exactly one geometric contour, namely $\mathbf{x} \in \mathscr{C}^{\mathrm{g}}_{\mathrm{P}}(\tau)$ where $\tau = \|\mathbf{F}^{\mathrm{g}}_{\mathrm{P}}(\mathbf{x})\|$. Therefore, we can relabel quantile regions so that, in the new labeling, the region associated with an order $\tau \in [0, 1)$ has P-probability content $\tau$. For this purpose, it is enough to observe that the map

$$\tau \mapsto \kappa_{\mathrm{P}}^{\mathrm{g}}(\tau) := \mathrm{P}[\mathscr{R}_{\mathrm{P}}^{\mathrm{g}}(\tau)], \quad \tau \in [0, 1),$$

from $[0, 1) \to [0, 1)$ is a continous and monotone increasing bijection. In fact, $\kappa_{\mathrm{P}}^{\mathrm{g}}$ is the univariate distribution function of the random variable $\|\mathbf{F}_{\mathrm{P}}^{\mathrm{g}}(\mathbf{X})\|$ with $\mathbf{X} \sim \mathrm{P}$. Consequently, we can define the *relabeled geometric distribution function* of P (of $\mathbf{X} \sim \mathrm{P}$) as

$$\mathbf{x} \mapsto \tilde{\mathbf{F}}_{\mathrm{P}}^{\mathrm{g}}(\mathbf{x}) := \mathbf{B}_{\mathrm{P}}^{\mathrm{g}}\big(\mathbf{F}_{\mathrm{P}}^{\mathrm{g}}(\mathbf{x})\big), \tag{11}$$

where

$$\mathbf{B}_{\mathrm{P}}^{\mathrm{g}}(\mathbf{x}) := \begin{cases} \kappa_{\mathrm{P}}^{\mathrm{g}}(\|\mathbf{x}\|)\frac{\mathbf{x}}{\|\mathbf{x}\|} & \text{if } \mathbf{x} \neq 0 \\ \mathbf{0} & \text{if } \mathbf{x} = 0 \end{cases}$$

is a homeomorphism. Consequently, the relabeled geometric distribution function $\tilde{\mathbf{F}}_{\mathrm{P}}^{\mathrm{g}}$ is also a homeomorphism, hence continuously invertible: define the *relabeled geometric quantile function* as its inverse $\tilde{\mathbf{Q}}_{\mathrm{P}}^{\mathrm{g}} := (\tilde{\mathbf{F}}_{\mathrm{P}}^{\mathrm{g}})^{-1}$. The collections of regions and contours associated with $\tilde{\mathbf{F}}_{\mathrm{P}}^{\mathrm{g}}$ are the same as those associated with $\mathbf{F}_{\mathrm{P}}^{\mathrm{g}}$—only the labels have changed. Call $\tilde{\mathscr{R}}_{\mathrm{P}}^{\mathrm{g}}(\tau) := \tilde{\mathbf{Q}}_{\mathrm{P}}^{\mathrm{g}}(\tau\mathbb{B}_d)$ the *relabeled geometric quantile region of order* $\tau$. Then, $\tilde{\mathscr{R}}_{\mathrm{P}}^{\mathrm{g}}(\tau)$ satisfies the essential property of a quantile region of order $\tau$:

(viii) the relabeled geometric quantile region $\tilde{\mathscr{R}}_{\mathrm{P}}^{\mathrm{g}}(\tau)$ has P-probability content $\tau$ for all $\tau \in [0, 1)$, irrespective of P.

We similarly define the *relabeled geometric quantile contours*

$$\tilde{\mathscr{C}}_{\mathrm{P}}^{\mathrm{g}}(\tau) := \tilde{\mathbf{Q}}_{\mathrm{P}}^{\mathrm{g}}(\tau\mathbb{S}^{d-1}), \quad \tau \in [0, 1).$$

One may wonder whether $\tilde{\mathbf{F}}_{\mathrm{P}}^{\mathrm{g}}$ now coincides with $\mathbf{F}_{\mathrm{P}}^{\pm}$ (hence, $\tilde{\mathbf{Q}}_{\mathrm{P}}^{\mathrm{g}}$ with $\mathbf{Q}_{\mathrm{P}}^{\pm}$). First, let us stress that, even though $\|\tilde{\mathbf{F}}_{\mathrm{P}}^{\mathrm{g}}(\mathbf{X})\|$, just like $\|\mathbf{F}_{\mathrm{P}}^{\pm}(\mathbf{X})\|$, is uniformly distributed over $[0, 1)$ when $\mathbf{X} \sim \mathrm{P}$, the behavior of $\mathbf{F}_{\mathrm{P}}^{\mathrm{g}}(\mathbf{X})/\|\mathbf{F}_{\mathrm{P}}^{\mathrm{g}}(\mathbf{X})\|$, to the best of our knowledge, remains unknown, and has not been investigated so far. Second, observe that $\mathbf{B}_{\mathrm{P}}^{\mathrm{g}}$ writes as the gradient $\nabla h_{\mathrm{P}}^{\mathrm{g}}$ of the convex function

$$\mathbf{x} \mapsto h_{\mathrm{P}}^{\mathrm{g}}(\mathbf{x}) := \int_0^{\|\mathbf{x}\|} \kappa_{\mathrm{P}}^{\mathrm{g}}(s)\ \mathrm{d}s, \quad \mathbf{x} \in \mathbb{R}^d.$$

It follows that $\tilde{\mathbf{F}}_{\mathrm{P}}^{\mathrm{g}} = \nabla h_{\mathrm{P}}^{\mathrm{g}} \circ \nabla \phi_{\mathrm{P}}^{\mathrm{g}}$. However, it is well known that the composition of two gradients of convex functions, in general, is not the gradient of a convex function. For these reasons, which do not constitute rigorous proofs, we believe that there is no obvious reason why the equality $\tilde{\mathbf{F}}_{\mathrm{P}}^{\mathrm{g}} = \mathbf{F}_{\mathrm{P}}^{\pm}$ should hold for general P. The simulations in Sect. 5) below—which do not constitute mathematical proofs—actually suggest that $\tilde{\mathscr{C}}_{\mathrm{P}}^{\mathrm{g}}(\tau)$ and $\mathscr{C}_{\mathrm{P}}^{\pm}(\tau)$, for some distributions P, very strongly differ.

### 4.2 Center-Outward Quantile Regions and Contours

Let $\mathrm{P} \in \mathcal{P}_d^+$. It is shown in [7] that

(i) the regions $\mathscr{R}_{\mathrm{P}}^{\pm}(\tau)$ are compact and arc-connected for all $\tau \in (0, 1)$;
(ii) the regions $\mathscr{R}_{\mathrm{P}}^{\pm}(\tau)$ are stricly nested as $\tau$ increases: $\mathscr{R}_{\mathrm{P}}^{\pm}(\tau_1) \subsetneq \mathscr{R}_{\mathrm{P}}^{\pm}(\tau_2)$ for any $0 < \tau_1 < \tau_2 < 1$;
(iii) the median region $\mathscr{R}_{\mathrm{P}}^{\pm}(0)$ is compact, convex, and has Hausdorff dimension at most $(d-1)$;
(iv) if, for some $k \in \mathbb{N}$ and $\alpha \in (0, 1)$, the density of P is locally of class $\mathcal{C}^{k,\alpha}$, then the contour $\mathscr{C}_{\mathrm{P}}^{\pm}(\tau)$ is a $(d-1)$-dimensional manifold of class $\mathcal{C}^{k+1,\alpha}$ for all $\tau \in (0, 1)$;
(v) for $d = 1$, $\mathbf{F}_{\mathrm{P}}^{\pm}$, $\mathbf{Q}_{\mathrm{P}}^{\pm}$, $\mathscr{R}_{\mathrm{P}}^{\pm}$, and $\mathscr{C}_{\mathrm{P}}^{\pm}$ reduce to $F_{\mathrm{P}}^{\pm}$, $Q_{\mathrm{P}}^{\pm}$, $\mathscr{R}_{\mathrm{P}}^{\pm}$, and $\mathscr{C}_{\mathrm{P}}^{\pm}$.

Note that (iv) was first established in [4] when the support of P is convex, then refined in [3]. By construction, we also have the following:

(vi) the P-probability content of $\mathscr{R}_{\mathrm{P}}^{\pm}(\tau)$ equals the $\mathrm{U}_d$-probability content of $\tau\mathbb{B}_d$, which is $\tau$: $\mathscr{R}_{\mathrm{P}}^{\pm}(\tau)$ and $\mathscr{C}_{\mathrm{P}}^{\pm}(\tau)$, thus, qualify as the center-outward quantile region and quantile contour *of order* $\tau$ of P;
(vii) if $\mathbf{X} \sim \mathrm{P}$, then $\mathbf{F}_{\mathrm{P}}^{\pm}(\mathbf{X}) \sim \mathrm{U}_d$; if $\mathbf{U} \sim \mathrm{U}_d$, then $\mathbf{Q}_{\mathrm{P}}^{\pm}(\mathbf{U}) \sim \mathrm{P}$: both $\mathbf{F}_{\mathrm{P}}^{\pm}$ and $\mathbf{Q}_{\mathrm{P}}^{\pm}$, thus, fully characterize P.[3]
(viii) the maps $\mathbf{F}_{\mathrm{P}}^{\pm}$ and $\mathbf{Q}_{\mathrm{P}}^{\pm}$ are cyclically monotone, hence monotone, over $\mathbb{R}^d$.

Finally, $\mathbf{F}_{\mathrm{P}}^{\pm}$ and $\mathbf{Q}_{\mathrm{P}}^{\pm}$ enjoy the following invariance/equivariance properties [8].

(ix) Let $\boldsymbol{\theta} \in \mathbb{R}^d$ and denote by $\mathbf{O}$ an arbitrary $d \times d$ orthogonal matrix. Letting $\mathbf{X} \sim \mathrm{P}$, write $\mathbf{F}_{\mathbf{X}}^{\pm}$, $\mathbf{Q}_{\mathbf{X}}^{\pm}$, etc. for $\mathbf{F}_{\mathrm{P}}^{\pm}$, $\mathbf{Q}_{\mathrm{P}}^{\pm}$, etc. Then,

$$\begin{aligned}
\mathbf{F}_{\boldsymbol{\theta}+\mathbf{OX}}^{\pm}(\boldsymbol{\theta} + \mathbf{Ox}) &= \mathbf{OF}_{\mathbf{X}}^{\pm}(\mathbf{x}) && \mathbf{x} \in \mathbb{R}^d, \\
\mathbf{Q}_{\boldsymbol{\theta}+\mathbf{OX}}^{\pm}(\mathbf{Ou}) &= \boldsymbol{\theta} + \mathbf{OQ}_{\mathbf{X}}^{\pm}(\mathbf{u}) && \mathbf{u} \in \mathbb{B}_d, \\
\mathscr{R}_{\boldsymbol{\theta}+\mathbf{OX}}^{\pm}(\tau) &= \boldsymbol{\theta} + \mathbf{O}\mathscr{R}_{\mathbf{X}}^{\pm}(\tau) && \tau \in [0, 1), \quad \text{and} \\
\mathscr{C}_{\boldsymbol{\theta}+\mathbf{OX}}^{\pm}(\tau) &= \boldsymbol{\theta} + \mathbf{O}\mathscr{C}_{\mathbf{X}}^{\pm}(\tau) && \tau \in [0, 1).
\end{aligned}$$

In particular, if P enjoys symmetry properties (rotational, central, or with respect to some hyperplanes), $\mathbf{F}_{\mathrm{P}}^{\pm}$, $\mathbf{Q}_{\mathrm{P}}^{\pm}$, $\mathscr{R}_{\mathrm{P}}^{\pm}(\tau)$, and $\mathscr{C}_{\mathrm{P}}^{\pm}(\tau)$ enjoy the same symmetry properties as P for all $\tau \in [0, 1)$. This is not the case for the distribution and quantile functions resulting from a transport to the Lebesgue uniform over the unit cube $[0, 1]^d$ which are favored by some authors.

---

[3] Note, however, that the collections of center-outward regions or center-outward contours alone (that is, without the mappings $\mathbf{F}_{\mathrm{P}}^{\pm}$ or $\mathbf{Q}_{\mathrm{P}}^{\pm}$ between $\mathbb{R}^d$ and $\mathbb{B}_d$) *do not* characterize P.

These properties, which for dimension $d = 1$ reduce to those of $F^{\pm}$ and $Q^{\pm}$ and constitute the essence of the concepts of distribution and quantile functions, fully justify the terminology *center-outward distribution and quantile functions* for $\mathbf{F}^{\pm}_{\mathrm{P}}$ and $\mathbf{Q}^{\pm}_{\mathrm{P}}$. The *only* difference, here, between the properties of $\mathbf{F}^{\pm}_{\mathrm{P}}$ and $\mathbf{Q}^{\pm}_{\mathrm{P}}$ and those of $\mathbf{F}^{\mathrm{g}}_{\mathrm{P}}$ and $\mathbf{Q}^{\mathrm{g}}_{\mathrm{P}}$ is (vi) above, and this difference disappears if the relabeled functions $\tilde{\mathbf{F}}^{\mathrm{g}}_{\mathrm{P}}$ and $\tilde{\mathbf{Q}}^{\mathrm{g}}_{\mathrm{P}}$ are substituted for $\mathbf{F}^{\mathrm{g}}_{\mathrm{P}}$ and $\mathbf{Q}^{\mathrm{g}}_{\mathrm{P}}$, respectively.

From this point of view, the relabeled geometric and the center-outward concepts, which, it seems, do not coincide, apparently both qualify as equally satisfactory multivariate concepts of distribution and quantile functions. Is there a way to break that tie and select one of them rather than the other? Since analytical expressions are impossible, one has to recur to numerical approximations—in statistical terms, simulations—in order to perform comparisons. This is what we are doing in the next section, where comparisons are based on simulations of the quantile contours $\mathcal{C}^{\pm}_{\mathrm{P}}(\tau)$ and $\mathcal{C}^{g}_{\mathrm{P}}(\tau)$.

# 5 Geometric and Center-Outward Quantile Contours: An Empirical Comparative Study

## *5.1 Numerical Computation of Contours*

Except for some very special distributions, such as the spherical ones, geometric and center-outward quantiles in dimension $d \geq 2$, unfortunately, are sharing the inconvenient absence of analytical expressions. Numerical solutions are possible, though, based on simulations with high numbers of replications. For all $N \in \mathbb{N}$, denote by $\mathrm{P}_N$ the empirical distribution of a random sample of size $N$ drawn from P. The accuracy of these numerical solutions relies on consistency results for the empirical counterparts $\mathbf{F}^{\mathrm{g}}_{\mathrm{P}_N}$ and $\mathbf{Q}^{\mathrm{g}}_{\mathrm{P}_N}$ in the geometric case (obtained by substituting $\mathrm{P}_N$ for P in (7)), for $\mathbf{F}^{\pm}_{\mathrm{P}_N}$ and $\mathbf{Q}^{\pm}_{\mathrm{P}_N}$ in the measure-transportation setting (see [7] for precise definitions and details).

In the geometric case, [14] established a Glivenko-Cantelli result of the form

$$\sup_{\mathbf{z}\in\mathbb{R}^d} \|\mathbf{F}^{\mathrm{g}}_{\mathrm{P}_N}(\mathbf{z}) - \mathbf{F}^{\mathrm{g}}_{\mathrm{P}}(\mathbf{z})\| \to 0 \tag{12}$$

with P-probability one as $N \to \infty$, provided that P admits a bounded density. In fact, one can show that the bounded density assumption is superfluous: similarly to distribution functions in dimension $d = 1$, the Glivenko-Cantelli result (12) is valid for an *arbitrary* probability measure. The same convergence holds for the sequence of relabeled counterparts $\tilde{\mathbf{F}}^{\mathrm{g}}_{\mathrm{P}}$ of $\mathbf{F}^{\mathrm{g}}_{\mathrm{P}}$ and $\tilde{\mathbf{F}}^{\mathrm{g}}_{\mathrm{P}_N}$ of $\mathbf{F}^{\mathrm{g}}_{\mathrm{P}_N}$; see [11].

In the measure transportation approach, [7] proved a similar Glivenko-Cantelli result for an interpolated version $\overline{\mathbf{F}}^{\pm}_{\mathrm{P}_N}$ of their empirical center-outward distribution functions $\mathbf{F}^{\pm}_{\mathrm{P}_N}$:

$$\sup_{\mathbf{z}\in\mathbb{R}^d} \|\overline{\mathbf{F}}^{\pm}_{\mathrm{P}_N}(\mathbf{z}) - \mathbf{F}^{\pm}_{\mathrm{P}}(\mathbf{z})\| \to 0$$

with P-probability one as $N \to \infty$, when $\mathrm{P} \in \mathcal{P}_d^+$ (see Sect. 2.2 for the definition of $\mathcal{P}_d^+$).

## 5.2 Numerical Comparisons of Quantile Contours

Figure 1 shows, for orders $\tau \in \{.25, .50, .75\}$, plots of the empirical contours $\tilde{\mathscr{C}}^{\mathrm{g}}_{\mathrm{P}_N}(\tau)$ and $\mathscr{C}^{\pm}_{\mathrm{P}_N}(\tau)$ based on random samples $\mathbb{X}_N = \{\mathbf{X}_1, \ldots, \mathbf{X}_N\}$ of size $N = 2,400$ from four different probability distributions on $\mathbb{R}^2$. Given the regular grid

$$\Big\{\mathbf{u}_k = \big(\cos(\theta_k), \sin(\theta_k)\big) : \theta_k = \frac{2k\pi}{70} \text{ and } 1 \le k \le \mathrm{K}\Big\}$$

of directions on $\mathbb{S}^1$, with $K = 70$, each contour $\tilde{\mathscr{C}}^{\mathrm{g}}_{\mathrm{P}_N}(\tau)$ is estimated by

$$\Big\{\tilde{\mathbf{Q}}^{\mathrm{g}}_{\mathrm{P}_N}(\tau\mathbf{u}_k) : 1 \le k \le K = 70\Big\},$$

where $\tilde{\mathbf{Q}}^{\mathrm{g}}_{\mathrm{P}_N}$ stands for the relabeled geometric quantile map defined in Sect. 3.1. In other words, to each level $\tau$ we first associate the level $\tau'$ such that

$$\tilde{\mathscr{C}}^{\mathrm{g}}_{\mathrm{P}_N}(\tau) = \mathscr{C}^{\mathrm{g}}_{\mathrm{P}_N}(\tau').$$

Then, the geometric contour $\mathscr{C}^{\mathrm{g}}_{\mathrm{P}_N}(\tau')$ is obtained as the collection $\{\mathbf{z}_1, \ldots, \mathbf{z}_K\}$ such that $\mathbf{z}_\mathbf{k}$ minimizes the empirical objective function $O^{\mathrm{P}_N}_{\tau',\mathbf{u}_\mathbf{k}}$.

Each measure-transportation-based contour is estimated by

$$\Big\{\mathbf{Q}^{\pm}_{\mathrm{P}_N}(\tau\mathbf{u}_k) : 1 \le k \le \mathrm{K}\Big\}.$$

To construct the map $\mathbf{Q}^{\pm}_{\mathrm{P}_N}$, we first fix a regular $n_R \times n_S$ grid $\mathfrak{G}_N$ over the punctured unit ball $(0, 1) \times \mathbb{S}^1$ such that $n_R n_S = N$;[4] our simulation relies on $N = 2,400$ replications, with $n_R = 40$ and $n_S = 60$. Then, we seek an optimal coupling between $\mathbb{X}_N$ and $\mathfrak{G}_N$, i.e., we associate to each $\mathbf{X}_k \in \mathbb{X}_N$ a unique $\mathbf{F}^{\pm}_{\mathrm{P}_N}(\mathbf{X}_k) \in \mathfrak{G}$ in such a way that

$$\sum_{k=1}^{N} \|\mathbf{X}_k - \mathbf{F}^{\pm}_{\mathrm{P}_N}(\mathbf{X}_k)\|^2 = \min_{\pi\in\Pi} \sum_{k=1}^{N} \|\mathbf{X}_{\pi(k)} - \mathbf{F}^{\pm}_{\mathrm{P}_N}(\mathbf{X}_k)\|^2,$$

[4] In the case of a genuine sample, $N$ is factorized into $N = n_R n_S + n_0$ with $n_0 < \min(n_R, n_S)$. We refer to [9] for details on the choice of $n_R$ and $n_S$. Here, however, we are dealing with simulations, and can choose $N$ such that $n_0 = 0$.

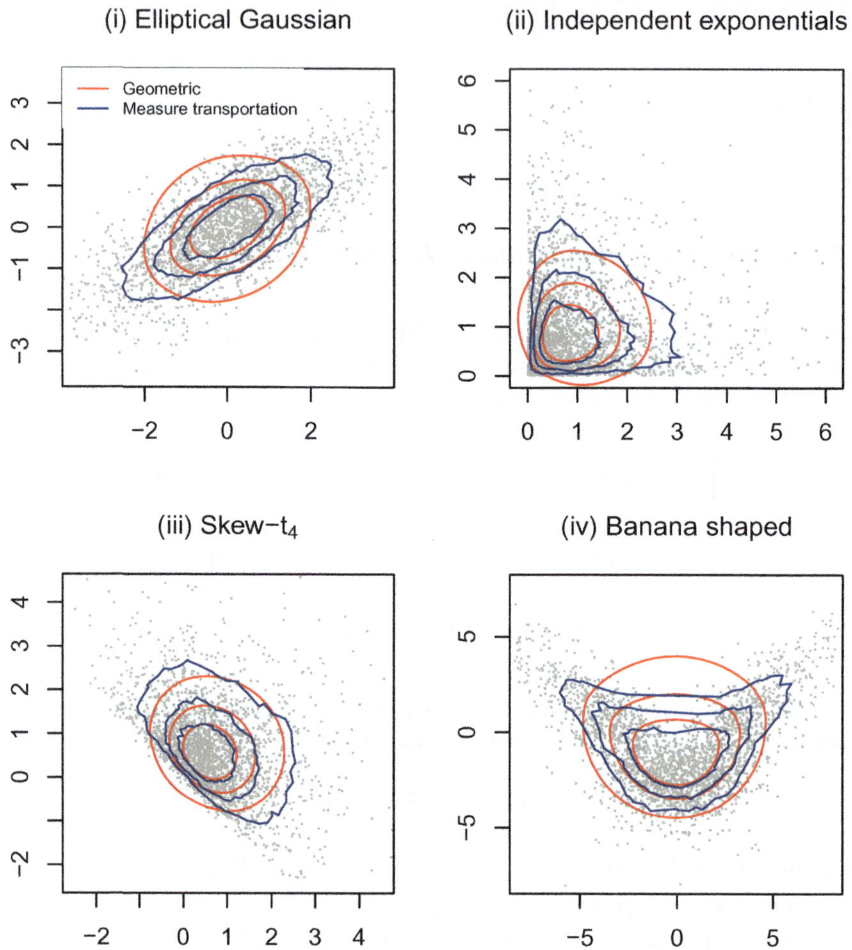


**Fig. 1** Geometric (red) and measure-transportation-based (blue) contours of levels $\tau = 0.25, 0.5, 0.75$ for samples of size $N = 2400$ drawn from four different distributions. Geometric contours are relabeled according to their probability content, and measure-transportation contours are obtained through the interpolation scheme (see [7]) of the optimal coupling between the observations and a $40 \times 60$ regular grid $\mathfrak{G}_N$ of the unit ball in $\mathbb{R}^2$; see Sect. 4.2 for details. Empirical quantiles, both geometric and center-outward, are connected by lines in the plots

where $\pi$ ranges over the set $\Pi$ of permutations of $\{1, \ldots, N\}$. A continuous map $\overline{\mathbf{F}}^{\pm}_{\mathrm{P}_N}$ then is obtained from the values $\mathbf{F}^{\pm}_{\mathrm{P}_N}(\mathbf{X}_k)$ of $\mathbf{F}^{\pm}_{\mathrm{P}_N}$ at $\mathbf{X}_k$ by interpolation techniques; see [7] for furter details.

These samples were generated from

(i) the centered bivariate normal distribution with covariance matrix $\Sigma = \begin{pmatrix} 2 & 1 \\ 1 & 1 \end{pmatrix}$;

(ii) the bivariate distribution with independent exponential marginals with mean one;
(iii) the standard skew-$t$ distribution with four degrees of freedom and slant vector $\alpha = (10, 10)$—see [1]—and
(iv) the banana-shaped distribution considered in [7], viz. the mixture

$$\frac{3}{8}\mathcal{N}(\mu_1, \Sigma_1) + \frac{3}{8}\mathcal{N}(\mu_2, \Sigma_2) + \frac{1}{4}\mathcal{N}(\mu_3, \Sigma_3),$$

of three bivariate normal distributions where

$$\mu_1 = \begin{pmatrix} -3 \\ 0 \end{pmatrix}, \quad \mu_2 = \begin{pmatrix} 3 \\ 0 \end{pmatrix}, \quad \mu_3 = \begin{pmatrix} 0 \\ -\frac{5}{2} \end{pmatrix},$$

and

$$\Sigma_1 = \begin{pmatrix} 5 & -4 \\ -4 & 5 \end{pmatrix}, \quad \Sigma_2 = \begin{pmatrix} 5 & 4 \\ 4 & 5 \end{pmatrix}, \quad \Sigma_3 = \begin{pmatrix} 4 & 0 \\ 0 & 1 \end{pmatrix}.$$

Inspection of Fig. 1 reveals the following facts.

(a) First, the contours obtained by measure transportation seem to better reflect the geometry of the underlying distributions. This is particularly obvious for the banana-shaped distribution (iv), where the concave shape of the mixture is well picked up by the center-outward contours but completely missed by the relabeled geometric ones. The same phenomenon also takes place, to a less spectacular extent, under distributions (i)-(iii): the relabeled geometric quantile region of order .75, under the exponential distribution (ii), largely extends beyond the exponential support, something the center-outward contours do not; the same geometric contours, for the skew-$t$ distribution (iii), contrary to the center-outward ones, hardly exhibit any skewness.
(b) Second, while relabeled geometric quantiles (as a direct consequence of their re-labelization rather than because of their $\mathrm{L}_1$ characterization) do provide the desired control over the probability content of their regions, hence yield, for each $\mathbf{X}_i \in \mathbb{X}_N$, a sound concept $R_i^{(N)} := \text{rank}(\|\mathbf{F}^{\mathrm{g}}_{\mathrm{P}_N}(\mathbf{X}_i)\|)$ of center-outward multivariate rank, they do not provide any clear notion of signs. Such signs, quite on the contrary, are well defined, in the measure-transportation-based approach, for the sample points $\mathbf{X}_i \in \mathbb{X}_n$, as the unit vectors $\mathbf{F}^{\pm}_{\mathrm{P}_N}(\mathbf{X}_i)/\|\mathbf{F}^{\pm}_{\mathrm{P}_N}(\mathbf{X}_i)\|$, $i = 1, \ldots, N$, which are i.i.d. (uniform over the unit vectors associated with the $N$ gridpoints in $\mathfrak{G}_N$ and independent of the ranks $R_1^{(N)}, \ldots, R_N^{(N)}$).
(c) On the other hand, geometric contours are more regular than their measure-transportation-based counterparts. This is a consequence of (iii) and (iv) in Sects. 3.1 and 3.2, respectively: geometric contours are continuous as soon as P is non-atomic and smooth (at least $C^{d-1}$) while the oscillations, hence also the variability, of the interpolated measure-transportation-based contours are further increased by the interpolation scheme.

While points (a) and (b) above are pleading for the measure-transportation-based concepts, the decisive argument in favor of the latter follows from a theoretical result by [6] on the "intriguing behavior" (a clear understatement) of high-order geometric quantile contours.

## 5.3 *Numerical Comparison of Extreme Contours*

Girard and Stupfler [6] establish the following disturbing result for extreme geometric quantiles: if P is a probability measure on $\mathbb{R}^d$ with finite second-order moments, i.e. if $\mathrm{E}_{\mathrm{P}}[\|\mathbf{X}\|^2] < \infty$ for $\mathbf{X} \sim \mathrm{P}$ with covariance matrix $\boldsymbol{\Sigma}$, then, denoting by $\mathrm{Tr}(\mathbf{A})$ the trace of a square matrix $\mathbf{A}$,

$$\|\mathbf{Q}^{\mathrm{g}}_{\mathrm{P}}(\tau\mathbf{u})\|^2(1-\tau) \rightarrow \frac{1}{2}\Big(\mathrm{Tr}\,\boldsymbol{\Sigma} - \mathbf{u}'\boldsymbol{\Sigma}\mathbf{u}\Big)$$

as $\tau \rightarrow 1$ for any $\mathbf{u} \in \mathbb{S}^{d-1}$. First, let us stress that the quantity $\mathrm{Tr}\,\boldsymbol{\Sigma} - \mathbf{u}'\boldsymbol{\Sigma}\mathbf{u}$ is always non-negative, and is strictly positive as soon as $\boldsymbol{\Sigma}$ is non-degenerate. In addition to providing the rate at which extreme geometric quantiles are "escaping to infinity," this result also provides valuable insights on the shape of the corresponding extreme contours. For instance, taking for $\mathbf{u}$ a unit eigenvector $\mathbf{u}_{\max}$ associated with $\boldsymbol{\Sigma}$'s *largest* eigenvalue yields a norm $\|\mathbf{Q}^{\mathrm{g}}_{\mathrm{P}}(\tau\mathbf{u}_{\max})\|$ for the contour $\mathscr{C}^{\mathrm{g}}_{\mathrm{P}}(\tau)$ of order $\tau$ in direction $\mathbf{u}$ which is *minimal* for $\tau \rightarrow 1$; the same contour, in the direction $\mathbf{u}_{\min}$ associated with the smallest eigenvalue of $\boldsymbol{\Sigma}$, asymptotically yields the *largest* norm $\|\mathbf{Q}^{\mathrm{g}}_{\mathrm{P}}(\tau\mathbf{u}_{\min})\|$. More generally, due to the fact that the shape factor $\mathrm{Tr}\,\boldsymbol{\Sigma} - \mathbf{u}'\boldsymbol{\Sigma}\mathbf{u}$ is a decreasing function of $\mathbf{u}'\boldsymbol{\Sigma}\mathbf{u}$, the asymptotic behavior of extreme quantiles—moving away fast from the center along the eigendirections of $\boldsymbol{\Sigma}$ associated with small eigenvalues, and slowly along the eigendirections of $\boldsymbol{\Sigma}$ associated with large eigenvalues—is exactly the opposite of what it is expected to be. This is highly counterintuitive and, actually, totally unacceptable.

Figure 2 provides a finite-sample empirical illustration of this theoretical fact. We generated a sample of size $N = 1,000$ from a centered bivariate Gaussian distribution $\mathcal{N}(\mathbf{0}, \boldsymbol{\Sigma})$ with covariance matrix $\boldsymbol{\Sigma} = \begin{pmatrix} 1/8 & 0 \\ 0 & 3/4 \end{pmatrix}$. Then, for $\tau \in \{0.90, 0.95, 0.99\}$, we computed the empirical relabeled geometric and transportation-based quantile contours of order $\tau$, as described in Sects. 5.1 and 5.2, with $K = 50$, $n_R = 20$, and $n_S = 50$ for the latter (see Sect. 5.2 for notation).

Figure 2 perfectly illustrates the pathological phenomenon described by [6]: the geometric contours are much wider in the (horizontal) directions associated with $\boldsymbol{\Sigma}$'s smaller eigenvalue, and narrower in the (vertical) directions associated with $\boldsymbol{\Sigma}$'s larger eigenvalue. This pathological behaviour is increasingly severe as $\tau$ approaches one: for $\tau = 0.90$, the contour is approximately circular. In sharp contrast, the center-outward contours, irrespective of $\tau$, are correctly accounting for the shape of the distribution and the relative magnitudes of $\boldsymbol{\Sigma}$'s eigenvalues.

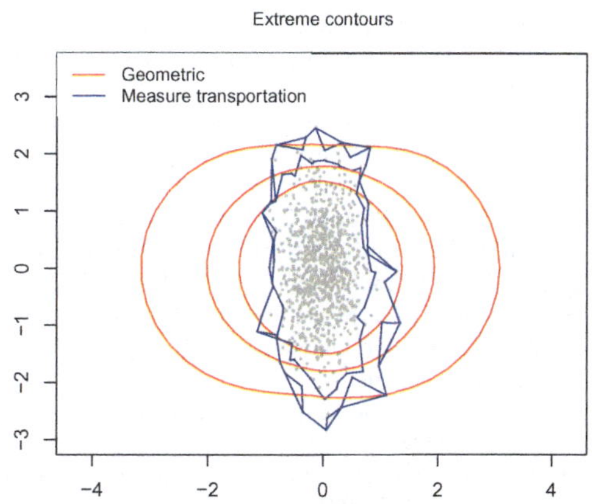


**Fig. 2** Geometric (red) and measure-transportation-based (blue) contours of order $\tau = 0.90, 0.95, 0.99$ for samples of size $N = 1,000$ drawn from a (nonspherical) bivariate Gaussian distribution on $\mathbb{R}^2$. Empirical quantiles, both geometric and center-outward, are connected by straight lines in the plots

## 6 Conclusions

Among all concepts of multivariate quantiles proposed in the literature, geometric quantiles (after due re-labelization) and measure-transportation-based center-outward quantiles are the most convincing ones: they both reduce to classical concepts in dimension one, and both satisfy most properties expected from quantiles, quantile regions, and quantile contours. Empirical comparisons and a highly pathological property of extreme geometric quantiles established by [6], however, are tilting the balance in favour of the measure-transportation-based concept.

**Acknowledgements** Marc Hallin acknowledges the support of the Czech Science Foundation grant GAČR22036365. Dimitri Konen is supported by a research fellowship from the Centre for Research in Statistical Methodology (CRiSM) of the University of Warwick. We thank Eustasio del Barrio and Alberto González-Sanz for kindly providing their codes for the computation of center-outward quantile functions and contours.

## References

1. Azzalini, A., Capitanio, A.: The Skew-Normal and Related Families. Cambridge University Press, IMS Monograph series (2014)
2. Chaudhuri, P.: On a geometric notion of quantiles for multivariate data. J. Am. Stat. Assoc. **91**, 862–872 (1996)
3. del Barrio, E., Gonzáles-Sanz, A.: Regularity of center-outward distribution functions in non-convex domains. Adv. Nonlinear Stud. (2024, to appear)
4. del Barrio, E., González-Sanz, A., Hallin, M.: A note on the regularity of optimal-transport-based center-outward distribution and quantile functions. J. Multivar. Anal. **180**, 104671 (2020)
5. Figalli, A.: On the continuity of center-outward distribution and quantile functions. Nonlinear Anal. **177**, 413–21 (2018)
6. Girard, S., Stupfler, G.: Intriguing properties of extreme geometric quantiles. REVSTAT **15**, 107–139 (2017)

7. Hallin, M., del Barrio, E., Cuesta-Albertos, J., Matrán, C.: Distribution and quantile functions, ranks and signs in dimension $d$: a measure transportation approach. Ann. Stat. **49**, 1139–1165 (2021)
8. Hallin, M., Hlubinka, D., Hudecová, Š.: Efficient fully distribution-free center-outward rank tests for multiple-output regression and MANOVA. J. Am. Stat. Assoc. **118**(543), 1923–1939 (2022)
9. Hallin, M., Mordant, G.: On the finite-sample performance of measure-transportation-based multivariate rank tests. In: Yi, M., Nordhausen, K. (eds.) Robust and Multivariate Statistical Methods: Festschrift in Honor of David E. Tyler, pp. 87–119. Springer (2023)
10. Koltchinski, V.: M-estimation, convexity and quantiles. Ann. Stat. **25**, 435–477 (1997)
11. Konen, D.: PDE characterization of geometric distribution functions and quantiles. Bernoulli (2024, to appear)
12. Konen, D., Paindaveine, D.: Multivariate $\rho$-quantiles: a spatial approach. Bernoulli **28**, 1912–1934 (2022)
13. McCann, R.J.: Existence and uniqueness of monotone measure-preserving maps. Duke Math. J. **80**, 309–323 (1995)
14. Mottonen, J., Oja, H., Tienari, J.: On the efficiency of multivariate spatial sign and rank tests. Ann. Stat. **25**, 542–552 (1997)
15. Paindaveine, D., Virta, J.: On the behavior of extreme $d$-dimensional spatial quantiles under minimal assumptions. In: Daouia, A., Ruiz-Gazen, A. (eds.) Advances in Contemporary Statistics and Econometrics, pp. 243–259. Springer (2020)
16. Rockafellar, R.T.: Convex Analysis. Princeton University Press, Princeton Mathematical Series (1970)

# Human Centered AI for Financial Decisions

**Polina Khrennikova**

**Abstract** We survey the state of the art of AI applications to financial expectations and the role quantum logic can play in further advancements of AI technologies. We discuss financial applications of such machine learning techniques as reinforcement learning and deep neural networks to the analysis of financial statements, algorithmic trading, portfolio management, and robo-advising. Next, we elaborate on the emergence and advancement of QML (quantum machine learning) and advocate for the wider exploration of the advantages of quantum inspired neural networks, steaming from the use of quantum logic that is able to capture agents' non- classical expectations and non expected utility decisions, also coined "bounded rationality". We would like to motivate to use human—like AI techniques that are centered on quantum, rather than classical logic to (i) represent the human brain type information processing; (ii) speed up the work of the AI algorithms; (iii) better operate in complex and uncertain environments.

**Keywords** Artificial intelligence · Machine learning algorithms · Deep neural networks · Quantum machine learning · Quantum logic · Bounded rationality · Financial expectations · Investments

## 1 Introduction

Artificial intelligence (AI) is by now a ubiquitous concept, frequently appearing in media, policy papers, and university curricula. AI encompasses all fields where human intellect can be substituted for automated processes. An algorithm performs some less (e.g., searching through a database) or more human-like tasks (executing trades against humans to maximise payoffs, analysing complex and incomplete datasets to arrive at predictions, making decisions). When talking about applications or problems that AI is solving in different scientific and economic domains, one

P. Khrennikova (✉)
University of Twente, Department of High-tech Business and Entrepreneurship, Financial Engineering Section, Hallenweg 17, 7522 NH Enschede, Netherlands
e-mail: p.khrennikova@utwente.nl



V. Kreinovich et al. (eds.), *Applications of Optimal Transport to Economics and Related Topics*, Studies in Systems, Decision and Control 556,
https://doi.org/10.1007/978-3-031-67770-0_7

notes specific algorithms to solve these problems as part of various machine learning techniques. In a nutshell, machine learning considers tasks of prediction that we can use in our decisions (which involve either prediction of the expected values of continuous variables or classification of variables with a finite number of realizations). An important role in contemporary economic and financial applications is played by the reinforcement learning (RL) technique, which entails the AI algorithm interacting with an outer environment. In such a dynamic context, the algorithm processes the feedback information to form expectations about the future states of the financial markets and ultimately make better investment decisions given a set of constraints. One central algorithm used as part of RL is neural networks, also known as deep neural networks (DNN).[1]

A computational neural network conceptualization can be traced back to the 1943 paper by [26] that paved the way to two avenues of research: (i) the biological structures that would be found in a human brain and (ii) the application of the artificial neural networks as part of what is now known as contemporary AI. The paper by [15] revolutionised the wider application of neural network structures to deep learning (DL). DL is by now an established technique that can be used in complex learning environments where adaptations to the constantly changing dynamic conditions are needed. One of such burgeoning areas are financial applications encompassing as algorithmic trading, financial advice, and risk management. AI algorithms learn how financial markets react to past signals and extract the missing information to form predictions, provide trading advice, or engage in active portfolio management.

In this contribution, we stipulate a motivation from the perspective of quantum logic, for a wider exploration and implementation of quantum machine learning techniques, notably quantum neural networks, which are getting an increased attention as more powerful tools to cope with complexity, non-linearity, unpredictability and high-dimensionality of learning environment, as discussed in the recent study by [1]. We draw upon the findings from behavioural economics and, particularly, from the concept of "bounded rationality" to explain the emergent non-rationality of human expectations and decisions in complex and unfamiliar environments. Importantly, from the perspective of rational expectations and expected utility maximisation in economics, 'non-rational' behaviour manifests itself in a contravention from the axiomatics of classical probability theory and Bayesian reasoning that are at the heart of expected utility maximisation under objective and subjective risks, [22, 35]. Against this backdrop, the so-called "biases" in human reasoning (that are often used as an umbrella for all non-expected utility-type behaviour) are gaining an increasing attention as part of responsible and explainable AI, see, e.g., a survey of possible biases in applications to business decisions by [12]. While some of the biases are indeed harmful for the decision-makers themselves and potentially for other stakeholders, we would like to focus this analysis on biassed information processing and

[1] Deep neural networks are in a sense an advancement of the earlier algorithms based on artificial neural networks, where the DNNs have a more complex structure, with multiple hidden layers between the input and output layers, aimed at mimicking the human brain information processing, see [30] for detailed discussion.

expectation formation under incomplete information. Hence, our aim is to advance AI support to closely resemble the human brain in terms of underlying logic and neurological processes, whence we draw upon the findings of an emergent field of "quantum cognition" that endeavours to capture human reasoning with the aid of the mathematics and logic from quantum physics, [9, 19, 32].

## 2 Expectation Formation of Classical AI Algorithms: Applications to Financial Decisions

One of the recent developments in AI algorithms is based on deep learning, which employs reinforcement learning algorithms to pursue the best (value-maximizing) courses of action when interacting with a dynamic financial market environment. This algorithm is best suited to assisting investors with financial predictions and decisions. Furthermore, there is a two-way exchange of information with the financial environment (prices, news, social media sentiment, etc.), which is very similar to how investors learn by trading and analysing news and market data. As a result, deep learning AI systems are frequently utilised in advisory as well as active roles in portfolio management (robo-advising, algorithmic stock trading, and customised chatbots). Financial organisations are increasingly using AI algorithms to predict and manage financial risks, such as value risk and probability of loss.[2]

In the context of active portfolio management, an AI algorithm can essentially replace an investor's or portfolio manager's decision to trade based on her stated preferences (e.g., to maximise the Sharpe ratio, avoid large losses, manage liquidity levels, etc.).

It's important to note that active portfolio management strategies make the implicit assumption that markets are inefficient and that excess returns can be realised from "hidden" or overlooked information. In light of this, it is crucial to emphasise that deep neural networks are learning from historical market data including financial asset prices that can deviate from the rational expectations equilibrium values. Additionally, these AI algorithms may be trained in noisy, highly uncertain contexts where 'black swan' events are possible, such as unforeseen volatility shocks. Therefore, predicting the expectation-driven trading of investors becomes equally important to processing the large datasets on past prices and trading volumes, especially in more complex financial environments (e.g., climate investments, cryptocurrencies, derivaties).

---

[2] For an introduction to AI algorithms in finance and investments, see, e.g., the primer by [31].

## 2.1 *Human Biases and AI: Do We Need Classical Rationality?*

One of the concerns in applications of AI to prediction and decision-making is that the learning of AI algorithms occurs from datasets produced by humans that could yield inaccurate or biassed predictions made by the trained AI algorithm.

As noted in Deloitte's report: *"As AI introduces new decision-making methods, there is potential for discrimination, testing societies' moral obligations if decisions adversely affect disparate groups."*[3]

When talking about deep learning and neural network applications in finance, the "input" data yielding the final decisions, e.g., credit approval, fraud detection, portfolio construction, and risk management, is difficult to trace back as the variety of data used is immense, with a high degree of uncertainty in relation to the input data. In this regard, the main ethical considerations concerning AI biases mainly revolve around the inaccuracy of historical training data and the amplification of existing biases in the process of AI algorithm training.

We would like to emphasise a slightly different manifestation of heuristics and biases, also coined "bounded rationality" in the seminal work by [34]. It has been subsequently shown that in many predominantly complex decision-making and learning contexts, people seek to ease the cognitive load and reason with shortcuts and approximations, [13]. Such cognitive mechanisms result in seemingly irrational human expectations and decisions from the perspective of classical logic, [7].

As shown by behavioural finance scholars, the non-Bayesian update of information by investors can lead to a suboptimal choice of portfolios and subsequently a miss-pricing of financial assets. At the same time, financial assets are valued by investors based on their future expectations and hence, deviations from fundamental prices can persist for longer periods.

It has been widely observed that investors tend to process macroeconomic and company-specific signals in a non-Bayesian fashion characterised by *underreaction* and a subsequent gradual and non-linear adjustment of future expectations about financial assets, as formalised by [4] and recently for high frequency trading in [17]. The literature on "sticky expectations" about e.g., macro and company-specific variables also considers a lack of belief adjustment as a form of underreaction to news that can explain the predictability of stock prices. As documented in [8], stocks of companies with high profitability ratios enjoy consistently higher returns on a risk-adjusted basis, caused by investors' non-Bayesian expectations of future profitability levels.

The above cognitive mechanisms behind non-classical expectation formation are more pronounced in unfamiliar environments, showing that investors engage in a form of *adaptive learning* that is at variance with the normative appeal of information update following the tenets of classical probability and rational expectations frameworks, [7, 35].

[3] From: "Banking on the bots: unintended bias in AI", [38].

The above discussion leads us to a consideration of a different model of human (and AI) expectation formation based on quantum logic, relaxing some of the mathematical axioms of classical probability theory. In the next section we will discuss the advances of quantum information in AI algorithms and provide brief notes on the distinguishing features of quantum logic due to [5].

## 3 Quantum Logic and Applications to AI

Quantum AI is closely associated with the advent of quantum computers. Richard Feynman advocated back in 1982 that quantum computation could be a superior solution to classical algorithms when dealing with high dimensionality and complexity. Henceforth, the field of quantum AI was closely associated with quantum computing and quantum algorithms, see [30] for a thorough treatment. The growth of AI applications to finance, economics, language and healthcare was tremendous in the last decade, yet developed in parallel with the studies on quantum computing and machine learning that were less associated with applications to human prediction and decision. For an in-depth review of the historical developments and main advances in quantum algorithms see e.g., [37].

Quantum information studies were again recognized as ground-breaking with the award of the Nobel Prize 2022 for experimental and theoretical studies on the behaviour of quantum systems. The worldwide recognition of quantum phenomena such as superposition and entanglement in micro- and macroscopic systems closely followed renowned experimental work carried out by Google corporation, showing an absolute advantage of quantum computers in terms of a tremendous speedup of computations, [2]. It was stated by the team that: "*The Google team estimates that simulating the full circuit would take 10,000 years even on a computer with one million processing units (equivalent to around 100,000 desktop computers). Sycamore [quantum computer name] took just 3 min and 20 s*". p. 462.

Against this backdrop, we can also mention the application of quantum calculus-based algorithms to information retrieval to overcome the constraints posed by classical AI algorithms based on symbolic logic, [27, 28]. The evidence is now vast that quantum computation and quantum algorithms are capable of: (i) speeding up computation and, (ii) providing more accurate solutions to prediction and classification of outcomes in various domains as documented in [6, 11, 25, 29].

Some practical applications of QML (quantum machine learning) and hybrid quantum-classical algorithms also revolutionized applications to domains outside physics. QML and quantum circuits were implemented to produce handwritten text, [16], and to enhance the quality of data processing within the context of ghost imaging, [36].

In this venue, quantum machine learning becomes a promising application along classical AI algorithms and deep reinforcement learning approaches, [32]. One fundamental distinction of quantum machine learning from classical supervised and unsupervised learning algorithms is its grounding in quantum, rather than classical logic. The core repercussions are between how the bits in classical versus quantum computers are constructed; while the classical bit is *either* 1 or 0, the qubit can be a *superposition* of these two states. The interference of quantum states has the potential to speed up algorithms tremendously, thus allowing them to process large volumes of information from a dynamic environment, e.g., such as high-frequency financial data. While quantum computing is still in its infancy, the fundamental idea of using quantum logic as part of quantum neural network has already established some promising practical applications.

## *3.1 Quantum Logic and Its Properties*

As discussed in section (1), there is a stream of research that aims to capture human reasoning and expectation formation with the aid of quantum probability calculus that obeys quantum logic, see among other analyses and reviews in [9, 10, 19, 21, 24, 32]. The field emerged following the ample evidence from the empirical findings of the behavioural economics school, e.g., [18, 21]. Quantum probability relaxes several axioms of classical probability theory, such as distributivity, commutativity, and consequently the Bayesian formula for conditional probability. Moreover, quantum probability is grounded on Hilbert space representation, where events are represented by vectors in that space. Operationally, the relaxation of the above axioms allows for the formalization of a co-existence of multiple expectations and choices at once when applied to human cognition. Put differently, the human brain can find itself in multiple (dynamically evolving) mental states simultaneously, [9, 10, 19, 21]. In a recent paper by [20] it was shown that the human cognitive system can also exhibit quantum properties of indeterminism on a micro-level as neurons can find themselves simultaneously in states of "firing" and "non-firing" that are associated with certain energy levels.

The phenomenon of indeterminism and coexistence is formally coined in quantum theory as *superposition* of states, and can well explain the 'boundedly rational' expectations about the future states of economy, expected dividends and prices of financial assets, see quantum probability frameworks of asset trading by [23, 24]. The observed asset prices can deviate from the fundamental values based on perfect information and perfect foresight, as postulated by the rational expectations hypothesis (REH). The deviations from the uniquely possible equilibrium price states in financial markets emerge from human expectation-driven trading decisions. As shown in [24], the quantum probability-based models can explain the trading behaviours of investors, who exhibit heterogeneous beliefs and hence can 'agree to disagree on the future states of financial variables'. It can be debated, whether non-classical reasoning is optimal for humans or should be corrected by AI algorithms, e.g., [3]. Here we would like to discuss AI advancements that could support investors, by mimicking their neural processes and yet, outperforming them in terms of computational ability and speed. We consider human cognitive processes being central to subsequently advancing "intelligent" AI systems as opposed to the normative rules of expectation formation based on "black box" approach, see also an analysis in [14].

### 3.1.1 Classical Versus Quantum Probability Calculus

Below we illustrate some examples on the structure of classical and quantum logic underpinning the functioning of AI algorithms, such as the above discussed deep neural networks and quantum neural networks.

The main distinguishing property of classical logic (and classical probability calculus) is that probabilistic measurements obey the rules of commutativity and distributivity. For some events, A, B and C, one can always get the expression:

$$A \cap (B \cup C) = (A \cap B) \cup (A \cap C) \tag{1}$$

In the quantum operational framework this proposition may not hold and this is exactly what occurs, when a so called "disjunction effect" in human decisions is observed, [33]. The other notable deviations from classical logic give raise to order effects (violation of commutativity) and contextual preferences—when the joint probability distribution cannot be observed. Commutativity and distributivity are required to derive the Bayes formula for conditional probability, as one can express the marginal probabilities via the "Law of total probability", where $P$ denotes probability of an objective or subjective nature and B&C are some disjoint complementary events.

$$P(A) = P(A \cap B) + P(A \cap C) \tag{2}$$

The same formula can be expressed via conditional probabilities:

$$P(A) = P(A|B)P(B) + P(A|C)P(C) \tag{3}$$

When the above equality in Eq. (2) does not hold, one deals with non-addivity of probabilities (or non- addivity of the subjective expectations of an economic agent). The non-commutativity emergence naturally yields a deviation from conditional expectation formation following Bayesian update.

It has been shown in experimental studies that non-distributivity and non-commutativity properties of human reasoning emerge when agents form and update beliefs in less familiar or complex environments.[4]

The violation of classical probability axioms in human reasoning is of a contextual nature that closely aligns with the earlier notion of human bounded rationality. In the quantum probability framework the non-distributive feature of human reasoning can be well captured by vector-based calculus, allowing for a superposition between

[4] The circumstances in which agents switch to non classical information processing is an intriguing question, yet to be fully explored, see [9] and a discussion in the setting of financial markets in [24].

expectations about events in question. The formula, (2), becomes extended with an interference term (IT), capturing the above mentioned non-additivity. Interference is a key feature of quantum probability that is based on information processing with complex probability amplitudes rather than absolute probability values.

$$P(A) = P(A \cap B) + P(A \cap C) + IT \tag{4}$$

We note that the interference term emerges due to a representation of probabilities with complex valued amplitudes and it can have both a negative and positive (constructive or destructive) interference.

We see that alike human neural processes a quantum logic based neural network would function by employing the superposition states (characterised by interference effects) between the quantum bits, whereby instead of a path approach as employed in Turing-type algorithms grounded on symbolic logic and functionalism a simultaneous propagation of information across multiple possible paths can be captured. Such quantum based neural network algorithm would form *non-deterministic* or *contextual* expectations that would be compatible with such human features as heterogeneity and bounded rationality.

## 4 Concluding Remarks

In this contribution, we aim to identify the necessary conditions to employ AI systems that use quantum-inspired information processing to naturally capture the human neural processes that affect trading decisions in the financial marketplace. We compared the normative frameworks by comparing normative frameworks for preference formation based on unlimited cognitive ability and time as postulated under RE and bounded rationality that relaxes the assumptions on how humans deal with existing knowledge and complex information. The use of quantum logic is well motivated and has been shown to better represent the nature of human reasoning in real-life decision-making (as widely known under the umbrella of "Quantum Cognition" studies).

Furthermore, we pointed out the fact that human neural processes can exhibit quantum features such as co-existence and strong correlations between these neural states. This is why quantum AI could be a natural advancement of classical AI algorithms to deliberate and form predictions in complex and uncertain environments, serving as a helpful aid for humans in various domains of financial decision-making, see also a report by [40] outlining the recent developments in financial applications. Against this backdrop, the core argument of this analysis is that AI could become a valuable human-like apparatus that allows us to capture the (boundedly rational) mechanism of human reasoning and equips us with better tools to deal with the high complexity and multidimensionality of financial data.

## References


1. Abbas, A., Sutter, D., Zoufal, C., et al.: The power of quantum neural networks. Nat Comput. Sci. **1**, 403–409 (2021)
2. Arute, F., Arya, K., Babbush, R., et al.: Quantum supremacy using a programmable superconducting processor. Nature **574**, 505–510 (2019). https://doi.org/10.1038/s41586-019-1666-5
3. Back, C., Morana, S., Spann, M.: When do robo-advisors make us better investors? The impact of social design elements on investor behavior. J. Behav. Exp. Econ. **103**, 101984 (2023)
4. Barberis, N., Shleifer, A., Vishny, R.: A model of investor sentiment. J. Financ. Econ. **49**, 307–343 (1998)
5. Birkhoff, G., von Neumann, J.: The logic of quantum mechanics. Ann. Math. **37**, 823–843 (1936)
6. Biamonte, J., et al.: Quantum machine learning. Nature **549**, 195–202 (2017)
7. Boole, G.: An Investigation of the Laws of Thought on Which are Founded the Mathematical Theories of Logic and Probabilities. Macmillan (1958[1854]). Reprinted with corrections, Dover Publications, New York, NY (reissued by Cambridge University Press, 2009)
8. Bouchard, J.P., Krüger, P., Landier, A., Thesmar, D.: Sticky expectations and the profitability anomaly. J. Financ. **74**, 639–674 (2019)
9. Busemeyer, J., Bruza, P.: Quantum Models of Cognition and Decision. Cambridge University Press (2012)
10. Busemeyer, J., Wang, Z.: What is quantum cognition, and how is it applied to psychology? Curr. Dir. Psychol. Sci. **24**(3), 163–169
11. Daley, A.J., et al.: Practical quantum advantage in quantum simulation. Nature **607**, 667–676 (2022)
12. Dr. Varsha, P.S.: How can we manage biases in artificial intelligence systems—A systematic literature review. Int. J. Inf. Manag. Data Insights **3**(1), 100–165 (2023)
13. Gigerenzer, G., Selten, R.: Bounded Rationality: The Adaptive Toolbox. Dahlem Workshop Reports (2001)
14. Glimcher, P., Fehr, E. (eds.): Neuroeconomics: Decision Making and the Brain. Academic Press, Elsevier (2014)
15. Hornik, K., Stinchcombe, M.B., White, H.L.: Multilayer feedforward networks are universal approximators. Neural Netw. **2**, 359–366 (1989)
16. Huang, H.L., et al.: Experimental quantum generative adversarial networks for image generation. Phys. Rev. Appl. **17**, 1013–1017 (2021)
17. Jiang, H., Zhengzi, L.S., Wang, H.: Pervasive underreaction: evidence from high-frequency data. J. Financ. Econ. **141**(2), 573–599 (2021)
18. Kahneman, D.: Thinking. Fast and Slow, Farrar Straus and Giroux (1994)
19. Khrennikov, A.: Open systems, quantum probability, and logic for quantum-like modeling in biology, cognition, and decision-making. Entropy **25**, 886 (2023). https://doi.org/10.3390/e25060886
20. Khrennikov, A., Basieva, I., Pothos, E.M., et al.: Quantum probability in decision making from quantum information representation of neuronal states. Sci. Rep. **8**, 16225 (2018)
21. Khrennikov, A., Haven, E.: Quantum mechanics and violations of the sure—thing principle: the use of probability interference and other concepts. J. Math. Psychol. **53**, 378–388 (2009)
22. Kreps, M.D.: Notes on the Theory of Choice. Westview Press (1988)
23. Khrennikova, P., Patra, S.: Asset Trading under non-classical ambiguity and heterogeneous beliefs. Phys. Stat. Mech. Appl. **521**, 562–577 (2019)
24. Khrennikova, P.: Application of quantum master equation for long term prognosis of asset prices. Phys. Stat. Mech. Appl. **450**, 253–263 (2016)
25. Liu, X., Yang, H., Gao, J., Wang, C.: FinRL: deep reinforcement learning framework to automate trading in quantitative finance. In: Proceedings of the Second ACM International Conference on AI (2021). https://doi.org/10.1145/3490354.3494366in Finance

26. McCulloch, W., Pitts, W.: A logical calculus of ideas immanent in nervous activity. Bull. Math. Biophys. **5**(4), 115–133 (1943)
27. Melucci, M.: Introduction to Information Retrieval and Quantum Mechanics. Springer, Berlin (2015)
28. Melucci, M.: Deriving a quantum information retrieval basis. Comput. J. **56**(11), 1279–1291 (2013)
29. Montanaro, A.: Quantum algorithms: an overview. Quantum Inf. **2**, 1–8 (2016)
30. Nielson, M.A., Chuang, I.L.: Quantum Computation and Quantum Information. Cambridge University Press (2010)
31. Osterrieder, J., GPT, C.: A Primer on Deep Reinforcement Learning for Finance (January 2, 2023). Available at SSRN: https://ssrn.com/abstract=4316650 or https://doi.org/10.2139/ssrn.4316650
32. Plotnitsky, A., Haven, E. (eds.) The Quantum -Like Revolution: A Festschrift for Andrei Khrennikov. Springer, Berlin/Heidelberg, Germany, New York, NY, USA (2023)
33. Shafir, E., Tversky, A.: Thinking through uncertainty: non-consequential reasoning and choice. Cogn. Psychol. **24**(4), 449–474 (1992)
34. Simon, H.A.: Behavioral model of rational choice. Q. J. Econ. **69**, 99–118 (1955)
35. Von Neumann, J., Morgenstern, O.: Theory of Games and Economic Behaviour, Princeton. Princeton University Press, NJ (1953)
36. Xiao, T., Zhai, X., Wu, X., et al.: Practical advantage of quantum machine learning in ghost imaging. Commun. Phys. **6**, 171 (2023). https://doi.org/10.1038/s42005-023-01290-1
37. Ying, M.: Quantum computation, quantum theory and AI. Artif. Intell. **174**(2), 162–176 (2010)
38. Deloitte UK., Banking on the bots: unintended bias in AI. https://www2.deloitte.com/uk/en/pages/financial-services/articles/banking-on-the-bots-unintended-bias-in-ai.html) (2023). Accessed on 7 Sept 2023
39. Financial Times: Advance of the AI adviser: machine learning's role in managing fortunes. https://on.ft.com/3ZfOELI (2023). Accessed on 31 Aug 2023
40. OECD: Artificial Intelligence, Machine Learning and Big Data in Finance: Opportunities, Challenges, and Implications for Policy Makers. https://www.oecd.org/finance/artificial-intelligence-machine-learningbig-data-in-finance.htm (2021). Accessed on 4 Sept 2023

# Digital Economy, Labor Mobility and Industrial Structure Optimization—Empirical Analysis Based on Mediating Effect and Threshold Effect

**Jianxu Liu, Zhidan Shen, and Yansong Li**

**Abstract** Using the provincial panel data in China from 2014 to 2021, this paper examines the effects and paths of digital economy and labor mobility on industrial structure optimization, based on the construction of industrial structure upgrading index and digital economy comprehensive development level index. The mediation effect model and the panel threshold model are used in the analysis process. The empirical results show that: ① Digital economy can promote the optimization and upgrading of industrial structure by facilitating the mobility of labor. ② Adopting the threshold model, this paper finds that the effect of digital economy on the optimization of industrial structure presents non-linear characteristics. ③ Both digital economy and industrial structure are characterized by heterogeneity in terms of regional distribution. Therefore, we can conclude that in order to realize the goal of high-quality development, it is necessary to actively promote the balanced development of the digital economy. We are supposed to further improve the relevant labor market policies to promote the coordinated development of the industrial structure and the digital economy.

**Keywords** Digital economy · Labor mobility · Industrial structure

## 1 Introduction

The digital economy is a broad concept, broadly defined to include all forms of economic activity that directly or indirectly use data to manage the role of resources and increase productivity [1]. In April 2023, China Academy of Information and Communications Technology released "Research Report on the Development of

J. Liu · Z. Shen · Y. Li (✉)
School of Economics, Shandong University of Finance and Economics, Jinan 250014, China
e-mail: 1364393958@qq.com

J. Liu
e-mail: liujianxu1984@163.com



V. Kreinovich et al. (eds.), *Applications of Optimal Transport to Economics and Related Topics*, Studies in Systems, Decision and Control 556,
https://doi.org/10.1007/978-3-031-67770-0_8

China's Digital Economy (2023)". According to the report: the scale of China's digital economy has reached 50.2 trillion yuan in 2022, with a nominal year-on-year growth of 10.3%, significantly higher than the nominal GDP growth rate in the same period for 11 consecutive years. The share of the digital economy in GDP has reached 41.5%, equivalent to the share of the secondary industry in the national economy. It can be seen that the digital economy has been and will be the core driving force for change in economic development in the longer term.

The digital economy has become a globally recognized driver of industrial upgrading. However, established studies have not reached a consensus on the impact mechanism. Based on the perspective of carbon emission, more scholars have tested and argued that the development of digital economy can promote the rationalization and upgrading of industrial structure through empirical analysis [2–4]. Zhao et al. [5] argued that the digital economy significantly promotes the technological progress and human capital upgrading of China's cities, which in turn promotes the upgrading of industrial structure. In addition, some studies have also demonstrated the impact of the digital economy on industrial structure from the perspective of industrial agglomeration [6]. However, there are few empirical analyses from the perspective of labor mobility.

Established studies have discussed the role of the digital economy as a driver of labor mobility. For example, Kawashima [7] found that the digitization of work has gradually enabled firms to outsource IT-supported service work to low-cost overseas production sites. However, this paper did not further analyze the relationship between labor mobility and industrial structure optimization. To fill this gap, this paper integrates digital economy, labor mobility, and industrial structure optimization into a unified analytical framework based on panel data from 31 provinces (autonomous regions and municipalities) in China from 2014 to 2021. Taking labor mobility as the starting point, this paper explores the interaction mechanism between digital economy and industrial structure optimization. In addition, this paper also analyzes the impact of different stages of digital economy development on industrial structure optimization. The empirical results show that: ① Digital economy obviously promotes industrial structure optimization in each province. ② Digital economy can influence the labor flow in each region, thus exerting the mediating effect to indirectly influence the adjustment of regional industrial structure. There is a single-threshold effect of digital economy on industrial structure upgrading.

The possible marginal contribution of this paper lies in two aspects: firstly, this paper adds to the literature on digital economy for industrial structure upgrading through the perspective of labor mobility; secondly, this paper analyses the impact of labor mobility on industrial structure upgrading, which helps to enrich the relevant research and is also of great practical significance for promoting reasonable labor mobility and achieving high-quality economic development.

## 2 Theoretical Analysis and Research Hypothesis

The continuous optimization of the industrial structure of market players has stimulated the endogenous momentum of development. It has enhanced the coordination and sustainability of China's economic development and further consolidated China's super-sized market advantage. Industrial upgrading based on the digital economy is becoming a new approach to high-quality economic development, laying a solid foundation for gaining new advantages in long-term development.

### *2.1 The Influence Mechanism of Digital Economy on Industrial Structure Optimization*

The digital economy is a new economic form carried by modern communication technology, using digitized information and knowledge as factors of production and integrating them with traditional industries. Its manifestations of openness, high resource mobility and value co-creation are important foundations for the enhancement of innovation capacity. At the same time, the digital economy boom has also brought about an impact on society and people's livelihood that can't be ignored, especially bringing a huge impact on the labor market. Established studies have shown that limited employment opportunities and low wage rates lead to labor migration out of the original region [8]. Therefore, abundant income and suitable job opportunities are the main evaluation criteria for workers to make mobility decisions. Due to the characteristics of non-competition, zero marginal cost and positive externality of digital information [9], the development of digital economy does not strongly depend on the initial economic development level of the region. Late-developed regions can directly introduce mature digital technology at low cost, and use the platform economy to reduce the cost of financial services, information access and logistics. At the same time, it can improve the circulation efficiency of production factors and market accessibility [10, 11], thus realizing the "frog-leaping" development of late-developed regions. The development of the digital economy creates a large number of employment opportunities and promotes intercity labor mobility.

Romer argues that long-term economic growth is contributed by technological progress, while short-term economic growth is contributed by increases in capital, labor, and other factor inputs [12]. Through labor mobility, labor factors are reallocated among different sectors and regions, and changes in the employment structure of the labor force bring about changes in factor productivity, which plays an important role in the economic growth of the whole country or region [13]. In addition, the labor force also plays an important role in industrial upgrading. Studies have shown that the steady accumulation of human capital inflow helps cities to attract a large number of high-quality workers under relatively unchanged trade costs. As a result, the urban consumer market will expand accordingly, and the regional market effect will expand. High value-added products will have the effect of selection into the market, while low value-added industries and industrial structure will change, and traditional manufacturers will be squeezed out [14, 15].

In summary, the development of digital economy may play a certain positive role in the optimization of regional industrial structure by using labor mobility as a mediating path. Therefore, this paper proposes research hypothesis 1: digital economy may have a positive effect on industrial structure optimization by improving labor mobility.

### 2.2 The Nonlinear Effect of Digital Economy on Industrial Structure Optimization

The application of digital technology poses a great challenge to the traditional economic model. To some extent, it effectively solves the inherent problems such as market information asymmetry [16]. In the early stage of digital economy development, the digital network penetration rate is low and cannot significantly reduce the cost of industrial sectors [17]. When the development of the digital economy reaches a certain level, the value of the Internet is fully utilized with the increase of participating users. The industrial sector realizes the trans-formation and upgrading at a lower cost and promotes the optimization of the regional industrial structure, while the optimization of the regional industrial structure generates diversified demand for the development of the digital economy, triggering a positive feedback mechanism [18].

Therefore, this paper proposes research hypothesis 2: the impact of digital economy on industrial structure optimization is non-linear.

## 3 Methodology and Data

### 3.1 Model Construction

**Basic model**

$$idc_{i,t} = \alpha_0 + \alpha_1 dige_{i,t} + \alpha_c Z_{i,t} + \mu_i + \delta_t + \varepsilon_{i,t} \quad (1)$$

In formula (1), $idc_{i,t}$ is the industrial structure of province (municipalities and counties) i in t period; $dige_{i,t}$ is the development level of digital economy of province (municipalities and counties) i in t period; vector $Z_{i,t}$ represents a set of control variables; $\mu_i$ represents the individual fixed effect of province (municipalities and counties) i that does not change over time; $\delta_t$ control time fixed effect; $\varepsilon_{i,t}$ represents a random disturbance term.

**Intermediary effect model**. In addition to the direct effect captured in Eq. (1), in order to discuss the possible mechanisms of action of the digital economy on industrial structure, this paper tests whether labor mobility is a mediating variable between the two. The specific steps are as follows: ① Confirm that the significance of the coefficient $\alpha_1$ of the linear regression model (1) of the digital economy development index *(dige)* on the industrial structure upgrading index *(idc)* passes the test ② Construct the linear regression equation of *(dige)* on the mediator variable of the net inflow rate of labor *(mig)* ③ Construct the regression equation of *(dige)* and the mediator variable of *(mig)* on *(idc)* ④ Use the significance of regression coefficients such as $\beta_1$, $\gamma_1$ and $\gamma_2$ to determine the existence of mediation effects.

$$mig_{i,t} = \beta_0 + \beta_1 dige_{i,t} + \beta_c Z_{i,t} + \mu_i + \delta_t + \varepsilon_{i,t} \tag{2}$$

$$idc_{i,t} = \gamma_0 + \gamma_1 dige_{i,t} + \gamma_2 mig_{i,t} + \gamma_c Z_{i,t} + \mu_i + \delta_t + \varepsilon_{i,t} \tag{3}$$

**Threshold model**. To empirically test the indirect transmission mechanism, in addition to the mediation effect model, the characteristics of different development stages of the digital economy should also be considered. The level of development of the digital economy may have a non-linear effect on the digital economy to promote the optimization of industrial structure. Therefore, set the panel threshold model as follows:

$$idc_{i,t} = \varphi_0 + \varphi_1 dige_{i,t} \times IAdj_{i,t} \leq \theta + \varphi_2 dige_{i,t} \times IAdj_{i,t} > \theta + \varphi_c Z_{i,t} + \mu_i + \varepsilon_{i,t} \tag{4}$$

In formula (4), $Adj_{i,t}$ are threshold variables such as digital economy and labor flow. I(·) is an indicator function with a value of 1 or 0, which is 1 if the conditions in parentheses are met, otherwise 0. Formula (4) considers the single threshold case, which can be extended to the multi-threshold case according to the steps such as measurement test of sample data.

## 3.2 Measurement and Description of Variables

**Explained variable**. Industrial structure upgrading index (*idc*). Learn from the method of Xu Min et al. [19], the Industrial Structure Upgrade Index (*idc*) is used to measure the industrial structure of each province. The calculation method is as follows (*Xi* is the share of industry *i* in the three industries):

$$idc = 1 \times X1 + 2 \times X2 + 3 \times X3 \tag{5}$$

**Core explanatory variable**. Digital economy comprehensive development index (*dige*). Considering that there are certain difficulties and deviations in directly estimating the scale of digital economy, this paper measures the scale of digital economy by constructing the digital economy development indicator system. Drawing on the ideas of Zhao Tao et al., this paper constructs the indicator system according to the reality of digital economy development and related concepts. The system takes the level of Internet development as the core of measurement, adding digital transaction-related measurements for supplementation and improvement. The system mainly contains five second level indicators and five third level indicators [20]. Table 1 shows the meaning of specific indicators. In this paper, the entropy value method is used to calculate the index of digital economy development level (*dige*) after giving weight to each indicator.

**Mediating variable**. Net inflow rate of labor force (*mig*). The net labor inflow rate is taken as 80% of the official net population inflow rate [21]

$$\begin{aligned} &Rate\ of\ net\ inflow\ of\ labor \\ &= \left[(Permanent\ population - Registered\ population)/Registered\ population\right] \\ &*80\% \end{aligned} \tag{6}$$

### Control variable

*Wage level(lnwage).* China's urban population is large and mobile, which to some extent determines the phenomenon of apparent wage differentials within Chinese cities. Therefore, this paper uses the average wage of employees to indicate the wage level of the labor market. Lichun Xiong et al. found that the average wage level of employees and the number of students enrolled in colleges and universities have a

**Table 1** Index system of comprehensive development level of digital economy

| First level indicator | Second level indicator | Third level indicator | Index attribute |
|---|---|---|---|
| Digital economy Comprehensive Development Index | Internet penetration rate | Internet users per 100 people | + |
| | Internet related practitioners | Percentage of employees in computer services and software | + |
| | Internet-related output | Total telecommunications services per capita | + |
| | Number of mobile Internet users | Number of mobile phone users per 100 people | + |
| | Inclusive development of digital finance | China Digital Financial Inclusion Index | + |

significant positive impact on the upgrading of the industrial structure of Beijing-Tianjin-Hebei at the (0.50–0.75) quantile [22]. Therefore, this paper expects that the increase in wage level will promote the upgrading of industrial structure.

*Urbanization rate (urban).* The degree of urbanization is measured by the ratio of the urban population to the total population. The development of urbanization is not only the transfer of the rural population to the city, but also accompanied by a change in the mode of production and living. Urbanization has an important impact on labor mobility, and the urbanization rate has a positive impact on industrial structure [23].

*Per capita fiscal expenditure (lngov).* Government support is conducive to easing bottlenecks in industrial development, thereby promoting the optimization of industrial structure. The main form of government support is government fiscal expenditure. This paper adopts the ratio of government fiscal expenditure to total population to express per capita fiscal expenditure. Therefore, this paper expects that an increase in per capita fiscal expenditure will promote the optimization of industrial structure.

*Degree of openness (lnopen).* It is an important indicator of the degree of openness of a country's economy. In this paper, it is expressed as the ratio of total regional imports and exports to total regional output. The study by Wei Wang et al. shows that the increase in foreign import and export trade will accelerate the transfer of industries from developed countries to developing countries. By taking over foreign industries, developing countries can skip the technological progress in optimizing the industrial structure of the equipment manufacturing industry. So that they can directly enter the stage of product upgrading [24]. Therefore, this paper expects that the increase in the degree of openness will have an impact on accelerating the optimization of industrial structure.

## *3.3 Data Sources and Descriptive Statistics*

The data in this study are panel data of 31 provinces (cities and counties) in China during 2014–2021. The data are mainly from "China Statistical Yearbook (2015–2022)" and "Provincial Statistical Yearbook (2015–2022)"; the "Digital Inclusive Finance Index" is derived from the "China Digital Inclusive Finance Index (2011–2021)" compiled by a joint research group formed by the Digital Fi-nance Research Center of Peking University and the Ant Technology Group Research Institute; the data of "Household population" are from the "China County Statistical Yearbook (2015–2022)" (Table 2).

**Table 2** Descriptive statistics of variables

| | Variable | Observed number | Mean value | Standard deviation | Minimum value | Maximum value |
|---|---|---|---|---|---|---|
| Explained variable | *idc* | 248 | 2.420 | 0.115 | 2.158 | 2.838 |
| Core explanatory variable | *dige* | 248 | 0.245 | 0.127 | 0.0546 | 0.684 |
| Mediating variable | *mig* | 248 | 0.0334 | 0.157 | −0.135 | 0.572 |
| Control variable | *lnwage* | 248 | 11.22 | 0.292 | 10.65 | 12.18 |
| | *urban* | 248 | 0.606 | 0.121 | 0.262 | 0.893 |
| | *lngov* | 248 | 9.576 | 0.427 | 8.740 | 11.01 |
| | *lnopen* | 248 | −1.861 | 0.963 | −4.875 | 0.126 |

**Table 3** Results of Levin-Lin-Chu unit root test

| Variable | t-value | p-value | Result |
|---|---|---|---|
| *idc* | −19.2947 | 0.0000 | Stationary |
| *dige* | −9.5174 | 0.0000 | Stationary |
| *mig* | −17.7385 | 0.0000 | Stationary |
| *lnwage* | −13.4193 | 0.0000 | Stationary |
| *urban* | −7.6065 | 0.0000 | Stationary |
| *lngov* | −11.2696 | 0.0000 | Stationary |
| *lnopen* | −36.3380 | 0.0000 | Stationary |

# 4 Empirical Results

## 4.1 Stationarity Test

Stationarity test Smoothness test means that the observed value has a relatively small error compared to the true value, i.e., the statistic has a certain characteristic or property on a certain distribution. In order to prevent pseudoregression, this paper performs the Levine-Lin-Chu unit root test on the selected variables [25]. Table 3 shows the results of the panel unit root test, and it can be seen that the variables have passed the stationarity test.

## 4.2 Regression Result Analysis

Stata16.0 software was used for all data processing in this paper.

## 4.3 Benchmark Regression

Table 4 reports the results of the linear estimation of the impact of the digital economy on industrial structure. In columns (1) and (2), the estimated coefficient of the core explanatory variable, the digital economy development index *(dige)*, is significantly positive, suggesting that the digital economy promotes the upgrading of the industrial structure in the provinces. In addition, column (2) is shown with control variables added: ① There is a non-significant positive correlation between the level of wage income *(lnwage)* and industrial structure, indicating that the wage level in the labor market has not been effectively improved with the optimization of industrial structure. ② The coefficient value of urbanization rate is positive but insignificant, indicating that urbanization does not significantly promote the upgrading of industrial structure in the region. This may be due to the fact that there is always a certain degree of deviation between the path of urbanization and the process of industrial structure upgrading, which results in China's industrialization being ahead of schedule and urbanization lagging behind [26]. ③ Per capita financial expenditure *(lngov)* and the degree of openness *(lnopen)* are positively correlated with the upgrading of industrial structure and remain significant at the 1% level. This shows the importance of government support and the degree of economic openness for the optimization and upgrading of industrial structure.

**Table 4** Benchmark regression results of the impact of digital economy on industrial structure

| Variable | *idc* | |
|---|---|---|
| | (1) | (2) |
| *dige* | 0.308***<br>(0.0195) | 0.111***<br>(0.0313) |
| *lnwage* | | 0.0163<br>(0.0219) |
| *urban* | | 0.135<br>(0.129) |
| *lngov* | | 0.0852***<br>(0.0259) |
| *lnopen* | | 0.0208***<br>(0.00521) |
| Constant term | 2.344***<br>(0.00503) | 1.350***<br>(0.202) |
| Province fixed | Yes | Yes |
| Year fixed | Yes | Yes |
| Period number | 8 | 8 |
| Number of provinces | 31 | 31 |
| *R-squared* | 0.536 | 0.637 |

*Note* ***, ** and *represent significance levels of 1, 5 and 10% respectively; values in parentheses are robust standard errors, as in the table below

**Intermediate effect model**

The previous paper theoretically analyzed the transmission mechanism of digital economy affecting industrial structure from the perspective of labor mobility. In order to verify the hypothesis of this mechanism of action, this paper chooses the mediation effect model for empirical testing. The regression results are shown in Table 5. Column (3) confirms that the digital economy has a cumulative effect on industrial structure, based on which, column (4) verifies whether the digital economy promotes labor mobility between provinces. The regression coefficients for the level of digital economy development in both are significantly positive. Finally, the mediating variable of labor mobility is reintroduced into the regression equation of digital economy's influence on industrial structure. By observing the coefficient values of the core explanatory variables and the change in significance, it can be concluded that the coefficient of the influence of digital economy on industrial structure in column (5) has decreased compared to that in column (3). This indicates that labor mobility is the mechanism of action of the digital economy in optimizing industrial structure. This empirical result supports Hypothesis 1.

**Nonlinear effect analysis**

This paper provides a theoretical elaboration of the existence of nonlinear effects on the impact of the digital economy on industrial structure optimization. To test hypothesis 2, a panel threshold regression model is used here for empirical testing. Before estimating the threshold model, a panel threshold existence test is first conducted based on Hansen's (1999) method [27]. After 1,000 times repeated sampling by the "bootstrap" method, the results show that: the threshold variables of the digital economy development index significantly passed the single threshold and failed the double and triple threshold tests; all other variables failed the single threshold test. A single-threshold regression model is built on this basis, and the regression results are shown in Table 6.

**Table 5** Test results of the effect mechanism of digital economy on industrial structure

| Variable | *idc* | *mig* | *idc* |
|---|---|---|---|
| | (3) | (4) | (5) |
| *dige* | 0.111***<br>(0.0313) | 0.0474**<br>(0.0224) | 0.0739*<br>(0.0435) |
| *mig* | | | 0.383***<br>(0.0443) |
| Control variable | Yes | Yes | Yes |
| Province fixed | Yes | Yes | Yes |
| Year fixed | Yes | Yes | Yes |
| Period number | 8 | 8 | 8 |
| Number of provinces | 31 | 31 | 31 |
| *R-squared* | 0.637 | 0.235 | 0.810 |

**Table 6** Regression results of the digital economy influence threshold model of industrial structure

| Variable | *idc* |
|---|---|
| | (6) |
| Threshold value $q$ | 0.157 |
| $dige*I\ (Th \leq q)$ | −0.0258<br>(0.0710) |
| $dige*I\ (Th > q)$ | 0.0873**<br>(0.0322) |
| Control variable | Yes |
| Period number | 8 |
| Number of provinces | 31 |
| *R-squared* | 0.676 |

From column (6), it is easy to see that as the value of digital economy development index increases, the impact of digital economy development on industrial structure shows non-linear characteristics. When the digital economy development index is lower than 0.157, the coefficient of *dige* is negative and insignificant. This may be due to the unbalanced development of the digital economy and the low penetration rate of the digital network in the regions at the initial stage of the development of the digital economy. Therefore, it cannot significantly reduce the cost of industrial sectors. Its impact on the industrial structure is uncertain. When the digital economy development index is higher than 0.157, the coefficient of *dige* is positive and significant at the 5% level, which indicates that the digital economy can restructure production relations. Thus, it can create new productivity and promote the optimization and upgrading of traditional industries. This is consistent with the findings of Sun Tianhao et al. [28]. Hypothesis 2 is supported.

**Regional heterogeneity**

Due to the different stages of development, both the level of development of the digital economy and the industrial structure are characterized by heterogeneity in regional distribution. Therefore, the impact of the digital economy on the industrial structure may also have regional heterogeneity. It is necessary to discuss this in depth. According to the principle of combining the level of economic and technological development and geographical location, China is divided into three major economic zones. From Table 7, we can see that the eastern region is ahead of the central and western regions in both industrial structure optimization and digital economy. This result lays the foundation for testing the regional heterogeneity of the digital economy, which affects the industrial structure.

Table 8 shows the regression analysis of regional heterogeneity. The results in columns (7), (8), and (9) show that digital economy has a significant effect on industrial structure optimization in the eastern and central regions, and the effect is not significant in the western region. This indicates that the positive effect of the digital economy on industrial structure optimization is stronger in the eastern and central regions when regional heterogeneity is taken into account. This result may be due

**Table 7** Differences in industrial structure and digital economy among different regions

| Industrial structure optimization (*idc*) | | | | | |
|---|---|---|---|---|---|
| Classification by region | Sample size | Mean value | Standard deviation | Minimum value | Maximum value |
| Eastern region | 80 | 2.513 | 0.145 | 2.293 | 2.838 |
| Central region | 64 | 2.364 | 0.0723 | 2.158 | 2.466 |
| Western region | 96 | 2.383 | 0.0502 | 2.268 | 2.491 |
| Digital economy development index (*dige*) | | | | | |
| Eastern region | 80 | 0.341 | 0.138 | 0.0885 | 0.684 |
| Central region | 64 | 0.216 | 0.0888 | 0.0877 | 0.445 |
| Western region | 96 | 0.190 | 0.0928 | 0.0546 | 0.449 |

**Table 8** Test of regional heterogeneity of industrial structure affected by digital economy

| Variable | Eastern region | Central region | Western region |
|---|---|---|---|
| | (7) | (8) | (9) |
| *dige* | 0.110***<br>(0.0355) | 0.190**<br>(0.0761) | 0.0862<br>(0.0559) |
| Control variable | Yes | Yes | Yes |
| Constant term | 1.649***<br>(0.187) | 1.750**<br>(0.732) | 2.080***<br>(0.526) |
| Province fixed | Yes | Yes | Yes |
| Year fixed | Yes | Yes | Yes |
| Period number | 8 | 8 | 8 |
| Number of provinces | 10 | 8 | 12 |
| *R-squared* | 0.838 | 0.772 | 0.503 |

to the fact that the digital economy has developed earlier and at a higher level in the eastern and central regions of China than in the western region, which makes the release of the digital economy dividend more appropriate.

## 5 Robustness Test

### *5.1 Alternate Explanatory Variable*

The digital financial inclusion index is used as an alternative core explanatory variable. The index is constructed from three dimensions of digital financial coverage breadth, usage depth and digitization level to construct the index system. It is both vertically and horizontally comparable, so it can more comprehensively reflect the

**Table 9** Robustness test results

| Variable | *idc* | |
|---|---|---|
| | (10) | (11) |
| *dige* | 0.153***<br>(0.0347) | 0.105***<br>(0.0325) |
| *lnwage* | −0.0858**<br>(0.0333) | 0.0115<br>(0.0234) |
| *urban* | −0.219<br>(0.149) | 0.0460<br>(0.141) |
| *lngov* | 0.0986***<br>(0.0254) | 0.103***<br>(0.0265) |
| *lnopen* | 0.0426<br>(0.0311) | 0.0238<br>(0.0312) |
| Constant term | 1.703***<br>(0.263) | 1.241***<br>(0.218) |
| Province fixed | Yes | Yes |
| Year fixed | Yes | Yes |
| Period number | 8 | 8 |
| Number of provinces | 31 | 31 |
| *R-squared* | 0.651 | 0.661 |

level of regional digital economy development [29]. The results in column (10) of Table 9 show that the regression coefficient of digital inclusive finance on industrial structure upgrading is positive at the 1% significance level. The results of the robustness test by replacing the explanatory variables are basically consistent with the results of the benchmark model.

## *5.2 Winsorization*

In the benchmark regression, there may be abnormally distributed values, which may bias the regression results. To ensure the accuracy of the regression results, the continuous variables are winsorized at the 1 and 99% levels [30]. The regression was then rerun. The results are presented in column (11) of Table 9. The regression results show that the promotion effect of digital economy on industrial structure optimization passes the significance test at 1% level. And the regression coefficients are similar to the benchmark regression, which means that the regression results are robust.

## 6 Conclusions and Policy Implication

In his report to the 20th Party Congress, General Secretary Xi Jinping emphasized: "Accelerate the development of the digital economy, promote the deep integration of the digital economy and the real economy, and build an internationally competitive digital industry cluster." The digital economy is an important embodiment of national comprehensive strength in the digital era. It is also the focus of a new round of international competition and a key area of play. Optimizing the industrial structure through the digital economy has become an important support for China's economic transformation. This paper cuts from the perspective of labor mobility. Utilizing China's province-level data from 2014 to 2021, this paper constructs an index of industrial structure upgrading and an index of the comprehensive development level of the digital economy. The impact of digital economy on industrial structure upgrading is empirically examined in various aspects by using fixed effect model, mediation effect model and panel threshold model.

The main conclusions are as follows: ① The digital economy obviously pro-motes the upgrading of the industrial structure of the provinces. The conclusion still holds through robustness tests after replacing the explanatory variables. In terms of regional heterogeneity, the eastern region enjoys a larger digital economy dividend than the central and western regions. ② The development of the digital economy can play a mediating effect by affecting the labor mobility in each region, which indirectly affects the adjustment of regional industrial structure. And the intermediary variable of labor mobility has a greater incentive effect on the optimization of industrial structure. ③ The nonlinear effect of digital economy in the process of industrial structure upgrading is also confirmed. There is a single threshold effect of digital economy on industrial structure upgrading. When the level of digital economy development is lower than the conditional threshold, the effect of digital economy on industrial structure upgrading is not significant; when the level of digital economy development slowly increases and crosses the single threshold value, digital economy will significantly affect industrial structure upgrading.

The findings of this paper also have the following policy implications: firstly, based on the reality that the digital economy can promote the upgrading of industrial structure, it is reasonable to improve the level of digital economy development. At the same time, focusing on R&D and human capital production can enhance the attractiveness of cities for labor. This can support the transformation and upgrading of the economic structure, and make the economy ascend to the high-quality aspect. Secondly, the positive impact of the digital economy on the upgrading of the industrial structure in the western region has yet to be deepened. This suggests the need to strengthen the construction of digital infrastructure in less economically developed regions. The rights of digitally disadvantaged groups should be guaranteed, and the "digital divide" should be continuously nar-rowed, so as to balance equity and efficiency in the process of economic development. Finally, the digital economy can serve as a mechanism to have a positive impact on the upgrading of industrial structure by improving labor mobility. This proves that high-quality economic development should be in line with China's national conditions. Because of China's large population base, it is necessary to improve the supply and demand system of the labor market, and to introduce ap-propriate policies. In this way, labor can move freely within each region, thus promoting industrial development and realizing industrial upgrading.

This study has the following areas for expansion: firstly, the study was con-ducted at the provincial level. In the future, the study can be expanded to the level of prefecture-level cities; secondly, there may also be spatial effects in the economic variables, which can be explored in the future.

Note: China's three economic belts are divided into: the eastern region includes 11 provinces (cities), including Beijing, Tianjin, Hebei, Liaoning, Shanghai, Jiangsu, Zhejiang, Fujian, Shandong, Guangdong and Hainan; The central region includes Shanxi, Jilin, Heilongjiang, Anhui, Jiangxi, Henan, Hubei and Hunan provinces. The western region includes 12 provinces (municipalities and counties) of Inner Mongolia, Guangxi, Chongqing, Sichuan, Guizhou, Yunnan, Tibet, Shaanxi, Gansu, Qinghai, Ningxia and Xinjiang.

## References

1. Williams, L.D.: Concepts of digital economy and industry 4.0 in intelligent and information systems. Int. J Intell. Netw. **2**, 122–129 (2021).
2. Gu, R., Li, C., Yang, Y., et al.: Impact of digital economy development on carbon emission intensity in the Beijing–Tianjin–Hebei region: a mechanism analysis based on industrial structure optimization and green innovation. Environ. Sci. Pollut. Res. **30**(14), 41644–41664 (2023)
3. Chen, S., Ding, D., Shi, G., et al.: Digital economy, industrial structure, and carbon emissions: an empirical study based on a provincial panel data set from China. Chines. J. Popul. Resour. Environ. **20**(4), 316–323 (2022)
4. Jiang, H., Chen, Z., Liang, Y., et al.: The impact of industrial structure upgrading and digital economy integration on China's urban carbon emissions. Front. Ecol. Evol. **11** (2023)
5. Zhao, S., Peng, D., Wen, H., et al.: Does the digital economy promote upgrading the industrial structure of Chinese cities. Sustainability **14**(16), 10235 (2023)
6. Chang, K., Zhang, H., Li, B.: The Impact of digital economy and industrial agglomeration on the changes of industrial structure in the Yangtze River delta. J. Knowl. Econ. 1–21 (2023)
7. Kawashima, K.: Service outsourcing and labour mobility in a digital age: transnational linkages between Japan and Dalian, China. Glob. Netw. **17**(4), 483–499 (2017)
8. Rahman, A.: Mobilitas Tenaga Kerja Dan Dampaknya Terhadap Kehidupan Sosial Budaya Pada Masyarakat Perdesaan. Jurnal Ekonomi dan Riset Pembangunan **1**(3), 53–63 (2023)
9. Arrow, K.J.: The economic implications of learning by doing. Rev. Econ. Stud. **29**(3), 155–173 (1962)
10. Chen, M.X., Wu, M.: The value of reputation in trade: evidence from Alibaba. Rev. Econ. Stat. **103**(5), 857–873 (2021)
11. Carballo, J., Chatruc, M.R., Santa, C.S., et al.: Online business platforms and international trade. J. Int. Econ. **137**, 103599 (2022)
12. Romer, D.: Advanced Macroeconomics. A Dzz~zszon of The McGraw. Hil1~ompante (1996)
13. Romer, P.M.: Endogenous technological change. J. Polit. Econ. **98**(5), 71–102 (1990)
14. Blundell, R., Bond, S.: Initial conditions and moment restrictions in dynamic panel data models. J. Econ. **87**(1), 115–143 (1998)

15. Anselin, L.: Spatial Econometrics: Methods and Models. Springer Science & Business Media (1988)
16. Khotamov, I.S.: World Tendencies and Development of the Digital Economy in Uzbekistan. Архив научных исследований (2019)
17. Chen, Y., Zhang, X.: The influence mechanism of digital economy development on the optimization of regional industrial structure–an empirical analysis based on provincial panel data from 2011–2019. Collected Essays Financ. Econ. **297**(04), 14–23 (2023). (Chinese)
18. Guo, B., Wang, Y., Zhang, H.: Digital economy, green technology innovation and industrial structure upgrading–empirical evidence from 282 cities in China. Lanzhou Acad. J. **2**, 58–73 (2022). (Chinese)
19. Xu, M., Jiang, Y.: Can China's industrial structure upgrade narrow the urban-rural consumption gap. J. Quant. Econ. **32**(03), 3–21 (2015). (Chinese)
20. Zhao, T., Zhang, Z., Liang, S.: Digital economy, entrepreneurial activity and high quality development–empirical evidence from Chinese Cities. Manag. World **36**(10), 65–76 (2020). (Chinese)
21. Card, D.: Immigrant inflows, native outflows, and the local labor market impacts of higher immigration. J. Law Econ. **19**(1), 22–64 (2001)
22. Xiong, L., Chen, B.: Fiscal decentralization, employed population and industrial structure optimization: an empirical analysis based on Beijing–Tianjin–Hebei Region. Modernization Manag. **38**(03), 31–34 (2018). (Chinese)
23. Han, F., Li, Y.: Urbanization and industrial structure optimization: a dynamic econometric analysis based on Hunan province. J. Nanjing Audit Univ. **7**(04), 8–15 (2010). (Chinese)
24. Wang, W., Qi, L.: Research on influential factors of regional equipment manufacturing industry structure optimization based on structural equation. Forum Sci. Technol. China **212**(12), 71–77 (2013). (Chinese)
25. Levin, A., Lin, C.F., Chu, C.S.J.: Unit root tests in panel data: asymptotic and finite–sample properties. J. Econ. **108**(1), 1–24 (2002)
26. Feng, S.: The road of urbanization with chinese characteristics and the upgrading of industrial structure. Jilin Univ. J. Soc. Sci. Edn. **5**, 130–134 (2005). (Chinese)
27. Hansen, B.E.: Threshold effects in non–dynamic panels: estimation, testing, and inference. J. Econ. **93**(2), 345–368 (1999)
28. Sun, T., Wang, Y.: Optimization of industrial structure under the perspective of digital economy: an empirical analysis based on the supply and demand equilibrium model. J. Stat. **4**(02), 1–16 (2023). (Chinese)
29. Xun, Z., Guanghua, W., Jiajia, Z., et al.: Digital economy, financial inclusion and inclusive growth. China Econ. **15**(3), 92–105 (2020)
30. Chen, C., Ye, F., Xiao, H., et al.: The digital economy, spatial spillovers and forestry green total factor productivity. J. Clean. Prod. **405**, 136890 (2023)

# Bayesian Model Selection Among Dispersed Integer-Valued Time Series Models

**Feng-Chi Liu, Cathy W. S. Chen, and Hsiao-Han Hsu**

**Abstract** This research evaluates model selection within a class of integer-valued time series models that feature overdispersion and extends these models to their generalized forms. The newly introduced models include: (1) dispersed integer-valued GARCH models incorporating negative binomial, double Poisson, or generalized Poisson distributions, and (2) a Double Log-form integer-valued GARCH model. The latter model avoids over-restrictions in the parameter space. We estimate parameters and select models within the Bayesian framework using adaptive Markov chain Monte Carlo (MCMC) sampling schemes, and employ the deviance information criterion (DIC) for model selection. We also design simulation studies to examine estimation accuracy and potential model misspecification. Using monthly crime counts in Bankstown, New South Wales, Australia, for an empirical illustration, our findings highlight the ability to select the most promising models among the competing ones based on DIC.

## 1 Introduction

Time series of count data are widely investigated in many empirical applications, such as numbers of crimes, traffic accidents, and disease cases. A recent review by [1] covers various aspects of the topic. Time series of counts typically feature autocorrelation and overdispersion, where the variance exceeds the mean. Studies in the literature explore autocorrelations in integer-valued time series, focusing on the autoregressive (AR) process in the mean function. Examples include the integer-valued autoregressive (INAR) model of [2, 3], the integer-valued generalized autoregressive conditional heteroskedasticity (INGARCH) model of [4], and the integer-valued autoregressive conditional heteroskedasticity (INARCH) models as discussed in [5, 6].

F.-C. Liu (✉) · C. W. S. Chen · H.-H. Hsu
Department of Statistics, Feng Chia University, Taichung 40724, Taiwan
e-mail: fcliu@mail.fcu.edu.tw

C. W. S. Chen
e-mail: chenws@mail.fcu.edu.tw

 
V. Kreinovich et al. (eds.), *Applications of Optimal Transport to Economics and Related Topics*, Studies in Systems, Decision and Control 556,
https://doi.org/10.1007/978-3-031-67770-0_9

The common assumption of a time series of counts being Poisson distributed is often violated due to overdispersion. Ferland et al. [4] propose a Poisson-INGARCH model. Despite the equal conditional mean and variance in a Poisson distribution, an INGARCH process typically exhibits overdispersion [5].

Some research suggests using a negative binomial (NB) distribution to handle overdispersion in various integer-valued models [7–11]. Zhu [7] recommends a NB-INGARCH model for potential extremes. Truong et al. [10] put forth a hysteretic Poisson INGARCH model, while [11] analyze temperature-crime correlation using Poisson and NB-INGARCH models with an exogenous variable. The generalized Poisson (GP) distribution includes an extra parameter (compared to the standard Poisson), which allows it to handle these dispersion situations. The additional parameter can adjust the variance independently from the mean, enabling a more flexible fit to the data. Many papers describe overdispersion using the GP distribution, for example, [12–14]. Chen and Lee [12] investigate GP autoregressive models to capture asymmetric and nonlinear responses, while [14] incorporate spatial-temporal patterns into multivariate integer-valued GARCH models with either a GP or a zero-inflated GP distribution.

Alternatively, one can use the double Poisson (DP) distribution to handle overdispersion. This distribution is a specific case of the double exponential family that [15] introduced. Xu et al. [8] suggest dispersed INARCH (DINARCH) models, by considering NB, DP, and GP distributions to analyze dengue cases in Singapore. With numerous models available for time series counts, this study seeks to identify the top performer among competing dispersion models. The research considers and extends a range of integer-valued time series models that account for overdispersion into generalized forms. The newly introduced models include: (1) dispersed INGARCH models incorporating NB, DP, or GP, and (2) a double log-form INGARCH model. The first model extends the work of [8], demonstrating that an extra parameter in INGARCH significantly impacts model fitting and prediction. The second model builds upon the idea of [9], applying it to the double log NB-INGARCH model, thus avoiding over-restrictions in the parameter space. Each model offers unique benefits when describing overdispersed time series counts. Therefore, choosing a promising model from the non-nested competitors for the target series poses a significant challenge.

Several related papers utilize maximum likelihood estimation (MLE) for inferences about integer-valued models [3, 4], whereas others employ Bayesian methods [10, 12, 16]. Notably, Bayesian methods can estimate unknown parameters in a complex model simultaneously, unlike the MLE method which relies on asymptotic theories. We target to provide parameter estimations and model selection within the Bayesian framework for all competing models, employ adaptive Markov chain Monte Carlo (MCMC) sampling schemes, and calculate the deviance information criterion (DIC) proposed by [17] for model selection.

DIC is an analogy of the Akaike information criterion (AIC), which combines the model's goodness of fit and complexity and exhibits broader applicability. It is computationally convenient because it can be calculated from the output of a MCMC simulation. The smallest DIC value is preferred as the best model. This study contributes by (1) developing new double log NB-INGARCH and dispersed INGARCH (DINGARCH) models to manage overdispersion; (2) employing the Bayesian approach for inferring all parameters; and (3) providing Bayesian model selection for all non-nested competing models. There are nine competing models in a row. The four newly developed models are NB-DINGARCH, DP-DINGARCH, GP-DINGARCH, and double log NB-INGARCH models. The other five competing models in the literature include Poisson-INARCH, Poisson-INGARCH, NB-DINARCH, DP-DINARCH and GP-DINARCH. Finally, we employ monthly crime counts in Bankstown, New South Wales, Australia for an empirical illustration to verify our proposed methods and models.

The structure of the rest of the paper is as follows. Section 2 introduces the proposed INGARCH models and the considered distributions. Section 3 illustrates the Bayesian MCMC method to estimate parameters and presents DIC to deal with model selection. Section 4 shows simulation results and compares them with those of the misspecified model. Section 5 provides some empirical analysis of New South Wales crime datasets. Section 6 presents concluding remarks.

## 2 Integer-Valued Time Series Models

As mentioned earlier, we generalize two types of integer-valued GARCH models. We now explain the proposed models.

### 2.1 *Double Log NB-INGARCH Model*

Let $Y_t$ be an integer-valued time series process characterized by autocorrelation and overdispersion. Chen et al. [9] develop an autoregressive conditional NB model to capture both the time-varying conditional mean and variance of counts. We improve upon the work of [9] by taking the logarithm of the conditional mean and variance. The advantage of our model is that it relaxes the restrictions on parameters for positive values. We refer to this model as a double log NB-INGARCH model. The term $\gamma_t$ represents the ratio of variance over mean. A key feature of this model is its ability to describe overdispersed count data, as $\gamma_t$ is greater than 1. The proposed model is presented as follows:

$$Y_t|\mathscr{F}_{t-1} \sim \text{NB}\left(\frac{\mu_t}{\gamma_t - 1}, \frac{1}{\gamma_t}\right), \ \gamma_t > 1, \tag{1}$$

$$\log(\mu_t) = \beta_0 + \beta_1 \log(Y_{t-1} + 1) + \beta_2 \log(\mu_{t-1}), \tag{2}$$

$$\log(\sigma_t^2) = \delta_0 + \delta_1 \log\left((Y_{t-1} - \mu_{t-1})^2 + 1\right) + \delta_2 \log(\sigma_{t-1}^2), \tag{3}$$

where $\mathscr{F}_t$ denotes all past information up to time $t$. The restrictions of the mean and variance parameters in Eqs. (2) and (3) satisfy the stationary conditions:

$$|\beta_2| < 1 \text{ and } |\delta_2| < 1. \tag{4}$$

Suppose that $Y_t \sim \text{NB}(r, p)$, and so the probability mass function of $Y_t$ is:

$$P(Y_t = y_t) = \frac{\Gamma(y_t + r)}{\Gamma(r)\Gamma(y_t + 1)} p^r (1 - p)^{y_t}. \tag{5}$$

Based on the NB set-up in Eq. (1), we provide the relations for this specified model:

(i) $E(Y_t|\mathscr{F}_{t-1}) = \left(\frac{\mu_t}{\gamma_t - 1}\right)\left(1 - \frac{1}{\gamma_t}\right)\left(\frac{1}{\gamma_t}\right)^{-1} = \mu_t$;

(ii) $\text{Var}(Y_t|\mathscr{F}_{t-1}) = \left(\frac{\mu_t}{\gamma_t - 1}\right)\left(1 - \frac{1}{\gamma_t}\right)\left(\frac{1}{\gamma_t}\right)^{-2} = \mu_t \gamma_t$;

(iii) $\dfrac{\text{Var}(Y_t|\mathscr{F}_{t-1})}{E(Y_t|\mathscr{F}_{t-1})} = \dfrac{\sigma_t^2}{\mu_t} = \gamma_t$.

The double log NB-INGARCH model is able to describe dynamic overdispersed phenomena of time series of counts.

## 2.2 NB-DINGARCH Model

An alternative approach to model overdispersed time series of counts, [8] introduce the dispersed INARCH (DINARCH) models with a dispersion parameter $\alpha$. In these models, the mean and variance differ and maintain a constant ratio. We generalize the DINARCH models to the dispersed INGARCH (DINGARCH) models and analyze three conditional distributions of $Y_t$ to account for overdispersion. Consequently, we present flexible versions of dispersed integer-valued models using NB, DP, and GP distributions.

We first consider a NB distribution for $Y_t$ given $\mathscr{F}_{t-1}$ with the nonnegative dispersion parameter $r_t = \mu_t/(\alpha - 1)$ and probability parameter $p = 1/\alpha$. The proposed NB-DINGARCH(1,1) model for $Y_t$ is as follows:

$$\begin{aligned} Y_t|\mathscr{F}_{t-1} &\sim \text{NB}\left(\frac{\mu_t}{\alpha - 1}, \frac{1}{\alpha}\right), \\ \mu_t &= \beta_0 + \beta_1 Y_{t-1} + \beta_2 \mu_{t-1}, \end{aligned} \tag{6}$$

where we restrict $\alpha > 1$ to present the overdispersion, $\beta_0 > 0$, $\beta_1, \beta_2 \geq 0$, and $\beta_1 + \beta_2 < 1$. For NB-DINGARCH$(1, 0)$, we call this model a NB-DINARCH model. Based on the NB set-up in Eq. (6), We provide various relations for this specified model:

(i) $E(Y_t|\mathscr{F}_{t-1}) = \frac{\mu_t}{\alpha-1}\left(1-\frac{1}{\alpha}\right)\left(\frac{1}{\alpha}\right)^{-1} = \mu_t$;
(ii) $\text{Var}(Y_t|\mathscr{F}_{t-1}) = \frac{\mu_t}{\alpha-1}\left(1-\frac{1}{\alpha}\right)\left(\frac{1}{\alpha}\right)^{-2} = \mu_t\alpha$;
(iii) $\dfrac{\text{Var}(Y_t|\mathscr{F}_{t-1})}{E(Y_t|\mathscr{F}_{t-1})} = \alpha.$

### 2.3 DP-DINGARCH Model

We alternatively consider a DP distribution for $Y_t$ given $\mathscr{F}_{t-1}$ with mean parameter $\mu_t$ and disperiosn parameter $\alpha$. The DP-DINGARCH$(1, 1)$ model is:

$$
\begin{aligned}
Y_t|\mathscr{F}_{t-1} &\sim \text{DP}\left(\mu_t, \mu_t\alpha\right), \\
\mu_t &= \beta_0 + \beta_1 Y_{t-1} + \beta_2\mu_{t-1},
\end{aligned} \tag{7}
$$

where we restrict $\alpha > 1$ to present the overdispersion, $\beta_0 > 0$, $\beta_1, \beta_2 \geq 0$, and $\beta_1 + \beta_2 < 1$. For DP-DINGARCH$(1, 0)$, we call this model a DP-DINARCH model. The probability mass function of a DP distribution for $Y_t$ is approximated by:

$$
\begin{aligned}
P\left(y_t \mid \mu_t, \alpha\right) &= C\left(\mu_t, \alpha\right) P_{\text{DP}}\left(y_t \mid \mu_t, \alpha\right), \\
C\left(\mu_t, \alpha\right) &= \left[1 + \frac{\alpha-1}{12\mu_t}\left(1+\frac{\alpha}{\mu_t}\right)\right]^{-1}, \\
P_{\text{DP}}\left(Y_t = y_t \mid \mu_t, \alpha\right) &= \left(\alpha^{-1/2}e^{-\mu_t/\alpha}\right)\left(\frac{e^{-y_t}y_t^{y_t}}{y_t!}\right)\left(\frac{e\mu_t}{y_t}\right)^{y_t/\alpha}.
\end{aligned} \tag{8}
$$

Efron [15] provides the first two moments for the DP distribution, which are as follows:

(i) $E(Y_t|\mathscr{F}_{t-1}) \doteq \mu_t$;
(ii) $\text{Var}(Y_t|\mathscr{F}_{t-1}) \doteq \mu_t\alpha$;
(iii) $\dfrac{\text{Var}(Y_t|\mathscr{F}_{t-1})}{E(Y_t|\mathscr{F}_{t-1})} = \alpha.$

Based on these moments, we conclude that the DP-DINGARCH model can effectively handle overdispersed count data.

### 2.4 GP-DINGARCH Model

We finally consider a GP distribution for $Y_t$ given $\mathscr{F}_{t-1}$ with parameters $\mu_t^*$ and $\kappa$, denoted by GP$(\mu_t^*, \kappa)$. We also define a GP-DINGARCH$(1, 1)$ model as:

$$
\begin{aligned}
Y_t|\mathscr{F}_{t-1} &\sim \text{GP}\left(\mu_t^*, \kappa\right), \\
\mu_t &= \beta_0 + \beta_1 Y_{t-1} + \beta_2\mu_{t-1},
\end{aligned} \tag{9}
$$

where $\mu_t^* = \mu_t/\sqrt{\alpha}$, $\kappa = 1 - 1/\sqrt{\alpha}$, $\beta_0 > 0$, $\beta_1, \beta_2 \geq 0$, and $\beta_1 + \beta_2 < 1$. For GP-DINGARCH(1, 0), we call this model a GP-DINARCH model. The probability mass function of a GP distribution for $Y_t$ is:

$$P_{\text{GP}}\left(Y_t = y_t \mid \mu_t^*, \kappa\right) = \frac{\mu_t^* \left(\mu_t^* + \kappa y_t\right)^{y_t - 1} e^{-\left(\mu_t^* + \kappa y_t\right)}}{y_t!}, \tag{10}$$

where $\mu_t^* > 0$, and we restrict $0 < \kappa < 1$ to present the overdispersion. The first and second moments for GP the distribution model are as follows.

(i) $E(Y_t|\mathscr{F}_{t-1}) = \dfrac{\mu_t^*}{1-\kappa} = \mu_t$;

(ii) $\text{Var}(Y_t|\mathscr{F}_{t-1}) = \dfrac{\mu_t^*}{(1-\kappa)^3} = \mu_t\alpha$;

(iii) $\dfrac{\text{Var}(Y_t|\mathscr{F}_{t-1})}{E(Y_t|\mathscr{F}_{t-1})} = \alpha$.

## 3 Bayesian Inference

To make inferences about the unknown parameters for the proposed models based on the Bayesian framework, we simulate samples from the posterior distributions. Each posterior distribution combines the prior distribution and the likelihood function. We introduce the details of the model likelihood, prior settings of parameters, MCMC sampling schemes and the DIC for model selection in the following subsections.

### 3.1 Priors and Model Likelihoods

Let $\boldsymbol{\theta}$ denote all parameters under the double log NB-INGARCH model, $\boldsymbol{\theta} = (\boldsymbol{\beta}', \boldsymbol{\delta}')'$, and $\boldsymbol{\beta} = (\beta_0, \beta_1, \beta_2)'$ and $\boldsymbol{\delta} = (\delta_0, \delta_1, \delta_2)'$ are parameters in Eqs. (2) and (3), respectively. The conditional likelihood function is:

$$\mathscr{L}(\boldsymbol{Y}|\boldsymbol{\theta}) = \prod_{t=2}^{n} \frac{\Gamma\left(\dfrac{\mu_t}{\gamma_t - 1} + Y_t\right)}{\Gamma\left(\dfrac{\mu_t}{\gamma_t - 1}\right)\Gamma\left(Y_t + 1\right)} \left(\frac{1}{\gamma_t}\right)^{\frac{\mu_t}{\gamma_t - 1}} \left(\frac{\gamma_t - 1}{\gamma_t}\right)^{Y_t},$$

where $\boldsymbol{Y} = (Y_2, \ldots, Y_n)'$, and $Y_1$ is prefixed. We make use of the following parameter blocks for the double log NB-INGARCH model: (1) $\boldsymbol{\beta}$ and (2) $\boldsymbol{\delta}$. We assume constrained uniform priors for $\boldsymbol{\beta}$ and $\boldsymbol{\delta}$ with restrictions in Eq. (4).

When we turn to a DP-DINGARCH model, $\boldsymbol{\theta}_{\text{DP}} = (\boldsymbol{\beta}', \alpha)$, $\boldsymbol{\beta} = (\beta_0, \beta_1, \beta_2)'$, and the conditional likelihood function of the model is:

$$\mathscr{L}\left(\boldsymbol{Y}|\boldsymbol{\theta}_{\mathrm{DP}}\right)=\prod_{t=2}^{n}\left[1+\frac{\alpha-1}{12\mu_t}\left(1+\frac{\alpha}{\mu_t}\right)\right]^{-1}\left(\alpha^{-1/2}e^{-\mu_t/\alpha}\right)\left(\frac{e^{-Y_t}Y_t^{Y_t}}{Y_t!}\right)\left(\frac{e\mu_t}{Y_t}\right)^{Y_t/\alpha}.$$

The parameter blocks for the DP-DINGARCH model are: (1) $\boldsymbol{\beta}$ and (2) $\alpha$. Similarly, we employ constrained uniform priors for $\boldsymbol{\beta}$ with restrictions. In other words, $\pi\ (\boldsymbol{\beta}) \propto I(A)$, where $A=\{\beta_0>0, \beta_1, \beta_2 \geq 0,\ \text{and}\ \beta_1+\beta_2<1\}$. Regarding $\alpha$, we either choose a truncated gamma prior or assume $\alpha^{-1} \sim$ Unif(0, 1).

If we consider a GP-DINGARCH model, $\boldsymbol{\theta}_{\mathrm{GP}}=(\boldsymbol{\beta}', \alpha)'$, $\boldsymbol{\beta}=(\beta_0, \beta_1, \beta_2)'$, then the conditional likelihood function of the model is:

$$\mathscr{L}\left(\boldsymbol{Y}|\boldsymbol{\theta}_{\mathrm{GP}}\right)=\prod_{t=2}^{n}\frac{\mu_t^*\left(\mu_t^*+\kappa Y_t\right)^{Y_t-1}e^{-\left(\mu_t^*+\kappa Y_t\right)}}{Y_t!}.$$

The parameter blocks for the GP-DINGARCH model and the prior settings are the same as the DP-DINGARCH model.

## 3.2 MCMC Sampling Schemes

The conditional posterior distribution is proportional to the likelihood function multiplied by the prior density for each parameter group. The posterior distribution for the parameter $\boldsymbol{\theta}_j$ appears as:

$$p(\boldsymbol{\theta}_j|\boldsymbol{Y}, \boldsymbol{\theta}_{-j}) \propto \mathscr{L}(\boldsymbol{Y}|\boldsymbol{\theta})\pi(\boldsymbol{\theta}_j|\boldsymbol{\theta}_{-j}), \tag{11}$$

where $\pi(\boldsymbol{\theta}_j)$ is its prior, and $\boldsymbol{\theta}_{-j}$ stands for the vector of $\boldsymbol{\theta}$ without the component of $\boldsymbol{\theta}_j$. Regarding the double log NB-INGARCH model, we employ the adaptive MCMC algorithm of [18] to sample MCMC samplers for the two parameter groups, $\boldsymbol{\beta}$ and $\boldsymbol{\delta}$, combining a random-walk (RW) Metropolis algorithm and an independent kernel Metropolis-Hastings (IK-MH) algorithm to accelerate convergence and to allow for optimal mixing. We conduct a total of $M$ iterations for MCMC sampling and $B$ iterations for burn-in samples. For the parameter $\boldsymbol{\beta}$, the procedure of the RW Metropolis algorithm runs as follows.

**Step 1:** At iteration $i$, generate a candidate $\boldsymbol{\beta}^*$ by

$$\boldsymbol{\beta}^*=\boldsymbol{\beta}^{[i-1]}+\boldsymbol{\varepsilon}, \quad \boldsymbol{\varepsilon} \sim N\left(\mathbf{0}, a\Sigma\right).$$

Here, $\boldsymbol{\beta}^{[i-1]}$ is the estimate of $\boldsymbol{\beta}$ at the $(i-1)$th iterate. Check the stationarity condition of $\boldsymbol{\beta}$. If the proposed $\boldsymbol{\beta}^*$ falls within the feasible range, proceed to Step 2; otherwise, revisit Step 1.

**Step 2:** Accept $\boldsymbol{\beta}^*$ as $\boldsymbol{\beta}^{[i]}$ when

$$p = \min\left\{\frac{p\left(\boldsymbol{\beta}^*\right)}{p\left(\boldsymbol{\beta}^{[i-1]}\right)}, 1\right\} > u, \quad \text{where} \quad u \sim U(0, 1),$$

where $p(\cdot)$ is the posterior distribution of $\boldsymbol{\beta}$ and $u$ is a probability drawn from the uniform distribution $U(0, 1)$. Otherwise, set $\boldsymbol{\beta}^{[i]} = \boldsymbol{\beta}^{[i-1]}$.
**Step 3:** Repeat Steps 1 and 2 for $B$ iterations.

The tuning parameter $a$ in Step 1 is used to achieve good convergence of the parameter estimates. For simplicity, we can set $\Sigma = \boldsymbol{I}$ as an identity matrix.

For the IK-MH algorithm, the procedure follows these steps:

**Step 1:** At iteration $i > B$, generate a candidate $\boldsymbol{\beta}^*$ by

$$\boldsymbol{\beta}^* = \boldsymbol{\mu}_\beta + \boldsymbol{\varepsilon}, \quad \boldsymbol{\varepsilon} \sim \boldsymbol{N}\left(\boldsymbol{0}, \Sigma_\beta\right),$$

where $\boldsymbol{\mu}_\beta$ and $\Sigma_\beta$ are the mean vector and covariance matrix from the burn-in samples. Check the stationarity condition of $\boldsymbol{\beta}$. If the proposed $\boldsymbol{\beta}^*$ falls within the feasible range, proceed to Step 2; otherwise, revisit Step 1.
**Step 2:** Accept $\boldsymbol{\beta}^*$ as $\boldsymbol{\beta}^{[i]}$ with

$$p = \min\left\{\frac{p\left(\boldsymbol{\beta}^*\right) g\left(\boldsymbol{\beta}^{[i-1]}\right)}{p\left(\boldsymbol{\beta}^{[i-1]}\right) g\left(\boldsymbol{\beta}^*\right)}, 1\right\} > u, \quad \text{where} \quad u \sim U(0, 1),$$

where $g(\cdot)$ is a Gaussian proposal density with mean $\boldsymbol{\mu}_\beta$ and covariance matrix $\Sigma_\beta$. Otherwise, set $\boldsymbol{\beta}^{[i]} = \boldsymbol{\beta}^{[i-1]}$.
**Step 3:** Repeat Steps 1 and 2 until iteration $i = M$.

The RW Metropolis and IK-MH algorithms for $\boldsymbol{\delta}$ follow similar procedures as for $\boldsymbol{\beta}$.

## 3.3 DIC for Model Selection

We further utilize the DIC, proposed by [17], as a Bayesian measure of fit or adequacy. We describe DIC as follows:

$$\begin{aligned}
\text{DIC} &= \overline{D(\boldsymbol{\theta})} + P_D, \\
\overline{D(\boldsymbol{\theta})} &= E[D(\boldsymbol{\theta})], \quad \text{and} \quad P_D = \overline{D(\boldsymbol{\theta})} - D(\bar{\boldsymbol{\theta}}), \\
D(\boldsymbol{\theta}) &= -2\ln \mathscr{L}(\boldsymbol{Y}|\boldsymbol{\theta}) + 2\ln p(\boldsymbol{Y}),
\end{aligned}$$

where $\bar{\boldsymbol{\theta}}$ is the posterior mean of $\boldsymbol{\theta}$, $\mathscr{L}(\boldsymbol{Y}|\boldsymbol{\theta})$ is the likelihood function, and $p(\boldsymbol{Y})$ is the integrated likelihood.

DIC can be divided into two parts: one is $\overline{D(\boldsymbol{\theta})}$, the posterior mean of the Bayesian deviance; and the other is $P_D$ which is the subtraction of $\overline{D(\boldsymbol{\theta})}$ and $D(\bar{\boldsymbol{\theta}})$. Since DIC is a by-product of the MCMC procedure, it can be swiftly procured during posterior sampling. Alternatively, one might consider using the posterior odds ratio or Bayes factor for pairwise comparisons. However, we would need to conduct numerous pairwise comparisons as the number of candidate models increases.

## 4 Simulation Study

In order to demonstrate the validity of the proposed Bayesian estimation, we conduct a simulation study on seven dispersed models, which are specified as follows.
**Model 1**: Double Log NB-INGARCH model

$$\begin{aligned}
&y_t \mid \mathscr{F}_{t-1} \sim \text{NB}\left(\frac{\mu_t}{\gamma_t - 1}, \frac{r_t - 1}{r_t}\right),\\
&\log(\mu_t) = 0.5 + 0.28\log(y_{t-1} + 1) + 0.3\log(\mu_{t-1}),\\
&\log\left(\sigma_t^2\right) = 1.64 + 0.3\log\left((y_{t-1} - \mu_{t-1})^2 + 1\right) + 0.5\log\left(\sigma_{t-1}^2\right).
\end{aligned}$$

**Model 2**: NB-DINARCH model

$$\begin{aligned}
&y_t \mid \mathscr{F}_{t-1} \sim \text{NB}\left(\frac{\mu_t}{\alpha - 1}, \frac{1}{\alpha}\right),\\
&\mu_t = 0.5 + 0.28y_{t-1}, \text{ and } \alpha = 1.5.
\end{aligned}$$

**Model 3**: DP-DINARCH model

$$\begin{aligned}
&y_t \mid \mathscr{F}_{t-1} \sim \text{DP}\left(\mu_t, \mu_t\alpha\right),\\
&\mu_t = 4.3 + 0.5y_{t-1}, \text{ and } \alpha = 5.0.
\end{aligned}$$

**Model 4**: GP-DINARCH model

$$\begin{aligned}
&y_t \mid \mathscr{F}_{t-1} \sim \text{GP}\left(\mu_t^*, \kappa\right),\\
&\mu_t = 0.5 + 0.3y_{t-1}, \ \mu_t^* = \frac{\mu_t}{\sqrt{\alpha}}, \text{ and } \kappa = 1 - \frac{1}{\sqrt{\alpha}}, \ \alpha = 2.0.
\end{aligned}$$

**Model 5**: NB-DINGARCH model

$$\begin{aligned}
&y_t \mid \mathscr{F}_{t-1} \sim \text{NB}\left(\frac{\mu_t}{\alpha - 1}, \frac{1}{\alpha}\right),\\
&\mu_t = 0.5 + 0.28y_{t-1} + 0.3\mu_{t-1}, \text{ and } \alpha = 1.5.
\end{aligned}$$

**Model 6**: DP-DINGARCH model

$$y_t \mid \mathscr{F}_{t-1} \sim \mathrm{DP}(\mu_t, \mu_t\alpha),$$
$$\mu_t = 0.5 + 0.28y_{t-1} + 0.6\mu_{t-1}, \text{ and } \alpha = 2.5.$$

**Model 7**: GP-DINGARCH model

$$y_t \mid \mathscr{F}_{t-1} \sim \mathrm{GP}\left(\mu_t^*, \kappa\right),$$
$$\mu_t = 0.5 + 0.3y_{t-1} + 0.28\mu_{t-1}, \ \mu_t^* = \frac{\mu_t}{\sqrt{\alpha}}, \text{ and } \kappa = 1 - \frac{1}{\sqrt{\alpha}}, \ \alpha = 2.0.$$

We simulate sample size $n = 1{,}000$ for each model and obtain parameter estimates with 100 replications. The total number of MCMC iterations is $M = 20{,}000$ with burn-in iterations $B = 8{,}000$. In order to speed up the convergence of MCMC sampling, we employ two sampling mechanisms. First, we use an adaptive MH method for MCMC sampling, which combines the RW-MH algorithm and the IK-MH algorithm. Second, due to autocorrelation, we choose every second sampler as thinning chain in MCMC output. The initial values for $\mu_1$ and $\sigma_1^2$ are set as the first observation $y_1$ and $\gamma_1\mu_1$, respectively, where $\gamma_1$ comes from a Gamma distribution, $G(1, 0.7)$.

Table 1 presents the true values of parameters, averages of posterior means, medians, standard deviations (Std.), and 95% Bayesian credible intervals ($\wp_{2.5}$ and $\wp_{97.5}$) for the seven models based on 100 replications. The results demonstrate that the proposed Bayesian method accurately estimates the model parameters, because the posterior means or medians are reasonably close to the true values, the posterior standard deviations are relatively small, and 95% Bayesian credible intervals include the true values inside the intervals.

To monitor the convergence of MCMC samples, we implement visual diagnostics, autocorrelation function (ACF) and trace plots, for all parameters with selected MCMC samples. We also employ the convergence diagnostic (CD) test of [19] and inefficiency factors (Ineff) of [20] to test the convergence of MCMC iterations. The CD test helps assess whether a given Markov chain has converged to its stationary distribution. The test involves splitting the Markov chain into two non-overlapping parts and comparing the means of these two segments. Geweke's CD test is essentially a Z-score test. In Bayesian analysis, the focus is on the direct interpretation of the Z-score. A lower inefficiency factor indicates that the chain has less autocorrelation and is more efficient, meaning that the samples are closer to being independent.

We present the average of absolute Z-scores of the CD and the Ineff number for each parameter in the last two columns of Table 1. The averages of the Z-scores in Table 1 are all less than 1.96, and all average Ineff values are significantly smaller than 20. This indicates that all parameter estimates have converged to equilibrium. The trace and ACF plots of MCMC samples also show the convergence and no dependence of parameter estimates. We omit the trace and ACF plots for saving the space.

**Table 1** Bayesian results for seven dispersed models based on 100 replications, with average of Z-scores for the CD test

| Parameter | True | Mean | Median | Std | $\wp_{2.5}$ | $\wp_{97.5}$ | Z-scores | Ineff |
|---|---|---|---|---|---|---|---|---|
| Double Log NB-INGARCH | | | | | | | | |
| $\beta_0$ | 0.50 | 0.471 | 0.463 | 0.121 | 0.258 | 0.720 | 1.830 | 5.532 |
| $\beta_1$ | 0.28 | 0.283 | 0.283 | 0.047 | 0.193 | 0.377 | 1.046 | 5.862 |
| $\beta_2$ | 0.30 | 0.308 | 0.310 | 0.112 | 0.090 | 0.521 | 0.431 | 4.543 |
| $\delta_0$ | 1.64 | 1.862 | 1.881 | 0.288 | 1.298 | 2.318 | 0.669 | 5.382 |
| $\delta_1$ | 0.30 | 0.314 | 0.313 | 0.062 | 0.197 | 0.439 | 1.578 | 5.605 |
| $\delta_2$ | 0.50 | 0.441 | 0.437 | 0.075 | 0.307 | 0.585 | 0.371 | 5.515 |
| NB-DINARCH | | | | | | | | |
| $\beta_0$ | 0.50 | 0.503 | 0.502 | 0.034 | 0.439 | 0.571 | 1.052 | 3.903 |
| $\beta_1$ | 0.28 | 0.278 | 0.278 | 0.038 | 0.207 | 0.353 | 0.135 | 3.812 |
| $\alpha$ | 1.50 | 1.513 | 1.508 | 0.089 | 1.353 | 1.701 | 0.126 | 3.929 |
| DP-DINARCH | | | | | | | | |
| $\beta_0$ | 4.30 | 4.610 | 4.607 | 0.277 | 4.074 | 5.161 | 0.032 | 3.302 |
| $\beta_1$ | 0.50 | 0.471 | 0.471 | 0.031 | 0.410 | 0.533 | 0.273 | 3.139 |
| $\alpha$ | 5.00 | 4.914 | 4.908 | 0.221 | 4.501 | 5.365 | 0.497 | 3.174 |
| GP-DINARCH | | | | | | | | |
| $\beta_0$ | 0.50 | 0.508 | 0.507 | 0.037 | 0.438 | 0.584 | 0.750 | 4.965 |
| $\beta_1$ | 0.30 | 0.298 | 0.297 | 0.039 | 0.223 | 0.376 | 0.500 | 3.938 |
| $\alpha$ | 2.00 | 2.022 | 2.011 | 0.154 | 1.753 | 2.353 | 1.143 | 7.694 |
| NB-DINGARCH | | | | | | | | |
| $\beta_0$ | 0.50 | 0.545 | 0.541 | 0.095 | 0.371 | 0.739 | 0.006 | 11.052 |
| $\beta_1$ | 0.28 | 0.278 | 0.277 | 0.035 | 0.212 | 0.347 | 0.358 | 5.973 |
| $\beta_2$ | 0.30 | 0.270 | 0.270 | 0.089 | 0.100 | 0.444 | 0.032 | 10.372 |
| $\alpha$ | 1.50 | 1.508 | 1.504 | 0.080 | 1.362 | 1.675 | 1.568 | 3.470 |
| DP-DINGARCH | | | | | | | | |
| $\beta_0$ | 0.50 | 0.656 | 0.646 | 0.139 | 0.411 | 0.959 | 1.820 | 3.866 |
| $\beta_1$ | 0.28 | 0.278 | 0.277 | 0.030 | 0.222 | 0.338 | 1.679 | 3.594 |
| $\beta_2$ | 0.60 | 0.568 | 0.570 | 0.051 | 0.463 | 0.661 | 1.658 | 3.666 |
| $\alpha$ | 2.50 | 2.538 | 2.535 | 0.114 | 2.324 | 2.772 | 1.179 | 3.323 |
| GP-DINGARCH | | | | | | | | |
| $\beta_0$ | 0.50 | 0.529 | 0.526 | 0.086 | 0.370 | 0.708 | 1.345 | 3.603 |
| $\beta_1$ | 0.30 | 0.305 | 0.304 | 0.036 | 0.236 | 0.377 | 0.627 | 4.448 |
| $\beta_2$ | 0.28 | 0.252 | 0.251 | 0.082 | 0.095 | 0.413 | 1.050 | 3.922 |
| $\alpha$ | 2.00 | 2.003 | 1.995 | 0.130 | 1.770 | 2.280 | 0.748 | 6.875 |

Z-scores: the average of absolute Z-scores of the CD test
Ineff.: inefficiency factors

**Table 2** Results of misspecifications for dispered INGARCH models based on 100 replications

| Parameter | True Model: Double Log NB-INGARCH | | | | | True Model: NB-DINGARCH | | | | |
|---|---|---|---|---|---|---|---|---|---|---|
| | True | Mean | $\wp_{2.5}$ | $\wp_{97.5}$ | $\text{DIC}_{\text{ave}}$ | True | Mean | $\wp_{2.5}$ | $\wp_{97.5}$ | $\text{DIC}_{\text{ave}}$ |
| Fitted Model: Double Log NB-INGARCH | | | | | | | | | | |
| $\beta_0$ | 0.50 | 0.496 | 0.274 | 0.748 | 5956.91 | | –0.078 | –0.161 | 0.006 | 7334.56 |
| $\beta_1$ | 0.28 | 0.279 | 0.190 | 0.367 | [99] | | 0.347 | 0.261 | 0.433 | [1] |
| $\beta_2$ | 0.30 | 0.282 | 0.063 | 0.502 | | | 0.234 | 0.035 | 0.417 | |
| $\delta_0$ | 1.64 | 1.907 | 1.365 | 2.315 | | | 0.479 | 0.230 | 0.810 | |
| $\delta_1$ | 0.30 | 0.313 | 0.199 | 0.432 | | | 0.137 | 0.050 | 0.240 | |
| $\delta_2$ | 0.50 | 0.432 | 0.311 | 0.569 | | | 0.169 | –0.180 | 0.472 | |
| Fitted Model: NB-DINGARCH | | | | | | | | | | |
| $\beta_0$ | | 2.110 | 1.270 | 2.903 | 6041.29 | 0.50 | 0.542 | 0.369 | 0.734 | 7267.97 |
| $\beta_1$ | | 0.067 | 0.017 | 0.139 | [1] | 0.28 | 0.280 | 0.214 | 0.349 | [62] |
| $\beta_2$ | | 0.163 | 0.023 | 0.457 | | 0.30 | 0.274 | 0.105 | 0.444 | |
| $\alpha$ | | 42.909 | 32.021 | 57.193 | | 1.50 | 1.516 | 1.369 | 1.683 | |
| Fitted Model: DP-DINGARCH | | | | | | | | | | |
| $\beta_0$ | | 1.610 | 1.036 | 2.121 | 16338.30 | | 0.541 | 0.366 | 0.739 | 7711.26 |
| $\beta_1$ | | 0.109 | 0.058 | 0.173 | [0] | | 0.280 | 0.212 | 0.350 | [0] |
| $\beta_2$ | | 0.262 | 0.094 | 0.499 | | | 0.274 | 0.100 | 0.447 | |
| $\alpha$ | | 11.723 | 10.737 | 12.799 | | | 1.511 | 1.385 | 1.649 | |
| Fitted Model: GP-DINGARCH | | | | | | | | | | |
| $\beta_0$ | | 0.529 | 0.370 | 0.708 | 7227.91 | | 0.541 | 0.370 | 0.733 | 7268.60 |
| $\beta_1$ | | 0.305 | 0.236 | 0.377 | [0] | | 0.280 | 0.215 | 0.349 | [37] |
| $\beta_2$ | | 0.252 | 0.095 | 0.413 | | | 0.274 | 0.105 | 0.443 | |
| $\alpha$ | | 2.003 | 1.770 | 2.280 | | | 1.518 | 1.365 | 1.696 | |
| Total | | | | | [100] | | | | | [100] |

$\text{DIC}_{\text{ave}}$: the average of DIC values over 100 replications
The numbers in square brackets represent the frequency with which the model is selected by DIC in 100 replications

To investigate the misspecification of the target integer-valued models and the utilized distribution, we conduct simulation studies on the double log NB-INGARCH, NB-DINGARCH, DP-DINGARCH, and GP-DINGARCH models. We evaluate the performance of DIC values in model or distribution misspecification cases. As indicated in Table 2, the true model is favored by DIC with 99% accuracy when it originates from the double log NB-INGARCH model. Conversely, the situation significantly changes when the true model is derived from the DINGARCH model with an NB distribution. Approximately 37% shift towards the DINGARCH model with a GP, while the remaining 62% favor the true model. This simulation result echoes [21] observation that the GP distribution is a mixture of Poisson distributions and that the fits to count data from both the GP and NB distributions are often analogous.

Our result is reasonable given that the NB-DINGARCH, DP-DINGARCH, and GP-DINGARCH models have the same parameter setup in intensity $\mu_t$. Consequently, the $\boldsymbol{\beta}$ and $\alpha$ estimates closely align with the true values for each fitted model.

Furthermore, our findings indicate that the DIC values are very close between the NB-DINGARCH and GP-DINGARCH models. In contrast, the DP-DINGARCH model does not yield similar DIC values when using an approximated probability function. A key insight from this experiment is the importance of identifying promising models, rather than solely focusing on the single best model.

## 5 Real Examples

We use monthly crime data from the Bureau of Crime Statistics and Research in New South Wales, spanning from January 1995 to June 2021, to demonstrate our proposed methods. There are three datasets with each sample size $n = 318$ considered in this study. We denote the data of dealing, trafficking in cocaine as Data 1, the data of dealing, trafficking in ecstasy as Data 2, and the data of dealing, trafficking in narcotics as Data 3. We consider the seven dispersed models as in the simulation study and two equal-dispersed models to investigate the problem of model selection for each dataset.

Figure 1 exhibits the time plots of Data 1, Data 2, and Data 3, respectively. The figures reveal that the three datasets have strong patterns of overdispersion and autocorrelation. We provide summary statistics and the dispersion ratio ($\sigma^2/\mu$) in Table 3. All dispersion ratios greater than six show a remarkable degree of overdispersion. Hence, it is worth to consider the integer-valued time series model with overdispersion. We employ the same MCMC method used in the simulation study to investigate three real examples. We monitor the trace and ACF plots to ensure the convergence of MCMC sampling in each model estimation. To save space, we do not provide diagnostic checking (the trace and ACF plots) for each model here.

To investigate the problem of model selection among dispersed integer-valued models, we calculate the DIC value for each considered model and select the best model by the smallest value of DIC. There are nine models considered for model selection, and we highlight the best two models selected by DIC for each dataset in Table 4. In Data 1, the models demonstrating the best fit, indicated by the lowest DIC values, are NB-DINGARCH and GP-DINGARCH. For Data 2, the GP-DINGARCH model and the Double Log NB-INGARCH model surpass others based on DIC values. When examining Data 3, the models with the lowest DIC values are the GP-DINARCH and the GP-DINGARCH. On the whole, the GP-DINGARCH model consistently ranks within the top two for all three data sets according to the DIC criterion.

Table 5 presents the Bayesian estimations, along with the CD test results and inefficiency factors, for the best models across three datasets. A smaller inefficiency factor is generally preferred because it indicates that fewer iterations are required to achieve a reliable estimation. The CD test and inefficiency factors confirm that all parameter estimates have achieved convergence. It is interesting to see that the estimates of $\alpha$ are away from 1. Focusing on the overdispersion of the three datasets, we obtain large values of $\alpha$, greater than 1, in GP-DINGARCH models for the three

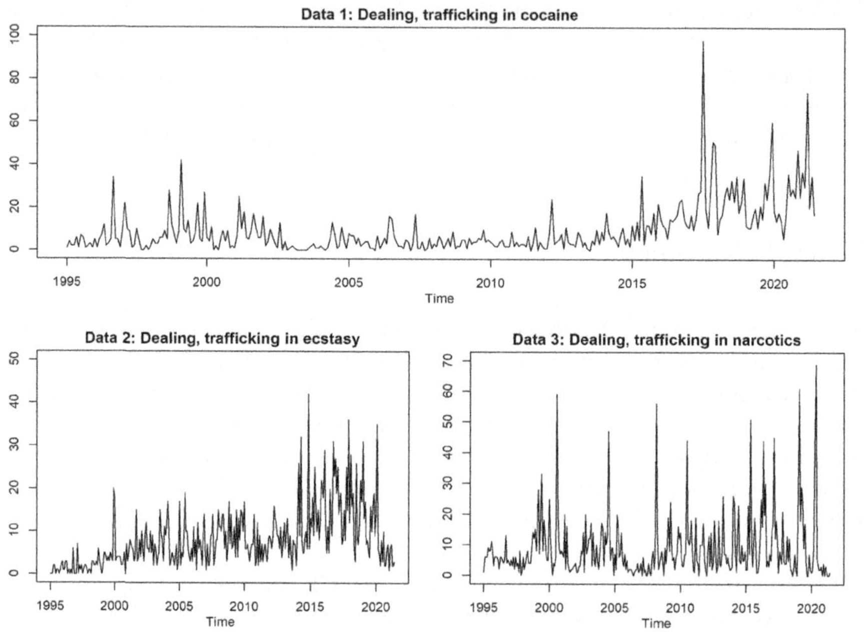


**Fig. 1** Time plots of Data 1, 2, and 3

**Table 3** Summary statistics of three datasets

| Crime data | Mean | Var. | Dispersion ratio | Min. | Median | Max. |
|---|---|---|---|---|---|---|
| Data 1: Dealing, trafficking in cocaine | 9.49 | 136.42 | 14.38 | 0 | 5 | 98 |
| Data 2: Dealing, trafficking in ecstasy | 8.03 | 48.86 | 6.09 | 0 | 6 | 42 |
| Data 3: Dealing, trafficking in narcotics | 8.84 | 105.68 | 11.96 | 0 | 6 | 69 |

datasets. For Data 1 and Data 2, the GP-DINGARCH models with large values of $\beta_2$ denote that a strong autocorrelation of $\mu_t$ exists. We conclude that the chosen models are effective for modeling both persistence and high overdispersion.

To perform the diagnostic checking of the selected models, we compute the standardized Pearson's residuals provided by [22]:

$$Z_t = \frac{y_t - \mathrm{E}\left(y_t \mid \mathscr{F}_{t-1}\right)}{\sqrt{\mathrm{Var}\left(y_t \mid \mathscr{F}_{t-1}\right)}}, \quad t = 2, \ldots, n,$$

where $\mathrm{E}\left(y_t \mid \mathscr{F}_{t-1}\right) = \mu_t$ and $\mathrm{Var}\left(y_t \mid \mathscr{F}_{t-1}\right) = \sigma_t^2$ or $(\mu_t\alpha)$ are respectively the conditional mean and conditional variance of the selected model (GP-DINGARCH,

**Table 4** DIC values of the nine competing models, highlighting the top two selected based on DIC

| Model | Data 1 | Data 2 | Data 3 |
|---|---|---|---|
| Poisson-INARCH | 7861.123 | 6169.199 | 8706.535 |
| Poisson-INGARCH | 7034.516 | 5355.561 | 8695.702 |
| NB-DINARCH | 4960.446 | 4702.650 | 5016.343 |
| DP-DINARCH | 5011.752 | 4728.950 | 5143.059 |
| GP-DINARCH | 4929.771 | 4691.894 | 4971.508 |
| NB-DINGARCH | 4804.711 | 4483.822 | 5014.263 |
| DP-DINGARCH | 4876.243 | 4513.576 | 5142.616 |
| GP-DINGARCH | 4772.471 | 4471.325 | 4969.055 |
| Double Log NB-INGARCH | 4857.153 | 4475.038 | 5042.514 |

The boxed values highlight the top two performers in terms of DIC

**Table 5** Bayesian estimation results of the best models for three datasets

| Parameter | Mean | Median | Std. | $\wp_{2.5}$ | $\wp_{97.5}$ | Z-scores | Ineff |
|---|---|---|---|---|---|---|---|
| DATA 1: GP-DINGARCH | | | | | | | |
| $\beta_0$ | 0.585 | 0.563 | 0.190 | 0.266 | 1.021 | 0.192 | 7.140 |
| $\beta_1$ | 0.220 | 0.218 | 0.039 | 0.150 | 0.301 | 0.684 | 6.067 |
| $\beta_2$ | 0.728 | 0.733 | 0.051 | 0.617 | 0.817 | 0.348 | 6.229 |
| $\alpha$ | 6.375 | 6.324 | 0.728 | 5.117 | 8.008 | 0.204 | 8.586 |
| DATA 2: GP-DINGARCH | | | | | | | |
| $\beta_0$ | 0.283 | 0.272 | 0.091 | 0.133 | 0.490 | 0.971 | 6.810 |
| $\beta_1$ | 0.151 | 0.148 | 0.037 | 0.088 | 0.232 | 0.814 | 5.820 |
| $\beta_2$ | 0.816 | 0.820 | 0.043 | 0.719 | 0.890 | 0.879 | 6.244 |
| $\alpha$ | 3.536 | 3.518 | 0.343 | 2.939 | 4.285 | 0.244 | 7.244 |
| DATA 3: GP-DINGARCH | | | | | | | |
| $\beta_0$ | 4.629 | 4.670 | 0.867 | 2.902 | 6.215 | 0.504 | 8.041 |
| $\beta_1$ | 0.309 | 0.307 | 0.053 | 0.209 | 0.420 | 0.943 | 6.081 |
| $\beta_2$ | 0.174 | 0.162 | 0.107 | 0.010 | 0.401 | 0.468 | 7.012 |
| $\alpha$ | 8.677 | 8.587 | 1.044 | 6.952 | 11.008 | 1.300 | 10.363 |

Z-scores: the absolute Z-scores of the CD test.
Ineff.: inefficiency factors

NB-DINGARCH or double log NB-INGARCH model). The mean of $Z_t$ should be 0, and no significant autocorrelation exists in the series of $Z_t$.

Figures 2, 3 and 4 present the plots of diagnostic checking, the time plots and ACF plots of standardized residuals from the best two models for the three datasets. The time plots reveal that the standardized residuals are distributed around 0, and few time points produce large residuals which exceed 3. Furthermore, there are no significant serial autocorrelations of the standardized residuals in each dataset. The results thus indicate that the best two models of each dataset are worth considering.

Focusing on the dynamic structure of the data, we estimate the dispersion parameter $\gamma_t$ from the double log NB-INGARCH model by following steps:

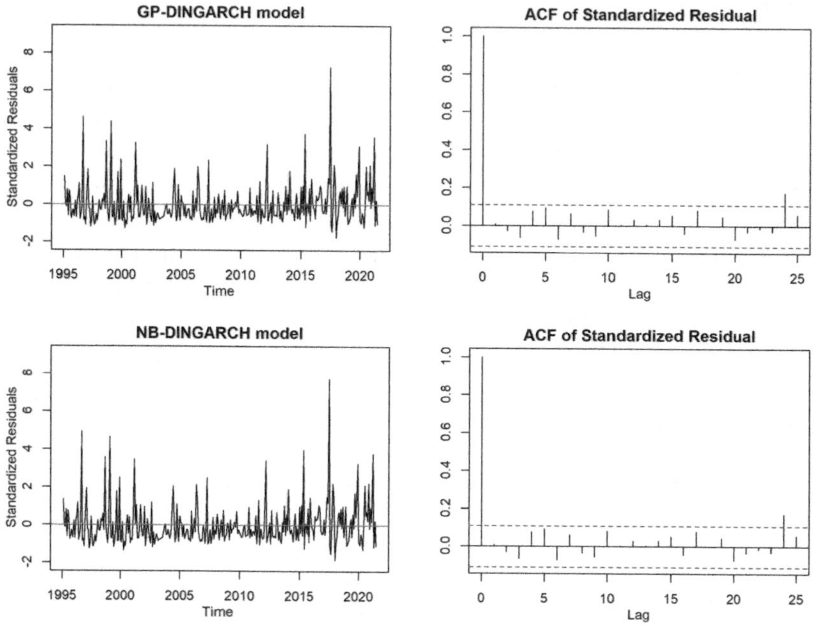


**Fig. 2** Diagnostic check for the best two models of Data 1: the GP-DINGARCH (upper panel) and NB-DINGARCH (lower panel) models

**Step 1:** During and after the burn-in MCMC iterations, we compute $\gamma_t^{[i]}$ for the $i$th iteration by

$$\gamma_t^{[i]} = \frac{\sigma_t^{2[i]}}{\mu_t^{[i]}}, \ i = (B+1), (B+2), \ldots, M.$$

**Step 2:** We estimate $\gamma_t$ by

$$\hat{\gamma}_t = \frac{1}{M-B} \sum_{i=B+1}^{M} \gamma_t^{[i]}, \ t = 1, 2, \ldots, n.$$

We present a time plot of $\hat{\gamma}_t$ based on the double log NB-INGARCH models for Data 2 in Fig. 5. The figure reveals a non-constant pattern of $\hat{\gamma}_t$ between 1 and 10, demonstrating dynamic overdispersed phenomena.

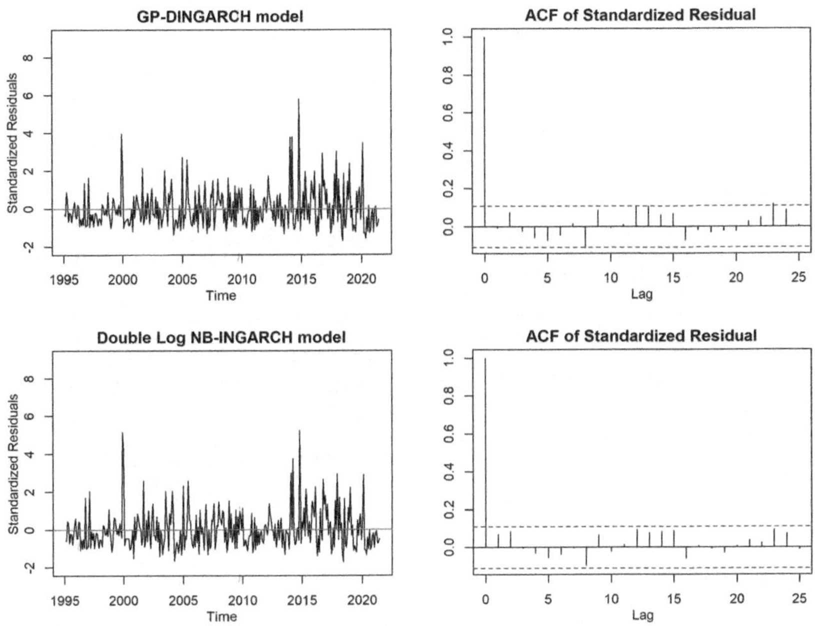


**Fig. 3** Diagnostic checking for the best two models of Data 2: GP-DINGARCH (upper panel) and double log NB-INGARCH (lower panel) models

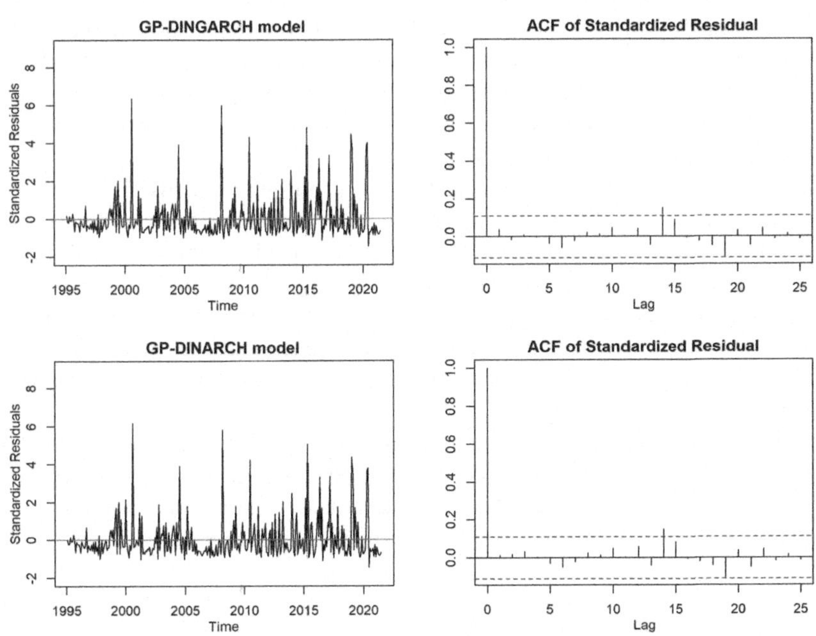


**Fig. 4** Diagnostic checking for the best two models of Data 3: GP-DINGARCH (upper panel) and GP-DINARCH (lower panel) models

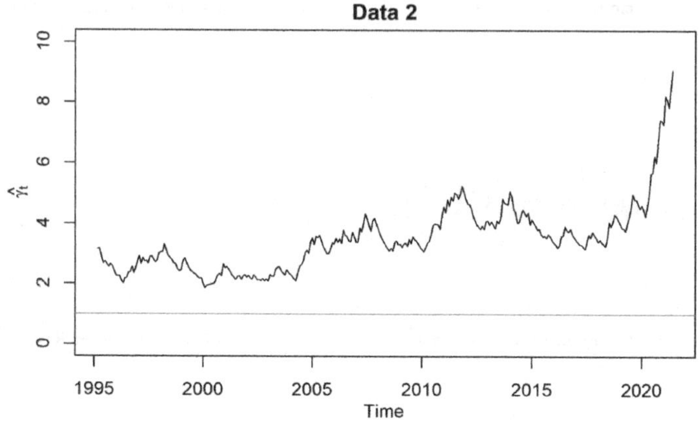


**Fig. 5** Time plots of $\hat{\gamma}_t$ based on the double log NB-INGARCH model for Data 2. The blue line is $\gamma_t = 1$

# 6 Conclusion

This study provides Bayesian inferences and model selection among a class of integer-valued time series models for describing overdispersions. The proposed models focus on various specifications for the GARCH process, including: Dispersed INGARCH models incorporating NB, DP, or GP distributions, and Double log-form INGARCH models. We successfully employ Bayesian inference using the MCMC sampling scheme to estimate model parameters and adopt the DIC as our model selection criterion. This approach enables us to identify the best model among the nine considered in the empirical application.

The simulation results demonstrate that the proposed methods effectively yield accurate estimates. Additionally, the model selection experiment underscores the importance of identifying promising models, rather than focusing solely on the single best model. Through the investigation of three datasets, we apply Bayesian methods to carry out parameter estimation and model selection among candidate models for describing overdispersed crime data. The results indicate that the dispersed INGARCH model with a GP distribution successfully captures the overdispersion characteristic in all three datasets.

In future work, we can address the issue of model selection for integer-valued time series data exhibiting a zero-inflated distribution, particularly when we encounter excess zeros in our count time series. We could also consider models incorporating exogenous variables when we obtain additional information to explain the target series.

**Acknowledgements** We thank the Editor and anonymous referee for their precious time and helpful comments, which have improved this paper. Feng-Chi Liu's research is funded by the National Science and Technology Council, Taiwan (NSTC111-2118-M-035-005- and NSTC112-2118-M-035-002-). Cathy W. S. Chen's research is funded by the National Science and Technology Council, Taiwan (NSTC109-2118-M-035-005-MY3 and NSTC112-2118-M-035 -001-MY3).

## References

1. Weiß, C.H.: An Introduction to Discrete-Valued Time Series. Wiley, Hoboken, NJ (2018)
2. Al-Osh, M.A., Alzaid, A.A.: First order integer-valued autoregressive (INAR(1)) process. J. Time Ser. Anal. **8**, 261–275 (1987)
3. Weiß, C.H., Homburg, A., Puig, P.: Testing for zero inflation and overdispersion in INAR(1) model. Stat. Pap. **60**, 823–848 (2019)
4. Ferland, R., Latour, A., Oraichi, D.: Integer-valued GARCH process. J. Time Ser. Anal. **27**, 923–942 (2006)
5. Weiß, C.H.: Modelling time series of counts with overdispersion. Stat. Methods Appl. **18**, 507–519 (2009)
6. Zhu, F., Li, Q., Wang, D.: A mixture integer-valued ARCH model. J. Stat. Plan. Inference **140**, 2025–2036 (2010)
7. Zhu, F.: A negative binomial integer-valued GARCH model. J. Time Ser. Anal. **32**, 54–67 (2011)
8. Xu, H.Y., Xie, M., Goh, T.N., Fu, X.: A model for integer-valued time series with conditional overdispersion. Comput. Stat. Data Anal. **56**, 4229–4242 (2012)
9. Chen, C.W.S., So, M.K.P., Li, J., Sriboonchitta, S.: Autoregressive conditional negative bionomial model applied to over-dispersed time series of counts. Stat. Methodol. **31**, 73–90 (2016)
10. Truong, B.C., Chen, C.W.S., Sriboonchitta, S.: Hysteretic Poisson INGARCH model for integer-valued time series. Stat. Model. **17**, 401–402 (2017)
11. Chen, C.W.S., Lee, S.: Bayesian causality test for integer-valued time series models with applications to climate and crime data. J. R. Stat. Soc. C **66**, 797–814 (2017)
12. Chen, C.W.S., Lee, S.: Generalized autoregressive conditional Poisson models for time series of counts. Comput. Stat. & Data Anal. **99**, 51–67 (2016)
13. Chen, C.W.S., Liu, F.C., Pingal, A.C.: Integer-valued transfer function models for counts that show zero inflation. Stat. Probab. Lett. **193**, 109701 (2023)
14. Chen, C.W.S., Chen, C.S., Hsiung, M.H.: Bayesian modeling of spatial integer-valued time series. Comput. Stat. & Data Anal. **188**, 107827 (2023)
15. Efron, B.: Double exponential families and their use in generalized linear regression. J. Am. Stat. Assoc. **81**, 709–721 (1986)
16. Pingal, A.C., Chen, C.W.S.: Bayesian modelling of integer-valued transfer function models. Stat. Model. (2022). https://doi.org/10.1177/1471082X221075477
17. Spiegelhalter, D.J., Best, N.G., Carlin, B.P., Van Der Linde, A.: Bayesian measures of model complexity and fit. J. R. Stat. Soc. B **64**, 583–639 (2002)
18. Chen, C.W.S., So, M.K.P.: On a threshold heteroscedastic model. Int. J. Forecast. **22**, 73–89 (2006)
19. Geweke, J.: Evaluating the accuracy of sampling-based approaches to the calculation of posterior moments (with discussion). In: Berger, J.O., Bernardo, J.M., Dawid, A.P., Smith, A.F.M. (eds.) Bayesian Statistics 4, pp. 169–193. Oxford University Press (1992)
20. Chib, S.: Markov chain Monte Carlo methods: computation and inference. In: J.J. Heckman, E. Leamer (eds.) Handbook of Econometrics, vol. 5, pp. 3569–3649. North Holland, Amsterdam (2001)

21. Joe, H., Zhu, R.: Generalized Poisson distribution: the property of mixture of Poisson and comparison with negative binomial distribution. Biom. J. **102**, 563–576 (2005)
22. Jung, R.C., Kukuk, M., Liesenfeld, R.: Time series of count data: modeling, estimation and diagnostics. Comput. Stat. & Data Anal. **51**, 2350–2364 (2006)

# Forecasting GDP with Many Predictors Using Sparse-Group LASSO MIDAS

**Wasin Siwasarit**

**Abstract** We conducted an investigation into four econometric models designed to handle mixed-frequency data. Our primary objective is to leverage a vast array of monthly macroeconomic variables to enhance the accuracy of forecasting quarterly Gross Domestic Product (GDP). To achieve this, we compared the following models: (1) The Autoregressive (AR) model, (2) The Mixed Data Sampling (MIDAS) model, which enables the combination of data at different frequencies, (3) The Lasso-MIDAS model, as proposed by [27], aimed at addressing issues related to inconsistent data frequencies and the curse of dimensionality arising from high-dimensional data, (4) the Sparse-group LASSO model, introduced by [3], which accommodates for the mixing processes and recognizes that the financial and the macroeconomic data may have heavier than exponential tails. An empirical analysis of forecasting GDP growth reveals that the sparse-group LASSO model consistently outperforms other models when forecasting four steps ahead, both before and after COVID-19 episodes. For short-term forecasting, both the MIDAS and sparse-group LASSO models exhibit favorable performance compared to alternative approaches. When comparing our findings before and after the COVID-19 episodes, it becomes evident that the MIDAS model significantly outperforms other models when incorporating COVID-19 data. Utilizing high-frequency data without any form of regularization appears to play a substantial role in improving forecasting performance, particularly during abrupt economic downturns. In essence, these two models can serve as an alternative "benchmark" for forecasting when sudden economic fluctuations occur, rendering conventional models like the AR model quickly outdated.

**Keywords** Forecasting · High-dimensional time series · Fat tail · MIDAS · LASSO · Sparse-group LASSO · Mixed frequency data

**JEL classification** C53 · C55 · E37

W. Siwasarit (✉)
Faculty of Economics, Thammasat University, Bangkok 10200, Thailand
e-mail: wasin@econ.tu.ac.th

 
V. Kreinovich et al. (eds.), *Applications of Optimal Transport to Economics and Related Topics*, Studies in Systems, Decision and Control 556,
https://doi.org/10.1007/978-3-031-67770-0_10

# 1 Introduction

Forecasting future economic growth presents a common challenge due to the need to work with economic and financial data reported at different frequencies. For example, Gross Domestic Product (GDP) data is typically reported quarterly, while business sentiment indices and stock return data are available at monthly and daily intervals, respectively.

Modern econometric solutions have emerged to extract insights from high-frequency data and apply them to forecast lower-frequency data. One straightforward approach is to align high-frequency data with lower-frequency data through a process known as time-aggregation. For instance, we can calculate a simple average over every three months to match quarterly data. However, this simple averaging method has its drawbacks, primarily concerning the inefficient and inconsistent estimation of parameters, as identified by [1].

To address this issue, [19] proposed a comprehensive framework called Mixed Data Sampling (MIDAS). This framework is particularly useful when forecasters aim to estimate a limited number of hyperparameters relative to the sampling frequency of higher-frequency variables, such as daily versus monthly data. Reference [19] employed distributed lag polynomials as weighting functions. Building upon this, [11] introduced a common factor to the MIDAS model, incorporating an autoregressive (AR) component. Their model demonstrated improved forecasting accuracy at short horizons, especially within-quarter forecasts, compared to benchmark AR or AR distributed-lag models.

Since the pioneering work of [11], a body of literature has emerged, exploring the application of MIDAS techniques in macroeconomic forecasting, both in large economies and small open economies.

For instance, [25] used monthly data to predict quarterly GDP in the euro-area. Their study compared the performance of the AR-MIDAS model by [10] with a vector autoregression (VAR) model. They found that the AR-MIDAS model outperformed the VAR model for one-quarter forecasts, while the VAR model excelled in three-quarter forecasts.

In the work of [4], they surveyed various methods for handling mixed-frequency data and observed that, in some cases, simple averaging of higher-frequency data didn't result in significant disadvantages. However, in other cases, explicitly modeling the data flow, as proposed by [19] using Mixed Data Sampling, proved to be more beneficial for forecasters, especially when constructing forecasts within specific time periods.

In addition to the challenge of dealing with varying frequencies during the estimation process, another issue that emerges is the curse of dimensionality resulting from a substantial number of variables. To address this problem, we can decide to choose a limited set of variables for forecasting. This approach is straightforward, yet it comes with a drawback as the information within each individual variable tends to fluctuate over time.

In our study, we focus on two approaches for extracting predictive insights from a large number of variables for GDP growth forecasting. The first involves pooling forecasts generated by many single indicator MIDAS models, while the second method, a promising alternative for forecasting with a vast mixed-frequency dataset, is to employ penalized regression techniques like Lasso or elastic net. While the lasso-midas model effectively addresses the issues related to the curse of dimensionality in high-dimensional data, it falls short in recognizing temporal dependencies among covariates at varying high-frequency lags. To tackle this issue, we can explore the sparse group lasso (sg-group) model, which combines elements of both lasso and group lasso. The sg-LASSO not only promotes sparsity within and across groups but also facilitates the extraction of predictive insights from each group. Therefore, we also compare sparse-group LASSO model as suggested by [3, 39] in this study as well.

The main contribution of this study is (1) Firstly, we utilized the basic MIDAS model proposed by [20] and the work of [1] which allows a set of high-frequency explanatory variables to explain low-frequency variables. (2) Second, in order to solve the multicollinearity problems that arise from a set of high-frequency explanatory variable, we add the lasso proposed by [38] which is a dimentionality reduction technique. Thus, the Lasso-MIDAS model is constructed by introducting the penalty variable selection method LASSO into the MIDAS model. (3) Third, we apply a new structured approach to high-dimensional regularized time regression problem as proposed by [3]. This framework allows for the mixing processes and recognizes that the financial and the macroeconomic data may have heavier than exponential tails.

The rest of the paper is organized as follows. The Sect. 2 presents the model specification. Section 3 explains the data and the forecasting strategy. In Sect. 4 presents our main finding and compares the performance from several models in forecasting the out-of-sample of GDP growth. Section 5 summarizes the conclusion and limitations of this study.

## 2 A Class of Forecasting Models

The purpose of this section is to introduce the standard econometric methodology for estimating the low-frequecy variables by utilizing the information that contains in the high-frequency data.

### 2.1 The Class of AR-MIDAS Models

To be able to have a better of the forcasting when the new high frequency data are available to use, we now can consider the model of Mixed-Data Sampling (MIDAS) Regression introduced by [19, 20]. For the traditional model of Autoregressive Distributed Lag Model -ADL($P_y$,$P_x$), it can present as:

$$y^Q_{t+1} = \mu + \alpha_1 y^Q_t + \cdots \alpha_P y^Q_{t-P_Y} + \beta_1 x^Q_t + \cdots + \beta_P x^Q_{t-P_X+1} + u_{t+1} \quad (1)$$

Based on Eq. (1) above, it becomes evident that the variable $y^Q_{t+1}$ depends on two distinct groups of explanatory variables. These groups consist of the autoregressive variables with $P_y$ lags and the variables represented by X with $P_x$ variables. This particular model is recognized as the conventional model utilized for forecasting purposes, as demonstrated in the work of [Stock and Watson 2003], for instance. However, it is important to note that a limitation of this model is that it necessitates the frequency of both Y and X to be identical.

For the case that explanatory variables are the higher-frequency data and we intend to use these inforecasting lower-frequency variables, one solution for this mismatch frequencies is to convert higher-frequency data to match the lower-frequency. Although the simplest method is to compute the simple average of the observations of X as shown in (2) that occur between samples of the lower-frequency variables but it discards any information about the timing of innovations to higher-frequency data.

$$y^Q_{t+1} = \mu + \beta y^Q_t + \beta \left[ \sum_{j=1}^{N_M} \frac{1}{N_M} x^M_{j,t} \right] + u_{t+1} \quad (2)$$

To allow the model to provide the unique slope coefficients for each of individual observations of X, we can alternatively add these data to the model directly as shown in (3):

$$y^Q_{t+1} = \mu + \beta y^Q_t + \sum_{j=1}^{N_M} \beta_j x^M_{j,t} + u_{t+1} \quad (3)$$

The number of parameters could become quite large, for example if we use the daily data with the lags $P_x = 4$ then we have to estimate the coefficients $22 * 4 = 88$ values. This situation could lead to overfitting and lead to the problem of Parameter Proliferation.

To solve the problem of overfitting while keeping some timing information, The MIDAS regression proposed by [19, 20] by applying hyper-parameterized weighting scheme to set the Autoregressive Distributed Lag Model -ADL($P_y$,$P_x$) as follow:

$$Y^Q_{t+1} = \mu + \sum_{j=0}^{p^Q_Y - 1} \beta_j Y^Q_{t-j} + \beta \sum_{j=0}^{p^M_X - 1} \sum_{i=1}^{N_M} \omega_k(\theta) X^M_{N_M-i,t-j} + u_{t+1} \quad (4)$$

From Eq. (11), we can see that when we apply hyper-parameterized weighting scheme to the model, the number of parameters then reduce to number of lags term $\beta$ and the set of parameters in the weighting scheme.

For more information about the weighting scheme used in AR-MIDAS model, see the work of [19, 20]

**Case 1** Normalized Exponential Almon Lag Polynomial

$$\omega_i^u = \omega_i(\theta_1, \theta_2) = \frac{e^{\theta_1 i + \theta_2 i^2}}{\sum_{i=1}^{N} e^{\theta_1 i + \theta_2 i^2}} \quad (5)$$

$$\omega_i^r = \omega_i(\theta_1, 0) \quad (6)$$

where N is the number of lags in MIDAS Polynomial where u presents the unrestricted equation and r presents the restricted equation.

**Case 2** Normalized Beta

$$\omega_i^u = \omega_i(\theta_1, \theta_2, 0) = \frac{x_i^{\theta_1 - 1}(1-x_i)^{\theta_2 - 1}}{\sum_{i=1}^{N} x_i^{\theta_1 - 1}(1-x_i)^{\theta_2 - 1}} \quad (7)$$

$$\omega_i^r = \omega_i(1, \theta_2, 0) \quad (8)$$

where $x_i = (i-1)/(N-1)$.

**Case 3** Almon Lag Polynomial Specification of Order P

$$\beta\omega_i(\theta_0, \ldots, \theta_P) = \sum_{p=0}^{P} \theta_p i^p \quad (9)$$

$$\begin{bmatrix} \omega_0 \\ \omega_1 \\ \omega_2 \\ \omega_3 \\ \vdots \\ \omega_N \end{bmatrix} = \begin{bmatrix} 1 & 0 & 0 & \cdots & 0 \\ 1 & 1 & 1 & \cdots & 1 \\ 1 & 2 & 2^2 & \cdots & 2^P \\ 1 & 3 & 3^2 & \cdots & 3^P \\ \vdots & \vdots & \vdots & \vdots & \vdots \\ 1 & N & N^2 & \cdots & N^P \end{bmatrix} \begin{bmatrix} \theta_0 \\ \theta_1 \\ \vdots \\ \theta_P \end{bmatrix} \quad (10)$$

**Case 4** Polynomail Specification with Step Function

$$\beta\omega_i(\theta_1, \ldots, \theta_P) = \theta_1 I_{i\in[a_0,a_1]} + \sum_{p=2}^{P} \theta_p I_{i\in(a_{p-1},a_p)}$$

$$a_0 = 1 < a_1 < \ldots < a_P = N$$

$$I_{i\in[a_{p-1},a_p]} = \begin{cases} 1, & a_{p-1} \le i \le a_p \\ 0, & otherwise \end{cases}$$

where $a_0 = 1 < a_1 < \ldots < a_P = N$.

## 2.2 The Class of LASSO-MIDAS Models

From the general form of ARDL-MIDAS, we can apply the technique proposed by [38] which use specifying properties of the $l_1$ norm to constrain the non-informative element $\beta_i$ to zero whereas retaining the crucial variables to reduce the dimentions of the explnanatory variables. We therefore use the LASSO-MIDAS by introducing the penalty term LASSO into the MIDAS model, which can be expressed as:

$$\min \sum_t \left( Y_{t+1}^Q - \mu - \sum_{j=0}^{p_Y^Q - 1} \beta_j Y_{t-j}^Q - \beta \sum_{j=0}^{p_X^M - 1} \sum_{i=1}^{N_M} \omega_k(\theta) X_{N_M - i, t-j}^M \right)^2 + \lambda \sum_{i=1}^{n} |\beta_i| \quad (11)$$

where $\lambda$ is the regularization parameters that can be used to control the strength of the penalty and variable selection.

## 2.3 The Class of Sparse-Group LASSO-MIDAS Models

While the lasso-midas model is effective in mitigating the challenges associated with the curse of dimensionality when dealing with high-dimensional data, it doesn't acknowledge the temporal relationships among covariates at different high-frequency lags. To address this concern, we can turn to the sparse group lasso (sg-group), which is a linear combination of both lasso and group lasso. The sg-LASSO not only concerns sparsity both within and between groups but also enables the extraction of predictive insights from each group. Therefore, the sparse -group LASSO estimator solve the following penalized least-squares problem:

$$\min \| y - Xb \|_T^2 + 2\lambda \Omega(b) \quad (12)$$

For a penalty function, it interpolates between the $l_1$ LASSO penalty and the group LASSO penalty:

$$\Omega(b) = \alpha |b|_1 + (1 - \alpha) \| b \|_{2,1},$$

where $\|b\|_{2,1} = \sum_{G \in \mathcal{G}} |b_G|_2$ is the group LASSO norm and $\mathcal{G}$ is a group structure.

From the Eq. (12), the parameter $\lambda$ is used to control the amount of penalization as same as in the Eq. (11). For the parameter $\alpha$, it is a weight parameter that determines the relative importance of the sparsity and the group structure. $\alpha \in [0, 1]$. By setting $\alpha = 1$, then we collapse the Eq. (12) back to the normal MIDAS-LASSO. In the Fig. 1a, it shows the geometry of the constrained set, $\{b \in \mathcal{R}^2 : \Omega(b) \leq 1\}$. As seen, the coefficients does not form a group where as if we allow the $\alpha = 0$ then, we have the geometry as depicted in (b) which is represented the case of the 1 group Lasso estimator. For the last two cases we allow to have the solutions that are both between

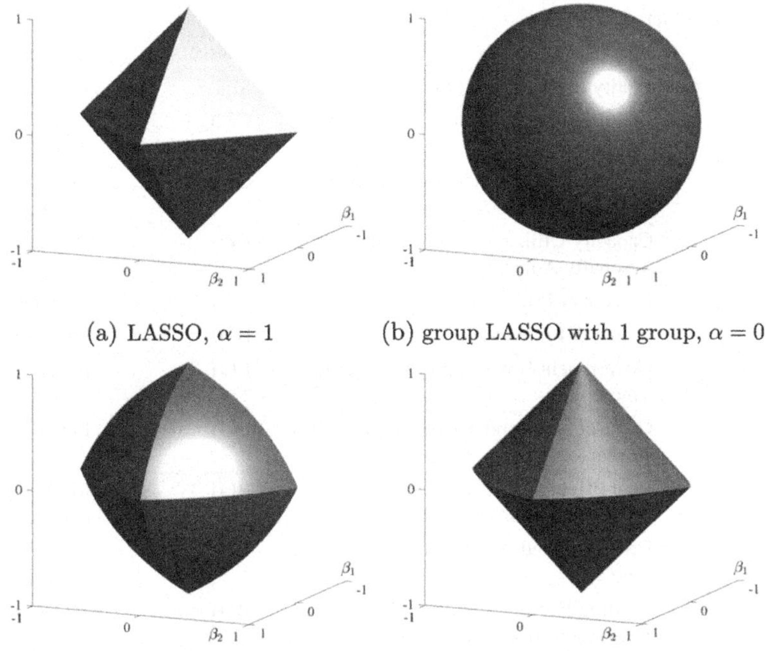


(a) LASSO, $\alpha = 1$ (b) group LASSO with 1 group, $\alpha = 0$

(c) sg-LASSO with 1 group, $\alpha = 0.5$ (d) sg-LASSO with 2 groups, $\alpha = 0.5$

**Fig. 1** Lasso, group lasso and ridge penalizations comparison *Source* [3]

and within group sparse. In (c) it shows the case one group with $\alpha = 0.5$ whereas (d) two groups with $\alpha = 0.5$. As suggested in the work of [3] the number of groups are defined by a particular problem and are specified by the econometrician, while $\alpha$ can be fixed or selected jointly with $\lambda$ using the cross-validation.

# 3 Data

## *3.1 GDP and Macroeoconomic Variables*

In this paper, to compare the four methods for handling mixed-frequency data, we design to use the monthly data to forecast quarterly data. In general, we compare forecasts of the quarterly GDP growth using the large scale dataset comprises 14 monthly indicators. The complete data listing is reported in Table 1. The summary of basic statistics are calculated and reported in Table 2. As expected, most of these macroeconomics and financial variables are heavier than exponential tails.

Figure 2 reports the time series for GDP growth movement in Thailand from 1994Q1 to 2023Q3. The economic growth movement tends to turn more volatile

**Table 1** Data description

| Parameter | Description | Source | Units |
|---|---|---|---|
| $X_1$ | Equity Market Index: SET | SET | Log-return % |
| $X_2$ | CEIC Leading Indicator: Smoothed: United States | CEIC | MOM % change |
| $X_3$ | CEIC Leading Indicator: Smoothed: China | CEIC | MOM % change |
| $X_4$ | Capacity Utilization Rate: 2016=100: Sesonally Adjusted | BOT | Ppt. change |
| $X_5$ | Consumer Price Index | BOT | Ppt. change |
| $X_6$ | Producer Price Index | BOT | Ppt. change |
| $X_7$ | Government Bond Yield: Average: BOT: 3 month | BOT | Level |
| $X_8$ | Government Bond Yield: Average: BOT: 10 year | BOT | Level |
| $X_9$ | Real Effective Exchange Rate Index: Trade Weight Broad 25 | BOT | MOM % change |
| $X_{10}$ | US$ Weighted Avg Interbank Exchange Rate | BOT | MOM % change |
| $X_{11}$ | Leading Economic Index (LEI): BOT: Seasonally Adjusted | BOT | MOM % change |
| $X_{12}$ | Business Sentiment Index | BOT | MOM % change |
| $X_{13}$ | Thai Industries Sentiment Index (TISI) | CEIC | MOM % change |
| $X_{14}$ | OPEC average crude oil price | CEIC | MOM % change |

*Notes*—Data description table (macro data)—The Series column gives a time-series name, which is given in the thrid column Source. The column Units denotes the data transformation applied to a time-series

during the recessions. This effect is particularly apparent for Asian crisis in 1997, the global financial crisis during 2007–2008 and the recent covid-19 pandermic during 2020–2021.

Although data for specific series can be traced back to before 2011Q1, our empirical analysis utilizes information from 2011Q1 to 2023Q3, a period during which we have data for the vast majority of the time series. The collected data have typically undergone seasonal adjustments, allowing for the necessary transformations to achieve stationarity in the series.

**Table 2** Descriptive statistics for all variables used in this study

| | X1 | X2 | X3 | X4 | X5 | X6 | X7 | X8 | X9 | X10 | X11 | X12 | X13 | X14 | Y |
|---|---|---|---|---|---|---|---|---|---|---|---|---|---|---|---|
| **Mean** | 0.353 | −0.017 | 0.030 | 0.504 | 66.881 | 1.630 | 1.614 | 2.711 | 0.016 | 0.111 | 0.179 | 0.181 | −0.068 | 0.405 | 2.261 |
| Median | 0.606 | −0.021 | −0.036 | 0.437 | 67.153 | 1.293 | 1.470 | 2.665 | 0.067 | −0.058 | 0.221 | 0.000 | −0.651 | 1.150 | 2.686 |
| Maximum | 17.855 | 9.545 | 6.048 | 4.406 | 76.919 | 7.859 | 3.460 | 4.210 | 7.170 | 3.922 | 2.559 | 24.359 | 7.086 | 42.288 | 15.468 |
| Minimum | −16.013 | −10.127 | −4.316 | −0.805 | 49.211 | −3.438 | 0.320 | 1.170 | −7.349 | −4.514 | −1.807 | −24.330 | −13.750 | −37.733 | −12.178 |
| Std. Dev. | 4.528 | 2.294 | 1.530 | 0.680 | 5.031 | 2.059 | 0.837 | 0.796 | 1.436 | 1.544 | 0.737 | 6.076 | 2.607 | 10.068 | 3.808 |
| Skewness | −0.074 | 0.147 | 0.797 | 1.657 | −0.489 | 0.695 | 0.344 | −0.083 | −0.211 | −0.062 | −0.056 | −0.441 | −1.272 | −0.107 | −0.642 |
| Kurtosis | 5.632 | 11.421 | 6.086 | 9.605 | 3.920 | 3.667 | 2.287 | 2.091 | 10.270 | 3.380 | 3.797 | 6.658 | 9.051 | 7.028 | 8.401 |
| Observations | 150 | 150 | 150 | 150 | 150 | 150 | 150 | 150 | 150 | 150 | 150 | 150 | 150 | 150 | 50 |

*Notes*—The descriptive statistics are calculated from the data that covered period from 2011Q1 to 2023Q3 for the GDP growth (Y). For all explanatory variables as listed in Table 1 covered period from 2011M1 to 2023M06

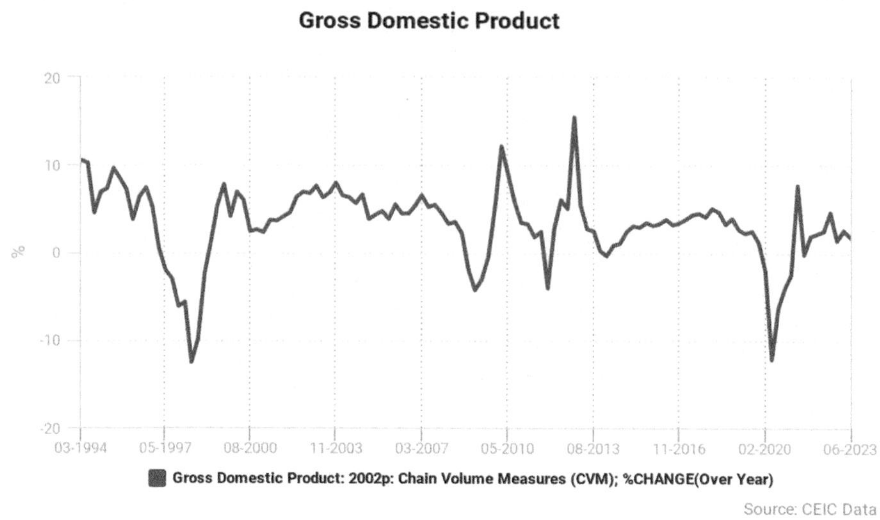


**Fig. 2** Gross domestic product

## *3.2 Forecasting Procedure*

The primary objective of this paper is to assess the suitability and effectiveness of the group LASSO-MIDAS model in forecasting GDP growth compared to conventional models such as the AR model, MIDAS model, and LASSO-MIDAS model. Our analysis focuses on two distinct scenarios: one covering the period before the COVID-19 pandemic and the other encompassing the period during the COVID-19 pandemic. The latter case represents instances where data structures exhibit greater volatility, making it more challenging to capture and filter information compared to typical conditions.

Specifically, in the first case, we divide the total sample into two parts: (1) the training set, where data covering the period from 2011Q1 to 2018Q3 are used for in-sample estimation, and (2) the test set, where data spanning from 2018Q4 to 2019Q3 are used to evaluate forecasting performance. We conduct one-step up to four-step ahead forecasting and evaluate the performance of each model using the mean absolute percentage error (MAPE) and root mean square error (RMSE).

In the second case, the training data is expanded to cover the period of the COVID-19 pandemic, spanning from 2011Q1 to 2022Q3. In this case, the testing set is from 2022Q4 to 2023Q3.

## 4 Empirical Analysis

In this section, we show the empirical results of using 4 different econometric models to forecast the quarterly economic growth using the large scale dataset comprises 14 monthly indicators. The complete data listing is reported in Table 1. The empirical results based on the MAPE and RMSE of the out-of sample forecasting as presented in Tables 3 and 4.

For the period preceding the COVID-19 pandemic, we present the forecasting results at various horizons from our end-of-period forecasting experiments across four models: (1) The Autoregressive (AR) model, (2) The Mixed Data Sampling (MIDAS) model, (3) The Lasso-MIDAS model, and (4) the Sparse-group LASSO model. We use the simple autoregressive model as a benchmark as it is considered to be a reasonable model for short-term GDP growth prediction. The optimal lag is selected based on the BIC criteria.

For the MIDAS model, we use a rolling window scheme to conduct each of single indicator MIDAS models. At the end of process, we then have 14 individual forcasts genarated from the MIDAS models. Given the findings in [2, 34, 35], we applied the Squared Discounted MSFE forecast combinations method to determine the forecast weights and calculate the forecast combination. Generally, we pool forecasts generated by many single indicator MIDAS models using the Squared Discounted MSFE forecast combinations. For the LASSO-MIDAS, we use the time-series cross-validation (TSCV) proposed by [23] for determining the size of $\lambda$. For the sg-LASSO-MIDAS, we tune sg-LASSO-MIDAS regularization parameters $\lambda$ and $\alpha$ using cross-validation as well.

As reported in Table 3, among the four models, the sparse-group LASSO model outperforms the others in terms of MAPE and RMSE. On average, the MAPE is less than 20% for all step-ahead forecasting. For one-step ahead forecasting, it appears that updated and useful information can help improve forecasting performance, as all of the models related to MIDAS perform better than the benchmark AR model. However, the sparse-group LASSO model stands out as it can simultaneously mitigate the issue of multicollinearity among predictors and consider temporal relationships among covariates at different high-frequency lags. The comparison of MAPE and RMSE of different models using the periods before COVID-19 is reported in Figs. 3 and 4, respectively. In general, the sparse-group LASSO can remove redundant variables and recognize time series data structures, allowing us to perform hierarchical model selection within and between groups.

In Table 4, we present findings from the second scenario in which the training data is extended to encompass the COVID-19 pandemic period, ranging from the 2011Q1 to the 2022Q3. The outcomes reveal that, for one-period ahead predictions, the MIDAS model exhibits a slight advantage over the sg-LASSO MIDAS model. However, it becomes apparent that the MIDAS model significantly outperforms the sg-LASSO MIDAS when forecasting for two- and three-periods ahead. In contrast, when looking at longer-term predictions, we observe significant enhancements in forecasting accuracy for the sg-LASSO MIDAS model as compared to MIDAS. As

**Table 3** Comparison of forecasts' RMSE and MAPE of different models using the periods before COVID-19

| AR | Actual | 4 | 3 | 2 | 1 |
|---|---|---|---|---|---|
| | 3.912 | 3.491 | | | |
| | 2.647 | 2.993 | 2.914 | | |
| | 2.226 | 2.815 | 2.731 | 2.819 | |
| | 2.458 | 3.198 | 3.111 | 3.202 | 3.186 |
| | **RMSE** | 0.581 | 0.584 | 0.744 | 0.728 |
| | **MAPE** | 23.208 | 24.623 | 28.434 | 29.602 |
| MIDAS | Actual | 4 | 3 | 2 | 1 |
| | 3.912 | 3.490 | | | |
| | 2.647 | 3.432 | 3.261 | | |
| | 2.226 | 2.999 | 3.209 | 3.193 | |
| | 2.458 | 2.846 | 2.808 | 2.942 | 2.983 |
| | **RMSE** | 0.674 | 0.738 | 0.484 | 0.525 |
| | **MAPE** | 26.712 | 29.191 | 31.545 | 21.371 |
| LASSO-MIDAS | Actual | 4 | 3 | 2 | 1 |
| | 3.912 | 3.718 | | | |
| | 2.647 | 2.889 | 2.945 | | |
| | 2.226 | 2.817 | 2.874 | 2.889 | |
| | 2.458 | 3.007 | 3.057 | 3.0655 | 3.055 |
| | **RMSE** | 1.734 | 1.749 | 1.751 | 1.748 |
| | **MAPE** | 19.342 | 26.740 | 27.253 | 24.272 |
| SG-MIDAS | Actual | 4 | 3 | 2 | 1 |
| | 3.912 | 3.236 | | | |
| | 2.647 | 2.876 | 2.918 | | |
| | 2.226 | 2.683 | 2.737 | 2.746 | |
| | 2.458 | 2.796 | 2.848 | 2.857 | 2.850 |
| | **RMSE** | 0.353 | 0.454 | 0.399 | 0.392 |
| | **MAPE** | 14.297 | 19.408 | 19.776 | 15.961 |

*Notes*—This table presents Root Mean Square Error (RMSE) and Mean Absolute Percentage Error (MAPE) values for four econometric models: (1) Autoregressive (AR) model, (2) MIDAS model with a rolling-window approach, (3) Lasso-MIDAS model, and (4) SG-MIDAS. The training dataset spans from March 2011 to September 2018, with data collected at a quarterly frequency. The explanatory variables used in these models consist of a monthly set of macroeconomic indicators

demonstrated in Table 4, the Mean Absolute Percentage Error (MAPE) for a four-period ahead forecast is approximately half as large as that of the MIDAS model. The comparison of MAPE and RMSE of different models using the periods after COVID-19 is reported in Figs. 5 and 6, respectively

When comparing our findings before and after the COVID-19 episodes, it becomes evident that the MIDAS model significantly outperforms other models when incor-

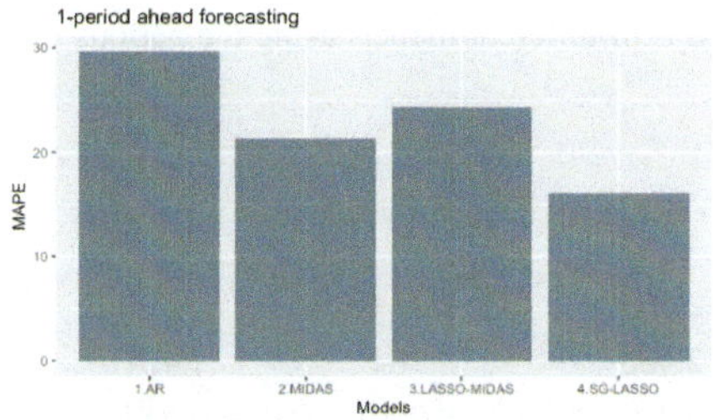


(a) 1-period ahead forecasting

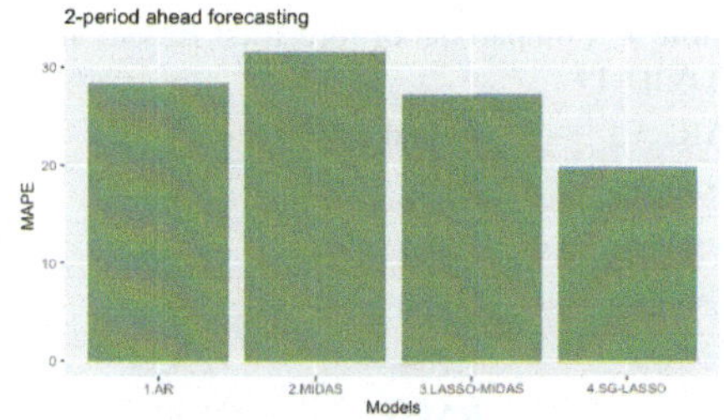


(b) 2-period ahead forecasting

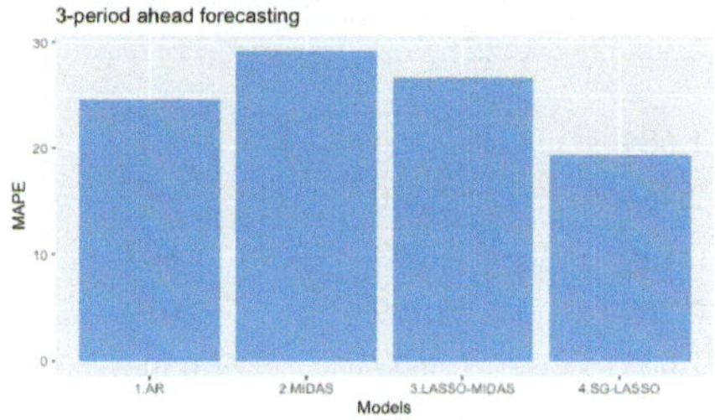


(c) 3-period ahead forecasting

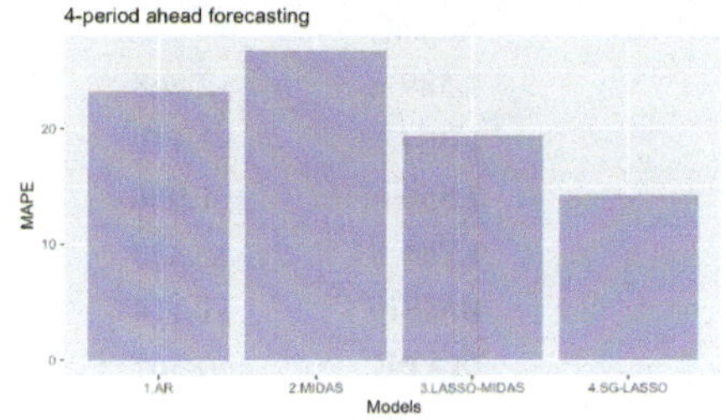


(d) 4-period ahead forecasting

**Fig. 3** Comparison of forecasts' MAPE of different models using the periods before COVID-19

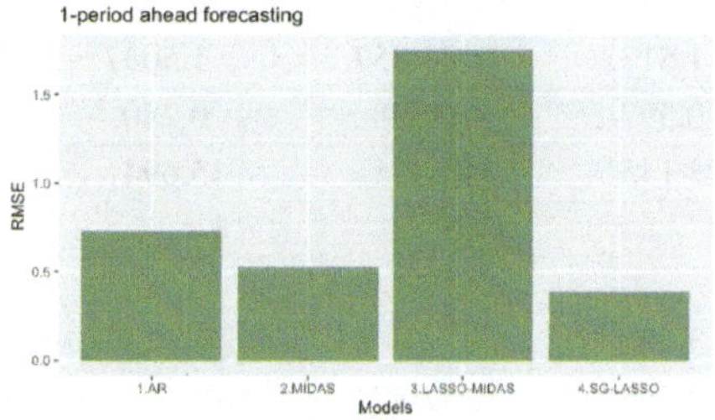


(a) 1-period ahead forecasting

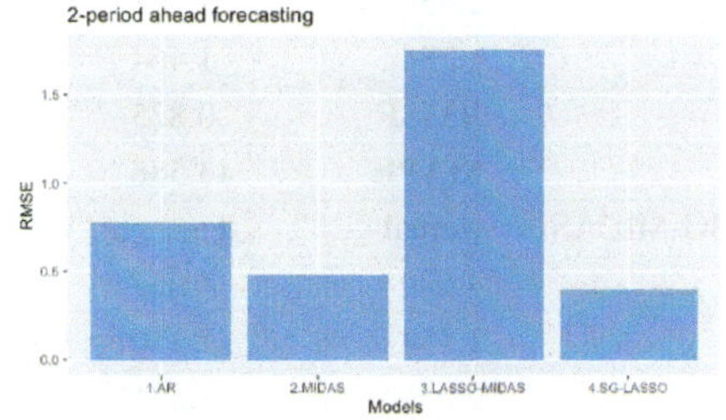


(b) 2-period ahead forecasting

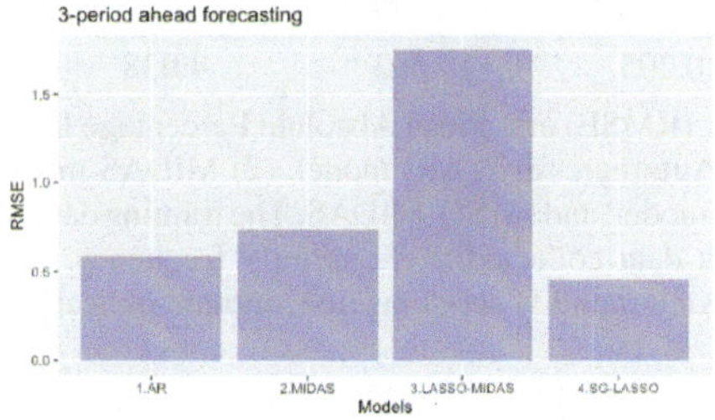


(c) 3-period ahead forecasting

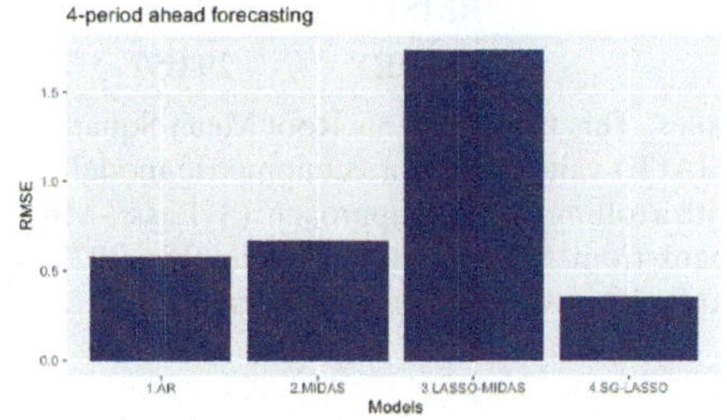


(d) 4-period ahead forecasting

**Fig. 4** Comparison of forecasts' RMSE of different models using the periods before COVID-19

**Table 4** Comparison of forecasts' RMSE and MAPE of different models using the periods after COVID-19

| AR | Actual | 4 | 3 | 2 | 1 |
|---|---|---|---|---|---|
| | 4.589 | 3.187 | | | |
| | 1.384 | 3.562 | 3.597 | | |
| | 2.556 | 2.470 | 2.505 | 2.474 | |
| | 1.784 | 2.433 | 2.468 | 2.384 | 2.421 |
| | **RMSE** | 1.313 | 0.486 | 0.600 | 0.637 |
| | **MAPE** | 65.723 | 20.205 | 18.431 | 35.719 |
| MIDAS | Actual | 4 | 3 | 2 | 1 |
| | 4.589 | 2.868 | | | |
| | 1.384 | 3.119 | 1.684 | | |
| | 2.556 | 1.906 | 2.464 | 2.530 | |
| | 1.784 | 2.323 | 2.146 | 1.893 | 1.803 |
| | **RMSE** | 1.114 | 0.264 | 0.109 | 0.019 |
| | **MAPE** | 60.361 | 11.973 | 3.567 | 1.059 |
| LASSO-MIDAS | Actual | 4 | 3 | 2 | 1 |
| | 4.589 | 2.599 | | | |
| | 1.384 | 2.481 | 2.5602 | | |
| | 2.556 | 1.698 | 1.7549 | 1.7346 | |
| | 1.784 | 1.464 | 1.5172 | 1.4857 | 1.5037 |
| | **RMSE** | 0.825 | 0.597 | 0.298 | 0.280 |
| | **MAPE** | 43.588 | 23.145 | 24.425 | 15.694 |
| SG-MIDAS | Actual | 4 | 3 | 2 | 1 |
| | 4.589 | 3.067 | | | |
| | 1.384 | 2.110 | 3.597 | | |
| | 2.556 | 1.831 | 2.505 | 1.847 | |
| | 1.784 | 1.671 | 2.468 | 1.651 | 1.712 |
| | **RMSE** | 0.596 | 0.486 | 0.133 | 0.072 |
| | **MAPE** | 29.052 | 20.205 | 17.593 | 4.038 |

Notes -This table presents Root Mean Square Error (RMSE) and Mean Absolute Percentage Error (MAPE) values for four econometric models: (1) Autoregressive (AR) model, (2) MIDAS model with a rolling-window approach, (3) Lasso-MIDAS model, and (4) SG-MIDAS. The training dataset spans from March 2011 to September 2022, with data collected at a quarterly frequency. The explanatory variables used in these models consist of a monthly set of macroeconomic indicators

porating COVID-19 data. Utilizing high-frequency data without any form of regularization appears to play a substantial role in improving forecasting performance, particularly during abrupt economic downturns. Nevertheless, for long-term forecasts, it is crucial to account for data structure control and address the noise stemming from the use of high-dimensional data.

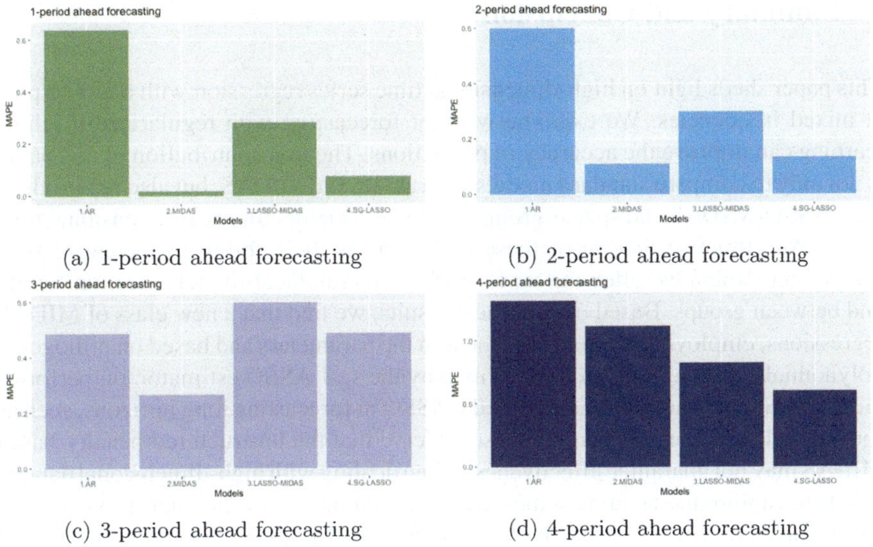


(a) 1-period ahead forecasting (b) 2-period ahead forecasting

(c) 3-period ahead forecasting (d) 4-period ahead forecasting

**Fig. 5** Comparison of forecasts' MAPE of different models using the periods after COVID-19

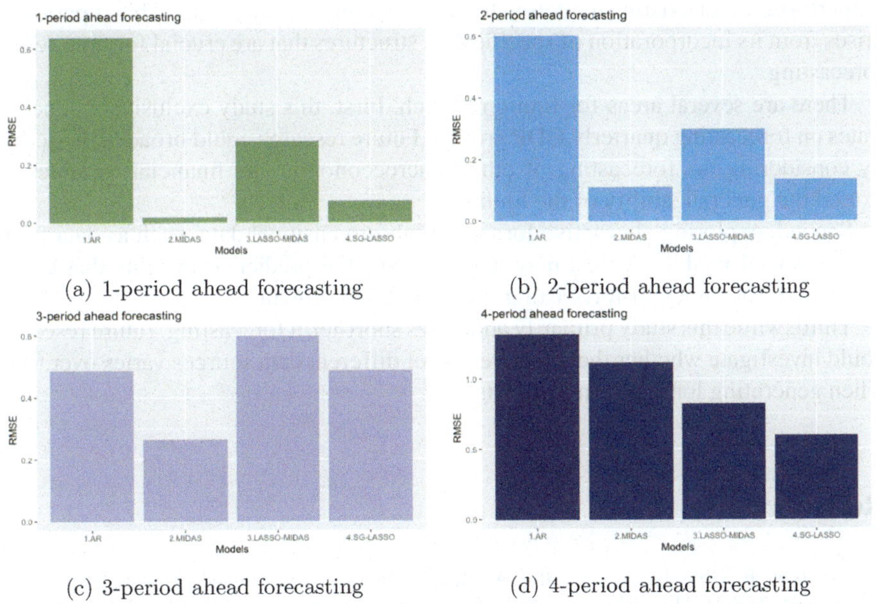


(a) 1-period ahead forecasting (b) 2-period ahead forecasting

(c) 3-period ahead forecasting (d) 4-period ahead forecasting

**Fig. 6** Comparison of forecasts' RMSE of different models using the periods after COVID-19

## 5 Summary and Conclusion

This paper sheds light on high-dimensional time series regression with data sampled at mixed frequencies. We examine whether forecasting with regularized machine learning can improve the accuracy of predictions. The first contribution of this paper is not only to compare standard models such as AR and MIDAS, but also to introduce the LASSO-MIDAS and sparse-group LASSO estimators for high-dimensional time series. An attractive feature of these estimators is their ability to recognize time series data structures, allowing us to perform hierarchical model selection within and between groups. Based on empirical results, we find that a new class of MIDAS regressions, employing dictionaries linear in the parameters and based on orthogonal polynomials with lag selection performed by the sg-LASSO estimator, outperforms the MIDAS model and the unstructured LASSO in forecasting long horizons, such as four-period ahead predictions. We also observe that the unstructured penalty-based MIDAS may not guarantee effectiveness when dealing with high-dimensional data, as it fails to capture the essential time-series stylized facts. For short-term forecasting, specifically one-period ahead, both MIDAS and sg-LASSO models perform well and do not exhibit significant differences. Overall, this study confirms that when dealing with data sampled at mixed frequencies, the new class of sg-LASSO MIDAS outperforms other existing models in forecasting long-term horizons. This superiority arises from its incorporation of specific data structures that are crucial for time series forecasting.

There are several areas for future research. First, this study exclusively concentrates on forecasting quarterly GDP growth. Future research could broaden the scope by considering the forecasting of other macroeconomic and financial variables to extend the generalizability of the findings.

Second, exploring alternative forms of high-dimensional data, such as data from new or social media sources, may uncover valuable predictive insights that could enhance the accuracy of forecasting quarterly GDP growth.

Third, while this study primarily addresses short-term forecasting, future research could investigate whether the effectiveness of different data sources varies over time when generating longer-term forecasts.

## References

1. Andreou, E., Ghysels, E., Kourtellos, A.: Regression models with mixed sampling frequencies. J. Econ. **158**(2), 246–261 (2010). Specification Analysis in Honor of Phoebus J. Dhrymes
2. Andreou, E., Ghysels, E., Kourtellos, A.: Should macroeconomic forecasters use daily financial data and how? J. Bus. Econ. Stat. **31**(2), 240–251 (2013)
3. Andrii Babii, E.G., Striaukas, J.: Machine learning time series regressions with an application to nowcasting. J. Bus. Econ. Stat. **40**(3), 1094–1106 (2022)
4. Armesto, M.T., Engemann, K., Owyang, M.: Forecasting with mixed frequencies. Review **92**, 521–536 (2010)

5. Aruoba, S.B., Diebold, F.X., Scotti, C.: Real-time measurement of business conditions. J. Bus. Econ. Stat. **27**(4), 417–427 (2009)
6. Bai, J., Ghysels, E., Wright, J.H.: State space models and MIDAS regressions. Economet. Rev. **32**(7), 779–813 (2013)
7. Birchenhall, C., Osborn, D., Sensier, M.: Predicting UK business-cycle regimes. Scott. J. Polit. Econ. **48**(2), 179–195 (2001)
8. Birchenhall, C., Osborn, D., Simpson, P.: Predicting US business-cycle regimes. J. Bus. Econ. Stat. **17**, 313–323 (1999)
9. Campbell, J.Y.: Asset prices, consumption, and the business cycle. Working paper 6485, National Bureau of Economic Research (1998)
10. Clements, M.P., Galvãso, A.B.: Macroeconomic forecasting with mixed-frequency data: forecasting output growth in the United States. J. Bus. Econ. Stat. **26**(4), 546–554 (2008)
11. Clements, M.P., Galvão, A.B.: Forecasting us output growth using leading indicators: an appraisal using MIDAS models. J. Appl. Economet. **24**(7), 1187–1206 (2009)
12. Doan, T., Litterman, R., Sims, C.: Forecasting and conditional projection using realistic prior distributions. Economet. Rev. **3**(1), 1–100 (1984)
13. Estrella, A., Mishkin, F.S.: Predicting US recessions: financial variables as leading indicators. Rev. Econ. Stat. **80**, 45–61 (1995)
14. Forni, M., Hallin, M., Lippi, M., Reichlin, L.: Do financial variables help forecasting inflation and real activity in the euro area? J. Monet. Econ. **50**(6), 1243–1255 (2003)
15. Francis, N., Ghysels, E., Owyang, M.: The low-frequency impact of daily monetary policy shocks. Discussion Paper Federal Reserve Bank of St. Louis Working Paper Series 2011-009B (2011)
16. Ghysels, E.: On the periodic structure of the business cycles. J. Bus. Econ. Stat. 289–298 (1994)
17. Ghysels, E.: Macroeconomics and the reality of mixed frequency data. UNC Discussion Paper (2012)
18. Ghysels, E.: Macroeconomics and the reality of mixed frequency data. J. Econ. **193**(2), 294–314 (2016). The Econometric Analysis of Mixed Frequency Data Sampling
19. Ghysels, E., Sinko, A., Valankov, R.: The MIDAS touch: mixed data sampling regression models. UNC Working Paper (2004)
20. Ghysels, E., Sinko, A., Valankov, R.: MIDAS regressions: further results and new directions. Econ. Rev. **26**(1) (2007)
21. Hamilton, J.: Time series analysis. Princeton University Press, Princeton (New Jersey) (1994)
22. Harvey, C.R.: Forecasts of economic growth from the bond and stock markets. Financ. Anal. J. **45**(5), 38–45 (1989)
23. Hyndman, R.J., Athanasopoulos, G.: Forecasting: Principles and Practice. OTEXTS (2018)
24. Kadiyala, K.R., Karlsson, S.: Numerical methods for estimation and inference in Bayesian VAR-models. J. Appl. Econ. **12**(2), 99–132 (1997)
25. Kuzin, V., Marcellino, M., Schumacher, C.: Midas versus mixed-frequency VAR: nowcasting GDP in the euro area. Int. J. Forecast. **27**(2), 529–542 (2011)
26. Litterman, R.B.: Forecasting with Bayesian vector auto regressions-five years of experience. J. Bus. Econ. Stat. **4**(1), 25–38 (1986)
27. Marsilli, C.: Variable selection in predictive MIDAS models. Working Papers 520, Banque de France (2014)
28. Mitchell, W.C., Burns, A.F.: Statistical Indicators of Cyclical Revivals. Princeton University Press, pp. 162–183 (1961)
29. Moore, G.H.: Business Cycle Indicators, vol. 1. Princeton University Press (1961)
30. Sims, C.A.: Macroeconomics and reality. Econometrica **48**(1), 1–48 (1980)
31. Sims, C.A., Zha, T.: Bayesian methods for dynamic multivariate models. Int. Econ. Rev. **39**(4), 949–968 (1998)
32. Stock, J.H., Watson, M.W.: New indexes of coincident and leading economic indicators. Working Paper 1380, National Bureau of Economic Research (1990)
33. Stock, J.H., Watson, M.W.: Business cycle fluctuations in US macroeconomic time series. Working Paper 6528, National Bureau of Economic Research (1998)

34. Stock, J.H., Watson, M.W.: Macroeconomic forecasting using diffusion indexes. J. Bus. Econ. Stat. **20**(2), 147–162 (2002)
35. Stock, J.H., Watson, M.W.: Combination forecasts of output growth in a seven-country data set. J. Forecast. **23**(6), 405–430 (2004)
36. Stock, J.H., Watson, M.W.: Forecasting output and inflation: the role of asset prices. J. Econ. Lit. **41**(3), 788–829 (2003)
37. Tay, A.S.: Mixing frequencies: stock returns as a predictor of real output growth. Macroeconomics Working Papers 22480, East Asian Bureau of Economic Research (2006)
38. Tibshirani, R.: Regression shrinkage and selection via the lasso: a retrospective. J. R. Stat. Soc. Ser. B: Stat. Methodol. **73**(3), 273–282 (2011)
39. Uematsu, Y., Tanaka, S.: High-dimensional macroeconomic forecasting and variable selection via penalized regression. Econ. J. **22**(1), 34–56 (2019)

# On Disintegration of Measures in Optimal Transport

**Hung T. Nguyen**

**Abstract** As part of a prelude to the applications of optimal transport theory to econometrics and machine learning, this tutorial paper focus on the notion of disintegration in measure theory in the analysis of Wassersrein metrics which are useful, in particular, for machine learning and financial risks. An elementary exposition on using disintegration to prove that Waserstein divergence is a bona fide metric is provided.

**Keywords** Disintegration · Financial risks · Machine learning · Optimal transport · Wasserstein metrics

## 1 Introduction

At present, Optimal Transport (OT) is a hot research topic in Mathematics, especially for applications in a variety of domains, such as Statistics, Econometrics, Machine Learning, Differential Geometry, Partial differential equations, etc. See Villani [4, 5] and Galichon [2]. These represent an ideal setting for Applied Mathematics.

Once an application problem revealed new mathematics, theoreticians can embark on "pure" mathematical research, such as Wasserstein metric spaces of probability measures, and Statistical Decision Theory, which, in turn, should contribute to advances of such applied domains (similar to the two-way street interaction between theoretical physics and mathematics).

The research on OT revealed, in particular, a new set of tools to compare probability measures (between models and observed data), namely Wasserstein divergences.

To establish that Wasserstein "divergences" are in fact bona fide metrics, we need to use disintegration to prove their triangle inequality property.

As such, this tutorial aims at elaborating a bit on disintegration of measures to point out several interesting things.

H. T. Nguyen (✉)
New Mexico State University (USA) & Chiang Mai Universi, Las Cruces, USA
e-mail: hunguyen@nmsu.edu

 
V. Kreinovich et al. (eds.), *Applications of Optimal Transport to Economics and Related Topics*, Studies in Systems, Decision and Control 556,
https://doi.org/10.1007/978-3-031-67770-0_11

## 2 What Is a Wasserstein Divergence?

Roughly speaking, a divergence, on a space of probability measures, is a "measure" of the difference between two probability measures. Such a divergence is used to compare probability measures, for example $D(\mu, \nu)$ is the difference between a model $\mu$ and a data $\nu$.

The most well known divergence is the Kullback-Leibler divergence on probability measures on $(\mathbb{R}, \mathscr{B}(\mathbb{R}))$:

$$KL(\mu//\nu) = \int_{\mathbb{R}} f(x) \log\left(\frac{f(x)}{g(x)}\right) d\gamma(x)$$

where $f(.), g(.)$ are probability density of $\mu, \nu$ respectivey (with respect to some dominating measure $\gamma$ on $\mathscr{B}(\mathbb{R})$). The $KL$ divergence ( also known as "relative entropy") is not a distance since it is not symmetric, but it does have analogous properties which could be used to substitute for a metric, such as the Total Variation metric

$$TV(\mu, \nu) = \sup\{|\mu(A) - \nu(A)| : A \in \mathscr{B}(\mathbb{R})\}$$

The $KL$ divergence appears in the model selection criterion $AIC$.

Metrics on spaces of probability measures are viewed as special divergences. Divergences abound. The choice of a divergence or a metric for comparing probability measures depends on its usefulness for the problem at hand. For example, the Kullback-Leibler divergence is used in $AIC$ because of its relation to Maximum Likelihood Estimation.

Here we are interested in a quantity arising from OT which turns out to be a new type of divergence (in fact, a new type of metric), especially suitable for Machine Learning as it takes into account the geometry of the underlying sample space.

The OT setting is this. Let $\mu, \nu$ be two probability measures on $\mathscr{X}, \mathscr{Y} \subseteq \mathbb{R}^n$, respectively. Let $\Pi(\mu, \nu)$ denote the set of joint probability measures on $\mathscr{X} \times \mathscr{Y}$ admitting $\mu, \nu$ as marginals. For $p \geq 1$, consider the convex optimization problem

$$\min\{\int_{\mathscr{X} \times \mathscr{Y}} ||x - y||^p d\lambda(x, y) : \lambda \in \Pi(\mu, \nu)\}$$

From OT, this minimization problem admits a minimizer, i.e., the infimum is attained. In fact, the solution of this OT problem is unique under suitable conditions.

In particular, when $\mathscr{X} = \mathscr{Y} \subseteq \mathbb{R}^n$ (or a polish space), we are interested in the following Wasserstein divergence (a priori) on the subset $\mathscr{P}_p(\mathscr{X})$ of the set $\mathscr{P}(\mathscr{X})$ of all (Borel) probability measures on $\mathscr{X}$, where

$$\mathscr{P}_p(\mathscr{X}) = \{\mu \in \mathscr{P}(\mathscr{X}) : \int_{\mathscr{X}} ||x||^p d\mu(x) < \infty\}$$

namely, $W_p(.,.) : \mathscr{P}_p(\mathscr{X}) \times \mathscr{P}_p(\mathscr{X}) \to [0, \infty)$

$$W_p(\mu, \nu) = \inf_{\lambda \in \Pi(\mu,\nu)} [\int_{\mathscr{X} \times \mathscr{X}} ||x - y||^p d\lambda(x, y)]^{\frac{1}{p}}$$

Specifically, we are going to show that the Wasserstein divergence $W_p(.,.)$ is in fact a bona fide metric on $\mathscr{P}_p(\mathscr{X})$, a well known fact in the literature.

Our main purpose of this talk is to carry out that proof with emphasis on the use of the notion of disintegration.

## 3 What Is Disintegration?

Disintegration is a process of extracting a conditional probability measure from a joint probability measure on a product space.

In various applications, the underlying spaces for which disintegrations exist, are very general, such as polish spaces (separable complete metric spaces) rather than just finitely dimensional Euclidean spaces. As such, we need to generalize the way we obtain classical *regular* conditional probability measures (or distributions).

To be concrete, let $\mathscr{X}, \mathscr{Y} \subseteq \mathbb{R}^n$, and $(\mathscr{X}, \mathscr{B}(\mathscr{X}))$, $(\mathscr{Y}, \mathscr{B}(\mathscr{Y}))$, $(\mathscr{X} \times \mathscr{Y}, \mathscr{B}(\mathscr{X} \times \mathscr{Y}))$, be (Borel) measurable spaces. We denote by $\mathscr{P}(\mathscr{X})$ , $\mathscr{P}(\mathscr{Y})$, $\mathscr{P}(\mathscr{X} \times \mathscr{Y})$ the set of all probability measures on these spaces.

For $\lambda \in \mathscr{P}(\mathscr{X} \times \mathscr{Y})$, its marginal probability measure on $\mathscr{X}$ is $\mu \in \mathscr{P}(\mathscr{X})$, defined as, for any $A \in \mathscr{B}(\mathscr{X})$, $\mu(A) = \lambda(A \times \mathscr{Y})$.

A disintegration of $\lambda$ with respect to $\mu$ is a family of probability measures $\nu_x \in \mathscr{P}(\mathscr{Y})$, for any $x \in \mathscr{X}$, such that, for $A \in \mathscr{B}(\mathscr{X})$ and $B \in \mathscr{B}(\mathscr{Y})$, we have

$$\lambda(A \times B) = \int_A \nu_x(B) d\mu(x)$$

Symbolically,

$$\lambda = \int_{\mathscr{X}} (\delta_x \otimes \nu_x) d\mu(x)$$

where $\delta_x$ is the Diract probability measure on $\mathscr{X}$, at $x \in \mathscr{X}$, and $\delta_x \otimes \nu_x$ denotes the product measure $(\delta_x \otimes \nu_x)(A \times B) = \delta_x(A)\nu_x(B)$, noting that $\delta_x(A) = 1$ or $0$, according to $x \in A$ or $x \notin A$.

The representation of $\lambda$ is so written since

$$\lambda(A \times B) = \int_{\mathscr{X}} (\delta_x \otimes \nu_x)(A \times B) d\mu(x) =$$

$$\int_{\mathscr{X}} (\delta_x(A)\nu_x)(B) d\mu(x) = \int_A \nu_x(B) d\mu(x)$$

Below is a tutorial on disintegration, just enough for using it in proving the triangle inequality for Wasserstein metrics. A reference could be Dudley [1] or Graf and Mauldin [3].

Let's us recap the situation on $\mathbb{R}^n$. Basically, we want to justify (rigorously) what do we mean by a conditional probability measure (or distribution) $P_{Y|X=x}$ when $P(X = x) = 0$? e.g., when $X$ is a continuous random variable, since then $P(X = x) = 0$ for all $x$.

Let $(X, Y) : (\Omega, \mathscr{A}, P) \to (\mathbb{R}^2, \mathscr{B}(\mathbb{R}^2))$ be a random vector, defined on its source of uncertainty $(\Omega, \mathscr{A}, P)$ (a probability space), with values in the measurable space $(\mathbb{R}^2, \mathscr{B}(\mathbb{R}^2))$, where $\mathscr{B}(\mathbb{R}^2)$ denotes the Borel $\sigma-$ field of $\mathbb{R}^2$.

The probability law of $(X, Y)$ is the probability measure $P_{(X,Y)}(.)$ on $\mathscr{B}(\mathbb{R}^2)$ which is the "push—forward" of $P$ by $(X, Y)$.

The marginal probability measures of $X$ and $Y$ are denoted as $\mu = P_X$ , and $\nu = P_Y$, respectively.

By Lebesgue-Stieltjes Theorem, $P_{(X,Y)}(.)$ is characterized by the joint distribution function $F_{(X,Y)}(.) : \mathbb{R}^2 \to [0, 1]$ :

$$F_{(X,Y)}(x, y) = P\{\omega \in \Omega : X(\omega) \le x, Y(\omega) \le y\} = P_{(X,Y)}((-\infty, x] \times (-\infty, y])$$

i.e., in this "Euclidean setting", and we write $P_{(X,Y)} = dF_{(X,Y)}$.

When $F_{(X,Y)}$ is known, its marginal distributions of $X$ and $Y$ are obtained as

$$F_X(x) = F_{(X,Y)}(x, \infty), \qquad , F_Y(y) = F_{(X,Y)}(\infty, y)$$

The question is: What is the conditional distribution function $y \to F_{Y|X}(y|x)$ of $Y$ given $X = x$ ? (we let $\nu_x = dF_{Y|X}(.|x)$).

Some concrete simple examples motivate our analysis.

(a) Suppose $(X, Y)$ is a discrete random vector, i.e., say, $X$ (resp. $Y$) takes values in a countable set $\mathscr{X}$ (resp.$\mathscr{Y}$), more specifically, in the measurable space $(\mathscr{X}, 2^{\mathscr{X}})$ (resp. $(\mathscr{Y}, 2^Y)$ .

For $x \in \mathscr{X}$ such that $\mu(\{x\}) = dF_X(\{x\}) = f_X(x) \neq 0$, we set, for any $B \subseteq \mathscr{Y}$ :

$$\nu_x(B) = P_{(X,Y)}\{(x, y) : y \in B\}/\mu(\{x\})$$

and for $x \in \mathscr{X}$ such that $\mu(\{x\}) = 0$, we let $\nu_x = \delta_{y_o}$ (the Dirac probability measure as a fixed point $y_o \in \mathscr{Y}$, i.e., $\nu_x(B) = 1_B(y_o)$).

Then, it can be checked that we obtain a family of bona fide probability measures $\{\nu_x : x \in \mathscr{X})$ on $(\mathscr{Y}, 2^Y)$.

This family of conditional probability measures on $\mathscr{Y}$, indexed by $\mathscr{X}$, is a "regular" conditional probability measure (distribution) which is called a disintegration of $P_{(X,Y)}$ with respect to the marginal $\mu = dF_X$.

(b) Consider now the case where $(X, Y)$ is a continuous random vector, i.e., say, $\mathscr{X} = \mathscr{Y} = \mathbb{R}$, so that $P_{(X,Y)} = dF_{(X,Y)}$ on $\mathscr{B}(\mathbb{R}^2)$.

In particular, suppose $P_{(X,Y)}$ is absolutely continuous with respect to Lebesgue measure $\lambda(dx, dy)$ on $\mathscr{B}(\mathbb{R}^2)$, i.e., $F_{(X,Y)}$ has a joint probability density function $f(x, y)$.

In this case, the marginal density of $X$ is obtained as $f_X(x) = \int_{-\infty}^{\infty} f(x, y)dy$.

If $f(x) \in (0, \infty)$, we let $f_{Y|X}(y, x) = \frac{f(x,y)}{f_X(x)}$, and $\nu_x$ be its corresponding probability measure.

If $f(x) = 0$ or $\infty$, we let $P_x$ the probability measure corresponding to some fixed density $g$ on $\mathbb{R}$.

Then $\nu_x, x \in \mathbb{R}$ are regular conditional probability measures for $P_{(X,Y)}$.
(a) and (b) justify the existence of regular conditional probability measures on $(\mathbb{R}, \mathscr{B}(\mathbb{R}))$.

*Remark* In the above settings, we "break" the joint probability measure $P_{(X,Y)}$ in two pieces: the marginal $P_X$ and the disintegration of $P_{(X,Y)}$ with respect to $\mu = P_X$, a process analogous to "fission" of particles in physics.

Of course, you ask "What's for?". The answer is: For representing integration with respect to $P_{(X,Y)}$ as an iterated integral with respect to $P_X$ and the disintegration of $P_{(X,Y)}$ with respect to $P_X$.

If $X$ and $Y$ are independent, so that $P_{(X,Y)} = P_X \otimes P_Y$ (product measure), then $\nu_x = \nu = P_Y$ for any $x \in \mathscr{X}$, and we have the Fubini Theorem: If $\varphi : \mathscr{X} \times \mathscr{Y} \to \mathbb{R}$ is $P_{(X,Y)}-$ integrable, then

$$\int_{\mathscr{X} \times \mathscr{Y}} \varphi(x, y)dP_{(X,Y)}(x, y) = \int_{\mathscr{X}} [\int_{\mathscr{Y}} \varphi(x, y)dP_Y(y)]dP_X(x)$$

Thus, disintegration generalizes Fubini to the dependence case of random variables (to be shown next).

The opposite of disintegration, in a simple setting, is this. Given distribution functions $F_X, F_Y$ on $\mathbb{R}$, how to "fuse" them to form a joint distribution function $F_{(X,Y)}$ on $\mathbb{R}^2$? (analogous to fusion of particles).

The answer is using copulas: By Sklar's theorem:

$$F_{(X,Y)}(x, y) = c(F_X(x), F_Y(y))$$

where $c(., .)$ is a copula.

We proceed now to establish the announced fact that disintegration (when it exists) is useful for representing an integral with respect to a joint probability measure as an iterated integral, namely:

**Theorem** *If the conditional probability measures $\nu_x, x \in \mathscr{X}$ exist, then for any $\varphi : \mathscr{X} \times \mathscr{Y} \to \mathbb{R}$ is $P_{(X,Y)}-$ integrable, we have*

$$\int_{\mathscr{X} \times \mathscr{Y}} \varphi(x, y)dP_{(X,Y)}(x, y) = \int_{\mathscr{X}} [\int_{\mathscr{Y}} \varphi(x, y)d\nu_x(y)]dP_X(x)$$

*Proof* By "the standard argument" of measure theory, it suffices to verify the representation for $\varphi(x, y) = 1_{A\times B}(x, y)$ for $A \in \mathscr{B}(\mathscr{X})$ and $B \in \mathscr{B}(\mathscr{Y})$, but that is obvious. QED.

Symbolically, for any $\lambda \in \Pi(\mu, \nu)$, there exist $\{\nu_x , x \in \mathscr{X}\}$, $\nu_x \in \mathscr{P}(\mathscr{Y})$ such that $\lambda = \int_{\mathscr{X}} (\delta_x \otimes \nu_x) d\mu(x)$.

Or, a disintegration of $\lambda$ is a representation $\lambda = \int_{\mathscr{X}} \nu_x d\mu(x)$, where $\nu_x \in \mathscr{P}(\mathscr{Y})$, the space of all probability measures on $\mathscr{Y}$.

Another interpretation of disintegration of $\lambda \in \mathscr{P}(\mathscr{X} \times \mathscr{Y})$ with marginal $\mu$ on $\mathscr{X}$ is this. There exists $\{\nu_x, x \in \mathscr{X}\}$, $\nu_x \in \mathscr{P}(\mathscr{Y})$, such that for any $A \in \mathscr{B}(\mathscr{X} \times \mathscr{Y})$, we have $\lambda(A) = \int_{\mathscr{X}} \nu_x(A_x) d\mu(x)$., where $A_x = \{y \in \mathscr{Y} : (x, y) \in A\}$.

A disintegration can be viewed as a map $T(.) : \mathscr{X} \to \mathscr{P}(\mathscr{Y})$ such that for any $x \in \mathscr{X}$, $T(x) \otimes \mu \in \Pi(\mu, \nu)$.

An interesting interpretation of Kantorovich's transport plans in OT is this.

As $\Pi(\mu, \nu)$ is the set of all possible transport plans, a disintegration of an $\lambda \in \Pi(\mu, \nu)$, with respect to $\nu$, means that we can split the mass $\mu$ to $\mu_x, x \in \mathscr{X}$ to transport it to $\nu$.

The transport plan $\mu \otimes \nu$ corresponds to the constant map $x \to \nu_x = \nu$.

**Important Remark**. Disintegration might not exist in general spaces.

However, disintegration does exist on polish spaces (separable complete metric spaces, e.g.. $\mathbb{R}^n$). For a proof, see a Text like Dudley [].

As we will see next, as we are going to prove that Wasserstein divergences are bona fide metrics by using disintegration, our underlying metric space $\mathscr{X}$ will be taken as a polish space.

## 4 Wasserstein Divergences are Metrics

Consider first the simplest setting ! Let $\mu, \nu$ be two probability measures on $(\mathbb{R}, \mathscr{B}(\mathbb{R}))$, and

$$W(\mu, \nu) = \inf\{\int_{\mathbb{R}^2} |x - y| d\lambda(x, y) : \lambda \in \Pi(\mu, \nu)\}$$

Then $W(\mu, \nu)$ is a bona fide metric on $\mathscr{P}(\mathbb{R})$. One "practical" way to see it is go ahead to compute it!

For this simplest setting, we proceed as follows.

If we let $X, Y : (\Omega, \mathscr{A}, P) \to (\mathbb{R}, \mathscr{B}(\mathbb{R}))$ be random variables with probability laws $\mu = dF$ , $\nu = dG$ , respectively, then

$$W(\mu, \nu) = W(dF, dG) = \inf\{E|X - Y| : X \sim \mu, Y \sim \nu\}$$

where the expectation is with respect to the joint law of $(X, Y)$ having $\mu, \nu$ as marginals. Then,

$$W(dF, dG) = \int_{-\infty}^{\infty} |F(t) - G(t)|dt$$

so that $W(.,.)$ is clearly a metric on $\mathscr{P}(\mathbb{R})$ !

*Proof* Since

$$|X - Y| = (X - Y)1_{(X \geq Y)} + (Y - X)1_{(X<Y)}$$

we let $\alpha = \max(X - Y, 0)$, and $\beta = \max(Y - X), 0)$, so that

$$E|X - Y| = E\alpha + E\beta$$

Since $\alpha \geq 0$, we have

$$E(\alpha|Y = y) = \int_0^{\infty} P(\alpha > z|Y = y)dz$$

Now, $E(\alpha) = EE(\alpha|Y)$, so that

$$E\alpha = \int_{-\infty}^{\infty} dG(y) \int_0^{\infty} P(X - Y \geq z|Y = y) =$$

$$\int_{-\infty}^{\infty} dG(y) \int_0^{\infty} P(X \geq y + z|Y = y)dz =$$

$$\int_{-\infty}^{\infty} dG(y) \int_y^{\infty} P(X \geq x|Y = y)dx =$$

$$\int\int_{(x,y);x>y} P(X \geq y, Y < y)dx = \int_{-\infty}^{\infty} P(X \geq y, Y < y)dy$$

Similarly,

$$E\beta = \int_{-\infty}^{\infty} P(Y \geq y, X < y)dy$$

Thus,

$$E|X - Y| = \int_{-\infty}^{\infty} P(X \geq y, Y < y)dy + \int_{-\infty}^{\infty} P(Y \geq y, X < y)dy =$$

$$\int_{-\infty}^{\infty} [P(X < y, Y \geq y) + P(Y < y, X \geq y)]dy$$

Now, look at the event $(X < y, Y \geq y)$.

Let $A = (X < y)$ and $B = (Y < y)$, then $(X < y, Y \geq y) = A \cap B^c$. But $A = (B^c \cap A) \cup (A \cap B)$, so that

$$P(X < y, Y \geq y) = P(A) - P(A \cap B) = P(X < y) - P(X < y, Y < y)$$

Thus,

$$E|X - Y| = \int_{-\infty}^{\infty} [P(X < y) + P(Y < y) - 2P(X < y, Y < y)]dy$$

Note that $P(X < y, Y < y)$ is the value of the joint distribution $H$ of the vector $(X, Y)$, and it is well known that $H(x, y) \leq F(x) \wedge G(y)$ (for example, using copula theory), so that

$$E|X - Y| \geq \int_{-\infty}^{\infty} [F(y) + G(y) - 2\min(F(y), G(y)]dy = \int_{-\infty}^{\infty} |F(y) - G(y)|dy$$

by noting that $|x - y| = x + y - 2(x \wedge y)$.

Therefore, $W(dF, dG) \geq \int_{-\infty}^{\infty} |F(y) - G(y)|dy$.

However, there are random variables $X \sim dF$, $Y \sim dG$ such that $E|X - Y|$ attains the bound $\int_{-\infty}^{\infty} |F(y) - G(y)|dy$, so that

$$W(dF, dG) = \int_{-\infty}^{\infty} |F(y) - G(y)|dy$$

Indeed, the quantile function of $F$ is $F^{[-1]}(.) : (0, 1) \to \mathbb{R}$, defined as $F^{[-1]}(u) = \inf\{x \in \mathbb{R} : F(x) \geq u\}$ (which is, in fact a minimum). Similarly for $G^{[-1]}$.

*Remark* If we "consider" random variables that take values in $\bar{\mathbb{R}} = [-\infty, +\infty]$, then $F^{[-1]}(.)$ is defined on $[0, 1]$ with $F^{[-1]}(0) = -\infty$, $F^{[-1]}(1) = +\infty$.

Since $F^{[-1]}(u) \leq x \iff F(x) \geq u$, it follows that the random variable $F^{[-1]}(U)$ has the same distribution $F$ of the random variable $X$, where $U$ is the random variable uniformly distributed on $(0, 1)$. Similarly, $G^{[-1]}(U) \overset{D}{=} Y \sim G$.

Thus,

$$E|X - Y| = E|F^{[-1]}(U) - G^{[-1]}(U)| = \int_0^1 |F^{[-1]}(u) - G^{[-1]}(u)|du$$

Now, $\int_{-\infty}^{\infty} |F(x) - G(x)|dx$ is the area between the graphs of $F$ and $G$, it is the same as $\int_0^1 |F^{[-1]}(u) - G^{[-1]}(u)|du$.

Thus, since $W(dF, dG) = \int_{-\infty}^{\infty} |F(x) - G(x)|dx$, it is obvious that $W(.,.)$ is a metric on $\mathscr{M}(\mathbb{R})$.

*Remark* (1) An "analytic" proof of

$$\int_0^1 |F^{[-1]}(u) - G^{[-1]}(u)|du = \int_{-\infty}^{\infty} |F(x) - G(x)|dx$$

is as follows. The following is justified by Fubini's theorem, namely if $\int_{A\times B} |f(x, y)|d(x, y) < \infty$, then

$$\int_{A\times B} |f(x, y)|d(x, y) = \int_A [\int_B f(x, y)dy]dx = \int_B [\int_A f(x, y)dx]dy$$

Now, for $u \in (0, 1)$, we have

$$|F^{[-1]}(u) - G^{[-1]}(u)| = [F^{[-1]}(u) - G^{[-1]}(u)]1_{\{u:F^{[-1]}(u)>G^{[-1]}(u)\}}(u) +$$

$$[G^{[-1]}(u) - F^{[-1]}(u)]1_{\{u:F^{[-1]}(u)\leq G^{[-1]}(u)\}}(u)$$

So let

$$A = \{u \in (0, 1) : F^{[-1]}(u) > G^{[-1]}(u)\}$$

$$A^c = \{u \in (0, 1) : F^{[-1]}(u) \leq G^{[-1]}(u)\}$$

We have

$$\int_0^1 |F^{[-1]}(u) - G^{[-1]}(u)|du =$$

$$\int_A |F^{[-1]}(u) - G^{[-1]}|(u)|du + \int_{A^c} |F^{[-1]}(u) - G^{[-1]}(u)|du$$

where we can write

$$\int_A |F^{[-1]}(u) - G^{[-1]}|(u)|du = \int_A [\int_{G^{[-1]}(u)}^{F^{[-1]}(u)} dx]du$$

Now, observe that, by definition of the quantile functions, we have $G^{[-1]}(u) \leq x \iff u \leq G(x)$ (and of course, $x < F^{[-1]}(u) \iff u > F(x)$) , so that

$$\int_A [\int_{G^{[-1]}(u)}^{F^{[-1]}(u)} dx]du = \int_{\mathbb{R}} [\int_{F(x)}^{G(x)} 1_A(u)1_{\{F(x)\leq G(x)\}}(x)du]dx$$

Similarly,

$$\int_{A^c} |F^{[-1]}(u) - G^{[-1]}(u)|du = \int_{\mathbb{R}} [\int_{G(x)}^{F(x)} 1_{A^c}(u) 1_{\{F(x)>G(x)\}}(x) du] dx$$

Hence,

$$\int_{\mathbb{R}} [\int_{F(x)}^{G(x)} 1_A(u) 1_{\{F(x)\leq G(x)\}}(x) du] dx + \int_{\mathbb{R}} [\int_{G(x)}^{F(x)} 1_{A^c}(u) 1_{\{F(x)>G(x)\}}(x) du] dx =$$

$$\int_{\mathbb{R}} |F(x) - G(x)|dx$$

(2) The infimum in the definition of $W(.,.)$ is attained (i.e., is a minimum) for $\mathbb{R}$ which is a separable and complete metric space, i.e., a Polish space. This fact is true in any Polish spaces. The metric space $(\mathscr{M}(\mathbb{R}), W)$ is called a Wasserstein space.

The $L_1$ distance $W(dF, dG) = \int_{-\infty}^{\infty} |F(y) - G(y)|dy$ (on the space of distributions with finite means) is also called the Kantorovich metric.

(3) The above calculations can also be obtained as follows.

Since $C(u, v) = u \wedge v$ is a copula, the joint distribution function $H(x, y) = F(x) \wedge G(y)$ corresponds to a joint probability measure $\pi^* = dH$ which is in $\Pi(\mu, \nu)$, where $F, G$ are distribution functions corresponding to $\mu, \nu$, respectively ($\mu = dF, \nu = dG$).

From the Optimal Transport theory (see Villani [1]) we know that $\pi^*$ is the optimal transport plan, i.e.

$$W(\mu, \nu) = E_{\pi^*}(|X - Y|) = \int_{\mathbb{R}^2} |x - y| d\pi^*(x, y)$$

Now, from copula theory, we know that if $(X, Y)$ has the joint distribution function $H(x, y) = F(x) \wedge G(y)$, then $X, Y$ are comonotonic and $(X, Y) \stackrel{D}{=} (F^{[-1]}(U), G^{[-1]}(U))$, where $U$ is uniform on $[0, 1]$.

This is seen as follows. Separately, we have $X \stackrel{D}{=} F^{[-1]}(U)$, and $Y \stackrel{D}{=} G^{[-1]}(U)$. Now, assume $H(x, y) = F(x) \wedge G(y)$, we have

$$P(F^{[-1]}(U) \leq x, G^{[-1]}(U) \leq y) = P(U \leq F(x), U \leq G(y)) =$$

$$P(U \leq F(x) \wedge G(y)) = F(x) \wedge G(y) = P(X \leq x, Y \leq y)$$

Thus,

$$W(\mu, \nu) = W(dF, dG) = \int_{\mathbb{R}^2} |x - y| d\pi^*(x, y) = \int_0^1 |F^{[-1]}(u) - G^{[-1]}(u)| du$$

by transfert theorem, noting that $\pi^* = (F^{[-1]} \times G^{[-1]})\#du$. But, as proved in (1),

$$\int_0^1 |F^{[-1]}(u) - G^{[-1]}(u)| du = \int_{-\infty}^{\infty} |F(x) - G(x)| dx$$

It looks like we arrive at a Wasserstein metric which is simply a $L^1$ metric on distribution functions of random variables, but in fact, a "hidden" factor is that it is defined in terms of a chosen metric on the underlying space $\mathbb{R}$, namely $|x - y|$. But even on $\mathbb{R}$, different metrics can be used, e.g., $|x - y|^2$, and more importantly, for $\mathbb{R}^n$ or general polish spaces, the associated Wasserstein metrics do take into account of the geometry of the underlying spaces, e.g., a manifold, and that is why Wasserstein metrics are suitable for, say, machine learning.

The above "motivating example" of Wasserstein distance can be placed in the context of Optimal Transport (OT) theory as follows.

Let $\mu, \nu$ be probability measures on $\mathscr{X}, \mathscr{Y} \subseteq \mathbb{R}^n$, say (or more generally, a polish space) respectively, and $c(., .) : \mathscr{X} \times \mathscr{Y} \to \mathbb{R}^+$, measurable, the OT problem is this.

Find the probability measure $\pi^*$ on $\mathscr{X} \times \mathscr{Y}$ such that

$$\pi^* = \arg\min\{\int_{\mathscr{X} \times \mathscr{Y}} c(x, y) d\pi(x, y)\}$$

where the minimum is taken over the set $\Pi(\mu, \nu)$ of joint probability measures with $\mu, \nu$ as marginals.

The metric side of OT is this. Consider the special case where $\mathscr{X} = \mathscr{Y}$, and $c = \rho$, a metric on $\mathscr{X}$. For example, with $\mathscr{X} = \mathbb{R}$, and $\rho(x, y) = |x - y|$, $\pi^*$ is the probability measure on $\mathbb{R}^2$ whose joint distribution function $H(x, y) = F(x) \wedge G(y)$, where $F, G$ are distribution functions corresponding to $\mu, \nu$, respectively. The (optimal) value

$$W(\mu, \nu) = W(dF, dG) = \min\{\int_{\mathscr{X} \times \mathscr{X}} \rho(x, y) d\pi(x, y) : \pi \in \Pi(\mu, \nu)\}$$

appears to be a distance between probability measures $\mu, \nu$ (of course, when they have finite first moments), i.e., $W(., .)$ is a metric on the set $\mathscr{M}(\mathbb{R})$ of (Borel) probability measures on $\mathbb{R}$ which have finite first moments, i.e., $\int_{\mathscr{X}} |x| d\mu(x) < \infty$.

For $\rho(x, y) = \frac{1}{2}|x - y|^2$, the above $\pi^*$ is also an optimal transport plan, and the optimal value is $\int_0^1 |F^{[-1[}(u) - G^{[-1]}(u)|^2 du$
so that

$$W_2(dF, dG) = [\int_0^1 |F^{[-1]}(u) - G^{[-1]}(u)|^2 du]^{\frac{1}{2}}$$

is a metric on the set of probability measures on $\mathbb{R}$ with finite second moments, i.e., $\int_{\mathscr{X}} |x|^2 d\mu(x) < \infty$.

In fact, for $p \geq 1$, the corresponding Wasserstein metric is

$$W_p(dF, dG) = [\int_0^1 |F^{[-1]}(u) - G^{[-1]}(u)|^p du]^{\frac{1}{p}}$$

*Remark* Except for $p = 1$, $W_p(dF, dG) \neq [\int_{\mathbb{R}} |F(x) - G(x)|^p dx]^{\frac{1}{p}}$, $p > 1$.

The proof, for $p \geq 1$, of

$$W_p^p(dF, dG) = \inf\{\int_{\mathbb{R}^2} ||x - y||^p d\pi(x, y) : \pi \in \Pi(dF, dG)\} =$$

$$\int_0^1 |F^{[-1]}(u) - G^{[-1]}(u)|^p du$$

can be obtained by observing that, in one hand, for $U$ uniform on (0, 1), the random variables $F^{[-1]}(U)$, $G^{[-1]}(U)$ are distributed as $dF, dG$, respectively, so that their joint measure is in $\Pi(dF, dG)$. But

$$E[|F^{[-1]}(U) - G^{[-1]}(U)|^p = \int_0^1 |F^{[-1]}(u) - G^{[-1]}(u)|^p du$$

so that

$$W_p(dF, dG) \leq [\int_0^1 |F^{[-1]}(u) - G^{[-1]}(u)|^p du]^{\frac{1}{p}}$$

on the other hand,

$$W_p^p(dF, dG) \geq \int_0^1 |F^{[-1]}(u) - G^{[-1]}(u)|^p du$$

by Fatou lemma.

Now, for $\mathscr{X} = \mathbb{R}^n$, with norm $||.||$, and $p \geq 1$, the $p$th-Wasserstein metric is

$$W_p^p(dF, dG) = \inf\{E_\pi ||X - Y||^p : X \sim dF, Y \sim dG, \pi \in \Pi(dF, dG)\}$$

where $F, G$ are $n-$ dimensional distribution functions of $X, Y$, respectively, and $\pi$ has $2n-$ dimensional distribution function $H$ with $F, G$ as marginals, i.e.,

$$H(x_1, ...x_n, \infty, ..., \infty) = F(x_1, ...x_n), \quad H(\infty, ..., \infty, x_1, ...x_n,) = G(x_1, ...x_n)$$

More generally, Wasserstein distance is a metric on spaces of probability measures. Let $(\mathscr{X}, \rho)$ be a metric space. Consider the situation where we are interested in probability measures governing the random evolution of random elements taking values in $\mathscr{X}$ (i.e., their "laws" operating on Borel $\sigma-$ field $\mathscr{B}(\mathscr{X})$ ). Comparisons of probability measures are standard concerns in applications, such as in the so-called empirical processes.

For $\mu, \nu$ two probability measures on $(\mathscr{X}, \mathscr{B}(\mathscr{X}))$, consider the nonnegative quantity

$$W(\mu, \nu) = \inf\{\int_{\mathscr{X}\times\mathscr{X}} \rho(x, y) d\pi(x, y)\} \leq +\infty$$

where the infimum is taken over all joint probability measure $\pi$ with marginals (projections) $\mu, \nu$.

We will denote by $\Pi(\mu, \nu)$ the set of probability measures $\pi$ on the product space $\mathscr{X} \times \mathscr{X}$ having $\mu, \nu$ as marginal measures, i.e., $\mu(.) = \pi(. \times \mathscr{X}), \nu(.) = \pi\ (\mathscr{X} \times .)$.

Note that the above quantity can be written as :

$$W(\mu, \nu) = \inf\{E\rho(X, Y) : X \sim \mu, Y \sim \nu\}$$

i.e., the infimum is taken over all random variables $X, Y$ with values in $(\mathscr{X}, \mathscr{B}(\mathscr{X}))$, and $X, Y$ are distributed as $\mu, \nu$, respectively.

On a subset of $\mathscr{P}(\mathscr{X})$ where $W(\mu, \nu) < \infty$, for $\mu, \nu$ in it, $W(., .)$ is a bona fide metric.

Specifically, if $(\mathscr{X}, \rho)$ is a Polish space, i.e., a separable and complete metric space (finitely dimensional or not, e.g. $\mathbb{R}^d$ , $C([0, 1], \mathbb{R}^d)$) then $W(., .)$ is a metric on the subspace of $\mathscr{P}(\mathscr{X})$ on which $W(\mu, \nu) < \infty$, and the convergence in Wasserstein distance is equivalent to the weak convergence of probability measures.

We come now to the main investigation of Wasserstein divergences (a priori) on a metric space $\mathscr{X}$ for which disintegration exists, such as $\mathbb{R}^n$ or a polish space.

This is a special OT setting where $\mathscr{X} = \mathscr{Y} \subseteq \mathbb{R}^n$ (for concreteness).

Let $\mathscr{P}(\mathscr{X})$ denotes the set of all (Borel) probability measures on $\mathscr{B}(\mathscr{X})$. For $p \geq 1$, let $\mathscr{P}_p(\mathscr{X}) \subseteq \mathscr{P}(\mathscr{X})$, be the subset of probability measures with finite $p-$ moment, i.e.,

$$\mathscr{P}_p(\mathscr{X}) = \{\mu \in \mathscr{P}(\mathscr{X}) : \int_{\mathscr{X}} ||x||^p d\mu(x) < \infty\}$$

where $||.||$ is the Euclidean norm of $\mathbb{R}^n$.

Consider the Wasserstein divergence on $\mathscr{P}_p(\mathscr{X})$: For $\mu, \nu \in \mathscr{P}_p(\mathscr{X})$, and $p \geq 1$, let

$$W_p(\mu, \nu) = \inf_{\lambda \in \Pi(\mu,\nu)} [\int_{\mathscr{X}\times\mathscr{X}} ||x - y||^p d\lambda(x, y)]^{\frac{1}{p}}$$

*Remark* Here for concreteness, we consider $(\mathscr{X}, D) = (\mathbb{R}^n, ||.||)$, but the analysis applies to more general situations, such as for a polish space $(S, D)$, where

$$W_p(\mu, \nu) = \inf_{\lambda \in \Pi(\mu,\nu)} [\int_{S\times S} D^p(x, y) d\lambda(x, y)]^{\frac{1}{p}}$$

This is just an exercise to verify that $W_p(., .)$ does satisfy the axioms of a metric, i.e., $W_p(., .) : \mathscr{P}_p(\mathscr{X}) \times \mathscr{P}_p(\mathscr{X}) \to \mathbb{R}^+ = [0, \infty)$ is such that
(i) $W_p(\mu, \nu) = W_p(\nu, \mu)$
(ii) $W_p(\mu, \nu) = 0 \Longleftrightarrow \mu = \nu$
(iii) For any $\mu, \nu, \gamma \in \mathscr{P}_p(\mathscr{X})$, $W_p(\mu, \nu) \leq W_p(\mu, \gamma) + W_p(\gamma, \nu)$

First, since, for $x, y \in \mathbb{R}^n$, $||x - y||^p \leq c(||x||^p + ||y||^p)$, so that , for $\mu, \nu \in \mathscr{P}_p(\mathbb{R}^n)$, we have

$$W_p(\mu, \nu) \leq c[\int_{\mathscr{X}} ||x||^p d\mu(x) + \int_{\mathscr{X}} ||x||^p d\nu(x)] < \infty$$

While (i) is obvious (since the function $(x, y) \to ||x - y||^p$ is symmetric, and $\Pi(\mu, \nu) \simeq \Pi(\nu, \mu)$) and (ii) can be seen as follows.

For $\nu = \mu$ , the optimal transport map $T : \mathscr{X} \to \mathscr{X}$ is the identity map $I(x) = x$, so that the optimal transport plan is $\lambda = (I, I)\#\mu$ concentrated on $\{(x, y) : x = y\}$ and hence

$$W_p(\mu, \mu) = \int_{\mathscr{X}\times\mathscr{X}} ||x - y||^p d\lambda(x, y) = 0$$

Conversely, if $W_p(\mu, \nu) = 0$, then, since

$$\inf_{\lambda \in \Pi(\mu,\nu)} [\int_{\mathscr{X}\times\mathscr{X}} ||x - y||^p d\lambda(x, y)]^{\frac{1}{p}}$$

is attained, there exists $\lambda \in \Pi(\mu, \nu)$ such that $\int_{\mathscr{X}\times\mathscr{X}} ||x - y||^p d\lambda(x, y) = 0$ so that $\lambda$ is concentrated on $\{(x, y) : x = y\}$ which, in turn, implies that, for any $A \in \mathscr{B}(\mathbb{R}^n)$,

$$\mu(A) = \lambda(A \times \mathbb{R}^n) = \lambda(A \times A) = \lambda(\mathbb{R}^n \times A) = \nu(A)$$

i.e., $\mu = \nu$.

*Remark* For $p \leq q$, we have $W_p(\mu, \nu) \leq W_q(\mu, \nu)$, since, by Jensen's inequality (with respect to the convex function $t \to t^{\frac{q}{p}}$), for any $\lambda \in \Pi(\mu, \nu)$,

$$W_p^q(\mu, \nu) \leq [\int_{\mathscr{X}\times\mathscr{X}} ||x - y||^p d\lambda(x, y)]^{\frac{q}{p}} \leq$$

$$[\int_{\mathscr{X}\times\mathscr{X}} ||x - y||^q d\lambda(x, y)] = W_q^q(\mu, \nu)$$

However, the triangle inequality (iii) is not so obvious!

Interestingly, it is the notion of disintegration which will provide a method to verify it.

We wish to show that, for any $\mu_j$, $j = 1, 2, 3$ in $\mathscr{P}_p(\mathbb{R}^n)$, for $p \geq 1$, with support $\mathscr{X}_j \subseteq \mathbb{R}^n$, $j = 1, 2, 3$, respectively, we should have

$$W_p(\mu_1, \mu_2) \leq W_p(\mu_1, \mu_3) + W_p(\mu_3, \mu_2)$$

For this, we first need a

**Lemma** *Let* $\mu_j$, $j = 1, 2, 3$ *in* $\mathscr{P}_p(\mathbb{R}^n)$, *for* $p \geq 1$, *with support* $\mathscr{X}_j \subseteq \mathbb{R}^n$, $j = 1, 2, 3$, *respectively. Let* $\lambda_{12} \in \Pi(\mu_1, \mu_2)$, *and* $\lambda_{23} \in \Pi(\mu_2, \mu_3)$.

Then there exists a probability measure $\lambda$ on $\mathscr{X}_1 \times \mathscr{X}_2 \times \mathscr{X}_3$ having marginals $\lambda_{12}$ and $\lambda_{23}$ on $\mathscr{X}_1 \times \mathscr{X}_2$, and $\mathscr{X}_2 \times \mathscr{X}_3$, respectively.

*Proof* Disintegrate both $\lambda_{12}$ and $\lambda_{23}$ with respect to their common marginal $\mu_2$, and denote their disintegrations as $\nu_{x_1}, \gamma_{x_3}$, respectively, so that

$$\lambda_{12} = \int_{\mathscr{X}_2} (\nu_{x_1} \otimes \delta_{x_2}) d\mu_2(x_2)$$

$$\lambda_{23} = \int_{\mathscr{X}_2} (\delta_{x_2} \otimes \gamma_{x_3}) d\mu_2(x_2)$$

Then $\lambda \in P(\mathscr{X}_1 \times \mathscr{X}_2 \times \mathscr{X}_3)$ constructed as

$$\lambda = \int_{\mathscr{X}_2} (\nu_{x_1} \otimes \delta_{x_2} \otimes \gamma_{x_3}) d\mu_2(x_2)$$

Then, for $A_1 \in \mathscr{B}(\mathscr{X}_1)$, $A_2 \in \mathscr{B}(\mathscr{X}_2)$, $A_3 \in \mathscr{B}(\mathscr{X}_3)$, we have

$$\lambda(A_1 \times A_2 \times \mathscr{X}_3) = \int_{\mathscr{X}_2} (\nu_{x_1} \otimes \delta_{x_2} \otimes \gamma_{x_3})(A_1 \times A_2 \times \mathscr{X}_3) d\mu_2(x_2) =$$

$$\int_{\mathscr{X}_2} (\nu_{x_1}(A_1) \delta_{x_2}(A_2) \gamma_{x_3}(\mathscr{X}_3) d\mu_2(x_2) = \int_{\mathscr{X}_2} (\nu_{x_1}(A_1) \delta_{x_2}(A_2) d\mu_2(x_2) =$$

$$\int_{\mathscr{X}_2} (\nu_{x_1} \otimes \delta_{x_2})(A_1 \times A_2) d\mu_2(x_2) = \lambda_{12}(A_1 \times A_2)$$

Similarly for

$$\lambda(\mathcal{X}_1 \times A_2 \times A_3) = \lambda_{23}(A_2 \times A_3)$$

QED.

Proof of triangle inequality for Wasserstein metrics.

Let $\mu_j, j = 1, 2, 3$ in $\mathscr{P}_p(\mathbb{R}^n)$, for $p \geq 1$, with support $\mathcal{X}_j \subseteq \mathbb{R}^n$, $j = 1, 2, 3$, respectively.

Note that, from OT theory (existence of solutions of Kantorovich's problem),

$$W_p(\mu_i, \mu_j) = \inf_{\lambda \in \Pi(\mu_i, \mu_j)} [\int_{\mathcal{X}_i \times \mathcal{X}_j} ||x_i - x_j||^p d\lambda(x_i, x_j)]^{\frac{1}{p}}$$

is attained with some *optimal* transport plan $\lambda_{ij} \in \Pi(\mu_i, \mu_j)$. Thus,

$$W_p(\mu_i, \mu_j) = [\int_{\mathcal{X}_i \times \mathcal{X}_j} ||x_i - x_j||^p d\lambda_{ij}(x_i, x_j)]^{\frac{1}{p}}$$

Now, let $\lambda$, in the above Lemma corresponding to $\mu_j$, $j = 1, 2, 3$, be the probability measure on $\mathcal{X}_1 \times \mathcal{X}_2 \times \mathcal{X}_3$ having marginals $\lambda_{12}$ and $\lambda_{23}$ on $\mathcal{X}_1 \times \mathcal{X}_2$, and $\mathcal{X}_2 \times \mathcal{X}_3$, respectively.

We then have

$$W_p(\mu_1, \mu_3) = [\int_{\mathcal{X}_1 \times \mathcal{X}_3} ||x_1 - x_3||^p d\lambda_{13}(x_1, x_3)]^{\frac{1}{p}} =$$

$$[\int_{\mathcal{X}_1 \times \mathcal{X}_3 \times \mathcal{X}_3} ||x_1 - x_3||^p d\lambda(x_1, x_2, x_3)]^{\frac{1}{p}} \leq$$

$$[\int_{\mathcal{X}_1 \times \mathcal{X}_3 \times \mathcal{X}_3} (||x_1 - x_2|| + ||x_2 - x_3||)^p d\lambda(x_1, x_2, x_3)]^{\frac{1}{p}} \leq$$

$$[\int_{\mathcal{X}_1 \times \mathcal{X}_3 \times \mathcal{X}_3} ||x_1 - x_2||^p d\lambda(x_1, x_2, x_3)]^{\frac{1}{p}} + [\int_{\mathcal{X}_1 \times \mathcal{X}_3 \times \mathcal{X}_3} ||x_2 - x_3||^p d\lambda(x_1, x_2, x_3)]^{\frac{1}{p}} =$$

$$[\int_{\mathcal{X}_1 \times \mathcal{X}_3} ||x_1 - x_2||^p d\lambda_{12}(x_1, x_2)]^{\frac{1}{p}} + [\int_{\mathcal{X}_3 \times \mathcal{X}_3} ||x_2 - x_3||^p d\lambda_{23}(x_1, x_2)]^{\frac{1}{p}} =$$

$$W_p(\mu_1, \mu_2) + W_p(\mu_2, \mu_3)$$

Q.E.D.

## References

1. Dudley, R.M.: Real Analysis and Probability. Cambridge University Press (2003)
2. Galichon, A.: Optimal Transport Methods in Economics. Princeton University Press (2016)
3. Graf, S., Mauldin, R.D.: A Classification of Disintegration of Measures, in Measure and Measurable Dynamics, No. 94, pp. 147–158. Contemp. Math, AMS (1989)
4. Villani, C.: Topics in Optimal Transportation. AMS (2003)
5. Villani, C.: Optimal Transport: Old and New. Springer (2009)

# Why Micro-funding? Why Small Businesses Are Important? Analysis Based on First Principles

**Hien D. Tran, Edwin Tomy George, and Vladik Kreinovich**

**Abstract** On the one hand, in economics, there is a well-known and well-studied economy of scale: when two smaller companies merge, it lowers their costs and thus, makes them more effective and therefore more competitive. At first glance, this advantage of big size would make economy dominated by big companies—but in reality, small business remain a significant and important economic sector. Similarly, it is well known and well studied that research collaboration enhances researchers' productivity—but still a significant portion of important results come from individual efforts. In several applications areas, there are area-specific explanations for this seemingly contradictory phenomenon. In this paper, we provide a general explanation based on first principles. Our reasoning also leads to a new explanation of the ubiquity of Zipf's Law—a law that describes, e.g., the distribution of companies by size.

## 1 Formulation of the Problem

**Economy of scale: well-known and well-studied phenomenon.** In economics, there is a known phenomenon of *economy of scale*, when a merger of two small companies helps lower the costs.

The same phenomenon is known in all kinds of activities. For example, when researchers collaborate, they can usually achieve much more than when they work on their own or in small groups.

H. D. Tran
Tan-Tao University, Ho Chi Minh City, Vietnam
e-mail: hien.tran@ttu.edu.vn

E. Tomy George · V. Kreinovich (✉)
Department of Computer Science, University of Texas at El Paso, 500 W. University, El Paso, TX 79968, USA
e-mail: vladik@utep.edu

E. Tomy George
e-mail: etomygeorg@miners.utep.edu

 
V. Kreinovich et al. (eds.), *Applications of Optimal Transport to Economics and Related Topics*, Studies in Systems, Decision and Control 556,
https://doi.org/10.1007/978-3-031-67770-0_12

Based on this logic, one would expect that this effectiveness leads to the dominance of big companies in economics and big well-funded projects in science.

**However, many small businesses are successful.** In practice, however, there is a stable and significant proportion of small businesses—which shows that there is economic benefit in having small businesses in addition to big companies.

**Micro-funding is useful.**

- Along the same lines, it has been empirically shown that the best way to stimulate economy is to provide funding both to big and small businesses, i.e., to combine macro-funding and micro-funding.
- Similarly, when supporting science, the best effect is achieved when usual-size grants are supplemented by micro-funding, i.e., by smaller-size grants; see, e.g., [2, 3, 7] and references therein.

**How can we explain this phenomenon?** In economics, in science sponsorship, and in other similar areas there are good explanations for this phenomenon.

**Remaining problem.** However, the current explanations are specific to each area, while the phenomenon is the same in all these areas. It is therefore desirable to look for a general explanation for this phenomenon.

**What we do in this paper.** In this paper, we provide such a general explanation.

*Comment.* Preliminary results from this paper first appeared in [8].

## 2 Analysis of the Problem and the Resulting Explanation

**Let us formulate the problem in precise terms.** In all such situations:

- we have a fixed amount of money, and
- we want to find the best way to distribute this amount.

Each distribution can be naturally described by a density function $f(m)$ for which the number of grants of sizes between $m$ and $m + \Delta m$ is equal to $f(m) \cdot \Delta m$. In these terms, the question is: What is the optimal function $f(m)$?

**What do we mean by optimal: towards a precise definition.** In the above formulation, we used the word "optimal". Usually, this means that we have an objective function—e.g.:

- profit for a company,
- research productivity for a group of researchers,

and we want to maximize the value of this function. However, in our case, we do not know the exact form of the objective function.

All we know is that:

- some distributions are more effective than others; we will denote this relation by $f(m) \succ g(m)$—and
- some distributions are of the same effectiveness as others; we will denote this relation by $f(m) \sim g(m)$.

**These relations must be consistent.** If $f$ is better than $g$ and $g$ is better than $h$, then we should be able to conclude that $f$ is better than $h$. This leads to the following definition.

**Definition 1** By an *optimality criterion* on a set $A$, we mean a pair of binary relations $\langle \succ, \sim \rangle$ that satisfy the following properties for all $a, b, c \in A$:

- if $a \succ b$ and $b \succ c$, then $a \succ c$;
- if $a \succ b$ and $b \sim c$, then $a \succ c$;
- if $a \sim b$ and $b \succ c$, then $a \succ c$;
- if $a \sim b$ and $b \sim c$, then $a \sim c$;
- if $a \sim b$, then $b \sim a$;
- $a \sim a$; and
- if $a \succ b$ then we cannot have $a \sim b$.

In these terms, an optimal alternative is an alternative which is better than—or at least of the same quality—as all other alternatives.

**Definition 2** We say that an element $a_0 \in A$ is *optimal* with respect to the optimality criterion $\langle \succ, \sim \rangle$ if for every $a \in A$, we have either $a_0 \succ a$ or $a_0 \sim a$.

**Optimality criterion must be final.** It is reasonable to require that there is one and only one optimal alternative. Indeed, if there are no optimal alternatives, then this criterion is useless. So, there must be at least one optimal alternative.

On the other hand, if there are two or more alternatives of equal quality, then we can use this non-uniqueness to optimize something else. For example, if a company has two alternatives with the same amount of expected profit, it is reasonable to select one with the lowest risk, etc. This means, in effect, that in this case our original optimality criterion $\succ$ was not final, and we add additional criterion $\succ_a$ to formulate a new criterion $\succ_n$ for which $a \succ_n b$ if and only if:

- either $a \succ b$ according to the original criterion,
- or $a \sim b$ according to the original criterion and $a \succ_a b$ according to the additional criterion.

If the new criterion still leads to several equally good optimal alternatives, this means that this new criterion is still not final: we can use this non-uniqueness to optimize something else. This process can continue until we reach the final optimality criterion, i.e., a criterion for which there is exactly one optimal alternative. This leads to the following definition.

**Definition 3** We say that the optimality criterion is *final* if there exists exactly one alternative that is optimal with respect to this criterion.

**Optimality criterion should be scale-invariant.** Previously, we talked about general properties of optimality criteria. Let us now consider our case, when alternatives are non-negative functions $f(m)$ defined for positive values $m$.

Each value $m$ describes the amount of funding. The numerical value of funding depends on what units we choose for counting money: we can use dollars, we can use Euros, we can use Mexican pesos, etc. The choice of a monetary unit is a matter of convenience. It is therefore reasonable to require that the relative quality of different functions should not depend on what unit we use for counting.

By definition of density, in the original units, for each pair of values $m$ and $\Delta m \ll m$, we have $f(m) \cdot \Delta m$ folks who receive grants with amounts from $m$ to $m + \Delta m$. If we replace one monetary unit with another monetary unit that is $\lambda$ times smaller, then all the numerical values will be multiplied by $\lambda$: $m \mapsto m' = \lambda \cdot m$ and $\Delta m \mapsto \Delta m' = \lambda \cdot \Delta m$. Let us denote by $f'(m')$ the probability density as described in the new units. In the new units, the same number of people $f(m) \cdot \Delta m$ is now equal to $f'(m') \cdot \Delta m'$. So,

$$f'(m') \cdot \Delta m' = f(m) \cdot \Delta m. \tag{1}$$

Here, $m = m'/\lambda$ and $\Delta m = \Delta m'/\lambda$. Substituting these expressions for $m$ and $\Delta m$ into the formula (1), we conclude that

$$f'(m') \cdot \Delta m' = f(m'/\lambda) \cdot \Delta m'/\lambda.$$

Dividing both sides by $\Delta m'$, we conclude that the density in the new units has the form $f'(m') = f(m'/\lambda)/\lambda$.

Thus, the dependence described, in the original unit, by a function $f(m)$ will now be described by a function $S_\lambda(f)$ for which $(S_\lambda(f))(m) \stackrel{\text{def}}{=} f(m/\lambda)/\lambda$. So, the requirement that the relative quality not depend on the unit takes the following form:

**Definition 4** Let $A$ be the set of all non-negative functions $f(m)$ defined for all positive values $m$. We say that an optimality criterion defined on this set is *scale-invariant* if for every two function $f, g \in S$, and for every $\lambda > 0$, the following two conditions hold:

- if $f \succ g$, then $S_\lambda(f) \succ S_\lambda(g)$, and
- if $f \sim g$, then $S_\lambda(f) \sim S_\lambda(g)$.

Now, we are ready to formulate our main result.

**Proposition.** *For every scale-invariant final optimality criterion, the optimal function has the form $f(m) = C/m$ for some value $C$.*

*Comment.* For readers' convenience, the proof of this result is placed in a special Proof section.

**Of course, our model is a simplification.** Of course, as usual in numerical analysis of real-life problems, our description is an idealization: we assume that the monetary amount can take any real values, but values which are too small or too large are not realistic:

- we cannot give a loan of 0.1 cents, and,
- for a different reason, we cannot give a loan of a thousand trillion dollars—no one has that much money.

However, similar simplifications work well in many other applications, so we believe that our result well describes what is happening for realistic value $m$.

This leads us to the following conclusion.

**Main conclusion.** The optimal function—as described by the above proposition—is everywhere positive. So in the optimal arrangement, we should always have some grants with small $m$.

This explains the ubiquity and effectiveness of micro-funding.

**This leads to a new explanation for the ubiquity of Zipf's Law.** The specific form of the optimal function $f(m)$ is known as Zipf's Law; see, e.g., [5] and references therein. The law is ubiquitous: in particular, it describes the distribution of the companies by size—one of the phenomena that we are trying to explain. Thus, our result provide one more explanation of Zipf's law; for other explanations, see, e.g., [1, 4].

## 3 Proof

We assumed that the optimality criterion is final. This means that there exists exactly one function that is optimal with respect to this criterion. Let us denote this optimal function by $F$.

Let us first prove that the function $F$ is itself scale-invariant, i.e., that for every $\lambda > 0$, we have $F = S_\lambda(F)$. Indeed, by definition of optimality, the fact that $F$ is optimal means that for every $a \in A$ we have either $F \succ a$ or $F \sim a$. In particular, for every function $a$, we have $F \succ S_{1/\lambda}(a)$ or $F \sim S_{1/\lambda}(a)$. By scale-invariance, we can conclude that $S_\lambda(F) \succ S_\lambda(S_{1/\lambda}(a))$ or $S_\lambda(F) \sim S_\lambda(S_{1/\lambda}(a))$.

One can easily see that $S_\lambda(S_{1/\lambda}(a)) = a$. Thus, for every $a \in A$, we have either $S_\lambda(F) \succ a$ or $S_\lambda(F) \sim a$. By definition of optimality, this means that the function $S_\lambda(F)$ is optimal. However, we assumed that the optimality criterion is final, which means that there is only one optimal function. Thus indeed $S_\lambda(F) = F$ for all $\lambda > 0$.

By definition of the expression $S_\lambda(F)$, thus means that for every $m > 0$ and every $\lambda > 0$, we have $F(m/\lambda)/\lambda = F(m)$. In particular, for $\lambda = m$, we get $F(m) = C/m$, where we denoted $C \stackrel{\text{def}}{=} F(1)$. The proposition is proven.

*Comment.* The main ideas behind this proof were first used in [1, 6].

**Acknowledgements** This work was supported in part by the National Science Foundation grants 1623190 (A Model of Change for Preparing a New Generation for Professional Practice in Computer Science), HRD-1834620 and HRD-2034030 (CAHSI Includes), EAR-2225395 (Center for Collective Impact in Earthquake Science C-CIES), and by the AT&T Fellowship in Information Technology.

It was also supported by the program of the development of the Scientific-Educational Mathematical Center of Volga Federal District No. 075-02-2020-1478, and by a grant from the Hungarian National Research, Development and Innovation Office (NRDI), and by the Center of Excellence in Econometrics, Faculty of Economics, Chiang Mai University, Thailand.

The authors are greatly thankful to all the participants of the NMSU/UTEP Workshop on Mathematics, Computer Science, and Computational Science (Las Cruces, New Mexico, April 1, 2023) for valuable discussions, and to Professor Hung T. Nguyen for his encouragement.

## References

1. Cervantes, D., Kosheleva, O., Kreinovich, V.: Why Zipf's Law: a symmetry-based explanation. Int. Math. Forum **13**(6), 255–258 (2018)
2. Flowers, R.M., Arrowsmith, J.R.: AGeS$^3$: microfunding an inclusive community grassroots effoirts to better understand the Earth system. GSA Today **32**, 52–53 (2022). https://doi.org/10.1130/GSATG549GW.1
3. Flowers, R., Arrowsmith, J.R., McConnell, V., Metcalf, J., Rittenour, T., Schoene, B.: The AGeS2 (Awards for Geochronology Student research 2) program: supporting community geochronology needs and interdiscioplinary science. GSA Today **29**(3) (2019). https://doi.org/10.1130/GSATG392GW.1
4. Kosheleva, O., Kreinovich, V., Autchariyapanikul, K.: Commonsense explanations of sparsity, Zipf Law, and Nash's bargaining solution. In: Thach, N.N., Ha, D.T., Trung, N.D., Kreinovich, V. (eds.) Prediction and Causality in Econometrics and Related Topics, pp. 67–74. Springer, Cham, Switzerland (2022)
5. Mandelbrot, B.B.: The Fractal Geometry of Nature. W. H. Freeman Publ, San Francisco, California (1983)
6. Nguyen, H.T., Kreinovich, V.: Applications of Continuous Mathematics to Computer Science. Kluwer, Dordrecht (1997)
7. Rappert, B.: Fostering data openness by enabling science: a proposal for micro-finding. Data Sci. J. **16** (2017). https://doi.org/10.5334/dsj-2017-044
8. Tomy George, E., Kreinovich, V.: Why micro-funding? Why small businesses are important? In: Abstracts of the NMSU/UTEP Workshop on Mathematics, Computer Science, and Computational Science, Las Cruces, New Mexico (2023)

# Local-Global Support for Earth Sciences: Economic Analysis

**Uyen Hoang Pham, Aaron Velasco, and Vladik Kreinovich**

**Abstract** Most funding for science comes from taxpayers. So, it is very important to be able to convince taxpayers that this funding is potentially beneficial for them. This task is easier in Earth sciences, e.g., in meteorology, where there are clear local benefits. The problem is that while many people support local studies focused on their region, they do not always have a good understanding of the fact that effective local benefits require also studying surrounding areas—and what should be the optimal balance between local and (more) global studies. In this paper, on a (somewhat) simplified model of the situation, we explain what is the appropriate balance. We hope that the corresponding methodology can (and will) be applied to more realistic—and thus, more complex—local-global models as well.

## 1 Formulation of the Problem

**Financial support is important for science.** While there arc few disciplines like theoretical mathematics or theoretical physics that mainly need brain power, most other sciences need funding. Even theoretical physics indirectly needs funding: it is great to come up with new theories and ideas, but their experimental testing requires funding—and sometimes very serious funding.

U. H. Pham
University of Economics and Law, Ho Chin Minh City, Vietnam
e-mail: uyenph@uel.edu.vn

A. Velasco
Department of Earth, Evironmental, and Resource Sciences, University of Texas at El Paso, 500 W. University, El Paso, TX 79968, USA
e-mail: aavelasco@utep.edu

V. Kreinovich (✉)
Department of Computer Science, University of Texas at El Paso, 500 W. University, El Paso, TX 79968, USA
e-mail: vladik@utep.edu



V. Kreinovich et al. (eds.), *Applications of Optimal Transport to Economics and Related Topics*, Studies in Systems, Decision and Control 556,
https://doi.org/10.1007/978-3-031-67770-0_13

**Who provides this financial support.** Some scientific research is done by companies, but the vast majority of scientific efforts are eventually supported by taxpayers:

- taxpayers directly support salaries of faculty at public universities,
- taxpayers indirectly support faculty at private universities—by supporting funding agencies that distribute research grants.

**How to maintain this support.** To maintain financial support of science, we need to continue convincing taxpayers that this support is beneficial to the society.

**Local-global aspects of seeking financial support.** In some areas—e.g., in fundamental physics—research promises to benefit humanity as a whole, irrespective of location. For example, when researchers come up with new ways to design computer chips that could lead to faster and/or less energy-consuming computers, everyone will benefit.

The situation is different with Earth science, be it geological sciences, environmental science, biological research, atmospheric and meteorological studies, etc. In these studies, many research efforts lead to clear local benefits; for example:

- geophysical studies of a given region can help better predict the probability of earthquakes of different strength, and thus, help protect the buildings against possible shocks without overstraining the budget;
- geophysical studies can help find underground water reservoirs in desert areas, they can find mineral deposts—which can boost local economies;
- meteorological studies help make short- and long-term weather predictions, which helps agriculture and helps better prepare for possible extreme events, etc.

**In Earth sciences, there is a local support, but this support is often too localized.** Because of the importance of the corresponding research efforts, people often support local research efforts. For example, in El Paso, Texas, where two of us live, recently, a doctoral student—now Doctor—Solymar Ayala performed a study of seismicity around the city. In many public and private places, including local schools, churches, and synagogues, he was very welcome to dig in corresponding equipment and perform measurements, even when this interfered with the usual activities. Local media usually emphasizes the importance of such activity, and thus, helps get community support for this research.

In short, population usually appreciates and supports local research efforts that have a potential local benefit. But there is not that much understanding of the fact that to make local predictions, we cannot restrict ourselves only the local area, we also need to perform studies in nearby—and sometimes faraway—areas. For example, to predict tomorrow's weather in El Paso, it is not sufficient to know today's temperature, wind, and humidity in the city itself, we need to also use measurements in the nearby locations that tell us whether, e.g., a cold front is coming. An extreme example is that many of El Paso geoscientists interested in El Paso seismicity perform measurements in Kenya—because many geological structures there are similar to Rio Grande Rift which goes through El Paso, and studies there have contributed (and continue to contribute) to a much better understanding of seismicity around El Paso.

**We are working on solving this problem.** There is a clear need to help local communities get a better understanding that local benefits come not only from strictly local research, but also from research in nearby areas. This community communications are one of the main objectives of the recently funded US-based Center for Collective Impact in Earthquake Science (C-CIES), a project sponsored by the US National Science Foundation, in which one of us (AV) is the Principle Investigator; see, e.g., [5].

**How should we balance local vs. global support: economic analysis is needed.** It is not enough just to convince people that support is needed. Resources are bounded, so we need to better understand what should be the balance between local and no-so-local research efforts that will maximize the research's impact on the local community.

**What we do in this paper.** In this paper, on a simplified model of local-global effects, we show how to find the optimal balance of efforts. We hope that the corresponding methodology can (and will) be applied to more realistic—and thus, more complex—local-global models as well.

## 2 Model of Local-Global Effects and the Resulting Optimization Problem

**Main task.** We want to predict—or determine—the value $y$ of some quantity at our location $x_0$. This may be tomorrow's temperature in our city, this may be the amount of water in an underground natural aquifer, etc. To estimate this value, we use our estimates of the values $f(x)$ of related quantities $f$ at different geographical locations $x$ around $x_0$. For this estimation, we have an algorithm $A$ that transforms estimates for the values $f(x)$ into an estimate for $y$: $y = A(f(x_1), f(x_2), \ldots)$.

**Need to take uncertainty into account.** Our information about the values $f(x)$ usually comes from measurements: either directly from measurements, or from processing results of measuring related quantities. Measurements are never absolutely accurate; see, e.g., [2]. Thus, the resulting estimates $\widetilde{f}(x)$ are, in general, different from the actual (unknown) values $f(x)$. Because of this, the estimate $\widetilde{y} = A(\widetilde{f}(x_1), \widetilde{f}(x_2), \ldots)$ that we obtain by using the estimates is, in general, different from the (almost) exact value $y = A(f(x_1), f(x_2), \ldots)$ that we would have gotten if we knew the exact values $f(x_i)$.

The purpose of additional measurements is to get more accurate estimates for the values $f(x_i)$ and thus, get a more accurate estimate for $y$.

**Possibility of linearization.** The estimates $\widetilde{f}(x)$ are usually reasonable, in the sense that the difference $\Delta f(x) \stackrel{\text{def}}{=} \widetilde{f}(x) - f(x)$ between each estimate $\widetilde{f}(x)$ and the corresponding actual value $f(x)$ is significantly smaller than the value $f(x)$ itself: if this difference was not smaller, it would have been a wild guess, not an estimate.

Thus, we have $f(x_i) = \widetilde{f}(x_i) - \Delta f(x_i)$, where $|\Delta f(x_i)| \ll |f(x_i)|$. So, the resulting estimation error $\Delta y \stackrel{\text{def}}{=} \widetilde{y} - y$ can be described as

$$\Delta y = A(\widetilde{f}(x_1), \widetilde{f}(x_2), \ldots) - A(f(x_1), f(x_2), \ldots) =$$

$$A(\widetilde{f}(x_1), \widetilde{f}(x_2), \ldots) - A(\widetilde{f}(x_1) - \Delta f(x_1), \widetilde{f}(x_2) - \Delta f(x_2), \ldots). \quad (1)$$

Since we have $|\Delta f(x_i)| \ll |f(x_i)|$, we can do what physicists usually do in such situations (see, e.g., [1, 4]): as the first approximation, expand the dependence on $\Delta f(x_i)$ in Taylor series and keep only linear terms in this expansion. In this approximation, $\Delta y$ becomes a linear function of the values $\Delta f(x_i)$:

$$\Delta y = \sum_i c_i \cdot \Delta f(x_i), \quad (2)$$

for some coefficients $c_i$.

In general, the closer the point $x_i$ to the location $x_0$, the more the corresponding term $f(x_i)$ affects the value $y$, i.e., the larger the value $|c_i|$.

**It is reasonable to assume that $\Delta y$ is normally distributed with 0 mean.** Measurements at different locations are independent, so the corresponding uncertainties $\Delta f(x_i)$ are independent random variables. According to the formula (2), the value $\Delta y$ is a linear combination of a large number of similar-size independent variables. According to the Central Limit Theorem (see, e.g., [3]), the distribution of such a linear combination is close to Gaussian (normal).

A normal distribution is characterized by two parameters: mean $\mu$ and variance $V$ (instead of the variance $V$, we can consider its square root $\sigma \stackrel{\text{def}}{=} \sqrt{V}$ known as *standard deviation*).

**The mean value is usually 0.** Measuring instruments are usually calibrated, so that if there is a *bias*—i.e., if the testing shows that the mean value of the measurement error is different from 0—we subtract this mean value from all the measurement results and thus, conclude that the mean value of each variable $\Delta f(x_i)$ is 0.

In general, the mean value of a linear combination is equal to the linear combination of the corresponding mean values. Since the mean value of each term $\Delta f(x_i)$ is 0, we conclude that the men value $\mu$ of their linear combination (2) is also 0.

Thus, the only characteristic of the uncertainty $\Delta y$ is its variable $V$.

**The formula for the variable $V$.** Since the variables $\Delta f(x_i)$ are independent, we have

$$V = \sum_i c_i^2 \cdot V_i, \quad (3)$$

where $V_i$ denotes the variance of the measurement error $\Delta f(x_i)$.

Let $V_0$ be the variance of a single measurement. According to statistics (see, e.g., [3]), if we repeat a measurement $k$ times, the variance of the resulting estimate becomes $k$ times smaller. Let $N$ denote the overall number of measurements that we can afford to make. Let $n_i$ be the number of measurements performed in the $i$th areas, so that

$$n_1 + n_2 + \ldots = N \tag{4}$$

Then, we have $V_i = V_0/n_i$ and thus, the formula (3) takes the form

$$V = \sum_i c_i^2 \cdot \frac{V_0}{n_i}. \tag{5}$$

**Resulting optimization problem.** We want to estimate $y$ with the smallest possible uncertainty. As we have mentioned, the uncertainty of $y$ can be gauged by its variance $V$. Thus, our objective is, among the values $n_i$ that satisfy the equality (4), to select the values that minimize the variance (5).

## 3 Solution to the Optimization Problem

**Let us use Lagrange multiplier method.** The usual way to solve constraint optimization problems is to use the *Lagrange multiplier method*, according to which optimizing a function $F(z)$ under constraint $G(z) = 0$ can be reduced to an unconstrained optimization of an auxiliary function $F(z) + \lambda \cdot G(z)$ for some parameter $\lambda$ known as *Lagrange multiplier*. Once we find the solutions corresponding to different $\lambda$, we select the value $\lambda$ for which the resulting solution satisfies the constraint $G(z) = 0$.

In our case, this method means that we maximize the following auxiliary function:

$$\sum_i c_i^2 \cdot \frac{V_0}{n_i} + \lambda \cdot \left( \sum_i n_i - N \right). \tag{6}$$

Differentiating this expression with respect to $n_i$ and equating the derivative to 0, we conclude that

$$-c_i^2 \cdot \frac{V_0}{n_i^2} + \lambda = 0,$$

hence

$$n_i^2 = c_i^2 \cdot \frac{V_0}{\lambda}$$

and thus,

$$n_i = c \cdot |c_i|, \tag{7}$$

where we denoted

$$c \stackrel{\text{def}}{=} \sqrt{\frac{V_0}{\lambda}}.$$

The value $c$ can be determined from the condition (4) that takes the form

$$c \cdot \sum_j |c_j| = N,$$

hence

$$c = \frac{N}{\sum_j |c_j|}.$$

For this value $c$, the optimal solution takes the following form.

**Optimal measurement strategy.** To get the smallest possible estimation error, we need to perform

$$n_i = \frac{N \cdot |c_i|}{\sum_j |c_j|} \tag{8}$$

measurements at each location $i$, where:

- the integer $N$ is the overall number of measurements that we can afford, and
- the value $c_i$ describes the degree to which $y$ depends on the estimate $f(x_i)$.

**Discussion.** In line with common sense, we need to perform not only local measurements, but also measurement at nearby points. The further away we are from our location $x_0$, the smaller the value $|c_i|$ and thus, the smaller amount of effort we should allocate to measuring the corresponding value $f(x_i)$.

*Comment.* This qualitative conclusion we could, of course, make without doing any calculations. We need calculations to determine how exactly to distribute the measurements.

**What is the resulting accuracy of estimating** $y$**: derivation.** Substituting the values (8) into the formula (5), we conclude that

$$V = \sum_i c_i^2 \cdot \frac{V_0}{n_i} = V_0 \cdot \sum_i \frac{c_i^2}{n_i} = \frac{V_0}{N} \cdot \left( \sum_j |c_j| \right) \cdot \sum_i \frac{c_i^2}{|c_i|}. \tag{9}$$

Here, $c_i^2 = |c_i|^2$, thus

$$\sum_i \frac{c_i^2}{|c_i|} = \sum_i \frac{|c_i|^2}{|c_i|} = \sum_i |c_i|,$$

and we get the following formula.

**What is the resulting accuracy $V$ of estimating $y$:the resulting formula.**

$$V = \frac{V_0}{N} \cdot \left(\sum_i |c_i|\right)^2. \tag{10}$$

**Acknowledgements** This work was supported in part by the National Science Foundation grants 1623190 (A Model of Change for Preparing a New Generation for Professional Practice in Computer Science), HRD-1834620 and HRD-2034030 (CAHSI Includes), EAR-2225395 (Center for Collective Impact in Earthquake Science C-CIES), and by the AT&T Fellowship in Information Technology.

It was also supported by the program of the development of the Scientific-Educational Mathematical Center of Volga Federal District No. 075-02-2020-1478, and by a grant from the Hungarian National Research, Development and Innovation Office (NRDI) and by the Center of Excellence in Econometrics, Faculty of Economics, Chiang Mai University, Thailand.

The authors are thankful to Hung T. Nguyen for his encouragement.

## References

1. Feynman, R., Leighton, R., Sands, M.: The Feynman Lectures on Physics. Addison Wesley, Boston, Massachusetts (2005)
2. Rabinovich, S.G.: Measurement Errors and Uncertainty: Theory and Practice. Springer, New York (2005)
3. Sheskin, D.J.: Handbook of Parametric and Nonparametric Statistical Procedures. Chapman and Hall/CRC, Boca Raton, Florida (2011)
4. Thorne, K.S., Blandford, R.D.: Modern Classical Physics: Optics, Fluids, Plasmas, Elasticity, Relativity, and Statistical Physics. Princeton University Press, Princeton, New Jersey (2021)
5. Velasco, A.A., Weidner, J., Karplus, M., Bilek, S., Brudzinski, M., Chandrasekhar, D., Ebel, J., Hobbs, T., Hurtado, J.M., Jaume, S., Jones, E., Kafka, A., Kreinovich, V., Lin, Y., Nuñez, A.-M., Pankow, K., Peng, Z., Savvaidis, A., Tien, I., Bolton Valencius, C., Vanacore, E., Alvillar, M.: Center for Collective Impact in Earthquake Science (C-CIES): building inclusive excellence, diversity, equity, and community into earthquake science. In: Abstracts of the 48th Annual Natural Hazards Research and Applications Workshop, Boulder, Colorado. Accessed from 9–13 July 2023 (2023)

# Approximate Stochastic Dominance Revisited

**Chon Van Le, Olga Kosheleva, and Vladik Kreinovich**

**Abstract** According to decision theory, in general, to recommend the best of possible actions, we need to know, for each possible action, the probabilities of different outcomes, and we also need to know the decision maker's utility function—that describes his/her preferences. For some pairs of probability distributions, however, we can make such a recommendation without knowing the exact form of the utility function—e.g., in financial applications, we only need to know that a larger amount is preferable to a smaller one. Such situations, when we can make decisions based only on the information about probabilities, are known as *stochastic dominance.* The usual analysis of such situations is based in the idealized assumption that any difference in utility, no matter how small, is important. In reality, very small changes in utility value are irrelevant. From this viewpoint, if the utility corresponding to the distribution $F_2(x)$ is always either larger or only slightly smaller than the utility corresponding to $F_1(x)$, then we can still conclude that the second action is better (or of the same quality) than the first action. In this paper, we show how to describe such approximate stochastic dominance in precise terms.

---

C. Van Le
International University, Vietnam National University, Ho Chin Minh City, Vietnam
e-mail: lvchon@hcmiu.edu.vn

O. Kosheleva
Department of Teacher Education, University of Texas at El Paso 500 W. University, El Paso, Texas 79968, USA
e-mail: olgak@utep.edu

V. Kreinovich (✉)
Department of Computer Science, University of Texas at El Paso 500 W. University, El Paso, Texas 79968, USA
e-mail: vladik@utep.edu

 
V. Kreinovich et al. (eds.), *Applications of Optimal Transport to Economics and Related Topics*, Studies in Systems, Decision and Control 556,
https://doi.org/10.1007/978-3-031-67770-0_14

## 1 Formulation of the Problem

**How rational people make decisions?** One of the main objectives of econometrics is to help people make decisions. For this purpose, it is necessary to know how people actually make decisions. This has been studied by *decision theory*; see, e.g., [2, 3, 6, 8–11]. To be more precise, decision theory describes how *rational* people make decisions.

Rational means, e.g., that if a person prefers $A$ to $B$, and prefers $B$ to $C$, then this person should also prefer $A$ to $C$. Real-life decision makers are not always rational in this sense; see, e.g., [5] and references therein. However, if we point out this inconsistency to the decision maker, he/she will correct it. From this viewpoint, it makes sense to restrict ourselves to rational decision makers.

How does decision theory describes decisions of such people? A decision means that the decision maker decides between several possible actions $a$. We cannot predict the outcome of each action with 100% accuracy—e.g., investing money in a certain stock:

- may increase the amount of money if this stock continues to rise,
- or it may lead to a loss of money if the corresponding company fails.

In general, each action has several possible outcomes $o_1, \ldots, o_n$. Based on prior experiences, we can usually estimate the probabilities $p_1, \ldots, p_n$ of different outcomes. In such situations, decisions of a rational decision maker can be described as follows:

- to each possible outcome $o$, we assign a numerical value $u(o)$ which is called the *utility* of $o$; the function that assigns, to each outcome, its utility, is (naturally) called *utility function*;
- to each action $a$ in which we get outcome $o_i$ with probability $p_i$, we assign the expected value $E[u]$ of the utility, i.e., the value

$$u(a) = E[u] \stackrel{\text{def}}{=} p_1 \cdot u(o_1) + \cdots + p_n \cdot u(o_n);$$

- then, out of several possible actions, the decision maker selects the action with the largest value of its utility $u(a)$.

The utility values can be elicited from the decision maker as follows. We select two ideal outcomes:

- an outcome $o_-$ which is worse than anything that we will actually encounter, and
- an outcome $o_+$ which is better than anything that we will actually encounter.

Then, for each value $p$ from the interval [0, 1], we can consider a "lottery"—that we will denote by $L(p)$—in which:

- we get the very good outcome $o_+$ with probability $p$, and
- we get the very bad outcome $o_-$ with the remaining probability $1 - p$.

For any actual outcome $o$:

- when the probability $p$ is small, the lottery $L(p)$ is close to the very bad outcome $a_-$ and is, thus, clearly worse than $o$; we will denote this by $L(p) < o$;
- when the probability $p$ is close to 1, the lottery $L(p)$ is close to the very good outcome $a_+$ and is, thus, clearly better than $o$: $o < L(p)$.

Of course, if $p < p'$, then the lottery $L(p)$ is worse than the lottery $L(p')$ in which the probability of the very good alternative is higher: $L(p) < L(p')$. Thus, if $p < p'$, then:

- $o < L(p)$ implies $o < L(p')$; and
- $L(p') < o$ implies $L(p) < o$.

So, for sufficiently small $p$, we have $L(p) < o$, and for sufficiently large $p$, we have $o < L(p)$. There is a threshold separating these two groups of values. This threshold is what is called the utility $u(o)$ of the outcome $o$. In this case:

- if $p < u(o)$, then $L(p) < o$, and
- if $u(o) < p$, then $o < L(p)$.

The very bad event $o_-$ is equivalent to the lottery $L(0)$, so $u(0_-) = 0$. Similarly, the very good event $o_+$ is equivalent to the lottery $o_+$, so $u(o_+) = 1$.

*Comment.* Thus defined utility is a probability, so it is always located on the interval [0, 1]. However, utility does not have to satisfy this restriction: one can easily check that if the function $u(o)$ describes the person's preferences, then, for each $a_0$ and $a_1 > 0$, the function $u'(o) \stackrel{\text{def}}{=} a_0 + a_1 \cdot u(o)$ describes the same preferences equally well—and for this new utility function, the values can be anywhere between $a_0$ and $a_0 + a_1$.

**Decision making: from general case to financial case**. As we have mentioned, in general, each action can be characterized by a probability distribution on the set of all possible outcomes. The decision maker prefers action 2 to action 1 if the expected value $E_2[u]$ of his/her utility under the action-2 distribution is larger than the expected value $E_1[u]$ under the action-1 distribution, i.e., when $E_1[u] \le E_2[u]$.

In financial situations, the outcome is usually the resulting amount of money $x$. So, the set of all possible outcomes is the real line, and each action can be characterized by a probability distribution on a real line—as described, e.g., by a cumulative probability distribution function $F(x) \stackrel{\text{def}}{=} \text{Prob}(X \le x)$.

**What if we do not know the utility function?** Often, we do not know the utility function $u(x)$ corresponding to a decision maker, all we know is that the more money, the better—i.e., that the utility function $u(x)$ is non-strictly increasing.

In such situations, sometimes, we can still decide which action is better. Thus situation—when one action is better than another based only on their probabilities, is known as *stochastic dominance.*

*Comments.*

- We will assume that the utility function $u(x)$ is *non-strictly* increasing, i.e., that $x < x'$ implies $u(x) \le u(x')$. The reason for this is that if a person has a huge amount of money, he/she may not get any additional opportunities if he/she gets some more money, so we have $u(x) = u(x')$ for some $x < x'$.
- In practice, the overall amount of money is bounded. Let us denote the largest possible amount of money by $M$. Corresponding, the largest amount of debt is also $M$, which corresponds to the amount of money $-M$.

  - When the value $x$ is equal to $-M$, the situation is as bad as it can be—which, as we have described, corresponds to the very bad outcome $o_-$, i.e., to utility 0.
  - When the amount $x$ is equal to $M$, the situation is as good as it can be—which, as we have described, corresponds to the very good outcome $o_+$ for which the utility is 1.

  Thus, if we restrict ourselves to utilities whose values are between 0 and 1, then we must have $u(-M) = 0$ and $u(M) = 1$.

**Stochastic dominance: known result.** Here is the known result about stochastic dominance.

**Proposition 1** *For every two probability distributions $F_1(x)$ and $F_2(x)$ on the real line, the following two conditions are equivalent to each other:*

- $F_1(x) \ge F_2(x)$ *for all* $x$, *and*
- $E_1[u] \le E_2[u]$ *for all possible non-strictly increasing functions* $u(x)$.

*Comment.* This is a known result, but we will nevertheless present its proof, since the proof of our result is based on this proof. For reader's convenience, all the proofs are placed in a special Proofs section.

**Let us reformulate the above known result.** In the above result, the inequalities in the two equivalent conditions are opposite to one another. We can make these inequalities going in the same direction if:

- instead of the cumulative distribution function $F(x) = \text{Prob}(X \le x)$,
- we consider the probabilities of the opposite event $S(x) \stackrel{\text{def}}{=} \text{Prob}(X > x)$.

The corresponding function has several names: it is known as *complementary distribution function*, *tail distribution*, *exceedance*, *survival function*, and *reliability function*. One can easily check that $S(x) = 1 - F(x)$, so $F_1(x) \ge F_2(x)$ is equivalent to $S_1(x) \le S_2(x)$. Thus, in terms of this function, Proposition 1 takes the following form:

**Proposition 1′.** *For every two survival functions $S_1(x) = 1 - F_1(x)$ and $S_2(x) = 1 - F_2(x)$ on the real line, the following two conditions are equivalent to each other:*

- $S_1(x) \le S_2(x)$ *for all* $x$*, and*
- $E_1[u] \le E_2[u]$ *for all possible non-strictly increasing functions* $u(x)$.

**Need for approximate stochastic dominance.** In the above text, we assumed that any difference in utility, no matter how small, is important. In reality, very small changes in utility value are irrelevant. For example, if two cars cost, correspondingly, \$30,000 and \$30,001, hardly anyone would take this difference seriously (although, as marketing shows, that difference between \$10.00 and \$9.99 does make a psychological difference for actual buyers—but we decide to focus on rational decisions makers.)

From this viewpoint, if the utility corresponding to the distribution $F_2(x)$ is always either larger then—or only slightly smaller than—the utility corresponding to $F_1(x)$, then we can still conclude that the second action is better—or of the same quality—as the first action. How can we describe such approximate stochastic dominance in precise terms?

**What is known.** Several results about approximate stochastic dominance are known; see, e.g., [1, 7] and references therein.

**What we do in this paper.** In this paper, we provide another general result about approximate stochastic dominance.

## 2 Main Result

**Definition 1** Let $M > 0$ be a real number. By a *0-1-M-utility function*, we mean a non-strictly increasing function $u(x)$ from the interval $[-M, M]$ to the interval $[0, 1]$ for which $u(-M) = 0$ and $u(M) = 1$.

**Definition 2** (*reminder*). A function $f(z)$ is called *concave* if for all real numbers $z_1, \ldots, z_n$ and for all non-negative numbers $c_1, \ldots, c_n$ whose sum is 1, we have

$$f(c_1 \cdot z_1 + \cdots + c_n \cdot z_n) \ge c_1 \cdot f(z_1) + \cdots + c_n \cdot f(z_n).$$

**Examples.** The following functions are concave:

- linear functions, e.g., function $f(z) = (1 + \varepsilon) \cdot z + \delta$ for some small $\varepsilon > 0$ and $\delta > 0$;
- a function $f(z) = z^a$ for $0 \le a \le 1$ as defined for non-negative $x$; e.g., the function $f(z) = z^{1-\varepsilon}$ for some small positive $\varepsilon$;
- the function $\ln(z)$, etc.

**Proposition 2** *Let $M > 0$ be a real number, and let $f(z)$ be a concave function. Then, for every two survival functions $S_1(x) = 1 - F_1(x)$ and $S_2(x) = 1 - F_2(x)$ on the interval $[-M, M]$, the following two conditions are equivalent to each other:*

- *$S_1(x) \le f(S_2(x))$ for all $x$, and*
- *$E_1[u] \le f(E_2[u])$ for all possible 0-1-$M$-utility functions $u(x)$.*

*Comments.*

- For $f(z) = z$, we have Proposition 1—restricted to 0-1-$M$-utility functions.
- For concave functions $f(z)$ corresponding to small $\varepsilon > 0$ and $\delta > 0$, we get exactly the desired result: that the utility values corresponding to the second distribution is always almost larger than the utility corresponding to $F_1$ if and only if the survival function corresponding to $F_2$ is always almost larger than the survival function correspond to $F_1(x)$.
- It is worth mentioning that some versions of stochastic dominance are related to Wasserstein metric and Optimal Transport Theory; see, e.g., [4]. It would be nice to extend our result to this case as well.

## 3 Proofs

**Proof of Proposition 1.** For convenience, we will prove the equivalent Proposition 1′.

1°. First, let us prove that if $E_1[u] \le E_2[u]$ for all $u$, then $S_1(x) \le S_2(x)$ for all $x$. Indeed, if we take a function $u(y) = \theta_x(y)$ that is:

- equal to 0 for all $y \le x$ and
- equal to 1 for all $y > x$,

then for all $i$, the expected value $E_i[u]$ will be exactly equal to $S_i(x)$.

2°. Let us now prove that if $S_1(x) \le S_2(x)$ for all $x$, then $E_1[u] \le E_2[u]$ for all $u$.

2.1°. Let us start with the piece-wise constant functions $u(x)$ for which there exist points

$$x_0 = -\infty < x_1 < \cdots < x_n < x_{n+1} = +\infty$$

and values $u_j = u(x_j)$ such that:

- for all $x \in (x_0, x_1]$, we have $u(x) = u(x_0)$,
- for all $x \in (x_1, x_2]$, we have $u(x) = u(x_1)$,
- ...
- for all $x \in (x_j, x_{j+1}]$, we have $u(x) = u(x_j)$,
- ...

- for all $x \in (x_{n-1}, x_n]$, we have $u(x) = u(x_{n-1})$, and
- for all $x \in (x_n, x_{n+1}]$, we have $u(x) = u(x_n)$.

It is easy to check that the resulting function $u(x)$ is a linear combination of the functions $\theta_{x_j}(x)$:

$$u(x) = u(x_0) + (u(x_1) - u(x_0)) \cdot \theta_{x_1}(x) + \cdots + (u(x_j) - u(x_{j-1})) \cdot \theta_{x_j}(x) + \cdots +$$

$$(u(x_n) - u(x_{n-1})) \cdot \theta_{x_n}(x).$$

The mean value of a linear combination is equal to the linear combination of mean values, so

$$E_i[u] = u(x_0) + (u(x_1) - u(x_0)) \cdot E_i[\theta_{x_1}] + \cdots + (u(x_j) - u(x_{j-1})) \cdot E_i[\theta_{x_j}] + \cdots +$$

$$(u(x_n) - u(x_{n-1})) \cdot E_i[\theta_{x_n}].$$

Here, for any $j$, we have $E_j[\theta_{x_j}(x)] = S_j(x_n)$, thus we have

$$E_i[u] = u(x_0) + (u(x_1) - u(x_0)) \cdot S_i(x_1) + \cdots + (u(x_j) - u(x_{j-1})) \cdot S_i(x_j) + \cdots +$$

$$(u(x_n) - u(x_{n-1})) \cdot S_i(x_n). \tag{1}$$

For each $j$, we have

$$S_1(x_j) \le S_2(x_j). \tag{2}$$

Since the utility function is non-strictly increasing and $x_{j-1} < x_j$, the difference $u(x_j) - u(x_{j-1})$ is non-negative. Thus, multiplying both sides of the inequality (1) by the non-negative difference $u(x_j) - u(x_{j-1})$, we conclude that

$$(u(x_j) - u(x_{j-1})) \cdot S_1(x_j) \le (u(x_j) - u(x_{j-1})) \cdot S_2(x_j). \tag{3}$$

Adding up all these inequalities, adding $u(x_0)$ to both sides, and taking into account the formula (1), we conclude that indeed

$$E_1[u] \le E_2[u]. \tag{4}$$

2.2°. For each non-strictly increasing function $u(x)$ and for each selection of values $x_1 < \cdots < x_n$, we can form a piece-wise constant function as described in Part 2.1 of this proof. As we take denser and denser set of values $x_j$ covering larger and larger domain, we will get functions that are closer and closer to $u(x)$. So in the limit, from the inequality (4) for these auxiliary functions, we get the same inequality for the original function $u(x)$.

The proposition is proven.

**Proof of Proposition 2.**

1°. Similarly to Part 1 of the previous proof, we can show that if $E_1[u] \le f(E_2[u])$ for all 0-1-$M$ utility functions $u$, then $S_1(x) \le f(S_2(x))$ for all $x$.

2°. Let us now prove that if $S_1(x) \le f(S_2(x))$ for all $x$, then $E_1[u] \le f(E_2[u])$ for all 0-1-$M$ utility functions $u$. Similarly to the previous proof, let us start with the case when the function $u(x)$ is piece-wise constant. In this case, we have $x_0 = -M$ and $x_{n+1} = M$. By definition of a 0-1-$M$ utility function, we have $u(x_0) = 0$ and $u(x_{n+1}) = 1$. Then, the formula (1) takes a somewhat simplified form

$$E_i[u] = u(x_1) \cdot S_i(x_1) + \cdots + (u(x_j) - u(x_{j-1})) \cdot S_i(x_j) + \cdots +$$
$$(u(x_n) - u(x_{n-1})) \cdot S_i(x_n) + (1 - u(x_n)) \cdot S_i(x_{n+1}). \quad (5)$$

From the inequalities $S_1(x_j) \le f(S_2(x_j))$, multiplying each of them by $u(x_j) - u(x_{j-1}) \ge 0$ and adding up the results, we conclude that

$$E_1[u] = u(x_1) \cdot S_1(x_1) + \cdots + (u(x_j) - u(x_{j-1})) \cdot S_1(x_j) + \cdots +$$
$$(u(x_n) - u(x_{n-1})) \cdot S_1(x_n) + (1 - u(x_n)) \cdot S_1(x_{n+1}) \le$$
$$u(x_1) \cdot f(S_2(x_1)) + \cdots + (u(x_j) - u(x_{j-1})) \cdot f(S_2(x_j)) + \cdots +$$
$$(u(x_n) - u(x_{n-1})) \cdot f(S_2(x_n)) + (1 - u(x_n)) \cdot f(S_2(x_{n+1})). \quad (6)$$

Since the function $f(z)$ is concave, and the non-negative coefficients $u(x_j) - u(x_{j-1})$ add up to 1, we conclude that

$$u(x_1) \cdot f(S_2(x_1)) + \cdots + (u(x_j) - u(x_{j-1})) \cdot f(S_2(x_j)) + \cdots +$$
$$(u(x_n) - u(x_{n-1})) \cdot f(S_2(x_n)) + (1 - u(x_n)) \cdot f(S_2(x_{n+1})) \le$$
$$f(u(x_1) \cdot S_2(x_1) + \cdots + (u(x_j) - u(x_{j-1})) \cdot S_2(x_j) + \cdots +$$
$$(u(x_n) - u(x_{n-1})) \cdot S_2(x_n) + (1 - u(x_n)) \cdot S_2(x_{n+1})) = f(E_2[u]). \quad (7)$$

From the inequalities (6) and (7), we conclude that $E_1[u] \le f(E_2[u])$ for piece-wise constant utility functions.

Similarly to the previous proof, we can then take the limit and thus prove that the desired inequality $E_1[u] \le f(E_2[u])$ holds for all 0-1-$M$ utility functions. The proposition is proven.

**Acknowledgements** This work was supported in part by the National Science Foundation grants 1623190 (A Model of Change for Preparing a New Generation for Professional Practice in Computer Science), HRD-1834620 and HRD-2034030 (CAHSI Includes), EAR-2225395 (Center for Collective Impact in Earthquake Science C-CIES), and by the AT&T Fellowship in Information Technology.

It was also supported by the program of the development of the Scientific-Educational Mathematical Center of Volga Federal District No. 075-02-2020-1478, by a grant from the Hungarian National Research, Development and Innovation Office (NRDI), and by the Center of Excellence in Econometrics, Faculty of Economics, Chiang Mai University, Thailand.
The authors are thankful to Hung T. Nguyen for his encouragement.

## References

1. Bruni, R., Cesarone, F., Scozzari, A., Tardella, F.: On exact and approximate stochastic dominance strategies for portfolio selection. Eur. J. Oper. Res. **259**, 322–329 (2017)
2. Fishburn, P.C.: Utility Theory for Decision Making. Wiley, New York (1969)
3. Fishburn, P.C.: Nonlinear Preference and Utility Theory. The John Hopkins Press, Baltimore, Maryland (1988)
4. Hosseini-Nodeh, Z., Khanjani-Shiraz, R., Pardalos, P.M.: Distributionally robust portfolio optimization with second-order stochastic dominance based on Wasserstein metric. Inf. Sci. **613**, 828–852 (2022)
5. Kahneman, D.: Thinking, Fast and Slow. Farrar, Straus, and Giroux, New York (2011)
6. Kreinovich, V.: Decision making under interval uncertainty (and beyond). In: Guo, P., Pedrycz, W. (eds.) Human-Centric Decision-Making Models for Social Sciences, pp. 163–193. Springer, Berlin (2014)
7. Kreinovich, V., Nguyen, H.T., Sriboonchitta, S.: What if we only have approximate stochastic dominance? In: Huynh, V.-N., Kreinovich, V., Sriboonchitta, S., Suriya, K. (eds.) Econometrics of Risk, pp. 53–61. Springer, Berlin, Heidelberg (2015)
8. Luce, R.D., Raiffa, R.: Games and Decisions: Introduction and Critical Survey. Dover, New York (1989)
9. Nguyen, H.T., Kosheleva, O., Kreinovich, V.: Decision making beyond Arrow's 'impossibility theorem', with the analysis of effects of collusion and mutual attraction. Int. J. Intell. Syst. **24**(1), 27–47 (2009)
10. Nguyen, H.T., Kreinovich, V., Wu, B., Xiang, G.: Computing Statistics under Interval and Fuzzy Uncertainty. Springer, Berlin, Heidelberg (2012)
11. Raiffa, H.: Decision Analysis. McGraw-Hill, Columbus, Ohio (1997)

# Practical Applications

# Border Crossing: Bayesian Gravity and Cross-Border Exports Between Myanmar and Its Neighbours

**Sunil Dash**

**Abstract** Using data from 18 border stations between 2018 and 2022, this paper analyzes the formal border exports between Myanmar and its neighbouring countries—Bangladesh, China, Laos, India, and Thailand. Firstly, it presents an overview of the export pattern that crosses international borders. Secondly, it applies a Bayesian gravity model to investigate the determinants of Myanmar's border exports. The posterior mean estimates give a clear prediction that the border exports are positively related to the market size of the destination and adversely related to the distance, which is in line with standard gravity literature. It was further found that (i) border infrastructure plays a significant role in driving border exports as border stations having relatively better infrastructure are predicted to have higher levels of exports than those without, (ii) documentation cost is predicted to be a significant negative predictor of border exports, and (iii) border exports to destinations that have free trade agreements with Myanmar are predicted to be higher than destinations without such agreements.

**Keywords** International trade · Cross-border trade · Myanmar · Bayesian inference and parameter estimation · Gravity model

## 1 Introduction

"Cross-border trade" facilitates the movement of goods through border stations across countries sharing land borders. Myanmar shares its borders with five countries, and border trade between Myanmar and its neighbours—China, Thailand, Bangladesh, India, and Laos have existed for several decades [1, 2]. Myanmar shares a border with China and Laos in the north and northeast, Bangladesh and the Indian states of Nagaland, Manipur, and Mizoram in the northwest and west, and Thailand

S. Dash (✉)
Faculty of Economics, Chulalongkorn University, Bangkok, Thailand
e-mail: sunil.dash@hotmail.com



V. Kreinovich et al. (eds.), *Applications of Optimal Transport to Economics and Related Topics*, Studies in Systems, Decision and Control 556,
https://doi.org/10.1007/978-3-031-67770-0_15

in the southeast.[1] With five countries sharing borders, the export of goods totaling around USD 5580 million crossed Myanmar's borders in 2022, while the average figure from 2018 to 2022 was USD 5916 million for the same [3]. Trade along the Indo-Myanmar, Lao-Myanmar, and Bangladesh-Myanmar borders is comparatively lower than the trade across Sino-Myanmar and Thai-Myanmar borders. Though Myanmar borders five countries, much of its cross-border trade has been with China and Thailand [4–7]. Myanmar, Laos, and Thailand are member states of the Association of Southeast Asian Nations (ASEAN), hence, parties to the ASEAN Trade in Goods Agreement. In 2019, ASEAN member states accounted for 23.8% of Myanmar's exports [7]. Trade between Myanmar and China has been governed under the ASEAN-Hong Kong, China, Free Trade Agreement (FTA) since June 2019 [7]. Myanmar signed the ASEAN-India FTA in 2009, which came into force in 2010; from India's "look east policy" in the 1990 s to the upgraded version of the "act east policy," which was launched in 2014, Myanmar has been in India's focus as India seeks to deepen its economic ties with the countries to its east [2, 6].

Agriculture (including forestry and fishing) is the backbone of Myanmar's economy, and it contributed 21.3% to the GDP and 24.3% of all exports in 2019 [7]. Myanmar is endowed with an abundance and diverse natural resources and is a net exporter of agricultural products. More recently, Myanmar has developed a relatively large manufacturing sector, and the two primary sub-sectors are the food and beverage industry and the textiles and clothing industry. The largest of these sectors is the textile and apparel sector, which accounts for about 14 percent of total exports. Myanmar's exports are less diverse since a small set of products makes up most of its export basket [8, 9].

Most of the existing research on Myanmar's border trade has been done to capture the bilateral trade patterns in the context of the Indo-Myanmar border [1, 2, 5, 6], and some literature captures the trade patterns of Sino-Myanmar border [4], and Thai-Myanmar border [10]. Most of these pieces of literature are comprehensive studies that explain the pattern of trade in goods, trade in services, cross-border investment, and infrastructure and identify the potential border trade linkages and constraints acting upon them. The following are some of the key findings of the recent literature:

(i) Given that there had been little evidence of industrialization until the late 2000s, the importance of the agricultural sector's share in the national economy was paramount. As non-agricultural sectors such as manufacturing and services have recently experienced significant growth, the agricultural sector's share has been declining [2, 4, 7–9].
(ii) Natural gas exports, especially those with product code HS 271,121, "*natural gas in the gaseous state*," have rapidly expanded in the last two decades, improving the state of the government sector and aiding in the restoration of macroeconomic stability [7–9].

[1] The political map of Myanmar showing all the neighbouring countries is included in the Appendices Section.

(iii) As Western economic sanctions became more severe in the 1990 s, Myanmar's economy steadily became more dependent on China. As a result, the international community's efforts to promote democratic changes were significantly hampered by China's increasing assistance to Myanmar [2, 5, 8].
(iv) Importantly, Myanmar is the largest trading partner of the Yunnan Province, China, which shares a long border with Myanmar, accounting for a majority of its total trade with China [4].
(v) The tightening of economic and political ties between Myanmar and China resulted in neighbouring nations taking steps towards closer ties with Myanmar to counterbalance and secure their influence due to Myanmar's geopolitical significance [2, 8].
(vi) To mitigate the impact of the pandemic, the authorities in Myanmar implemented several trade and investment measures, including changes in import and export licensing requirements [7].

So far, no empirical research has been conducted from the perspective of Myanmar's cross-border trade with all its neighbouring countries. For Myanmar, a border-station-specific study was carried out by [11] that explores the characteristics of cross-border trade; however, the study is limited to only one township, i.e., Myawaddy Township in Kayin State [11]. The authors do not examine the determinants of cross-border exports, and no gravity is employed as well. In this study, we analyze the data of 18 border stations across the Myanmar borders using the most recent data made available from 2018 to 2022 by the Ministry of Commerce, Myanmar [3]. These border stations are gateways for the transit of goods between Myanmar and its neighbouring countries. We contribute to the border trade literature (i) by examining Myanmar's border export patterns using multiple border-station-specific data, (ii) by employing a Bayesian gravity model to examine the determinants of Myanmar's border exports, (iii) by showing that better border infrastructure is a positive predictor and documentation cost at the border is a negative predictor of Myanmar's border exports, and (iv) by predicting that border exports to destinations that have free trade agreement with Myanmar are likely to be higher than destinations without such agreements.

The paper is structured as follows. In the following section, we provide an overview of Myanmar's cross-border exports with its neighbouring countries at an aggregate level and sector level as well. In Sect. 3, we explain the gravity model for border exports, followed by the motivation of Bayesian inference in Sect. 4. In Sect. 5, we then discuss the empirical findings, and policy implications, and outline our model selection process. Section 6 concludes.

## 2 Overview of Cross-Border Exports

The graphical summary of the aggregate border export flows of Myanmar from 2018 to 2022 is provided in Fig. 1. Cross-border exports accounted for roughly 40% of Myanmar's total exports in the last 5 years, on average. This highlights that border crossing is important for Myanmar as far as trade in goods is concerned. The highest percentage of border exports was in 2020, whereas the lowest was in 2021 when compared to the total volume of exports. Due to the second wave of COVID-19, border operations were disrupted significantly; hence, 2021 saw a decrease in the share, most likely.

Figure 2 illustrates the sector-wise border export pattern from 2018 to 2022. Looking at shares of cross-border exports disaggregated by product, some notable tendencies emerge. The share of cross-border exports tends to be high for agricultural, marine, and manufacturing products but low for forest and mineral products. The percentage of border exports is roughly half for agricultural, marine, and manufacturing products compared to the total volume of exports of each of the sectors, respectively. It appears that most animal products are exported via borders. For forest and mineral products, exports by the border are significantly low compared to other products. The share of agricultural and animal products significantly dropped in 2021 (possibly due to COVID-19 restrictions); however, they seem to have increased in 2022.

In total, there are 18 border stations in Myanmar, with 7 on the Thai-Myanmar border, 5 on the Sino-Myanmar border, and 2 on each of the borders with Bangladesh, India, and Laos. The locations of the border stations are presented in Fig. 3.

The exports that passed through these border stations from 2018 to 2022 are summarized and presented in Table 1. Analyzing the cross-border export pattern over the years, it is evident that Muse on the Sino-Myanmar border and Htee Khee on the Thai-Myanmar border are the most important border stations—the Lao-Myanmar

**Fig. 1** Border exports pattern of Myanmar, 2018–2022. *Source* [3]

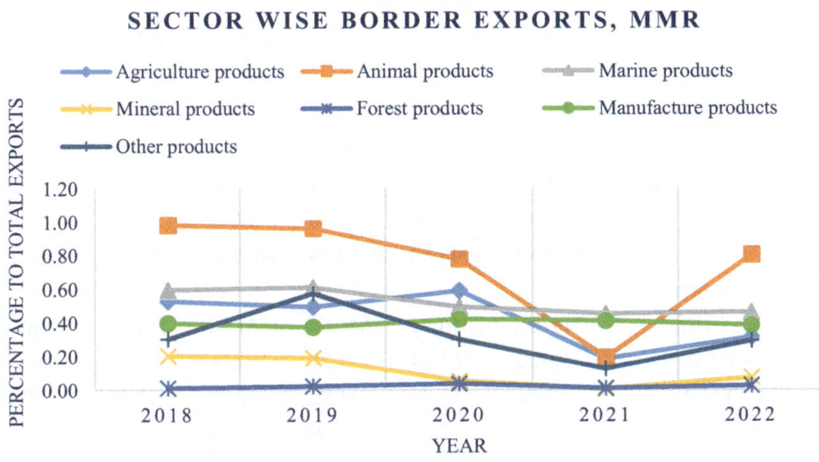


**Fig. 2** Sector-wise border exports pattern. *Source* [3]

**Fig. 3** Location of border stations in Myanmar. *Source* (Google Maps)

border account for the smallest volume of exports out of all the crossings. On the Lao-Myanmar border, the Kenglap border station accounted for a small volume of exports in 2018 and 2019, whereas the Htan Ta Lan border station has seen no exports in these five years. Table 1 indicates in bold the border stations having exports above

**Table 1** Year-wise exports across border stations, MMR (2018–2022)

| SL | Border stations | 2018 | 2019 | 2020 | 2021 | 2022 | TOTAL |
|---|---|---|---|---|---|---|---|
| Sino-Myanmar border | | | | | | | |
| 1 | CHIN SHWEHAW | **460.36** | **424.95** | 349.62 | 73.27 | 145.56 | 1453.76 |
| 2 | KANPITETEE | 264.11 | 262.34 | 266.45 | 62.21 | 73.84 | 928.96 |
| 3 | KYAING TONG | 6.99 | 2.89 | 12.51 | 9.48 | 7.02 | 38.89 |
| 4 | LWEJEL | 122.84 | 112.76 | 173.10 | 53.98 | 87.38 | 550.05 |
| 5 | MUSE | **3156.41** | **3049.88** | **2919.34** | **1482.26** | **1784.09** | **12,392.01** |
| Thai-Myanmar border | | | | | | | |
| 6 | HTEE KHEE | **2467.54** | **2010.91** | **1646.66** | **1551.62** | **2191.91** | **9868.63** |
| 7 | KAWTHAUNG | 175.22 | 265.74 | 234.71 | 194.26 | 174.98 | 1044.92 |
| 8 | MAWTAUNG | 12.08 | 12.79 | 11.82 | 10.28 | 12.61 | 59.58 |
| 9 | MESE | 1.13 | 2.48 | 0.00 | 0.00 | 0.00 | 3.61 |
| 10 | MYAWADDY | 211.48 | 385.64 | **871.74** | **948.34** | **639.27** | **3056.48** |
| 11 | MYEIK | 121.48 | 55.32 | 49.76 | 36.37 | 25.32 | 288.25 |
| 12 | TARCHILEIK | 20.11 | 68.21 | 158.22 | 35.81 | 42.83 | 325.17 |
| Lao-Myanmar border | | | | | | | |
| 13 | KENGLAP | 0.02 | 0.41 | 0.00 | 0.00 | 0.00 | 0.43 |
| 14 | HTAN TA LAN | 0.00 | 0.00 | 0.00 | 0.00 | 0.00 | 0.00 |
| Bangladesh-Myanmar border | | | | | | | |
| 15 | MAUNG DAW | 8.90 | 11.49 | 4.86 | 12.70 | 15.15 | 53.11 |
| 16 | SITTWE | 12.98 | **432.55** | 12.12 | 13.80 | 62.99 | 534.43 |
| Indo-Myanmar border | | | | | | | |
| 17 | RHI | 82.11 | 48.86 | 164.41 | 38.84 | 38.84 | 373.07 |
| 18 | TAMU | 95.39 | 40.73 | 33.56 | 6.03 | 9.35 | 185.06 |

*Note* Export figures are in USD Millions
*Source* [3]

the mean level for each of the years (Column 3 to Column 7) and for the entire period (Column 8).

## 3 The Gravity Model of Cross-Border Exports

Motivated by the theoretical foundations of the gravity model, in this section, we describe the empirical specification we adopt for modeling the cross-border exports where the data consists of 18 border stations of Myanmar. We aim to examine the determinants of border exports in Myanmar. Beyond market size and distance, we augment the gravity model by adding further variables that may influence border exports, such as border infrastructure, free trade agreements, and documentation costs.

The simplest notion of the gravity model in trade is that the trade flows between two countries are positively influenced by their respective market sizes and negatively by the distance between them [12–15]. In the basic form of the model, the relationship between trade flow and its determinants is typically estimated in the gravity equation, a log-linear specification relating export flows to the Gross Domestic Product (GDP) of trading partners (as a measure of market size) and the geographical distance (a measure of trade friction) between them.

A basic gravity equation capturing trade flows between Country A and Country B has the following form (Eq. (1)):

$$Trade_{AB} = constant \times \frac{GDP_A \times GDP_B}{Distance_{AB}} \tag{1}$$

Eventually, researchers augmented the gravity model and allowed the barriers between trading partners to take many forms, including distance, tariffs, non-tariff barriers (like administrative or documentation costs), etc. [16–18] So, the augmented gravity model can be expressed as (Eq. (2)):

$$Trade_{AB} = constant \times \frac{GDP_A \times GDP_B}{Trade\ Barriers_{AB}} \tag{2}$$

Other factors related to familiarity, like language, a historical connection like colonial ties, and geography (like contiguity and landlock status) are also considered as additional variables in gravity equations. Familiarity and contiguity are expected to improve trade, while landlocked countries are expected to dampen trade flows compared with countries that are not, as they do not have access to marine transport at their disposal.

For capturing market size, the real gross domestic product of the destination country is proxied. The information on the destination GDP (in US dollars) comes from the World Bank's *World Development Indicators Database*. The sign of the variable is expected to be positive. The gravity model generally includes both importer and exporter market sizes as explanatory variables: however, as the export in our data originates from a single source country, i.e., Myanmar, its market size will be captured in the regression constant [19].

In the gravity model, distance is frequently proxied for transportation costs because (a) transportation costs increase as distance increases and (b) an increase in transportation costs results in less trade [15, 17, 20–23]. The relationship between distance and border exports is expected to be negative. In this paper, instead of the great circle distance between the origin country and destination country, we use the bilateral distance between the state capital where the border station is located to the capital city of the state/province of the destination country on the other side of each of the borders. This record has been extracted from *Google Maps*, and the measure is in kilometers.

Supplementing the distance proxy for trade costs of exporting to the foreign destination, we add an additional variable capturing documentation costs incurred at the border. The administrative complexity of the documentation process while exporting through the border stations, as measured by the number of documents that need to be completed, is considered. The database of the *Ministry of Commerce, Myanmar*, is the source of this information [3]. We expect a negative association between this variable and exports.

Furthermore, we consider the "border infrastructure" as an explanatory variable. The absence of proper border infrastructure or its poorly developed state can be a major reason for high trade costs. Hence, we hypothesize that border stations where bilateral infrastructure development projects are relatively better developed are more likely to have a bigger volume of export flows compared to the stations where the border infrastructure is relatively inferior. We briefly explain some of the bilateral projects Myanmar has developed by partnering with its neighbouring countries here. The cities of Myawaddy in Kayin State, Myanmar, and Mae Sot in Tak Province, Thailand are connected by two Thai–Myanmar friendship bridges. The first bridge opened on August 15, 1997, whereas on October 30, 2019, the second bridge began operations [24]. Cross-border operations, however, suffered substantially during the Covid-19 period due to restrictions.

As part of the *Belt and Roads Initiative*, Myanmar has developed three economic zones since 2018 intending to foster infrastructure development along the border regions as well as economic integration between the southern border of China and Myanmar. These areas are established in Kachin State's Kanpiketi Township, Shan State's Laukkai Township, and Shan State's Muse Township [25]. On the Indo-Myanmar border, a friendship road is constructed with Indian assistance and runs from Moreh/Tamu to Kyigone and Kalewa. The TKK road, as it is known, was handed over to Myanmar in 2009. The road has made it easier to travel between Tamu and Kalewa in the Indian bordering Sagaing area of Myanmar [6]. A friendship bridge, which was built on February 16, 2013, connects Laungnamtha Province in Laos with Tachilek in the Eastern Shan State of Myanmar. To capture the border infrastructure, we use a dummy variable that takes the value 1 if any joint infrastructure development project is completed near the border station at time $t$ and 0, otherwise. We expect the coefficient of this variable to be positive.

Finally, we add a dummy variable capturing the free trade agreements between Myanmar and the destinations. Having free trade agreements in place allows better bilateral trade and economic relations. In addition to lowering and eliminating tariffs, free trade agreements also assist in removing behind-the-border obstacles that would otherwise prevent the free movement of goods. The dummy variable FTA takes the value 1 if there is an FTA in place between Myanmar and the destination at time $t$ and 0; otherwise. We expect the sign of this variable to be positive.

The gravity model adopted for the analysis takes the form (Eq. (3)):

$$\begin{aligned} lnEXPORTS_{id,t} = \alpha_1 + \alpha_2 lnGEODIST_{id} + \alpha_3 lnMSIZE_{d,t} + \alpha_4 BORDERINFRA_{id,t} \\ + \alpha_5 FTA_{id,t} + \alpha_6 lnDOCCOST_{id,t} + \psi_i + \varepsilon_{id,t} \end{aligned} \quad (3)$$

where

| | |
|---|---|
| $\ln EXPORTS_{id,t}$ | Exports (In USD Millions) crossing through the border station *i* to destination *d* at time *t* (expressed in natural logarithmic form). |
| $\ln GEODIST_{id}$ | Bilateral geographic distance (In Kilometers) between the state capital where border station *i* is located and the state/provincial capital of the destination *d* across the border (expressed in natural logarithmic form). |
| $\ln MSIZE_{d,t}$ | Real Gross Domestic Product (In USD) of the destination *d* at time *t* (expressed in natural logarithmic form). |
| $BORDERINFRA_{i,t}$ | A dummy variable takes the value 1 if any joint infrastructure between Myanmar and the neighbouring country is developed and/or operating near the border station *i*, such as highways, friendship bridges, etc., at time *t* and 0, otherwise. |
| $FTA_{id,t}$ | A dummy variable takes the value 1 if there is an FTA between Myanmar and its neighbouring country in place at time *t* and 0; otherwise. |
| $\ln DOCCOST_{id,t}$ | Number of documents required to let the exports cross the border station *i* to destination *d* at time *t* (expressed in natural logarithmic form). |

The summary statistics are presented in Table 2.

**Table 2** Summary statistics

| Variable | Description | Obs | Mean | Std. dev. |
|---|---|---|---|---|
| EXPORTS | Exports (In USD Millions) | 90 | 346.18 | 719.33 |
| GEODIST | Geographic distance (In Kilometers) | 90 | 744.11 | 538.27 |
| MSIZE | GDP of the destination country (In USD) | 90 | 4.60E + 12 | 6.34E + 12 |
| BORDERINFRA | A dummy variable that captures joint border infrastructures between Myanmar and its neighbouring countries | 90 | 0.38 | 0.49 |
| FTA | A dummy variable that captures FTAs between Myanmar and its neighbouring countries | 90 | 0.88 | 0.31 |
| DOCCOST | Documents required for border clearance (In Numbers) | 90 | 11.80 | 1.47 |

*Source* (Author)

## 4 Motivations for Bayesian Inference

Traditionally, gravity models are estimated using the p-value technique. More recently, the p-value technique has come under scrutiny, and researchers identify the technique to be providing limited information about the data [26, 27]. Bayesian analysis is argued to provide an alternative to the traditional p-value technique, which has been frequently used by researchers more recently [26, 28, 29]. In this paper, we model the gravity based on the Bayesian inference and parameter estimation procedure. The foundation of Bayesian inference is built on the premise that new data can gradually update our understanding of a given system and the uncertainty that goes along with it [30–32]. The Bayes' rule is the fundamental of the Bayesian inference, which is expressed for two random vectors, A and B, such as (Eq. (4)) [33, 34]:

$$p(B|A) = \frac{p(A|B)p(B)}{p(A)} \tag{4}$$

Competing hypotheses are transformed into a set of statistical models and parameters in the Bayesian system of analysis. Each parameter's potential values are given (i.e., the prior) according to a particular statistical distribution based on existing information about the parameters. By evaluating the significance of the input parameters, a likelihood function (i.e., the likelihood) is used to determine the odds of any specific output. The updated information (i.e., the posterior) is then computed to infer a) the likelihood that a particular model generated the existing evidence; and b) the relative explanatory power of a collection of competing models [32, 34]. The posterior distribution is computed by combining the likelihood and the prior with the Bayes' rule (Eq. (5)) [33]:

$$Posterior \propto Likelihood \times Prior \tag{5}$$

There are multiple observations per border station in our data, indicated by subscript $i$ in the gravity Eq. (3). We fitted a Bayesian panel-data linear model to estimate random (panel) effects together with all other model parameters. As a result, each random effect (RE) is represented by an entire Markov Chain Monte Carlo (MCMC) sample from its posterior distribution. Using default normal priors for regression coefficients and default inverse-gamma priors for the error variance and the variance of random intercepts, we executed the empirical procedure in STATA-17. It is noteworthy that in this instance, we fitted a linear random-effects panel-data model to the outcome variable, i.e., export value with several predictors and "border station" as the panel or group identifier id. Essentially, we specified the standard deviation of normal priors for regression coefficients by bayesopts normalprior (100) and shape and scale of inverse-gamma prior for variance components by igammaprior (0.01 0.01).

## 5 Empirical Findings

The output of the Bayesian panel regression is presented in Table 3. The results for total exports show, as expected, a significant negative relationship between exports and distance and a significant positive coefficient on the market size of the destination. The distance coefficient on exports is −2.101, which is higher than the average distance elasticity of −0.9 found in the meta-analysis of general gravity model papers by [15]. The documentation cost of the border exporting process, as measured by the number of documents that need to be completed, is negatively associated with exports. Authorities in Myanmar should aim to reduce the administrative complexity at the border, which will lessen the burden of paperwork and related costs resulting in improved cross-border export flows.

As expected, the variable BORDERINFRA has a positive relationship with the exports. This empirical result suggests that border stations with better infrastructure are likely to witness a greater volume of export flows than border stations with subpar infrastructure. One of the major obstacles, coupled with security, is the subpar infrastructure on the Indo-Myanmar border, as existing literature documents. Hence,

**Table 3** Results of bayesian RE panel regression

| Variables | Mean | Std. dev. | MCSE*** | Median | [95% cred. interval] | |
|---|---|---|---|---|---|---|
| GEODIST | –2.101 | 0.752 | 0.112 | −2.108 | −3.613 | −0.676 |
| MSIZE | 1.323 | 0.370 | 0.039 | 1.318 | 0.615 | 2.053 |
| BORDERINFRA | 0.583 | 0.888 | 0.114 | 0.557 | −1.094 | 2.310 |
| FTA | 1.834 | 1.910 | 0.237 | 1.789 | 1.766 | 5.692 |
| DOCCOST | −1.689 | 0.710 | 0.016 | −1.698 | −3.087 | −0.297 |
| Constant | −14.286 | 7.067 | 0.741 | −14.219 | −28.362 | −1.099 |
| MCMC iterations | 12,500 | 12,500 | 12,500 | 12,500 | 12,500 | 12,500 |
| Burn-in | 2,500 | 2,500 | 2,500 | 2,500 | 2,500 | 2,500 |
| MCMC sample size | 10,000 | 10,000 | 10,000 | 10,000 | 10,000 | 10,000 |
| Number of obs | 79 | 79 | 79 | 79 | 79 | 79 |
| Number of groups | 17 | 17 | 17 | 17 | 17 | 17 |
| Obs. per group: min | 2 | 2 | 2 | 2 | 2 | 2 |
| Obs. per group: avg | 4.6 | 4.6 | 4.6 | 4.6 | 4.6 | 4.6 |
| Obs. per group: max | 5 | 5 | 5 | 5 | 5 | 5 |
| Acceptance rate | 0.823 | 0.823 | 0.823 | 0.823 | 0.823 | 0.823 |
| Efficiency: min | 0.004 | 0.004 | 0.004 | 0.004 | 0.004 | 0.004 |
| Efficiency: avg | 0.084 | 0.084 | 0.084 | 0.084 | 0.084 | 0.084 |
| Efficiency: max | 0.400 | 0.400 | 0.400 | 0.400 | 0.400 | 0.400 |

*Notes* Gibbs sampling is used for regression coefficients and variance components; the Group variable is "border station"; *** Monte-Carlo Standard Error (MCSE)
*Source* (Author)

policymakers in Myanmar might enhance communication with Indian authorities to jointly upgrade the Indo-Myanmar border infrastructure, which would enable Myanmar to capitalize on the opportunity to expand into a bigger market like India. The results on variable FTA indicate that FTAs significantly influence Myanmar's border exports since more export is anticipated to be flowing to the neighbouring destinations with FTAs compared to destinations without such agreements.

Furthermore, we would like to explain the model selection process adopted in this paper. The posterior distributions for parameter values depending on a certain model $M_j$ are given to us by the Bayesian parameter estimation. When several models are in use, Bayes' rule can be extended, and the data can be used to update the relative possibility of each candidate model from the collection of competing models [35]. In the case of the two models, $M_0$ and $M_1$, Eqs. (6) and (7) show how the prior model odds (i.e., the relative possibility of $M_0$ and $M_1$ before observing the data) are updated to posterior model odds (i.e., the relative possibility of $M_0$ and $M_1$ after observing the data). The change from prior odds to posterior odds is given by the Bayes Factor [33, 34], which indicates the models' relative predictive performance for the data observed (Eqs. (6) and (7)):

$$\frac{p(M_1|Data)}{p(M_0|Data)} = \frac{p(M_1)}{p(M_0)} \times \frac{p(Data|M_1)}{p(Data|M_0)} \tag{6}$$

$$Posterior\ model\ odds = Prior\ model\ odds \times Bayes\ Factor\ (BF_{10}) \tag{7}$$

When the Bayes factor $BF_{10}$ is 5, this indicates that the data are 5 times more likely under $M_1$ than $M_0$. The Bayes factor subscripts indicate which model is in the numerator and denominator; for instance, if $BF_{10} = 0.20$, then $\frac{1}{BF_{10}} = BF_{01} = 5$, which means that the data are 5 times more likely under $M_0$ than under $M_1$.

Table 4 outlines the eight best gravity models from the Bayesian regressions. Model specification is presented in Column (1), prior model probabilities are presented in Column (2), and the posterior model probabilities after observing the data are presented in Column (3). The change from prior odds to posterior odds is presented in Column (4). The Bayes factor $BF_{01}$, which quantifies a model's relative predictive performance is outlined in Column (5) in terms of the best model. We put the best model in the first row and compare the subsequent models with the best model, i.e., Model (1). The explained variance of respective models is presented in Column (6).

As the results suggest, all eight best models contain the market size and geographic distance variables, indicating that these predictors are important to account for the border exports. This is consistent with the theoretical foundations of the gravity model which assumes that trade is positively related to market size and negatively related to distance. The relative predictive performance of Model (1) is marginally higher than that of Model (8); hence, confirms that our selection of Model (1) is appropriate based on the data.

**Table 4** Model comparison—lnExports

| Models | P(M) | P(M\|data) | $BF_M$ | $BF_{01}$ | $R^2$ |
|---|---|---|---|---|---|
| (1) msize + geodist + fta + borderinfra + doccost | 0.167 | 0.222 | 1.423 | 1.000 | 0.250 |
| (2) msize + geodist + borderinfra | 0.017 | 0.189 | 13.739 | 0.117 | 0.249 |
| (3) msize + geodist | 0.017 | 0.158 | 11.095 | 0.140 | 0.216 |
| (4) msize + geodist + fta + borderinfra | 0.033 | 0.125 | 4.139 | 0.355 | 0.250 |
| (5) msize + geodist + borderinfra + doccost | 0.033 | 0.120 | 3.938 | 0.371 | 0.249 |
| (6) msize + geodist + fta | 0.017 | 0.050 | 3.084 | 0.446 | 0.218 |
| (7) msize + geodist + doccost | 0.017 | 0.045 | 2.804 | 0.488 | 0.216 |
| (8) msize + geodist + fta + doccost | 0.033 | 0.033 | 0.995 | 1.336 | 0.218 |

*Notes* The variables in Column (1) are written in lowercase. The top 8 models shown out of 32 visited
*Model enumeration* $2^5 = 32$ considered models
*Source* (Author)

# 6 Conclusion

Using export data from 18 border stations between 2018 and 2022, this paper examines the formal cross-border exports between Myanmar and its neighbouring countries—Bangladesh, China, Laos, India, and Thailand. Firstly, it presents an overview of the export pattern that crosses the international borders of Myanmar. Within the Bayesian framework, a Bayesian gravity analysis was conducted to examine the determinants of border exports. In addition to the standard gravity variables of market size and distance, this paper considers further variables that may influence border exports, such as border infrastructure, free trade agreements, and border documentation costs. The posterior estimates show that better border infrastructure is a positive predictor, and documentation cost at the border is a negative predictor of Myanmar's border exports. Finally, the model predicts that border exports to destinations having free trade agreements with Myanmar are likely to be higher than destinations without such agreements.

## Appendix I: Political Map of Myanmar

Source: (Political Map of Myanmar - Nations Online Project)

## References


1. Singh, T.B.: India's border trade with its neighbouring countries with special reference to Myanmar. Margin: J. Appl. Econ. Res. **1**(4), 359–382 (2007)
2. Das RU, Enhancing India-Myanmar Border Trade: Policy and Implementation Measures 2016, p. 110. Department of Commerce, Ministry of Commerce and Industry, Government of India, New Delhi
3. Commerce Ministry, Border Trade Data, Republic of the Union of Myanmar, Editor. 2023: Nay Pyi Taw
4. Koji, K.: Myanmar's Cross-Border Trade with China: Beyond Informal Trade, p. 23. Institute of Developing Economies, JETRO, Chiba, Japan (2016)
5. Shahriar, S.: India's economic relations with Myanmar: a study of border trade. J. Borderlands Stud. **37**(3), 599–621 (2022)
6. Taneja, N., et al.: India-Myanmar Border Trade, p. 42. Indian Council for Research on International Economic Relations (2019)
7. WTO, T.: Trade Policy Review: Myanmar, p. 117. Geneva (2020)
8. Odaka, K.: The Myanmar Economy: Its Past, Present, and Prospects. Springer, Berlin (2015)
9. Khandelwal, A.K., Teachout, M.: Special Economic Zones for Myanmar, I.G. Centre, Editor, p. 41. London School of Economics and Political Science, London (2016)
10. Krainara, C., Routray, J.K.: Cross-border trade and commerce between Thailand and neighbouring countries: policy implications for establishing special border economic zones. J. Borderlands Stud. **30**(3), 345–363 (2015)
11. Naing, Y.Y., Chulasai, L., Panthamit, N.: Characteristics of cross border trade in Myawaddy township kayin state, Myanmar. Chiang Mai Univ. J. Econ. **14**(2), 100–122 (2010)
12. Anderson, J.E.: A Theoretical Foundation for the Gravity Equation. American Economic Review **69**(1), 106–116 (1979)
13. Bergstrand, J.H.: The gravity equation in international trade: some microeconomic foundations and empirical evidence. Rev. Econ. Stat. **67**(3), 474–481 (1985)
14. Yotov, Y.V., et al.: An Advanced Guide to Trade Policy Analysis: The Structural Gravity Model, p. 142. The WTO, Geneva (2016)
15. Disdier, A.-C., Head, K.: The puzzling persistence of the distance effect on bilateral trade. Rev. Econ. Stat. **90**(1), 37–48 (2008)
16. Deardorff, A.: Determinants of bilateral trade: does gravity work in a neoclassical world? In: The Regionalization of the World Economy, pp. 7–32. University of Chicago Press (1998)
17. Chaney, T.: Distorted gravity: the intensive and extensive margins of international trade. Am. Econ. Rev. **98**(4), 1707–1721 (2008)
18. Feenstra, R.C.: Advanced International Trade: Theory and Evidence. Princeton University Press (20150
19. Lawless, M.: Deconstructing gravity: trade costs and extensive and intensive margins. Can. J. Econ. **43**(4), 1149–1172 (2010)
20. Anderson, J.E., Van Wincoop, E.: Gravity with gravitas: a solution to the border puzzle. Am. Econ. Rev. **93**(1), 170–192 (2003)
21. Bernard, A.B., et al.: Firms in international trade. J. Econ. Perspect. **21**(3), 105–130 (2007)
22. Cheong, J., Kwak, D.W., Tang, K.K.: The distance effects on the intensive and extensive margins of trade over time. Empirical Econ. **50**(2), 253–278 (2016)
23. Crozet, M., Koening, P.: Structural gravity equation with extensive and intensive margins. Can. J. Econ./Revue canadienne d'économique **43**(1), 41–62 (2010)
24. Authority, T.: Thai-Myanmar Friendship Bridge (2023). Available from: https://www.tourismthailand.org/Attraction/thai-myanmar-friendship-bridge
25. ASEANBriefing.: Myanmar Plans Economic Zones on Chinese Border (2018)
26. Held, L., Ott, M.: On p-values and Bayes factors. Ann. Rev. Stat. Appl. **5**(1), 593–419 (2018)
27. Halsey, L.G.: The reign of the p-value is over: what alternative analyses could we employ to fill the power vacuum? Biol. Lett. **15**(5), 8 (2019)

28. Maneejuk, P., Yamaka, W.: Significance test for linear regression: how to test without P-values? J. Appl. Stat. **48**(5), 827–845 (2021)
29. Tran, H.D., et al.: An alternative to p-values in hypothesis testing with applications in model selection of stock price data. In: Robustness in Econometrics, pp. 305–319. Springer (2017)
30. Etz, A., Vandekerckhove, J.: Introduction to bayesian inference for psychology. Psychon. Bull. Rev. **25**(1), 5–34 (2018)
31. Kosheleva, O., Kreinovich, V., Autchariyapanitkul, K.: Why Beta Priors: Invariance-Based Explanation. In: Behavioral Predictive Modeling in Economics, pp. 141–144. Springer (2021)
32. Morey, R.D., Romeijn, J.-W., Rouder, J.N.: The philosophy of Bayes factors and the quantification of statistical evidence. J. Math. Psychol. **72**, 6–18 (2016)
33. Jeffreys, H.: Theory of probability. Oxford University Press, Oxford (1961)
34. Berrar, D.: Bayes' Theorem and Naive Bayes Classifier. In: Encyclopedia of Bioinformatics and Computational Biology: ABC of Bioinformatics, pp. 403–412. Elsevier Inc., Amsterdam (2019)
35. Bergh, D.V.d., et al.: A tutorial on Bayesian multi-model linear regression with BAS and JASP. Behav. Res. Methods 1–21 (2021)

# Investigating the Environmental Kuznets Curve Hypothesis: Empirical Study on the Relationships Among Carbon Emissions, International Crude Oil Prices, Total Factor Productivity, Research and Development, International Trade Balance in US

**Nunnapat Tawong and Worrawat Saijai**

**Abstract** Our research investigated the relationships among carbon emissions, crude oil prices, total factor productivity (TFP), research and development (R&D), and international trade balance in the US. Using the autoregressive distributed lag (ARDL) model, we analyzed annual data from 1970 to 2021. Our findings revealed a two-way relationship between carbon emissions and TFP, and identified oil prices as causes of carbon emissions, while carbon emissions influenced trade balance. In the long term, carbon emissions had minimal effects on TFP and R&D but significantly impacted trade balances. TFP negatively affected carbon emissions. Moreover, increasing TFP and investment in R&D are crucial for mitigating emissions. In the short term, the impact of these variables on carbon emissions was lower. Changes in TFP and R&D influenced by carbon emissions had persistent negative effects, while oil prices and trade balance had modest, long-term positive impacts on emission volatility. These findings emphasize that insufficient growth in TFP hampers efforts to reduce environmental degradation and highlight the importance of enhancing total factor productivity and investing in research and development to mitigate carbon emissions in the long run.

**Keywords** ARDL · Total factor productivity · $CO_2$ emissions · Environmental kuznets curve

N. Tawong (✉) · W. Saijai
Faculty of Economics, Chiang Mai University, Chiang Mai 50200, Thailand
e-mail: cadent141@gmail.com

W. Saijai
e-mail: worrawat.s@cmu.ac.th

 
V. Kreinovich et al. (eds.), *Applications of Optimal Transport to Economics and Related Topics*, Studies in Systems, Decision and Control 556,
https://doi.org/10.1007/978-3-031-67770-0_16

## 1 Introduction

Global warming caused by the greenhouse effect has become one of the major problems people around the world are facing today. A significant contributor to this issue is the massive use of fossil fuels, particularly crude oil, in modern industries [1]. The greenhouse effect resulting from the combustion of fossil fuels leads to the emission of greenhouse gases, which in turn contributes to climate change.

Crude oil, being a widely traded and extensively utilized fossil fuel, plays a crucial role in various economic activities. The lower cost of energy consumption associated with crude oil enables the manufacturing sector to produce goods and services more efficiently, thereby positively impacting the state of the economy as a whole.

When assessing the state of the economy and its growth, several indicators come into play. Traditional indicators like gross domestic product (GDP) and gross national product (GNP) are commonly used. However, GDP primarily focuses on the increase or decrease in inputs such as labor and capital, neglecting the aspect of improving production efficiency or quality. Hence, GDP has its limitations [2].

In the present era, many countries are grappling with aging populations, leading to a decline in the workforce. Additionally, the utilization of capital inputs, such as machinery, has decreased and become more challenging [2]. In light of this, we aim to employ the Total Factors Productivity as an alternative measure to demonstrate economic progress rather than relying solely on GDP growth.

TFP provides insight into economic progress by capturing changes in technological advancements in both production processes and resource allocation [3]. It takes into account both input improvements and the increased quantity of input. Therefore, TFP proves to be a more appropriate measure for evaluating economic progress than solely relying on GDP growth.

Given its status as one of the largest economies globally, the United States plays a significant role in global environmental efforts. Analyzing carbon emissions and economic growth in the U.S. helps in assessing sustainable development pathways that promote both economic prosperity and environmental sustainability. Although countries like China emit the highest carbon emissions overall, when considering emissions per capita, developed nations like the United States rank highest [4].

Figure 1 demonstrates the correlation between GDP and carbon emissions in the United States from 1970 to 2021. Notably, after 2007, carbon emissions started to decline, coinciding with the introduction of carbon credits trading. This trading system provides organizations with incentives to reduce their carbon emissions annually, leading to changes in product pricing, international trade dynamics, social costs, and the development of carbon-reduction innovations and technologies.

In the long run, the reduction of carbon emissions in the United States would align with the Environmental Kuznets Curve (EKC) hypothesis, which proposes an inverted U-shaped relationship between environmental degradation and economic development. According to this hypothesis, as per capita income or economic growth

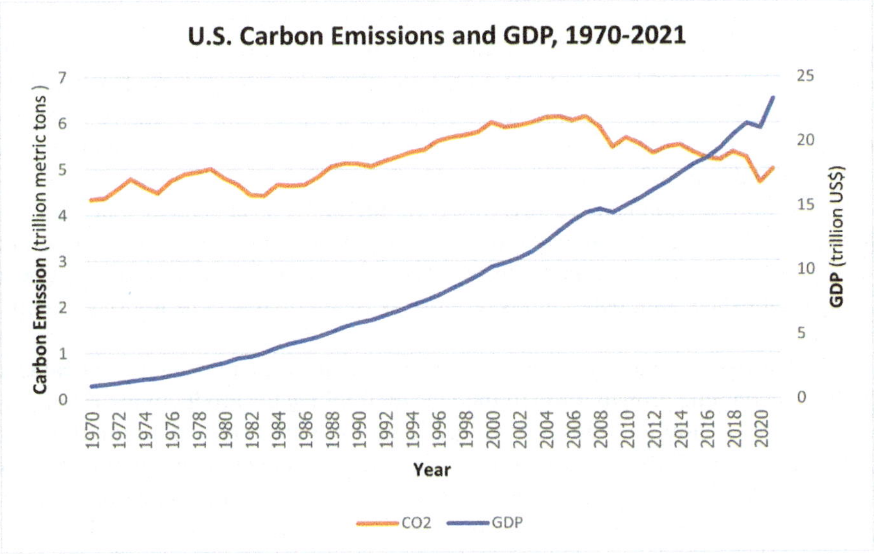


**Fig. 1** U.S. carbon emissions and GDP during 1970–2021

increases, environmental degradation initially worsens but eventually reaches a turning point where it starts to improve. This relationship is closely tied to both short-term and long-term economic growth [5].

Despite a reduction in carbon emissions, the United States remains the developed country with the highest carbon emissions. This underscores the need for advanced technology and innovation to address carbon removal as much as possible [5]. Consequently, total factor productivity and research and development become vital variables that impact carbon emissions and can also influence international trade dynamics.

The literature examining the relationship between energy consumption, economic growth, and carbon emissions consists of numerous studies that employ various indicators of economic growth and research methods, such as econometric models and machine learning techniques [1, 6–11]. These studies yield different results due to variations in the research methodology, sample composition, and testing tools utilized. Additionally, the extensive use of TFP in research has not yet been firmly established, despite its potential as another indicator of economic growth [3].

The relationship between TFP, energy consumption, and carbon emissions has been investigated in various studies. Reference [12] found that more open trade is conducive to higher TFP levels. Currently, the majority of research in this area focuses on China, where increasing carbon emissions have prompted studies on enhancing TFP. To raise TFP levels, China has prioritized investment in research and development. Therefore, research and development and international trade dynamics are also considered in this study [13, 14].

The empirical results from the mentioned research exhibit conflicting outcomes. Some studies validate the EKC assumption [6, 9, 12, 15, 16], while others reject its validity [1, 7, 8, 10, 11, 13, 14]. The diverse outcomes can be attributed to variations in study methods, samples, and testing tools employed. Moreover, the extensive adoption of TFP in research is yet to be fully established, despite its potential as an additional indicator of economic growth [3].

Based on a comprehensive review of the literature, it is evident that different studies yield divergent results due to variations in research methods, samples, and testing tools. Furthermore, the widespread use of TFP in research has not yet been firmly established, although it holds promise as an alternative indicator of economic growth [3].

Considering the conclusions drawn from the literature, there is significant value in studying economic growth through the lens of Total Factor Productivity. Economic growth is inherently intertwined with economic activity, energy consumption, and carbon emissions. This research aims to explore the impact of international crude oil prices, TFP, research and development, and international trade balance on short-term and long-term carbon emissions in the United States. To achieve this, the ARDL model will be utilized to test whether energy consumption and economic development in the United States align with the hypothesized environmental curve. The ARDL model is advantageous for our topic as it addresses non-stationarity, captures cointegration, enables dynamic analysis, and facilitates policy implications for short-run and long-run terms. Its application allows us to gain a comprehensive understanding of the relationships among variables and their effects on carbon emissions and economic growth.

To further elucidate the results of this study, the Impulse Response Function (IRF) and Granger Causality Tests will be incorporated. The IRF based on the Vector Autoregressive (VAR) model helps demonstrate the magnitude of each variable's interaction over short time and its eventual return to stability and equilibrium. Additionally, the Granger Causality Test will be employed to explore the lead-lag relationships between the variables. This analysis will provide insights into which variables should be prioritized and developed first to support the progress of other variables.

The main objectives of this study can be summarized as follows: (1) To test the Environmental Kuznets Curve hypothesis by examining the relationship and estimating the short-run and long-run connections between carbon dioxide emissions, international crude oil prices, total factor productivity, research and development, and international trade balance. (2) To investigate the Granger causality between the variables, thereby identifying key variables that have a significant impact on the others and should be prioritized for development.

The remainder of this paper is organized as follows: Sect. 2 provides an overview of the methodology employed, while Sect. 3 provides a detailed description of the data utilized. The empirical findings of our study are presented in Sect. 4, followed by the conclusions in Sect. 5.

## 2 Methodology

We analyze the short-term and long-term relationships between 5 variables using the ARDL model. Furthermore, we employ the Impulse Response Function to uncover the short-term dynamic relationships and conduct Granger Causality Tests to investigate the lead and lag variables.

### 2.1 ARDL Model

We employ the Autoregressive Distributed Lag model to assess the validity of the Environmental Kuznets Curve (EKC) hypothesis. The ARDL model is particularly suitable for this analysis as it allows for the inclusion of variables with different levels of stationarity, capturing both $I(0)$ and $I(1)$ processes. Additionally, the ARDL model enables the examination of both short-run and long-run relationships within the same framework.

To ensure the robustness of our analysis, we logarithmically transform all variables to mitigate potential issues related to heteroscedasticity. The determination of appropriate lags for the ARDL model is based on the results of the stationary tests outlined in Sect. 2.2.1. By incorporating these considerations, we construct the long-run ARDL model, which can be represented as Eq. 1.

$$
\begin{aligned}
\Delta \ln Y_t = & \alpha_0 + \sum_{i=1}^{p_1} \alpha_{1i} \Delta \ln Y_{t-i} + \sum_{i=1}^{p_2} \alpha_{2i} \Delta \ln X_{1,t-i} + \sum_{i=1}^{p_3} \alpha_{3i} \Delta \ln X_{2,t-i} \\
& + \sum_{i=1}^{p_4} \alpha_{4i} \Delta \ln X_{3,t-i} + \sum_{i=1}^{p_5} \alpha_{5i} \Delta \ln X_{4,t-i} + \beta_1 \ln Y_{t-i} + \beta_2 \ln X_{1,t-i} \\
& + \beta_3 \ln X_{2,t-i} + \beta_4 \ln X_{3,t-i} + \beta_5 \ln X_{4,t-i} + \omega_t \qquad (1)
\end{aligned}
$$

The long-run parameters ($\beta$) represents the relationship between variables in the ARDL model. To estimate the long-run ARDL model, we utilized the Ordinary Least Squares (OLS) method. We then employed F-Test statistics to assess the significance of the parameter values and determine the presence of a long-term relationship among the variables. In the F-Test, the null hypothesis ($H_0$) assumes no long-term relationship between the variables. Conversely, the alternative hypothesis ($H_1$) suggests the existence of a long-term relationship. The appropriate lag values for all variables are selected using the Bayesian Information Criterion (BIC). We employ different lag values to determine the most suitable lag structure for our analysis.

To uncover the short-term associations between the variables, we utilize the F-Test with the lag length determined using the BIC value, following a similar approach used for examining long-term relationships.

In the short-run equation, we introduce an error correction term (ECT) that represents the speed of adjustment to equilibrium. This term captures the dynamics of how the variables converge back to their long-run relationship in response to short-term shocks or deviations from equilibrium. The short-run ARDL model can be expressed as Eq. 2, which incorporates the error correction term and the relevant lagged variables.

$$\begin{aligned}\Delta \ln Y_t =& \alpha_1 + \sum_{i=1}^{m_1} \gamma_{1i} \Delta \ln Y_{t-i} + \sum_{i=1}^{m_2} \gamma_{2i} \Delta \ln X_{1,t-i} + \sum_{i=1}^{m_3} \gamma_{3i} \Delta \ln X_{2,t-i} \\ &+ \sum_{i=1}^{m_4} \gamma_{4i} \Delta X_{3,t-i} + \sum_{i=1}^{m_5} \gamma_{5i} \Delta X_{4,t-i} + \theta_1 ECT_{t-1} + u_t \end{aligned} \quad (2)$$

In the short-run ARDL model, the short-run parameter ($\gamma$) represents the relationship between the variables in the short term. The error correction term captures the speed of adjustment towards the long-run equilibrium. To assess the existence of short-term relationships between the variables, we again utilize the F-Test. The null hypothesis ($H_0$) assumes no short-term relationship.

## 2.2 *Impulse Response Function*

By examining the Impulse Response Function, we can observe the magnitude and duration of the effects that occur over time following the introduction of a specific shock. The IRF provides insights into the short-term dynamic relationships among variables, illustrating how changes in one variable impact others within the system.

In the Vector autoregression system, the error terms are referred to as impulses or innovations. These error terms capture the unexplained or residual components of the variables in the system. Shocks or impulses to endogenous variables are typically introduced through these error terms.

Suppose we consider a scenario where there is a one standard deviation shock to Carbon Emission. This shock will have an immediate impact on income as depicted in the first equation of the model. However, the consequences of this shock will manifest in subsequent time periods for the other endogenous variables in the system.

VARs are written in the form of Vector Moving Average (VMA) of [17, 18], can be expressed in equation as follows.

$$\begin{bmatrix} Y_t \\ X_{1,t} \\ X_{2,t} \\ X_{3,t} \\ X_{4,t} \end{bmatrix}_{5\times 1} = \begin{bmatrix} A_{11} & A_{12} & A_{13} & A_{14} & A_{15} \\ A_{12} & A_{22} & A_{23} & A_{24} & A_{25} \\ A_{13} & A_{32} & A_{33} & A_{34} & A_{35} \\ A_{14} & A_{42} & A_{43} & A_{44} & A_{45} \\ A_{15} & A_{52} & A_{53} & A_{54} & A_{55} \end{bmatrix}_{5\times 5} \times \begin{bmatrix} Y_{t-1} \\ X_{1,t-1} \\ X_{2,t-1} \\ X_{3,t-1} \\ X_{4,t-1} \end{bmatrix}_{5\times 1} + \begin{bmatrix} u_{1t} \\ u_{2t} \\ u_{3t} \\ u_{4t} \\ u_{5t} \end{bmatrix}_{5\times 1} \quad (3)$$

$A_{ij}$ is Matrix parameters which indicates immediate effects of the change in the tolerance with that variable over a period.

## 2.3 Granger Causality Test

To find the causal relationship, we employ Granger Causality Test, which its equation can be expressed as follows.

$$Y_t = c_1 + \sum_{i=1}^{q} \phi_{1,i} Y_{1,t-i} + \sum_{i=1}^{q} \phi_{2,i} X_{t-i} \tag{4}$$

$$X_t = c_2 + \sum_{i=1}^{r} \delta_{1,i} X_{t-i} + \sum_{i=1}^{r} \delta_{2,i} Y_{t-i}, \tag{5}$$

We utilize this approach with two variables to examine the reciprocal cause-and-effect connection. For example, in (4) when the optimal lag is q, which the best lag is defined by the BIC value. If the values of these coefficients $\varphi_{2,1} = \varphi_{2,2} = \cdots = \varphi_{2,q} = 0$, it suggests that there is no causal relationship between $X$ and $Y$, otherwise, it has causal relationship which $X$ is the cause of $Y$. Conversely, if at least one of the coefficients are non-zero, it implies the presence of a causal relationship, where $X$ can be considered the cause of $Y$. By analyzing the results of the Granger Causality Test, we can determine whether a causal relationship exists between the variables under investigation. We consider the same when $Y$ is independent variable, but $X$ is dependent (5). Suppose that the optimal lag is $r$, if the values of these coefficients $\delta_{2,1} = \delta_{2,2} = \ldots = \delta_{2,r} = 0$, it suggests that there is no causal relationship between $X$ and $Y$.

# 3 Data

This study aims to examine the hypothesis of the Environmental Kuznets Curve in the United States by exploring the relationship between energy consumption, economic activities, and economic growth. To conduct this analysis, we gathered annual time series data spanning from 1970 to 2021, resulting in a total of 52 observations. The data consists of carbon emissions (CO) as the dependent variable, international crude oil prices (OP), total factor productivity (TFP), research and development (RD), and international trade balance (TR) as factors influencing economic growth. The data used for this study was obtained from the Energy Information Administration and the Federal Reserve Bank of St. Louis (FRED). In order to ensure comparability and statistical robustness, all variables were transformed into natural logarithm.

**Table 1** Data description

| | lnCO | lnOP | lnTFP | lnRD | lnTR |
|---|---|---|---|---|---|
| Mean | 0.716300 | 1.355622 | 0.555692 | 0.539439 | 1.322735 |
| Median | 0.718499 | 1.345814 | 0.591059 | 0.573993 | 1.349661 |
| Maximum | 0.788168 | 1.982226 | 0.946943 | 0.991226 | 1.489149 |
| Minimum | 0.637490 | 0.502427 | −0.154902 | −0.397940 | 1.031698 |
| Std. Dev. | 0.044033 | 0.397265 | 0.233349 | 0.247280 | 0.115208 |
| Skewness | −0.015357 | −0.277784 | −1.012484 | −1.011287 | −0.701504 |
| Kurtosis | 1.913444 | 2.479971 | 3.950066 | 5.261373 | 3.105242 |
| Jarque-Bera | 2.560020 | 1.254688 | 10.84010 | 19.94333 | 4.288930 |
| (MBF) | (0.817) | (1.570) | (0.000) | (0.013) | (0.344) |
| ADF test (level) | −0.679838 | 2.356989 | 4.697619*** | −4.285532*** | −2.589831 |
| (MBF) | (2.848) | (1.167) | (0.006) | (0.020) | (0.842) |
| ADF test (1st diff) | −3.730391** | −6.427989*** | −5.594061*** | −9.821553** | −6.913868*** |
| (MBF) | 0.087 | (0.000) | (0.001) | (0.000) | (0.000) |
| Integrated order | I(1) | I(1) | I(0) | I(0) | I(1) |

*Note* "**" and "***" denote moderate and strong support of the null hypothesis, respectively, according to Minimum Bayes Factor (MBF) (see, [19])

The test results in Table 1 shows the descriptive statistics of all variables in this study. In skewness, all variables are negative. For Kurtosis, lnTFP, lnRD, and lnTR have high Kurtosis. According to the skewness and kurtosis for our variables, the Jarque-Bera normality suggests that lnCO, lnOP, lnRD and lnTR are not normally distributed because the Maximum Bayes Factor (MBF) values are higher than 0.10.

In the stationary test conducted for the variables, it is found that not all variables exhibit stationarity at order I(0). In the I(1), the test statistic indicates that all variables are first-order difference stationary. This indicates that our variables are not stationary at the same level. The ARDL model is well-suited for handling such mixed order of integration in the variables. It can effectively capture the relationships between variables with different levels of stationarity, allowing for the analysis of both short-run and long-run dynamics (Table 2).

## 4 Estimation Results

### *4.1 Granger Causality*

According to Table 4, which provides the results of the Granger causality tests, a two-way relationship exists between carbon emissions and total factor productivity. Moreover, it was revealed that oil prices do serve as Granger causes for carbon

**Table 2** Granger causality tests

| | F-Statistic | Prob. | The null hypothesis | Causal relationship |
|---|---|---|---|---|
| *Dependent variable: lnCO* | | | | |
| lnOP | 4.24517 | 0.0448** | Refuse | Yes |
| lnTFP | 4.4366 | 0.0411** | Refuse | Yes |
| *Dependent variable: InTFP* | | | | |
| | F-Statistic | Prob. | The null hypothesis | Causal relationship |
| lnCO | 8.55012 | 0.0053*** | Refuse | Yes |
| lnTR | 5.81505 | 0.0198 | Refuse | Yes |
| *Dependent variable: InRD* | | | | |
| | F-Statistic | Prob. | The null hypothesis | Causal relationship |
| lnTFP | 11.6384 | 0.0013*** | Refuse | Yes |
| *Dependent variable: lnTR* | | | | |
| | F-Statistic | Prob. | The null hypothesis | Causal relationship |
| lnCO | 4.16990 | 0.0367** | Refuse | Yes |

*Note* "**" and "***" denote moderate and strong support of the null hypothesis, respectively, according to Minimum Bayes Factor (MBF) (see, [19])

emissions, while carbon emissions are Granger causes for international trade balance. Additionally, international trade balance acts as a Granger cause for TFP, and TFP serves as a Granger cause for research and development. This insight highlights the significant contribution of the Granger causes variables, which not only enhance explanatory power but also prove valuable in short-term predictions of the dependent variable.

## 4.2 Long and Short Run Analysis

The test results in Table 3, in long-run analysis shows that energy consumption in term of oil price does not affect carbon emissions, but all economic indicator variables affect carbon emissions. It suggests that other Factors, such as research and development and international trade balance, play a more crucial role in influencing carbon emissions in the long term and certain types of R&D activities, particularly in industries heavily reliant on fossil fuels or energy-intensive production processes, such as Automotive and Aerospace, Chemicals and Materials, and vice versa, could potentially lead to increased carbon emissions. International trade balance has the greatest effect in same direction on the emissions as same as research and development. Increasing international trade balance and research and development expenditures will result in more quantity of carbon emissions, but increasing the amount of

**Table 3** Estimation result of the long run ARDL (3, 4, 4, 4, 2) and short run ARDL (3, 4, 4, 4, 2)

| Variables | Coefficient | Standard error | T-statistic | MBF |
|---|---|---|---|---|
| *Long-run analysis* | | | | |
| Constant | 0.011418 | 0.187860 | 0.060782 | 2.799 |
| lnOP | −0.050834 | 0.049700 | −1.022815 | 0.928 |
| lnTFP | −0.214287** | 0.089221 | −2.401743 | 0.070 |
| lnRD | 0.118199** | 0.056831 | 2.079853 | 0.140 |
| lnTR | 0.611956*** | 0.150456 | 4.067349 | 0.000 |
| R-squared:0.96519 | | | | |
| AIC: −5.985608 | | | | |
| BIC: −5.127974 | | | | |
| *Short-run analysis* | | | | |
| lnOP | −0.031569** | 0.014996 | −2.105120 | 0.133 |
| lnTFP | −0.046171*** | 0.012644 | −3.651464 | 0.004 |
| lnRD | 0.029696** | 0.008205 | 3.619350 | 0.0238 |
| lnTR | −0.244516** | 0.090491 | −2.702099 | 0.035 |
| ECT (−1) | 0.303153*** | 0.068368 | 4.434146 | 0.000 |
| R-squared: 0.714878 | | | | |
| AIC: −6.193941 | | | | |
| BIC: −5.531224 | | | | |

*Note* "**" and "***" denote moderate and strong support of the null hypothesis, respectively, according to Minimum Bayes Factor (MBF) (see, [19])

total factor productivity can help to reduction quantity of carbon emissions. In other words, it can be said that although research and development is one of the factors that affect total factor productivity, but the results of the study show that the two variables affect carbon emissions in different directions. Nevertheless, there emerges a plausible scenario where the impact of research and development on carbon emissions continues to decline. It will ultimately be the negative impact on carbon emissions depending on the ratio of R&D expenditure to GDP.

In short-run analysis, all variables have a significant negative impact on carbon dioxide emissions, except R&D, but the obtained parameters values were lower than the parameters obtained in the long term besides international trade balance. Especially, the total factor productivity value that has negative impact on carbon emissions lower than the long term. Therefore, the economic growth of the United States using total factor productivity as an indicator Still inconsistent with the Kuznets environmental curve hypothesis.

According to Congressional Research Service [20], most of the research and development expenditure in proportion of GDP is invested in the business (Technology and Electronics, Automotive and Aerospace, Healthcare and Medical Devices, and vice versa) and education sectors, resulting in technological advances that are expressed through the size of the total factor productivity did not increase in size as expected. In the prospect of the future, heightened government investments in research and development pertaining to environmental domains, including Energy and Renewable Resources, and Emission Reduction Technologies, are anticipated to yield a commensurate reduction in carbon emissions. Therefore, US total factor productivity level is not high enough to cause the inversion point to reduce environmental degradation, while the increasing quantity of carbon emission will result in more research and development expenditure.

## 4.3 *Impulse Response*

Figure 2a shows that changes in total factor productivity affected by carbon emissions have a negative impact in the first cycle and reach to the lowest point in the second cycle. Figure 2b shows that the changes in research and development impacted by carbon emissions peaked negatively in the second cycle and reached to the lowest point in the third cycle. This indicates that carbon emissions have a persistent negative effect on the long-term fluctuations of TFP and R&D. These variables are unable to adjust back to equilibrium in response to changes in carbon emissions.

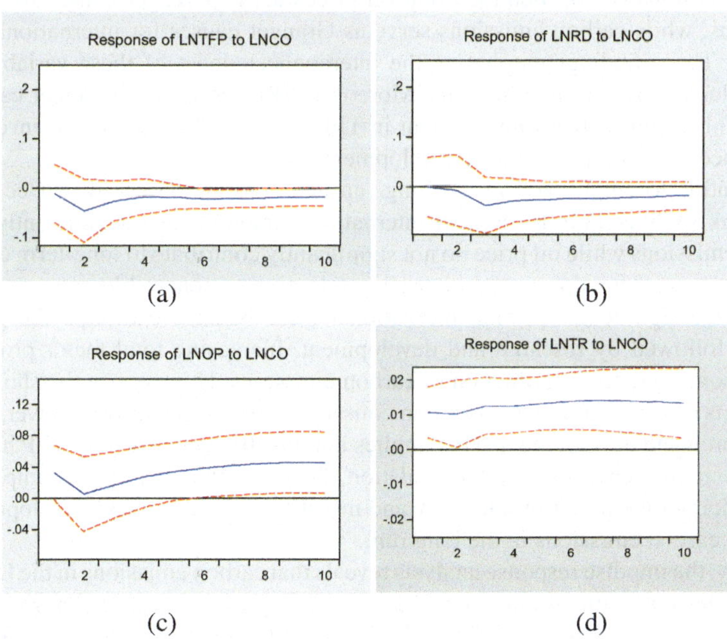


**Fig. 2** Impulse response

Figure 2c shows that changes in oil prices affected by carbon emissions have a negative impact in the first cycle and cannot adjust to stable level, while Fig. 2d shows that the changes in the international trade balance affect carbon emissions and change its course. This suggests that carbon emissions have a modest, long-term positive impact on the volatility of oil prices and international trade.

## 5 Conclusion

Our research delved into the investigation of the environmental Kuznets curve hypothesis and explored the relationships among carbon emissions, international crude oil prices, total factor productivity, research and development, and international trade balance in the United States. By employing the Autoregressive Distributed Lag model, we aimed to understand how these variables impact short-term and long-term carbon emissions to the annual time series data spanning from 1970 to 2021, resulting in a total of 52 observations.

Our findings shed light on several important aspects. Firstly, our analysis of Granger causality uncovered a two-way relationship between carbon emissions and TFP, indicating that there are many economic factors that can impact carbon emissions. Additionally, we identified that oil prices act as Granger causes for carbon emissions, while carbon emissions serve as Granger causes for international trade balance. These findings emphasize the interconnectedness of these variables and their influence on carbon emissions. Moreover, TFP serves as a Granger cause for R&D. This suggests that improvement in TFP might leads to increased investment or advancements in research and development.

Secondly, to emphasize, in the long-run analysis, we observed that economic indicators such as TFP, R&D, and international trade balance significantly affect carbon emissions while oil price do not significantly contribute to long-term changes in carbon emissions due to the increasing adoption of renewable energy sources. Among the significant factors, international trade balance exhibited the greatest impact, followed by research and development. Increasing total factor productivity can contribute to the reduction of carbon emissions. However, in the short term, the effects of these variables on carbon emissions were comparatively lower, except for research and development. This implies that insufficient growth in TFP hampers efforts to reduce environmental degradation. These results highlight the importance of enhancing total factor productivity and investing in research and development to mitigate carbon emissions in the long run.

Lastly, the impulse response analysis reveals that carbon emissions in the US were found to have a minimal effect on TFP and R&D growth because changes in TFP and R&D have persistent negative effects in the long term. Conversely, changes in oil prices and international trade balance have modest, long-term positive impacts on the volatility of carbon emissions.

These findings have implications for policymakers and stakeholders involved in sustainable development, emphasizing the importance of promoting economic growth while simultaneously prioritizing environmental sustainability through measures such as enhancing total factor productivity, fostering research and development, and fostering balanced international trade. By understanding these relationships, we can work towards achieving a more sustainable and environmentally conscious future.

## References

1. Ren, X., Liu, Z., Jin, C., Lin, R.: Oil price uncertainty and enterprise total factor productivity: evidence from China. Int. Rev. Econ. Financ. **83**, 201–218 (2023)
2. Berkman, L.F., Börsch-Supan, A., Avendano, M.: Labor-force participation, policies & practices in an aging America: adaptation essential for a healthy & resilient population. Daedalus **144**(2), 41–54 (2015)
3. Ladu, M.G., Meleddu, M.: Is there any relationship between energy and TFP (total factor productivity)? a panel cointegration approach for Italian regions. Energy **75**, 560–567 (2014)
4. Crippa, M., Guizzardi, D., Banja, M., Solazzo, E., Muntean, M., Schaaf, E., Pagani, F., Monforti-Ferrario, F., Olivier, J., Quadrelli, R., et al.: $CO_2$ emissions of all world countries. JRC Sci. Policy Rep., Eur. Comm., EUR 31182 (2022)
5. Nguyen, T.T., Pham, T.A.T., Tram, H.T.X.: Role of information and communication technologies and innovation in driving carbon emissions and economic growth in selected g-20 countries. J. Environ. Manage. **261**, 110162 (2020)
6. Soytas, U., Sari, R., Ewing, B.T.: Energy consumption, income, and carbon emissions in the united states. Ecol. Econ. **62**(3–4), 482–489 (2007)
7. Sadorsky, P.: Renewable energy consumption, $CO_2$ emissions and oil prices in the G7 countries. Energy Econ. **31**(3), 456–462 (2009)
8. Hammoudeh, S., Nguyen, D.K., Sousa, R.M.: Energy prices and $CO_2$ emission allowance prices: a quantile regression approach. Energy Policy **70**, 201–206 (2014)
9. Heidari, H., Katircioğlu, S.T., Saeidpour, L.: Economic growth, $CO_2$ emissions, and energy consumption in the five ASEAN countries. Int. J. Electr. Power Energy Syst. **64**, 785–791 (2015)
10. Amri, F.: Carbon dioxide emissions, total factor productivity, ICT, trade, financial development, and energy consumption: testing environmental Kuznets curve hypothesis for Tunisia. Environ. Sci. Pollut. Res. **25**(33), 33691–33701 (2018)
11. Mardani, A., Streimikiene, D., Nilashi, M., Aranda, D.A., Loganathan, N., Jusoh, A.: Energy consumption, economic growth, and $CO_2$ emissions in G20 countries: application of adaptive neuro-fuzzy inference system. Energies **11**(10), 2771
12. Miller, S.M., Upadhyay, M.P.: The effects of openness, trade orientation, and human capital on total factor productivity. J. Dev. Econ. **63**(2), 399–423 (2000)
13. Chen, H., Guo, W., Feng, X., Wei, W., Liu, H., Feng, Y., Gong, W.: The impact of low-carbon city pilot policy on the total factor productivity of listed enterprises in China. Resour. Conserv. Recycl. **169**, 105457 (2021)
14. Gao, Y., Zhang, M., Zheng, J.: Accounting and determinants analysis of china's provincial total factor productivity considering carbon emissions. China Econ. Rev. **65**, 101576 (2021)
15. Zou, X.: VECM model analysis of carbon emissions, GDP, and international crude oil prices. Discret. Dyn. Nat. Soc. **2018**, 1–11 (2018)
16. Balashova, S.: Total factor productivity growth and oil price volatility. Int. J. Trade Glob. Mark. **14**(6), 548 (2021)
17. Lütkepohl, H.: Impulse response function (2010)

18. Simbachawene, M.: Taxes, financing and spending synchronization in Tanzania: ARDL bound test and impulse response function. Int. J. Financ. Account. **7**(3), 67–75 (2018)
19. Maneejuk, P., Yamaka, W.: Significance test for linear regression: how to test without P-values? J. Appl. Stat. **48**(5), 827–845 (2021)
20. Sargent Jr, J.F., Lipiec, E., Croft, G.K., Morgan, D., Gallo, M.E., Sekar, K., Harris, L.A., Yen, J.H.: Federal research and development (R&D) funding: Fy2022. CRS Report, vol. R46869 (2022)

# Do Green Taxes and Renewables Energy Matter for Environmental Quality in OECD Countries? Evidence from Monte Carlo Simulations

**Bui Hoang Ngoc, Nguyen Ngoc Thach, and Nguyen Huynh Mai Tram**

**Abstract** The governments of the Organization for Economic Cooperation and Development (OECD) countries have issued several strict policies to improve environmental quality. Thereby, analysis of factors that affect environmental matters in OECD members might provide helpful lessons for emerging countries. The study aims to investigate the role of tax and renewable energy consumption on environmental issues prevention in OECD countries from 1994 to 2018. By applying Monte Carlo simulations, the findings reveal that renewable energy consumption is a good policy, while tax regulation only has a medium negative impact on environmental quality. The study also finds that regulatory quality does not support environmental quality improvement in these countries. Based on findings, the study suggests some helpful implications for administrators in planning ecological protection policies and encouraging renewable energy consumption.

**Keywords** Environmental pollution · Renewable energy · Green taxes · Bayesian inference · OECD countries

B. H. Ngoc
Ho Chi Minh City University of Industry and Trade, 140 Le Trong Tan Street, Tan Phu District, Ho Chi Minh City 70000, Vietnam
e-mail: ngocbh@huit.edu.vn

N. N. Thach
Banking University of Ho Chi Minh City, 39 Ham Nghi Street, District 1, Ho Chi Minh City 70000, Vietnam
e-mail: thachnn@buh.edu.vn

N. H. M. Tram (✉)
Ho Chi Minh City Open University, 97 Vo Van Tan Street, District 3, Ho Chi Minh City 70000, Vietnam
e-mail: tramnhm.20ae@ou.edu.vn

 
V. Kreinovich et al. (eds.), *Applications of Optimal Transport to Economics and Related Topics*, Studies in Systems, Decision and Control 556,
https://doi.org/10.1007/978-3-031-67770-0_17

# 1 Introduction

In contemporary times, enhancing energy efficiency and addressing environmental challenges via the implementation of environmental regulations and taxation are regarded as key factors that propel climate change policies forward [23, 39]. The relationship between green taxes, renewable energy, and environmental sustainability has garnered significant attention in recent years [26]. As concerns over climate change and environmental degradation continue to mount, policymakers and researchers have sought effective strategies to promote sustainable development and mitigate harmful environmental impacts.

However, the findings on the relationship between green taxes, renewable energy, and environmental sustainability in the OECD countries have been varied and sometimes contradictory. Some studies have highlighted the positive impact of green taxes and renewable energy on environmental sustainability, while others have raised concerns about their effectiveness or identified specific regional variations. Green taxes, also known as environmental taxes or eco-taxes, are fiscal measures imposed on activities or products that have negative environmental effects [31]. These taxes are intended to internalize the environmental costs associated with such activities and incentivize more sustainable practices [19, 33]. On the one hand, green taxes can contribute to achieving cleaner environmental objectives, but on the other hand, they can sometimes result in unintended negative environmental outcomes [14, 16]. As the global energy demand continues to rise, coupled with the alarming consequences of climate change, the need for clean, renewable energy sources has never been more pressing. Renewable energy, derived from sources such as solar, wind, hydro, and geothermal power, offers the potential to reduce greenhouse gas emissions, mitigate environmental degradation, and promote a more sustainable energy system [13, 47].

In the OECD context, where economic growth and industrialization have often come at the expense of the environment, the implementation of green taxes and the promotion of renewable energy have emerged as crucial policy measures. These measures aim to strike a balance between economic development and environmental protection, ultimately fostering long-term sustainability. The effectiveness of green taxes and renewable energy in achieving environmental sustainability in OECD countries is a topic of interest and research. Therefore, this research aims to contribute to the existing body of knowledge by examining the complex relationship between green taxes, renewable energy, and environmental sustainability within the OECD context between 1994 and 2018. By analyzing empirical data and employing a nonparametric Bayesian approach, we elucidate a few of the following contributions:

(i) The available studies have usually used carbon dioxide emissions as proxied environmental quality. However, Dogan et al. [12] noted that it is not a comprehensive indicator because it has ignored several perspectives of economic activities on the environment, such as biodiversity and forests required to absorb $CO_2$ emissions. Nowadays, OECD countries have effectiveness controlled the $CO_2$ emissions. Thus, this study used ecological footprint index is a proxy for environmental quality since it is is a comprehensive measure of pressure on the ecosystem due to various economic activities in terms of six categories of land: cropland, fishing ground, grazing land, forests (timber and fuelwood), forests required to absorb $CO_2$ emissions due (carbon footprint), and build-up land (infrastructure footprint).

(ii) A further contribution of this study is the employment of an advanced econometric technique. The frequentist inference have provided unknown estimated coefficients, but fixed. In the case of coefficient is not significant, it is difficult to believe on suggested policy implications. In this study, we adopt the Bayesian inference with the integrated Markov chain Monte-Carlo (MCMC) sampler to provide probabilistic interpretations of model uncertainty and varying effects of key macroeconomic indicators on environmental quality in OECD members. That means we further shed light on the importance of each investigated variable in the model in view of probability, and thus policymakers can more effectively allocate available resources to improve environmental quality.

The structure of the remaining sections is presented as follows: Sect. 2 is a literature review of the topic, Sect. 3 exhibits the data and methodology, Sect. 4 discusses the empirical results, and the last section is the conclusion and some policy implications.

## 2 Literature Review

### *2.1 Green Taxes and Environmental Sustainability*

The relationship between green taxes and environmental sustainability refers to how the imposition of taxes on environmentally harmful activities or products can contribute to the long-term preservation and well-being of the natural environment. Green taxes are designed to internalize the costs of environmental externalities by making polluting activities or products more expensive [10, 31, 32]. The primary objective of these taxes is to incentivize individuals, businesses, and industries to reduce their negative impact on the environment and promote more sustainable practices. However, previous studies on the nexus between green tax and environmental sustainability are still controversial.

On the one hand, environmental tax is seen as an effective market-based mitigation tool to promote green growth. Green taxes can alter the behavior of individuals and businesses by making environmentally harmful activities more costly [28, 29]. By increasing the price of polluting goods or activities, taxes can encourage a shift towards cleaner and more sustainable alternatives [27]. This can lead to reduced resource consumption, lower emissions, and more sustainable use of natural resources. Moreover, the revenue generated from green taxes can be used to support environmental initiatives, such as conservation projects, renewable energy development, or environmental research [1, 52]. By allocating funds towards these initiatives, environmental taxes can help promote sustainable practices and technologies [20]. Green taxes can be integrated into broader policy frameworks to promote environmental sustainability. They can complement other policy measures, such as regulations, subsidies, or voluntary programs, to create a comprehensive approach towards sustainable development. [30] evaluated the effects of green taxes on pollution in European countries between 1995 and 2006. By employing the two-step GMM approach, the study showed that the implementation of these taxes had a detrimental impact on reducing carbon emissions ($CO_2$). For OECD countries, [16] emphasized that a 1% rise in per capita revenue from environmental taxes leads to a 0.03% decrease in $CO_2$. [15] examined the impact of green taxes on fostering environmental sustainability in G7 countries over the period 1991–2017. The research used the Cross-Sectionally Augmented Auto-regressive Distributive lag (CS-ARDL) model and revealed that environmental taxes have a significant negative effect on $CO_2$ emissions. Furthermore, [18] utilized an innovative dynamic simulated ARDL (DYARDL) model to illustrate the positive impact of green taxes and renewable energy on improving the environmental quality in Italy. These measures were found to effectively reduce the ecological footprint, thus indicating their contribution to environmental sustainability. And the emphasis that environmental-related taxes have a substantial negative impact on the long-term ecological deficit is also found in [35] (29 OECD countries), [48] (top-seven green economies), [51] (China).

On the other hand, [24] employed the difference-in-difference approach to assess the impact of a carbon tax on $CO_2$ emissions in five countries in northern Europe. The results of their analysis revealed that the implementation of an environmental tax led to a noteworthy decrease in emissions only in Finland, while no significant effect was observed in the remaining countries (Denmark, Sweden and Netherlands, Norway). The effects of the energy tax were analyzed in [42] using China as a case study. The results showed that the energy tax had different impacts on different components of the ecological footprint. For the actual land use component, the energy tax increased the export of the ecological footprint but decreased its import. The net exports of China decreased during the period from 2015 to 2020 under both low and high tax intensities. For the virtual land use component, the energy tax led to a decrease in both export and import. The average net import of this component would increase under both low and high tax intensities. Finally, for the total ecological footprint, the energy tax reduced both export and import, resulting in a decrease in the net import under both low and high tax intensities. Telatar and Birinci [43] conducted a non-linear investigation in Turkey to investigate the impact of an environmental tax on the ecological footprint and $CO_2$ emissions between the years 1994 and 2019. The estimation results indicated that the environmental taxes do not have any long-term effects on the ecological footprint and $CO_2$ emissions. Consequently, it can be inferred that the environmental taxes implemented in Turkey do not contribute to the prevention of environmental degradation. Meanwhile, [50] found out the effectiveness of environmental tax and renewable energy in reducing $CO_2$ emissions across 18 countries in Latin America and the Caribbean (LAC) during the period of 1994–2018. The findings presented that the impact of environmental tax and renewable energy on $CO_2$ emissions varies across countries, with a significant negative effect observed in higher-emitting countries but an insignificant effect in lower-emitting countries. And this argument also received consensus from [40], given the context of top renewable energy countries.

## 2.2 Renewable Energy and Environmental Sustainability

Numerous studies have examined the relationship between renewable energy consumption and environmental sustainability; however, the findings have been inconclusive and contradictory [3, 7]. Renewable energy sources such as solar, wind, hydropower, and biomass are generally considered cleaner and have lower greenhouse gas emissions compared to fossil fuels [6, 36]. We can significantly mitigate climate change and reduce air pollution by shifting to renewable energy. Renewable energy also diversifies the energy mix and reduces dependence on non-renewable resources like coal, oil, and natural gas [4]. This reduces the environmental risks associated with fossil fuel extraction, such as oil spills and destructive mining practices. Sarkodie et al. [38] displayed that the immediate rise in renewable energy, environmental sustainability, and income level leads to a decreasing impact on emissions and environmental degradation. Integrating renewable energy sources into the energy mix results in reduced emissions by 0.38% and environmental degradation by 0.21% in China. In contrast, the utilization of fossil fuel energy contributes to a 1.93% increase in environmental degradation and a 1.58% rise in emissions.

Unlike finite fossil fuel reserves, renewable energy sources are abundant and virtually inexhaustible [8, 37]. They do not deplete natural resources, ensuring long-term sustainability and minimizing environmental degradation associated with resource extraction. Usman et al. [49] highlighted that renewable energy exhibit a positive association with the ecological footprint in the short run, based on quarterly data from 1985:Q1 to 2014:Q4 in the United States. However, in the long term, renewable energy was found to harm the ecological footprint. Kirikkaleli and Adebayo [21] sought to investigate the long-term relationship and causal impact of renewable energy consumption on environmental sustainability, taking into account technological innovation and economic growth on a global scale. To achieve this objective, various statistical techniques, including fully modified OLS (FMOLS), dynamic OLS (DOLS), and canonical cointegrating regression (CCR), were utilized. The findings of this study demonstrated that, in the long run, there is a significant positive association between global renewable energy consumption and environmental sustainability. Abbasi et al. [2] investigated the effects of fossil fuel energy, renewable energy, and GDP on $CO_2$ emissions in China from 1980 to 2018, employing dynamic ARDL simulations and Frequency Domain Causality (FDC) models. The findings exhibited that although renewable energy usage initially has a negative impact on $CO_2$ emissions in the short term, it plays a crucial role in achieving long-term sustainable environmental objectives by discouraging the use of fossil fuels in the energy mix.

In contrast, other studies reported renewable energy consumption to have a harmful bearing on environmental quality. For instance, [5] showed that renewable energy consumption exhibits a substantial and adverse influence on pollution levels in Central and Eastern Europe, Western Europe, East Asia and the Pacific, South Asia, and the Americas. Nonetheless, the findings indicated that renewable energy consumption does not have a significant impact on pollution levels in the Middle East and North Africa, as well as Sub-Saharan Africa over the period 1980 to 2010. Ben Jebli and Ben Youssef [9] confirmed negative relationship between renewable energy and environmental sustainability in five North African countries over the period 1980 to 2011. Meanwhile, [25] in China and [34] in Turkey opined that renewable energy consumption has no impact on environmental degradation.

## 3 Methodology

Since the end of the XX century, a sharp rise in the utilization of Bayesian statistical methods has taken place in many areas (e.g., [17, 44, 45, 46), due to enormous developments in computer software packages and some superiorities of the Bayesian paradigm over the traditional frequentist statistics. Bayesian methods are able to solve models that are too complex or too demanding for frequentist estimators. A Bayesian approach can interpret estimation results in a straightforward way. A simple probability rule, Bayes' law is applicable to all parametric models. Let us mention one more superior property of Bayesian methods over frequentist estimation: the former does not rely on asymptotic theory. Prior knowledge combined with data produces a posterior distribution that is representative for a distribution function regardless of data sample size. Prior distributions incorporate background information about parameters prior to seeing the data. In the case of information unavailability, non-informative priors are often specified. In this case, nevertheless, the Bayesian framework will lose its major advantages. Smid et al. [41] emphasized three types of Bayesian inference: naïve, thoughtful, and data-driven, arguing that research using naïve or data-driven priors is justifiable under some circumstances; they are considered as thoughtful priors.

In case information about model parameters lacks, [11] recommended data-driven priors. [22] concluded that Bayesian estimation with the use of data-driven priors outperforms frequentist and naïve Bayesian estimations. Our Bayesian model is specified as follows:

$$EF_{it} = \beta_0 + \beta_1.TAX_{it} + \beta_2.REN_{it} + \beta_3.REG_{it} + u_{it} \quad (1)$$

where, $u$ is the errors, $i$ is the countries, and $t$ is time. A panel dataset was constructed to cover 29 OECD countries for the period of 1994–2018. The dependent variable (EF) is proxied by environmental quality, and collected from Footprint Global Networks. The tax variable (TAX) presents environmental tax regulation of OECD countries, while the regulatory quality (REG) variable denotes institutional quality, and the renewable variable (REN) is proxied by renewable energy consumption. Three variables are collected from the databases of OECD statistics at https://stats.oecd.org/.

# 4 Empirical Results and Discussion

## *4.1 Empirical Results*

To specify a data-dependent prior of the parameter of interest, renewable, we run an OLS regression. The coefficient estimate of variable renewable derived from the OLS regression will serve as the prior mean of the normal distribution for variable renewable in Bayesian estimation. To examine how prior specifications influence the posterior conclusion, we conduct a sensitivity analysis. To this end, we vary the prior variance of the normal distribution in a range from 10,000 to 1 (Table 1). The results reported in Tables 1 and 3 show that, first, the posterior conclusions are robust to various prior specifications as the posterior mean estimates of all the predictors, in particular, the parameter renewable, do not change under the different prior variances; second, the narrower the prior variance (that is the more informative the prior distribution for the slope coefficient of variable renewable), the better the Bayesian simulations perform. Based on the estimates of the log marginal likelihood, the model with the prior variance of 1 performs the best. So, we can use this model summary for further analysis (Table 2).

Prior to inferences, we should check MCMC convergence for all the parameters in the selected model. There are several convergence diagnostics. A Gelman-Rubin diagnostic is used in the study. The results of the test are demonstrated in Table 3, where the convergence rule is satisfied for all the parameters. Hence, there is no anomalies concerning MCMC convergence.

As Table 3 shows, the 95% credibility intervals for variables renewable and regulatory quality do not contain zero, meaning that two these variables exert strong effects

**Table 1** Variance specifications and log marginal—likelihood with the use of a data-driven prior in Bayesian simulation

| Prior variance specification | 10,000 | 1000 | 100 | 50 |
|---|---|---|---|---|
| Log marginal—likelihood | −1236.4 | −1235.2 | −1233.9 | −1233.7 |

| 30 | 10 | 5 | 3 | 1 |
|---|---|---|---|---|
| −1233.4 | −1232.8 | −1232.5 | −1232.3 | −1231.7 |

**Table 2** Gelman-Rubin convergence test

| Parameters | Rc |
|---|---|
| Dependent variable: EF | |
| Tax | 0.999975 |
| Renewable | 1.000352 |
| Regulatory quality | 1.000118 |
| Intercept | 1.000059 |
| Variance | 0.999967 |

**Table 3** Simulation results with the N (−0.04, 1) prior for the mean coefficient of variable renewable

| Parameters | Mean | Std. dev. | Monte Carlo SE | Median | Credibility interval |
|---|---|---|---|---|---|
| Dependent variable: EF | | | | | |
| Tax | −0.066 | 0.097 | 0.001 | −0.067 | [−0.257, 0.124] |
| Renewable | −0.036 | 0.008 | 0.000 | −0.036 | [−0.051, −0.021] |
| Regulatory quality | 3.086 | 0.179 | 0.002 | 3.087 | [2.738, 3.438] |
| Intercept | 2.391 | 0.324 | 0.003 | 2.387 | [1.755, 3.033] |
| Variance | 3.926 | 0.235 | 0.002 | 3.916 | [3.494, 4.412] |

on response EF; variable renewable is negatively related to the response whereas variable regulatory quality is positively associated with it. Besides, variable tax has a weaker negative effect on the response with a probability of over 70%.

## 4.2 *Discussion*

The finding of study reveal that adopting renewable energy consumption could serve as a viable pathway to address environmental degradation in OECD countries, where governance is not characterized by dictatorial regimes. This conclusion is in line with [2, 9, 38]. Some reasons are given to explain this finding as: (i) Most renewable energy sources, such as solar, wind, hydro, and geothermal, produce little to no greenhouse gas emissions during power generation. Unlike fossil fuels, which release large amounts of carbon dioxide and other pollutants, renewables help mitigate climate change and reduce air pollution, improving overall air quality [4], (ii) Renewable energy installations, such as solar farms and wind turbines, can be built on existing infrastructure or marginal lands, minimizing the need for deforestation and habitat destruction. This allows ecosystems to remain intact and supports wildlife conservation efforts; (iii) Renewable energy sources are considered sustainable as they are naturally replenished. Unlike fossil fuels, which are finite resources, renewable energy can provide a reliable and continuous energy supply without depleting the Earth's resources [21].

The study also confirms that taxes regulation has a negative impact on environmental degradation in OECD countries, implying that green taxes provide powerful economic incentives to reduce environmental harm, promote sustainable practices, and support the transition to a more environmentally friendly society. This conclusion is similar to [28, 29]. According to [10], some reasons why green taxes can help improve environmental quality as: (i) Green taxes impose charges on activities that have a negative environmental impact. By making polluting or resource-intensive activities more expensive, individuals and businesses are incentivized to adopt cleaner and more sustainable practices to reduce their tax burden. This can lead to a shift towards greener technologies and behaviors; (ii) Taxes on the extraction or use of natural resources, such as water or minerals, can incentivize efficient use and conservation; (iii) Green taxes on landfill waste or non-recyclable materials encourage waste reduction and recycling efforts. Such taxes create an economic incentive to minimize waste generation and promote the development of more sustainable waste management practices. Additionally, revenue generated from green taxes can be used to fund environmental projects and initiatives. This includes investments in renewable energy infrastructure, conservation efforts, and research and development of sustainable technologies.

## 5 Conclusion and Policy Implications

The study performed a Bayesian analysis of the effects of green taxes, renewable energy consumption, and regulatory quality on environmental quality measured by the ecological footprint index for 29 OECD countries during the period 1994–2018. The use of data-driven priors for the parameters yielded meaningful results. Importantly, a sensitivity analysis indicated that the more information is included in the model, the better the model performs. Moreover, the posterior conclusion is valid to different prior specifications. The findings of study are summarized:

(i) An increase in renewable energy consumption leads to decreased environmental degradation;
(ii) Taxes regulation has a medium negative impact on environmental quality;
(iii) Regulatory quality exerts strong effects on environmental matters. However, it does not support environmental quality improvement in OECD countries.

Based on the findings, some policies are suggested for governments of OECD countries:

(i) Firstly, governments should establish a mandate that requires enterprises to use a certain percentage of their energy from renewable sources. This policy encourages the growth of renewable energy generation and ensures a steady increase in renewable energy consumption over time.
(ii) Secondly, credit policies for individuals and businesses that invest in renewable energy systems, such as solar panels or wind turbines, should be considered. These incentives reduce operation costs. Thereby, it encourages enterprises' greater adoption of renewable technologies.
(iii) Thirdly, governments should establish a carbon pricing mechanism, such as a carbon tax or cap-and-trade system, to put a price on carbon emissions. This incentivizes businesses to reduce their carbon footprint and promotes the transition to renewable energy sources.

(iv) Fourthly, governments should invest in public awareness campaigns and educational programs to inform citizens about the benefits of renewable energy consumption and how they can prevent environmental degradation.

Finally, even though significant empirical evidence has been explored in this study, we agree that it still has some limits. In general, many macroeconomic variables can affect environmental pollution. Thus, further related economic variables should be considered, such as economic growth, globalization, foreign direct investment, and trade policies. Additionally, future research might apply our model to other developing countries to help policy-makers systematically understand the role of tax regulation and renewable energy consumption in environmental matters. We also suggest that several novel econometric techniques, such as the spatial regression approach and method of moments quantile regression, should be applied to support policy-makers with comprehensive policy guidelines.

## References

1. Abbas, J., Wang, L., Ben Belgacem, S., Pawar, P.S., Najam, H., Abbas, J.: Investment in renewable energy and electricity output: Role of green finance, environmental tax, and geopolitical risk: Empirical evidence from China. Energy **269**, 126683 (2023). https://doi.org/10.1016/j.energy.2023.126683
2. Abbasi, K.R., Shahbaz, M., Zhang, J., Irfan, M., Alvarado, R.: Analyze the environmental sustainability factors of China: The role of fossil fuel energy and renewable energy. Renewable Energy **187**, 390–402 (2022). https://doi.org/10.1016/j.renene.2022.01.066
3. Adams, S., Nsiah, C.: Reducing carbon dioxide emissions; Does renewable energy matter? Sci. Total. Environ. **693**, 133288 (2019). https://doi.org/10.1016/j.scitotenv.2019.07.094
4. Ajide, K.B., Mesagan, E.P.: Heterogeneous analysis of pollution abatement via renewable and non-renewable energy: lessons from investment in G20 nations. Environ. Sci. Pollut. Res. **29**(24), 36533–36546 (2022). https://doi.org/10.1007/s11356-022-18771-5
5. Al-Mulali, U., Ozturk, I., Solarin, S.A.: Investigating the environmental Kuznets curve hypothesis in seven regions: The role of renewable energy. Ecol. Ind. **67**, 267–282 (2016). https://doi.org/10.1016/j.ecolind.2016.02.059
6. Amponsah, N.Y., Troldborg, M., Kington, B., Aalders, I., Hough, R.L.: Greenhouse gas emissions from renewable energy sources: A review of lifecycle considerations. Renew. Sustain. Energy Rev. **39**, 461–475 (2014). https://doi.org/10.1016/j.rser.2014.07.087
7. Apergis, N.: Renewable Energy and its Finance as a Solution to the Environmental Degradation. In: Özcan, B., Öztürk, I. (eds.) Environmental Kuznets Curve (EKC), pp. 55–63. Academic Press (2019)
8. Ashraf Chaudhry, M., Raza, R., Hayat, S.A.: Renewable energy technologies in Pakistan: Prospects and challenges. Renew. Sustain. Energy Rev. **13**(6), 1657–1662 (2009). https://doi.org/10.1016/j.rser.2008.09.025
9. Ben Jebli, M., Ben Youssef, S.: The role of renewable energy and agriculture in reducing $CO_2$ emissions: Evidence for North Africa countries. Ecol. Ind. **74**, 295–301 (2017). https://doi.org/10.1016/j.ecolind.2016.11.032
10. Cheng, B., Qiu, B., Chan, K.C., Zhang, H.: Does a green tax impact a heavy-polluting firm's green investments? Appl. Econ. **54**(2), 189–205 (2022). https://doi.org/10.1080/00036846.2021.1963663
11. Darnieder, W.F. (2011). Bayesian methods for data-dependent priors (Doctoral dissertation). The Ohio State University.

12. Dogan, E., Ulucak, R., Kocak, E., Isik, C.: The use of ecological footprint in estimating the environmental Kuznets curve hypothesis for BRICST by considering cross-section dependence and heterogeneity. Sci. Total Environ. **723**, 138063 (2020)
13. Elum, Z.A., Momodu, A.S.: Climate change mitigation and renewable energy for sustainable development in Nigeria: A discourse approach. Renew. Sustain. Energy Rev. **76**, 72–80 (2017). https://doi.org/10.1016/j.rser.2017.03.040
14. Guo, B., Wang, Y., Feng, Y., Liang, C., Tang, L., Yao, X., Hu, F.: The effects of environmental tax reform on urban air pollution: A quasi-natural experiment based on the Environmental Protection Tax Law. Front. Public Health **10**, 967524 (2022). https://doi.org/10.3389/fpubh.2022.967524
15. Hao, L.-N., Umar, M., Khan, Z., Ali, W.: Green growth and low carbon emission in G7 countries: How critical the network of environmental taxes, renewable energy and human capital is? Sci. Total. Environ. **752**, 141853 (2021). https://doi.org/10.1016/j.scitotenv.2020.141853
16. Hashmi, R., Alam, K.: Dynamic relationship among environmental regulation, innovation, $CO_2$ emissions, population, and economic growth in OECD countries: A panel investigation. J. Clean. Prod. **231**, 1100–1109 (2019). https://doi.org/10.1016/j.jclepro.2019.05.325
17. Hung T. Nguyen, Trung N.D., & Thach N.N. (2019). Beyond Traditional Probabilistic Methods in Econometrics. In: Kreinovich V., Thach N., Trung N., Van Thanh D. (eds), Beyond Traditional Probabilistic Methods in Economics. ECONVN 2019. Studies in Computational Intelligence, vol 809. Springer, Cham. https://doi.org/10.1007/978-3-030-04200-4_1
18. Javed, A., Rapposelli, A., Khan, F., Javed, A.: The impact of green technology innovation, environmental taxes, and renewable energy consumption on ecological footprint in Italy: Fresh evidence from novel dynamic ARDL simulations. Technol. Forecast. Soc. Chang. **191**, 122534 (2023). https://doi.org/10.1016/j.techfore.2023.122534
19. Jiang, X., Li, G., & Fan, X. (2023). Environmental Protection Fee-to-Tax and Corporate Environmental Social Responsibility: A Test Based on Corporate Life Cycle Theory. Sustainability, 15(3), 2128. Retrieved from https://www.mdpi.com/2071-1050/15/3/2128
20. Karmaker, S.C., Hosan, S., Chapman, A.J., Saha, B.B.: The role of environmental taxes on technological innovation. Energy **232**, 121052 (2021). https://doi.org/10.1016/j.energy.2021.121052
21. Kirikkaleli, D., Adebayo, T.S.: Do renewable energy consumption and financial development matter for environmental sustainability? New global evidence. Sustainable Development **29**(4), 583–594 (2021). https://doi.org/10.1002/sd.2159
22. Lee, S.Y., Song, X.Y.: Evaluation of the Bayesian and maximum likelihood approaches in analyzing structural equation models with small sample sizes. Multivar. Behav. Res. **39**(4), 653–686 (2004)
23. Li, J., Wang, X.: Energy and climate policy in China's twelfth five-year plan: A paradigm shift. Energy Policy **41**, 519–528 (2012). https://doi.org/10.1016/j.enpol.2011.11.012
24. Lin, B., Li, X.: The effect of carbon tax on per capita $CO_2$ emissions. Energy Policy **39**(9), 5137–5146 (2011). https://doi.org/10.1016/j.enpol.2011.05.050
25. Lin, B., Moubarak, M.: Renewable energy consumption – Economic growth nexus for China. Renew. Sustain. Energy Rev. **40**, 111–117 (2014). https://doi.org/10.1016/j.rser.2014.07.128
26. Lin, Z., Alvarez-Otero, S., Belgacem, S. B., & Fu, Q. (2022). Role of sustainable finance, geopolitical risk and economic growth in renewable energy investment: Empirical evidence from China. Geological Journal, n/a(n/a), 1–9. https://doi.org/10.1002/gj.4654
27. Liu, H., Zafar, M. W., Sinha, A., & Khan, I. (2023). The path to sustainable environment: Do environmental taxes and governance matter? Sustainable Development, n/a(n/a). https://doi.org/10.1002/sd.2505
28. Maitre-Ekern, E., Dalhammar, C.: Towards a hierarchy of consumption behaviour in the circular economy. Maastricht Journal of European and Comparative Law **26**(3), 394–420 (2019). https://doi.org/10.1177/1023263X19840943
29. McLaughlin, C., Elamer, A.A., Glen, T., AlHares, A., Gaber, H.R.: Accounting society's acceptability of carbon taxes: Expectations and reality. Energy Policy **131**, 302–311 (2019). https://doi.org/10.1016/j.enpol.2019.05.008

30. Morley, B.: Empirical evidence on the effectiveness of environmental taxes. Appl. Econ. Lett. **19**(18), 1817–1820 (2012). https://doi.org/10.1080/13504851.2011.650324
31. Mpofu, F. Y. (2022). Green Taxes in Africa: Opportunities and Challenges for Environmental Protection, Sustainability, and the Attainment of Sustainable Development Goals. Sustainability, 14(16), 10239. Retrieved from https://www.mdpi.com/2071-1050/14/16/10239
32. Nguyen, T.L.T., Laratte, B., Guillaume, B., Hua, A.: Quantifying environmental externalities with a view to internalizing them in the price of products, using different monetization models. Resour. Conserv. Recycl. **109**, 13–23 (2016). https://doi.org/10.1016/j.resconrec.2016.01.018
33. Osório, A., Zhang, M.: Using contests to design emission tax mechanisms. Sustainable Production and Consumption **31**, 762–776 (2022). https://doi.org/10.1016/j.spc.2022.03.033
34. Pata, U.K.: Renewable energy consumption, urbanization, financial development, income and $CO_2$ emissions in Turkey: Testing EKC hypothesis with structural breaks. J. Clean. Prod. **187**, 770–779 (2018). https://doi.org/10.1016/j.jclepro.2018.03.236
35. Rafique, M.Z., Fareed, Z., Ferraz, D., Ikram, M., Huang, S.: Exploring the heterogenous impacts of environmental taxes on environmental footprints: An empirical assessment from developed economies. Energy **238**, 121753 (2022). https://doi.org/10.1016/j.energy.2021.121753
36. Rahman, A., Farrok, O., Haque, M.M.: Environmental impact of renewable energy source based electrical power plants: Solar, wind, hydroelectric, biomass, geothermal, tidal, ocean, and osmotic. Renew. Sustain. Energy Rev. **161**, 112279 (2022). https://doi.org/10.1016/j.rser.2022.112279
37. Rahman, M.A., Kim, J.-H., Hossain, S.: Recent advances of energy storage technologies for grid: A comprehensive review. Energy Storage **4**(6), e322 (2022). https://doi.org/10.1002/est2.322
38. Sarkodie, S.A., Adams, S., Owusu, P.A., Leirvik, T., Ozturk, I.: Mitigating degradation and emissions in China: The role of environmental sustainability, human capital and renewable energy. Sci. Total. Environ. **719**, 137530 (2020). https://doi.org/10.1016/j.scitotenv.2020.137530
39. Shahzad, U.: Environmental taxes, energy consumption, and environmental quality: Theoretical survey with policy implications. Environ. Sci. Pollut. Res. **27**(20), 24848–24862 (2020). https://doi.org/10.1007/s11356-020-08349-4
40. Shayanmehr, S., Radmehr, R., Ali, E.B., Ofori, E.K., Adebayo, T.S., Gyamfi, B.A.: How do environmental tax and renewable energy contribute to ecological sustainability? New evidence from top renewable energy countries. Int. J. Sustain. Dev. World Ecol. 1–21 (2023). https://doi.org/10.1080/13504509.2023.2186961
41. Smid, S.C., McNeish, D., Miočević, M., van de Schoot, R.: Bayesian versus frequentist estimation for structural equation models in small sample contexts: a systematic review. Struct. Equ. Model.: Multi. J. **27**(1), 131–161 (2019). https://doi.org/10.1080/10705511.2019.1577140
42. Sun, Y., Mao, X., Liu, G., Yin, X., Zhao, Y.: Modelling the effects of energy taxes on ecological footprint transfers in China's foreign trade. Ecol. Model. **431**, 109200 (2020). https://doi.org/10.1016/j.ecolmodel.2020.109200
43. Telatar, O.M., Birinci, N.: The effects of environmental tax on Ecological Footprint and Carbon dioxide emissions: a nonlinear cointegration analysis on Turkey. Environ. Sci. Pollut. Res. **29**(29), 44335–44347 (2022). https://doi.org/10.1007/s11356-022-18740-y
44. Thach N.N., Ha D.T., Trung N.D. & Kreinovich V. (2022a). Prediction and Causality in Econometrics and Related Topics. Cham: Springer, vol. 983. https://doi.org/10.1007/978-3-030-77094-5
45. Thach N.N., Kreinovich V., Ha D.T. & Trung N.D. (2022b). Financial Econometrics: Bayesian Analysis, Quantum Uncertainty, and Related Topics. Cham: Springer, vol. 427. https://doi.org/10.1007/978-3-030-98689-6
46. Thu, P.X., Thach, N.N., Van Diep, N., Tram, N.H.M.: Globalization, economic freedom, and the shadow economy in ASEAN countries: insights from bayesian modeling. In: Partial Identification in Econometrics and Related Topics, pp. 363–382. Springer Nature Switzerland, Cham (2024)

47. Tripathi, L., Mishra, A.K., Dubey, A.K., Tripathi, C.B., Baredar, P.: Renewable energy: An overview on its contribution in current energy scenario of India. Renew. Sustain. Energy Rev. **60**, 226–233 (2016). https://doi.org/10.1016/j.rser.2016.01.047
48. Ullah, S., Luo, R., Adebayo, T.S., Kartal, M.T.: Dynamics between environmental taxes and ecological sustainability: Evidence from top-seven green economies by novel quantile approaches. Sustain. Dev. **31**(2), 825–839 (2023). https://doi.org/10.1002/sd.2423
49. Usman, O., Akadiri, S.S., Adeshola, I.: Role of renewable energy and globalization on ecological footprint in the USA: implications for environmental sustainability. Environ. Sci. Pollut. Res. **27**(24), 30681–30693 (2020). https://doi.org/10.1007/s11356-020-09170-9
50. Wolde-Rufael, Y., Mulat-Weldemeskel, E.: The moderating role of environmental tax and renewable energy in $CO_2$ emissions in Latin America and Caribbean countries: Evidence from method of moments quantile regression. Environmental Challenges **6**, 100412 (2022). https://doi.org/10.1016/j.envc.2021.100412
51. Xiong, Z., Li, H.: Ecological deficit tax: A tax design and simulation of compensation for ecosystem service value based on ecological footprint in China. J. Clean. Prod. **230**, 1128–1137 (2019). https://doi.org/10.1016/j.jclepro.2019.05.172
52. Zhang, Q., Zhang, Y., Liao, Q., Guo, X.: Effect of green taxation on pollution emissions under ESG concept. Environ. Sci. Pollut. Res.Pollut. Res. **30**(21), 60196–60211 (2023). https://doi.org/10.1007/s11356-023-26699-7

# Can Monetary Policy Uncertainty Predict Exchange Rate Volatility? New Evidence from Hybrid Neural Network–GARCH Model

**Parevee Maneejuk, Terdthiti Chitkasame, Chaiwat Klinlampu, and Pichayakone Rakpho**

**Abstract** This paper aims to examine the capacity of the US Monetary Policy Uncertainty (MPU) Index in accurately forecasting the volatility of foreign exchange rates for the Dollar Index, Euro/US$, and Yen/US$. To achieve this, we introduce several hybrid Artificial Neural Networks (ANN)−GARCH models, namely GARCH, ANN−EGARCH, and ANN−GJR−GARCH, which incorporate MPU as the exogenous variable (X). In addition to that, a significant challenge in ANN modeling is choosing the appropriate activation function. Therefore, we consider and compare various forms of activation functions, including logistic, Gompertz, Tanh, ReLU, and leakyReLU. Our results demonstrate that incorporating MPU improves the forecasting performance of the benchmark ANN−GARCH−type models both in- and out-of-sample. In particular, we find that incorporating MPU into the ANN−EGARCH model yields the largest forecasting gains compared to all other variants of the ANN−GARCH−type models. Additionally, our findings reveal that ReLU is the best activation function for predicting Dollar and Yen volatility, while Gompertz performs the best for predicting Euro volatility.

**Keywords** Activation functions · Foreign exchange rates · Hybrid Artificial Neural Networks (ANN)−GARCH models · Volatility forecasting

P. Maneejuk · C. Klinlampu · P. Rakpho (✉)
Faculty of Economics, Center of Excellence in Econometrics, Chiang Mai University, Chiang Mai, Thailand
e-mail: pichayakone@gmail.com

P. Maneejuk · T. Chitkasame
Faculty of Economics, Chiang Mai University, Chiang Mai, Thailand
e-mail: terdthiti_c@cmu.ac.th

C. Klinlampu · P. Rakpho
Office of Research Administration, Chiang Mai University, Chiang Mai 50200, Thailand



V. Kreinovich et al. (eds.), *Applications of Optimal Transport to Economics and Related Topics*, Studies in Systems, Decision and Control 556,
https://doi.org/10.1007/978-3-031-67770-0_18

# 1 Introduction

The global economy is currently facing a multitude of challenges, including inflation, geopolitical tensions resulting from Russia's armed conflict in Ukraine, and the resurgence of COVID−19 cases in China. These events have had a significant impact on global economic activity in 2022, and their effects are expected to continue in 2023. To address these challenges and ensure economic stability, the US Federal Reserve has employed a range of monetary policy tools, such as regulating the banking system, controlling the money supply, and adjusting short-term interest rates. However, the future trajectory of monetary policy is inherently uncertain, as unforeseen macroeconomic events can necessitate adjustments to the central bank's intended course of action. Consequently, it is essential to acknowledge that forward guidance is contingent upon incoming economic data, as unexpected developments may warrant a change in the policy stance [25].

This study aims to investigate the influence of US monetary policy uncertainty (MPU) on the volatility forecasts of the three largest currencies in the world, namely the Dollar Index, Euro/US$, and Yen/US$. The MPU indices are constructed as scaled frequency counts of newspaper articles that discuss monetary policy uncertainty. The impact of monetary policy on exchange rates has been the subject of much debate in the literature, with numerous studies highlighting its importance [11, 12]. It is well-established that higher uncertainty surrounding monetary policy can lead to greater volatility in financial markets, including foreign exchange markets. This can create fluctuations in exchange rates that can impact the global economy [9]. Furthermore, heightened MPU can signal to investors that there is a greater likelihood of unexpected changes in interest rates or other monetary policy tools. As a result, market participants may adjust their positions in response to changing expectations about future economic conditions, leading to changes in exchange rates [12]. Moreover, MPU can impact the risk premium that investors demand when investing in a particular currency. Higher MPU can increase the risk premium, resulting in higher borrowing costs in that currency and potentially causing the exchange rate to depreciate. Conversely, lower MPU can lead to a lower risk premium, lower borrowing costs, and an appreciation of the exchange rate [21].

Forecasting foreign exchange rate volatility has become increasingly crucial for businesses, investors, and policy-makers as the volatility poses significant risks to international trade and financial stability. Businesses engaged in foreign trade and investment rely heavily on accurate exchange rate forecasts to plan their operations and manage risks. Accurate forecasting helps businesses prepare for future uncertainties, respond to changes confidently, control business operations, and make strategic decisions that drive future growth. Similarly, investors need reliable exchange rate forecasts to make informed investment decisions and minimize risks associated with currency fluctuations. Policy-makers also rely on exchange rate forecasts to manage macroeconomic stability, balance of payments, and inflationary pressures [2, 8, 14]. Incorrect predictions of exchange rates can result in financial losses, missed chances, and instability in the economy. The recent global economic challenges, such as inflation, geopolitical tensions, and the COVID-19 pandemic, have made the accurate forecasting of exchange rate volatility more crucial than ever. Therefore, there is a growing need for advanced analytical tools and models to predict exchange rate volatility accurately.

Various methods have been proposed in the literature to predict volatility, with the GARCH model being the most widely used [3, 6, 10, 23, 24]. However, the asymmetric nature of exchange rate volatility which has been highlighted by several studies Rakpho et al., [23] and Ramasamy and Munisamy [24] may not have been captured by the original GARCH model, leading to inaccurate conclusions. To address this issue, we employ the EGARCH and the GJR-GARCH models developed by Nelson [22] and Glosten et al. [13], respectively. These models have been shown to account for asymmetry, where negative unexpected returns have a stronger impact on future volatility than positive unexpected returns, as well as leverage, which refers to the negative correlation between shocks and subsequent shocks to volatility.

Despite the popularity of the GARCH−type models in predicting exchange rate volatility, Kristjanpoller and Minutolo [16] have shown that these models still suffer from substantial errors in their predictions. To address this issue, Liao et al. [18] and Rakpho et al. [23] have suggested combining artificial neural networks (ANNs) with the GARCH−type models to improve their forecasting accuracy. ANNs can model complex non-linear volatility structures and uncover non-linear correlations between input and output, making them a valuable tool for forecasting financial markets. In fact, Liao et al. [18] successfully utilized a hybrid ANN and GARCH−type model to predict exchange rates, showing significant improvements over the benchmark GARCH−type models. This hybrid approach has the potential to be applied to foreign exchange rate forecasting and may provide more accurate predictions than traditional GARCH−type models alone.

The main objective of this study is to investigate the predictive power of the US Monetary Policy Uncertainty (MPU) Index in forecasting exchange rate volatility. While the Economic Policy Uncertainty (EPU) Index has been used in previous literature as a measure of policy implementation [1, 4, 18] it encompasses a broader range of government economic policies and does not solely focus on monetary policy [15]. Husted et al. [15] introduced the MPU Index as a new measure of uncertainty surrounding Federal Reserve policy actions and their effects. The Index is constructed by searching major newspapers for keywords related to monetary policy uncertainty, similar to the methodology used in constructing the EPU Index. Therefore, our contribution is to assess the effectiveness of index the MPU Index in predicting exchange rate volatility, which has not been explored in previous literature. We examine the predictive power of the MPU Index alone, rather than in conjunction with other predictors like the EPU Index. By doing so, we aim to provide a more accurate and precise measure of the impact of monetary policy uncertainty on exchange rate volatility, which could lead to better risk management strategies for businesses and investors [7, 20].

The research methods used in this work are arranged as follows. The methods used in this study are described in Sect. 2 of this report. Data description is provided in Sect. 3. The study's findings are discussed in Sect. 4, and the study's conclusion is presented in Sect. 5.

# 2 Methodology

The hybrid ANNs−GARCH−X−types model are employed to predict exchange rate volatility in this study. In this section, we firstly describe the three GARCH−X type models, then the ANN structure is briefly explained. Finally, we present the hybrid ANN−GARCH−X−type models.

## 2.1 GARCH−X−type Models

### 2.1.1 GARCH−X Model

The Generalized Autoregressive Conditional Heteroscedasticity with exogenous variable (GARCH−X) model is a widely used method for estimating the volatility of financial data, as it allows for the inclusion of additional explanatory variables or exogenous variables that may influence the conditional variance of the data. In this study, we focus on the Monetary Policy Uncertainty (MPU) Index as the exogenous variable, as it has been found to have a significant impact on asset price volatility [7].

$$\begin{aligned} y_t &= \mu + \varepsilon_t \\ \varepsilon_t &= \sigma_t \cdot v_t, \\ \sigma_t^2 &= \omega + \alpha\varepsilon_{t-1}^2 + \beta\sigma_{t-1}^2 + \delta MPU \end{aligned} \tag{1}$$

where $\omega, \alpha, \beta \geq 0$are the coefficient of parameter; $\alpha + \beta \leq 1$. $\varepsilon_t$ is the error term and variance is $\sigma_t^2$. $v_t$ is white noise. $\delta$ is the coefficient of MPU.

### 2.1.2 EGARCH−X Model

The EGARCH−X model is an extension of the GARCH model that can capture the asymmetry in the volatility response of a time series variable to positive and negative shocks, while also incorporating the impact of exogenous variables on the conditional variance. This is important in financial markets because negative shocks often have a larger impact on volatility than positive shocks, which can lead to significant differences in risk management and asset pricing Nelson [22]. This model can be written as follows:

$$\ln(\sigma_t^2) = \omega + \gamma \left| \frac{\varepsilon_{t-1}}{\sqrt{\sigma_{t-1}^2}} \right| + \alpha \left( \frac{\varepsilon_{t-1}}{\sqrt{h_{t-1}}} \right) + \beta \ln\left(\sigma_{t-1}^2\right) + \delta MPU, \quad (2)$$

where the volatility leverage impact is described by the asymmetric leverage coefficient ($\gamma$). The utilization of this model provides greater flexibility in specifying the conditional variance, leading to enhanced precision in forecasting volatility when compared to alternative models.

### 2.1.3 GJR–GARCH–X Model

The GJR GARCH with exogenous Variable (GJR–GARCH–X) model is proposed by Glosten et al. [13]. This model allows for asymmetric volatility, which means that positive and negative shocks to returns may affect volatility differently, resulting in an asymmetric leverage volatility effect. The model can be expressed as:

$$\sigma_t^2 = \omega + (\alpha + \gamma I_{t-1})\,\varepsilon_{t-1}^2 + \beta\sigma_{t-1}^2 + \delta MPU \quad (3)$$

where

$$I_{t-1} = \begin{cases} 0 \; if \; \varepsilon_{t-1} \geq 0, \\ 1 \; if \; \varepsilon_{t-1} < 0. \end{cases} \quad (4)$$

and $\gamma$ denote an asymmetric leverage effect.

## 2.2 *Artificial Neural Network (ANN)*

An artificial neural network (ANN) is a computational model that draws inspiration from the neural structure and functioning of the human brain. It comprises interconnected nodes or neurons organized in layers, with each layer dedicated to processing specific types of information. This connects input and output signals in a way that is comparable to how biological neurons in the human brain work. Figure 1 shows ANN's working principle:

To simplify the estimation process in this work, we take into account the multilayer perceptron, which has just one input layer, one hidden layer, and one out-put layer. The input layer is represented by vector $X = (x_1, x_2, \ldots, x_d)'$ and the output layer is a vector $y$. The ANN model can be determined as follows:

$$y = \tilde{\phi}(\widehat{\phi}(dw^I + b^I)w^O + b^O), \quad (5)$$

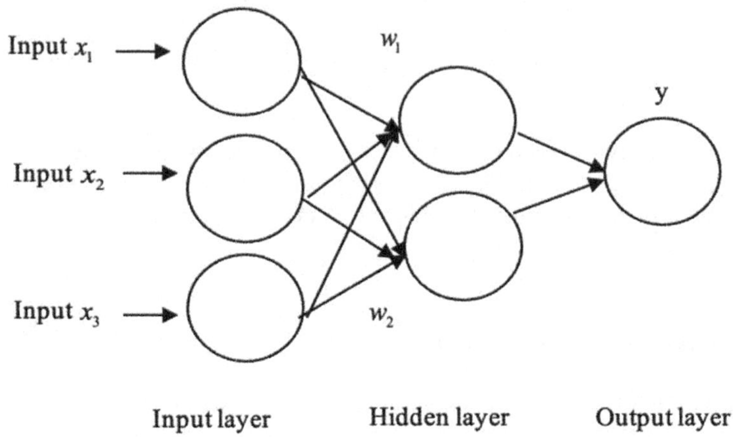


**Fig. 1** Multilayer feed-forward back propagation neural network

where $\tilde{\phi}$ is the input activation and $\widehat{\phi}$ is the output activation function. $b^I$ is the bias term of input and $b^O$ is the bias term of output. $w^I$ is the weight vector of the hidden layer. $w^O$ is the weight vector of the output layer.

## 2.3 *Hybrid Models*

In this study, we explore the integration of the Artificial Neural Network (ANN) model with the GARCH−type−X models. Specifically, we focus on incorporating the conditional variance processes of three commonly used models: GARCH, EGARCH, and GJR-GARCH, in conjunction with the ANN model. We propose three-different models, namely the ANN−GARCH(1,1)−X, ANN−EGARCH(1,1)−X and ANN−GJR−GARCH(1,1)−X. This is the combination ANN model with the GARCH different-type model, incorporating exogenous variables to capture the conditional volatility of each variable.

- ANN−GARCH(1,1)−X model

$$\sigma_t^2 = \omega + \alpha\varepsilon_{t-1}^2 + \beta\sigma_{t-1}^2 + \delta MPU_t + \theta\left(\psi\left(z_t\Upsilon\right)\right) \tag{6}$$

- ANN−EGARCH(1,1)−X model

$$\ln(\sigma_t^2) = \omega + \gamma\left|\frac{\varepsilon_{t-1}}{\sqrt{\sigma_{t-1}^2}}\right| + \alpha\left(\frac{\varepsilon_{t-1}}{\sqrt{h_{t-1}}}\right) + \beta\ln\left(\sigma_{t-1}^2\right) + \delta MPU_t + \theta\left(\psi\left(z_t\Upsilon\right)\right), \tag{7}$$

- ANN−GJR−GARCH(1,1)−X model

$$\sigma_t^2 = \omega + (\alpha + \gamma I_{t-1}) \varepsilon_{t-1}^2 + \beta\sigma_{t-1}^2 + \delta MPU_t + \theta (\psi (z_t \Upsilon)), \tag{8}$$

where $\theta$ denotes the additional output weight of the neural network component, while the activation function of the output layer of the ANN model, $\psi (z_t \Upsilon)$ is the sigmoid activation function of the output layer of the ANN part, and $z_{t-1} = (\varepsilon_{t-1} - E(\varepsilon_t)) / \sqrt{E(\varepsilon_t^2)}$ is the standardized residual used in the activation function. To simplify the computational process, we adopt a straightforward approach of selecting the values using a uniform random number generator within the range of -1 to 1.

## 2.4 Activation Functions

Selecting an activation function to link the input, hidden, and output layers is crucial. This study considers five activation functions: Sigmoid (logistic), Gompertz, Tanh, ReLU, and Leaky ReLU, which are presented as follows [26].

### 2.4.1 Sigmoid (logistic)

The sigmoid function takes an input value and squashes it into a range between 0 and 1. They are commonly used in the output layer of a binary classification problem and can also be used in the hidden layers of the network to introduce non-linearity and help the model capture complex patterns in the data. This function is defined as

$$f(x) = \frac{1}{1 + e^{-x}}, \tag{9}$$

where $x$ as the input to the sigmoid function, and e represents Euler's number, a mathematical constant. This function always produces an output between 0 and 1, making it beneficial for applications that involve modeling binary outcomes or probabilities

### 2.4.2 Gompertz

The Gompertz function has a sigmoidal shape, similar to the sigmoid function, but with a steeper initial slope and a more gradual flattening. The function can be written as

$$N(t) = N(0) \cdot \exp^{-b \cdot \exp^{-c \cdot t}} \tag{10}$$

where $N(0)$ is the initial number of cells/organisms at time zero. In short, the Gompertz activation function has demonstrated proficiency in handling complex nonlinear patterns exhibited in data for certain applications. Nonetheless, its efficacy is reliant on the properties of the data being modeled and the specific task requirements.

### 2.4.3 Tanh

The Tanh function has a similar shape to the sigmoid function but its output ranges from $-1$ to 1. The output of the Tanh function is always between $-1$ and 1, and it is symmetrical around the origin. it introduces non-linearity into the model and can help capture complex patterns in the data. It is often preferred over the sigmoid activation function because it has a steeper gradient and is, therefore, more sensitive to small changes in the input value, which can help accelerate the learning process during training, which can be displayed as follows:

$$f(x) = \tanh(x) = \frac{(e^x - e^{-x})}{(e^x + e^{-x})} \tag{11}$$

### 2.4.4 Rectified Linear Unit (ReLU)

ReLU is computationally efficient and helps to avoid the vanishing gradient problem during training. The vanishing gradient problem can occur when the gradient of the activation function becomes too small, making it difficult to train deep neural networks effectively. It is defined as:

$$f(x) = \begin{cases} 0, \ if \ x \le 0 \\ x, \ if \ x > 0 \end{cases} \tag{12}$$

### 2.4.5 Leaky ReLU

The Leaky ReLU function is preferred over the standard ReLU function in certain situations because it helps to address the "dying ReLU" problem, which can occur when a large fraction of the inputs to a neuron are negative, causing the neuron to become permanently inactive. The Leaky ReLU function allows some information to flow through the neuron, even if the inputs are negative, which can help avoid this problem.

We select the following activation function:

$$f(x) = \begin{cases} 0.01x \ for \ x < 0 \\ x \ for \ x \ge 0 \end{cases} \tag{13}$$

## 2.5 Comparison Criteria

In this work, the loss function for our forecasting models is the Root Mean Square Error (RMSE). It determines the discrepancy between volatility as anticipated and as actual. The RMSE equation is:

$$RMSE = \sqrt{\frac{\sum_{t=1}^{N} \left(\widehat{\sigma}_t^2 - \sigma_t^2\right)^2}{N}} \tag{14}$$

where $\widehat{\sigma}_t^2$ is the forecasted foreign exchange rates volatility. $\sigma_t^2$ is the realized volatility of forecast foreign exchange rates and $N$ is the number of forecasting data. Note that the realized volatility $\sigma_t^2$ can be computed as the sample variance of log returns in a 21 days' window to the future (approximately one month of transactions),

$$\sigma_t^2 = \frac{1}{21} \sum_{i=t+1}^{t+21} (r_i - \bar{r}_t)^2 \tag{15}$$

where $r_i$ is the return of exchange rate and $\bar{r}_t$ is the average return of exchange rate.

# 3 Data Specification

Our study aims to forecast foreign exchange rates volatility in three-different currency using the hybrid ANN–GARCH type models with MPU as the exogenous variable. The selected exchange rates include the Dollar Index (US–INDEX), Euro/US$ (EUR/USD), and Yen/US$ (JPY/USD), which are considered major global currencies. The Dollar Index is a measure of the value of the United States dollar relative to a basket of foreign currencies. We obtained monthly data from the Thomson Reuters database covering the period from 16/04/1993 to 16/04/2023, with a total of 359 observations. The data was transformed using log-return. Table 1 contains descriptive statistics. It reveals that all exchange rates have kurtosis values larger than 3, which means that the data series do not follow a normal distribution. Additionally, for every exchange rate series, the Jarque–Bera test has strongly rejected the null hypothesis of normal distribution. Furthermore, we tested the processed time series using the augmented Dickey–Fuller (ADF) unit root test, which showed that the return of all variables has a stationary condition with decisive evidence (see MBF–Unit root).

**Table 1** Descriptive statistics

| | MPU | US–INDEX | EUR/USD | JPY/USD |
|---|---|---|---|---|
| Mean | −7.05E-05 | 0.00011 | −0.00013 | −0.00016 |
| Median | −0.0079 | 0.00067 | −0.00056 | −0.00146 |
| Maximum | 0.5204 | 0.02978 | 0.03556 | 0.06874 |
| Minimum | −0.353 | −0.03747 | −0.03663 | −0.04664 |
| Std. Dev. | 0.1123 | 0.00998 | 0.01158 | 0.01397 |
| Skewness | 0.6164 | −0.2218 | 0.05041 | 0.50667 |
| Kurtosis | 4.8782 | 3.66811 | 3.50432 | 4.84781 |
| Jarque-Bera | 75.5063 | 9.62181 | 3.95657 | 66.4344 |
| MBF-JB | 0.0000 | 0.0000 | 0.0004 | 0.0000 |
| Unit-root | −14.397 | −18.403 | −17.707 | −19.655 |
| MBF-Unit root | 0.0000 | 0.0000 | 0.0000 | 0.0000 |
| Observations | 359 | 359 | 359 | 359 |

*Note* MBF is the Minimum Bayes factor

# 4 Estimated Results

## *4.1 Performance Results for the Forecasting*

Table 2 displays the performance of the models based on their RMSE values. To determine the most precise model for forecasting exchange rate return volatility, we assess both in-sample and out-of-sample predictions. The in-sample dataset, which spans 21 years from April 1993 to April 2014, comprises 70% of the data. The remaining 30% of the data is reserved for out-of-sample evaluation of the model, covering the period from April 2015 to April 2023.

The findings firstly indicate that the hybrid ANN models with MPU exhibit lower RMSEs compared to the all-hybrid models without the MPU. These results suggest that incorporating MPU as exogenous variable in the ANN-GARCH-type models can enhance the accuracy of volatility forecasting. Secondly, the study evaluated different types of GARCH models and ANN models with various activation functions.

The results indicate that the ANN–EGARCH–X model with ReLU is the most appropriate for forecasting the volatility of US-Index and JPY/USD exchange rate returns, with an RMSE of 0.00094. On the other hand, for forecasting the volatility of EUR/USD exchange rate returns, the ANN–EGARCH–X model with Gompertz is the most suitable, with an RMSE of 0.00220.

**Table 2** Results of the in-sample performance of hybrid GARCH−type models and hybrid GARCH−type with exogenous (X) models with various activation functions

| Activation function | US-INDEX | | |
|---|---|---|---|
| | ANN−GARCH | ANN−EGARCH | ANN−GJR−GARCH |
| | 0.2532 | 0.4523 | 0.2753 |
| | ANN−GARCH−X | ANN−EGARCH−X | ANN−GJR−GARCH−X |
| Logistic | 0.0246 | 0.0021 | 0.1464 |
| Gompertz | 0.1376 | 0.0370 | 0.00099 |
| Tanh | 0.0564 | 0.0010 | 0.01816 |
| ReLU | 0.0010 | 0.0009 | 0.00109 |
| LeakyReLU | 0.0206 | 0.443 | 0.04374 |
| Activation | EUR/USD | | |
| Function | ANN−GARCH | ANN−EGARCH | ANN−GJR−GARCH |
| | 0.1852 | 0.5312 | 0.0912 |
| | ANN−GARCH−X | ANN−EGARCH−X | ANN−GJR−GARCH−X |
| Logistic | 0.0022 | 0.0502 | 0.0551 |
| Gompertz | 0.0249 | 0.0022 | 0.0808 |
| Tanh | 0.0022 | 0.3468 | 0.0024 |
| ReLU | 0.0339 | 0.0022 | 0.0026 |
| LeakyReLU | 0.0342 | 0.0022 | 0.0089 |
| Activation function | JPY/USD | | |
| | ANN−GARCH | ANN−EGARCH | ANN−GJR−GARCH |
| | 0.5021 | 0.1000 | 0.7012 |
| | ANN−GARCH−X | ANN−EGARCH−X | ANN−GJR−GARCH−X |
| Logistic | 0.2041 | 0.0022 | 0.14644 |
| Gompertz | 0.0022 | 0.037 | 0.0010 |
| Tanh | 0.0319 | 0.0010 | 0.0181 |
| ReLU | 0.0039 | 0.0009 | 0.001 |
| LeakyReLU | 0.0435 | 0.0416 | 0.0437 |

## 4.2 *Estimation Results of the ANN−EGARCH−X model*

After that, we have analyzed the volatility persistence of exchange rate calculated by the sum of the estimated parameter $(\alpha + \beta)$ from GARCH process reported in Table 3. When the total summation is close to 1, there is a tendency for high levels of volatility to persist in the future, based on their past values. The results show the volatility persistence value of US-INDEX to be 0.9599 and that of JPY/USD to be 0.8963 indicating a high level of volatility which is much higher than the case of JPY/USD (0.1975). Furthermore, we discover that the output weight of the neural network $(\theta)$ for the EGARCH−X models is significant, proving that conditional volatility can be predicted using the ANN term. Moreover, the exchange rates of

**Table 3** Estimation results of the ANN−EGARCH−X model

ANN−EGARCH-X (1,1)

| Parameter | US-INDEX | | | EUR/USD | | | JPY /USD | | |
|---|---|---|---|---|---|---|---|---|---|
| | Coef. | t-statistic | MBF | Coef. | t-statistic | MBF | Coef. | t-statistic | MBF |
| $\omega$ | −0.289 | −1396.7 | 0.00 | −5.752 | −3.26 | 0.005 | −0.42 | −1847.1 | 0.00 |
| $\alpha$ | 0.004 | 1677.8 | 0.00 | 0.043 | 4.57 | 0.00 | −0.04 | −41.34 | 0.00 |
| $\beta$ | 0.959 | 2514.7 | 0.00 | 0.154 | 5.49 | 0.00 | 0.942 | 1995.1 | 0.00 |
| $\gamma$ | −0.157 | −2198.6 | 0.00 | 0.457 | 3.17 | 0.007 | −0.19 | −1075.7 | 0.00 |
| $\delta$ | 0.449 | 2550.1 | 0.00 | −0.461 | −0.17 | 0.984 | 0.496 | 6524 | 0.00 |
| $\theta$ | 0.002 | 2354.1 | 0.00 | 0.002 | 5410 | 0.00 | 0.042 | 1841.1 | 0.00 |
| $(\alpha+\beta)$ | 0.9599 | | | 0.8963 | | | 0.1975 | | |

*Note* MBF is the Minimum Bayes factor

the US-INDEX and JPY/USD have negative values as seen from the asymmetric leverage effect's parameters $\gamma$ of ANN−EGARCH (1, 1). Negative shocks exert a greater influence on volatility compared to positive shocks. This means that the exchange rate is more sensitive to bad news than to good news. On the other hand, the EUR/USD case has a positive value suggesting that the positive shocks have a larger impact on volatility than the negative shocks or that the EUR/USD exchange rate is more sensitive to good news than to bad news. Furthermore, the findings indicate that the incorporation of the exogenous variable, US Monetary Policy Uncertainty (MPU) Index, in the ANN−GARCH−type models enhances the accuracy of exchange rate volatility forecasting, particularly for the US Index and JPY/USD exchange rate. This suggests that policy uncertainty, as captured by the MPU Index, plays a significant role in exchange rate volatility. This is consistent with a previous research work that showed the impact of economic policy uncertainty on exchange rate volatility [26]. Moreover, the finding that MPU has a significant positive impact on exchange rate volatility can be interpreted as indicating that when there is uncertainty regarding the Federal Reserve's monetary policy decisions, there is a corresponding increase in the MPU Index, and this, in turn, leads to a higher exchange rate volatility. This underscores the importance of clarity and transparency in monetary policy decisions in order to reduce uncertainty and volatility in financial markets.

## 5 Conclusion

The purpose of this study had been to investigate the capacity of an exogenous variable, specifically the US Monetary Policy Uncertainty (MPU) Index, on the estimation of exchange rate volatility and to explore ways to improve forecasting models by incorporating this variable as the key predictor. Our study findings have had several important consequences for policymakers, investors, and financial institutions. Firstly, the study has shown that MPU has positive impact on the volatility of the

US-INDEX and JPY/USD exchange rates. This suggests that the uncertainty surrounding US monetary policy decsions, like changes in the Federal Funds Rate, can generate negative attitudes among stakeholders towards policy implementation, leading to increased volatility in financial markets. Thus, it is imperative to underscore the importance of transparency and clarity in monetary policy decisions in order to reduce uncertainty and volatility in financial markets and promote stability and growth. Secondly, the study has demonstrated that the ANN−EGARCH−X models with appropriate activation functions can improve the accuracy of exchange rate volatility forecasting. The ReLU activation function was found to be more appropriate for capturing the non-linear patterns in the data and predicting the volatility of US-INDEX and JPY/USD exchange rate, while the Gompertz activation function performed better for forecasting the volatility of EUR/USD exchange rate. These findings suggest that incorporating the ANN-EGARCH-X models with appropriate activation functions in exchange rate volatility forecasting can improve the predictive accuracy and provide useful information for decision-making. Based on these findings, this recommend that policymakers, investors, and financial institutions are recommended to consider the relevance of MPU in ex-change rate volatility and incorporate this variable in exchange rate volatility forecasting models to improve their accuracy. Furthermore, policymakers should prioritize transparency and clarity in monetary policy decisions to reduce uncertainty and promote stability in financial markets. Finally, we suggest that future research attempts explore the influence of other exogenous variables on exchange rate volatility and evaluate the performance of different ANN−GARCH−X models with various activation functions for exchange rate volatility forecasting.

**Acknowledgements** This research work was partially supported by Chiang Mai University and the authors are grateful to the Centre of Excellence in Econometrics for the suggestion support.

## References

1. Arouri, M., Estay, C., Rault, C., Roubaud, D.: Economic policy uncertainty and stock markets: long-run evidence from the US. Finance Res. Lett. **18**, 136–141 (2016)
2. Asseery, A., Peel, D.A.: The effects of exchange rate volatility on exports: some new estimates. Econ. Lett. **37**(2), 173–177 (1991)
3. Bala, D.A., Asemota, J.O.: Exchange-rates volatility in Nigeria: application of GARCH models with exogenous break. CBN J. Appl. Stat. **4**(1), 89–116 (2013)
4. Balcilar, M., Gupta, R., Kyei, C., Wohar, M.E.: Does economic policy uncer-tainty predict exchange rate returns and volatility? Evidence from a nonparametric cau-sality-in-quantiles test. Open Econ. Rev. **27**, 229–250 (2016)
5. Beckmann, J., Czudaj, R.: Exchange rate expectations and economic policy uncertainty. European J. Polit. Econ. **47**, 148–162 (2017)
6. Clement, A., Samuel, A.: Empirical modeling of Nigerian exchange rate vola-tility. Math. Theory Model. **1**(3), 8–15 (2011)
7. Chiang, T.C.: Spillovers of US market volatility and monetary policy uncertainty to global stock markets. North Am. J. Econ. Finance **58**, 101523 (2021)

8. De Grauwe, P.: Exchange rate variability and the slowdown in growth of international trade. Staff Pap. **35**(1), 63–84 (1988)
9. Devereux, M.B., Lane, P.R., Xu, J.: Exchange rates and monetary policy in emerging market economies. Econ. J. **116**(511), 478–506 (2006)
10. Dhamija, A.K., Bhalla, V.K.: Financial time series forecasting: comparison of various arch models. Glob. J. Financ. Manag. **2**(1), 159–172 (2010)
11. Dornbusch, R.: Exchange rate expectations and monetary policy. J. Int. Econ. **6**(3), 231–244 (1976)
12. Gali, J., Monacelli, T.: Monetary policy and exchange rate volatility in a small open economy. Rev. Econ. Stud. **72**(3), 707–734 (2005)
13. Glosten, L.R., Jagannathan, R., Runkle, D.E.: On the relation between the expected value and the volatility of the nominal excess return on stocks. J. Financ. **48**(5), 1779–1801 (1993)
14. Hakkio, C.S.: Exchange rate volatility and federal reserve policy. Econ. Rev. **69**(Jul), 18–*31 (1984)
15. Husted, L., Rogers, J., Sun, B.: Monetary policy uncertainty. J. Monetary Econ. **115**, 20–36 (2020)
16. Kristjanpoller, W., Minutolo, M.C.: Gold price volatility: a forecasting approach using the Artificial Neural Network-GARCH model. Exp. Syst. Appl. **42**(20), 7245–7251 (2015)
17. Lastrapes, W.D.: Exchange rate volatility and US monetary policy: an ARCH application. J. Money Credit Bank. **21**(1), 66–77 (1989)
18. Liao, R., Yamaka, W., Sriboonchitta, S.: Exchange rate volatility forecasting by hybrid neural network Markov switching Beta-t-EGARCH. IEEE Access **8**, 207563–207574 (2020)
19. Liu, L., Zhang, T.: Economic policy uncertainty and stock market volatility. Financ. Res. Lett. **15**, 99–105 (2015)
20. Li, Xiafei, Wei, Yu., Chen, Xiaodan, Ma, Feng, Liang, Chao, Chen, Wang: Which un-certainty is powerful to forecast crude oil market volatility? New evidence. Inte. J. Financ. Econ. **27**(4), 4279–4297 (2022)
21. Mueller, P., Tahbaz-Salehi, A., Vedolin, A.: Exchange rates and monetary policy uncertainty. J. Financ. **72**(3), 1213–1252 (2017)
22. Nelson, D.B.: Conditional heteroskedasticity in asset returns: a new approach. Econometrica: J Econom. Soc. 347–370 (1991)
23. Rakpho, P., Yamaka, W., Phadkantha, R.: Predicting energy price volatility using hybrid artificial neural networks with GARCH-Type models. In: Integrated Un-certainty in Knowledge Modelling and Decision Making: 9th International Symposium, IUKM 2022, Ishikawa, Japan, March 18–19, 2022, Proceedings, pp. 317–328. Springer International Publishing, Cham (2022)
24. Ramasamy, R., Munisamy, S.: Predictive accuracy of GARCH, GJR and EGARCH models select exchange rates application. Glob. J. Manag. Bus. Res. **12**(15), 89–100 (2012)
25. Siklos, P.L.: Central Banks into the Breach: from Triumph to Crisis and the Road Ahead. Oxford University Press (2017)
26. Zurada, J.: Introduction to Artificial Neural Systems. West Publishing Co (1992)

# Factors Affect Happiness and the Risk of Unhappiness in Thailand

**Kunsuda Nimanussornkul and Chaiwat Nimanussornkul**

**Abstract** In this research, we analyze the factors that contribute to happiness and unhappiness in Thailand. We use linear and diminishing relationships with Gross Province Product per capita to determine these factors. To select model variables, we use Lasso regression, and Ordinary Least Square tests the Easterlin Paradox and hedonic treadmill. We base our study on data collected from 77 provinces in Thailand in the year 2020. Our findings confirm the Easterlin Paradox and indicate that happiness and unhappiness are influenced by an inherent force. Furthermore, our results reveal that the number of criminal suspects has a significant impact on happiness and reduces the risk of unhappiness.

**Keywords** Happiness · Risk of Unhappiness · Thailand · Easterlin Paradox

## 1 Introduction

Happiness is a universal human pursuit, while pain is something everyone seeks to avoid. There are three main schools of thought when it comes to the philosophy of happiness: Hedonism, Desire Theory, and Objective List Theory. Hedonism Theory explains that happiness maximizes feeling or showing pleasure and minimizes pain. Desire theory defines happiness as getting what you want [18]. This theory can better explain why we prefer certain things over others. Objective List theory suggests that happiness is achieved by accomplishing certain things from a list of worthwhile goals [32, 36, 37]. Psychological theories of happiness are categorized into two approaches: Hedonic and Eudaimonic [22, 23, 45]. The Hedonic approach centers on fulfilling desires and seeking pleasure while minimizing pain. Eudaimonic approaches involve being in a state of good spirits [35].

In the field of Economics, the concept of the economics of happiness involves evaluating well-being by combining principles from both economics and psychology.

K. Nimanussornkul (✉) · C. Nimanussornkul
Faculty of Economics, Chiang Mai University, Chiang Mai 50200, Thailand
e-mail: kunsudan@gmail.com

 
V. Kreinovich et al. (eds.), *Applications of Optimal Transport to Economics and Related Topics*, Studies in Systems, Decision and Control 556,
https://doi.org/10.1007/978-3-031-67770-0_19

Richard Easterlin was the first modern economist to re-explore this concept in the early 1970s [11, 13]. Other researchers have contributed to this area of study e.g. Blanchflower and Oswald [3], Clark and Oswald [8], Frey and Stutzer [15], Graham and Pettinato [16], and Graham [17].

The topic of happiness is of great importance to many countries. The first World Happiness Report was introduced on April 2nd, 2012, and since then, the International Day of Happiness is celebrated each year on March 20th. The rankings of happiness are determined by the scores received in response to the life evaluation question, which is rated on a scale of 0–10 [48]. The World Happiness Report of 2021 compared the rankings of happiness from 2017–2019 to those from 2018–2020 and found a high correlation of 0.99 for the 95 countries with 2020 data. This suggests that overall rankings have only modestly changed due to the COVID-19 pandemic. In Thailand, the happiness ranking for 2020 was 48 with a score of 5.885, which is lower than the ranking from 2017 to 2019 of 44 with a score of 5.999 [20]. Furthermore, the ranking of happiness for Thailand based on a three-year average from 2020–2022 is 60 out of 137 countries with a score of 5.843, indicating a trend of decreasing happiness rankings [21].

Numerous studies on happiness economics have been conducted in Thailand, primarily focusing on individual data and specific groups such as the elderly and organizations. These studies e.g. Tansuchat and Khamkaew [41], Nikhamyang and Noikhamyang [29], Lapha, Waiwitriyakul and Thowannang [25], Wannapong and Hong-ngam [47] and Teeranupattana [42]. However, Senasu [38] conducted a study on the entire country that measured income in logarithmic form but did not factor in environmental and life safety aspects. Additionally, no studies have explored the causes of unhappiness, which could lead to depression if left unaddressed for a prolonged period [14].

The purpose of this study is to examine the factors that contribute to happiness and the risk of unhappiness in Thailand. We measured the risk of unhappiness based on the percentage of individuals with lower mental health scores compared to the general population. We examine both diminishing and linear relationships with Gross Province Product per capita, we use the Lasso (Least Absolute Shrinkage and Selection Operator) linear regression to select variables for the model. After selecting the variables with the Lasso, we estimate the model using Ordinary Least Square to test the Easterlin Paradox and the hedonic treadmill from the model. By understanding the factors that impact happiness and the risks of unhappiness, we can provide recommendations to government officials and policymakers. The paper is structured as follows: Factors Affecting Happiness are presented in Sect. 2, followed by the methodology in Sect. 3. Section 4 contains the empirical results and Sect. 5 is the conclusion.

# 2 Literature Review

## *2.1 Factors Affecting Happiness*

The definition of happiness is not unique. Subjective well-being (SWB) is often used to represent happiness and is studied in the field of behavioral science [10]. An individual's level of happiness can be determined by their balance of positive and negative feelings, as explained by Ng [27]. Numerous studies have been conducted in various countries to explore the factors that influence happiness.

Many traditional economists believe that people are rational and intelligent, which leads them to predict that income buys happiness. This is because they assume that people make decisions based on maximizing their wealth and income and that this will lead to more happiness. The idea is that when people have more money, they can buy more things and have more opportunities to improve their quality of life, which in turn makes them happier. However, this relationship between income and happiness is not a simple one-to-one correlation, as traditional economics recognizes the concept of diminishing returns. This means that people with higher income or wealth tend to be less happy with income/wealth increases compared to those with lower income or wealth. This is why the relationship between income and happiness is a nonlinear positive one. However, behavioral economists like Kahneman and Tversky have a different view. They argue that people often make mistakes and fail to maximize their happiness because they don't always know what they actually like. This means that having better information and understanding that information can help people correct these mistakes and make better choices that lead to greater happiness [1].

Although income is often seen as a key factor in determining happiness, there are other important considerations. These include mental and physical health, religious beliefs, and participation in political organizations, educational level, job satisfaction, and personal characteristics such as age, gender, and marital status [26, 28].

Empirical research on factors affecting subjective well-being or happiness typically uses simple additive functions, such as this equation:

$$SWB_i = \alpha + \sum_n \beta_n X_{n,i} + u_i \tag{1}$$

where $i$ refers to the individual or country, $SWB$ is a measure of subjective well-being or Happiness, $X$ is a set of explanatory variables, such as income and other determinants of happiness, $\beta$ are the parameters and $u$ is an error term [9].

### 2.2 Easterlin Paradox

In 1974, Richard Easterlin first challenged the correlation between income and happiness, which is now known as the Easterlin Paradox. The paradox states that although happiness and income are positively related among and within nations at any given point, long-term income growth does not necessarily lead to increased happiness [11, 12]. There are three main points to consider from the Easterlin Paradox. The first is the positive correlation between individual income and happiness. The second is the positive correlation between a country's GDP per capita and happiness. The third point is that there is no correlation between the growth of happiness and the growth of GDP per capita. This paper focuses only on the second point, which is the positive relationship between happiness and GDP per capita among countries but uses data from each province in Thailand.

The previous studies found a positive relationship between happiness and GDP per capita e.g. Tella et al. [44], Tella and MacCulloch [43], Proto and Rustichini [33]. Some studies used logarithm form e.g. Stevenson and Wolfers [40], Angeles [2], and Slag et al. [39].

### 2.3 The Hedonic Treadmill

The hedonic treadmill, also known as hedonic adaptation, is a theory proposed by Brickman and Campbell [6]. It explains that people tend to quickly return to a neutral happiness set point after experiencing changes or positive/negative events in life. As a result, people return to their baseline level of happiness after experiencing happiness or unhappiness. This phenomenon has been observed in various fields of psychology and economics, such as lottery windfalls [7] and losing a limb to cancer [46]. However, Wundera and Schwarzec [49] and Bottan and Truglia [5] did not find hedonic adaptation. They found that happiness has a positive autoregressive coefficient, resulting in a positive correlation between current and previous happiness, suggesting that happiness has an inertial force.

## 3 Methodology

We use Lasso linear regression to select variables and then estimate the model with Ordinary Least Squares to test the Easterlin Paradox and the hedonic treadmill.

Lasso regression (Least Absolute Shrinkage and Selection Operator) was used in several papers e.g. Boonyakunakorn, Nunti and Yamaka [4] and Kubin et al. [24]. The Lasso regression offers a trade-off between lower forecasting error or lower variance by using a biased model. Its estimated results should provide less variance than OLS when variables are correlated. The details of the Lasso linear regression model used in this paper are as follows:

Lasso linear regression estimates the parameter by shrinking the parameters with respect to the norm. The objective is to minimize $\sum_{i=1}^{n}\left(y_i - \beta_0 - \sum_{j=1}^{p}\beta_j x_{ij}\right)^2$ with subject to $\|\beta\|_1 \le t$ then we can set the Lagrangian function of Lasso linear regression as follows

$$\min_{\beta_0,\beta_j}\left[\sum_{i=1}^{n}\left(y_i - \beta_0 - \sum_{j=1}^{p}\beta_j x_{ij}\right)^2 + \lambda\|\beta\|_1\right] \tag{2}$$

where $|\beta\|_1$ is the sum of absolute values of the betas.

The paper proposes four models: Model 1 in equation (3) shows a diminishing relation between mental health grade point average ($Happiness$) and Gross Province Product per capita ($GPPper$), Model 2 in equation (4) depicts a linear relation between $Happiness$ and $GPPper$, Model 3 in equation (5) reveals a diminishing relation between the percentage of individuals with a lower mental health grade point compared to the general population ($Lower$) and $GPPper$, and Model 4 in equation (6) shows a linear relation between $Lower$ and $GPPper$. Four models can be expressed as follows:

**Model 1**

$$\begin{aligned} Happiness_{it} = f(&lnGPPper_{it}, Happiness_{it-1}, Expenditure_{it}, Deaths_{it}, Poor_{it},\\ &Medicalstaff_{it}, Man_{it}, Waste_{it}, Divorce_{it}, Suspect_{it},\\ &Transparency_{it}, Employ_{it}, Inflation_{it}) \end{aligned} \tag{3}$$

**Model 2**

$$\begin{aligned} Happiness_{it} = f(&GPPper_{it}, Happiness_{it-1}, Expenditure_{it}, Deaths_{it}, Poor_{it},\\ &Medicalstaff_{it}, Man_{it}, Waste_{it}, Divorce_{it}, Suspect_{it},\\ &Transparency_{it}, Employ_{it}, Inflation_{it}) \end{aligned} \tag{4}$$

**Model 3**

$$\begin{aligned} Lower_{it} = f(&lnGPPper_{it}, Lower_{it-1}, Expenditure_{it}, Deaths_{it}, Poor_{it},\\ &Medicalstaff_{it}, Man_{it}, Waste_{it}, Divorce_{it}, Suspect_{it},\\ &Transparency_{it}, Employ_{it}, Inflation_{it}) \end{aligned} \tag{5}$$

**Model 4**

$$
\begin{aligned}
Lower_{it} = f(&GPPper_{it}, lower_{it-1}, Expenditure_{it}, Deaths_{it}, Poor_{it},\\
&Medicalstaff_{it}, Man_{it}, Waste_{it}, Divorce_{it}, Suspect_{it},\\
&Transparency_{it}, Employ_{it}, Inflation_{it})
\end{aligned} \tag{6}
$$

where

| | |
|---|---|
| $Happiness_{it}$ | = Mental health grade point average in Province $i$ of the year 2020 |
| $lnGPPper_{it}$ | = Gross Province Product per capita in Province $i$ of the year 2020 in logarithm form |
| $Happiness_{it-1}$ | = Mental health grade point average in Province $i$ of the year 2018 (because this data did not collect every year) |
| $GPPper_{it}$ | = Gross Province Product per capita in Province $i$ of the year 2020 |
| $Lower_{it}$ | = the percentage of individuals with a lower mental health grade point compared to the general population in Province $i$ of the year 2020 |
| $Lower_{it-1}$ | = the percentage of individuals with a lower mental health grade point compared to the general population in Province $i$ of the year 2018 (because this data did not collect every year) |
| $Expenditure_{it}$ | = Average expenditure per month of household in Province $i$ of the year 2020 |
| $Deaths_{it}$ | = Number of death in Province $i$ of the year 2020 |
| $Poor_{it}$ | = Number of poor people in Province $i$ of the year 2020 |
| $Medicalstaff_{it}$ | = Number of medical staff in Province $i$ of the year 2020 |
| $Man_{it}$ | = Number of the male population in Province $i$ of the year 2020 |
| $Waste_{it}$ | = Number of waste in Province $i$ of the year 2020 |
| $Divorce_{it}$ | = Number of divorce registration in Province $i$ of the year 2020 |
| $Suspect_{it}$ | = Number of criminal suspects in Province $i$ of the year 2020 |
| $Transparency_{it}$ | = Rating score of Integrity and Transparency in Province $i$ of the year 2020 |
| $Employ_{it}$ | = Number of employed people in Province $i$ of the year 2020 |
| $Inflation_{it}$ | = Inflation rate in Province $i$ of the year 2020 |

We test for the Easterlin Paradox by the null hypothesis ($H_0$) as the coefficient of $GPPper_{it}$ or $lnGPPper_{it} = 0$ implying the Easterlin Paradox is not hold. Then, we test the hedonic treadmill using the null hypothesis ($H_0$), which states that the coefficient of $Happiness_{it-1}$ is equal to 0. This suggests that the hedonic treadmill is present. However, if we reject the null hypothesis, it implies that happiness is inertial. Then we used $-eplogp$ calibration, which $p$ is the p-value to apply the Minimum Bayes Factor (MBF) approach. The MBF is interpreted as follows in Table 1.

**Table 1** Interpret from Minimum Bayes Factor

| Bayes Factor (MBF) | Strength of Evidence for alternative hypothesis (or Evidence against a null hypothesis) |
|---|---|
| 1 to 1/3 | Weak |
| 1/3 to 1/10 | Moderate |
| 1/10 to 1/30 | Substantial |
| 1/30 to 1/100 | Strong |
| 1/100 to 1/300 | Very strong |
| < 1/300 | Decisive |

Remark: Adapted from Held and Ott [19]

# 4 Data and Variables

We gathered information on the mental health grade point average and the percentage of people with a lower mental health grade point than the general population in 77 provinces. This data was collected from 2018 and 2020. For all other variables, we only considered data from 2020. The data obtained from the National Statistic Organization of Thailand includes the Integrity and Transparency rating score from the National Anti-Corruption Commission of Thailand [30], the Mental Health Grade Point Average, and the percentage of individuals with lower mental health grades compared to the general population in 2018, which are sourced from Prukkanone et al. [34].

Both the National Statistic Organization of Thailand and Prukkanone et al. [34] use the same method to observe the Mental Health Grade Point Average and the percentage of individuals with lower mental health grades compared to the general population. They gather data by interviewing households of individuals aged 15 and above from all provinces in Thailand and evaluating their mental health. The mental health questionnaire consists of 15 questions divided into four parts. The first part comprises six questions and 18 points, which evaluate the individual's current mental state. The questions are divided into two groups, one assessing positive feelings and the other negative feelings, with each group consisting of three questions worth nine points in total. The second section comprises of three questions worth nine points related to mental capacity, while the third section consists of three questions worth nine points regarding mental quality. The fourth section includes three questions worth nine points concerning supporting factors. Each equation consists of four scales, ranging from 0 (none) to 3 (very much), resulting in a total of 45 points. The National Statistical Organization of Thailand has categorized mental health grades into three groups based on total points. The first group comprises individuals with a lower mental health grade point than the general population if the score falls within the range of 0.00–27.00 points. The second group includes individuals with an equal mental health grade point to the general population if the score falls within the range of 27.01–34.00 points. The third group comprises individuals with a higher mental health grade point than the general population if the score falls within the range of

34.01–45.00 points. After categorizing the grades, the National Statistic Organization of Thailand calculated the percentage of individuals with a lower mental health grade point than the general population in each province of Thailand [31]. The descriptive statistics can be found in Table 2 .

Table 2 shows the mental health grade point averages ($Happiness$) in different regions of Thailand. Maha Sarakham has the highest Happiness, while Samut Prakan has the lowest due to air pollution and waste issues from factories. Additionally, Sukhothai has the highest percentage of individuals with lower mental health grades ($Lower$) compared to the general population, while Narathiwat has the lowest.

## 5 Empirical Results

Table 3 shows the number of nonzero coefficients for each model. The variables selected by Lasso for model 1 are $lnGPPper_{it}$, $Happiness_{it-1}$, $Inflation_{it}$, and $Suspect_{it}$, while for model 2 they are $GPPper_{it}$, $Happiness_{it-1}$, $Inflation_{it}$, $Suspect_{it}$, and $Expenditure_{it}$. The results of estimating models 1 and 2 using OLS are shown in Tables 4 and 5, respectively. Similarly, Table 3 shows that models 3 and 4 have 2 and 5 nonzero coefficients, respectively. The variables selected by Lasso for model 3 are $lnGPPper_{it}$ and $Lower_{it-1}$, and for model 4 they are $GPPper_{it}$, $Lower_{it-1}$, $Inflation_{it}$, $Suspect_{it}$, and $Expenditure_{it}$. The results of estimating models 3 and 4 using OLS are shown in Tables 6 and 7, respectively.

The results of factors affecting happiness show a diminishing relation and linear relation to Gross Province Product per capita ($GPPper$) in Tables 4 and 5. The Minimum Bayes Factor (MBF) of $lnGPPper_{it}$ provides weak evidence, while the MBF of $GPPper_{it}$ provides moderate evidence, meaning there is no Easterlin Paradox in the case of a diminishing relationship, but it holds in the case of a linear relationship. The MBF of $Happiness_{it-1}$ in both cases provides decisive evidence, implying that happiness has an autoregressive or an inertial force. $Inflation_{it}$ and $Suspect_{it}$ affect happiness in both cases, with the MBF of $Inflation_{it}$ and $Suspect_{it}$ providing moderate and strong evidence. The coefficient of $Inflation_{it}$ and $Suspect_{it}$ show that an increase in inflation in the province leads to an increase in happiness, while an increase in the number of criminal suspects in the province leads to a decrease in happiness. Additionally, Table 5 has Expenditure in the model following Lasso selection, but the MBF of $Expenditure_{it}$ provides weak evidence. This means that Expenditure does not significantly affect happiness.

Tables 6 and 7 show the results of factors that affect the relationship between $Lower$ and $GPPper$. The Minimum Bayes factor (MBF) for $lnGPPper_{it}$ indicates weak evidence, while the MBF for $GPPper_{it}$ suggests strong evidence. This indicates that there is only a linear correlation between the percentage of individuals with a lower mental health grade point compared to the general population ($Lower$) and $GPPper$. Furthermore, in both cases (diminishing and linear relation), the MBF for $Lower_{it-1}$ provides strong evidence, indicating that Lower has autoregressive properties.

**Table 2** Data Description

| | Happiness | Lower | GPPper | Expenditure | Deaths | Poor | Medical Staff | Man |
|---|---|---|---|---|---|---|---|---|
| Max | 37.39 | 24.30 | 115.60 | 33,823.84 | 44,645.00 | 531.55 | 54,305.00 | 2,625,938.00 |
| Mean | 33.92 | 10.05 | 18.12 | 19,173.31 | 6,512.18 | 61.66 | 3,111.35 | 420,461.45 |
| Min | 30.57 | 0.50 | 5.02 | 11,532.11 | 1,034.00 | 0.00 | 611.00 | 91,825.00 |
| Observations | 77 | 77 | 77 | 77 | 77 | 77 | 77 | 77 |
| | Waste | Divorce | Suspect | Transparency | Employ | Inflation | | |
| Max | 12,281.00 | 15,567.00 | 2,314.00 | 85.91 | 5,287,678.31 | 0.67 | | |
| Mean | 900.29 | 1,571.57 | 219.86 | 68.33 | 492,558.45 | −0.86 | | |
| Min | 140.00 | 231.00 | 57.00 | 57.81 | 106,473.73 | −2.57 | | |
| Observations | 77 | 77 | 77 | 77 | 77 | 77 | | |

**Table 3** Result from LASSO

| Dependent | Equation | Lambda | Number of nonzero coefficient | Out-of-sample R-squared | CV mean prediction error |
|---|---|---|---|---|---|
| Happiness | Model 1 | 0.11964 | 4 | 0.1168 | 2.24042 |
| | Model 2 | 0.10045 | 5 | 0.1778 | 2.08550 |
| Lower | Model 3 | 0.87431 | 2 | 0.0131 | 32.76349 |
| | Model 4 | 0.51570 | 5 | 0.0813 | 30.50044 |

**Table 4** Results of factor affecting to Happiness in diminishing relation (Model 1)

| | Coefficient | MBF | Interpretation |
|---|---|---|---|
| $lnGPPper_{it}$ | 0.07462 | 0.96156 | Weak |
| $Happiness_{it-1}$ | 0.60678 | 0.00034 | Decisive |
| $Inflation_{it}$ | 0.46975 | 0.24385 | Moderate |
| $Suspect_{it}$ | −0.00100 | 0.03232 | Strong |
| $Constant$ | 15.21774 | 0.01015 | Strong |

**Table 5** Results of factor affecting to Happiness in linear relation (Model 2)

| | Coefficient | MBF | Interpretation |
|---|---|---|---|
| $GPPper_{it}$ | 0.01165 | 0.26059 | Moderate |
| $Happiness_{it-1}$ | 0.62830 | 0.00013 | Decisive |
| $Inflation_{it}$ | 0.42333 | 0.32465 | Moderate |
| $Suspect_{it}$ | −0.00112 | 0.02683 | Strong |
| $Expenditure_{it}$ | −0.00003 | 0.74345 | Weak |
| $Constant$ | 15.14997 | 0.00746 | Very strong |

**Table 6** Results of factor affecting to Lower in diminishing relation (Model 3)

| | Coefficient | MBF | Interpretation |
|---|---|---|---|
| $lnGPPper_{it}$ | 0.22497 | 0.97390 | Weak |
| $Lower_{it-1}$ | 0.30167 | 0.01581 | Strong |
| $Constant$ | 4.48327 | 0.09090 | Substantial |

In the selected model following Lasso selection, Table 7 includes $Inflation_{it}$, $Suspect_{it}$, and $Expenditure_{it}$. However, the MBF of $Inflationit_{it}$ shows weak evidence, indicating that it does not have a significant effect on $Lower$. On the other hand, the MBF of $Suspect_{it}$ and $Expenditure_{it}$ provide substantial and moderate evidence, respectively. Both $Suspect_{it}$ and $Expenditure_{it}$ show a positive correlation with $Lower_{it}$, meaning if the number of criminal suspects and average household

**Table 7** Results of factor affecting to Lower in linear relation (Model 4)

| | Coefficient | MBF | Interpretation |
|---|---|---|---|
| $GPPper_{it}$ | −0.07332 | 0.02474 | Strong |
| $Lower_{it-1}$ | 0.30579 | 0.01145 | Strong |
| $Inflation_{it}$ | −0.66581 | 0.83023 | Weak |
| $Suspect_{it}$ | 0.00399 | 0.04394 | Substantial |
| $Expenditure_{it}$ | 0.00028 | 0.31499 | Moderate |
| $Constant$ | −0.44832 | 0.99025 | Weak |

expenditure per month in the province increase, the percentage of individuals with a lower mental health grade point compared to the general population ($Lower$) also increases.

## 6 Conclusion

The goal of this study is to explore the various factors that contribute to happiness and unhappiness in Thailand. The level of unhappiness is determined by calculating the percentage of individuals with a lower mental health grade point compared to the general population. This study analyzes both linear and diminishing relationships with Gross Province Product per capita. To select variables for the model, the Lasso linear regression technique is used. After selecting the variables, the study utilizes Ordinary Least Square to test both the Easterlin Paradox and the hedonic treadmill. In our study, we analyzed mental health grade point averages and the percentage of individuals with lower mental health grades compared to the general population in 77 provinces. The data was collected between 2018 and 2020, while all other variables were only based on data from 2020.

The result shows that the relationship between Gross Province Product per capita and happiness is only observable in a linear regression form, confirming the Easterlin Paradox in a linear relationship. This finding is consistent with earlier studies conducted by Tella, MacCulloch and Oswald [44], Tella and MacCulloch [43] and Proto and Rustichini [33]. Thus, these results provide definitive support for the Easterlin Paradox. It was found that there is a negative and statistically significant relationship between Gross Province Product (GPP) per capita and the risk of unhappiness, but only when analyzed through linear regression.

The results of the hedonic treadmill test indicate a link between past levels of happiness and current levels of happiness, as well as a connection between past risk of unhappiness and current risk of unhappiness. It is suggested that happiness and unhappiness are influenced by an inherent force, and the concept of the hedonic treadmill may not be applicable here. This supports previous research by Wundera and Schwarzec [49] and Bottan and Truglia [5] which found a positive and significant coefficient on lagged happiness.

In addition, there are several factors that influence happiness and the risk of unhappiness, but they do so in opposite ways. These factors include the number of criminal suspects, the inflation rate, and the average monthly household expenditure. However, only the number of criminal suspects has a statistically significant impact on happiness and the risk of unhappiness. The findings presented in this statement are consistent with the results of a previous study conducted by Tella and MacCulloch [43]. Their research also demonstrated that there is a negative correlation between crime rates and levels of happiness. It is clear from this information that it is essential for governments and other relevant authorities to prioritize law enforcement in order to maintain the safety and security of citizens. By doing so, they can help to promote a sense of happiness and well-being among the population, while also reducing the risk of unhappiness.

**Acknowledgements** This research was partially supported by Chiang Mai University.

## References

1. Altman, M.: Behavioral Economics for Dummies. Canada. Wiley Canada, Ltd., 360 pages (2012)
2. Angeles, L.: A closer look at the Easterlin Paradox. J. Socio-Econ. **40**, 67–73 (2011)
3. Blanchflower, D.G., Oswald, A.J.: Well-Being over Time in Britain and the USA. J. Publ. Econ. **88**, 1359–1387 (2004)
4. Boonyakunakorn, P., Nunti, C., Yamaka, W.: Forecasting of Thailand's rice exports price: based on ridge and lasso regression. In: Proceedings of the 2nd International Conference on Big Data Technologies, pp. 354–357 (2019)
5. Bottan, N.L., Truglia, R.P.: Deconstructing the hedonic treadmill: is happiness autoregressive? J. Socio-Econ. **40**, 224–236 (2011)
6. Brickman, P., Campbell, D.T.: Hedonic relativism and planning the good society. In: Appley, M.H. (ed.), Adaptation-level Theory. Academic Press, New York (1971)
7. Brickman, P., Coates, D., Janoff-Bulman, R.: Lottery winners and accident victims: is happiness relative? J. Pers. Soc. Psychol. **36**(8), 917–927 (1978)
8. Clark, A.E., Oswald, A.J.: Unhappiness and unemployment. Econ. J. **104**(424), 648–659 (1994)
9. Cordero, J.M., Salinas-Jiménez, J., Salinas-Jiménez, M.M.: Exploring factors affecting the level of happiness across countries: a conditional robust nonparametric frontier analysis. European J. Oper. Res. **256**, 663–672 (2017)
10. Diener, E., Scollon, C.N., Lucas, R.E.: The evolving concept of subjective wellbeing: the multifaceted nature of happiness. Adv. Cell Aging Gerontol. **15**, 187–219 (2003)
11. Easterlin, R.A. Does economic growth improve the human lot? Some empirical evidence. Nations and Households in Economic Growth, pp. 89–125 (1974)
12. Easterlin, R.A.: Will raising the incomes of all increase the happiness of all? J. Econ. Behav. Organ. **27**(1), 35–48 (1995)
13. Easterlin, R.A.: Explaining happiness. Proc. Natl. Acad. Sci. **100**(19), 11176–83 (2003)
14. Fitzgerald, J.: The difference between depression and sadness. Medically reviewed by Timothy J. Legg. Available via DIALOG (2019). https://www.medicalnewstoday.com/articles/314418

15. Frey, B.S., Stutzer, A.: Happiness and Economics: How the Economy and Institutions Affect Human Well-Being. Forthcoming with Princeton University Press (2002)
16. Graham, C., Pettinato, S.: Happiness and Hardship: Opportunity and Insecurity in New Market Economies. The Brookings Institution Press, Washington, D.C. (2002)
17. Graham, C.: The economics of happiness: insights on globalization from a novel approach. World Econ. **6**(3), 41–55 (2005)
18. Griffin, J.: Well-being: its Meaning, Measurement, and Moral Importance. Clarendon Press, Oxford, England (1986)
19. Held, L., Ott, M.: On p-Values and Bayes Factors. Ann. Rev. Stat. Appl. **5**, 393–419 (2018)
20. Helliwell, J.F., Layard, R., Sachs, J.D., Aknin, L.B., De Neve, J.-E., Wang, S. (eds.), World Happiness Report 2021, (9th ed.) (2021)
21. Helliwell, J.F., Layard, R., Sachs, J.D., Aknin, L.B., De Neve, J.-E., Wang, S. (eds.), World Happiness Report 2023 (11th ed.) (2023)
22. Huta, V., Waterman, A.S.: Eudaimonia and its distinction from hedonia: developing a classification and terminology for understanding conceptual and operational definitions. J. Happiness Stud. **15**, 1425–1456 (2014)
23. Intelisano, S., Krasko, J., Luhmann, M.: Integrating philosophical and psychological accounts of happiness and well-being. J. Happiness Stud. **21**, 161–200 (2020)
24. Kubin, W., Xie, Y., Bokati, L., Kreinovich, V., Autchariyapanitkul, K.: In: Sriboonchitta, S (ed.), Why Geometric Progression in Selecting the LASSO Parameter: A Theoretical Explanation (2022)
25. Lapha, R., Waiwitriyakul, N., Thowannang, R.: Happiness and associated factors among elderly people living in the urban community of Udonthani Province. Community Health Dev. Q. Khon Kaen Univ. **6**(3), 411–422 (2018). (In Thai)
26. Mehrdadi, A., Shahnaz, S., Direkvand-Moghadam, A., Hashemian, A.: Factors affecting happiness: a cross-sectional study in the Iranian Youth. J. Clin. Diagn. Res. **10**(5), VC01–VC03 (2016)
27. Ng, Y-K.: Happiness-Concept, Measurement and Promotion. Springer e-book (2022). https://doi.org/10.1007/978-981-33-4972-8
28. Nguyen, P.V., Nguyen, P.T., Huynh, V.D.B., Nguyen, Q.L.H.T. T.: Critical Factors Affecting the Happiness: A Vietnamese Perspective. MPRA Paper No. 96587, Available via DIALOG. https://mpra.ub.uni-muenchen.de/96587/
29. Nikhamyang, C., Noikhamyang, P.: Factors Affecting Work Happiness Index of the Library Staff at the Central Library, Srinakharinwirot University. (In Thai) (2012)
30. National Anti-Corruption Commission of Thailand. Integrity and Transparency Assessment Report 2020 (2020). Available via DIALOG. http://www.stopcorruption.moph.go.th/ (In Thai)
31. National Statistic Organization of Thailand, Summary of Key Results of the Mental Health Survey (Happiness) of People in Thailand 2020, 50 pages (In Thai) (2021)
32. Nussbaum, M.: Human functioning and social justice: in defense of Aristotelian essentialism. Polit. Theory **20**, 202–246 (1992)
33. Proto, E., Rustichini, A.: A reassessment of the relationship between GDP and life satisfaction. PLOS ONE **11**(8), 1–10 (2013)
34. Prukkanone, B., Kittirattanapaiboon, P., Tanaree, A., Supreda, B.: Happiness in Thai Population: a 2018 National Survey Analysis at Provincial Level. J. Mental Health Thailand **30**(1), 75–86 (2022). (In Thai)
35. Schaffner, A.K.: Hedonic versus Eudaimonic Wellbeing: How to Reach Happiness (2023). Available via DIALOG. https://positivepsychology.com/hedonic-vs-eudaimonic-wellbeing/
36. Seligman, M.E.P., Royzman, E.: Happiness: The Three Traditional Theories. The University of Pennsylvania (2003). Available via DIALOG. https://www.authentichappiness.sas.upenn.edu/newsletters/authentichappiness/happiness
37. Sen, A.: Commodities and Capabilities. North-Holland, Amsterdam (1985)
38. Senasu, K.: Factors Affect Happiness in Thailand. Research report. National Institute of Development Administration (NIDA) (2016). Abstract from (In Thai)

39. Slag, M., Burger, M.J., Veenhoven, R.: Did the Easterlin Paradox apply in South Korea between 1980 and 2015? A case study. Int. Rev. Econ. **66**, 325–351 (2019)
40. Stevenson, B., Wolfers, J.: Economic Growth and Subjective Well-being: Reassessing the Easterlin Paradox. Nber Working paper Series. Working Paper 14282, 79 pages (2008)
41. Tansuchat, R., Khamkaew, T.: Happiness and well-being index of community members of the Royal Project Foundation. Chiang Mai Univ. J. Econ. **13**(2), 1–27 (2009). (In Thai)
42. Teeranupattana, C.: Factors Affecting Happiness: Comparison between Residents of a Community and Thai Rotary Club Members in Bangkok. Master of Science (Human Resource and Organization Development) School of Human Resource Development, National Institute of Development Administration (NIDA), 121 pages, (In Thai) (2020)
43. Tella, R.D., MacCulloch, R.: Gross national happiness as an answer to the Easterlin Paradox? J. Dev. Econ. **86**, 22–42 (2008)
44. Tella, R.D., MacCulloch, R., Oswald, A.J.: The macroeconomics of happiness. Rev. Econ. Stat. **85**(4), 809–827 (2003)
45. Tov, W.: Well-being concepts and components. In: Diener, E., Oishi, S., Tay, L. (eds.) Handbook of Well-Being. DEF Publishers, Salt Lake City, UT (2018)
46. Tyc, V.L.: Psychosocial adaptation of children and adolescents with limb deficiencies. Clin. Psychol. Rev. **12**, 275–291 (1992)
47. Wannapong, S., Hong-ngam, J.: Factors affecting life happiness of elderly people: an empirical study in Amphoe Muang, Khon Kaen Province. J. Econ. Manag. Strat. **5**(2), 21–32 (2018). (In Thai)
48. World Happiness Report (2023). About. Available via DIALOG. https://worldhappiness.report/about/
49. Wundera, C., Schwarzec, J.: What (if anything) do satisfaction scores tell us about the intertemporal change in living conditions? SOEPpapers on Multidisciplinary Panel Data Research 306, DIW Berlin, The German Socio-Economic Panel (SOEP) (2010)

# Financial Development and Economic Growth: A Case of Developing Countries

**Diem Thi Thuy Pham and Hoai Trong Nguyen**

**Abstract** This study analyses the effect of financial development on economic growth in 64 developing countries during the period 2003–2017 using a Bayesian model averaging approach. The results indicate that financial development has a significant U-shaped effect on economic growth, providing new insight concerning the relationship between financial development and economic growth. We also demonstrate that investment to gross domestic product ratio and foreign direct investment have significant positive effects on economic growth, whereas population growth has a significant negative impact on economic growth.

**Keywords** Financial development · Economic growth · Developing countries · Bayesian model averaging

**JEL Classifications** C11 · N2 · O1 · O4

## 1 Introduction

The relationship between financial development and economic growth has traditionally been a central issue in development economics research over the past few decades [5, 23, 36, 56, 69, 76]. Most empirical studies conclude that financial systems influence savings and investment decisions, leading to economic growth in developed and developing countries [68, 103]. Specifically, Levine [68, p. 869] suggests that the overall functions of financial systems include the reduction of information, enforcement, and transaction costs through five channels of (i) producing information ex-ante regarding possible investments and allocation of capital,(ii) monitoring investments and corporate governance following financial provision,(iii) facilitating the trade,

D. T. T. Pham (✉)
Ho Chi Minh University of Banking, Ho Chi Minh City, Vietnam
e-mail: diemptt@hub.edu.vn; diempham.edu.vn@gmail.com

D. T. T. Pham · H. T. Nguyen
University of Economics Ho Chi Minh City, Ho Chi Minh City, Vietnam

 
V. Kreinovich et al. (eds.), *Applications of Optimal Transport to Economics and Related Topics*, Studies in Systems, Decision and Control 556,
https://doi.org/10.1007/978-3-031-67770-0_20

diversification, and management of risk; (iv) mobilizing and pooling savings; and (v) easing the exchange of goods and services. Financial system efficiency refers to the effective performance of the five basic functions, and financial development implies improving financial systems' efficiency.

Over the last two decades, developing economies have experienced a wave of financial sector reforms facilitated by the elimination of government restrictions on the financial system in terms of interest rates, high reserve requirements, and quantitative restrictions on credit allocation to promote economic growth [75, 93]. However, economic growth in developing economies has been unsteady, suffering reversals at certain times during the period 2003–2017. Relatively rapid economic growth was experienced during 2003–2007, a period of relatively high financial development. Economic growth suddenly decreased during 2007–2009, dramatically reversing in 2009, followed by stagnation in economic growth that prevailed from the latter period 2010–2017 (Fig. 1). In order to get a better understanding of this growth stagnation, this study aims to provide new evidence on the impact of financial development on economic growth in developing countries over the period 2003–2017.

Our analysis extends the literature in six respects. First, research on the relationship between financial development and economic growth that considers openness factors (both trade and financial openness) has received considerably less attention, particularly for developing countries. The present study aims to close this gap in the literature by investigating whether and how financial development and

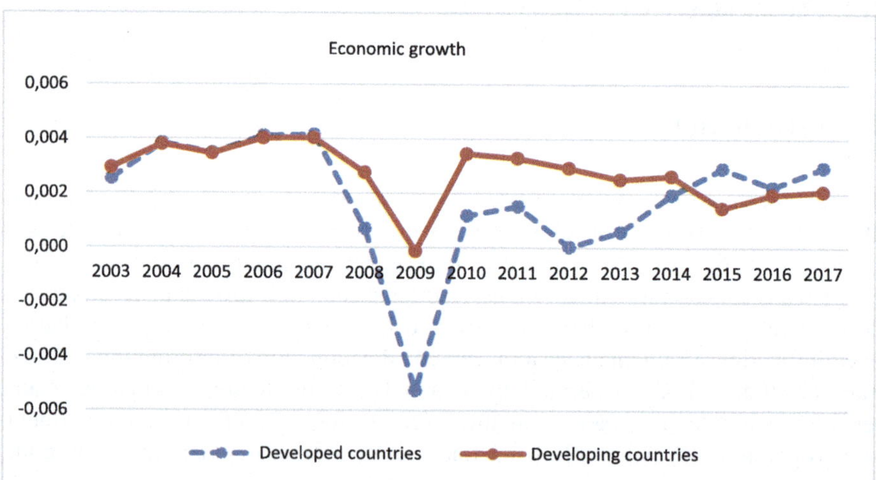


Note: Economic growth was transformed into natural logarit of (1+Real GDP per capita growth)

Source: World Development Indicators (WDI) and authors'calculations

**Fig. 1** Economic growth in developing countries over the period 2003–2017

openness influence financial development in developing countries. Second, previous empirical studies primarily focus on panel data methods based on traditional or frequentist statistics, whereas no studies examine the impact of financial development on economic growth using approaches based on Bayesian statistics. Hence, the study applies the Bayesian model averaging (BMA) approach proposed by several seminal studies, such as [44, 85–87], to consider model uncertainty. Third, there are several controversies regarding the measurement of financial development. The ratio of private credit to gross domestic product (GDP) is widely used as a proxy for financial development [13, 46] and has often been assumed as the most suitable indicator for the characteristics of developing countries. Fourth, there is a lack of research to investigate the impact of financial development on economic growth for developing economies examining legal origins variables as a crucial determinant of the protection of corporate shareholders and creditors in a financial system. Fifth, the study examines whether the interaction between trade openness and financial openness and the interaction between trade openness and institutional quality has a significant influence on economic growth. Finally, the inclusion of regional dummy variables as control variables increases the explanatory power of levels of economic growth across regions.

This study is organized as follows. The next section, Sect. 2, presents the related literature. Section 3 defines the variables, outlines the methodology, and describes the dataset. Section 4 depicts a discussion of the empirical findings. Finally, Sect. 5 provides the concluding remarks.

## 2 Literature Review

First, some studies demonstrate that a country with a higher growth-enhancing effect that is closer to a well-developed financial market provides strong support to the proposition that more finance leads to more economic growth. The first division of researchers focuses on both developed and developing countries. One of the earliest studies of this type is by [57]. They investigate the relationship between financial depth and economic growth by examining the data of 77 countries over the period 1960–1989 and revealing a statistically significant positive relationship between financial development and economic growth. Levine and Zervos [65] use a two-stage least squares analysis for 47 countries over the period 1976–1993 to examine the relationship between long-term stock market development, banking development, and economic growth, determining that banking development and stock market liquidity are both positive and correlated with output growth, capital accumulation, and productivity growth. Levine et al. [69] use generalized method of moments (GMM) estimators to elicit international evidence on the impact of financial intermediation development on economic growth using a dataset of 71 countries during the 1960–1995 period. The results indicate a positive link between financial development and economic growth. Calderón and Liu [18] examine the direction of causality between financial development and economic growth using the Geweke decomposition test on the panel data of 109 developing and industrial countries for the period 1960–1994. The findings demonstrate that financial development leads to real GDP per capita growth as a proxy for economic growth. Furthermore, a causal relationship is revealed between financial development and economic growth. Using data from 40 countries for the period 1976–1998, Beck and Levine [68] apply GMM estimators to new data to re-examine the relationship between stock market development, banking development, and economic growth. The results indicate that stock market and banking development positively influence economic growth. Using data from 125 countries from 1990 to 2008, [35] employ the least squares dummy variable model to investigate the relationship between financial development and economic growth in developing countries in Asia. The findings indicate that bank and stock market development and overall financial development exert a significant positive effect on economic growth in both developed and developing countries. Bangake and Eggoh [8] employ cointegration tests and panel cointegration analysis for 71 developed and developing countries for the period 1960–2004 to estimate the long-term relationship between financial development and economic growth. The evidence suggests a single long-run equilibrium between financial development, GDP per capita, and control variables, clearly supporting the existence of bidirectional causality between finance and economic growth. Bittencourt [16] investigates the role of financial development in generating economic growth in four Latin American countries for the period 1980–2007. Based on panel time-series analysis, the results support the Schumpeterian prediction, indicating that financial development authorises entrepreneurs to invest in productive activities and improves economic growth. Caporale et al. [19] investigate the relationship between financial development and economic growth using a dynamic panel model for ten new European Union members from 1994 to 2007. The evidence indicates that the effects of stock and credit markets on economic growth are limited due to a lack of financial depth, however, a more efficient banking sector could improve economic growth. A Granger causality test shows that causality runs from financial development to economic growth but not in the opposite direction. Valickova et al. [97] analyse 1,334 estimates from 67 studies to examine the relationship between financial development and economic growth using a meta-analysis. The findings demonstrate that stock markets accelerate economic growth faster than other financial intermediaries.

The second division of studies focuses on developing economies, predominantly on finding a positive relationship between financial development and economic growth. For example, Al-Yousif [4] uses a Granger causality test within an error correction framework to examine the causality between financial development and economic growth. Their dataset comprises 30 developing countries for the period 1977–1999. The results suggest that the causality between financial development and economic growth is bidirectional, however, the findings are country-specific and depend on the proxies used to estimate financial development. Using the autoregressive distributed lag (ARDL) approach, [48] investigate the relationship between financial development and economic growth in Sierra Leone from 1970 to 2008. The results suggest that financial development generally spurs economic growth through the investment channel. Using data from 286 Chinese cities for the period 2001–2006, [102] examine an empirical relationship between financial development and economic growth at the city level. Based on system GMM estimators for dynamic panel data, the results demonstrate that financial development is positively associated with economic growth. Based on a Cobb–Douglas production, [96] investigate the relationship between financial development and economic growth in Kenya from 1971 to 2011. Using the autoregressive distributed lag (ARDL) bounds test approach and the error correction model, the results indicate that the development of the financial sector positively affects economic growth. Adu et al. [3] examine the long-run growth effects of financial development in Ghana over the period 1961–2010 using the autoregressive distributed lag model (ARDL). The findings reveal that the positive or negative impact of financial development on economic growth depends on the indicator used to proxy for financial development.

Second, several other studies purport the existence of a threshold impact of financial development on economic growth. For instance, Cecchetti and Kharroubi [20] suggest an inverted U-shaped relationship between financial sector size and productivity growth. It implies that there is a point at which the further enlargement of financial development may reduce real economic growth. In addition, financial sector growth may become a drag on productivity growth which can be explained because the financial sector competes with the rest of the economy for scarce resources. In general, financial booms are not growth-enhancing. Arcand et al. [7] also suggest that a negative finance–growth link, wherein financial development begins to have a negative effect when credit to the private sector reaches 100% of the GDP, occurs in high-income countries. These authors indicate that their findings support the "vanishing effect" hypothesis of financial development. This phenomenon is not driven by low institutional quality, output volatility, banking crises, or differences in the banking regulatory and supervisory environment. Rioja and Valev [90] suggest a threshold effect in the relationship between financial development and economic growth. In particular, financial development has a significant positive effect on economic growth only if it has reached a certain level of financial development. These authors assert that financial development from low- to high-level has a significant influence in shaping the impact of finance on economic growth. Financial development has a large positive effect on economic growth in economies at an intermediate level, but this effect is smaller in economies at high levels of financial development. However, financial development is found to be insignificant in promoting economic growth in economies with a low level of financial development. Shen and Lee [94] also find an inverse U-shaped link between financial development and economic growth. Applying an innovative dynamic panel threshold technique to new data, [62] analyse the relationship between finance and economic growth for 87 developed and developing countries from 1980 to 2010, demonstrating a finance threshold in the finance–growth nexus.

Third, previous studies indicate differences in the relationship between financial development and economic growth with varying income levels. For instance, Rioja and Valev [91] provide evidence of an insignificant link between financial development and economic growth in low-income economies, whereas a significant positive relationship is found for middle-income economies; however, this relationship is weakly significant in high-income economies. Conversely, [28, 45] demonstrate that the positive effect on economic growth from financial development is much less significant in high-income economies than in low- and middle-income economies. There is a contradiction between these results on the financial development and economic growth link at different income levels, and the nonlinear link between financial development and economic growth. Estrada et al. [36] use the Arellano–Bond GMM approach to examine the impact of financial development on economic growth for 108 countries (of which 20 are developing Asian economies) from 1977 to 2011, finding that the development of financial systems contributes to overall economic growth. The positive impact of financial development on economic growth is more substantial in developing countries than in advanced economies.

Lastly, some studies do not support the relationship between financial development and economic growth. Lucas [72, p. 6] asserted, "the importance of financial matters is very badly over-stressed in popular and even much professional discussion." Using data from 15 Middle East and North African countries for the period 1980–2007, [47] investigate the direction of causality between financial development and economic growth. Based on the seemingly unrelated regressions and panel causality testing approach developed by [59], the results demonstrate no evidence of bidirectional causality between financial development and economic growth for all measurements of financial development. Menyah et al. [76] apply the panel bootstrapped approach to Granger causality to investigate data from 21 African countries from 1965 to 2008, finding no evidence that either financial development or trade liberalization contributes to economic growth. Moreover, they do not find any causality between financial development and real GDP per capita.

Overall, multiple studies have examined the impact of financial development on economic growth, adopting a variety of econometric techniques, such as cross-sectional, time-series, panel data, and international- and country-level studies [11, 15, 56, 57, 66, 68, 69, 88], however, the findings are mixed. Moreover, the multidimensional nature of financial development is one of the most critical explanations for the inconsistent impact of financial development on economic growth.

## 3 Variables, Methodology, and Data

### *3.1 Construction of Variables*

**Financial development**. In the context of developing countries, we employ the ratio of private credit to GDP as a measure of financial development, which is the most common indicator of the nature of financial development in these countries. The indicator shows the overall growth of private banking markets by excluding credit provided to the public sector and credit issued by the central bank and development banks [13, 46]. This method is more appropriate for measuring financial activity in bank-based financial systems because it specifically focuses on the role of financial intermediaries and financial markets in funding the private entrepreneurial sector [28].

**Economic growth**. According to Levine [66], several popular indicators for economic growth include real per capita GDP growth, productivity growth, and average per capita capital stock growth. Therefore, this study employs real per capita GDP growth, as suggested in previous studies [16, 18, 19, 65, 68, 69, 76, 90, 91, 102].

**Control variables**. The study also includes several control variables to account for other factors affecting economic growth as follows.

*The ratio of investment to GDP*. Levine [68] suggests that financial institutions and markets can promote economic growth through several channels, including investment as a proxy for capital accumulation, one of the most relevant determinants of the growth process. Including the ratio of investment to GDP as a conditioning factor allows us to assess the channels through which financial institutions and markets can stimulate economic growth by boosting productivity or boosting savings resources [68]. Hence, this indicator is the most relevant factor of the growth process as a proxy for capital accumulation.

*Inflation*. The inflation rate is an explanatory variable as it might distort decision-making. More specifically, moderate-to-high levels of inflation may discourage financial intermediation and stimulate saving for real assets. Therefore, it may act as a proxy variable for macroeconomic stability [17, 32, 46]. Here, we use consumer prices (annual percentage) as a proxy for inflation.

*Government consumption*. This study uses government expenditures as a share of GDP to control government size. Keynes [52] asserts that government expenditure may be used to boost aggregate demand to bring an economy out of recession,therefore, this measure is also considered an indicator of macroeconomic stability [39, 69].

*Human capital*. To measure human capital, the average years of secondary schooling in the population above 25, referencing [9], is employed as a proxy variable for the national endowment of human capital [95]. This indicator is expected to have a positive effect on economic growth by influencing productivity [19].

*Trade openness*. As for the trade openness indicator, we follow [6, 27, 56, 57, 64, 68, 69, 89], and others. Trade openness is measured by the sum of exports and imports of goods and services over GDP. Trade openness illustrates the extent of actual exposure to trade interactions and accounts for the effective level of integration [53–55]. More openness to trade is related to greater economic growth [98, 99].

*Financial openness*. While there is a wealth of evidence from studies to support the view that trade openness leads to faster economic growth [98, 99], there are few arguments to support the positive impact of financial liberalization on economic growth [5, 42, 67]. Conversely, for the case of financial openness, the link between financial openness and economic growth reveals mixed results, and there is a lack of evidence on this issue for developing countries in particular [60, 80]. Therefore, to measure financial openness, we use the [24].

*Legal origins.* According to [67], the law and finance theories stress two channels via which legal origin influences financial development: (i) the rights of private property owners, (ii) the legal adaptability channel. Therefore, legal origin plays a crucial role in shaping international differences in financial development. Following this line of reasoning, [89] demonstrate that only a portion of financial development is dependent on the inherited legal systems, revealing significant variation in the degree of financial development in the last century. Based on data availability, the legal origins include British, French, and Socialist legal systems.

*Foreign direct investment.* Theoretically, foreign direct investment (FDI) may help promote economic growth in recipient countries based on the benefits of enhancing technological change via the spillover effects of knowledge and new capital goods, which we collectively refer to as the process of technological diffusion. However, the magnitude of FDI contributions depends on the overall business climate in recipient countries [21].

*Institutional quality.* "Institutions are the humanly devised constraints that structure political, economic, and social interaction. They consist of both informal constraints (sanctions, taboos, customs, traditions, and codes of conduct), and formal rules (constitutions, laws, property rights)" [78, p. 97]. Acemoglu et al. [1] acknowledge that economic institutions are the fundamental source of long-term growth. Moreover, researchers have argued that the proposition of "better finance, more growth" is more accurate than "more finance, more growth." Therefore, the development of financial markets, along with a sound institutional framework, is asserted to be more important to economic growth [29, 30, 79]. We measure institutional quality by the mean of the six dimensions of governance referencing [50].

*Population growth.* Rapid population growth is likely to contribute to economic sluggishness in less-developed countries. According to the Malthusian theory of population, rapid population growth impedes economic growth [74]. Conversely, alternative views suggest that rapid labour force growth can be economically beneficial [51, 61]. Therefore, evidence for the relationship between population growth and economic growth has shown mixed results.

*The square of financial development.* As evidence suggests, we also explore the potential financial development threshold in the finance–growth relationship [2, 20, 92, 94].

*Other variables.* We control for institutions and their interactions with trade and financial openness to determine whether variations in institutional quality interacted with trade and financial openness influencing economic growth [58]. Moreover, the assessment of the interactive effects of trade openness and financial openness on economic growth is used for evaluating the simultaneity of openness, as argued by [89]. In contrast, some previous studies have examined the effects of trade openness and financial openness separately, e.g., in [3, 36, 76], among others. To investigate whether the effects of trade/financial openness on economic growth depend on the country's level of development, this study includes interactive terms between trade/financial openness and a logarithm of real GDP per capita [33]. Finally, the inclusion of regional dummies and time trends allows us to explain differences among regions and over time. For more details on the variables, Table 3 (see Appendix 1) shows brief definitions of the variables used in the study.

## 3.2 Methodology

The issue of model uncertainty has received considerable attention in both statistics and econometrics literature for decades [22, 26, 31, 82, 83, 101]. However, classical methods based on traditional or frequentist statistics provide little guidance on managing model uncertainty [25, 40, 44, 85–87]. Although several studies attempt to explore the impact of financial development on economic growth, almost all of them use popular techniques for panel data based on standard statistics practice, such as ordinary least squares [20, 94], the GMM estimator, the system GMM estimator, and/or the GMM estimator for dynamic panel data [19, 36, 68, 69, 90, 91, 102], two-stage least squares and/or the three-stage least squares [34, 57, 65], and/or autoregressive distributed lag [3, 48, 96]. Notably, these studies fail to consider model uncertainty, meaning that there is a persistent scarcity of empirical work considering the issue of model uncertainty in economic growth research, particularly in developing countries. Therefore, this study employs a BMA technique suggested by several seminal works, such as [44, 85–87], to consider model uncertainty for these models in the next section. Our approach closely follows [37].

Hoeting et al. [44], Fernandez et al. [37, 38], Zeugner and Feldkircher [100], Ley and Steel [70], among others, use the BMA approach in the context of linear regression. Therefore, we begin with the following linear regression model.

$$y = \gamma + \pi X_i + \psi Z_i + \upsilon_i$$

where $y$ is the real GDP per capita growth over a particular period as a proxy measure for economic growth; $\gamma$ is the constant term in the regression; $X_i$ is the ratio of private credit to GDP as a proxy measure for financial development; $Z_i$ is a set of additional control variables; and $\upsilon_i$ is the error term, assumed to have homoscedastic, normally distributed disturbances. In this context, we have $N = 64$ developing countries and a set of $K = 19$ potential explanatory variables. Particularly, Sect. 3.1 constructs all the selected variables concerning the impact of financial development on economic growth in the context of developing countries.

BMA creates a weighted average of many models that include all potential subsets of explanatory variables. Because it aggregates several different meta-regression models, BMA can be regarded as a meta-analysis of meta-analyses. Posterior model probabilities are used as weights in this process as a goodness-of-fit measure in models, similar to information criteria or R-squared (adjusted), wherein the best-fitting models have the highest posterior probability model, and vice versa. Next, the posterior inclusion probability, which is the total posterior probability of all true models, can then be computed for each explanatory variable as an indicator of its likelihood. Finally, we can obtain the posterior coefficient distribution across all "true" regressions for each explanatory variable. Based on the posterior coefficient distribution, we can deduce the posterior mean and the posterior standard deviation.

Because there are 19 explanatory variables to examine, it is not physically viable to list all $2^{19}$ conceivable combinations; this would require several years on a basic personal computer. In such cases, we use Markov Chain Monte Carlo Model Composition (MC3) proposed by [73] to investigate the model space stochastically with one million burn-ins and three million draws of the MC3 sampler. We employ the BMS (Bayesian model sampling) package for R developed by [100] for the computation. For selecting the best-fit model, we use three criteria for parameter prior, including Unit Information Prior (UIP) [38, 49], benchmark prior (BRIC) [38], and a local empirical Bayes approach (LEB) prior [41, 71]. We discuss posterior estimates under uniform model priors in the context of their application in Sect. 4.

### *3.3 Data Sources*

The study uses panel data from 64 developing countries and an updated period analysis of 2003–2017. The total number of observations is the most significant sample compared with previous empirical studies of economic growth research, such as [4, 20, 36]. The list of developing countries is based on the Human Development Index of the United Nations Development Programme (Table 1). The data utilized in this study were drawn from a variety of databases. Specifically, the ratio of private credit to GDP was collected from the Global Financial Development Database. Real GDP per capita growth, the ratio of total trade (exports and imports) to GDP, consumer prices (annual percentage), general government final consumption expenditure (per cent of GDP), the ratio of FDI to GDP, and population growth (annual percentage) were obtained from the World Development Reports. In addition, we used the data availability for the KAOPEN index of Chinn and Ito [24] based on the IMF's Annual Report on Exchange Arrangements and Exchange Restrictions (AREAER 2018). Legal origins were collected from the data provided by Porta et al. [84]. We estimate institutional quality by six governance dimensions obtained from Worldwide Governance Indicators. The ratio of investment to GDP and the human capital index was collected from the Penn World Table version 6.3.

## 4 Results and Discussion

The findings from the BMA approach are reported in Table 2 in the context of the cross-country economic growth regression model. The dependent variable is the growth rate of real GDP per capita (Y) as a proxy for economic growth over the entire period of 2003–2017. We use 19 variables from 64 developing economies, which means that our model space contains 219 (= 524,288) different models. The estimations are derived from the Markov Chain of 20,000,000 recorded draws

**Table 1** The list of selected developing countries

| |
|---|
| **Africa** (*27 developing countries*): Algeria, Angola, Benin, Botswana, Burkina Faso, Burundi, Cameroon, Central African Republic, Egypt Arab Rep., Gabon, Ghana, Kenya, Malawi, Mali, Mauritius, Morocco, Namibia, Niger, Rwanda, Senegal, Sierra Leone, South Africa, Togo, Tunisia, Uganda, United Republic of Tanzania, and Zambia |
| **Asia** (*21 developing countries*): Bangladesh, Cambodia, China, India, Indonesia, Iran (Islamic Rep.), Israel, Jordan, Kuwait, Malaysia, Mongolia, Myanmar, Nepal, Pakistan, Philippines, Singapore, Sri Lanka, Thailand, Turkey, United Arab Emirates, and Viet Nam |
| **Latin America and the Caribbean** (*16 developing countries*): Barbados, Bolivia, Brazil, Chile, Colombia, Costa Rica, Dominican Republic, Ecuador, El Salvador, Guatemala, Honduras, Jamaica, Nicaragua, Paraguay, Peru, and Uruguay |

after 10,000,000 burn-ins and 3,962,772 models are visited. The posterior expected model size is 13.465 (i.e., the average number of included regressors). Following the assumption of uninformative priors over the parameters within each model, Table 2 presents the unit information prior (UIP), the benchmark prior (BRIC), and the local empirical Bayes (LEB) approach with posterior inclusion probabilities (PIPs), posterior means (Post means), and posterior standard deviations (Post SD).

Following the literature cited above, only variables with PIPs above 0.5 are considered robust determinants of economic growth. As shown in Table 2, the ratio of private credit to GDP in measuring financial development and the square of private credit to GDP ratio have PIPs exceeding 0.5 under all three alternative prior structures on parameters; hence, these variables appear to influence economic growth significantly. More specifically, the results indicate that financial development negatively influences economic growth; however, financial development has a significant threshold effect on economic growth. Therefore, it implies that higher levels of financial development only lead to economic growth if a specific threshold of financial development has been attained. Figure 2 shows the marginal densities of financial development from BRIC.

In addition to financial development, Table 2 presents the results of estimates where the independent variables demonstrate an important role in economic growth with PIPs exceeding 0.5 under all three alternative prior structures on parameters, including population growth, investment to GDP ratio, the Socialist legal origin dummy, time trend, the Latin America dummy, FDI, and the Africa dummy. Specifically, investment to GDP ratio, Socialist legal origin dummy, and FDI have negative effects on economic growth, whereas population growth, time trend, the Latin America dummy, and the Africa dummy have positive associations with economic growth. Regarding the interaction between trade and financial openness, this interacted variable is shown to have an insignificant effect on economic growth in developing countries. Conversely, the remaining variables have PIPs of less than 0.5, including inflation, trade openness, institutional quality, government consumption, human capital, and financial openness, suggesting that these independent variables are not linked to economic growth.

**Table 2** The impact of financial development on economic growth: posterior estimates under uniform model priors

| Variable | g = BRIC | | | g = UIP | | | g = LEB | | |
|---|---|---|---|---|---|---|---|---|---|
| | PIP | Post mean | Post SD | PIP | Post mean | Post SD | PIP | Post mean | Post SD |
| Population growth | 1.000 | −1.046 | 0.092 | 1.000 | −1.046 | 0.092 | 1.000 | −1.014 | 0.091 |
| Investment | 1.000 | 0.019 | 0.004 | 1.000 | 0.019 | 0.004 | 1.000 | 0.019 | 0.004 |
| Socialist legal origin dummy | 1.000 | 0.022 | 0.005 | 0.998 | 0.022 | 0.005 | 1.000 | 0.022 | 0.005 |
| Time trend | 0.994 | −0.001 | 0.000 | 0.994 | −0.001 | 0.000 | 0.993 | −0.001 | 0.000 |
| Latin America and The Caribbean dummy | 0.982 | −0.014 | 0.004 | 0.981 | −0.014 | 0.004 | 0.999 | −0.015 | 0.004 |
| Foreign direct investment | 0.972 | 0.106 | 0.034 | 0.972 | 0.106 | 0.034 | 0.992 | 0.104 | 0.031 |
| Africa dummy | 0.898 | −0.010 | 0.005 | 0.897 | −0.010 | 0.005 | 0.950 | −0.011 | 0.004 |
| Financial development | 0.502 | −0.495 | 5.326 | 0.501 | −0.499 | 5.352 | 0.553 | −2.354 | 11.454 |
| The square of financial development | 0.502 | 0.243 | 2.663 | 0.502 | 0.245 | 2.676 | 0.553 | 1.172 | 5.727 |
| Inflation | 0.334 | −0.015 | 0.024 | 0.334 | −0.015 | 0.024 | 0.631 | −0.026 | 0.025 |
| Trade openness | 0.195 | −0.001 | 0.002 | 0.196 | −0.001 | 0.002 | 0.572 | −0.003 | 0.003 |
| Institutional quality | 0.168 | 0.001 | 0.002 | 0.168 | 0.001 | 0.002 | 0.714 | 0.005 | 0.004 |
| Government consumption | 0.149 | −0.001 | 0.003 | 0.150 | −0.001 | 0.003 | 0.462 | −0.003 | 0.004 |
| Human capital | 0.110 | −0.001 | 0.002 | 0.110 | −0.001 | 0.002 | 0.579 | 0.004 | 0.004 |
| Interaction between financial openness and institutional quality | 0.076 | 0.000 | 0.001 | 0.076 | 0.000 | 0.001 | 0.395 | −0.001 | 0.001 |
| Financial openness | 0.076 | 0.000 | 0.000 | 0.076 | 0.000 | 0.000 | 0.297 | 0.000 | 0.001 |

(continued)

**Table 2** (continued)

| Variable | g = BRIC | | | g = UIP | | | g = LEB | | |
|---|---|---|---|---|---|---|---|---|---|
| | PIP | Post mean | Post SD | PIP | Post mean | Post SD | PIP | Post mean | Post SD |
| Interaction between trade openness and institutional quality | 0.055 | 0.000 | 0.001 | 0.055 | 0.000 | 0.001 | 0.272 | 0.000 | 0.002 |
| Interaction between trade openness and financial openness | 0.042 | 0.000 | 0.000 | 0.042 | 0.000 | 0.000 | 0.240 | 0.000 | 0.001 |
| British legal origin dummy | 0.040 | 0.000 | 0.001 | 0.040 | 0.000 | 0.001 | 0.228 | 0.000 | 0.002 |

*Notes* BRIC denotes benchmark prior; UIP denotes unit information prior; LEB denotes local empirical Bayes approach; and PIP denotes posterior inclusion probability
*Source* Authors' analysis

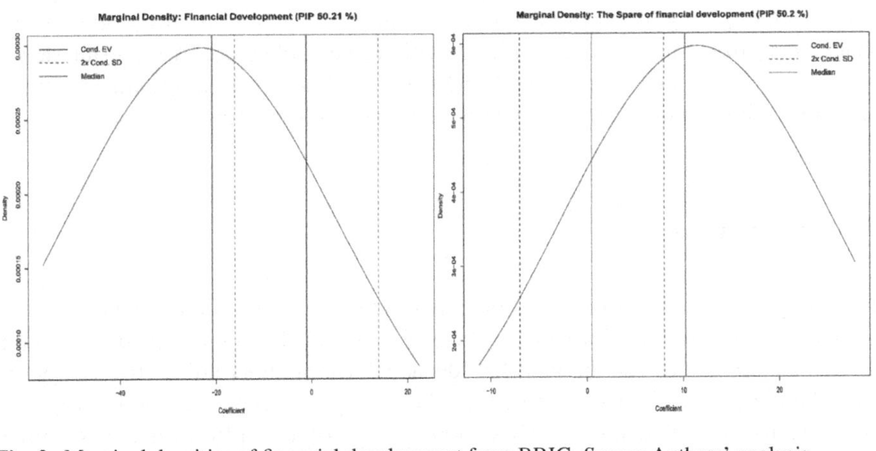


**Fig. 2** Marginal densities of financial development from BRIC. *Source* Authors' analysis

Previous studies have examined the influence of financial development on economic growth, while other potential crucial determinants of economic growth, such as macro-characteristics and institutional quality, were controlled. This approach is supported by [1, 58], among others. The results reveal a U-shaped relationship between finance and economic growth, wherein financial development (proxied by the ratio of private credit to GDP) is negative only up to a certain point, after which it begins to have a positive effect on output growth. This finding supports the proposition that superior finance may mean more growth in developing countries. Our findings are inconsistent with [7, 20, 62, 94], who suggested an inverted U-shaped curve between financial development and economic growth. These findings may be explained by the fact that, in developing countries, the impact of financial development on economic growth depends on the investment behaviour of the private sector than the public sector in the economy. That is, an increase in private lending leads to lower growth due to the private sector insufficiently developing to transform available resources into profitable investments. On the other hand, up to a certain level of financial development is required, financial development stimulates economic growth, which may be explained by active and innovative entrepreneurial activity.

The other variables are also positively associated with economic growth in three ways. First, the investment to GDP ratio has positive effects on economic growth, which is remarkably close to [19, 48, 62, 68]. Second, the results of this study provide strong support for Socialist legal origin in boosting economic growth compared to French legal origin, as argued by [84, 89]. Third, another important finding is the statistically and economically significant positive impact of FDI on economic growth, as argued in [43, 63, 81]. However, this finding is inconsistent with [102].

Conversely, our findings suggest that population growth has a significant negative impact on economic growth, which is consistent with the Malthusian theory of population but does not support [51, 61]. Moreover, this study finds a significant negative time trend, implying that factors which are common to all developing countries, but change over time, diminish economic growth. Finally, compared with Asian developing countries, Latin America and The Caribbean, and Africa dummies mitigate the positive impact of financial development on economic growth.

## 5 Conclusions

This study endeavours to investigate the impact of financial development on economic growth, which has long remained a central issue in the agenda of development economics research over the past few decades. The authors are motivated by the lack of empirical work investigating the relationship between financial development and economic growth using the BMA approach proposed by [38], which was applied in the case of 64 developing countries covering the period 2003–2017. This study used the ratio of private credit to GDP as a proxy for financial development and real per capita GDP growth as a proxy for economic growth in developing countries.

The main findings indicate a U-shaped relationship between financial development and economic growth wherein financial development is negative only up to a certain point, after which it positively affects output growth. This study provides strong evidence that as developing economies experience decreases in economic growth at a certain level, increases in the ratio of private credit to GDP lead to higher economic growth and vice versa. Our results do not support previous studies, indicating that this relationship is an inverse U-shaped curve, such as [7, 20, 62, 94].

Regarding control variables, the study suggests that the investment to GDP ratio has positive effects on economic growth, which is remarkably similar to [19, 48, 62, 68]. Concerning legal origins, the findings provide strong support for the Socialist legal origin in economic growth compared to the French legal origin, as argued by [84, 89]. The study also demonstrates that FDI has a statistically and economically significant positive impact on economic growth, consistent with [43, 63, 81], and inconsistent with [102]. The study also provides evidence supporting the Malthusian theory of population that population growth has a significant negative impact on economic growth, which does not align with [51, 61]. Finally, the study demonstrates that time trend, Latin America and The Caribbean, and Africa dummies diminish the positive impact of financial development on economic growth in developing countries.

As argued above, this study provides three notable insights. (i) This study provides new evidence on the impact of financial development (proxied by the ratio of private credit to GDP) on economic growth in developing countries, presented as a U-shaped curve. (ii) The results provide strong support for the Socialist legal origin in boosting economic growth compared to the French legal origin. This finding has not been addressed previously in economic growth research in developing countries. (iii) The study confirms that population control policy could be one of the most important policies for promoting economic growth in developing countries.

Although this study contributes to narrowing some of the existing gaps in the literature, it has three limitations. First, although the BMA approach is useful for considering model uncertainty, it does not provide insights regarding the causality relationship between financial development and economic growth. There is a need for further research to use advanced methods to resolve this limitation. Second, due to scarce data on developing countries, some other variables were not addressed in this study, such as black-market premiums, and state-owned entities/total fixed asset investments. Finally, due to data availability, this study covers the period of 2003–2017, which is somewhat out-of-date. Further research should be expanded by including other potential variables and more recent data.

**Acknowledgements** We would like to thank Vladik Kreinovich (the editor), Hung Trung Nguyen, and Woraphon Yamaka, as well as the TES 2024 Conference Organising Committee, for their helpful support. Any remaining errors are the authors' own responsibility. This research is partly funded by the University of Economics Ho Chi Minh City, Vietnam.

**Declaration of Competing Interest** The authors declare no conflicts of interest. Besides, I hereby declare that this paper is my own work, which has been performed as a part of my Ph.D. thesis at the University of Economics Ho Chi Minh City, Vietnam.

# Appendix

See Table 3

**Table 3** Information on variables

| Variable | Signs of proxies | Description | Unit of measurement | Source | Expected signs |
|---|---|---|---|---|---|
| Economic growth | $Y^{(**)}$ | Real GDP per capita growth | % | World Development Indicators | |
| Financial development | $PrivCredit^{(*)}$ | The ratio of private credit to GDP | % | Global Financial Development Database | (±) |
| Trade openness | $TradeOpen^{(*)}$ | The ratio of total trade (exports and imports) to GDP | % | World Development Indicators | (+) |
| Financial openness | $KAOPEN^{(1)}$ | The capital account openness index (KAOPEN) is based on four binary dummy variables associated with the financial account in AREAER, including the presence of multiple exchange rates, restrictions on current account transactions, restrictions on capital account transactions, and the requirement of the surrender of export proceeds. | Scale from − 1.92 to 2.35 | AREAER; Chinn and Ito [24] | (±) |
| Investment | $Investment^{(*)}$ | The ratio of investment to GDP | % | Penn World Table version 6.3 | (+) |
| Inflation | $Inflation^{(**)}$ | Inflation, consumer prices (annual %) | % | World Development Indicators | (−) |

(continued)

**Table 3** (continued)

| Variable | Signs of proxies | Description | Unit of measurement | Source | Expected signs |
|---|---|---|---|---|---|
| Government consumption | GovCons$^{(*)}$ | General government final consumption expenditure (percent of GDP) | % | World Development Indicators | (+) |
| Human capital | HCapital | Human capital index, based on years of schooling and returns to education (see Human capital in the Penn World Table version 9.0) | | Penn World Table version 9.1 | (+) |
| Foreign direct investment | FDI$^{(**)}$ | The ratio of FDI to GDP | % | World Development Indicators | (+) |
| Legal origins | LegalOrigin | British legal origin dummy (legor_uk), Socialist legal origin dummy (legor_so) or French legal origin dummy (legor_fr) | 1 or 0 | Porta et al. [84] | (±) |
| Institutional quality | InsQuality | Average of six dimensions of governance: (i) voice and external accountability; (ii) political stability and lack of violence; (iii) government effectiveness; (iv) lack of regulatory burden; (v) rule of law; (vi) control of corruption (see Kaufmann, Kraay, and Mastruzzi, 2005) | Scale from −2.5 to 2.5 | Worldwide Governance Indicators | (+) |
| Population growth | PopGrowth$^{(**)}$ | Population growth (annual %) | % | World Development Indicators | (±) |

(continued)

**Table 3** (continued)

| Variable | Signs of proxies | Description | Unit of measurement | Source | Expected signs |
|---|---|---|---|---|---|
| Regional dummies | Regional dummies | Africa dummy, Asia dummy, Latin America and The Caribbean dummy | 1 or 0 | United Nations | (±) |
| Time trend | TimeTrend | The time trend from 2003 to 2017 | From 1 to 15 | | (+) |

*Notes*[(1)] To get the data relative to *KAOPEN* for this thesis, we used the updated index of Chinn and Ito [24]. The dataset is available at http://web.pdx.edu/~ito/Chinn-Ito_website.htm.,[(*)]The above variables were transformed into natural logarithms; [(**)]The above variables were transformed into natural logarithms of (1 + variable)
*Source* Authors'summary from the literature review

## References


1. Acemoglu, D., Johnson, S., Robinson, J.A.: Institutions as a fundamental cause of long-run growth. In: Aghion, P., Durlauf, S. (eds.) Handbook of Economic Growth, vol. 1, pp. 385–472 (2005). https://doi.org/10.1016/S1574-0684(05)01006-3
2. Adeniyi, O., Oyinlola, A., Omisakin, O., Egwaikhide, F.O.: Financial development and economic growth in Nigeria: evidence from threshold modelling. Econ. Anal. **47**, 11–21 (2015). https://doi.org/10.1016/j.eap.2015.06.003
3. Adu, G., Marbuah, G., Mensah, J.T.: Financial development and economic growth in Ghana: Does the measure of financial development matter? Rev. Dev. Financ. **3**(4), 192–203 (2013). https://doi.org/10.1016/j.rdf.2013.11.001
4. Al-Yousif, Y.K.: Financial development and economic growth: another look at the evidence from developing countries. Rev. Financ. Econ. **11**(2), 131–150 (2002). https://doi.org/10.1016/S1058-3300(02)00039-3
5. Andersen, T.B., Finn, T.: Financial liberalization, financial development and economic growth in LDCs. J. Int. Dev. **15**(2), 189–209 (2003). https://doi.org/10.1002/jid.971
6. Antweiler, W., Copeland, B.R., Scott Taylor, M.: Is free trade good for the environment? Am. Econ. Rev. **91**(4), 877–908 (2001). https://doi.org/10.1257/aer.91.4.877
7. Arcand, J.L., Berkes, E., Panizza, U.: Too much finance? J. Econ. Growth **20**(2), 105–148 (2015). https://doi.org/10.1007/s10887-015-9115-2
8. Bangake, C., Eggoh, J.C.: Further evidence on finance-growth causality: a panel data analysis. Econ. Syst. **35**(2), 176–188 (2011). https://doi.org/10.1016/j.ecosys.2010.07.001
9. Barro, R.J., Lee, J.W.: International measures of schooling years and schooling quality. Am. Econ. Rev. **86**(2), 218–223 (1996). https://www.jstor.org/stable/2118126
10. Beck, T., Levine, R.: Stock markets, banks, and growth: panel evidence. J. Bank. Financ. **28**(3), 423–442 (2004). https://doi.org/10.1016/S0378-4266(02)00408-9
11. Beck, T., Levine, R.: Legal institutions and financial development. In: Menard, Shirley (eds.) Handbook of New Institutional Economics, pp. 251–278. Springer, Boston (2005). https://doi.org/10.1007/0-387-25092-1_12
12. Beck, T., Demirgüç-Kunt, A., Levine, R.: Legal theories of financial development. Oxf. Rev. Econ. Policy **17**(4), 483–501 (2001). https://doi.org/10.1093/oxrep/17.4.483
13. Beck, T., Demirgüç-Kunt, A., Levine, R.: Finance, inequality and the poor. J. Econ. Growth **12**(1), 27–49 (2007). https://doi.org/10.1007/s10887-007-9010-6
14. Beck, T., Demirgüç-Kunt, A., Maksimovic, V.: Financial and legal constraints to growth: does firm size matter? J. Financ. **60**(1), 137–177 (2005). https://doi.org/10.1111/j.1540-6261.2005.00727.x

15. Beck, T., Levine, R., Loayza, N.: Finance and the sources of growth. J. Financ. Econ. **58**(1–2), 261–300 (2000). https://doi.org/10.1016/S0304-405X(00)00072-6
16. Bittencourt, M.: Financial development and economic growth in Latin America: is Schumpeter right? J. Policy Model. **34**(3), 341–355 (2012). https://doi.org/10.1016/j.jpolmod.2012.01.012
17. Boyd, J.H., Levine, R., Smith, B.D.: The impact of inflation on financial sector performance. J. Monet. Econ. **47**(2), 221–248 (2001). https://doi.org/10.1016/S0304-3932(01)00049-6
18. Calderón, C., Liu, L.: The direction of causality between financial development and economic growth. J. Dev. Econ. **72**(1), 321–334 (2003). https://doi.org/10.1016/S0304-3878(03)00079-8
19. Caporale, G.M., Rault, C., Sova, A.D., Sova, R.: Financial development and economic growth: evidence from 10 new European Union members. Int. J. Financ. Econ. **20**(1), 48–60 (2014). https://doi.org/10.1002/ijfe.1498
20. Cecchetti, S.G., Kharroubi, E.: Reassessing the impact of finance on growth. BIS Working Paper no. 381. Bank for International Settlements (2012). https://www.bis.org/publ/work381.pdf
21. Chamarbagwala, R., Ramaswamy, S., Wunnava, P.V.: The role of foreign capital in domestic manufacturing productivity: empirical evidence from Asian economies. Appl. Econ. **32**(4), 393–398 (2000). https://doi.org/10.1080/000368400322561
22. Chatfield, C.: Model uncertainty, data mining and statistical inference. J. R. Stat. Soc. A. Stat. Soc. **158**(3), 419–444 (1995). https://doi.org/10.2307/2983440
23. Chinn, M.D., Ito, H.: Capital account liberalization, institutions and financial development: cross country evidence. In: NBER Working Paper no. 8967. National Bureau of Economic Research, Cambridge, MA (2002). https://doi.org/10.3386/w8967
24. Chinn, M.D., Ito, H.: The Chinn-Ito Index–A de jure measure of financial openness (2019). Available at: http://web.pdx.edu/~ito/Chinn-Ito_website.htm. Accessed 20 May 2022
25. Chipman, H., George, E.I., McCulloch, R.E., Clyde, M., Foster, D.P., Stine, R.A.: The practical implementation of Bayesian model selection. Lect. Notes-Monogr. Ser. 65–134 (2001). https://doi.org/10.1214/lnms/1215540964
26. Clyde, M., George, E.I.: Model uncertainty. Stat. Sci. **19**(1), 81–94 (2004). https://doi.org/10.1214/088342304000000035
27. Cole, M.A., Elliott, R.J.R.: Determining the trade–environment composition effect: the role of capital, labor and environmental regulations. J. Environ. Econ. Manag. **46**(3), 363–383 (2003). https://doi.org/10.1016/S0095-0696(03)00021-4
28. De Gregorio, J., Guidotti, P.E.: Financial development and economic growth. World Dev. **23**(3), 433–448 (1995). https://doi.org/10.1016/0305-750X(94)00132-I
29. Demetriades, P., Law, S.H.: Finance, institutions and economic development. Int. J. Financ. Econ. **11**(3), 245–260 (2006). https://doi.org/10.1002/ijfe.296
30. Demetriades, P., Andrianova, S.: Finance and growth: what we know and what we need to know. In: Goodhart C.A.E (ed.) British Association for the Advancement of Science. Palgrave Macmillan, London. https://doi.org/10.1057/9780230374270_2
31. Draper, D.: Assessment and propagation of model uncertainty. J. R. Stat. Soc. B **57**(1), 45–70 (1995). https://doi.org/10.1111/j.2517-6161.1995.tb02015.x
32. Easterly, W., Rebelo, S.: Fiscal policy and economic growth. J. Monet. Econ. **32**(3), 417–458 (1993). https://doi.org/10.1016/0304-3932(93)90025-B
33. Edwards, S.: Capital mobility and economic performance: are emerging economies different? In: NBER Working Paper no. 8076. National Bureau of Economic Research, Cambridge, MA (2001). https://doi.org/10.3386/w8076
34. Ergungor, O.E.: Financial system structure and economic growth: structure matters. Int. Rev. Econ. Financ. **17**(2), 292–305 (2008). https://doi.org/10.1016/j.iref.2006.09.001
35. Estrada, G.B., Park, D., Ramayandi, A.: Financial development and economic growth in developing Asia. In: ADB Economics Working Paper no. 233. Asian Development Bank (2010). Available at: https://www.adb.org/publications/financial-development-and-economic-growth-developing-asia

36. Estrada, G.B., Park, D., Ramayandi, A.: Financial development, financial openness, and economic growth. In: ADB Economics Working Paper no. 442. Asian Development Bank (2015). Available at: https://www.adb.org/publications/financial-development-financial-openness-and-economic-growth
37. Fernandez, C., Ley, E., Steel, M.F.J.: Model uncertainty in cross-country growth regressions. J. Appl. Economet. **16**(5), 563–576 (2001). https://doi.org/10.1002/jae.623
38. Fernandez, C., Ley, E., Steel, M.F.J.: Benchmark priors for Bayesian model averaging. J. Appl. Economet. **100**(2), 381–427 (2001). https://doi.org/10.1016/S0304-4076(00)00076-2
39. Fischer, S.: The role of macroeconomic factors in growth. J. Monet. Econ. **32**(3), 485–512 (1993). https://doi.org/10.1016/0304-3932(93)90027-D
40. Fragoso, T.M., Bertoli, W., Louzada, F.: Bayesian model averaging: a systematic review and conceptual classification. Int. Stat. Rev. **86**(1), 1–28 (2018). https://doi.org/10.1111/insr.12243
41. George, EdwardI., Foster, D.P.: Calibration and empirical Bayes variable selection. Biometrika **87**(4), 731–747 (2000). https://doi.org/10.1093/biomet/87.4.731
42. Gibson, H.D., Tsakalotos, E.: The scope and limits of financial liberalisation in developing countries: a critical survey. J. Dev. Stud. **30**(3), 578–628 (1994). https://doi.org/10.1080/00220389408422329
43. Hermes, N., Lensink, R.: Foreign direct investment, financial development and economic growth. J. Dev. Stud. **40**(1), 142–163 (2003). https://doi.org/10.1080/00220380412331293707
44. Hoeting, J.A., Madigan, D., Raftery, A.E., Volinsky, C.T.: Bayesian model averaging: a tutorial. Stat. Sci. **14**(14), 382–417 (1999). https://doi.org/10.1214/ss/1009212519
45. Huang, H.-C., Lin, S.-C.: Non-linear finance–growth nexus: a threshold with instrumental variable approach. Econ. Transit. **17**(3), 439–466 (2009). https://doi.org/10.1111/j.1468-0351.2009.00360.x
46. Ito, H.: Financial development and financial liberalization in Asia: thresholds, institutions and the sequence of liberalization. North Am. J. Econ. Financ. **17**(3), 303–327 (2006). https://doi.org/10.1016/j.najef.2006.06.008
47. Kar, M., Nazlıoğlu, Ş, Ağır, H.: Financial development and economic growth nexus in the MENA countries: bootstrap panel granger causality analysis. Econ. Model. **28**(1–2), 685–693 (2011). https://doi.org/10.1016/j.econmod.2010.05.015
48. Kargbo, S.M., Adamu, P.A.: Financial development and economic growth in Sierra Leone. J. Monet. Econ. Integr. **9**(2), 30–61 (2009)
49. Kass, R.E., Wasserman, L.: A reference Bayesian test for nested hypotheses and its relationship to the Schwarz criterion. J. Am. Stat. Assoc. **90**(431), 928–934 (1995). https://doi.org/10.1080/01621459.1995.10476592
50. Kaufmann, D., Kraay, A., Mastruzzi, M.: Governance matters IV: governance indicators for 1996–2004. In: Policy Research Working Paper no. 3630. The World Bank, Washington, DC (2005). https://openknowledge.worldbank.org/handle/10986/8221
51. Kelley, A.C., Schmidt, R.M.: Economic and demographic change: a synthesis of models, findings, and perspectives. In: Duke Economics Working Paper No. 99–01 (1999). Available at: https://ssrn.com/abstract=152888
52. Keynes, J.M.: The General Theory of Employment, Interest and Money. Harcourt, Brace and Co, New York (1936)
53. Kim, D.-H., Lin, S.-C., Suen, Y.-B.: Dynamic effects of trade openness on financial development. Econ. Model. **27**(1), 254–261 (2010). https://doi.org/10.1016/j.econmod.2009.09.005
54. Kim, D.-H., Lin, S.-C., Suen, Y.-B.: Are financial development and trade openness complements or substitutes? South. Econ. J. **76**(3), 827–845 (2010). https://doi.org/10.4284/sej.2010.76.3.827
55. Kim, D.-H., Lin, S.-C., Suen, Y.-B.: The simultaneous evolution of economic growth, financial development, and trade openness. J. Int. Trade Econ. Dev. **21**(4), 513–537 (2012). https://doi.org/10.1080/09638199.2010.497933

56. King, R.G., Levine, R.: Finance and growth: schumpeter might be Right. Q. J. Econ. **108**(3), 717–737 (1993a). https://doi.org/10.2307/2118406
57. King, R.G., Levine, R.: Finance, entrepreneurship and growth: theory and evidence. J. Monet. Econ. **32**(3), 513–542 (1993b). https://doi.org/10.1016/0304-3932(93)90028-E
58. Klein, M.W.K., Michael, W.: Capital account liberalization, institutional quality and economic growth: theory and evidence. In: NBER Working Paper no. 11112. National Bureau of Economic Research, Cambridge, MA (2005). https://doi.org/10.3386/w11112
59. Kónya, L.K., László: Exports and growth: ganger causality analysis on OECD countries with a panel data approach. Econ. Model. **23**(6), 978–992 (2006). https://doi.org/10.1016/j.econmod.2006.04.008
60. Kose, M.A., Prasad, E., Rogoff, K., Wei, S.-J.: Financial globalization: a reappraisal. IMF Staff. Pap. **56**(1), 8–62 (2009). https://doi.org/10.3386/w12484
61. Kuznets, S.: Population and economic growth. Proc. Am. Philos. Soc. **111**(3), 170–193 (1967). https://www.jstor.org/stable/985714
62. Law, S.H., Singh, N.: Does too much finance harm economic growth? J. Bank. Financ. **41**, 36–44 (2014). https://doi.org/10.1016/j.jbankfin.2013.12.020
63. Lee, C.-C., Chang, C.-P.: FDI, financial development, and economic growth: international evidence. J. Appl. Econ. **12**(2), 249–271 (2009). https://doi.org/10.1016/S1514-0326(09)60015-5
64. Levine, R., Renelt, D.: A sensitivity analysis of cross-country growth regressions. Am. Econ. Rev. 942–963 (1992). https://www.jstor.org/stable/2117352
65. Levine, R., Zervos, S.: Stock markets, banks, and economic growth. Am. Econ. Rev. **88**(3), 537–558 (1998). https://www.jstor.org/stable/116848
66. Levine, R.: Financial development and economic growth: views and agenda. In: Policy Research Working Paper no. 178. The World Bank, Washington, DC (1996)
67. Levine, R.: International financial liberalization and economic growth. Rev. Int. Econ. **9**(4), 688–702 (2001). https://doi.org/10.1111/1467-9396.00307
68. Levine, R.: Finance and growth: theory and evidence. In: NBER Working Paper no. 10766. National Bureau of Economic Research, Cambridge, MA (2004). https://doi.org/10.3386/w10766
69. Levine, R., Loayza, N., Beck, T.: Financial intermediation and growth: causality and causes. J. Monet. Econ. **46**(1), 31–77 (2000). https://doi.org/10.1016/S0304-3932(00)00017-9
70. Ley, E., Steel, M.F.J.: On the effect of prior assumptions in Bayesian model averaging with applications to growth regression. J. Appl. Economet. **24**(4), 651–674 (2009). https://doi.org/10.1002/jae.1057
71. Liang, F., Paulo, R., Molina, G., Clyde, M.A., Berger, J.O.: Mixtures of g priors for Bayesian variable selection. J. Am. Stat. Assoc. **103**(481), 410–423 (2008). https://doi.org/10.1198/016214507000001337
72. Lucas, R.E.: On the mechanics of economic development. J. Monet. Econ. **22**(1), 3–42 (1988). https://doi.org/10.1016/0304-3932(88)90168-7
73. Madigan, D., York, J., Allard, D.: Bayesian graphical models for discrete data. Int. Stat. Rev./Revue Internationale de Statistique **6**(3), 215–232 (1995). https://doi.org/10.2307/1403615
74. Malthus, T.R.: An Essay on the principle of population, as it Affects the Future Improvement of Society, with Remarks on the Speculations of Mr Godwin, M. Condorcet, and other Writers. Printed for J. Johnson, In St. Paul's Church-Yard, London (1798)
75. McKinnon, R.I.: Money and Capital in Economic Development. Brookings Institution Press, Washington (1973)
76. Menyah, K., Nazlioglu, S., Wolde-Rufael, Y.: Financial development, trade openness and economic growth in African countries: new insights from a panel causality approach. Econ. Model. **37**, 386–394 (2014). https://doi.org/10.1016/j.econmod.2013.11.044
77. NBER Working Paper no. 10766. National Bureau of Economic Research, Cambridge, MA
78. North, D.C.: Institutions. J. Econ. Perspect. **5**(1), 97–112 (1991). https://www.jstor.org/stable/1942704

79. North, D.C., Weingast, B.R.: Constitutions and commitment: the evolution of institutions governing public choice in seventeenth-century England. J. Econ. Hist. **49**(4), 803–832 (1989). https://doi.org/10.1017/S0022050700009451
80. Obstfeld, M.M.: International finance and growth in developing countries: what have we learned? IMF Staff Pap. **56**(1), 63–111 (2009). https://doi.org/10.1057/IMFSP.2008.32
81. Omran, M., Bolbol, A.: Foreign direct investment, financial development, and economic growth: evidence from the Arab countries. Rev. Middle East Econ. Financ. **1**(3), 37–55 (2003). https://doi.org/10.1080/1475368032000158232
82. Onatski, A., Williams, N.: Modeling model uncertainty. J. Eur. Econ. Assoc. **1**(5), 1087–1122 (2003). https://doi.org/10.3386/w9566
83. Park, I., Amarchinta, H.K., Grandhi, R.V.: A Bayesian approach for quantification of model uncertainty. Reliab. Eng. Syst. Saf. **95**(7), 777–785 (2010). https://doi.org/10.1016/j.ress.2010.02.015
84. La Porta, R., Lopez-de-Silanes, F., Shleifer, A., Vishny, R.W.: Law and finance. J. Polit. Econ. **106**(6), 1113–1155 (1998). https://doi.org/10.1086/250042
85. Raftery, A.E.: Bayesian model selection in structural equation models. Sage Focus Edn. 154: 163–180 (1993). https://sites.stat.washington.edu/raftery/Research/PDF/bollen1993.pdf
86. Raftery, A.E., Madigan, D., Hoeting, J.A.: Bayesian model averaging for linear regression models. J. Am. Stat. Assoc. **92**(437), 179–191 (1997). https://doi.org/10.1080/01621459.1997.10473615
87. Raftery, A.E., Gneiting, T., Balabdaoui, F., Polakowski, M.: Using Bayesian model averaging to calibrate forecast ensembles. Mon. Weather Rev. **133**(5), 1155–1174 (2005). https://doi.org/10.1175/MWR2906.1
88. Rajan, R.G., Zingales, L.: Financial dependence and growth. In: NBER Working Paper no. 5758. National Bureau of Economic Research, Cambridge, MA (1996). https://doi.org/10.3386/w5758
89. Rajan, R.G., Zingales, L.: The great reversals: the politics of financial development in the twentieth century. J. Financ. Econ. **69**(1), 5–50 (2003). https://doi.org/10.1016/S0304-405X(03)00125-9
90. Rioja, F., Valev, N.: Does one size fit all?: a reexamination of the finance and growth relationship. J. Dev. Econ. **74**(2), 429–447 (2004). https://doi.org/10.1016/j.jdeveco.2003.06.006
91. Rioja, F., Valev, N.: Finance and the sources of growth at various stages of economic development. Econ. Inq. **42**(1), 127–140 (2004). https://doi.org/10.2139/ssrn.438261
92. Samargandi, N., Fidrmuc, J., Ghosh, S.: Is the relationship between financial development and economic growth monotonic? evidence from a sample of middle-income countries. World Dev. **68**, 66–81 (2015). https://doi.org/10.1016/j.worlddev.2014.11.010
93. Shaw, E.S.: Financial Deepening in Economic Development. Oxford University Press, New York (1973)
94. Shen, C.-H., Lee, C.-C.: Same financial development yet different economic growth: why? J. Money, Credit, Bank. **38**(7), 1907–1944 (2006). https://doi.org/10.1353/mcb.2006.0095
95. Svaleryd, H., Vlachos, J.: Financial markets, the pattern of industrial specialization and comparative advantage: evidence from OECD countries. Eur. Econ. Rev. **49**(1), 113–144 (2005). https://doi.org/10.1016/S0014-2921(03)00030-8
96. Uddin, G.S., Sjö, B., Shahbaz, M.: The causal nexus between financial development and economic growth in Kenya. Econ. Model. **35**, 701–707 (2013). https://doi.org/10.1016/j.econmod.2013.08.031
97. Valickova, P., Havranek, T., Horvath, R.: Financial development and economic growth: a meta-analysis. J. Econ. Surv. **29**(3), 506–526 (2015). https://doi.org/10.1111/joes.12068
98. Vamvakidis, A.: Regional trade agreements or broad liberalization: which path leads to faster growth? IMF Staff Pap. **46**(1), 42–68 (1999). https://www.jstor.org/stable/3867634
99. Wacziarg, R.: Measuring the dynamic gains from trade. World Bank Econ. Rev. **15**(3), 393–429 (2001). https://doi.org/10.1093/wber/15.3.393

100. Zeugner, S., Feldkircher, M.: Benchmark priors revisited: on adaptive shrinkage and the supermodel effect in Bayesian model averaging. In: Working Paper no. 2009/202. International Monetary Fund, Washington, DC (2009)
101. Wit, E., van den Heuvel, E., Romeijn, J.-W.: 'All models are wrong...': an introduction to model uncertainty. Stat. Neerl. **66**(3), 217–236 (2012). https://doi.org/10.1111/j.1467-9574.2012.00530.x
102. Zhang, J., Wang, L., Wang, S.: Financial development and economic growth: recent evidence from China. J. Comp. Econ. **40**(3), 393–412 (2012). https://doi.org/10.1016/j.jce.2012.01.001
103. Zhuang, J., Gunatilake, H.M., Niimi, Y., Khan, M.E., Yi, J., Rana, H., Niny, K., Martin, A.L., Bracey, P., Biao, H.: Financial sector development, economic growth, and poverty reduction: A literature review. In: ADB Economics Working Paper no. 173. Asian Development Bank (2009). Available at: https://papers.ssrn.com/sol3/papers.cfm?abstract_id=1617022.

[illegible]

# Estimation Methods for the Coefficient of Variation of the Three-Parameter Lognormal Distribution: Their Application to Wind Speed Data Indicating Soil Erosion Effects

**Patcharee Maneerat, Pisit Nakjai, Sa-Aat Niwitpong, and Noppadon Yosboonruang**

**Abstract** Highly right-skewed data can often be modeled by using the three-parameter lognormal distribution. Meanwhile, the performances of extant confidence interval estimates are poor when the variance is large. Herein, methods to estimate the confidence interval for the coefficient of variation of the three-parameter lognormal distribution using the likelihood-based, generalized confidence interval, and method of variance estimates recovery methods are presented. The results of a numerical simulation study reveal that the likelihood-based confidence interval estimate performed well in terms of the actual coverage percentage and average length, even with a large coefficient of variation. The methods' applicability is demonstrated through their application to wind speed data from Chiang Mai, Thailand, aiming to indicate soil erosion effects.

## 1 Introduction

The three-parameter lognormal distribution involving the threshold ($\beta$), mean ($\mu_X$), and variance ($\sigma_X^2$) is an important model for analyzing highly right-skewed data that cannot be modeled by using the better-known two-parameter lognormal

P. Maneerat · P. Nakjai
Uttaradit Rajabhat University, Uttaradit 53000, Thailand
e-mail: m.patcharee@uru.ac.th

P. Nakjai
e-mail: pisit.nak@uru.ac.th

S.-A. Niwitpong
King Mongkut's University of Technology North Bangkok, Bangkok 10800, Thailand
e-mail: sa-aat.n@sci.kmutnb.ac.th

N. Yosboonruang (✉)
University of Phayao, Phayao 56000, Thailand
e-mail: noppadon.yo@up.ac.th

 
V. Kreinovich et al. (eds.), *Applications of Optimal Transport to Economics and Related Topics*, Studies in Systems, Decision and Control 556,
https://doi.org/10.1007/978-3-031-67770-0_21

distribution [14]. When fitting a population to a three-parameter lognormal distribution, the probability density function of $X$ becomes

$$h(x; \beta, \omega, \sigma_Y) = \frac{1}{(x-\beta)\sqrt{2\pi\sigma_Y^2}} \exp\left\{-(2\sigma_Y^2)^{-1}\left[\ln^2\left(\frac{x-\beta}{\omega}\right)\right]\right\} \tag{1}$$

Having considered $X$, it is reasonable that random variable $X - \beta$ can be transformed to be log-normally distributed if $Y = \ln(X - \beta)$ is normally distributed. Note that the two- and three-parameter lognormal distributions are identical when $\beta = 0$. This is the special case of the three-parameter lognormal distribution. Expanding the parameters of the three-parameter lognormal distribution provides the mean $\mathrm{E}(X) = \mu_X = \beta + \omega \exp(\sigma_Y^2)$, the median (related to the threshold) $\mathrm{M}(X) = \beta + \omega$, and the variance $\mathrm{Var}(X) = \sigma_X^2 = \omega^2 \exp(\sigma_Y^2)\left[\exp(\sigma_Y^2) - 1\right]$, where $\omega = \exp(\mu_Y)$.

The three-parameter distribution has been applied in several studies. In operational hydrology, Charbeneau [4] applied it for in-stream flow synthesis, while Cohen and Whitten [8] and Chen and Miao [6] used it to determine the maximum flood level from the Susquehanna river in Harrisburg, Pennsylvania. In industry, Chen [5] and Chen and Miao [6] employed it to examine plastic laminate strength in a manufacturing process, and it has also been utilized in fatigue life analysis [8, 9, 13].

Several statisticians have estimated the parameters of the three-parameter lognormal distribution. Burges et al. [2] used the sample mean, median, and standard deviation to estimate the threshold parameter and found that these three common measures produced a larger bias than the approximated threshold. Cohen et al. [9] modified the moment estimates for all three parameters; the results of their numerical analysis revealed that the moment estimates performed well when the skewness approached zero but less satisfactorily than the local maximum likelihood (ML) estimate in terms of the mean-squared error and bias. Later, Cohen and Whitten [8] revised the moment and local ML estimates for the three related parameters of the three-parameter lognormal distribution by considering the third standard moment for each of them.

Some scholars have constructed confidence interval estimates for the parameters of the three-parameter lognormal distribution. For example, Griffiths [11] estimated the confidence intervals for all three based on their likelihood functions by considering the tangents to the contour enclosing the 95% confidence level region. Next, Chen and Miao [6] constructed the exact confidence interval and the exact upper confidence interval limit for the threshold parameter. Recently, Maneerat et al. [12] estimated the confidence interval for the median of the three-parameter lognormal distribution by using bootstrap, normal approximation, and generalized pivotal quantity methods. However, none of them performed well with a large variance. The coefficient of variation is a statistical measure of the relative dispersion of data points in a data series around the mean. The relative precision can be used to evaluate inequality in a population because it is more sensitive to individual data points in a positively skewed distribution than the Gini coefficient [1].

Motivated by Maneerat et al. [12], we have identified a special case involving a large variance in the three-parameter lognormal population, which presents a previously unfilled research gap. Notably, within this context, there's a need to estimate its coefficient of variation, which is then employed to approximate the variability of wind speeds as an indicator of soil erosion effects. Therefore, we used existing methods to estimate the confidence interval for the coefficient of variation of the three-parameter lognormal distribution using the likelihood-based, generalized confidence interval, and method of variance estimates recovery (MOVER) methods, which are covered in the next section. The actual coverage percentage (ACP), average length (AL), and relative average length were the performance measures used to evaluate their efficacies both numerically and with real wind speed data. The paper concludes with a brief discussion and summary of findings.

## 2 Intervals Formed on the Coefficient of Variation

From the Eq. (1), the coefficient of variation of $X$ is defined as

$$\delta = \frac{\sigma_X}{\mu_X} = \frac{\omega\left\{\exp(\sigma_Y^2)\left[\exp(\sigma_Y^2) - 1\right]\right\}^{1/2}}{\beta + \omega\sqrt{\exp(\sigma_Y^2)}} \tag{2}$$

which is log-transformed as $\kappa = \ln\left(\frac{\sigma_X}{\mu_X}\right) = \ln\omega + \frac{1}{2}\left[2\sigma_Y^2 + \ln\left(1 - \frac{1}{\exp(\sigma_Y^2)}\right)\right] - \ln\left[\beta + \omega\sqrt{\exp(\sigma_Y^2)}\right]$. For large $\sigma_Y^2$, it can lead to $\lim_{\sigma^2\to\infty}\ln\left(1 - \frac{1}{\exp(\sigma_Y^2)}\right) = 0$, and then

$$\delta \approx \ln\omega + \sigma_Y^2 - \left[\ln\beta + \ln\left(\frac{\omega\sqrt{\exp(\sigma_Y^2)}}{\beta} + 1\right)\right] \tag{3}$$

Using this as the starting point, the parameter $\delta$ is needed to estimate by using the methods as follows.

### 2.1 Likelihood-Based Method

The likelihood and log-likelihood functions of $\boldsymbol{\theta} = (\beta, \omega, \sigma_Y)'$ are

$$L(\boldsymbol{\theta}) = \left(2\pi\sigma_Y^2\right)^{-n/2} \exp\left\{\sum_{i=1}^{n} -(2\sigma_Y^2)^{-1}\left[\ln^2\left(\frac{x_i - \beta}{\omega}\right)\right]\right\}\prod_{i=1}^{n}(x_i - \beta)^{-1} \tag{4}$$

$$l(\boldsymbol{\theta}) = -\frac{n}{2}\ln(2\pi\sigma_Y^2) - (2\sigma_Y^2)^{-1}\sum_{i=1}^{n}\left[\ln^2\left(\frac{x_i - \beta}{\omega}\right)\right] + \sum_{i=1}^{x}\frac{1}{\ln(x_i - \beta)} \tag{5}$$

The log-likelihood $l(\boldsymbol{\theta})$ is differentiated and equaling to zero as follows:

$$\left(\frac{\partial l(\boldsymbol{\theta})}{\partial \beta}, \frac{\partial l(\boldsymbol{\theta})}{\partial \omega}, \frac{\partial l(\boldsymbol{\theta})}{\partial \sigma_Y}\right) = \frac{1}{\sigma_Y^2}\left(\sum_{i=1}^{n}(x_i-\beta)^{-1}\ln\left[\frac{x_i-\beta}{\omega}\right] + \sum_{i=1}^{n}(x_i-\beta)^{-1}, \right.$$
$$\left. \frac{1}{\omega}\sum_{i=1}^{n}\ln\left[\frac{x_i-\beta}{\omega}\right], \frac{1}{\sigma_Y}\sum_{i=1}^{n}\ln^2\left[\frac{x_i-\beta}{\omega}\right] - \frac{n}{\sigma_Y}\right) \tag{6}$$

The second derivatives become

$$\left(\frac{\partial l^2(\boldsymbol{\theta})}{\partial \beta^2}, \frac{\partial l^2(\boldsymbol{\theta})}{\partial \omega^2}, \frac{\partial l^2(\boldsymbol{\theta})}{\partial \sigma_Y^2}\right) = \frac{1}{\sigma_Y^2}\left(\sum_{i=1}^{n}\left[(x_i-\beta)^{-2}\left\{\ln\left[\frac{x_i-\beta}{\omega}\right]-1\right\}\right]\right.$$
$$+\sum_{i=1}^{n}(x_i-\beta)^{-2}, -\frac{1}{\omega^2}\sum_{i=1}^{n}\ln(x_i-\beta) - n\left[\frac{1-\ln\omega}{\omega^2}\right],$$
$$\left.-\frac{3}{\sigma_Y^2}\sum_{i=1}^{n}\ln^2\left[\frac{x_i-\beta}{\omega}\right]+n\right) \tag{7}$$

and

$$\left(\frac{\partial l^2(\boldsymbol{\theta})}{\partial \omega \partial \beta}, \frac{\partial l^2(\boldsymbol{\theta})}{\partial \sigma_Y \partial \beta}, \frac{\partial l^2(\boldsymbol{\theta})}{\partial \sigma_Y \partial \omega}\right) = \frac{1}{\sigma_Y^2}\left(-\frac{1}{\omega}\sum_{i=1}^{n}(x_i-\beta)^{-1}, -2\sum_{i=1}^{n}(x_i-\beta)^{-1}\right. \tag{8}$$
$$\left.\ln\left[\frac{x_i-\beta}{\omega}\right], -\frac{2}{\omega}\sum_{i=1}^{n}\ln\left[\frac{x_i-\beta}{\omega}\right]\right)$$

Next, the Fisher information matrix is expressed as

$$I(\boldsymbol{\theta}) = -E\begin{bmatrix} \frac{\partial l^2(\boldsymbol{\theta})}{\partial \beta^2} & \frac{\partial l^2(\boldsymbol{\theta})}{\partial \beta \partial \omega} & \frac{\partial l^2(\boldsymbol{\theta})}{\partial \beta \partial \sigma_Y} \\ \frac{\partial l^2(\boldsymbol{\theta})}{\partial \omega \partial \beta} & \frac{\partial l^2(\boldsymbol{\theta})}{\partial \omega^2} & \frac{\partial l^2(\boldsymbol{\theta})}{\partial \omega \partial \sigma_Y} \\ \frac{\partial l^2(\boldsymbol{\theta})}{\partial \sigma_Y \partial \beta} & \frac{\partial l^2(\boldsymbol{\theta})}{\partial \sigma_Y \partial \omega} & \frac{\partial l^2(\boldsymbol{\theta})}{\partial \sigma_Y^2} \end{bmatrix} = \begin{bmatrix} \frac{(1+\sigma_Y^2)\tau}{\omega^2\sigma_Y^2} & \frac{\sqrt{\tau}}{\omega^2\sigma_Y^2} & \frac{-2\sqrt{\tau}}{\omega^2\sigma_Y} \\ \frac{\sqrt{\tau}}{\omega^2\sigma_Y^2} & \frac{1}{\omega^2\sigma_Y^2} & 0 \\ \frac{-2\sqrt{\tau}}{\omega^2\sigma_Y} & 0 & \frac{2}{\sigma_Y^2} \end{bmatrix} \tag{9}$$

which is derived by Cohen [7]; $\tau = \exp(\sigma_Y^2)$. Then, we obtain the asymptotic distribution of $\hat{\boldsymbol{\theta}}$ which is a multivariate normal with mean vector $\boldsymbol{\theta}$ and variance-covariance matrix $\boldsymbol{V}$, that is

$$V = \frac{I(\boldsymbol{\theta})^{-1}}{n} = \begin{bmatrix} \frac{\omega^2\sigma_Y^2}{n\tau}\mathrm{B}^{-1} & \frac{\omega\sigma_Y^2}{n\sqrt{\tau}}\mathrm{B}^{-1} & \frac{\omega\sigma_Y^3}{n\sqrt{\tau}}\mathrm{B}^{-1} \\ \frac{\omega\sigma_Y^2}{n\sqrt{\tau}}\mathrm{B}^{-1} & \frac{\omega^2\sigma_Y^2}{n}[1+\mathrm{B}^{-1}] & \frac{\omega\sigma_Y^3}{n}\mathrm{B}^{-1} \\ \frac{\omega\sigma_Y^3}{n\sqrt{\tau}}\mathrm{B}^{-1} & \frac{\omega\sigma_Y^3}{n}\mathrm{B}^{-1} & \frac{\sigma_Y^2}{n}[1+2\sigma_Y^2\mathrm{B}^{-1}] \end{bmatrix} \tag{10}$$

where $\mathrm{B} = \tau(1+\sigma_Y^2) - 2\sigma_Y^2 - 1$. Since $\delta$ is the function of $\boldsymbol{\theta}$ such that $\hat{\delta}$ is asymptotic normal distribution with mean $\delta$ and variance $V_a$ as $\hat{\delta} \sim \mathrm{N}(\delta, V_a)$;

$$V_a = \left(\frac{\partial\delta}{\partial\boldsymbol{\theta}}\right)' \boldsymbol{V} \left(\frac{\partial\delta}{\partial\boldsymbol{\theta}}\right) \tag{11}$$

where $\left(\frac{\partial\delta}{\partial\boldsymbol{\theta}}\right)' = \left(\frac{\partial\delta}{\partial\beta}; \frac{\partial\delta}{\partial\omega}; \frac{\partial\delta}{\partial\sigma_Y}\right) = \left(-\frac{1}{\mu_X}; \frac{1}{\omega} - \frac{\sqrt{\tau}}{\mu_X}; \sigma_Y\left[2 - \frac{\omega\sqrt{\tau}}{\mu_X}\right]\right)'$. Equation (11) is derived from the variance behavior theorem presented in Casella and Berger [3]. Therefore, we can construct the $100(1-\alpha)\%$ likelihood-based confidence interval for $\delta$, expressed as

$$l_{\mathrm{L}} = \hat{\delta} + W_{\alpha/2}\sqrt{V_a} \tag{12}$$

$$u_{\mathrm{L}} = \hat{\delta} + W_{1-\alpha/2}\sqrt{V_a} \tag{13}$$

where $P(W_{\alpha/2} \le W \le W_{1-\alpha/2}) = 1-\alpha$ and $W_\alpha$ denotes the $\alpha$th quantile of standard normal distribution: $N(0, 1)$.

## 2.2 Generalized Confidence Interval Method

In accordance with Cohen [7], we obtain the asymptotic variance of the ML estimate of $\beta$ derived from the Eq. (10). Using the central limit theorem, it leads to gain a new random variable $T_\beta$ has a standard normal. Given $\hat{\beta}$, we also produce a new random variables $T_\omega$, while a random variable $T_{\sigma_Y}$ is derived in form of $T_{\sigma_Y^2}$ having chi-square distribution. Thus, the three random variables are written as

$$\left(T_\beta, T_\omega, T_{\sigma_Y}\right) = \left(\frac{\hat{\beta}-\beta}{\sqrt{V(\hat{\beta})}}, \exp\left(\frac{\ln\hat{\omega} - \ln\omega}{\sigma_Y/\sqrt{n}}\right), \sqrt{\frac{(n-1)\hat{\sigma}_Y^2}{\sigma_Y^2}}\right) \tag{14}$$

After that the estimates $(\hat{\beta}, \hat{\omega}, \hat{\sigma}_Y)$ can be expressed as the functions of $(T_\beta, T_\omega, T_{\sigma_Y})$ which is

$$(\hat{\beta}, \hat{\omega}, \hat{\sigma}_Y) = \left(\beta + T_\beta\sqrt{V(\hat{\beta})},\ \exp\left(\ln\omega + T_\omega\sigma_Y/\sqrt{n}\right),\ T_{\sigma_Y}\sqrt{\sigma_Y^2/(n-1)}\right) \tag{15}$$

Given an observation $(\hat{\beta}^{[o]}, \hat{\omega}^{[o]}, \hat{\sigma}_Y^{[o]})$ and $(T_\beta^{[o]}, T_\omega^{[o]}, T_{\sigma_Y}^{[o]})$, the Eq. (15) has the unique solution as follows.

$$(\beta, \omega, \sigma_Y) = \left(\hat{\beta}^{[o]} - T_\beta^{[o]}\sqrt{V(\hat{\beta}^{[o]})},\ \exp\left[\ln\omega^{[o]} - T_\omega^{[o]}T_{\sigma_Y^{[o]}}^{-1}\sqrt{(n-1)\hat{\sigma}_Y^{2[o]}}/\sqrt{n}\right],\right.$$
$$\left. T_{\sigma_Y^{[o]}}^{-1}\sqrt{(n-1)\hat{\sigma}_Y^{2[o]}}\right) \tag{16}$$

which has the properties of the generalized pivotal quantity for $(\beta, \omega, \sigma_Y)$, denoted by $(G_\beta, G_\omega, G_{\sigma_Y})$. As a consequence, the $100(1-\alpha)\%$ generalized confidence interval for $\delta$ is formulated by

$$l_G = G_\delta(\alpha/2) \tag{17}$$
$$u_G = G_\delta(1-\alpha/2) \tag{18}$$

where $G_\delta = \ln G_\omega + G_{\sigma_Y}^2 - \left[\ln G_\beta + \ln\left(\frac{G_\omega\sqrt{G_\tau}}{G_\beta} + 1\right)\right]$; $G_\tau = \exp(G_{\sigma_Y}^2)$. The $G_\delta(\alpha)$ denotes the $\alpha$th quantile of $G_\delta$.

## 2.3 *MOVER Method*

According to Donner and Zou [10], the MOVER can be formulated from a combination of an individual confidence interval for each parameter. Here the MOVER-based confidence interval for $\delta$ is needed to construct for comparing other methods in terms of the performance measures. The individual confidence intervals for the parameters $\beta, \omega, \sigma_Y$ are formulated by using the normal approximation as follows:

$$[l_\beta, u_\beta] = \left[\hat{\beta} \mp C_{\beta,1-\alpha/2}\sqrt{\widehat{V}_{11}}\right] \tag{19}$$
$$[l_\omega, u_\omega] = \left[\hat{\omega} \mp C_{\omega,1-\alpha/2}\sqrt{\widehat{V}_{22}}\right] \tag{20}$$
$$[l_{\sigma_Y}, u_{\sigma_Y}] = \left[\hat{\sigma}_Y \mp C_{\sigma_Y,1-\alpha/2}\sqrt{\widehat{V}_{33}}\right] \tag{21}$$

where $(C_{\beta,\alpha}, C_{\omega,\alpha}, C_{\sigma_Y,\alpha})$ denote the $\alpha^{\text{th}}$ quantile of $(C_\beta, C_\omega, C_{\sigma_Y})$ having standard normal distributions. The asymptotic variances of $(\hat{\beta}, \hat{\omega}, \hat{\sigma}_Y)$ have been solved in the Eq. (10), while their estimated variances $(\widehat{V}_{11}, \widehat{V}_{22}, \widehat{V}_{33})$ are obtained by substituting their ML estimates, respectively. Recall that $\delta \approx \ln\omega + \ln\tau -$

$\left[\ln\beta + \ln\left(\frac{\omega\sqrt{\tau}}{\beta} + 1\right)\right]$ such that we let $\gamma = \ln\left[\exp(\eta) + 1\right]$. The $100(1-\alpha)\%$ MOVER confidence interval for $\eta = \ln\left(\frac{\omega\sqrt{\tau}}{\beta}\right) = \ln\omega + \ln\sqrt{\tau} - \ln\beta$ is expressed as

$$l_\eta = \hat{\eta} - \sqrt{\left[\ln\hat{\omega} + \ln\sqrt{\hat{\tau}} - l_{\ln\omega+\ln\sqrt{\tau}}\right]^2 + \left[u_{\ln\beta} - \ln\hat{\beta}\right]^2} \quad (22)$$

$$u_\eta = \hat{\eta} + \sqrt{\left[u_{\ln\omega+\ln\sqrt{\tau}} - (\ln\hat{\omega} + \ln\sqrt{\hat{\tau}})\right]^2 + \left[\ln\hat{\beta} - l_{\ln\beta}\right]^2} \quad (23)$$

where

$$l_{\ln\omega+\ln\sqrt{\tau}} = \ln\hat{\omega} + \ln\sqrt{\hat{\tau}} - \sqrt{\left[\ln\hat{\omega} - l_{\ln\omega}\right]^2 + \left[\ln\sqrt{\hat{\tau}} - l_{\ln\sqrt{\tau}}\right]^2} \quad (24)$$

$$u_{\ln\omega+\ln\sqrt{\tau}} = \ln\hat{\omega} + \ln\sqrt{\hat{\tau}} + \sqrt{\left[u_{\ln\omega} - \ln\hat{\omega}\right]^2 + \left[u_{\ln\sqrt{\tau}} - \ln\sqrt{\hat{\tau}}\right]^2} \quad (25)$$

and

$$[l_{\ln\omega}, u_{\ln\omega}] = \ln\,[l_\omega, u_\omega] \quad (26)$$

$$\left[l_{\ln\beta}, u_{\ln\beta}\right] = \ln\left[l_\beta, u_\beta\right] \quad (27)$$

$$\left[l_{\ln\sqrt{\tau}}, u_{\ln\sqrt{\tau}}\right] = \frac{1}{2}\left[l_{\sigma_Y}, u_{\sigma_Y}\right]^2 \quad (28)$$

This leads to obtain the $100(1-\alpha)\%$ MOVER for $\gamma$ is

$$l_\gamma = \ln\left(\exp l_\eta + 1\right) \quad (29)$$

$$u_\gamma = \ln\left(\exp u_\eta + 1\right) \quad (30)$$

Next, the $100(1-\alpha)\%$ MOVER confidence interval for $\ln\omega + \ln\tau - \ln\beta$ becomes

$$l_{\ln\omega+\ln\tau-\ln\beta} = \ln\hat{\omega} + \ln\hat{\tau} - \ln\hat{\beta} - \sqrt{\left[\ln\omega + \ln\tau - l_{\ln\omega+\ln\tau}\right]^2 + \left[u_{\ln\beta} - \ln\hat{\beta}\right]^2} \quad (31)$$

$$u_{\ln\omega+\ln\tau-\ln\beta} = \ln\hat{\omega} + \ln\hat{\tau} - \ln\hat{\beta} + \sqrt{\left[u_{\ln\omega+\ln\tau} - (\ln\hat{\omega} + \ln\hat{\tau})\right]^2 + \left[\ln\hat{\beta} - l_{\ln\beta}\right]^2} \quad (32)$$

where

$$[l_{\ln\tau}, u_{\ln\tau}] = \left[l_{\sigma_Y}, u_{\sigma_Y}\right]^2 \tag{33}$$

$$l_{\ln\omega+\ln\tau} = \ln\hat{\omega} + \ln\hat{\tau} - \sqrt{\left[\ln\hat{\omega} - l_{\ln\omega}\right] + \left[\ln\hat{\tau} - l_{\ln\tau}\right]} \tag{34}$$

$$u_{\ln\omega+\ln\tau} = \ln\hat{\omega} + \ln\hat{\tau} + \sqrt{\left[u_{\ln\omega} - \ln\hat{\omega}\right] + \left[u_{\ln\tau} - \ln\hat{\tau}\right]} \tag{35}$$

Finally, the $100(1-\alpha)\%$ MOVER confidence interval for $\delta$ can be established as

$$l_M = \hat{\delta} - \sqrt{\left[\ln\hat{\omega} + \ln\hat{\tau} - \ln\hat{\beta} - l_{\ln\omega+\ln\tau-\ln\beta}\right]^2 + \left[u_\gamma - \hat{\gamma}\right]^2} \tag{36}$$

$$u_M = \hat{\delta} + \sqrt{\left[u_{\ln\omega+\ln\tau-\ln\beta} - \left(\ln\hat{\omega} + \ln\hat{\tau} - \ln\hat{\beta}\right)\right]^2 + \left[\hat{\gamma} - l_\gamma\right]^2} \tag{37}$$

## 3 Numerical Results

To verify the theoretical results in Sect. 2, we conducted a Monte Carlo simulation by using R statistical software to assess the methods for estimating the confidence interval for the coefficient of variation of the three-parameter lognormal distribution in terms of ACP, AL and RAL.

For the 95% nominal coverage level, the mean ($\mu_Y$) was fixed at 2.5; the lower bound (threshold: $\beta$) was varied as 3, 5, 10, and 20; the sample sizes ($n$) was varied as 25, 50, 75, and 100; and a large variance ($\sigma_Y^2$) was set as 2.50 and 3.25. For each parameter combination, the data were independently generated for a three-parameter lognormal distribution $M = 5000$ times. For each run, we generated $m = 5000$ generalized pivotal quantities for the generalized confidence interval method. The ACPs were calculated as the percentage of times the confidence interval for the coefficient of variation falling was within the 95% confidence level for the 5000 simulations. For each scenario, the best-performing confidence interval estimate provided an ACP equal or greater than the 95% confidence level and the shortest AL considered from the smallest RAL. Note that the RAL was computed as the ratio between its AL and that using the MOVER method. The numerical results reported in Table 1 and Figs. 1 and 2 led us to the conclusions as follows:

(1) The performance of the likelihood-based confidence interval was a fairly good coverage percentage, while it also provided the shortest AL even when the variance was large.
(2) Regardless of the values of $\beta$, the generalized confidence interval could not provide an ACP close to the 95% nominal coverage level for any scenario.
(3) Considering all of the scenarios, the MOVER provided the highest ACPs, meanwhile its ALs were wider than the other methods.

**Table 1** Numerical results of 95% CIs for $\delta$

| $n$ | $\beta$ | Methods | $\sigma_Y^2 = 2.50$ | | | $\sigma_Y^2 = 3.25$ | | |
|---|---|---|---|---|---|---|---|---|
| | | | ACP | AL | RAL | ACP | AL | RAL |
| 25 | 20 | LCI | 96.86 | 3.242 | 0.577 | 97.60 | 3.123 | 0.528 |
| | | GCI | 80.44 | 2.510 | 0.447 | 83.60 | 2.381 | 0.403 |
| | | MCI | 97.14 | 5.615 | | 99.72 | 5.914 | |
| | 10 | LCI | 96.80 | 2.954 | 0.514 | 97.68 | 2.889 | 0.479 |
| | | GCI | 79.54 | 2.320 | 0.404 | 83.26 | 2.225 | 0.369 |
| | | MCI | 97.42 | 5.749 | | 99.64 | 6.027 | |
| | 5 | LCI | 96.56 | 2.773 | 0.471 | 97.92 | 2.759 | 0.448 |
| | | GCI | 77.66 | 2.205 | 0.375 | 81.94 | 2.140 | 0.347 |
| | | MCI | 96.94 | 5.886 | | 99.84 | 6.163 | |
| | 3 | LCI | 96.48 | 2.687 | 0.450 | 97.24 | 2.656 | 0.435 |
| | | GCI | 77.30 | 2.151 | 0.360 | 82.76 | 2.064 | 0.338 |
| | | MCI | 95.58 | 5.967 | | 99.68 | 6.106 | |
| 50 | 20 | LCI | 98.60 | 2.131 | 0.579 | 98.20 | 2.515 | 0.538 |
| | | GCI | 85.12 | 1.528 | 0.415 | 86.16 | 1.856 | 0.397 |
| | | MCI | 99.90 | 3.681 | | 99.92 | 4.676 | |
| | 10 | LCI | 98.40 | 1.912 | 0.511 | 98.34 | 2.312 | 0.486 |
| | | GCI | 85.56 | 1.377 | 0.368 | 85.28 | 1.716 | 0.361 |
| | | MCI | 99.99 | 3.745 | | 99.92 | 4.757 | |
| | 5 | LCI | 98.36 | 1.782 | 0.465 | 97.98 | 2.178 | 0.454 |
| | | GCI | 83.62 | 1.291 | 0.337 | 85.76 | 1.624 | 0.338 |
| | | MCI | 99.98 | 3.831 | | 99.98 | 4.801 | |
| | 3 | LCI | 97.96 | 1.716 | 0.444 | 97.88 | 2.136 | 0.439 |
| | | GCI | 82.84 | 1.247 | 0.323 | 84.80 | 1.597 | 0.328 |
| | | MCI | 99.88 | 3.865 | | 99.99 | 4.866 | |
| 75 | 20 | LCI | 98.72 | 1.690 | 0.587 | 98.44 | 2.000 | 0.546 |
| | | GCI | 87.58 | 1.182 | 0.410 | 87.78 | 1.440 | 0.393 |
| | | MCI | 99.90 | 2.882 | | 99.96 | 3.665 | |
| | 10 | LCI | 98.58 | 1.518 | 0.515 | 98.78 | 1.836 | 0.490 |
| | | GCI | 87.06 | 1.064 | 0.361 | 88.12 | 1.327 | 0.354 |
| | | MCI | 99.99 | 2.946 | | 99.99 | 3.745 | |
| | 5 | LCI | 98.40 | 1.414 | 0.469 | 98.42 | 1.730 | 0.457 |
| | | GCI | 85.20 | 0.994 | 0.330 | 87.62 | 1.253 | 0.331 |
| | | MCI | 99.99 | 3.015 | | 99.99 | 3.782 | |
| | 3 | LCI | 99.06 | 1.362 | 0.452 | 98.26 | 1.681 | 0.443 |
| | | GCI | 85.14 | 0.960 | 0.318 | 87.14 | 1.220 | 0.321 |
| | | MCI | 99.99 | 3.043 | | 99.99 | 3.798 | |

**Table 1** (continued)

| $n$ | $\beta$ | Methods | $\sigma_Y^2 = 2.50$ | | | $\sigma_Y^2 = 3.25$ | | |
|---|---|---|---|---|---|---|---|---|
| | | | ACP | AL | RAL | ACP | AL | RAL |
| 100 | 20 | LCI | 99.16 | 1.450 | 0.591 | 98.58 | 1.713 | 0.549 |
| | | GCI | 88.34 | 1.003 | 0.409 | 88.98 | 1.220 | 0.391 |
| | | MCI | 99.99 | 2.455 | | 99.99 | 3.122 | |
| | 10 | LCI | 98.92 | 1.304 | 0.517 | 98.94 | 1.569 | 0.493 |
| | | GCI | 87.04 | 0.902 | 0.358 | 88.26 | 1.119 | 0.351 |
| | | MCI | 99.99 | 2.523 | | 99.96 | 3.186 | |
| | 5 | LCI | 98.86 | 1.207 | 0.470 | 98.72 | 1.480 | 0.459 |
| | | GCI | 85.90 | 0.836 | 0.326 | 88.42 | 1.058 | 0.328 |
| | | MCI | 99.99 | 2.565 | | 99.99 | 3.226 | |
| | 3 | LCI | 98.82 | 1.167 | 0.449 | 98.66 | 1.440 | 0.444 |
| | | GCI | 85.68 | 0.810 | 0.312 | 88.20 | 1.030 | 0.318 |
| | | MCI | 99.99 | 2.599 | | 99.99 | 3.242 | |

*Note* LCI, likelihood-based confidence interval;
GCI, generalized confidence interval;
MCI, MOVER-based confidence interval

Overall, the likelihood-based method generated ACPs closer to the 95% nominal coverage level, and it produced shorter ALs than the generalized confidence interval and MOVER-based methods, even for extreme situations involving large variance and highly right-skewed data.

## 4 The Efficacies of the Methods Using Real Data

The wind speed data for Chiang Mai during January–March 2021 obtained from the Automatic Weather System of the Thai Meteorological Department [15] are highly right-skewed. The dataset can be fitted to a three-parameter lognormal distribution (see Fig. 3), which was checked by using Akaike and Bayesian Information Criteria [12]. Without loss of generality, we excluded the covariates that affect the wind speed data. For the dataset, the coefficient of variation is 0.9319 and the ML estimates of $\beta$, $\mu_Y$, and $\sigma_Y$ were $\hat{\beta} = 1.9356$, $\hat{\mu}_Y = 2.6544$, and $\hat{\sigma}_Y = 0.3197$, respectively. The 95% confidence interval for the coefficient of variation of the wind speed data computed by using the likelihood-based, generalized confidence interval, and MOVER methods were [0.6483,1.3394], [0.6547,1.6011], and [0.1502,4.1998] with ALs of 0.6911, 0.9464, and 4.0496, respectively. The results can be interpreted as indicating stable wind speeds in Chiang Mai during the period from January to March 2021. Additionally, this stability in wind speed suggests that it might not significantly contribute to soil erosion.

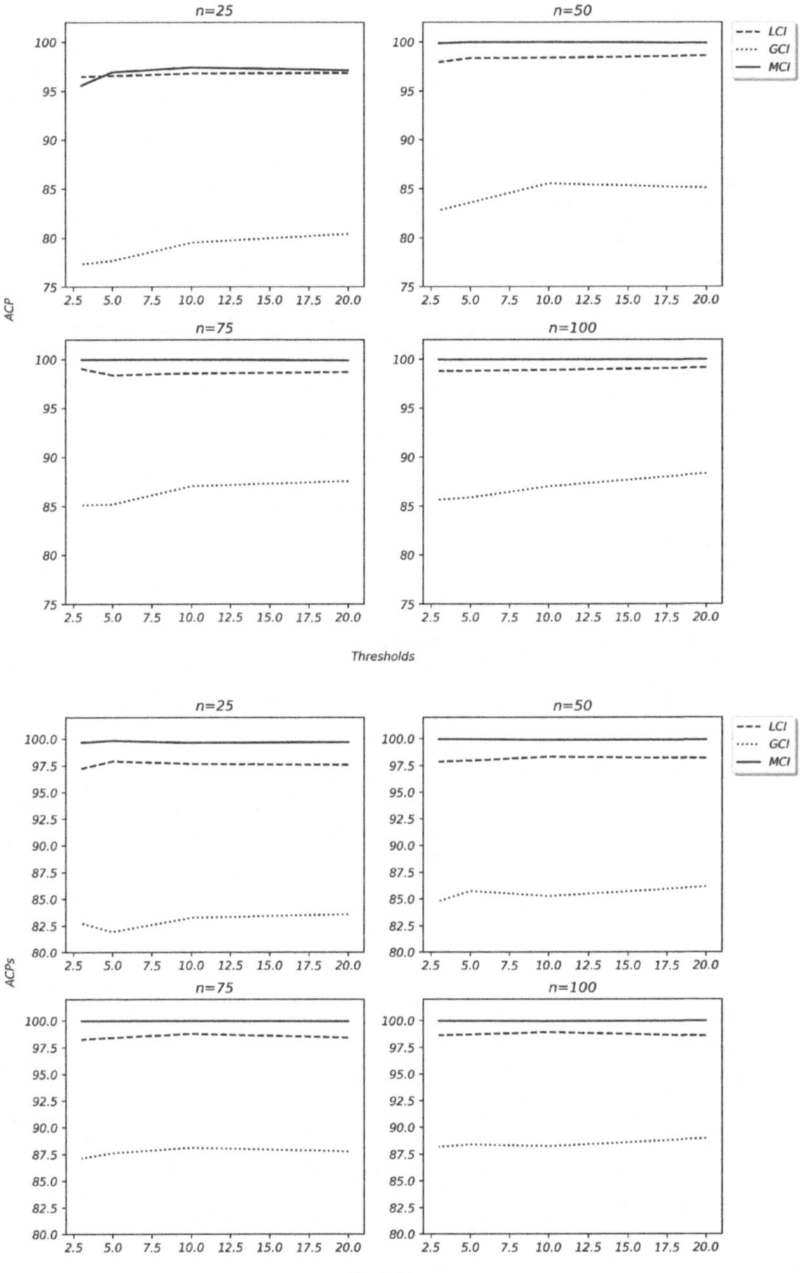


**Fig. 1** ACP performances of 95% CIs for $\delta$: **a** $\sigma_Y^2 = 2.50$ and **b** $\sigma_Y^2 = 3.25$

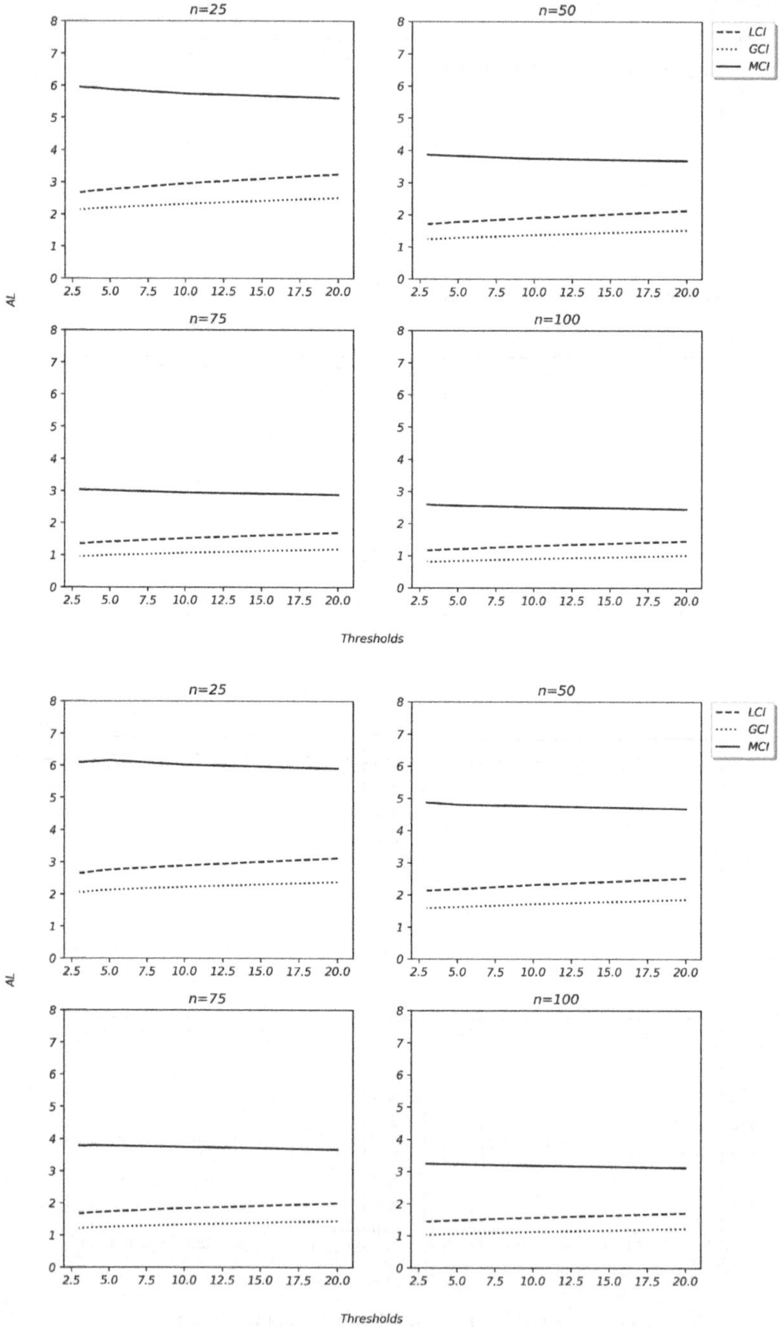


**Fig. 2** AL performances of 95% CIs for $\delta$: **a** $\sigma_Y^2 = 2.50$ and **b** $\sigma_Y^2 = 3.25$

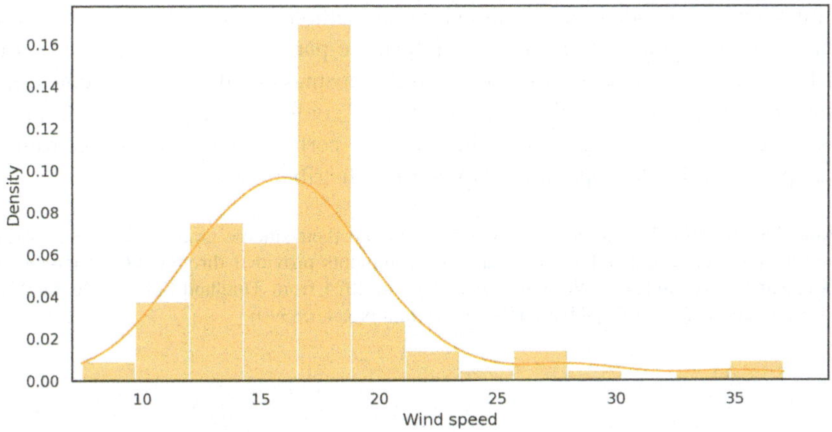


**Fig. 3** Density plot of wind speed data (km/hr) in Chiang Mai during January–March 2021

It can be seen from the above analysis that the likelihood-based method is the most suitable for estimating the confidence interval for the coefficient of variation of a three-parameter lognormal distribution because it produced the narrowest AL with an appropriate ACP. Moreover, the performances of all methods when applied to wind speed data agree with the numerical results in the previous section.

## 5 Discussion and Conclusions

Herein, we present three existing methods for the coefficient of variation of the three-parameter lognormal distribution based on the likelihood-based, generalized confidence interval, and MOVER approaches. When constructing these confidence interval estimators, the problem that needs to be overcome is how to handle the threshold limit ($\beta$). In the first method, the Fisher information matrix was used to obtain approximate large-sample variances and covariances for the ML estimate of $\beta$. The second method uses the generalized pivotal quantity for $\beta$ from a normal approximation using its ML estimate and variance-based Fisher information matrix. This method tends to result in an underestimate in terms of ACI. It might be caused by the parameter $\delta$ being reparameterized to examine for a large variation especially. Moreover, the generalized pivotal quantity of $\beta$ is formulated by the ML estimate $\hat{\beta}$ affecting to estimate $\hat{\sigma}_Y^2$ as well. Then, it is possible that it performs poorly in these situations.

For the last method, individual confidence interval estimates are produced for the mean, variance, and threshold from a normal approximation. The results of a numerical simulation study reveal that the likelihood-based method outperformed the other two, even with highly right-skewed data with a large variance. Therefore, we highly recommended the likelihood-based method for estimating the confidence interval for the coefficient of variation of the three-parameter lognormal distribution. Further research is required to evaluate the robustness of the methods concerning point estimation, as opposed to the coverage of confidence intervals. Additionally, it is important to investigate how the methods perform when departing from the assumptions of the three-parameter lognormal distribution.

**Acknowledgements** The authors would like to extend their sincere gratitude to the editor and reviewers for their expert advice and valuable comments provided throughout the manuscript. This research was funded by the Fundamental Fund 2024 from Thailand Science Research and Innovation, and Uttaradit Rajabhat University (Grant No. 4708390).

## References

1. Bendel, R.B., Higgins, S.S., Teberg, J.E., Pyke, D.A.: Comparison of skewness coefficient, coefficient of variation, and Gini coefficient as inequality measures within populations. Oecologia **78**(3), 394–400 (1989)
2. Burges, S.J., Lettenmaier, D.P., Bates, C.L.: Properties of the three-parameter log normal probability distribution. Water Res. Res. **11**(2), 229–235 (1975)
3. Casella, G., Berger, R.L.: Statistical Inference, 2nd ed. Duxbury (2002)
4. Charbeneau, R.J.: Comparison of the two- and three-parameter log normal distributions used in streamflow synthesis. Water Res. Res. **14**(1), 149–150 (1978)
5. Chen, C.: Tests of fit for the three-parameter lognormal distribution. Comput. Stat. Data Anal. **50**(6), 1418–1440 (2006)
6. Chen, Z., Miao, F.: Interval and point estimators for the location parameter of the three-parameter lognormal distribution. Int. J. Quality Stat. Reliab. **2012**(370), 1–6 (2012)
7. Cohen, A.C.: Estimating parameters of logarithmic-normal distributions by maximum likelihood. J. Am. Statist. Assoc. **46**(254), 206–212 (1951)
8. Cohen, A.C., Whitten, B.J.: Estimation in the three-parameter lognormal distribution. J. Am. Statist. Assoc. **75**(370), 399–404 (1980)
9. Cohen, A.C., Whitten, B.J., Ding, Y.: Modified moment estimation for the three-parameter lognormal distribution. J. Qual. Technol. **17**(2), 92–99 (1985)
10. Donner, A., Zou, G.Y.: Closed-form confidence intervals for functions of the normal mean and standard deviation. Stat. Methods Med. Res. **21**(4), 347–359 (2010)
11. Griffiths, D.A.: Interval estimation for the three-parameter lognormal distribution via the likelihood function. Appl. Stat. **29**(1), 58–68 (1980)
12. Maneerat, P., Niwitpong, S.-A., Nakjai, P.: Estimating wind speed by using confidence intervals for the median in a three-parameter lognormal model. In: Honda, K., Entani, T., Ubukata, S., Huynh, V.-N., Inuiguchi, M. eds., Integrated Uncertainty in Knowledge Modelling and Decision Making. Lecture Notes in Computer Science, pp. 329–341. Springer International Publishing, Cham (2022)
13. McCool, J.I.: Inferential Techniques for Weibull Populations. Technical Report (1975). https://apps.dtic.mil/sti/citations/ADA009645
14. Singh, V., Cruise, J., Ma, M.: A comparative evaluation of the estimators of the three-parameter lognormal distribution by Monte Carlo simulation. Comput. Stat. Data Anal. **10**(1), 71–85 (1990)
15. Thai Meteorological Department, Daily Weather Summary, Technical Report (2021). https://www.tmd.go.th/en/climate.php?FileID=1

# The Determinants of Carbon Dioxide Emissions in ASEAN-7: Evidence from Bayesian Model Averaging Approach

**Tho M. Nguyen, Pathairat Pastpipatkul, and Worrawat Saijai**

**Abstract** The study uses the Bayesian model averaging (BMA) approach to identify the determinants of carbon dioxide emissions in selected ASEAN countries for the period 1990–2020. The environmental Kuznets curve (EKC) hypothesis is re-examined during the analysis. Applying both non-informative and informative priors, the results show that energy factors are the main causes affecting $CO_2$ emission levels. Based on the obtained models, some recommendations are proposed for each group of countries to control the generation of $CO_2$ outputs.

**Keywords** Carbon dioxide ($CO_2$) emissions · Bayesian model averaging · ASEAN

## 1 Introduction

Along with economic growth, global warming and climate change have been known to be critical worldwide issues. In 1997, the Kyoto Protocol was adopted for emission reduction goals. It officially entered into force in 2005. More recently, the Paris Climate Agreement (aka Paris Agreement or COP21) was adopted in 2015 to require all countries including major carbon emitters that did not join the Kyoto Protocol to limit and reduce the emissions of $CO_2$. The demand for identifying the robust determinants of $CO_2$ emissions has been increasing. Though lots of work on $CO_2$ emission and its determinants have been analyzed, the nexus between environmental quality and economic factors is still an ongoing debate. This study is going to answer the following questions: What are the factors of $CO_2$ emissions in ASEAN countries?

T. M. Nguyen (✉) · P. Pastpipatkul · W. Saijai
Faculty of Economics, Chiang Mai University, Chiang Mai, Thailand
e-mail: tho.nm@ou.edu.vn

P. Pastpipatkul
e-mail: pathairat.past@cmu.ac.th

W. Saijai
e-mail: worrawat.s@cmu.ac.th



V. Kreinovich et al. (eds.), *Applications of Optimal Transport to Economics and Related Topics*, Studies in Systems, Decision and Control 556,
https://doi.org/10.1007/978-3-031-67770-0_22

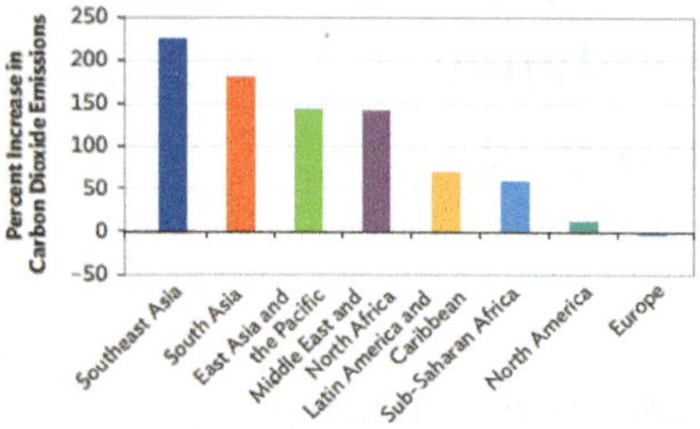


Source: Asian Development Bank (2015)

(a) Increases in total $CO_2$ in world regions 1990-2010

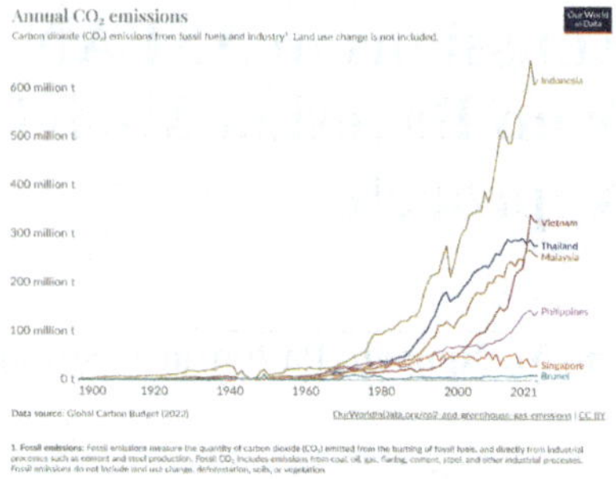


Source: Our World in Data (2023)

(b) Annual $CO_2$ in selected countries

**Fig. 1** Increase of $CO_2$ in ASEAN countries

Is there a non-linear relationship between the growth of economics and the quality of the environment?

Instead of focusing on a specific country, this study identifies the determinants of $CO_2$ emissions in the Association of Southeast Asian Nations (ASEAN) countries. Since the Asian region is pointed as the world's largest energy consumer all over the world [4, 19]. The $CO_2$ emissions from consuming energy in developing countries since the 1990s would be much more than in industrialized countries [15]. Figure 1 implies that the ASEAN region has been the most growing emitter and the most vulnerable region globally.

This study contributes to the existing literature in some points, as follows: (1) few studies analyze the case in ASEAN by employing the Bayesian approach, and most of them are implemented with the frequentist approach. As the model is uncertain, the more explanatory variables the more reasonable the analysis is. However, following the frequentist analysis, a small, predefined set of variables is often concerned. It can lead to the problem of omitted variable, pre-testing bias, and the risk of p-hacking [3, 7, 12], (2) adding more candidate determinant that has not been concerned before.

The remainder is organized as follows. Section 2 is the literature review of existing studies. The data and econometric strategy are mentioned in Sect. 3. Section 4 reports in detail the results and discussions. The paper is summed up by the conclusions in Sect. 5.

## 2 Literature Review

Shortly, the EKC hypothesis was first introduced by [9]. It can be illustrated by an inverted U-shaped relationship between environmental quality and the growth of economic. Many researches have been focused on the hypothesis but the results are

**Table 1** Summary of related literature

| Region | Authors | Variables | Techniques | Findings |
|---|---|---|---|---|
| ASEAN-5 1990–2006 | [17] | $CO_2$, electricity consumption, GDP, $GDP^2$ | Panel VECM & Granger causality tests | Support EKC |
| ASEAN-5 1971–2009 | [22] | $CO_2$, economic growth, energy consumption | Autoregressive Distributive Lag (ARDL) & Granger causality test based on VECM | EKC only in SGP, THA |
| ASEAN-5 1980–2008 | [10] | $CO_2$, GDP, energy consumption | Panel smooth transition regression model | Support EKC |
| ASEAN+3[a] 1991–2010 | [1] | $CO_2$, GDP, energy consumption, urbanization, trade & transportation | Cointegration test, the Granger causality based on VECM | GDP, EC, trade are main factors |
| ASEAN-5 1971–2013 | [16] | $CO_2$, GDP, energy use & trade openness | ARDL, Pooled Mean Group (PMG) | EKC only in THA |
| ASEAN-5 1980–2016 | [20] | $CO_2$, GDP, energy consumption | Bootstrapped panel cointegration, Granger non-causality | Support EKC |
| ASEAN-9 1990–2018 | [21] | Trade, RE, non-RE, GDP, $GDP^2$ | ARDL, mean group (MG), PMG | Support EKC |
| ASEAN-6[b] 1985–2012 | [2] | $CH_4$ (methane), GDP, energy consumption, trade openness | Dynamic panel data estimations: MG & PMG | Support EKC |
| ASEAN-4[c] 1980–2015 | [6] | $CO_2$, hydropower consumption, FDI, manufacturing performance | ARDL Bound test approach | Significant effects |

[a]All members of ASEAN and 3 other countries (China, Japan, and South Korea)
[b]Indonesia, Malaysia, the Philippines, Singapore, Thailand and Vietnam
[c]Indonesia, Malaysia, the Philippines and Thailand

still debatable even within the same sample. For details, Table 1 summarizes only the works that are related to the ASEAN. The similarity of previous studies is the highly positive interrelation between $CO_2$ emissions and energy consumption. Regarding the EKC hypothesis, it is an inconclusive result.

Reference [6] use hydropower consumption as a proxy for energy consumption. They find that besides GDP per capita and electricity consumption, many more important factors significantly affect $CO_2$ emissions such as foreign direct investment (FDI), or manufacturing value added. Reference [1] include the impacts of urbanization, trade openness, and transportation in the investigation. The empirical result only supports three main factors (economic growth, energy consumption, and trade openness). Reference [21] investigate the determinants of $CO_2$ emissions in ASEAN-9 and point out that trade and renewable energies can highly mitigate the emissions but non-renewable energies make the worse environment.

As far as we are concerned, there is no research using the Bayesian approach (specifically Bayesian model averaging, BMA technique) to define first the true model, then the robust indicators of $CO_2$ in ASEAN. Among the existing studies, many of them employ Granger causality tests. There is a restriction that this technique can only investigate the relationship between two factors, pair-by-pair. Hence, the number of variables is normally small. Another limitation of previous studies is criticized by the application of panel unit root and cointegration test with the assumption that the error terms are not correlated. It results in the biased findings [10, 20].

The spirit of the BMA approach is that when dealing with model uncertainty, it can be used to estimate all possible combinations of the features and take a weighted average over all the candidate models. In which, the weights are defined as the probabilities that the prospective model is the true one [3]. From the Bayesian perspective, the prior distribution is important as it can significantly affect the posterior model probabilities. According to [8], the specification of prior is even much harder to identify. As a result of this, this study is going to employ both non-informative and informative priors.

## 3 Data and Methodology

We collect the data of ASEAN-7[1] for the period 1990–2020, the rest of ASEAN members (Cambodia, Laos, and Myanmar) are excluded due to the lack of data. The variable of interest is $CO_2$ outflows (metric tons per capita) including the emissions during the consumption of solid, liquid, gas fuels, and gas flaring. The set of potential predictors is listed in Table 2. All variables are obtained from the World Bank unless total energy supply (TES) and electricity consumption (EC) are from the International Energy Agency (IEA).

The transformation of data into the natural logarithm form is necessary to induce the stationary process, and to reduce the possibility of heteroscedasticity as well as autocorrelation to exist. Given that the effects of the factors of $CO_2$ emissions can depend on the level of income in a country [3], we separate the selected ASEAN

[1] The selected countries include Brunei, Indonesia, Malaysia, Philippines, Singapore, Thailand, and Vietnam.

**Table 2** List of potential determinants of $CO_2$ emissions

| Variables | Descriptions | Natural logarithm |
|---|---|---|
| GDP | Per capita values for gross domestic product (GDP) expressed in current international dollars converted by purchasing power parity | Yes |
| $GDP^2$ | To check the existence of the EKC hypothesis (apply ln, then square) | Yes |
| POP | Total population | Yes |
| IND | Industry value added (% of GDP) | No |
| AFF | Agriculture, forestry, and fishing value added (% of GDP) | No |
| FDI | Foreign direct investment, net inflows (% of GDP) | No |
| Trade | Trade openness, measured by the sum of exports and imports (% of GDP) | Yes |
| Forest | Forest area (squared km) | Yes |
| URB | Urban population (% of total population) | No |
| RE | Renewable energy consumption (% of total final energy consumption) | No |
| Rents | Total natural resources rents (% of GDP), the sum of oil, natural gas, coal, mineral, and forest rents | No |
| AVIA | The number of registered carrier departures worldwide which are domestic takeoffs and takeoffs abroad of air carriers registered in a country, as a proxy for transportation | Yes |
| TES | Total energy supply per capita | Yes |
| EC | Electricity consumption per capita | Yes |

countries into three groups based on income level, according to the World Bank's classification:

1. the high-income group (Brunei and Singapore),
2. the upper-middle-income group (Malaysia and Thailand), and
3. the lower-middle-income group (Indonesia, Philippines, and Vietnam)

Regarding time-effect, the dataset is separated into two periods (1990–2005 and 2006–2020). The break-time point is the year when the Kyoto Protocol was entered into force. Two distinct variables are added to the investigation: (1) AVIA as the proxy for transportation effect, and (2) the presence of total natural resource rents. These factors are expected to have positive impacts on the environment. In other words, the more either flights or natural resource rents, the more $CO_2$ emissions are released.

The aim is to specify the determinants of $CO_2$ emissions. The research model is defined as follows:

$$CO_2 = f(X_1, X_2, \ldots, X_{14}) \tag{1}$$

$$\ln(CO_{2_{i,t}}) = \alpha_i + X^{'}_{i,t}\theta + \varepsilon_{i,t} \quad , \quad \varepsilon_{i,t} \sim \mathcal{N}(0, \sigma^2) \tag{2}$$

where $CO_{2i,t}$ is the level of $CO_2$ emissions measured in metric tonnes per capita in group $i$ and year $t$; $X^{'}_{i,t}$ is the set of potential triggers of $CO_2$ for group $i$ at year $t$. The intercept $\alpha_i$ captures the time-invariant unobservable factors in group $i$; $\theta$ is the set of coefficients. Finally, $\varepsilon_{i,t}$ is a normal i.i.d error term with variance $\sigma^2$.

We use the Bayesian inference instead of the classical inference for the following reasons: The pivotal restriction of the frequentist approach is that it does not condition on the observed data [14]. Meanwhile, Bayes rule prescribes how the observed data update *prior* beliefs for $\theta$ (i.e., $p(\theta)$) to *posterior* beliefs (i.e., the probability distribution after combining prior and available data $p(\theta|\text{data})$). Reference [5] appraises that under classical inference, the sample data are taken as random while population parameters $\theta$ are taken as *'fixed'*. In the Bayesian perspective, parameters are treated as *'random'* and follow a probability distribution themselves, prior. This advantage is clearly shown in the context of time series, the Bayesian approach outperforms in error correction and stochastic volatility (the full parameter density is estimated and results in the posterior). Moreover, the Bayesian analysis can simplify estimation under the condition of non-standard distributions or non-linear regression that makes the results more realistic; or it may facilitate the models without the stationary requirement.

As the model of the nexus between $CO_2$ emissions and its triggers is uncertain, according to [11, 13], the BMA approach is particularly useful. In this context, multiple models are the nuisance factor but may result in significant differences in prediction and estimation. With the posterior beliefs about the potential models, BMA can eliminate the nuisance factor to produce the best predictions or parameter estimates. The mechanism of BMA can be explained by estimating all plausible combinations of the factors and taking a weighted average over all the proposed models. In which, the weights are defined by applying Bayes rule [3]. The posterior model distribution of model $p$, $M_p$ is the true model given the data $y$ can be specified as follows:

$$P\left(M_p \mid y, X\right) = \frac{P\left(M_p\right) \times P\left(y \mid M_p, X\right)}{P(y \mid X)} = \frac{P\left(M_p\right) \times P\left(y \mid M_p, X\right)}{\sum_{q=1}^{2k} P\left(M_q\right) \times P\left(y \mid M_q, X\right)} \tag{3}$$

In the middle term, $P(y \mid X)$ is the marginal likelihood which is constant over all models. Hence, the posterior model probability (PMP) is proportional to the marginal likelihood $P\left(y \mid M_p, X\right)$. Reference [25] point out the difference between $P(y \mid X)$ and $P\left(y \mid M_p, X\right)$ is that integration is once over the model space (the former) and once for a given model over the parameter space (the latter). In the last term, $k$ is the number of concerned predictors, and the number of possible linear models is $2^k$. By re-normalization of the above product, we can infer the model weighted posterior distribution for the set of unknown parameters $\theta$ as follows:

$$P(\theta \mid y, X) = \sum_{p=1}^{2^k} P\left(\theta \mid M_p, y, X\right) \frac{P\left(M_p\right) \times P\left(M_p \mid y, X\right)}{\sum_{q=1}^{2^k} P\left(M_q\right) \times P\left(M_q \mid y, X\right)} \tag{4}$$

Regarding the priors, we first apply the uniform distribution on the model space, with hyper-g prior = BRIC[2] (Zellner's g-prior[3] as in [3, 8]). Then, to check the robustness of the uniform results, we implement different model priors to re-examine the most potential predictors of $CO_2$ emissions. The sensitivity analysis will show the results with two alternative model priors: the *random theta* prior by binomial-beta hyper prior [18] and *fixed common* prior inclusion probabilities for each factor as suggested in [23].

# 4 Results

## *4.1 Bayesian Model Averaging with Fixed Prior*

In this part, we employ the BMA approach with non-informative priors (uniform probability for each of $2^k$ models) as the lack of prior knowledge, but we assign hyper-g=BRIC.

The results show the estimates of all determinants' posterior inclusion probability (PIP) that can be interpreted as the probability in which a specific factor belongs to the true model. In other words, PIP defines the importance of the explanatory variables in explaining the variable of interest. The next column, Posterior mean (Post Mean) displays the coefficients averaged over all possible models, including the models in which the coefficient is zero. The coefficients' posterior standard deviations (Post SD) provide the magnitude of the coefficients' sign. All predictors are ranked by their PIP values.

If taking PIP = 0.5 to perform the robustness of the determinants, in Brunei and Singapore, total energy supply (TES) and urbanization rate (URB) have significantly negative effects on environmental quality. It is common sense as these two countries mostly depend on natural gas and oil to create 90% of the total energy supply (IEA). These resources release high amounts of $CO_2$ into the atmosphere. In the aspect of URB, the more people move into and live in the cities, the more requirements for energy consumption. In the meanwhile, this explains the reason why EC contributes to the increase of $CO_2$. Interestingly, the population variable shows the positive impacts on air pollution. It means that more people in these high-income countries will help to mitigate the generation of $CO_2$. To give a better explanation for this relationship, more data on behavioral actions, awareness, development and

[2] BRIC stands for Bayesian Ridge, Lasso, and Elastic Net. It is a specification of Zellner's g-prior that extends BMA application to regularization techniques.

[3] It is introduced by Zellner (1986) and is the informative prior distribution for the coefficient parameters [24]

**Table 3** Determinants of $CO_2$ emissions in the high-income ASEAN countries

| | PIP | Post mean | Post SD |
|---|---|---|---|
| TES | 0.9920 | 0.3846 | 0.1030 |
| POP | 0.8399 | (0.4342) | 0.2617 |
| URB | 0.7994 | 0.0287 | 0.0183 |
| RE | 0.6767 | (0.1754) | 0.1470 |
| EC | 0.3402 | 0.0815 | 0.1539 |
| IND | 0.3067 | 0.0017 | 0.0036 |
| AFF | 0.2976 | (0.0556) | 0.1108 |
| $GDP^2$ | 0.2357 | (0.0322) | 0.2777 |
| GDP | 0.2306 | 0.0359 | 0.5480 |
| Rents | 0.1260 | 0.0001 | 0.0016 |
| Forest | 0.1258 | (0.0018) | 0.0617 |
| AVIA | 0.0999 | (0.0010) | 0.0181 |
| Trade | 0.0953 | (0.0058) | 0.0408 |
| FDI | 0.0719 | (0.0000) | 0.0006 |

**Table 4** Determinants of $CO_2$ emissions in the upper-middle-income countries

| | PIP | Post mean | Post SD |
|---|---|---|---|
| RE | 0.9998 | (0.0160) | 0.0022 |
| Forest | 0.9841 | (1.4956) | 0.3641 |
| TES | 0.9546 | 0.5910 | 0.2086 |
| $GDP^2$ | 0.2964 | 0.0421 | 0.1591 |
| IND | 0.2781 | 0.0018 | 0.0035 |
| GDP | 0.2765 | 0.0306 | 0.3151 |
| EC | 0.1362 | 0.0127 | 0.0461 |
| Trade | 0.1196 | (0.0067) | 0.0303 |
| URB | 0.1033 | 0.0001 | 0.0008 |
| Rents | 0.0883 | (0.0002) | 0.0011 |
| POP | 0.0871 | 0.0021 | 0.0247 |
| AVIA | 0.0787 | (0.0007) | 0.0076 |
| FDI | 0.0745 | (0.0000) | 0.0011 |
| AFF | 0.0724 | (0.0001) | 0.0013 |

improvement of infrastructure (such as the efficiency of public transportation) should be considered. The other variables with PIP values that are lower than 0.3 can be considered to have less influence on $CO_2$ output. Regarding the EKC relationship, the results in Table 3 support the hypothesis.

From Table 4, the main determinants of $CO_2$ in Malaysia and Thailand include RE (the amount of renewable energy consumption), the area of forest, and the total energy

**Table 5** Determinants of $CO_2$ emissions in the lower-middle-income countries

| | PIP | Post mean | Post SD |
|---|---|---|---|
| TES | 1.0000 | 1.0773 | 0.0858 |
| RE | 0.9996 | (0.0117) | 0.0025 |
| URB | 0.9976 | (0.0125) | 0.0021 |
| EC | 0.9853 | 0.1727 | 0.0480 |
| IND | 0.7179 | 0.0042 | 0.0033 |
| AFF | 0.3101 | 0.0019 | 0.0034 |
| AVIA | 0.2649 | 0.0069 | 0.0142 |
| Forest | 0.2222 | 0.0111 | 0.0301 |
| Rents | 0.2196 | (0.0010) | 0.0024 |
| POP | 0.1440 | (0.0104) | 0.0525 |
| FDI | 0.1123 | 0.0004 | 0.0016 |
| Trade | 0.1054 | (0.0004) | 0.0231 |
| $GDP^2$ | 0.1053 | 0.0000 | 0.0725 |
| GDP | 0.1049 | 0.0006 | 0.1442 |

supply. Similar to the previous group, the resources of energy supply are not only from coal, and natural gas but also highly from oil. However, upper-middle-income countries have more potential abilities to reduce the emissions of $CO_2$ by focusing on increases in renewable energy consumption instead of other sources, more and more afforestation, or reforestation projects. The results of GDP and $GDP^2$ imply that there is no evidence of the EKC correlation in this group for the period 1990–2020.

From Table 5, it is shown that TES is revealed to have absolute effects on $CO_2$ emissions. The similarity of TES in this group is that in addition to the mentioned resources to generate energy, these countries still likely depend on biofuels and waste. However, renewable energy consumption can be found as a core solution for alleviating $CO_2$ exhalations. These three countries should foster their ability in renewable energy supply (from wind, solar, hydro, etc.) (Fig. 2)

Electricity final consumption (EC) plays an important role that affects $CO_2$ emissions (increase by nearly 810%, 312%, and 3.387% in Indonesia, the Philippines, and Vietnam respectively). Once again, the urbanization rate (URB) shows prominent reductions in pollution as in the results of high-income ASEAN countries. The manufacturing sector (IND) is closely linked to amplifying the worsening environment. From the bottom of Table 5, we can see that the EKC hypothesis might not exist in lower-middle-income ASEAN countries. It implies the lack of investment and efficiency in renewable energy. The governments should pay severe attention to this area as they have more potential for renewable energy supply than other members. Nevertheless, they may consume them ineffectively.

From Tables 3 to 5, all anticipated regressors are kept. In other words, they all affect the variable of interest to some extent. In making the final inference, it is prudent to be informed by all predictors [11]. Among all explanatory variables, total

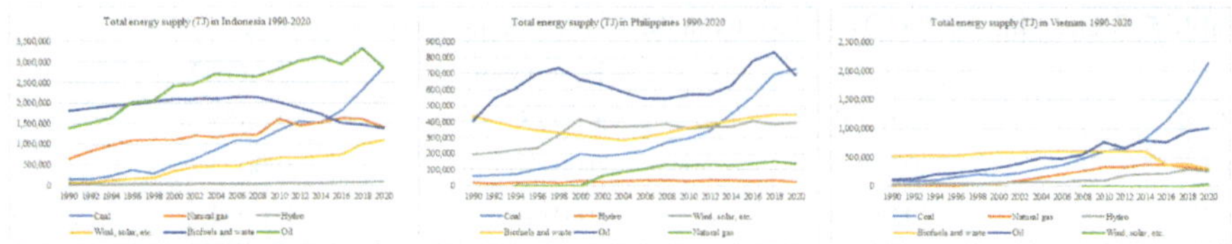


**Fig. 2** Total energy supply (TES, TJ) *Source* IEA (2023)

energy supply (TES) and renewable energy consumption (RE) are the key factors that have significant impacts on $CO_2$ emissions. It implies a strong potential solution for reducing pollution via utilizing renewable resources to generate energy in ASEAN countries. In the next section, we will present the true model by changing the priors to ensure the robustness of the above results.

## 4.2 Sensitivity Analysis

The purpose of the sensitivity analysis is to double-check the reasonableness of the above results by using alternative priors. As proposed in [3, 18, 23], we employ "random theta" prior (binomial-beta hyperprior on the prior inclusion probability) and "fixed common" prior inclusion probabilities for each predictor. The comparisons of three different priors are summarized in the following graphs (Fig. 3).

Generally, these two additional priors perform the same results in Sect. 4.1. Regarding the changes in model priors, the determinants of $CO_2$ emissions are specified in the same order.

(a) True model in high-income countries

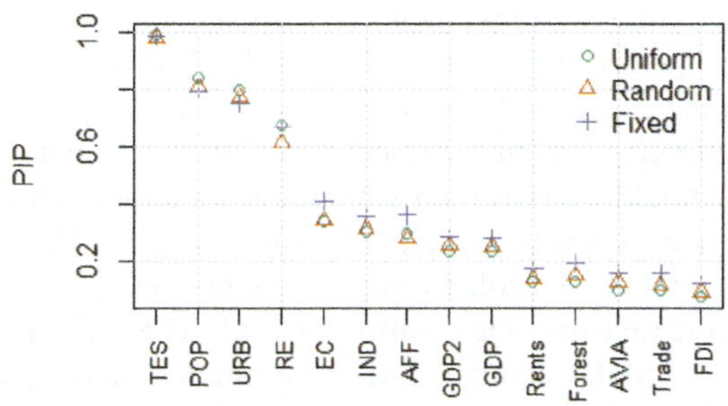


(b) True model in upper-middle-income countries

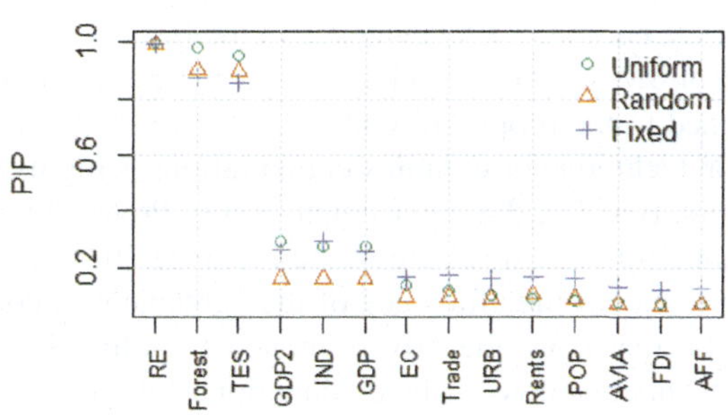


(c) True model in lower-middle-income countries

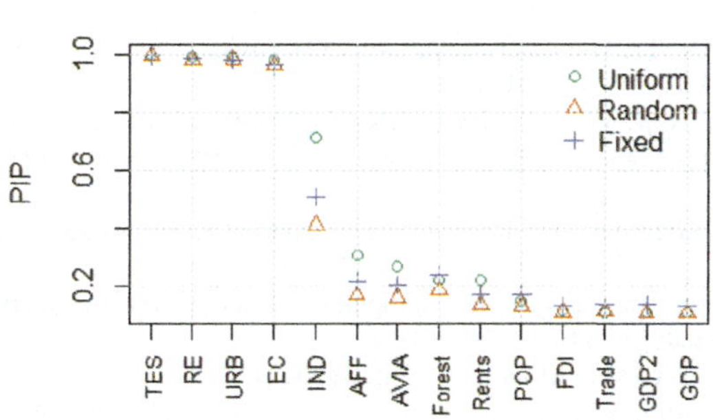


**Fig. 3** True determinants of $CO_2$ emissions by three different priors

# 5 Conclusion

Due to the model uncertainty, we implement the Bayesian model averaging to identify the most important factors of $CO_2$ emissions in ASEAN countries for the period 1990–2020. Based on the classification of income levels, it is found that each group has diverse features affecting the environmental quality. In Brunei and Singapore, the key determinants should be governed which are the total energy supply, population, urbanization rate, renewable energy consumption, etc. Renewable energy sources (such as hydro, wind, and solar systems) can be considered for investment and development in these high-income countries. Beside TES and RE, the forest is a substantial factor that the Malaysian and Thai governments should pay more attention for the goal of net zero. The last group (Indonesia, the Philippines, and Vietnam) emphasizes the importance of electricity consumption. The records of this variable have been changed sharply during the examined period. Some suggestions can be raised to handle the situation such as improving the awareness of saving electricity consumption, replacement of saving energy appliances, or incorporating renewable energy supply as much as possible. In the aspect of the EKC hypothesis, it may exist in high-income countries, but there is no evidence in the rest.

The main challenge of BMA is the dependence on the prior probabilities (including g-prior) that are assigned to the proposed models. In future research, the combination of cross-validation technique or complex computations is highly suggested for optimal choice of priors. Besides, the replacement Kyoto Protocol by Paris Agreement (2016), more than 190 countries have ratified it since 2021. The next studies can update more data to assess the efficiency of the agreement in controlling $CO_2$ emission magnitude. Furthermore, the drivers related to political institutions and government have not been considered. The conjunction of these factors will assess the importance of governmental management to the environment.

# References

1. Ab-Rahim, R., Xin-Di, T.: The determinants of $CO_2$ emissions in ASEAN+ 3 countries. J. Entrepr. Bus. (JEB) **4**(1), 38–49 (2016)
2. Adeel-Farooq, R.M., Raji, J.O., Adeleye, B.N.: Economic growth and methane emission: testing the EKC hypothesis in ASEAN economies. Manag. Environ. Quality: An Int. J. **32**(2):277–289 (2021)
3. Aller, C., Ductor, L., Grechyna, D.: Robust determinants of $CO_2$ emissions. Energy Econ **96**, 105154 (2021)
4. Aruga, K.: Investigating the energy-environmental Kuznets Curve hypothesis for the Asia-Pacific region. Sustainability **11**(8), 2395 (2019)
5. Congdon, P.: Applied Bayesian Modelling. Wiley (2014)
6. Danmaraya, I.A., Danlami, A.H.: Impact of hydropower consumption, foreign direct investment and manufacturing performance on $CO_2$ emissions in the ASEAN-4 countries. Int. J. Energy Sect. Manag. **16**(5), 856–875 (2022)

7. Fakher, H.-A., Ahmadian, M., Abedi, Z., Shaygani, B.: Bayesian econometrics approach in determining of effecting factors on pollution in developing countries (based on environmental performance index). Pollution **4**(3), 447–457 (2018)
8. Fernandez, C., Ley, E., Steel, M.F.J.: Benchmark priors for Bayesian model averaging. J. Econom. **100**(2):381–427 (2001)
9. Grossman, G.M., Krueger, A.B.: Environmental Impacts of a North American Free Trade Agreement (1991)
10. Heidari, H., Katircioğlu, S.T., Saeidpour, L.: Economic growth, $CO_2$ emissions, and energy consumption in the five ASEAN countries. Int. J. Electr. Power Energy Syst. **64**, 785–791 (2015)
11. Hinne, M., Gronau, Q.F., van den Bergh, D., Wagenmakers, E.-J.: A conceptual introduction to bayesian model averaging. Adv. Methods Pract. Psychol. Sci. **3**(2), 200–215 (2020)
12. Hoang, C.C.: Examining the relationship between electricity consumption, financial development and economic growth in ASEAN countries: evidence from a Bayesian analysis. Int. J. Energy Econ. Policy (2021)
13. Hoeting, J.A., Madigan, D., Raftery, A.E., Volinsky, C.T.: Bayesian model averaging: a tutorial (with comments by Clyde, M., Draper, D., George, E.I. and a rejoinder by the authors). Stat. Sci. **14**(4), 382–417 (1999)
14. Hoijtink, H., Klugkist, I., Boelen, P.A.: Bayesian Evaluation of Informative Hypotheses, vol. 361. Springer (2008)
15. Adnan Kasman and Yavuz Selman Duman: $CO_2$ emissions, economic growth, energy consumption, trade and urbanization in new EU member and candidate countries: a panel data analysis. Econ. Model. **44**, 97–103 (2015)
16. Kisswani, K.M., Harraf, A., Kisswani, A.M.: Revisiting the environmental Kuznets curve hypothesis: evidence from the ASEAN-5 countries with structural breaks. Appl. Econ. **51**(17), 1855–1868 (2019)
17. Hooi Hooi Lean and Russell Smyth: $CO_2$ emissions, electricity consumption and output in ASEAN. Appl. Energy **87**(6), 1858–1864 (2010)
18. Ley, E., Steel, M.F.J.: On the effect of prior assumptions in bayesian model averaging with applications to growth regression. J. Appl. Econ. **24**(4), 651–674 (2009)
19. Mundaca, L., Markandya, A.: Assessing regional progress towards a 'green energy economy'. Appl. Energy **179**, 1372–1394 (2016)
20. Munir, Q., Lean, H.H., Smyth, R.: $CO_2$ emissions, energy consumption and economic growth in the ASEAN-5 countries: a cross-sectional dependence approach. Energy Econ. **85**, 104571 (2020)
21. Nawab, T., Aamir, M.F.: Determinants of $CO_2$ emissions in the ASEAN economies: the role of renewable and non-renewable energy. iRASD J. Econ. **2**(1), 13–24 (2020)
22. Saboori, B., Sulaiman, J.: $Co_2$ emissions, energy consumption and economic growth in association of southeast asian nations (asean) countries: a cointegration approach. Energy **55**, 813–822 (2013)
23. Martin, X.S.-I., Doppelhofer, G., Miller, R.I.: Determinants of long-term growth: a Bayesian averaging of classical estimates (BACE) approach. Am. Econ. Rev. **94**(4), 813–835 (2004)
24. Wang, M.: Assigning g in Zellner's g prior for Bayesian variable selection. Ph.D. thesis (2015)
25. Zeugner, S., Feldkircher, M.: Bayesian model averaging employing fixed and flexible priors: the BMS package for R. J. Stat. Soft. **68**, 1–37 (2015)

# Data Analytics-Based Algorithm for Returning Intention to Rural Tourism Destinations of Tourists

**Nguyen Ha Thach and Bui Huy Khoi**

**Abstract** The paper seeks to evaluate the decision criteria in the returning intention to rural tourism destinations of tourists in Vietnam. This research sample comprises tourists. This chapter used the quantitative technique of PLS-SEM. The PLS Algorithm shows that the empirical model has a measurement of determination of 0.876, with 87.6% variation of intention, perceived value, attitude, travel motivation, subjective norm, and perceived behavior control. The factor that has the strongest influence on the intention to return to a rural tourist destination (INT) is the perceived value (PV) of tourists (0.367), followed by the second strongest factor on the returning intention to rural tourism destinations is the subjective norm factor (SN) of tourists (0.340). Finally, managers need suggestions to raise the returning intention to rural tourism destinations of tourists: evidence from Vietnam.

**Keywords** Algorithm · Returning intention · Rural tourism destinations

## 1 Introduction

Vietnam currently has nearly 65.6% of the population living in rural areas, which not only has a peaceful life, and vivid natural scenery but are also a place where a lot of cultural and historical values are kept by the nation, Vietnam has strengths in developing tourism associated with rural areas. Therefore, rural tourism is considered a trend in the tourism industry in the coming time [1]. Rural tourism is defined as a type of tourism taking place in rural areas, with a small business scale, open space, direct contact, and immersion in nature, associated with typical characteristics in rural areas, socio-cultural heritages, and traditional culture in villages show the diverse characteristics of the environment, economy, history, and location of each rural area. Rural tourism is a series of activities, services, and facilities provided in rural areas to exploit values in rural areas as a resource to meet the needs of

N. H. Thach · B. H. Khoi (✉)
Industrial University of Ho Chi Minh City, Ho Chi Minh City, Vietnam
e-mail: buihuykhoi@iuh.edu.vn



V. Kreinovich et al. (eds.), *Applications of Optimal Transport to Economics and Related Topics*, Studies in Systems, Decision and Control 556,
https://doi.org/10.1007/978-3-031-67770-0_23

tourists with experiences of non-rural residents or in other rural areas. The resource values in rural areas are typical of the culture and traditional village lifestyles associated with agricultural production. Rural tourism has many forms in Vietnam. It can be classified into three basic types: community tourism, agricultural tourism, and ecotourism. Which: Ecotourism is travel centered on the natural world and indigenous cultures that promote environmental awareness and involve the local population in conservation and sustainable development initiatives. In relatively untouched natural areas, ecotourism refers to the responsible enjoyment of nature and the accompanying cultural values of the past and present. It encourages conservation, has a small adverse effect on the environment, and has positive effects and socioeconomic poles for nearby communities. Almost every community offers some form of ecotourism [2]. Rural tourism is understood as a series of activities, services, and amenities provided in rural areas to exploit the strengths of the countryside as a resource that can meet the needs of tourists. Rural tourism uses material and technical facilities of rural communities (villages) with village cultural facilities (communes, temples, pagodas, shrines), traditional production facilities, combined roads, electricity, water, telecommunications services etc., associated with the natural environment, cultural identity, daily life and agricultural activities of the residential community. According to data from the World Tourism Organization, 10% of all tourists engage in rural and ecological tourism, which generates around USD 30 billion in revenue annually. Vietnamese travelers, both domestic and foreign, are increasingly choosing to visit rural areas, particularly during and after the Covid-19 pandemic's breakout to spend time in the peaceful, natural spaces of the countryside and escape the busy, noisy cities [1]. This study aims to find out the variables influencing travelers' intentions to return to their rural vacation spot in Ho Chi Minh City and to suggest some governance implications for agencies, organizations, and businesses to attract these visitors in the future.

## 2 Method

Table 1 demonstrated that 95 questionnaires were accepted out of the total samples gathered using two sampling techniques [3, 4], to achieve the required sample size for the element examination procedure. The survey was coded and entered into statistical software for data handling and analysis. The statistics of sample characteristics are described in Table 1.

**Table 1** Statistics of sample

| Characteristics | | Amount | Percent (%) |
|---|---|---|---|
| Sex | Male | 48 | 50.5 |
| | Female | 47 | 49.5 |
| Traveling to the countryside with whom? | Alone | 11 | 11.6 |
| | Family, Relatives | 30 | 31.6 |
| | Friends, Colleagues | 27 | 28.4 |
| | Lover | 27 | 28.4 |

# 3 Results

## *3.1 Measurement Model*

Algorithm was used to test the model and hypotheses that were presented [5, 6]. The measurement model was validated by assessing the internal consistency, convergent, and discriminant validity, as well as the dependability of the various indicators through their loadings (Table 2). It should be mentioned that due to multicollinearity, some components in various constructs have to be eliminated from the measurement model [7].

From Table 2, it's acknowledged that the outer loadings of the questionnaire objects on this research thing are greater than 0.7, which meets the verification preferred [15].

## *3.2 Reliability and Validity*

Internal consistency and convergent validity performance of composite reliability (CR) are superior (Table 3). 0.7 was suggested by Nunnally and Bernstein [16] as a reasonable reliability threshold. Everything is welcome.

All constructs fulfill the AVE criterion placed forth by using Fornell and Larcker [17] in terms of convergent validity by exceeding the cut-off of 0.5; every assembled debt for a minimum of 50% of the variance of the furnished indicators [18]. The primary diagonal values show that the correlations among any assemble and some other constructs are more than the rectangular root of the variance among the assemblers and its indicators (AVE), demonstrating discriminant validity (the rest of the matrix).

The estimation and validation of the structural version are proven in Fig. 1. After figuring out that there has been no multicollinearity (VIF 5 for all indicators), the parameters were calculated with the usage of a 5000-pattern bootstrap method. All the parameters had been extensive, except for the route between the parameters [19].

**Table 2** Standard loadings (OL)

| Factors and items | | Source | OL |
|---|---|---|---|
| SN1 | Most of my family members are in favor of my trip to the countryside | [8] | 0.769 |
| SN2 | Most of the people in the area I know will go to rural tourism as a vacation destination | New | 0.894 |
| SN3 | When I went back to rural tourism, I was influenced by the locals | [9] | 0.899 |
| MO1 | Because of good infrastructure | [10] | 0.742 |
| MO2 | Because there is a historical, cultural, and religious attraction | New | 0.821 |
| MO3 | Because of the beautiful natural scenery in the countryside | [10] | 0.816 |
| MO4 | Because I want to experience new things and travel to areas, I've never been to before | [11] | 0.822 |
| PV1 | Compared to other tourist spots I have visited, the countryside is a good value for money tourism | [12] | 0.648 |
| PV2 | Rural areas' tourism service systems and related services are of adequate quality | [13] | 0.837 |
| PV3 | Rural tourist industry personnel exhibit good customer service attitudes | [14] | 0.873 |
| PV4 | The surrounding scenery appears much more impressive after traveling through it | | 0.896 |
| BE1 | I think the countryside destination is one of my favorites | New | 0.883 |
| BE2 | I think the countryside is an attractive destination | | 0.891 |
| BE3 | The countryside is a great place to unwind with loved ones and friends | | 0.907 |
| AT1 | I think the countryside destination is one of my favorites | New | 0.908 |
| AT2 | I think the countryside is an attractive destination | | 0.888 |
| AT3 | With family and friends, you may unwind in the countryside | | 0.879 |
| INT1 | In the future, I'll think about moving back to the country | New | 0.868 |
| INT2 | In the future, I intend to return to the countryside to rest | | 0.903 |
| INT3 | In the future, I might think about moving back to the country | | 0.861 |

**Table 3** Measurement model

| Reliability and validity | | | Fornell-Larcker criterion | | | | | |
|---|---|---|---|---|---|---|---|---|
| | CR | AVE | AT | BE | INT | MO | PV | SN |
| AT | 0.921 | 0.796 | 0.892 | | | | | |
| BE | 0.923 | 0.799 | 0.652 | 0.894 | | | | |
| INT | 0.909 | 0.770 | 0.756 | 0.805 | 0.877 | | | |
| MO | 0.877 | 0.641 | 0.658 | 0.678 | 0.820 | 0.801 | | |
| PV | 0.890 | 0.671 | 0.805 | 0.783 | 0.909 | 0.886 | 0.819 | |
| SN | 0.891 | 0.733 | 0.695 | 0.822 | 0.881 | 0.732 | 0.850 | 0.856 |

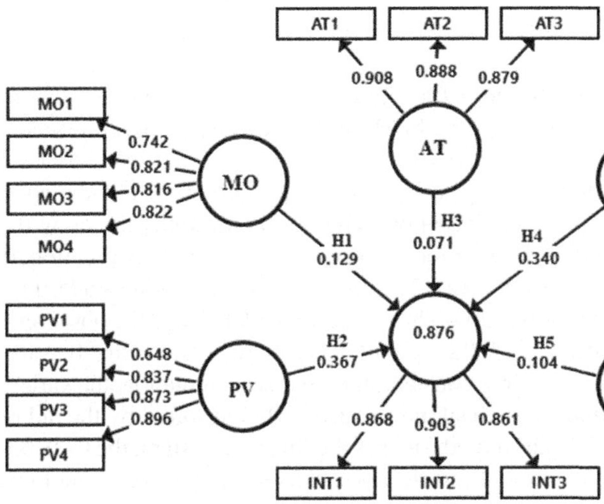


**Fig. 1** PLS algorithm

The PLS Algorithm, which takes five factors into account, is used to decide whether visitors will return to rural tourism destinations in about 87.6% of cases.

## *3.3 Hypothesis Testing*

Tourists' intentions to return to rural tourism sites are positively affected by two factors (PV [0.045:0.619], SN [0.111: 0.546]) and are unaffected by three elements (MO [−0.032: 0.406], AT [−0.062: 0.186], BE [−0.027: 0.294]) (Table 4). Given that there is statistical significance for the three paths, two hypotheses can be unaccepted to be based on the bootstrapped confidence intervals.

**Table 4** Path

| Path | β | SE | 2.5% | 97.5% | Decision |
|---|---|---|---|---|---|
| MO - > INT (H1) | 0.129 | 0.116 | −0.032 | 0.406 | Rejected |
| PV - > INT (H2) | 0.367 | 0.144 | 0.045 | 0.619 | Accepted |
| AT - > INT (H3) | 0.071 | 0.065 | −0.062 | 0.186 | Rejected |
| SN - > INT (H4) | 0.340 | 0.109 | 0.111 | 0.546 | Accepted |
| BE - > INT (H5) | 0.104 | 0.082 | −0.027 | 0.294 | Rejected |

## 4 Conclusions

This look aims to discover elements affecting the goal of returning to the rural vacationer vacation spot of vacationers in Ho Chi Minh City. Losing thorough research in this location can also affect domestic vacationers' intentions to go to rural tourism websites once more. Strong high-quality dating, but, can provide an excellent picture of the influencing variables primarily based on home traffic favorably affecting the protection and boom of lengthy-time period relationships with foreign vacationers and tourism business owners in rural areas. The regression check findings additionally display that the research version's hypotheses are not violated, and the observer's R2 is legitimate (87.6%). Even though R2 is small, it can be understood that because of the effects of the elements taken into consideration in this examination, the variety of domestic tourists' intentions to go back might also still be inspired by the aid of different factors that have not been studied, such as the financial system, the COVID-19 epidemic's results, the influence of lifestyle and society, the impact of technology, the impact of fee. The research on the difficulty can serve as a manual for similar research that includes similar material.

**Limitations and Further Research**

Although the desired research results have been achieved and the research objectives are satisfied, the results of this scientific research still have some limitations:

First, this study was conducted over five months. The time constraint has more or less affected choosing the research model and building the most suitable scale, and also affecting the research results of factors.

Second, in carrying out the research, the author also encountered certain difficulties such as having difficulty accessing leading experts, limited meeting time as well as not being able to fully comprehend all the relevant topics' opinions and certain knowledge. Collecting secondary data also faces many difficulties because some agencies in Vietnam have not disclosed data transparently.

To complete the research model and achieve better results, further research topics on the return intention of domestic tourists to rural tourism destinations need to: Conduct research with many tourists larger sample to achieve a better representation of the population; Research on a larger scale in Ho Chi Minh City, with many domestic tourists to increase the diversity of survey subjects; It is possible to focus on studying each domestic tourist to devise a specific strategy for each different group of tourists.

In addition, the research model has also proposed many new observed variables to reflect the current status of rural tourism destinations, as well as to measure the level of visitor satisfaction with service quality, characteristics, and service quality of the tourism industry in rural tourism destinations. In particular, the field of research on the intention of returning domestic tourists to a rural tourist destination is still quite new, creating a good basis for further studies to contribute necessary actual solutions for the development of rural tourism.

## References


1. https://nhandan.vn/du-lich-nong-thon-xu-huong-dang-khuyen-khich-post688350.html
2. https://vietnamtourism.gov.vn/index.php/items/37368
3. Worthington, R.L., Whittaker, T.A.: Scale development research: a content analysis and recommendations for best practices. Couns. Psychol. **34**(6), 806–838 (2006)
4. Tabachnick, B.G., Fidell, L.S., Ullman, J.B.: Using multivariate statistics. Pearson Boston, MA (2007)
5. Khoi, B.H., An, P.T.H., Van Tuan, N.: Applying the PLS-SEM model for the loyalty of domestic travelers. In: Book Applying the PLS-SEM Model for the Loyalty of Domestic Travelers, pp. 392–400. Springer, Berlin (2021)
6. Mai, D., Hai, P., Cuong, D., Khoi, B.: PLS-SEM algorithm for the decision to purchase durian milk with seeds. In: Book PLS-SEM algorithm for the decision to purchase durian milk with seeds, pp. 012013. IOP Publishing (2021)
7. Ngan, N.T., Khoi, B.H.: Partial least squares algorithm for deciding on buying innisfree cosmetics: evidence from Vietnam. In: Book Partial Least Squares Algorithm for Deciding on Buying Innisfree Cosmetics: Evidence from Vietnam, pp. 508–516. Springer, Berlin (2022)
8. Campón-Cerro, A.M., Folgado-Fernández, J.A., Hernández-Mogollón, J.M.: Rural destination development based on olive oil tourism: the impact of residents' community attachment and quality of life on their support for tourism development. Sustainability **9**(9), 1624 (2017)
9. Crawley, H., Hagen-Zanker, J.: Deciding where to go: Policies, people and perceptions shaping destination preferences. Int. Migr. **57**(1), 20–35 (2019)
10. Kanwal, S., Rasheed, M.I., Pitafi, A.H., Pitafi, A., Ren, M.: Road and transport infrastructure development and community support for tourism: the role of perceived benefits, and community satisfaction. Tour. Manag. **77**, 104014 (2020)
11. Jang, S.S., Wu, C.-M.E.: Seniors' travel motivation and the influential factors: an examination of Taiwanese seniors. Tour. Manag. **27**(2), 306–316 (2006)
12. Moon, H., Han, H.: Tourist experience quality and loyalty to an island destination: the moderating impact of destination image. J. Travel Tour. Mark. **36**(1), 43–59 (2019)
13. Derek, M.: Nature on a plate: Linking food and tourism within the ecosystem services framework. Sustainability **13**(4), 1687 (2021)
14. Zhang, Y., Wang, L.: Influence of sustainable development by tourists place emotion: analysis of the multiply mediating effect of attitude. Sustainability **11**(5), 1384 (2019)
15. Henseler, J., Ringle, C.M., Sarstedt, M.: A new criterion for assessing discriminant validity in variance-based structural equation modeling. J. Acad. Mark. Sci. **43**(1), 115–135 (2015)
16. Nunnally, J.C., Bernstein, I.: The assessment of reliability. Psychom. Theory **3**(1), 248–292 (1994)
17. Fornell, C., Larcker, D.F.: Structural equation models with unobservable variables and measurement error: algebra and statistics. J. Market. Res. 382–388 (1981)
18. Khoi, B.H., An, P.T.H., Van Tuan, N.: Applying the PLS-SEM model for the loyalty of domestic travelers. In: Prediction and Causality in Econometrics and Related Topics, pp. 392–400 (2022)
19. Ngan, N.T., Khoi, B.H.: Using PLS-SEM algorithm for choice of university in Vietnam. In: Book Using PLS-SEM algorithm for Choice of University in Vietnam, pp. 012066. IOP Publishing (2021)

## References

[illegible]

# Determination Weights for Models with a Large Number of Criteria, Applications for Evaluating the Quality of Life

**Jakub Hanousek**

**Abstract** This article deals with determination weights for models with many criteria. The application part is on the evaluation of the quality of life in the regional cities of the Czech Republic. The criteria are divided into main criteria and sub-criteria. There are 7 main criteria and 33 sub-criteria in this article. Each main criterion has several sub-criteria. The criterion weights are determined using the DEMATEL—ANP method. The advantage of this method is that it can display complex relationships between criteria, which is certainly a desired benefit.

**Keywords** Weights · DEMATEL-ANP · MCDA

## 1 Introduction

This article deals with a multi-criteria evaluation of the quality of life in the regional cities of the Czech Republic. The Czech Republic is a country in Central Europe. The Czech Republic has an area of 78,867 square kilometers with less than 11 million inhabitants. The Czech Republic is divided into 14 regions. Each region has its own capital city, except for one region that does not have a capital and shares a capital with a neighboring region. That is why the Czech Republic has only 13 regional cities. These 13 regional cities are the subject of analysis in this article. The evaluation is carried out using several criteria. Determining the weights of criteria is an essential step in multi-criteria decision-making models. Changing the weights of individual criteria can fundamentally affect the entire analysis. In this article there is a great emphasis on obtaining criteria weights. The criterion weights are obtained using the DEMATEL-ANP method.

---

J. Hanousek (✉)
nám., Winstona Churchilla 1938/4, 120 00 Praha 3-Žižkov, Prague, Czechia
e-mail: xhanj52@vse.cz

 
V. Kreinovich et al. (eds.), *Applications of Optimal Transport to Economics and Related Topics*, Studies in Systems, Decision and Control 556,
https://doi.org/10.1007/978-3-031-67770-0_24

## 2 Methods

The expression of the decision maker's preferences fundamentally affects the entire analysis. We may have the same data, but a different expression of preference is likely to completely change the outcome. The decision maker expresses preferences in terms of variants (it is also possible to express preferences between variants in terms of criteria) that are important to him.

### 2.1 *Dematel*

The DEMATEL method is used to map complex relationships between criteria. This method can be used to determine which criterion is dominant and influences other criteria, or which criterion is influenced by another criterion [7]. The DEMATEL method can be divided into six basic steps, which is graphically displayed in Fig. 1.

**Step 1. Collecting data from experts and calculating the average matrix A**

There are $p$ experts and $n$ criteria. Each expert is asked to determine the degree of influence between each pair of criteria how does he think criterion $i$ affects criterion $j$, or how does criterion $j$ affect criterion $i$. He does this on the basis of a pairwise comparison. The size of the direct influence is determined on a scale from 0 to 4:

*0—no influence*

*1—little influence*

*2—medium influence*

*3—strong influence*

*4—very strong influence*

The result will be the creation of a non-negative matrix $\mathbf{X}^k$ size of matrix is $n \cdot n$ from every expert. There will be 0 on the diagonal because one criterion cannot be evaluated against each other,

$$\mathbf{X}^k = \begin{bmatrix} 0 & \cdots & x_{1j}^k & \cdots & x_{1n}^k \\ \vdots & \ddots & \vdots & & \vdots \\ x_{i1}^k & \cdots & 0 & \cdots & x_{in}^k \\ \vdots & & \cdots & \ddots & \vdots \\ x_{n1}^k & \cdots & x_{nj}^k & \cdots & 0 \end{bmatrix}, \tag{1}$$

The letter $k$ indicates a specific expert who expressed his opinion. Must apply

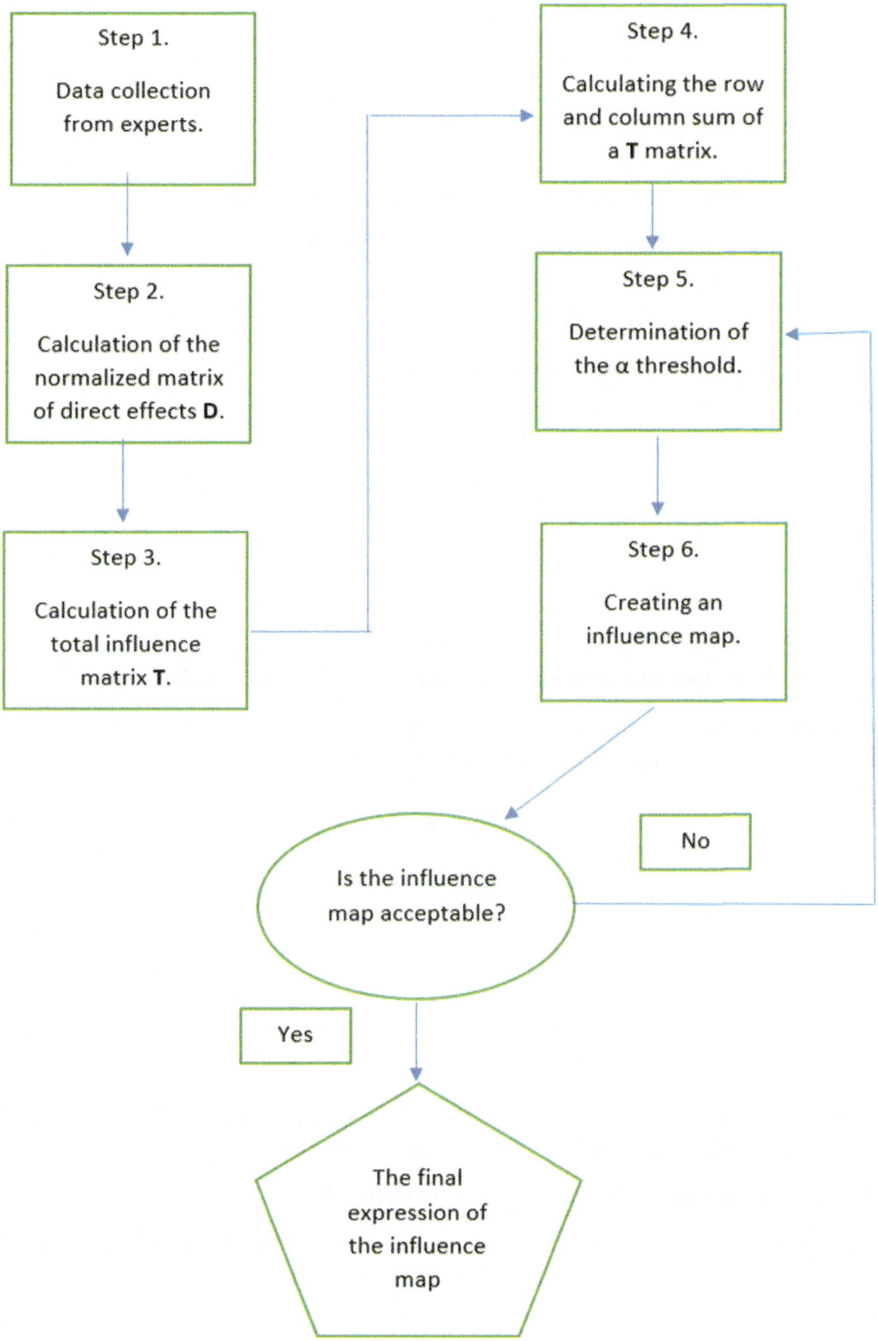


**Fig. 1** Basic steps of the DEMATEL method

$$1 \le k \le p$$

$$x_{ii}^k = 0, \text{ for}, k = 1, 2, ...p \text{ a } i = 1, 2, ...n.$$

At the end of this step, the **A** matrix is calculated. We calculate the matrix **A** as an arithmetic (geometric can also be used) average of the matrices $\mathbf{X}^k$.

$$a_{ij} = \frac{1}{p} \sum_{k=1}^{p} x_{ij}^k. \tag{2}$$

Matrix **A** is called the primary matrix of direct effects.

$$\mathbf{A} = \begin{bmatrix} 0 & \cdots & a_{1j} & \cdots & a_{1n} \\ \vdots & \ddots & \vdots & & \vdots \\ a_{i1} & \cdots & 0 & \cdots & a_{in} \\ \vdots & & \vdots & \ddots & \vdots \\ a_{n1} & \cdots & a_{nj} & \cdots & 0 \end{bmatrix} \tag{3}$$

**Step 2. Calculation of the normalized matrix of direct effects D**

We calculate the normalized matrix of direct effects **D**, the elements of which are denoted by $d_{ij}$, $d_{ij} \in \langle 0, 1 \rangle$, according to the relationship

$$\mathbf{D} = \lambda \cdot \mathbf{A} \tag{4}$$

where

$$\lambda = \min\left( \frac{1}{\max\limits_{1 \le i \le n} \sum_{j=1}^{n} |a_{ij}|}, \frac{1}{\max\limits_{1 \le j \le n} \sum_{i=1}^{n} |a_{ij}|} \right). \tag{5}$$

We then multiply each element of the matrix **A** by the constant λ, this operation gives us the normalized primary matrix of direct effects **D**.

**Step 3. Calculation of the total influence matrix T**

The matrix **T** expresses the overall relationships between each pair of criteria *i* and *j*. The elements of the matrix $t_{ij}$ express the direct and indirect relationships between criteria *i* and *j*.

$$\lim_{m \to \infty} \mathbf{D}^{\mathbf{m}} = 0, \tag{6}$$

0 is the zero matrix with dimensions $n \cdot n$. We can then calculate the matrix **T** in the following way

$$\begin{aligned}
\mathbf{T} &= \lim_{m\to\infty}\left(\mathbf{D}+\mathbf{D}^2+\cdots+\mathbf{D}^{\mathbf{m}}\right)\\
&= \sum_{m=1}^{\infty}\mathbf{D}^{\mathbf{i}}\\
&= \mathbf{D}+\mathbf{D}^2+\cdots+\mathbf{D}^{\mathbf{m}}\\
&= \mathbf{D}\left(\mathbf{I}+\mathbf{D}+\mathbf{D}^2+\cdots\mathbf{D}^{\mathbf{m}-1}\right)\\
&= \mathbf{D}(\mathbf{I}-\mathbf{D})^{-1}(\mathbf{I}-\mathbf{D})\left(\mathbf{I}+\mathbf{D}+\mathbf{D}^2+\cdots+\mathbf{D}^{\mathbf{m}-1}\right)\\
&= (\mathbf{I}-\mathbf{D})^{-1}\left(\mathbf{I}-\mathbf{D}^{\mathbf{m}}\right)\\
&= \mathbf{D}(\mathbf{I}-\mathbf{D})^{-1},
\end{aligned} \tag{7}$$

**I** is an $n \cdot n$ identity matrix.

**Step 4. Calculating the column and row sum of a T matrix**

The sum of the rows of the matrix **T** is denoted by the vector **r** and the sum of the columns of the matrix **T** is denoted by the vector **c**.

$$\mathbf{r} = [r_i]_{n\cdot 1} = \left(\sum_{j=1}^{n} t_{ij}\right)_{n\cdot 1} \tag{8}$$

$$\mathbf{c} = \left[c_j\right]_{1\cdot n} = \left(\sum_{i=1}^{n} t_{ij}\right)_{1\cdot n}, \tag{9}$$

$r_i$ expresses the total influence (direct and indirect) of the criteria $i$ and on all other criteria, and $c_j$, it expresses the total influence (direct and indirect) by which criterion $j$ is influenced by all other criteria. For $i = j$, we are interested in the overall effect of the criterion in the system, whose value we calculate as $r_{i\,+}\ c_j$.

**Step 5. Determining the threshold value $\alpha$**

For a better orientation in the **T** matrix, it is good to remove influences that are marginal, and instead focus on those influences that are significant. We achieve this with the threshold value $\alpha$. The threshold value $\alpha$ can be determined in several ways. It can be determined directly by experts, or it can be calculated in several ways. A simple and often used calculation of the threshold value $\alpha$ is using the averages of all values in the matrix **T**

$$\alpha = \frac{\sum_{i=1}^{n}\sum_{j=1}^{n} t_{ij}}{n^2}. \tag{10}$$

Influence values that exceed the threshold value $\alpha$ in the matrix **T** are marked as significant, on the contrary, those that do not exceed the threshold value are excluded from the matrix as insignificant.

$$t_{ij}^{\alpha} = 0, \ if \ t_{ij} < \alpha, \text{otherwise} \ t_{ij}^{\alpha} = t_{ij}.$$

**Step 6. Creating an influence map**

To create an influence map, it is necessary to calculate the coordinates $[X, Y]$ for each criterion, where

$$X = r_i + c_i, \tag{11}$$

$$Y = r_i - c_i. \tag{12}$$

Only those influences that exceeded the threshold value $\alpha$ are included in the influence map. The decision-maker can reduce or increase the threshold value $\alpha$ so that the influence map is acceptable for him.

Weights for individual criteria can also be obtained using the DEMATEL method, this is an extension of the DEMATEL method known as DEMATEL–ANP. DEMATEL–ANP will be used in the application part to determine the weights of criteria, because the method also takes into account relationships between criteria that are not completely obvious at first glance. DEMATEL–ANP is a combination of the DEMATEL method and the ANP method. The ANP method is based on the AHP method, which is based on pairwise comparison. AHP and ANP methods are briefly described below.

## *2.2 Analytic Hierarchy Process*

The method is used for setting priorities in hierarchical multi-criteria systems. Preference modeling is based on a sequence of pairwise comparisons of appropriately determined parts of the system. The AHP method consists of three steps [9]:

**Step 1. Creating a hierarchical structure of goals**

**Step 2. Pairwise comparison**

**Step 3. Determination of weights**

## 2.3 Analytic Network Process

The ANP method is a network generalization of the AHP method. The main difference is that in the AHP method, the dependence between levels was unidirectional. With the ANP method, dependencies can be bidirectional, dependencies on themselves are also allowed [10].

Individual parts of the system are called clusters. Elements with similar properties are grouped into clusters. A necessary condition is that the clusters are connected, or that at least one element from the cluster is connected to an element from another cluster. The arrangement of elements into clusters and their interconnection is called a network.

We construct a supermatrix **W**, which has dimensions n x n and displays the dependencies of individual elements on others [10]. Dependencies are expressed using pairwise comparisons.

We denote the clusters by $C_h$, where $h = 1,2,\ldots,N$. There are $n_h$ elements in each cluster $(e_{h1}, e_{h2},\ldots,e_{hn})$.

The supermatrix $\mathbf{W_{ij}}$ consists of blocks that compare the elements of cluster *i* with the elements of cluster *j*. If there is a link, the column sum is equal to 1. If there are no links between an element of cluster *i* and *j*, the matrix is zero. The matrix **W** is column stochastic. The sum of the columns in the matrix is equal to the number of observed clusters. The supermatrix **W** is not weighted, so that we can calculate the resulting weights, it is necessary to transform this matrix into a weighted matrix, which we denote by $\mathbf{W_w}$.

It is necessary that the supermatrix $\mathbf{W_w}$ also remains column stochastic. We therefore multiply the relevant blocks of the unweighted supermatrix **W** by a matrix that contains pairwise comparisons of individual clusters. Subsequently, it is only sufficient to multiply the $\mathbf{W_w}$ matrix to a sufficiently large number and thus obtain a limit supermatrix. In this limit supermatrix, the columns express the global priorities of the elements of the decision-making system, so we obtained weights.

## 2.4 Dematel–ANP

Using the DEMATEL method, we obtained the resulting influence map, which we can describe as a network representation of the decision problem. The next step will be the use of the matrix **T**, which displays the overall effects of individual clusters [11]. We will use the matrix **T** to normalize the super matrix **W**.

The matrix **T** needs to be transformed into the matrix $\mathbf{T}_\alpha$. We perform the transformation using the threshold value $\alpha$, which we obtained in the 2nd step of the DEMATEL method. We set the elements of the matrix $\mathbf{T}_\alpha$ that have a smaller value than $\alpha$ equal to 0. This means that cluster *i* has a negligible influence on cluster *j*.

Now we calculate the values of $d_i$ as the column sums of the elements of the matrix $\mathbf{T}_\alpha$

$$d_i = \sum_{j=1}^{n} t_{ij}^{\alpha}, j = 1, 2, .., n. \tag{13}$$

We then divide each element of the matrix $\mathbf{T}_\alpha$ by the corresponding value $d_i$. With this step, we obtain a normalized matrix $\mathbf{T_s}$, whose column sums will be equal to 1

$$\mathbf{T_s} = \begin{bmatrix} \frac{t_{11}^{\alpha}}{d_1} & \cdots & \frac{t_{1j}^{\alpha}}{d_j} & \cdots & \frac{t_{1n}^{\alpha}}{d_n} \\ \vdots & \ddots & \vdots & & \vdots \\ \frac{t_{i1}^{\alpha}}{d_1} & \cdots & \frac{t_{ij}^{\alpha}}{d_j} & \cdots & \frac{t_{in}^{\alpha}}{d_n} \\ \vdots & & \vdots & \ddots & \vdots \\ \frac{t_{n1}^{\alpha}}{d_1} & \cdots & \frac{t_{nj}^{\alpha}}{d_j} & \cdots & \frac{t_{nn}^{\alpha}}{d_n} \end{bmatrix} = \begin{bmatrix} t_{11}^{s} & \cdots & t_{1j}^{s} & \cdots & t_{1n}^{s} \\ \vdots & \ddots & \vdots & & \vdots \\ t_{i1}^{s} & \cdots & t_{ij}^{s} & \cdots & t_{in}^{s} \\ \vdots & & \vdots & \ddots & \vdots \\ t_{n1}^{s} & \cdots & t_{nj}^{s} & \cdots & t_{nn}^{s} \end{bmatrix}. \tag{14}$$

Using the matrix $\mathbf{T_s}$, we obtain the weighted super matrix $\mathbf{W_w}$. We multiply the individual blocks of the super matrix $\mathbf{W}$ by each element of the matrix $\mathbf{T_s}$

$$\mathbf{W_w} = \begin{bmatrix} t_{11}^{s} \cdot W_{11} & \cdots & t_{1j}^{s} \cdot W_{12} & \cdots & t_{1n}^{s} \cdot W_{1n} \\ \vdots & & \vdots & & \vdots \\ t_{i1}^{s} \cdot W_{i1} & \vdots & t_{ij}^{s} \cdot W_{ij} & \cdots & t_{in}^{s} \cdot W_{in} \\ \vdots & & & & \vdots \\ t_{n1}^{s} \cdot W_{n1} & \cdots & t_{nj}^{s} \cdot W_{nj} & \cdots & t_{nn}^{s} \cdot W_{nn} \end{bmatrix}. \tag{15}$$

The last step is to raise the matrix $\mathbf{W_w}$ to a sufficiently large number, thus obtaining the limit super matrix $\mathbf{W_l}$. The elements of the limit super matrix $\mathbf{W_l}$ represent the resulting weights of the individual criteria.

## 3 Application

In this section we analyze the quality of life in 14 regional cities of the Czech Republic. Regional cities of the Czech Republic are variants in this analysis (Table 1).

The evaluation criteria are in Table 2. Criteria are divided into main criteria and sub-criteria. Each main criterion consists of several subcriteria.

**Health**:

- Number of premature deaths
- Life expectancy for men
- Life expectancy for woman
- Number of physicians.

**Table 1** Regional cities in the Czech Republic

| Number | Regional city |
|---|---|
| 1 | Praha |
| 2 | Brno |
| 3 | Ostrava |
| 4 | Plzen |
| 5 | Ceske Budejovice |
| 6 | Olomouc |
| 7 | Zlin |
| 8 | Liberec |
| 9 | Usti nad Labem |
| 10 | Hradec Kralove |
| 11 | Pardubice |
| 12 | Jihlava |
| 13 | Karlovi Vary |

**Table 2** Main criteria

| Number | Main criteria |
|---|---|
| 1 | Health |
| 2 | Housing |
| 3 | Safety |
| 4 | Infrastructure and amenities of the village |
| 5 | Employment and work |
| 6 | Income and wealth |
| 7 | Education |
| 8 | Environment |

**Housing**:

- Time to save to buy real estate
- Price per $m^2$ of real estate
- Price per $m^2$ of land
- Price per $m^2$ of rent.

**Safety**:

- Accident rate
- Criminality
- Fires
- Flood risk.

**Infrastructure and Amenities of the Village**:

- Store availability

- ATM availability
- Availability of petrol station
- Density of public transport stops
- Mobile signal coverage.

**Employment and work**:

- Unemployment
- Persons in execution
- Vacancies.

**Income and wealth**:

- Investment purchases of the city
- Average salary.

**Education**:

- Kindergarten capacity
- Primary school capacity
- Secondary school offer
- Availability of secondary schools
- Residents with a university degree
- Residents with primary education
- Residents with high school education with high school diploma.

**Environment**:

- Greenery in the village
- Natural habitats
- Species
- Air quality.

The main criteria were presented to 15 experts who are actively interested in the quality of life. The experts were asked to fill in the degree of influence between each pair of criteria. They could express the degree of influence on the scale:

*0—no influence*

*1—little influence*

*2—medium influence*

*3—strong influence*

*4—very strong influence*

Applying the DEMATEL-ANP method, we obtained the following weights (Table 3):

**Table 3** Weight criteria

| Criteria | Weights |
|---|---|
| Number of premature deaths | 0.0313 |
| Life expectancy for men | 0.0469 |
| Life expectancy for woman | 0.0469 |
| Number of physicians | 0.0273 |
| Time to save to buy real estate | 0.0234 |
| Price per $m^2$ of real estate | 0.0273 |
| Price per $m^2$ of land | 0.0273 |
| Price per $m^2$ of rent | 0.0273 |
| Accident rate | 0.0156 |
| Criminality | 0.0430 |
| Fires | 0.0156 |
| Flood risk | 0.0469 |
| Store availability | 0.0508 |
| ATM availability | 0.0195 |
| Availability of petrol station | 0.0156 |
| Density of public transport stops | 0.0352 |
| Mobile signal coverage | 0.0430 |
| Unemployment | 0.0469 |
| Persons in execution | 0.0117 |
| Vacancies | 0.0156 |
| Investment purchases of the city | 0.0078 |
| Average salary | 0.0117 |
| Kindergarten capacity | 0.0195 |
| Primary school capacity | 0.0195 |
| Secondary school offer | 0.0352 |
| Availability of secondary schools | 0.0156 |
| Residents with a university degree | 0.0273 |
| Residents with primary education | 0.0195 |
| Residents with high school education with high school diploma | 0.0234 |
| Greenery in the village | 0.0547 |
| Natural habitats | 0.0469 |
| Species | 0.0391 |
| Air quality | 0.0625 |

## 4 Conclusion

This paper focuses to determine the weights of the individual sub-criteria. We had 33 sub-criteria. These 33 sub-criteria were assigned to clusters. We can call the cluster the main criteria. Criteria weights were obtained from 15 experts using the DEMATEL-ANP method. These obtained weights will be used in further research.

**Acknowledgements** This work was supported by the project no. F4/42/2021 of the Internal Grant Agency, Faculty of Informatics and Statistics, University of Economics, Prague.

## References

1. Anuntavoranich, P.: Dematel method. J. Eng. Manag. Appl. Sci. Technol. **4**(2), 25–33
2. Bouška, J., Černý, M., Gluckaufová, D.: Interaktivní postupy rozhodování ( Interactive decision making procedures ). Academica, Praha (1984)
3. BROŽOVÁ, H., HOUŠKA, M., ŠUBRT, T.: *Modely pro vícekriteriální rozhodování (Models for multicriteria decision making).* Praha: ČZU, 2009.
4. Fiala, P., Jablonský, J., Maňas, M.: Vícekriteriální rozhodování (Multi-criteria decision making). Vysoká škola ekonomická v Praze (University of Economics and Business in Prague) (1994)
5. Fiala, P., Jablonský, J., Maňas, M.: Vícekriteriální rozhodování (Multi-criteria decision making). Vysoká škola ekonomická v Praze (University of Economics in Prague) (1997)
6. Fiala, P.: Modely a metody rozhodování (Decision-making models and methods). Oeconomica, Praha (2013)
7. Gabus, A., Fontela, E.: World Problems, An Invitation to Further Thought within the Framework of DEMATEL. Battelle Geneva Research Centre, Geneva (1972)
8. Hwang, C.L., Yoon, K.: Multiple Attribute Decision Making. Springer, Methods and Application. Berlin (1981)
9. Saaty, T.L.: The Analytic Hierarchy Process: Planning, Priority Setting, Resource allocation. McGraw-Hill, New York (1980)
10. Saaty, T.L., Vargas, L.G.: Decision Making with the Analytic Network Process: Economic, Political, Social and Techonological Application with Benefits, Opportunities, Costs and Risks. Springer Science + Business Media, New York (2006). ISBN 0-387-33859-4
11. Yang, Y.O., Shieh, H., Leu, J., Tzeng, G.: A novel hybrid MCDM model combined with DEMATEL and ANP with applications. Int. J. Oper. Res. **5**(3), 56–73 (2008)

# The Empirical Analysis of the Effects of Sudden GDP and Commodity Price Changes on Inflation Using Markov-Switching Model: Insights from Six Southeast Asian Countries

**Worrawat Saijai, Somsak Chanaim, and Sukrit Thongkairat**

**Abstract** This study investigates the impact of abrupt changes in GDP growth and global commodity prices in six Southeast Asian countries: Indonesia, Malaysia, The Philippines, Singapore, Thailand, and Vietnam. Utilizing the Markov-switching model, we analyze the relationship between these variables and inflation, recognizing that each country has unique characteristics. The monthly data of each country is collected from January 2010 to May 2023, comprising 161 observations. Our findings reveal that global oil price fluctuations significantly influence inflation during higher inflation periods in most countries, while the impact diminishes during lower inflation periods, particularly in Singapore, Indonesia, and Vietnam. The effect of global wheat prices on inflation varies across countries, with a stronger impact observed in the higher inflation regime of the Philippines and the lower inflation regime of Indonesia. Global sugar prices play a substantial role in driving inflation dynamics in both higher and lower inflation periods in most countries, with exceptions in Thailand and the Philippines. Furthermore, GDP growth exhibits a significant positive relationship with inflation in both higher and lower regimes for most countries, with varying magnitudes of influence. Notably, negative GDP growth is associated with stagflation in the higher inflation regime. These insights have vital implications for policymakers, enabling tailored strategies to address specific challenges in each country. During high volatility periods, measures can mitigate inflationary effects from oil price shocks, while lower inflation periods require managing potential deflationary pressures from negative GDP shocks.

**Keywords** Commodity price · GDP growth · Inflation · Markov-switching · Nonlinear model

---

W. Saijai · S. Thongkairat (✉)
Center of Excellence in Econometrics, Faculty of Economics, Chiang Mai University, Chiang Mai 50200, Thailand
e-mail: sukrit415@gmail.com

S. Chanaim
International College of Digital Innovation (ICDI), Chiang Mai University, Chiang Mai, Thailand
e-mail: somsak.chanaim@cmu.ac.th



V. Kreinovich et al. (eds.), *Applications of Optimal Transport to Economics and Related Topics*, Studies in Systems, Decision and Control 556,
https://doi.org/10.1007/978-3-031-67770-0_25

## 1 Introduction

Inflation, a multifaceted economic phenomenon, holds paramount importance for policymakers and economists on a global scale. Its implications are particularly critical for Southeast Asian countries, which face distinctive challenges in their pursuit of sustainable growth amidst a rapidly evolving economic landscape. To ensure effective policymaking and maintain economic stability in the region, it is imperative to gain a comprehensive understanding of the intricate interplay between inflation and various socioeconomic factors.

Southeast Asia, comprising nations such as Indonesia, Malaysia, The Philippines, Thailand, Singapore, and Vietnam, has witnessed remarkable economic growth and development in recent decades. However, this progress has been accompanied by a host of challenges, with inflation management emerging as a top priority for these countries. The complex relationship between inflation and socioeconomic factors has been further compounded by rapid economic expansion, deeper integration into global markets, and shifting consumption patterns.

The link between commodity prices and inflation has been the subject of extensive research, yielding diverse outcomes. For instance, a study examined the relationship between economic growth, oil prices, exchange rates, and correlated inflation in Saudi Arabia from 1969 to 2014. By employing Morlet's wavelet method, the study demonstrated the dynamic evolution of associations between variables across time and frequency, revealing heterogeneous relationships within different time domains [1]. Referent [2] investigated the non-linear impact of oil prices on economic growth in Indonesia from 2000 to 2019, discovering that palm oil price volatility exerted a greater influence on economic growth compared to crude oil prices. Moreover, another identified a significant short-term effect of oil prices on economic growth across 31 provinces in China, employing a Nonlinear Autoregressive Distributed Lag methodology [3]. Similar findings were reported by [4], highlighting the asymmetrical effects of oil price volatility on economic growth.

For instance, Aloui et al. [1] examined the relationship between economic growth, oil prices, exchange rates, and correlated inflation in Saudi Arabia from 1969 to 2014. By employing Morlet's wavelet method, the study demonstrated the dynamic evolution of associations between variables across time and frequency, revealing heterogeneous relationships within different time domains. Prabhesh and Laila [2] investigated the non-linear impact of oil prices on economic growth in Indonesia from 2000 to 2019, discovering that palm oil price volatility exerted a greater influence on economic growth compared to crude oil prices. Baek et al. [3] identified a significant short-term effect of oil prices on economic growth across 31 provinces in China, employing a Nonlinear Autoregressive Distributed Lag methodology. Similar findings were reported by Baek and Young [4], highlighting the asymmetrical effects of oil price volatility on economic growth.

Cevik et al. [5] conducted an analysis of oil prices and stock market returns spanning 1990 to 2017, utilizing the EGARCH model. Their estimates revealed a substantial influence of crude oil prices on stock market returns, measured through the

Brent benchmark. However, no significant spillover effects were observed. Collectively, these studies shed light on the intricate interplay between commodity prices, economic growth, inflation, and related indicators.

Apart from the relationship between commodity prices and inflation, extensive research has explored the association between inflation and gross domestic product (GDP). Studies by Fischer [6], Barro [7], Sarel [8], Bruno and Easterly [9], Ghosh and Phillips [10], Sepehri and Moshiri [11], Mubarik [12], Lee and Wong [13], and others have investigated this relationship using diverse methodologies and datasets. Fischer [6] was among the first to identify a negative relationship between inflation and economic growth across developing and industrialized countries. Sarel [8] utilized panel data to highlight the impact of low levels of inflation on economic growth, while Bruno and Easterly [9] focused on countries that experienced inflation crises, revealing varying effects on economic growth. Lee and Wong [13] examined core inflation levels for Taiwan and Japan, revealing thresholds beyond which inflation became detrimental to economic growth. These studies underscore the necessity of investigating the effects of GDP growth and inflation on developed and developing countries. The contextual differences in time and the presence of asymmetrical effects warrant further examination, as they contribute to varied effects on inflation.

This academic study aims to provide comprehensive insights into the significance of inflation in Southeast Asian countries, encompassing an analysis of its multifaceted nature and its impact on various socioeconomic factors. By thoroughly analyzing the unique characteristics of these countries, identifying the causes and consequences of inflation, and evaluating policy responses, this study seeks to offer a nuanced understanding of the importance of inflation for Southeast Asian economies. Additionally, the study examines the drivers and determinants of inflation, considering both external factors such as global commodity prices and domestic factors like GDP growth. Moreover, food inflation in Southeast Asia can be significantly influenced by global wheat and sugar prices. Wheat and sugar are essential commodities in the region, and any fluctuations in their prices on the global market can have a substantial impact on the cost of food products. By comprehending the underlying causes of inflation, policymakers can develop targeted strategies to effectively manage and mitigate its adverse effects. While previous research has explored factors contributing to inflation, many studies have primarily relied on linear models, often neglecting evolving economic structures and volatilities.

To address this academic gap, this study proposes the utilization of a non-linear Markov-switching Model (MS model) with two regimes to analyze the effects of commodity prices and GDP growth on inflation while accounting for economic volatility. This modeling approach offers enhanced effectiveness and flexibility compared to single-regime models that assume no dynamics in parameters. By examining data from six Southeast Asian countries, this research endeavors to provide a more accurate understanding of the complex dynamics between commodity prices, inflation, and economic volatility.

The insights derived from this study are expected to deepen our comprehension of the factors influencing inflation and the consequences of economic volatility, thus providing valuable guidance for policymakers in their pursuit of effective and sustainable economic management in Southeast Asia.

The structure of this paper consists of several sections. Section 2, titled "Methods and Procedures", primarily focuses on the MS model. In Sect. 3, the data and preliminary analysis are presented. The empirical reports and findings are discussed in Section 4. Finally, Section 5 provides the conclusions drawn from the study.

## 2 Methods and Procedures

This research aims to investigate the effects of changes in global commodity prices and gross domestic product on inflation within the targeted countries individually. Then, we employ time-series analysis in our studies, assuming that each country has different characteristic. The global commodities are including oil (OIL), wheat (WHT), sugar (SUG), and gold (GOLD). The study will analyze both the positive and negative impacts of these changes of GDP (GR) and changes in commodity prices on inflation, utilizing the Markov-switching Model analysis proposed by Chang-Jin Kim et al. [15]. We assume that positive and negative GDP growth cast different impact to inflation, and these variables can represent the positive and negative shock respectively. Then, we intend to put positive GDP shock ($GR^+$) as a variable, ($GR^+ = MAX(\Delta GDP, 0)$), to capture the abrupt change effect in positive GDP that affect Inflation. Moreover, we try to capture the impact of a sudden change in negative GDP growth by another variable, $GR^-$, where $GR^- = MIN(\Delta GDP, 0)$. Moreover, we augment Markov-switching to see the difference structure in different state, e.g. lower and higher volatility of inflation. The research seeks to provide a comprehensive understanding of the relationship between commodity price returns, GDP growth, and inflation, specifically within the Asian region.

### 2.1 *Markov Switching Model*

The Markov permutation model can be written as follows

$$y_t = x_t'\beta_{S_t} + \sigma_{S_t}\varepsilon_t, \tag{1}$$

where $y_t$ and $x_t$ represent the dependent and precursor variables, respectively. The parameter $\beta_{S_t}$ is the coefficient of the initial variable at state $S_t$, $\sigma_{S_t}$ is the variance at state $S_t$, and $\varepsilon_t$ is the assumed confounding variable. The normal distribution has mean equal to 0 and variance equal to 1.

The parameters $\beta_{S_t}$ and $\sigma_{S_t}$ are determined by the state variable $S_t$, where $S_t = i$ and $i = 1, 2, \ldots, N$ are latent variables or the state of the economy itself. In this study, it is assumed that the model in Equation 1 can explain the study results in two states ($N = 2$), that is, a high-inflation state and a low-inflation state.

The variable $S_t$ is an unobservable variable, and is predicted according to the first-order derivative of the Markov chain whose probability change is:

$$P(S_t = i \mid S_{t-1} = j, z_t) = P_{ij}(z_t) \tag{2}$$

In equation (2), the change in probability is influenced by a $(q \times 1)$ vector of variances of the external variable or the previously defined $z_t$ variable, where $z_t$ may be an element of $x_t$ and $\sigma_{S_t}$ included in the Markov chain. To create this probability, it is assumed that $S_t$ contains a normal cumulative distribution function, for which the probability equation can be written as follows:

$$S_t = \left\{ \frac{1 \text{ if } \eta_t < \beta_{S_{t-1}}}{2 \text{ if } \eta_t \geq x_t' \beta_{S_{t-1}}} \right\}, \tag{3}$$

thus, the probabilities for each economic state can be estimated from

$$p_{1j}(z_t) = P\left(\eta_t < x_t' \beta_{S_t=j}\right) = \Phi\left(x_t' \beta_{S_t=j}\right) \tag{4}$$

$$p_{2j}(z_t) = P\left(\eta_t \geq z_t \beta_{S_t=j}\right) = 1 - \Phi\left(z_t \beta_{S_t=j}\right) \tag{5}$$

where $\Phi$ is the normal cumulative distribution function.

## 2.2 Smooth Probabilities

Smooth probabilities are crucial for understanding the likelihood of being in a particular state at different points in time, given the observed data. These probabilities are computed using filtering and smoothing algorithms in the context of Markov-switching models.

The smoothing probabilities $\Pr(S_t = i \mid \Omega_{t+1}, \zeta_t; \theta)$ can be estimated from the filtering probabilities and the likelihood function:

$$\begin{aligned} &\Pr(S_t = i \mid \Omega_{t+1}, \zeta_t; \theta) \\ &= \frac{1}{f(y_t \mid \Omega_t, \zeta_{t-1}; \theta)} \sum_j f(y_t \mid S_t = i, S_{t-1} = j, \Omega_t, \zeta_{t-1}; \theta) \times \\ &\Pr(S_t = i, S_{t-1} = j \mid \Omega_t, \zeta_{t-1}; \theta) \end{aligned} \tag{6}$$

## 2.3 Maximum Likelihood Estimation

Let $\Omega_t = \left(x_t', x_{t-1}', \ldots, x_1'\right)'$ and $\zeta_t = (y_t, y_{t-1}, \ldots, y_1)'$ is an observation vector from where t and $\theta = \left(\beta_{S_t}, \sigma_{S_t}\right)$ are vectors of the parametric model. In addition, the condition of the likelihood function for the observation $\zeta_t$ is $L(\theta) = \Pi_{t=1}^{T} f\,(y_t \mid \Omega_t, \zeta_{t-1}; \theta)$ where

$$f\,(y_t \mid \Omega_t, \zeta_{t-1}; \theta) = \sum_i \sum_j f\,(y_t \mid S_t = i, S_{t-1} = j, \Omega_t, \zeta_{t-1}; \theta) \\ \Pr\,(S_t = i, S_{t-1} = j \mid \Omega_t, \zeta_{t-1}; \theta) \tag{7}$$

Probability weights in Eq. (6) are calculated using the Bayes' theorem.

$$\begin{aligned} &\Pr\,(S_t = i, S_{t-1} = j \mid \Omega_t, \zeta_{t-1}; \theta) = P_{ij}\,(z_t) \Pr\,(S_{t-1} = j \mid \Omega_t, \zeta_{t-1}; \theta)\,, \\ &\Pr\,(S_t = i \mid \Omega_{t+1}, \zeta_t; \theta) = \Pr\,(S_t = i \mid \Omega_t, \zeta_t; \theta) \\ &= \frac{1}{f\,(y_t \mid \Omega_t, \zeta_{t-1}; \theta)} \sum_j f\,(y_t \mid S_t = i, S_{t-1} = j, \Omega_t, \zeta_{t-1}; \theta) \times \\ &\Pr\,(S_t = i, S_{t-1} = j \mid \Omega_t, \zeta_{t-1}; \theta) \end{aligned} \tag{8}$$

which, when the likelihood function can be approximated, all parameters are estimated. A model that maximizes the likelihood function.

## 3 Data

We investigate the relationship between inflation and the changes in world commodity prices, including oil price, wheat, and sugar, on inflation. We apply log transformation to all the data. The log-transformed variables include West Texas Intermediate crude oil log(OIL), global wheat price log(WHT), global sugar price log(SUG), global oil price log(OIL). Furthermore, to capture the asymmetrical effect of abrupt positive and negative changes of GDP on inflation, we include positive and negative GDP growth ($GR_i^+$ and $GR_i^-$) into our study. While the positive and negative GDP growth of each country are not in log-transformed values, the inflation ($INF_i$) is also in non-log-transformed values due to its negative values, where $i$ = Singapore (*SG*), Malaysia (*ML*), Thailand (*TH*), Indonesia (*IN*), Vietnam (*VN*), and The Philippines (*PH*). All data are collected by Yahoo Finance.com, except ($GR_i$) and ($INF_i$) that are obtained from Thomson Reuter DataStream. The monthly sample covers the period from January 2010 to May 2023 with 161 observations. We include this range because it include. Table 1. displays a presence of positive skewness and low kurtosis for log(WHT), log(GOLD), and log(SUG), while log(OIL) has negative skewness and low kurtosis. Additionally, a significant amount of the $GR_i$ and $INF_i$ exhibit both positive/negative skewness and high kurtosis, suggesting that the variables are not follow a normal distribution.

**Table 1** Descriptive statistics of log-transformed returns

| | Variables | | | | | | | |
|---|---|---|---|---|---|---|---|---|
| | log(OIL) | log(WHT) | log(SUG) | log(GOLD) | $GR_{SI}$ | $GR_{ML}$ | $GR_{TH}$ | $GR_{IN}$ |
| Mean | 3.80 | 0.02 | 2.47 | 4.94 | 1.02 | 1.12 | 0.62 | 1.71 |
| Median | 3.79 | −0.12 | 2.31 | 4.89 | 0.89 | 1.29 | 0.76 | 1.25 |
| Minimum | 2.84 | −0.63 | 1.69 | 4.65 | 9.17 | 18.23 | 9.38 | 3.5 |
| Maximum | 4.48 | 1.04 | 3.34 | 5.28 | −12.19 | −14.72 | −9.17 | −6.26 |
| Std. Dev. | 0.42 | 0.47 | 0.45 | 0.17 | 2.03 | 2.48 | 1.81 | 0.91 |
| Skewness | −0.13 | 0.50 | 0.37 | 0.23 | −1.24 | 0.01 | −0.59 | −4.91 |
| Kurtosis | −1.36 | −1.10 | −1.06 | −1.35 | 16.15 | 28.05 | 13.55 | 35.34 |
| | Variables | | | | | | | |
| | $GR_{VN}$ | $GR_{PH}$ | $INF_{SI}$ | $INF_{ML}$ | $INF_{TH}$ | $INF_{IN}$ | $INF_{VN}$ | $INF_{PH}$ |
| Mean | 1.38 | 1.3 | 2.03 | 2.05 | 1.77 | 4.23 | 5.26 | 3.39 |
| Median | 1.48 | 1.58 | 1.04 | 2.09 | 1.49 | 3.84 | 3.52 | 3.24 |
| Minimum | 10.58 | 7.29 | 7.48 | 4.92 | 7.86 | 8.36 | 23.02 | 8.68 |
| Maximum | −5.42 | −14.2 | −1.57 | −2.89 | −3.44 | 1.32 | −0.97 | −0.37 |
| Std. Dev. | 1.46 | 2.13 | 2.35 | 1.51 | 2.03 | 1.81 | 4.89 | 1.77 |
| Skewness | 1.02 | −4.27 | 0.62 | −0.81 | 0.51 | 0.49 | 1.98 | 0.55 |
| Kurtosis | 15.95 | 29.53 | 2.08 | 3.83 | 3.45 | 2.47 | 6.72 | 3.61 |

# 4 Empirical Results

Before interpreting the results of the MS model. We examine the goodness of fit between the traditional OLS and MS model. Where the equation of interest from the traditional and the MS model are as follows:

$$
\begin{aligned}
\mathrm{INF_{i,t}} = \mathrm{c}_i &+ \beta_{\log(\mathrm{OIL}),i} \log(\mathrm{OIL})_{i,t} + \beta_{\log(\mathrm{WHT}),i} \log(\mathrm{WHT})_{i,t} + \beta_{\log(\mathrm{SUG}),i} \log(\mathrm{SUG})_{i,t} \\
&+ \beta_{\log(\mathrm{GOLD}),i} \log(\mathrm{GOLD})_{i,t} + \beta_{\mathrm{GR^+,i}} \mathrm{GR^+, i} + \beta_{\mathrm{GR^-,i}} \mathrm{GR^-, i}
\end{aligned} \tag{9}
$$

and

$$
\begin{aligned}
INF_{i,\ \mathrm{S_{i,t}},t} = c_{i,s_{i,t}} &+ \beta_{\log(OIL),i,s_{i,t}} \log(OIL)_{i,t} + \beta_{\log(WHT),i,s_{i,t}} \log(WHT)_{i,t} \\
&+ \beta_{\log(SUG),i,s_{i,t}} \log(SUG)_{i,t} + \beta_{\log(GOLD),i,s_{i,t}} \log(GOLD)_{i,t} \\
&+ \beta_{GR^+,i,s_{i,t}} GR^+{}_{i,t} + \beta_{GR^-,i,s_{i,t}} GR^-{}_{i,t} + \sigma_{i,s_{i,t}} \varepsilon_{i,t}.
\end{aligned} \tag{10}
$$

Table 2 represents the value of Bayesian information criterion (BIC), showing that the MS model with two regimes outperforms the single regime model in all cases (countries) by lower BIC value. This indicates that using our Markov-switching model has higher ability to manipulate and explain about the property of the historical data. Then, we decide to use the parameters generated by the MS model to explain in the next step.

**Table 2** BIC estimates for the single regime model and the two-regime-MS model

| Country | BIC value | |
|---|---|---|
| | Single regime model | Two-regime-MS model |
| Singapore | 524.77 | 435.65 |
| Malaysia | 522.70 | 477.88 |
| Thailand | 547.47 | 449.81 |
| Indonesia | 537.85 | 398.76 |
| Philippines | 568.61 | 486.32 |
| Vietnam | 844.41 | 692.90 |

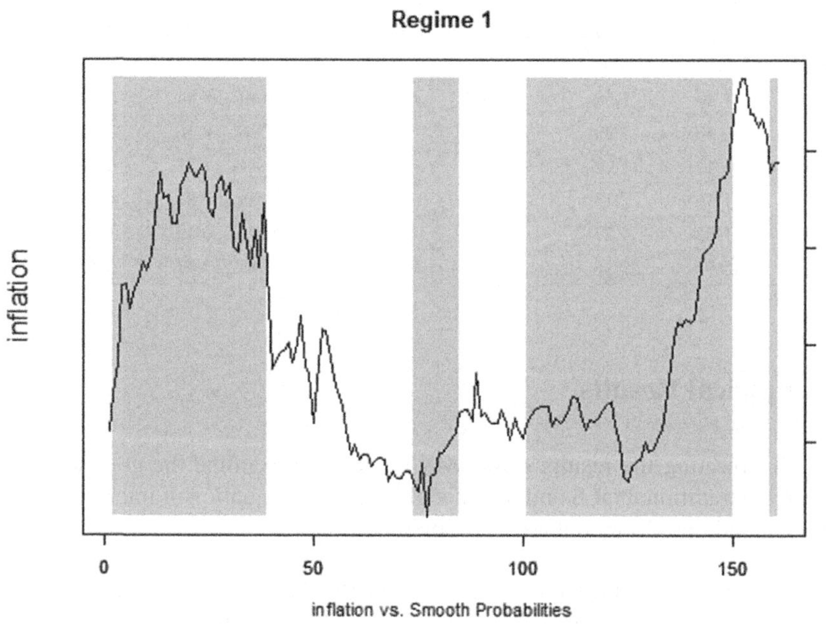


**Fig. 1** Singapore's inflation smooth probabilities

Based on Figs. 1, 2, 3, 4, 5 and 6, a noticeable difference can be observed when comparing inflation generated with smooth probabilities. The shaded period, representing the first regime, predominantly covers higher inflation periods, while the unshaded period corresponds to lower inflation. Consequently, we label the first regime as the "higher inflation regime", and the subsequent periods as the "lower inflation regime". However, it is important to note that in the smooth probabilities chart of Vietnam, although the unshaded period covers the very beginning period, the rest of the shaded periods exhibit higher inflation on average. This suggests that Vietnam's inflation dynamics might exhibit some variations in comparison to the other countries being analyzed.

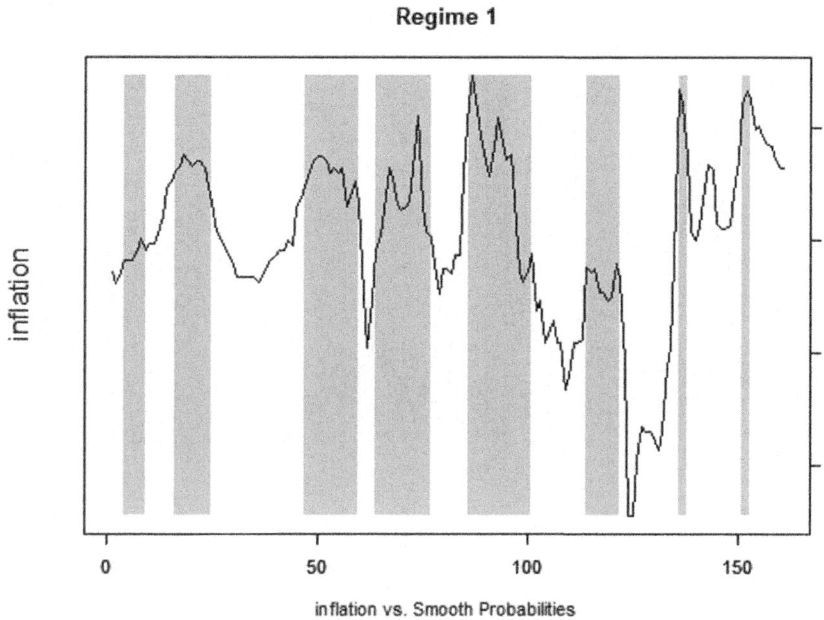


**Fig. 2** Malaysia's inflation smooth probabilities

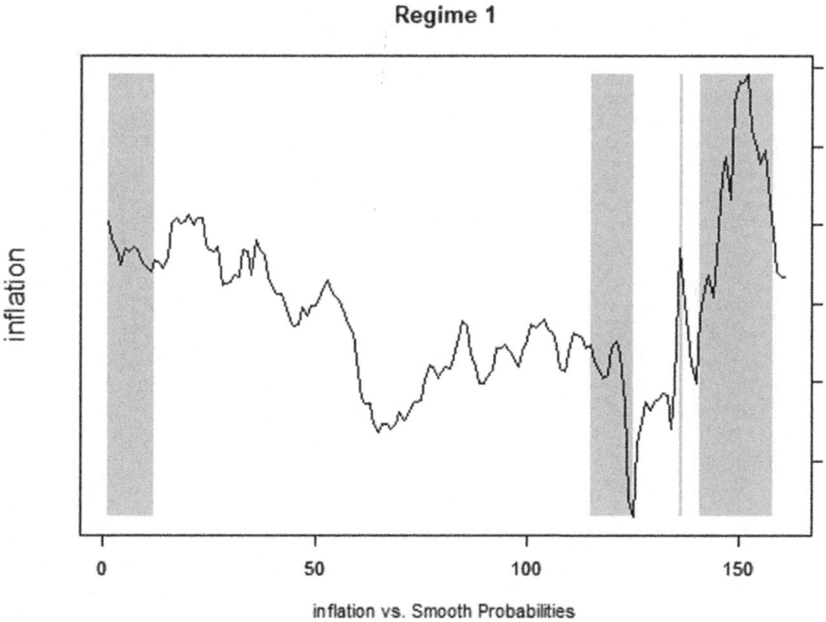


**Fig. 3** Thailand's inflation smooth probabilities

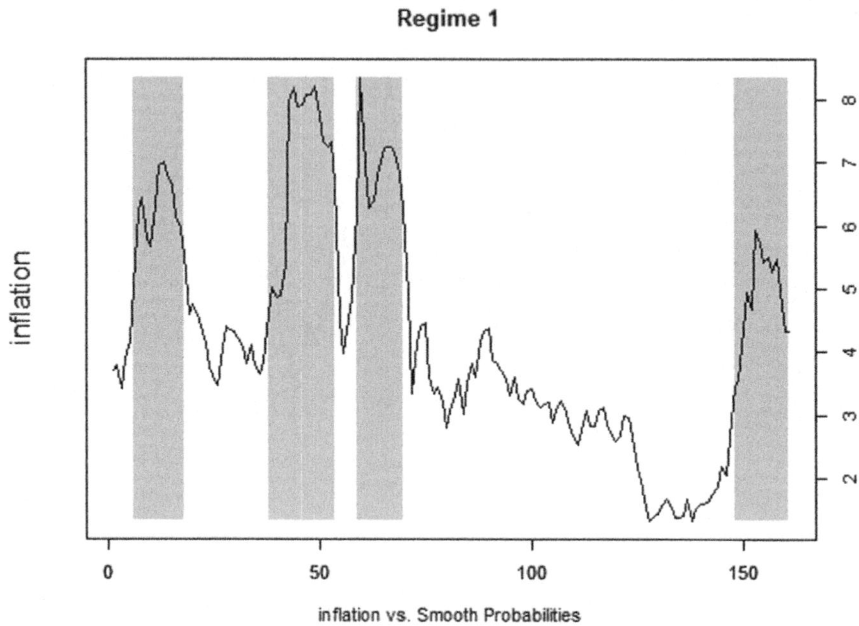


**Fig. 4** Indonesia's inflation smooth probabilities

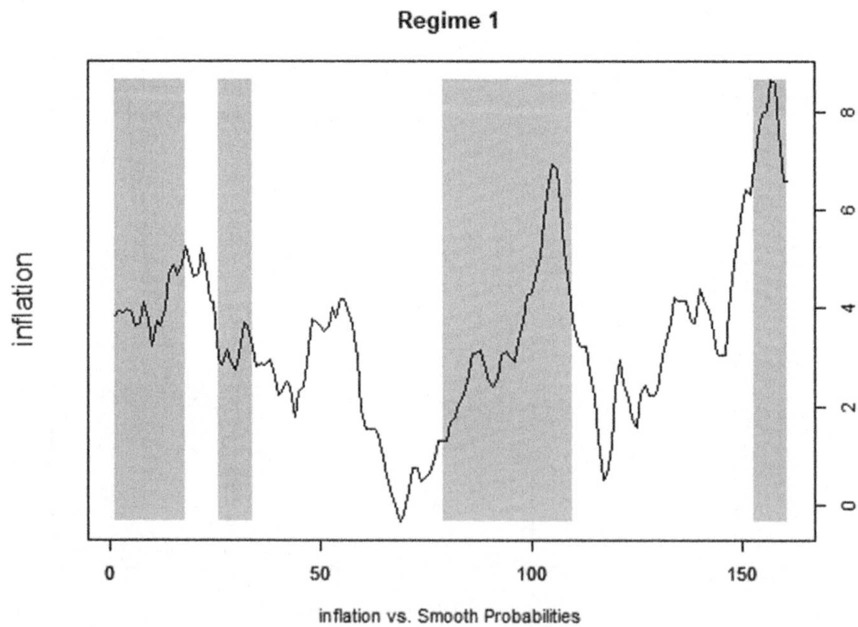


**Fig. 5** Philippines's inflation smooth probabilities

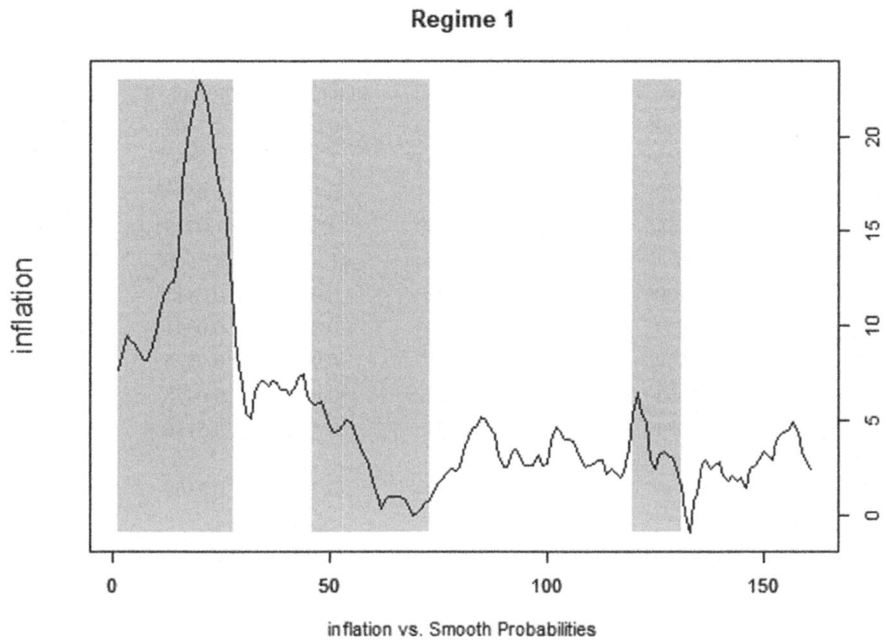


**Fig. 6** Vietnam's inflation smooth probabilities

According to Table 3, the persistence probability in the higher inflation regime, $P_{11}$, is lower than the probability observed in the lower inflation regime, $P_{22}$, for five out of six countries, except Singapore. This suggests that the higher inflation period have a shorter duration compared to the lower volatility regime. Furthermore, the Multiple $R^2$ of our MS models is notably high, ranging from 0.704 to 0.959, indicating strong explanatory power and a good fit of the model to the data of the MS model. Next, we will provide a detailed explanation of the results obtained from the MS-model for each country.

Upon examining the overall scenario, it becomes evident that there is a significantly positive relationship between inflation and global oil prices in most countries during the higher inflation periods, except for Malaysia. This suggests that fluctuations in global oil prices can significantly impact inflation, leading to either an increase or decrease in these countries during periods of higher inflation. However, the results in the lower inflation regime show a positive relationship between global oil prices and inflation in Malaysia, Thailand, Philippines, and Vietnam. Additionally, the magnitudes of the oil price parameters in the lower regime are lower than in the higher regime, confirming the varying degrees of influence during different time periods characterized by fluctuating inflation rates. This indicates that the impact of global oil prices on inflation may vary based on the prevailing inflationary conditions in each country.

**Table 3** Results of the MS model

*Higher inflation regime*

| | Singapore | Malaysia | Thailand | Indonesia | Philippines | Vietnam |
|---|---|---|---|---|---|---|
| $c_{S_1}$ | −34.6262[a] | 17.7902[a] | −1.1603 | 50.0486[a] | 4.2108 | −7.2882 |
| $\beta_{\log(OIL),S_1}$ | 4.6842[a] | −1.0361[a] | 7.0047[a] | 1.2108[a] | 5.5563[a] | 4.5880[a] |
| $\beta_{\log(WHT),S_1}$ | 0.0473 | −1.2146[a] | 0.4861 | −0.1083 | 1.8639[a] | −1.5471[a] |
| $\beta_{\log(SUG),S_1}$ | 0.7757[a] | 4.0054[a] | −0.8789 | 2.8153[a] | −1.0749[a] | 2.1009[a] |
| $\beta_{\log(GOLD),S_1}$ | 3.1627[a] | −4.3906[a] | −4.8684 | −11.8148[a] | −4.7560[a] | −2.1448[a] |
| $\beta_{GR^+,S_1}$ | 0.0120[a] | 0.0041 | 0.0223 | 0.0211[a] | 0.0411[a] | −0.0020 |
| $\beta_{GR^-,S_1}$ | −0.0039 | −0.0700[a] | −0.0503 | −0.0184 | 0.0341[a] | 0.0893[a] |
| $P_{11}$ | 0.9598 | 0.8914 | 0.9267 | 0.9301 | 0.9525 | 0.9666 |
| $P_{12}$ | 0.0402 | 0.1086 | 0.0732 | 0.0699 | 0.0475 | 0.0333 |
| S.E. of residual | 0.5584 | 0.5533 | 0.7745 | 0.5864 | 0.5116 | 0.7495 |
| Multiple $R^2$ | 0.9396 | 0.7037 | 0.9084 | 0.7983 | 0.9159 | 0.8312 |

*Lower inflation regime*

| | Singapore | Malaysia | Thailand | Indonesia | Philippines | Vietnam |
|---|---|---|---|---|---|---|
| $c_{S_2}$ | −24.4962[a] | 3.1323 | −38.4024[a] | 7.7362[a] | −26.1007[a] | −144.938[a] |
| $\beta_{\log(OIL),S_2}$ | 0.1710 | 1.9552[a] | 2.1592[a] | −0.1282 | 2.5266[a] | −3.2766[a] |
| $\beta_{\log(WHT),S_2}$ | 1.0254 | −0.4099 | −0.9129[a] | 0.8484[a] | −3.0957[a] | 0.3857 |
| $\beta_{\log(SUG),S_2}$ | 2.9365[a] | 3.4364[a] | 1.4604[a] | 0.4429[a] | 2.8896[a] | 1.0931 |
| $\beta_{\log(GOLD),S_2}$ | 2.7591[a] | −4.1744[a] | 5.5822[a] | −0.9963[a] | 2.2597[a] | 34.3548[a] |
| $\beta_{GR^+,S_2}$ | 0.0338[a] | 0.0238[a] | 0.0288[a] | 0.0018 | 0.0055[a] | −0.0714[a] |
| $\beta_{GR^-,S_2}$ | −0.0999[a] | −0.0118 | 0.1105[a] | 0.0928[a] | −0.0236[a] | 1.7038[a] |
| $P_{21}$ | 0.9451 | 0.9026 | 0.9617 | 0.9701 | 0.9651 | 0.9706 |
| $P_{22}$ | 0.0548 | 0.0974 | 0.0382 | 0.0298 | 0.0349 | 0.0293 |
| S.E. of residual | 0.4789 | 0.5120 | 0.4824 | 0.4001 | 0.7177 | 2.254 |
| Multiple $R^2$ | 0.959 | 0.8858 | 0.8977 | 0.8292 | 0.7762 | 0.8807 |

[a] denotes a strong rejection or higher of the null hypothesis, according to Minimum Bayes Factor (MBF) (see, Maneejuk and Yamaka, [14])

However, for Singapore, Indonesia, and Vietnam, variations in the global oil price do not exert a substantial influence on either elevating or reducing inflation in the lower regime. In contrast, changes in oil prices are found to drive inflation during the lower inflation period in Malaysia. This suggests that the impact of changes in global oil prices on inflation varies among these specific countries. Taking these findings into consideration. we will now move on to another section to provide more detailed insights into inflation resulting from a different variable.

When examining the impact of global wheat prices on inflation, we find a positive relationship between changes in global wheat prices and inflation, specifically observed in the higher inflation regime of the Philippines and the lower inflation regime of Indonesia. However, this relationship is not evident in both regimes of Singapore, Thailand, and Vietnam. These findings suggest that inflationary or deflationary pressures in the Six-Asian do not respond promptly to fluctuations in global wheat prices, indicating a weaker price transmission mechanism during shorter time periods. In other words, the influence of global wheat price changes on inflation in these Asian countries is less pronounced or possibly delayed, potentially due to various factors affecting the transmission of price effects in their respective economies. Further analysis is required to delve into the underlying factors that contribute to the varying impact of global wheat prices on inflation across the region and to gain a comprehensive understanding of the dynamics at play.

The analysis demonstrates a notably significant positive relationship between global sugar prices and inflation during higher inflation periods in most countries, except for Thailand and the Philippines. Moreover, in the lower inflation regime, all countries exhibit a significant positive relationship, with the exception of Vietnam. These findings indicate that global sugar prices play a substantial role in driving deflation or inflation in the higher and lower regimes of these economies. In simpler terms, fluctuations in global sugar prices have a considerable impact on inflation dynamics in most Southeast Asian countries, especially during periods of lower inflation. However, Thailand and the Philippines appear to be less sensitive to such changes in global sugar prices during their higher inflation periods.

The impact of global gold prices on inflation differs across countries. The MS model shows a significantly negative relationship between global gold prices and inflation in the higher inflation regime of four out of six countries, excluding Singapore and Thailand. Conversely, a positive relationship is found in the lower inflation regime in most countries, except Malaysia and Indonesia. These findings suggest that the inflation levels in these countries are less responsive to changes in global gold prices. The fluctuations in global gold prices do not seem to play a major role in driving inflation dynamics during these specific economic conditions in any of the countries in our study.

The analysis reveals a significant positive relationship between GDP growth and inflation in both the higher and lower regimes for most countries. However, the magnitude of the positive GDP parameter differs between the higher and lower inflation periods, particularly in the comparatively lower-income countries like Indonesia and the Philippines. In these countries, the impact of positive GDP growth on inflation is greater during the higher inflation period compared to the lower inflation regime, indicating that positive GDP growth play more important role in the higher regime. On the other hand, the higher-income countries, such as Singapore, Malaysia, and Thailand, show a different pattern. Here, the effect of positive GDP growth on inflation is more pronounced during the lower inflation regime rather than the higher inflation period.

Furthermore, it is important to note that negative GDP growth has a negative relationship on inflation in the higher inflation regime, indicating the presence of stagflation. Stagflation is characterized by lower output and higher inflation in the lower regime. This suggests that during economic downturns or periods of slower growth, inflation and a decline in output occur simultaneously.

The persistence probability in the higher volatility regime, $P_{11}$, is lower than the probability observed in the lower volatility regime, $P_{22}$, for four out of six countries. This suggests that highly volatile inflation tends to have a shorter duration compared to the lower volatility regime, especially for higher income nations, where the volatility of inflation in the wealthier countries is generally lower than the developing countries.

## 5 Conclusion

This study aims to study the impact of abrupt changes in GDP growth and global commodity prices in six Southeast Asian countries, including Indonesia, Malaysia, The Philippines, Singapore, Thailand, and Vietnam. From the point of view that every country is unique, we expect that the relationship between the global commodity prices and GDP shock on inflation should not be the same, and it should vary through time; thus, we employ Markov-switching model to capture the impact of the explained variables during high and low volatility of inflation. The monthly data employed in this study was collected from January 2010, to May 2023, with all 161 observations, and we will apply log transformation to all the data. The global market prices include NYMEX West Texas Intermediate crude oil, global wheat, sugar, and Gold Price of COMEX. While the country indicators are including positive and negative GDP growth and inflation, where the GDP growth represents shock to GDP. The model selection result indicates that the Markov-switching model outperforms the conventional-single regime one.

In conclusion, our analysis has shed light on the dynamics of inflation in Southeast Asian countries and its relationship with various global economic factors. We found that fluctuations in global oil prices significantly impact inflation during higher inflation periods in most countries, while the impact is more limited during lower inflation periods, particularly in Singapore, Indonesia, and Vietnam. The influence of global wheat prices on inflation varies across countries, with a stronger effect observed in the higher inflation regime of the Philippines and the lower inflation regime of Indonesia.

Furthermore, global sugar prices play a substantial role in driving inflation dynamics in both higher and lower inflation periods in most countries, with exceptions in Thailand and the Philippines. Global gold prices have varying effects on inflation, showing a significantly negative relationship in the higher inflation regime of some countries and a positive relationship in the lower inflation regime of others.

Additionally, GDP growth exhibits a significant positive relationship with inflation in both higher and lower regimes for most countries, with varying magnitudes of influence. Notably, negative GDP growth is associated with stagflation in the higher inflation regime.

These findings hold crucial implications for policymakers, empowering them to design targeted strategies to tackle the unique challenges encountered by each country. During high volatility periods, appropriate measures can be implemented to counteract inflationary effects resulting from oil price shocks. For instance, policymakers may consider lowering domestic energy prices, reducing gasoline taxes, or implementing measures to stabilize food prices, depending on the specific characteristics of each country's economy. On the other hand, during lower inflation periods, policymakers should concentrate on managing potential deflationary pressures arising from negative GDP shocks. In response, they may take measures to stimulate job demand and boost economic activity. This proactive approach will help mitigate the negative impacts of economic downturns, enhancing stability in the economy during such phases. By adopting the strategies based on these findings, policymakers can effectively navigate their respective countries through different economic conditions, fostering sustainable growth and stability.

**Acknowledgements** Supported by CMU Junior Research Fellowship Program.

## References

1. Aloui, C., Hkiri, B., Hammoudeh, S., Shahbaz, M.: A multiple and partial wavelet analysis of the oil price, inflation, exchange rate, and economic growth nexus in Saudi Arabia. Emerg. Markets Finance Trade **54**(4), 935–956 (2018)
2. Prabheesh, K.P., Laila, N.: Asymmetric effect of crude oil and palm oil prices on economic growth: evidence from Indonesia. Buletin Ekonomi Moneter Dan Perbankan **23**(2), 253–268 (2020)
3. Baek, J., Lu, G., Nam, S.: On the asymmetric effects of changes in crude oil prices on economic growth: new evidence from China's 31 provinces. Aust. Econo. Pap. **60**(2), 328–360 (2021)
4. Baek, J., Young, T.B.: A new look at the crude oil prices and economic growth nexus: asymmetric evidence from Alaska. Miner. Econ. **34**, 339–344 (2021)
5. Cevik, N.K., Cevik, E.I., Dibooglu, S.: Oil prices, stock market returns and volatility spillovers: evidence from Turkey. J. Policy Modeling **42**(3), 597–614 (2020)
6. Fischer, S.: The role of macroeconomic factors in growth. J. Monet. Econ. **32**(3), 485–512 (1993)
7. Barro, R.J.: Inflation and economic growth. Bank England Quart. Bull. **35**(4), 166–176 (1995)
8. Sarel, M.: Nonlinear effects of inflation on economic growth. Staff Pap. **43**(1), 199–215 (1996)
9. Bruno, M., Easterly, W.: Inflation crises and long-run growth. J. Monet. Econ. **41**(1), 3–26 (1998)
10. Ghosh, A.R., Phillips, S.: Warning: inflation may be harmful to your growth. Staff Pap. **45**(4), 672–710 (1998)
11. Sepehri, A., Moshiri, S.: Inflation-growth profiles across countries: evidence from developing and developed countries. Int. Rev. Appl. Econ. **18**(2), 191–207 (2004)
12. Mubarik, Y.A., Riazuddin, R.: Inflation and growth: an estimate of the threshold level of inflation in Pakistan. State Bank of Pakistan. Res. Bull. **1**(1), 71–84 (2005)
13. Lee, C., Wong, S.Y.: Inflationary threshold effects in the relationship between financial development and economic growth: evidence from Taiwan and Japan. J. Econ. Dev. **30**(1), 49 (2005)
14. Maneejuk, P., Yamaka, W.: Significance test for linear regression: how to test without P-values? J. Appl. Stat. **48**(5), 827–845 (2021)
15. Kim, C.J., Nelson, C.R.: State-Space Models with Regime Switching: Classical and Gibbs-Sampling Approaches with Applications, vol. 1. MIT Press Books (1999)

# Technical Efficiency of Agricultural Production in the North East of Thailand: An Empirical Study with a Stochastic Frontier Analysis

**Teerawut Teetranont and Kittawit Autchariyapanitkul**

**Abstract** This research paper utilized a conventional stochastic frontier approach, employing maximum likelihood estimation, to investigate empirical production data pertaining to agricultural products, namely tobacco, rice, and chilli. The application of this rigorous methodology focused on analyzing agricultural production data gathered from Wiengkuk Subdistrict in Muang Nong Khai, Thailand. The outcomes of this study yielded significant findings, unveiling a striking reality: a mere average of 60% of the agricultural products under scrutiny were effectively produced. These findings serve as a testament to the pivotal role played by government policies, specifically agricultural promotion, in shaping the overall quantity of agricultural output.

**Keywords** SFA · Agricultural production · Technical efficiency · Land used

## 1 Introduction

The analysis of production efficiency holds paramount importance within the realm of research for several compelling reasons. Firstly, it serves as a catalyst for identifying untapped areas with the potential for productivity improvement, which in turn leads to cost savings, increased output, and enhanced profitability. Secondly, it enables researchers to meticulously evaluate the effectiveness of their methods and processes, facilitating the refinement and optimization of workflows. Lastly, the analysis of production efficiency provides invaluable insights into the overall health and sustainability of research projects or organizations, contributing to their long-term success [10].

T. Teetranont
Faculty of Interdisciplinary Studies, Khon Kaen University, Khon Kaen, Thailand
e-mail: teerate@kku.ac.th

K. Autchariyapanitkul (✉)
Faculty of Economics, Maejo University, Chiang Mai 52000, Thailand
e-mail: Kittawit_a@mju.ac.th



V. Kreinovich et al. (eds.), *Applications of Optimal Transport to Economics and Related Topics*, Studies in Systems, Decision and Control 556,
https://doi.org/10.1007/978-3-031-67770-0_26

A primary benefit of scrutinizing production efficiency within this research context is the ability to identify specific areas where productivity can be improved. By thoroughly examining the various stages of a research project, ranging from data collection and analysis to publication and dissemination, researchers can pinpoint bottlenecks and inefficiencies that hinder progress [15]. For instance, they may discover tasks taking longer than anticipated or identify unnecessary delays in the review process. Addressing these issues allows researchers to streamline their workflows, thereby increasing output without compromising quality.

In addition to enhancing productivity, the analysis of production efficiency empowers researchers to evaluate the effectiveness of their methods and processes. By closely monitoring key performance indicators (KPIs) such as time-to-completion, error rates, and resource utilization, researchers can identify areas where their methods fall short and make necessary adjustments. For example, they may uncover inaccuracies stemming from a particular data analysis technique or identify underperforming tools or software programs. Addressing these issues enables researchers to improve the accuracy and reliability of their work.

In agricultural communities, farmers can sustain their livelihoods by effectively utilizing available resources and directly selling their agricultural products. A new agricultural theory emphasizes the efficient utilization of farmers' existing resources, as well as those provided within their communities, for agricultural production, consumption, and trading. This theory aligns with the Sustainable Development Goals (SDGs) set by the United Nations, particularly the goals of poverty eradication, achieving food security, improving nutrition, and promoting sustainable agriculture. These goals encompass various aspects, including ensuring food safety, access to nutrition sources, addressing food shortages, increasing productivity, providing resource access, and creating opportunities for small-scale farmers. For farmers, these goals signify self-sufficiency, utilizing community resources, and generating income through agricultural production and sales.

Given the global COVID-19 pandemic, many countries implemented isolation and quarantine measures, closed borders, and suspended domestic and international flights as preventive measures to curb the spread of the virus. Such measures were necessary to safeguard public health worldwide, including in Thailand. In 2020, Thailand experienced its first wave of the rapidly spreading virus, followed by an escalation in infection and fatality rates in 2021. Consequently, the government imposed stringent measures to contain the spread and save lives. These measures resulted in the suspension of domestic and international trade, compelling everyone to adapt to the new "normal" of primarily conducting activities online.

Nong Khai was one of the provinces significantly affected by the government's imposed shutdown. After the closure of Thailand's borders with the Lao People's Democratic Republic, trading activities within the province and between the two countries came to a standstill, significantly impacting the tourism sector. However, amidst these challenges, local farmers in the Wiengkuk Sub-district of Muang Nong Khai demonstrated resilience by effectively utilizing natural and community resources to sustain agricultural production for local consumption and trade.

By embracing new agricultural and sustainable agricultural theories, these farmers managed to maintain their livelihoods even during the COVID-19 pandemic.

This paper employs stochastic frontier analysis to examine the production of specific agricultural products, including tobacco, rice, and chilli. Stochastic frontier analysis is a statistical model used to estimate the production function of firms or industries, considering both random and systematic factors that influence the production process, such as technical inefficiency, measurement errors, and external shocks. This widely utilized econometric model facilitates the analysis of productivity and efficiency in various domains.

The underlying principle of the stochastic frontier model involves decomposing the observed output of a firm or industry into two components. The first component represents the deterministic aspect, indicating the maximum attainable output based on available inputs and technology. The second component represents the stochastic aspect, encompassing random deviations from the maximum output due to factors like technical inefficiency and measurement errors. The stochastic component is assumed to follow a normal distribution with a zero mean and constant variance.

Estimating the stochastic frontier model involves employing various econometric techniques, including maximum likelihood estimation, Maximum Entropy, and Bayesian methods [3]. The estimated model parameters facilitate the measurement of the technical efficiency of firms or industries, which represents the ratio of actual output to the maximum potential output based on available inputs and technology. Leveraging these technical efficiency measurements enables the identification of best practices and benchmarking opportunities for enhancing productivity and reducing costs.

The stochastic frontier model finds application in diverse fields such as agriculture, finance, healthcare, manufacturing, and transportation. It enables the analysis of factors influencing productivity and efficiency, including technology adoption, regulatory influences, competition, and environmental considerations. Furthermore, it allows for comparisons of firm or industry performance across different countries and regions.

The remainder of the paper is structured as follows: Sect. 2 provides a comprehensive review of Stochastic Frontier Analysis, encompassing relevant research. Section 3 outlines the Stochastic Frontier Analysis method. Section 4 presents the application of the method to survey data and showcases the results. Finally, Sect. 5 concludes the paper.

## 2 Reviews on Stochastic Frontier Analysis

Stochastic frontier analysis (SFA) is a widely used econometric technique that allows researchers to estimate the production efficiency of firms or organizations. Here are some literature reviews on stochastic frontier by various authors. Reference [4] proposes a model for estimating technical inefficiency effects in a stochastic frontier production function for panel data. The authors demonstrate the usefulness of their approach through an application to a panel dataset of Australian dairy farms. Reference [10] provides a comprehensive overview of the recent developments in stochastic frontier analysis (SFA), including new estimation techniques, model specifications, and applications. The authors also discuss the challenges associated with SFA and suggest future research directions. Reference [2] reviews provides an overview of the empirical applications of stochastic frontier analysis (SFA) in agriculture and food markets. The authors discuss the advantages and limitations of SFA in this context and provide recommendations for future research. Reference [7] focuses on the application of stochastic frontier analysis (SFA) in health economics and outcomes research. The authors provide an overview of the methodological issues associated with SFA in this context, including model specification, data requirements, and interpretation of results. Reference [11] provides an overview of recent developments in stochastic frontier models (SFM), including new estimation methods, model specifications, and applications. The authors also discuss the advantages and limitations of SFM and provide recommendations for future research.

## 3 Methods and Formulas

### *3.1 Stochastic Frontier Analysis*

The stochastic frontier is a statistical model used to estimate the efficiency of production processes. It was first introduced by Aigner, Lovell, and Schmidt in their seminal paper "Formulation and Estimation of Stochastic Frontier Production Function Models" published in 1977 [1].

The stochastic frontier model is based on the assumption that observed output is produced using a combination of inputs, some of which are not fully under the control of the producer. These uncontrollable factors are referred to as inefficiency or random error. The model estimates the level of inefficiency and separates it from the productive efficiency of the production process. Mathematically, suppose we have a production function $f(X_i, \alpha)$. In a state without error or inefficiency, the $i_{th}$ agent would generate

$$Q_i = f(X_i, \alpha) \tag{1}$$

SFA assumes that each agent potentially produces less than this due to a degree of inefficiency. More specifically,

$$Q_i = f(X_i, \alpha)\pi_i \tag{2}$$

where $\pi_i$ is the degree of efficiency for agent $i$; $\pi_i$ must be in the range (0, 1]. If $\pi_i = 1$, the agent is succeeding the optimal output with the technology consolidated in the production function $Q_i = f(X_i, \alpha)$. When $\pi_i < 1$, the firm is not making the most of the inputs $X_i$ given the technology embodies in the production function

$f(X_i, \alpha)$. Consequently, the output must be strictly positive ($Q_i > 0$) and the level of technical efficiency is supposed to be strictly positive ($\pi_i > 0$). Output is also assumed to be subject to random shocks, implying that

$$Q_i = f(X_i, \alpha)\pi_i \exp(v_i) \tag{3}$$

Taking the natural log of both sides yields

$$\ln Q_i = \ln f(X_i, \alpha) + \ln \pi_i + v_i \tag{4}$$

Assuming the there are $k$ inputs and that the production function is linear in logs, defining $u_i = -\ln \pi_i$ yields

$$\ln Q_i = \alpha_0 + \sum_{j=1}^{k} \alpha_j \ln X_{ji} + v_i - u_i \tag{5}$$

Because $u_i$ is subtracted from $\ln(q_i)$, restricting $u_i \geq 0$ implies that $0 < \pi \leq 1$, as specified above.

## 3.2 Technical Efficiency

Consider the following equation

$$Q_i = X_i\alpha + v_i - u_i \tag{6}$$

where $Q_i$ is the output variable, $X_i$ is $1 \times k$ vector of observations on the explanatory variables included as indent covariates, $\alpha$ is $k \times 1$ vector of coefficients. The log-likelihood function for Normal/half-normal model is as follows [5].

$$\ln L = \sum_{i=1}^{N} \left( \frac{1}{2} \ln \left( \frac{2}{\pi} \right) - \ln \sigma_S + \ln \Phi \left( -\frac{\varepsilon_i \lambda}{\sigma_S} \right) - \frac{\varepsilon_i^2}{2\sigma_S^2} \right) \tag{7}$$

where $\sigma_S = (\sigma_u^2 + \sigma_v^2)$, $\lambda = \frac{\sigma_u}{\sigma_v}$, $\varepsilon_i = Q_i - X_i\alpha$, and $\Phi()$ is the cumulative distribution function of the standard normal distribution.

To obtain estimation for $u_i$, we use the mean of the conditional distribution $f(u|\varepsilon)$ [14, 16].

$$E(u_i|\varepsilon_i) = \mu_{*i} + \sigma_* \left[ \frac{\phi(\frac{-\mu_{*i}}{\sigma_*})}{\Phi(\frac{\mu_{*i}}{\sigma_*})} \right] \tag{8}$$

Then the technical efficiency will be estimated by [9, 13]

$$TE_i = E\{\exp(-u_i)|\varepsilon_i\} = \left\{\frac{1 - \Phi(\sigma_* - \mu_{*i}/\sigma_*)}{1 - \Phi(-\mu_{*i}/\sigma_*)}\right\} \exp\left(-\mu_{*i} + \frac{1}{2}\sigma_*^2\right) \tag{9}$$

where $\mu_{*i}$ and $\sigma_*$ are defined for the normal/half-normal model as

$$\mu_{*i} = -\varepsilon_i \sigma_u^2/\sigma_S^2 \tag{10}$$

$$\sigma_* = \sigma_u \sigma_v/\sigma_S \tag{11}$$

Thus, the half-normal distribution ($N^+(0; \sigma_u^2)$) is a fold at the mean of an ordinary normal distribution with mean zero.

## 4 Application On The Survey Data and Formulation of Model

### 4.1 Data

The study gathered empirical data from farmers residing in the Wiengkuk Subdistrict of Muang Nong Khai, encompassing a comprehensive sample size of 247 observations collected over the period between June and August 2022. The principal focus of the investigation centered on the total production of agricultural products, designated as the dependent variable, while concurrently considering a multitude of factors that exert influence on the aggregate output as covariates. These covariates encompass Agricultural Products (AP), serving as a quantitative measure denoting the cumulative weight of all agricultural commodities cultivated within the farmers' operational domain, quantified in kilograms. Agricultural Land (AL) captures the expanse of cultivated land, measured in rai, occupied by the farmers for their agricultural pursuits. Labor (LA) signifies the count of laborers embedded within the farmers' immediate families. Cost (CO) signifies the capital investment incurred by the farmers for the purpose of agricultural cultivation within their respective areas.

In addition to the aforementioned covariates, the analysis incorporates a series of dummy variables, namely Community water resources (CW), intended to depict the availability of water sources localized within the farmers' communal domain. Organic fertilizer (OF) is utilized to delineate the utilization of organic fertilizers in the cultivation and nourishment of agricultural produce. Saving (SA) encapsulates the accumulated savings attributed to the families of the farming community. Local natural resources (LN) serve as an indicator denoting the presence of natural resources indigenous to the immediate community. Public land in the community (PL) elucidates the accessibility of communal land designated for public utilization. Native plants in the community (NP) signify the existence of indigenous plant species within the farmers' community. Livestock in the community (LS) serves as an indicator to portray the prevalence of livestock within the farming community. Local wisdom (LW) conveys the utilization and preservation of indigenous knowledge systems within the farmers' community. Lastly, local cultural sites (LC) delineate the presence of culturally significant sites within the farmers' community.

## 4.2 A Formulation of Model

The agricultural production frontier is specified as

$$\ln AP_i = \alpha_0 + \alpha_1 \ln AL_i + \alpha_2 \ln LA_i + \alpha_3 \ln CO_i + \alpha_4 CW_i + \alpha_5 OF_i + \alpha_6 SA_i + \alpha_7 LN_i + \alpha_8 PL_i + \alpha_9 NP_i + \alpha_{10} LS_i + \alpha_{11} LW_i + \alpha_{12} LC_i + v_i - u_i \tag{12}$$

where $AP_i$ is the agricultural output, $AL_i$, $LA_i$, $CO_i$, $CW_i$, $OF_i$, $SA_i$, $LN_i$, $PL_i$, $NP_i$, $LS_i$, $LW_i$ and $LC_i$ are the corresponding variable as mentioned in Sect. 4.1, respectively.

# 5 Results

In this section, we show the estimated parameters and the technical inefficiency from conventional SFM. We made the usual assumptions of half-normal and normal distributions for one-sided and two-sided errors, respectively. Table 1 shows parameter estimated from SFM as below.

In Table 1, we present the parameter estimates for the SFM, shedding light on the key factors influencing it, namely, the labor force and production costs. This conclusion gains robust support through evidence from the MBF goodness-of-fit test. The estimated SFM parameters consistently align with our expectations. Specifically, a one percentage increase in both land use and production costs corresponds to a 0.427% and 0.462% increase in agricultural product quantity, respectively.

However, when it comes to variables like LA, CW, OF, SA, PL, and LN, the evidence supporting their impact on production is weak. Despite the theoretically significant role of the labor force (LA) in production, the results hint at its diminished importance due to the widespread use of farmers' own members for production.

Finally, factors that represent local knowledge among farmers, such as NP, LS, LW, LC, show substantial evidence supporting their influence on production. This underlines the importance of considering these elements when assessing the dynamics of agricultural production.

**Table 1** Estimated parameters of SFM for agricultural products in the North East of Thailand

| Input variables | Estimated parameters | Std err | MBF | Strength of evidence [8] |
|---|---|---|---|---|
| Constant | 3.313 | 0.591 | 0.000 | Decisive |
| ln AL | 0.427 | 0.082 | 0.000 | Decisive |
| ln CO | 0.462 | 0.054 | 0.000 | Decisive |
| ln LA | −0.127 | 0.136 | 0.642 | Weak |
| CW | 0.227 | 0.239 | 0.636 | Weak |
| OF | −0.177 | 0.140 | 0.446 | Weak |
| SA | 0.039 | 0.147 | 0.964 | Weak |
| PL | 0.092 | 0.138 | 0.798 | Weak |
| LN | 0.119 | 0.181 | 0.804 | Weak |
| NP | −0.820 | 0.278 | 0.017 | Substantial |
| LS | −0.775 | 0.336 | 0.069 | Substantial |
| LW | 0.481 | 0.178 | 0.026 | Substantial |
| LC | −0.417 | 0.191 | 0.093 | Substantial |
| MSE | 4.760 | | | |
| Obs. | 247 | | | |

MBF is Minimum Bayes Factors

**Fig. 1** The scatter plot of TE and average product

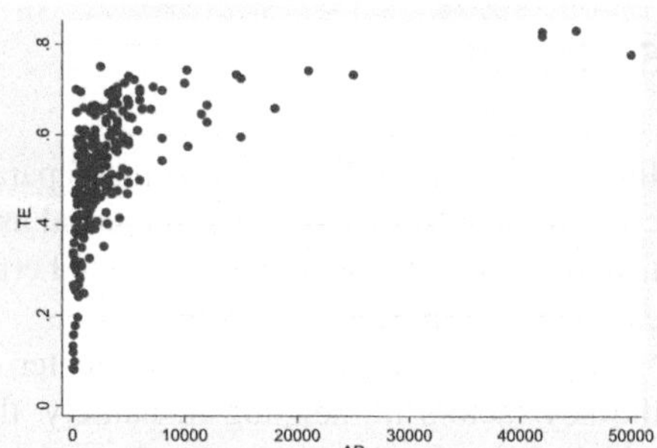


Figure 1 visually represents the connection between technical efficiency (TE) and average production. The outcomes highlight that farmers with TE values below 0.6 generally produce agricultural yields below 0.625 kg/m$^2$, whereas those with TE exceeding 0.8 exhibit production levels surpassing 25 kg/m$^2$. Figure 2, a scatter plot showcases the relationship between TE and community water resources, revealing that a significant portion of farmers operates within the sphere of community water resources. However, it also hints at the relatively modest impact of water supply on agricultural production.

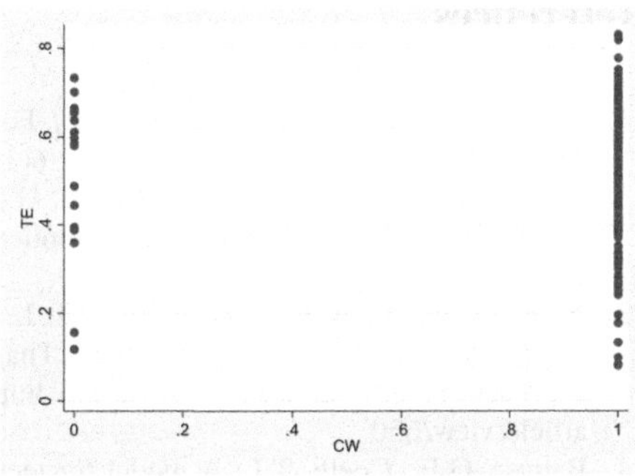


**Fig. 2** The scatter plot of TE and community water resources (CW)

These findings harmonize with various prior studies, fortifying the coherence and dependability of the results across diverse research undertakings. This consistency contributes to a broader understanding of the interplay between technical efficiency, agricultural production, and the role of water resources in farming contexts [3, 6, 12].

## 6 Conclusions

In conclusion, the meticulous analysis of production efficiency assumes utmost significance for research endeavors and organizational endeavors alike, as they strive to optimize productivity, streamline workflows, and ensure enduring prosperity The In the context of our study. We embraced the widely acclaimed Stochastic Frontier Method (SFM) to meticulously scrutinize agricultural products. Our comprehensive investigation encompassed not only quantitative determinants such as labor, production costs, and land utilization but also qualitative factors including CW, OF, SA, LN, PL, NP, LS, LW, and LC. To gauge the strength of evidential support for each covariate, we judiciously employed the Minimal Bayes Factor (MBF) as a dependable metric of goodness of fit.

The study shows that, on average, a mere 60% of agricultural products achieved successful production. This revelation underscores a compelling and imperative demand for the government to diligently enhance production techniques. By methodically identifying areas with the potential for productivity enhancement, meticulously evaluating the effectiveness of methods and processes, and rigorously maintaining consistent monitoring of outcomes, researchers can judiciously make data-driven decisions. Such informed decisions serve as a catalyst for fostering superior outcomes, propelling advancements in agricultural productivity.

**Acknowledgements** This research is funded by the Young Researcher Development Project of Khon Kaen University Year 2022.

## References

1. Aigner, D., Lovell, C.K., Schmidt, P.: Formulation and estimation of stochastic frontier production function models. J. Econom. **6**(1), 21–37 (1977)
2. Alemu, B.A., Nuppenau, E.A., Boland, H.: Technical efficiency of farming systems across agro-ecological zones in Ethiopia: an application of stochastic frontier analysis. Agric. J. **4**(4), 202–207 (2009)
3. Autchariyapanitkul, K., Srisirisakulchai, J., Kunasri, K., Ayusuk, A.: Technical efficiency in rice production at farm level in northern Thailand: a stochastic frontier with maximum entropy approach. Thai J. Math. 121–132 (2017). https://thaijmath2.in.cmu.ac.th/index.php/thaijmath/article/view/650
4. Battese, G.E., Coelli, T.J.: A model for technical inefficiency effects in a stochastic frontier production function for panel data. Empir. Econ. **20**, 325–332 (1995)
5. Belotti, F., Daidone, S., Ilardi, G., Atella, V.: Stochastic frontier analysis using Stata. Stata J. **13**, 719–758 (2013)
6. Chaovanapoonphol, Y., Singvejsakul, J., Sriboonchitta, S.: Technical efficiency of rice production in the upper north of thailand: clustering copula-based stochastic frontier analysis. Agriculture **12**(10), 1585 (2022). https://doi.org/10.3390/agriculture12101585
7. Hamidi, S., Akinci, F.: Measuring efficiency of health Systems of the Middle East and North Africa (MENA) region using stochastic frontier analysis. Appl. Health Econ. Health Pol. **14**, 337–347 (2016)
8. Held, L., Ott, M.: On p-values and Bayes factors. Ann. Rev. Stat. Appl. **5**, 393–419 (2018)
9. Karakaplan, M.U.: Fitting endogenous stochastic frontier models in Stata. Stata J. **17**, 39–55 (2017)
10. Kumbhakar, S.C., Lovell, C.K.: Stochastic Frontier Analysis. Cambridge University Press (2000)
11. Post, T., Potì, V.: Stochastic efficiency analysis using relative entropy and empirical likelihood. Available at SSRN 2489889 (2014)
12. Shanmugam, K.R., Venkataramani, A.: Technical efficiency in agricultural production and its determinants: an exploratory study at the district level. Indian J. Agr. Econ. **61**(902-2016-68012) (2006)
13. Stevenson, R.E.: Likelihood functions for generalized stochastic frontier estimation. J. Econom. **13**, 57–66 (1980)
14. Tauchmann, H.: Partial frontier efficiency analysis. Stata J. **12**, 461–478 (2012)
15. Worthington, A.C.: Frontier efficiency measurement in health care: a review of empirical techniques and selected applications. Med. Care Res. Rev. **61**(2), 135–170 (2004)
16. Zellner, A., Revankar, N.S.: Generalized production functions. Rev. Econ. Stud. **36**, 241–250 (1969)

# Determinants of Fertility Intention and Behavior Under Economic, Health and Policy Constraints in China

**Shiqi Zhou, Supanika Leurcharusmee, Piyaluk Buddhawongsa, and Paravee Maneejuk**

**Abstract** This study investigates the individual characteristics associated with fertility constraints. Specifically, this study uses data from the China Labor Force Dynamic Survey (CLDS2018), which includes information on individuals' number of children and their ideal number of children. This allows us to assess whether individuals' fertility choices align with their actual number of children or if they are constrained by factors such as health, economic circumstances, and government policies. Employing Probit regression analysis, the research identifies factors contributing to the ideal-behavior gap, representing the overall constraint on fertility. The findings reveal that individuals with gender preferences, higher education levels, residing in developed regions, and employed in government are more likely to experience a substantial ideal-behavior gap, implying constraints on their fertility behaviors due to factors such as health, economics, or policies. In addition, the study identifies individual characteristics specifically associated with facing restrictions imposed by the two-child policy. Focusing on the policy constraint, the findings reveal that older females and individuals with agricultural hukou, particularly those who prefer daughters, are less likely to be constrained by the two-child policy.

**Keywords** Ideal children number · Fertility intention · Fertility behavior · Family planning policy · Two-child policy

---

S. Zhou (✉)
Faculty of Economics, Chiang Mai University, Chiang Mai, Thailand
e-mail: z1768610879@gmail.com

S. Leurcharusmee · P. Buddhawongsa
Faculty of Economics, Center of Human Resource and Public Health Economics, Chiang Mai University, Chiang Mai, Thailand
e-mail: supanika.l@cmu.ac.th

P. Buddhawongsa
e-mail: piyaluk.b@cmu.ac.th

P. Maneejuk
Faculty of Economics, Center of Excellence in Econometrics, Chiang Mai University, Chiang Mai, Thailand
e-mail: paravee.m@cmu.ac.th

 
V. Kreinovich et al. (eds.), *Applications of Optimal Transport to Economics and Related Topics*, Studies in Systems, Decision and Control 556,
https://doi.org/10.1007/978-3-031-67770-0_27

## 1 Introduction

Three weeks after the statistics of China's seventh census were released, China's government introduced the three-child policy to accelerate population growth which is only five years after the universal two-child policy. Based on the background of the policy and the heated debate about the effect of family planning policy liberalization in contemporary society, the purpose of this paper is to explore what kinds of people are more affected by policy constraints and who are subjected to total constraint effect which is considering policy economy and healthy conditions together when making fertility decisions in the special social environment of China today. For the policy, the decade-long Cultural Revolution that brought down China's politics and economy after the founding of the People's Republic of China after World War II, the tremendous population growth impeded the enhancement and growth of people's living standards [3]. In order to rapidly reduce the population to solve the problem of feeding the people and improving their living quality, in 1979, the Chinese Communist Party began to propose the Family Planning policy, also known as the one-child policy.

However, after the 1990s, as a result of the country's long-standing one-child policy, China entered an era of low fertility [11, 12]. In order to get away from the aging population and stimulate the rise of total fertility, the selective two-child policy was introduced in 2011, and the universal two-child policy was standardized in 2015 after 35 years of the one-child policy.

Regarding The Seventh National Population Census of the People's Republic of China in 2020, the number of births on the Chinese mainland was only 12 million, the lowest amount since 1960. The proportion of individuals aged 0–14, 15–59, and 60 and above has increased by 1.35%, decreased by 6.79%, and increased by 5.44%, respectively, as compared to the 2010 sixth census [1]. The aging problem and the declining labor force are two significant demographic challenges that China is facing. The latest census results reveal that these challenges are becoming increasingly severe, and they affect all age groups across the population. It is evident that China's aging problem is becoming increasingly grave, and the labor force is declining significantly, which poses significant threats to the country's economic growth, social welfare, and the healthcare system.

The vast population-based demographic dividend in China is a foundation for the country's continued fast economic development in the last 20 years. The aging population and low natural growth rate are directly slowing down China's economic development. Since 2010, however, the trend in the changing age structure of the population has reversed, and there has been a labor shortage, a slowdown in the rate of quality improvement, a fall in the rate of return on investment, and a narrowing of the scope for resource reallocation [13]. Accordingly, the average growth rate of China's GDP has been decreasing year after year since 2010, falling to an average annual rate of 2.2% in 2020 from 10.6% in 2010 [2]. In response to this discernible economic deceleration, the Chinese government has undertaken a strategic initiative with the implementation of the three-child policy in 2021.

Undoubtedly, the announcement of this regulation has sparked a great deal of discussion among Chinese citizens, and opinions vary, especially regarding policy constraint's effect on fertility growth nowadays. This study's purpose is to explore the effects of economic and health constraints as well as the family planning policy, on an individual's fertility decisions. And the advantage of the study is to propose feasible policy recommendations on the three-child policy, with a focus on child-rearing cost-cutting measures and social welfare policies, especially protection and children's well-being.

## 2 Literature Review

According to Miller and Pasta, the reproductive behavior of humans is a complex process that spans both temporal and quantitative dimensions [4]. This comprehensive fertility process includes fertility motivation, fertility intention, the formulation of a birth plan, and subsequent fertility behavior, and ultimately culminates in the determination of the fertility rate. Similarly, in 2012, Zheng categorized the reproductive considerations of Chinese families into four levels, namely, ideal situation, fertility intention, birth plan, and fertility outcome [5]. These levels provide a framework for understanding the various factors that influence individuals' fertility decisions and highlight the importance of considering the different stages of the fertility process. Understanding this process is crucial for policymakers and researchers seeking to develop effective interventions for promoting fertility in China.

### *2.1 Ideal Children Number, Fertility Intention and Fertility Behavior*

Chinese scholars have defined the "Ideal Children Number" as individuals' willingness and demand to have children, as reflected in their expectations regarding the number, timing, gender, and quality of children. Additionally, people's attitudes towards their perceptions of fertility behavior are also seen as part of this concept. "Fertility Intention" refers to people's expectations regarding the number of children they wish to have over the course of their lifespan, given particular social, economic, and cultural conditions [14]. After considering their expectations and the actual situation, individuals make their fertility decision, which is reflected in their actual number of children, and is referred to as "fertility behavior". After a thorough analysis of the average ideal number of children, the intended number of children, and the actual fertility rate in the EU15 and the US between 1995 and 2000, Hagewen and Morgan discovered something that caught their attention: in both countries, the ideal number of children is greater than the intended number of children, and the intention is greater than the actual number of children [6].

## 2.2 Gender Preference

Gender preference has been shown to be an important influence on women's fertility intentions in China in several articles. Despite China's fast-declining birth rate, the sex ratio at birth has continued to rise in recent years [15]. Gender preferences can be divided into three categories: 1. Having boy 2. Having girl 3. Being a parent to both boy and girl. Numerous studies conducted in China have revealed that the high real fertility rate of women is influenced by gender preference. Couples of childbearing age who fail to meet the desired gender composition may have extra children if they do not already have the appropriate number of children.

According to Morgan, a significant preference for sons in China would increase the total fertility rate [16]. The traditional Chinese culture of preference for sons "preference for sons", "male supremacy over female", and "transmission of sons" derives from the fact that the birth of a boy signifies an increase in the family's labor force especially in rural areas. Meanwhile, some publications contend that son preference would reduce the number of children and, consequently, the fertility rate in China by causing sex-selective abortion [17–19]. Compared to childbearing-age women without a son preference, women with a son preference are less likely to have a second child, particularly if their first child is a boy [7].

## 2.3 Other Explanatory Factors of Fertility Intention and Behavior

Age, gender, marital status, and income are frequently discussed among the factors that influence people's decisions to have children. There are two other popular topics related to people's income: education level and region which also have significant effects in the field of reproductive research. In general, those with a higher level of education have greater financial resources and are able to cover more childcare expenses. Yuan and Jianqi mentioned that the lower the levels of education, the lower the desire to have another child compared to those with a university degree or more [20]. Contrary to the aforementioned opinions, Wei and Yongai think as women's education increases, they become more independent and have more liberal attitudes toward fertility, whereas the pursuit of education and a career tends to reduce the period of fertility Consequently, both the ideal and actual numbers of children tend to decrease [21]. The level of development of people's regions is inextricably linked to their personal and family income, and their attitudes toward reproduction, and both have direct and indirect effects on their fertility behavior. Various studies have looked into the geographic implications of splitting China into eastern, western, and central regions, with the eastern side typically being more developed by cities closer are closer to the sea. Ideal children number and fertility in rural areas have always been much greater than in urban areas [8], and there is a strong correlation between the Hukou makeup of a population and fertility [9].

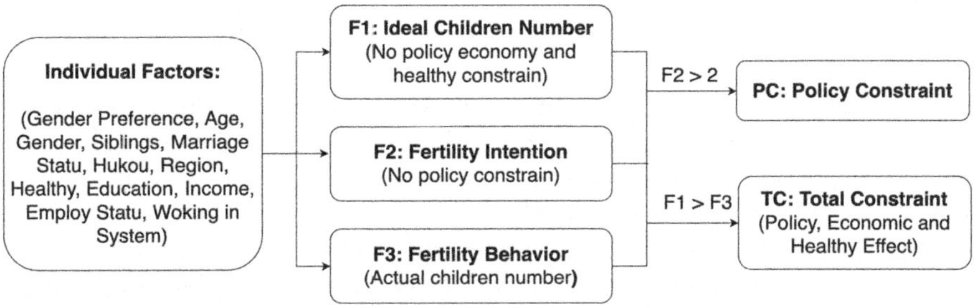


**Fig. 1** Conceptual framework

In China, factors such as the presence of siblings, the work sector, and parental support are often considered when individuals make decisions about their fertility. As noted in Jianxin and Yunliang's research, the absence of siblings among husbands has a facilitating influence on real fertility due to the substitution effect [11]. This is because couples without siblings may be more likely to have additional children to provide companionship and support for their existing child. Understanding the factors that influence fertility decisions is crucial for policymakers and researchers seeking to develop effective interventions for promoting fertility in China. By identifying the factors that are most strongly associated with fertility behavior, policymakers can design targeted interventions that address the unique circumstances of different groups of people and promote higher fertility rates.

The family planning policy is stricter for those who work in the system, such as government employees, primarily government agencies, public institutions, and state-owned businesses. They risk losing their jobs and political careers if they breach the family planning policy [10]. However, employment within the system affords better treatment, stability (almost no termination), social standing, and social security than outside the system. Due to their employment supporting national policy or low opportunity cost, system or government personnel may be more eager to procreate under an open birth policy (Fig. 1).

## 3 Research Methodology

There are two steps to examine the factors that affect the policy constraint effect and total constraint effect about having children. The first is done by using math calculations to sort data subjects with different constraints into distinct categories. The following prediction is using the generalized linear model. In this study, dichotomous and continuous explanatory variables collected in the first stage will be analyzed using probit regression.

## 3.1 Data

The secondary data used in this study were obtained from the China Labor Force Dynamics Survey (CLDS) conducted in 2018. The CLDS is a nationally representative survey that covers a large sample of individuals and workers aged 15–64 in 29 provinces across China. The survey collects a vast amount of information on various aspects of labor force dynamics, including employment status, wages, education, and family demographics. A total of 16,538 recordings were obtained, after the data cleaning procedure, which involved removing samples with missing or extreme values, 5,933 samples were retrieved and used for analysis in this study.

## 3.2 Descriptive Statistics

## 3.3 Methodology

### 3.3.1 Measurement of the Fertility Constraint Variables

To measure the effects of economic and health constraints, and the family planning policy, we generate two dummy variables. The dummy variable PC measures the effect of the two-child policy constraint, where PC equals 1 when the individual is constrained by the two-child policy and 0 otherwise.

Dummy variable TC measures the effect of policy, economic, and health constraints. When TC equals 1, it means the individual is constrained by all three conditions.

$$PC = \begin{cases} 1, & if\ F_2 > 2 \\ 0, & Otherwise \end{cases}$$

$$TC = \begin{cases} 1, & if\ F_1 > F_3 \\ 0, & Otherwise \end{cases}$$

### 3.3.2 Factors Associated with the Fertility Constraints

To examine factors associated with fertility constraints, this study uses the Probit Regression as follows.

$$P_r\{PC = 1\} = \mathscr{L}(X\beta_{PC})$$

$$P_r\{TC = 1\} = \mathscr{L}(X\beta_{TC})$$

where PC is the effect of the two-child policy constraint and TC is the Total constraint which is the policy, economic, and healthy effect together. $\mathscr{L}$ represents the standard normal distribution function, $\boldsymbol{\beta}$ is the parameters of the model, and X is a vector of individual and household factors determining the ideal children number, fertility intention, and fertility behavior.

To determine whether each explanatory variable significantly determines the outcome, the Bayes factor upper bound (BFUB) is used instead of the traditional p-value. As stated in Halsey [22], the BFUB is the upper bound of the odds in favor of the alternative hypothesis ($H_a$: $\boldsymbol{\beta}, \boldsymbol{\gamma} \neq 0$) relative to the null hypothesis ($H_0$: $\boldsymbol{\beta}, \boldsymbol{\gamma} = 0$). In particular, the BFUB can be written as a function of the p-value (p) as follows:

$$BFUB = \frac{-1}{e \cdot p \cdot ln(p)}$$

It is pertinent to observe that within the context of this investigation, BFUB values surpassing 999 are denoted as 999. A BFUB value of 999 signifies that the odds favoring the alternative hypothesis $H_a$ over the null hypothesis $H_o$ are at a ratio of 999 to 1. This substantial evidentiary magnitude underscores a compelling basis to assert that the explanatory variable significantly contributes to the determination of the outcome.

## 4 Results and Analysis

The descriptive findings presented in Table 1 indicate that the average values for the three fertility-related variables do not exceed 2, with even the mean number of children reported by the interviewed sample being only 1.26. The results indicate that the policy constraint effect and total constraint effect have a stronger influence on individuals' fertility decisions than the gender preference effect. Additionally, the two constraint effects are found to be more influenced by the total constraint effect, while gender preference does not exert a significant influence. One surprising finding of the study is that the average age of the sample is relatively high, at 42.88 years. This could potentially impact the subsequent regression analysis. It is possible that this advanced age is due to the absence of family planning policies during their birth, resulting in having more siblings. Moreover, the limited availability of schools during their youth could have contributed to their overall lower level of education. These factors may have contributed to the observed low fertility rates in the sample.

Table 2 presents empirical results subject to Policy Constraint (PC) and Total Constraint (TC) based on a set of defined criteria. Individuals with fertility intentions greater than 2 in 2018 are considered to be subject to PC, while those who disregard policy, economy, and health conditions and only consider their ideal number of children compared to their actual number of children are deemed to be subject to TC. The comparison of the results of Policy Constraint (PC) and Total Constraint (TC) in Table 2 that the economy and health have a greater impact on fertility decisions than policy. Gender preference, age, marital status, and type of hukou play a significant influence on the fertility decisions of people who are constrained by two separate

**Table 1** Variable description and sample statistics

| Variables | Description | Mean | MD | N |
|---|---|---|---|---|
| F1 | Ideal children number (Not considering policy, economy, and healthy conditions) | 1.98 | 0.530 | 5933 |
| F2 | Fertility intention (no family planning policy constraint) | 1.93 | 0.988 | 5933 |
| F3 | Fertility behavior (Actual Children Number) | 1.26 | 0.736 | 5933 |
| PC | policy constraint effect (=1 if F2>2) | 0.16 | 0.365 | 5933 |
| TC | Total constraint effect (Policy, Economic and Healthy constrain effect) (=1 if F1>F3) | 0.53 | 0.499 | 5933 |
| Son | =1 if the individual prefers to have a son | 0.16 | 0.368 | 933 |
| Daug | =1 if the individual prefers to have a daughter | 0.13 | 0.331 | 5933 |
| i_Age | Age | 42.88 | 8.908 | 5933 |
| i_Female | =1 if the individual is a female | 0.49 | 0.500 | 5933 |
| i_Married | =1 if the individual is in marriage | 0.83 | 0.378 | 5933 |
| i_Siblings | Number of individual's siblings | 2.39 | 1.818 | 5933 |
| i_Agriculture | =1 if the individual has Agriculture hukou | 0.77 | 0.418 | 5933 |
| i_Healthy | How healthy the individual think they are | 3.83 | 0.896 | 5933 |
| i_East | =1 if the individual from east of China | 0.50 | 0.500 | 5933 |
| i_Mid | =1 if the individual from middle of China | 0.21 | 0.411 | 5933 |
| i_Highschool | =1 if the highest education of individual member is high school | 0.16 | 0.371 | 5933 |
| i_University | =1 if the highest education of individual member is more than highschool | 0.24 | 0.426 | 5933 |
| i_CurEmploy | =1 if the individual is currently hired | 0.92 | 0.278 | 5933 |
| i_Gover | =1 if the individual work in system | 0.16 | 0.364 | 5933 |
| i_Gover | =1 if the individual work in system | 0.16 | 0.364 | 5933 |
| i_Gover | =1 if the individual work in system | 0.16 | 0.364 | 5933 |
| i_ToincLY | =log total income last year | 9.83 | 2.550 | 5933 |

conditions. For individuals who consider policy, economic, and health constraints, education level, and working sector also influence their fertility decisions.

Married or female individuals are subject to fewer policy constraints, which decrease with age. This is because the average age of the study participants is around 42, indicating that they are past childbearing age and have stable family structures. Additionally, individuals who prefer daughters receive fewer policy constraints because parents who prefer girls tend to do so out of preference, unlike son lovers who desire more sons. Consequently, those who prefer daughters face fewer policy constraints as they are not as eager to have another child. Similarly, people in rural areas with agricultural hukou have more children irrespective of family planning policies, as they view newborns as labor.

**Table 2** Probit regression results

| Variables | (1) Policy constraint | (2) Total constraint |
|---|---|---|
| Son | 0.101<br>(1.6) | 0.242<br>(999.0) |
| Daug | −0.148<br>(3.4) | 0.198<br>(999.0) |
| i_Age | −0.0510<br>(999.0) | −0.0187<br>(999.0) |
| i_Female | −0.212<br>(999.0) | −0.0566<br>(1.7) |
| i_Married | −0.813<br>(999.0) | −1.019<br>(999.0) |
| i_Sinblings | 0.0123<br>(1.0) | −0.0119<br>(1.0) |
| i_Agriculture | −0.138<br>(4.9) | −0.300<br>(999.0) |
| i_Healthy | −0.0178<br>(1.0) | −0.0531<br>(10.6) |
| i_East | 0.0777<br>(1.3) | 0.230<br>(999.0) |
| i_Mid | 0.0228<br>(1.7) | −0.0778<br>(1.5) |
| i_Highschool | 0.0379<br>(1.1) | 0.221<br>(999.0) |
| i_University | 0.0699<br>(1.1) | 0.272<br>(999.0) |
| i_CurEmploy | 0.0454<br>(1.2 | −0.126<br>(2.4) |
| i_Gover | 0.0217<br>(1.7) | 0.192<br>(53.2) |
| i_ToincLY | −0.00657<br>(1.0) | 0.00371<br>(1.2) |

*Note* For BFUB higher than 999, the value 999 is reported

Those who had the total constraint effect, regardless of their gender preference, tend to consider more realistic conditions when making fertility decisions, possibly due to their specific preferences. However, people with agricultural hukou are less likely to be constrained by health and economic factors and may have a greater desire to have more children, irrespective of policy factors. Compared to the aforementioned group, in eastern China, which is a more developed area with a higher level of education and a larger number of government employees, the TC effect on fertility behavior is greater.

## 5 Conclusion

The research aims to examine the individual-level factors that are associated with fertility constraints, with a particular focus on the influence of China's two-child policy on individuals' fertility decisions. In order to achieve this objective, the study draws on the China Labor Force Dynamic Survey(CLDS2018) as its primary data source. Based on the descriptive analysis, it was revealed that family planning policy is no longer the predominant fertility constraint for individuals, as economic and health considerations have reduced their fertility behavior. The Probit regression models for policy and total constraints indicate that older females and individuals with agricultural hukou, particularly those who prefer daughters, are less constrained by two-child policy constraints. Moreover, the analysis of total constraint reveals that individuals with gender preferences, those residing in developed regions, and those with higher education or working for the government are more likely to be constrained by health, economic, or the two-child policy conditions.

China has experienced significant economic growth over the past two decades due to its demographic dividend. For the purpose of sustaining this growth in the later stages of development, a large population base will continue to be a critical factor. To stimulate people's fertility intentions and behaviors, it is recommended to introduce policies beyond the three-child policy that address economic and health considerations. For instance, improving medical and maternity care systems, as well as social insurance systems, would enable older women to feel more confident about giving birth or having another child. Additionally, reducing the cost of education and providing child-rearing subsidies would alleviate the financial burden on young couples, especially those who are already pregnant. Lastly, increasing publicity in government agencies and universities would help to reduce people's fear of getting married and having children.

## References

1. Jizhe, N.: Main data results of the seventh national population census. National Bureau of Statistics (2021). https://www.stats.gov.cn/sj/zxfb/202302/t20230203_1901089.html
2. GDP Growth (Annual %) - China (1961–2022). The World Bank. https://data.worldbank.org/indicator/NY.GDP.MKTP.KD.ZG?locations=CN
3. Faxiang, Y.: Historical research on family planning of contemporary China (2003)
4. Miller, W.B., Pasta, D.J.: Behavioral intentions: which ones predict fertility behavior in married couples? J. Appl. Soc. Psychol. **25**(6), 530–555 (1995)
5. Bao, L., Chen, F., Zheng, Z.: Transition in second birth intention in a low fertility context: the case of Jiangsu. China. Asian Popul. Stud. **13**(2), 198–222 (2017)
6. Hagewen, K.J., Morgan, S.P.: Intended and ideal family size in the United States, 1970–2002. Popul. Dev. Rev. **31**(3), 507–527 (2005)
7. Jiang, Q., Li, Y., Sánchez-Barricarte, J.J.: Fertility intention, son preference, and second childbirth: survey findings from Shaanxi Province of China. Soc. Indic. Res. **125**, 935–953 (2016)

8. Valentina, S., Arjan, T.: Determinant factors for fertility: case of Albania. Stud. Business & Econ. **6**(1) (2011)
9. Cai, Y.: China's below-replacement fertility: Government policy or socioeconomic development? Popul. Dev. Rev. **36**(3), 419–440 (2010)
10. Chuanhui, L., Xingbang, H.: Identity in system, child bearing choice and optimization of overall two-child policy. J. Sichuan Normal Univ. (Social Sciences Edition) **43** (2016)
11. Jianxin, L., Yunliang, P.: The influentials of actual birth level in low desired fertility condition in our country: using Bongaarts low fertility model. Popul. & Econ. **4**(4.9), 10–27 (2018)
12. Mao, Z., Hao, L.: Difference between fertility intention and fertility behavior for women subject to the two-children policy: an empirical study based on the theory of planned behavior. Popul. Res. **37**(4.5), 84–93 (2013)
13. Fang, C.: The second demographic dividend as driver of China's growth. Universidade Federal De Juiz De Fora Curso **6**(5), 1–12 (2010)
14. Zhou, F.: Research on fertility intention and its measurement indicators (2005)
15. Wei, C.: Sex preference and fertility behaviors of Chinese women, p. 7.2 (2002)
16. Morgan, S.P., Guo, Z., Hayford, S.R.: China's below-replacement fertility: recent trends and future prospects. Popul. Dev. Rev. **35**(3), 605–629 (2009)
17. Yang, S., Wang, G.: Fertility decline and imbalance of sex ratio under birth control. Market & Demograph. Anal. **12**(4), 18–28 (2006)
18. Guo, Z.: China's low fertility and its determinants. Popul. Res. **32**(4), 1–12 (2008)
19. Cai, Y.: China's lowest low fertility in the global context: opportunities and challenges. In: Wang, F., Peng, X., Gu, B. (eds.) Globalization and Low Fertility: China's Choice. Fudan University Press, Shanghai (2011)
20. Zhang, Y., Chen, J.: The transition of fertility desire and its policy implications: an empirical research on the determinants of procreation desire in China: 1991–2011. J. Guizhou Univ. Finance Econ. **3**, 79–91 (2015)
21. Chen, W., Jin, Y.A.: Gaps between fertility intensions and fertility outcomes in China: patterns and determinants. Popul. J. **33**(2), 3–13 (2011)
22. Halsey, L.G.: The reign of the p-value is over: What alternative analyses could we employ to fill the power vacuum? Biol. Lett. **15**(5), 20190174 (2019)

# Bayesian Fixed-Effects Panel Kink Regression with Unknown Threshold with Application to the Impact of Economic Freedom on Economic Growth

**Emmanuel Mensaklo, Chukiat Chaiboonsri, Kanchana Chokethaworn, and Songsak Sriboonchitta**

**Abstract** This paper employs a fixed-effects Bayesian panel kink regression framework to examine the dynamics of the impact of economic freedom on economic growth in five selected Sub-Saharan African countries. Our novel findings have empirically established that economic freedom has a nonlinear impact on economic growth with a kink point value of approximately 59%. The results show that economic freedom positively impacts economic growth in the lower region, but negatively in the higher region. Specifically, with a 95% posterior probability, a 1% point rise in the total economic freedom score will lead to about 0.14% point increase in economic growth in the lower region, whilst it will be about 0.31% point decline in growth in the higher regime with a 95% posterior probability, ceteris paribus. This finding starkly highlights the fact that, in addition to the nonlinearity of the impact of economic freedom on growth, the impact in the high region is more than twice larger than it is in the lower region. This finding is very relevant for policy choices and implementations that focus on economic freedom and economic growth.

**Keywords** Bayesian panel kink regression · Economic growth · Economic freedom

E. Mensaklo (✉) · C. Chaiboonsri · K. Chokethaworn · S. Sriboonchitta
Faculty of Economics, Chiang Mai University, Chiang Mai, Thailand
e-mail: emmanuel_mensaklo@cmu.ac.th

C. Chaiboonsri
e-mail: chukiat.chai@cmu.ac.th

K. Chokethaworn
e-mail: kanchana.ch@cmu.ac.th

S. Sriboonchitta
e-mail: songsak.s@cmu.ac.th



V. Kreinovich et al. (eds.), *Applications of Optimal Transport to Economics and Related Topics*, Studies in Systems, Decision and Control 556,
https://doi.org/10.1007/978-3-031-67770-0_28

# 1 Introduction

Since the revolutionary work by [1] on regression kink designs, which work is a generalization of the pioneering work on threshold regression by [2], kink regression, with its many variants, has received a considerable attention from both theoretical and empirical researchers. On the theoretical front, several extensions have been made, some of which specifically focused on classical framework (of estimation, testing, and inference methods) using parametric, semi-parametric, and nonparametric modelling approaches, whilst some few employed Bayesian framework of modelling. The pattern in the empirical literature domain follows that of the theoretical one. Hansen [3] provides a catalogue of literature references on some of the main theoretical and empirical extensions made within the classical framework, whilst [4] succinctly provide a comparative synopsis on the broad categories of parameter estimation and testing within the classical approach and Bayesian paradigm.

Although there exist several studies that examined the impact of economic freedom on economic growth and other productivity measures, there is no evidence of any existing literature on theoretical and empirical applications of kink regression approaches to study the existence of possible non-linear impacts of economic freedom on growth and, particularly where the explanatory variables are bounded indexes with policy choices. Our suspicion is that the long-established economic growth-economic freedom nexus that posits a linear positive impact of the latter on the former has escaped econometric scrutiny. Since all macroeconomic policy and individual choices are always subject to constraints and, as a result, have opportunity costs, it will be of significant economic importance to commence research in this regard. We undertake to fill this gap.

To accomplish the above, this study aims to investigate the economic freedom-economic growth nexus in the top five-performing economies in Sub-Saharan Africa by employing fixed-effects Bayesian panel kink regression approach. With this framework, this research seeks to provide a comprehensive understanding of the dynamics of the impact of economic freedom on economic growth, while accounting for potential nonlinearities, which may be characterized by thresholds or kink points. By incorporating this nonlinear element, the study aims to capture any potential nonlinear impact of economic freedom on economic growth. By combining fixed effects with kink modelling to assess country-specific effects, this study will be able to account for heterogeneity across the selected countries to provide more precise estimates, which may have significant policy implications, compared to traditional panel regression models. As is evident in the literature, the Bayesian framework allows for robust estimation, efficient parameter inference, and model flexibility, making it an appropriate tool for our panel data for this study.

The selection of the top five-performing economies in Sub-Saharan Africa is motivated by their sustained economic growth rates, favorable investment climates, and overall economic resilience. These economies, which have showcased notable economic transformations, include Benin, Cote d'Ivoire, Ethiopia, Rwanda, and Tanzania. African Economic Outlook [5] ranks these economies as the top-5 best performing economies in Sub-Saharan Africa prior to and after the peak of the Covid-19 pandemic. By focusing on these economies, this study aims to provide empirical insights into the specific dynamics of the economic freedom-economic growth nexus within this subset of high-performing Sub-Saharan countries, which may differ from the broader regional patterns. The findings and insights from this study could serve as valuable reference points for home-bred policy models for the rest of the other economies within the Sub-Saharan region to emulate. The rest of the paper is organized as follows. Section 2 presents a brief literature review of some theoretical postulates, paradigms, and empirical studies. Section 3 covers data and the methodology used in the study, while Sect. 4 covers results and discussions. The last section, Sect. 5, presents the conclusions.

In conclusion, by contributing to the existing literature on Sub-Saharan Africa, this research seeks to advance our understanding of the economic freedom-economic growth nexus in the top five-performing economies in Sub-Saharan Africa. It is anticipated that the results will inform policymakers about the importance of economic freedom and its potential nonlinear effects on economic growth, thus aiding in the formulation of effective policies to promote sustainable economic growth and development within the sub-region.

## 2 Review of Some Related Literature

Several theoretical propositions have been propounded to explain the channels or conduits through which various proxies of economic freedom affect economic outcomes, including productivity and real growth rates (see [6–11]). Recently, some other theorists have emphasized how economic freedom could positively affect output measures through international trade (see, for example, [12–15]).

Whilst all the above theoretical perspectives posit a positive impact of economic freedom on output measures, there are some others who postulate opposing views. For instance, some argue that excessive economic freedom that liberalizes international trade and permits unbounded trade openness may be counterproductive to developing economies in many practical perspectives (see, for example, [16–19]).

Empirically, some previous studies have found strong evidence that economic freedom affects economic outcomes positively. For example, [11] provides empirical evidence supporting the idea that economic freedom and free trade contribute to efficiency gains, productivity improvements, and overall economic growth. In a study by [20], security of property rights, quality of monetary policy, freedom to trade and regulatory policies have been found to be positively related to economic growth.

However, some studies found contrasting evidence. For example, [21] found that economic freedom has a negative impact on the long-run economic growth of Latin American and Caribbean countries. This finding is in consonance with that of [22] who earlier found that some categories of economic freedom indicators have a negative relationship with growth. Panahi et al. [23] also found that not all economic freedom indicators have a statistically and economically significant impact on growth.

Based on the above varied views in the literature on the impact of economic freedom on economic growth, we propose the application of a Bayesian panel kink regression to provide empirical evidence on the impact of the former on the later.

## 3 Data and Methodology

### 3.1 Data

The study used freely obtained balanced panel data of the selected economies, covering a period of twenty-eight years, which is from 1995 to 2022. Data on economic growth, our dependent variable, measured by real economic growth rate, were obtained from the [24], which are augmented by data from [5] for missing values. Following the neoclassical growth theoretical propositions, we included the following as regressors in our model: capital, economic freedom, foreign direct investment (FDI), labor, and population. These variables are, respectively, proxied by the annual growth rate of gross fixed capital formation (fixed capital), the overall index of economic freedom (economic freedom), the net inflows of FDI as a percent of GDP (FDI), labor force participation rate (labor force) for youths and adults between the ages of 15–64, and the real annual population growth rate (population). The first and third were sourced from the website of the [25] World Development Indicators, whilst, respectively, economic freedom, labor, and population variables were sourced from [26], The International Labor Organization, and United Nation Conference on Trade and Development.

Data on economic freedom, proxied by index of economic freedom, require a brief discussion. Various authors have defined economic freedom in different ways. We adopt the definition by [27]. There are 12 main dimensions of economic freedom measured by the index, and these are grouped under four broad categories, which include the rule of law (judicial effectiveness, property rights, and government integrity), government size (fiscal health, tax burden, and government spending), regulatory efficiency (business freedom, labor freedom, and monetary freedom), and market openness (trade freedom, investment freedom, and financial freedom). The overall score is an average of all 12 dimensions with equal weights (see [27] for a detailed definition of the variables).

## 3.2 The Model

Kink regression modeling has witnessed several variants of specifications, largely motivated by the main research goals at stake. For purposes of this current study, we integrate the model specifications by [4, 28] with some extensions to account for our panel data and country-specific unobserved heterogeneities. Our model is given by

$$y_{it} = \beta_0 + \beta_1 (X_{1,it} - \delta)_- + \beta_2 (X_{1,it} - \delta)_+ + \beta_j^{\top} X_{j,it} + \omega_{i-1} + u_{it}, \tag{1}$$

where $i = 1, \ldots, N$ are the countries, or the cross-sectional units, which are the selected Sub-Saharan African countries in our study and $t = 1, \ldots, T$ is the time index. $X_1$ is our independent variable of interest, whose kink point we want to determine, $X_{\mathrm{j},it}$ is a $d \times 1$ vector of regressors, which serve as control variables, and $\omega_i$ represents the unobserved heterogeneity of the $i$th country which can be correlated with $X_{it} = \left(X_{1,it}, X'_{j,it}\right)'$ with $j = 2, \ldots, J$. $\left(\beta_0, \beta_1, \beta_2, \beta_{\mathrm{j}}, \delta\right)$ are parameters to be estimated. For $k = X_{1,it} - \delta$, $(k)_- = \min(k, 0)$ and $(k)_+ = \max(k, 0)$. For values of $X_{1,it}$ for which $X_{1,it} \leq 0$, the slope is given by $\beta_1$, otherwise the slope is given by $\beta_2$. It is important to highlight the fact that all regressors are continuous variables. However, in exception of all other variables' parameters, the parameters at the kink point are discontinuous. For concreteness, the entire regression function

$$\left(\beta_0 + \beta_1 (X_{1,it} - \delta)_- + \beta_2 (X_{1,it} - \delta)_+ + \beta_j^{\top} X_{j,it} + \omega_{i-1}\right)$$

is continuous as a function of $X_{1,it}$ and $X_{\mathrm{j},it}$. However, there exists a discontinuity at the threshold, or the kink, point $\delta$, regarding the slope function with respect to $X_1$. Our kink regressor uniquely has a compact and bounded domain ranging from zero to hundred and, as a result, the validity of our results will largely depend on whether the kink point lies within the entire support of the kink variable. This requires that we determine whether $\beta_1$ and $\beta_2$ are both individually identified, without which our best model will be to fit a linear model.

Our next step is to undertake a test to confirm the existence of a kink point. Reparametrizing Eq. (1) to preserve an implicit linear regression form, equation:

$$y_{it} = b_o + b_1 X_{1,it} + b_2 (X_{1,it} - \delta)_+ + \beta_j^T X_{j,it} + \omega_{i-1} + u_{it}, \tag{2}$$

where $b_0 = \beta_0 - \beta_1\delta$, $b_1 = \beta_1$, and $b_2 = \beta_2 - \beta_1$. It is important to note that $b_2$ seeks to capture two separate domains of $X_1$, at which point the kink point value is either greater or less than $X_{1,it}$. From $b_2 = \beta_2 - \beta_1$, it follows that $b_2 = 0$ if and only if $\beta_2 = \beta_1$, in which case Eq. (2) reduces to the usual linear regression model. If, on the contrary, $\beta_2 \neq \beta_1$, then $b_2 \neq 0$. For this later case, the slope of $X_{1,it}$ in Eq. (2) will be $b_1 + b_2$ for $X_{1,it} > \delta$ and $b_1$ for $X_{1,it} < \delta$. Following [28], who proposed sub-gradients statistics of residuals based on CUSUM, our null and alternative hypotheses for this test are, respectively, $H_0 : b_2 = 0$ for all $\delta \in \theta$ and $H_1 : b_2 \neq 0$ for some $\delta \in \theta$, where $\theta$ is the domain of the kink-dependent independent variable which is a bounded and compact support for the kink point. As has been noted by [28], the relevance of the application of a kink regression largely hinges on the identification of parameter $b_2$, otherwise, the ideal model choice should be a linear regression if we fail to reject the null. The sub-gradients statistics of the residual based CUSUM test is given by

$$R_n\left(\delta \middle| \hat{b}_1, \hat{B}_j, \hat{\omega}_{i-1}\right) = N^{-\frac{1}{2}} \sum_{t=1}^{N} \left(y_{it} - \hat{b}_1 X_{1,it} - \hat{B}_j^{\top} X_{j,it} - \omega_{i-1}\right) \times \left(X_{1,it} - \delta\right) I\left(X_{1,it} \leq \delta\right). \tag{3}$$

The panel structure of the data becomes immaterial after the sub-gradient statistics are computed, where now we have $t = 1, 2, \ldots., NT$. Since no prior is specified on the kink point value, [28] proposed a grid search for its candidates using a uniform distribution over the compact domain of $X_{1,it}$. The associated test statistic is given by

$$K_{NT}(\delta) = \sup_{\delta \in \theta} \left| R_{NT}\left(\delta \middle| \hat{b}_1, \hat{B}_j, \hat{\omega}_{i-1}\right)\right. \tag{4}$$

tends to be rejected when $K_{NT}(\delta)$ is statistically large. $|\cdot|$ denotes $L_1$ norm. This test statistic is simulated through residual bootstrap algorithm which yields p-values for $K_{NT}(\delta)$ which have asymptotic validity, since $K_{NT}(\delta)$ has a nonstandard limiting distribution. Following Li et al., the bootstrap procedures to test for the kink effect are outlined below. Given $\hat{u}_t = y_t - \hat{b}_1 X_{1,t} - \hat{B}_j^{\top} X_{j,t} - \omega_{i-1}$,

1. Generate iid $e_1, \ldots, e_{NT}$ from $N(0, 1)$ distribution.
2. Set $\tilde{y}_t = \hat{b}_1 X_{1,t} + \hat{B}_j^{\top} X_{j,t} + \hat{\omega}_{i-1} + \hat{u}_t e_t$ for $t = 1, \ldots, NT$.
3. Calculate the test statistic $\hat{K}_{NT}(\delta) = \sup_{\delta \in \theta} \left| R_{NT}\left(\delta \middle| \hat{b}_1, \hat{B}_j, \hat{\omega}_{i-1}\right)\right.$ by using the observations $\left\{\tilde{y}_t, X_{1,t}, X_{j,t}\right\}_{t=1}^{NT}$. Estimate $\hat{b}_1, \hat{B}_j, \hat{\omega}_{i-1}$ by minimizing $\frac{1}{NT} \sum_{t=1}^{NT} \left(\hat{y}_t - \hat{b}_1 X_{1,t} - \hat{B}_j^{\top} X_{j,t} - \hat{\omega}_{i-1}\right)^2$.
4. Repeat steps 1–3 B times to obtain a set of simulated statistics $\hat{K}_{NT}^{(1)}(\delta), \ldots, \hat{K}_{NT}^{(B)}(\delta)$. Calculate the p-value using $\hat{p}_{NT} = B^{-1} \sum_{b=1}^{B} I\left(\tilde{K}_{NT}^{(b)}(\delta) \geq K_{NT}(\delta)\right)$, which is the proportion of simulated statistics exceeding the actual value.

*Bayesian Estimation*

We implement an MCMC sampling process to estimate all unknown model parameters using a Bayesian estimator, which concurrently provides us with the benefits of the associated parameter inferences. Following the normal practice in Bayesian literature, we formulate the prior distributions for our unknown parameters, where we make use of $(m_0, V_0, c_0, d_0, a, b)$ as hyperparameters. We then combine the joint density to construct the posterior distribution. Following [29], we make use of a non-informative prior for the coefficient parameters. The prior for β is specified using the $b_0$ and $V_o$ arguments, which are the prior mean and prior precision, respectively, and we set the prior as an independent multivariate normal distribution, expressed as.

$$b \sim \mathcal{N}_K\left(m_0, V_0^{-1}\right)$$

The prior for the nuisance parameter $\sigma^2$ is predicated on two main parameters, $c_0$ and $d_0$. We assume $u_i$ has zero mean and variance $\sigma^2$, and the nuisance parameter $\sigma^2$ is inverse Gamma (IG) distributed with $\sigma^2 \sim IG\left(\frac{c_0}{2}, \frac{d_0}{2}\right)$, which requires that we use a conjugate prior for it. We set the hyperpapramеters a = 5% and b = 95% as our lower and upper quantiles, respectively, to handle edge effects of the drawing. We obtain the conditional posterior distribution for each parameter by the product of their respective priors and their Gaussian likelihood functions as expressed below:

The conditional posterior distribution for $b$ is $N\left(m_*, V_*^{-1}\right)$, where

$$m_* = \left[m_0 V_0 + \frac{\sum_{t=1}^{NT} X_{1,t} y_t}{\sigma^2}\right]\left[V_0 + \frac{\sum_{t=1}^{NT} X_{1,t} X_{1,t}^{\top}}{\sigma^2}\right]^{-1}$$

and

$$V_* = \left[V_0 + \frac{\sum_{t=1}^{NT} X_{1,t} X_{1,t}^{\top}}{\sigma^2}\right]$$

The conditional posterior distribution for $\sigma^2$ is $IG\left(\frac{c_0+NT}{2}, \frac{d_0+NTs^2}{2}\right)$, where

$$s^2 = NT^{-1}\sum_{t=1}^{NT}\left[y_t - b^{\top} X_{1,t}\right]^2.$$

The posterior distribution of the kink parameter is non-standard. Its density is given by

$$f\left(\delta \middle| b, \sigma^2\right) \propto \exp\left\{-\sum_{t=1}^{NT}\frac{1}{2\sigma^2}\left[y_t - b^{\top} x_t(\delta)\right]^2\right\} I_{(a,b)}(\delta)\sigma^{-NT}.$$

Gibbs sampling algorithm has the capability of handling small and larger parameter sizes. Based on this, we use the Gibbs sampler to draw our MCMC iterates. Marginal likelihood functions will be computed using the method of Chib. For the kink parameter, we follow [28] and use the standard Gaussian random walk Metropolis algorithm which is nested within the Gibbs sampler to obtain the Markov chains, since its posterior is nonstandard. We follow the Gibbs sampling procedure proposed by [28], which is outlined below, for the regression kink model.

5. Initialize the parameter values $\left(b_{(0)}, \delta_{(0)}, \sigma^2_{(0)}\right)$.
6. Generate iteratively from the conditional posteriors distribution for $\left(b_{(k)}, \delta_{(k)}, \sigma^2_{(k)}\right)$.

   i. Draw $b_{(k)}|\delta_{(k-1)}, \sigma^2_{(k-1)}$;
   ii. Draw $\sigma^2_{(k)}|b, \delta_{(k-1)}$;
   iii. Draw $\delta_{(k)}|b_{(k)}, \sigma^2_{(k)}$.

We implement the algorithm by initializing $\delta$. We set it equal to the median of $X_t = \delta_{(0)}$. We then estimate $\hat{b}(\delta_0)$ conditional on $\delta_0$ to obtain the initial values $b_{(0)}$. We will need to do the iterations for several runs to get stationary simulated chains to guarantee consistency of the posterior distributions. We will undertake diagnostic assessments using trace and density plots to determine whether the Markov chains have attained convergence to their stationary distributions. Our empirical model for this study is given below. All variables are as defined earlier.

$$\begin{aligned}&economic\ growth =\\&f(economic freedom,\ fixed\ capital,\ population,\ labour force,\ FDI, \omega_i)\end{aligned}$$

For a comparative analysis of our estimates, we will estimate a fixed effects Bayesian panel model without a kink.

## 4 Results and Discussions

### 4.1 *Descriptive Statistics, Panel Unit Root Tests, and Scatter Plots*

Table 1 below presents the descriptive statistics of both the dependent and independent variables used in our study.

Since our study employs panel data, it is important to determine the panel stationarity of our data by conducting a panel unit root test. Table 2 below presents results of the Levin, Lin and Chu first-generation homogeneous panel unit test of Levin et al. The results show that, at level, economic growth, economic freedom index, and fixed capital have strong significance at 0.01 level, when the intercept term is included in the test, whilst Population, labor force, and FDI have a strongly significance at 0.05. Without the intercept term, economic freedom, fixed capital, labor force and FDI have strong significance at 0.01 level. The rest of the variables have strong significance at 0.05 level. This shows that our variables are stationary at level and, therefore, we can include them in our estimations and analysis without producing spurious results. We next present some plots to informally assess the possibility of

**Table 1** Descriptive statistics of variables

| Statistic | Economic growth | Economic freedom | Fixed capital | Population growth | Labor force | Net FDI |
|---|---|---|---|---|---|---|
| Mean | 6.1643 | 55.5556 | 10.2762 | 3.1170 | 72.1469 | 1.9041 |
| Median | 6.2000 | 55.8500 | 8.6251 | 2.7439 | 71.7715 | 1.6253 |
| Maximum | 24.500 | 71.1000 | 155.7850 | 36.2064 | 88.3500 | 5.6637 |
| Minimum | −4.9000 | 38.3000 | −19.8464 | 1.1924 | 52.8400 | −0.6586 |
| Std. Dev | 3.8664 | 5.9111 | 16.9309 | 3.2024 | 9.7729 | 1.4468 |
| Skewness | 0.3546 | −0.2994 | 4.6661 | 8.7251 | 0.1740 | 0.5772 |
| Kurtosis | 6.2899 | 3.8821 | 40.4774 | 85.6140 | 1.6039 | 2.7577 |

**Table 2** Levin, Lin and Chu panel unit root test on variables

| Variable name | [1]Statistic | [1]Prob | [2]Statistic | [2]Prob |
|---|---|---|---|---|
| Economic growth | −3.9666 | 0.0000 | −1.7085 | 0.0438 |
| Economic freedom | −4.1323 | 0.0000 | −5.7631 | 0.0000 |
| Fixed capital | −4.3539 | 0.0000 | −3.6441 | 0.0001 |
| Population | −1.3716 | 0.0451 | −1.1849 | 0.0021 |
| Labor force | −0.6303 | 0.0301 | −3.2592 | 0.0000 |
| FDI | −3.1020 | 0.0164 | −4.1011 | 0.0000 |

*Note* [1]Panel unit root test with the intercept term; [2]Panel unit root test without the intercept term. Null: Unit root (assumes common unit root process)

applying a kink regression or, otherwise, to capture the impact of economic freedom on growth.

Figures 1 and 2 below, respectively, present scatter plots of economic growth and economic freedom for individual countries and stacked cross-sections of the countries. Pictorially, all plots suggest a kind of non-linear relationship between both variables. This offers a good reason to validate this important observation by testing for the presence of a kink effect. By using the test proposed by [28], our p-value obtained from bootstrap is 0.016. This result implies that we fail to accept the null of the existence of linearity, and therefore, we fit a kink model.

## 4.2 Estimation Results and Discussion

With a sample size per chain of 10,000, 10,000 iterations, and a thinning interval of 1, clearly, we observe that trace plots for all parameters exhibit good mixing, stable behaviour, and absence of any unique trends. Additionally, all chains indicate convergence to their stationary distributions. Table 3 below presents the results of the Bayesian panel kink estimation. The estimated posterior mean value of the kink

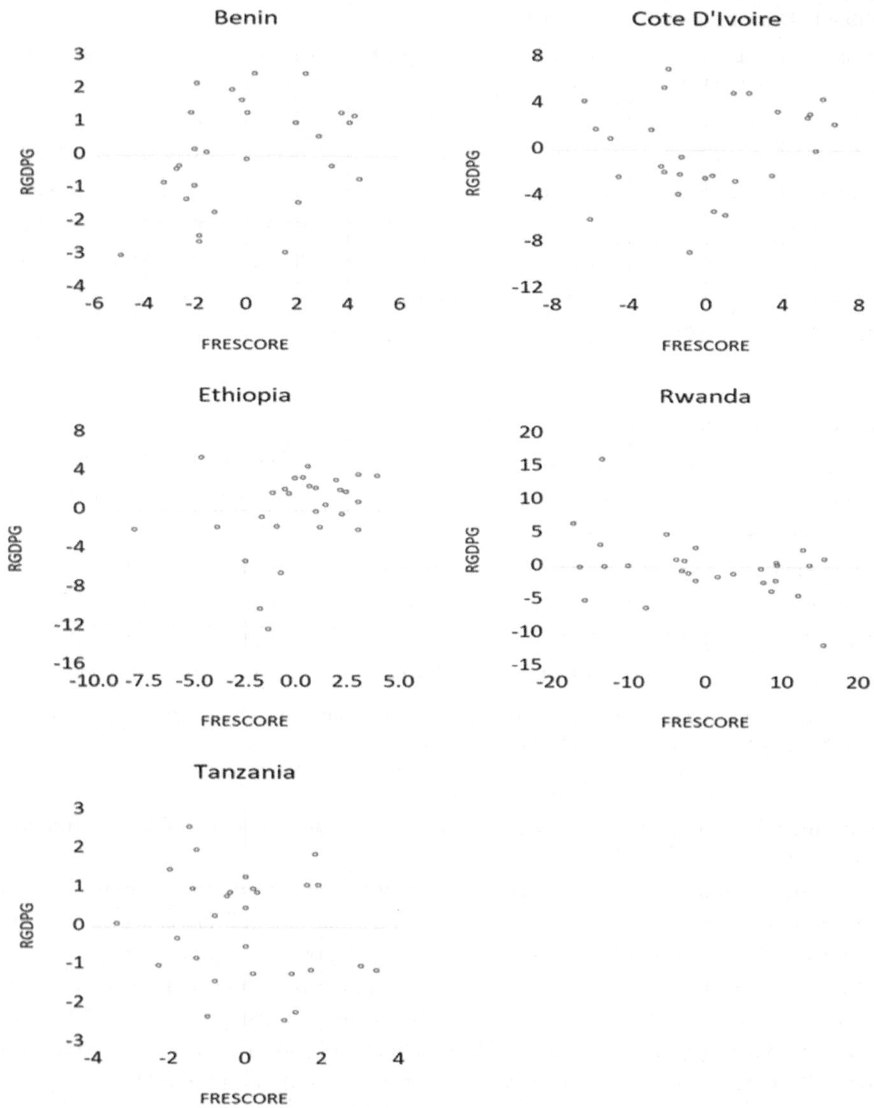


**Fig. 1** Scatter plots of Economic growth versus Economic freedom for individual Countries

parameter is 58.9195, with a 95% credible interval of (58.3894, 59.2149). The results show that, relative to the kink point, economic freedom has a non-linear impact on growth. With a 95% posterior probability, a 1 percentage point rise in the total economic freedom score leads to a 0.1424% point increase in economic growth in the lower region, ceteris paribus. However, beyond the threshold or the kink point, an increase in the freedom score activates adverse impacts on growth. Specifically,

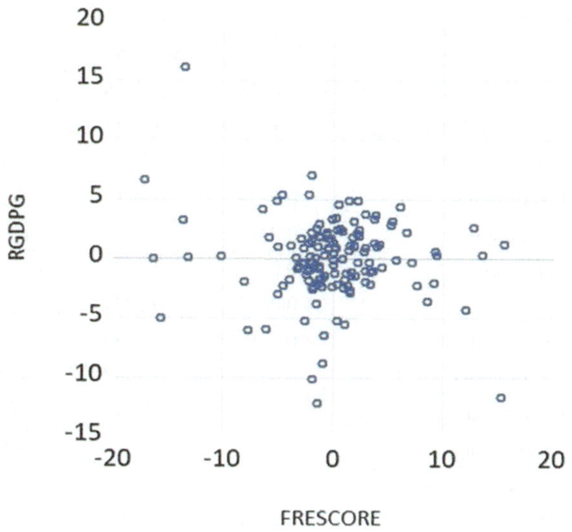


**Fig. 2** A scatter plot of Economic growth versus Economic freedom of stacked cross-sections of the countries

a 1% point increase in economic freedom score triggers a 0.3050% point decline in growth with a 95% posterior probability. This represents approximately 114.19% basis point change in absolute terms. Even after adequately accounting for country-specific heterogeneities, these estimates are still valid. This finding is diametrically contrary to the findings of earlier studies, some of which found either only a positive impact or a negative impact only. For example, [11, 20] provide empirical evidence for only positive impacts whilst [21, 22] provide evidence that some categories of economic freedom indicators have a negative impact on economic growth.

Evidently, the above contrast obtains since the estimation frameworks applied in all these previous studies excluded kink regression approaches. Our findings are also strongly at variance with several propositions of many mainstream growth theories that extol economic freedom as a core game changer in the pursuit of economic growth in less developed economies without setting baselines. Rather, our findings appear to lend some support to emerging views that hold that pursuing unrestrained economic freedom in Sub-Saharan Africa will be counterproductive and growth debilitating, given the general economic landscape and dynamics of the sub-region. Some authors emphasize that any forms of economic freedom that enlarge the private sector uncontrollably and reduce government size can hamper public investment in some critical areas (for example human and physical capital investments) of the macro economy [16–19].

The parameter estimates of the other explanatory variables have the a priori signs, apart from net inflows of capital as a percentage of GDP, which turns out be to negative, with a posterior mean of $-0.2842$. This indicates that a 1 percentage point rise in it will reduce real growth rate by 0.2842 with a 95% posterior probability, however with quite a wide credible interval. Fixed capital, proxied by the annual growth rate of gross fixed capital formation has posterior mean of 0.0818 at a 95% probability,

**Table 3** Estimated results from the Bayesian panel kink regression

| | Mean | Median | SD | Lower | Upper |
|---|---|---|---|---|---|
| Intercept | 2.8178 | 2.7735 | 6.2237 | −9.2899 | 15.1052 |
| Economic freedom$_{\mathbf{LR}}$ | 0.1424 | 0.1425 | 0.0938 | −0.0437 | 0.3289 |
| $^{UR}$Economic freedom | −0.3050 | −0.3015 | 0.2340 | −0.7631 | 0.1537 |
| Fixed capital | 0.0818 | 0.0817 | 0.0207 | 0.0418 | 0.1229 |
| Population | 0.2436 | 0.2442 | 0.1165 | 0.0111 | 0.4671 |
| Labor force | 0.0157 | 0.0168 | 0.0877 | −0.1599 | 0.1847 |
| FDI | −0.2842 | −0.2844 | 0.2216 | −0.7154 | 0.1522 |
| $\omega_2$: Cote d'Ivoire | −0.1489 | −0.1439 | 0.9272 | −1.9619 | 1.6717 |
| $\omega_3$: Ethiopia | 4.4771 | 4.4592 | 1.4280 | 1.7578 | 7.2778 |
| $\omega_4$: Rwanda | 3.8200 | 3.8318 | 1.3141 | 1.2223 | 6.3774 |
| $\omega_5$: Tanzania | 1.4001 | 1.3944 | 1.6161 | −1.7122 | 4.6563 |
| $\delta$ : Kink parameter | 58.9195 | 58.9484 | 0.2340 | 58.3894 | 59.2149 |
| sigma2 | 8.9089 | 8.8067 | 1.1356 | 6.9456 | 11.3522 |

$^{UR}$Positive part of the kink variable; $^{LR}$Negative part of the kink variable
Upper: 97.5% quantile simulated chain; Lower: 2.5% quantiles of each simulated chain.

whilst the real annual population growth rate and labor force participation rate for youths and adults between the ages of 15–64, respectively have posterior means of 0.2436 and 0.0157. These results imply that, all other things being equal, individually, a 1 percentage point increase in these variables will induce approximately 0.2436 and 0.0157 rise in real economic growth rate. Country-specific heterogeneity estimates are revealing, and they appear to be strongly in consonance with economic reports from various issues of Economic Outlook of leading international economic institutions like the World Bank, IMF, and AfDB. With respect to Benin, Ethiopia is more efficient, leading with a difference of 4. 4771, and followed by Rwanda, 3.82, and Tanzania, 1.4001. However, La Cote d'Ivoire is less efficient by a margin of 0.1489, with a posterior probability of 95%.

Table 4 above presents the estimated outputs from the Bayesian panel estimation without the kink point. Our key regressor, economic freedom, unsurprisingly, has a positive impact on growth. The results show that a 1 percentage point rise in total freedom score, all other things being equal, will lead to a rise in economic growth by 0.0443 percentage points with a posterior probability of 95%. This finding is in line with previous studies that found a positive impact. However, clearly, this outcome is partly inaccurate since our test for the existence of a kink effect has strongly failed to accept a linear relationship between economic growth and economic freedom.

**Table 4** The estimated results from the Bayesian panel regression with no kink

| | Mean | Median | SD | Lower | Upper |
|---|---|---|---|---|---|
| Intercept | −1.6737 | −1.6934 | 7.2672 | −15.7233 | 12.6807 |
| Economic freedom | 0.0443 | 0.0444 | 0.0550 | −0.0639 | 0.1516 |
| Fixed capital | 0.0846 | 0.0845 | 0.0205 | 0.0449 | 0.1258 |
| Population | 0.2135 | 0.2142 | 0.1142 | −0.0071 | 0.4405 |
| Labor force | 0.0386 | 0.0391 | 0.0852 | −0.1333 | 0.2048 |
| FDI | −0.2520 | −0.2504 | 0.2253 | −0.6922 | 0.1824 |
| $\omega_2$:Cote d'Ivoire | −0.1883 | −0.1898 | 0.9334 | −2.0121 | 1.6347 |
| $\omega_3$:Ethiopia | 3.6840 | 3.6931 | 1.3073 | 1.0779 | 6.2142 |
| $\omega_4$:Rwanda | 3.0962 | 3.1043 | 1.1871 | 0.7212 | 5.4409 |
| $\omega_5$:Tanzania | 1.1124 | 1.1167 | 1.5948 | −2.0449 | 4.1767 |
| sigma2 | 8.9727 | 8.8836 | 1.1257 | 6.9992 | 11.4065 |

Upper: 97.5% quantile; Lower: 2.5% quantiles

## 5 Conclusions

We explored the implications of the presence of a kink-point in economic freedom for economic growth within the context of a Bayesian panel kink regression modelling, and we further examined the dynamics of the kick point on economic growth. Our motivation derives from the current conflicting conjecture in the economic growth-economic freedom literature, fueled by the current remarkable showcasing of economic resilience by the selected Sub-Saharan African economies after the receding devastating impacts of covid-19 pandemic. Whereas some researchers found that economic freedom only has a positive impact on growth, some others found the direct converse. However, our findings have empirically showed that, first, economic growth has a significant and a substantial kink effect on economic growth. This result has conclusively indicated that there is a non-linear relationship between both variables. Using the proposed bootstrap procedures of [28], we found that the kink point of economic freedom occurs at the index score of approximately 59% with a very close credible interval of (58.3894, 59.2149) at a 95% posterior probability. In the lower region of the kink, economic freedom has a positive impact on economic growth. The upper region has the converse.

The above finding has quite a significant repercussion for policy designs and program initiatives that focus on economic freedom-economic growth within Sub-Saharan African countries. For concreteness, the key policy areas captured by [27] in its index of economic freedom, which include property rights, judicial effectiveness, government integrity, tax burden, government spending, fiscal health, business freedom, labor freedom, monetary freedom, trade freedom, investment freedom, and financial freedom require a cautious policy crafting and implementation that must consider the existence of kink-points or thresholds. It is important to remark that this is a novel finding in the economic growth-economic freedom literature. in addition, our findings have established that fixed capital, population, and labor force variables have positive impacts on economic growth, whereas, quite surprisingly, FDI has a negative impact, all with very modest credible intervals with 95% posterior probabilities.

There are numerous extension possibilities to our study that deserve future consideration. Further research is recommended to include all countries in Sub-Saharan Africa, conditional on data availability. Including more factors such the political environment proxies, resource rents, among others, is also recommended. We also suggest that further research can focus on multiple kink-point estimations.

## References

1. David, C., David, L., Zhuan, P., Andrea, W.: Nonlinear policy rules and the identification and estimation of causal effects in a generalized regression kind design". No 18564. In: NBER Working Papers, National Bureau of Economic Research, Inc. (2012)
2. Tong, H.: Threshold Models in Non-linear Time Series Analysis. Springer, Berlin (1983)
3. Hansen, B.E.: Regression kink with an unknown threshold. J. Bus. Econ. Stat. **35**(2), 228–240 (2017). https://doi.org/10.1080/07350015.2015.1073595
4. Zhang, Y., Zhou, Q., Jiang, L.: Panel kink regression with an unknown threshold. Econ. Lett. **157**, 116–121 (2017). https://doi.org/10.1016/j.econlet.2017.05.033
5. AfDB: African Economic Outlook (2023). www.afdb.org
6. Schumpeter Joseph A.: The Theory of Economic Development An Inquiry into Profits, Capital, Credit, Interest, and the Business Cycle (1934)
7. Solow, R.M.: Technical change and the aggregate production function. Rev. Econ. Stat. **39**(3), 312 (1957). https://doi.org/10.2307/1926047
8. Solow, R.M.: A contribution to the theory of economic growth. Q. J. Econ. **70**(1), 65 (1956). https://doi.org/10.2307/1884513
9. Swan, T.W.: Economic growth and capital accumulation. Econ. Record **32**(2), 334–361 (1956). https://doi.org/10.1111/j.1475-4932.1956.tb00434.x
10. Buchanan, M.J.M., Tullock, G.: The Calculus of Consent: Logical Foundations of Constitutional Democracy, vol. 3, pp. 21–34. University of Michigan Press (1962)
11. Krueger, A.: Why trade liberalisation is good for growth. Econ. J. **108**(450), 1513–1522 (1998). https://doi.org/10.1111/1468-0297.00358
12. Hall, J.C., Lawson, R.A.: Economic freedom of the world: an accounting of the literature. Contemp. Econ. Policy **32**(1), 1–19 (2014). https://doi.org/10.1111/coep.12010
13. Feenstra, C.R.: Advanced International Trade: Theory and Evidence, 2nd edn. Princeton University Press (2015)
14. Mitra, A., Sharma, C., Véganzonès-Varoudakis, M.-A.: Trade liberalization, technology transfer, and firms' productive performance: the case of Indian manufacturing. J. Asian Econ. **33**, 1–15 (2014). https://doi.org/10.1016/j.asieco.2014.04.001
15. Doucouliagos, C., Ulubasoglu, M.A.: Economic freedom and economic growth: does specification make a difference? Eur. J. Polit. Econ. **22**(1), 60–81 (2006). https://doi.org/10.1016/j.ejpoleco.2005.06.003
16. Zahonogo, P.: Trade and economic growth in developing countries: evidence from sub-Saharan Africa. J. African Trade **3**(1–2), 41 (2017). https://doi.org/10.1016/j.joat.2017.02.001
17. Bergh, A., Nilsson, T.: Do liberalization and globalization increase income inequality? Eur. J. Polit. Econ. **26**(4), 488–505 (2010). https://doi.org/10.1016/j.ejpoleco.2010.03.002
18. Carter, J.R.: An empirical note on economic freedom and income inequality. Public Choice **130**(1–2), 163–177 (2007). https://doi.org/10.1007/s11127-006-9078-0

19. Kneller, R., Bleaney, M.F., Gemmell, N.: Fiscal policy and growth: evidence from OECD countries. J. Public Econ. **74**(2), 171–190 (1999). https://doi.org/10.1016/S0047-2727(99)00022-5
20. Kacprzyk, A.: Economic freedom–growth nexus in European Union countries. Appl. Econ. Lett. **23**(7), 494–497 (2016). https://doi.org/10.1080/13504851.2015.1083076
21. Santiago, R., Fuinhas, J.A., Marques, A.C.: The impact of globalization and economic freedom on economic growth: the case of the Latin America and Caribbean countries. Econ. Chang. Restruct. **53**(1), 61–85 (2020). https://doi.org/10.1007/s10644-018-9239-4
22. Justesen, M.K.: The effect of economic freedom on growth revisited: new evidence on causality from a panel of countries 1970–1999. Eur. J. Polit. Econ. **24**(3), 642–660 (2008). https://doi.org/10.1016/j.ejpoleco.2008.06.003
23. Panahi, H., Assadzadeh, A., Refaei, R.: Economic freedom and economic growth in mena countries. Asian Econ. Financ. Rev. **4**, 105–116 (2014)
24. IMF (2023). https://www.imf.org/en/Data
25. World Bank: World Development Indicators (2023). https://databank.worldbank.org/source/world-development-indicators
26. The Heritage Foundation: Index of Economic Freedom (2023). https://www.heritage.org/index/pdf/2023/book/04_2023_IndexOfEconomicFreedom_PREFACE.pdf
27. Heritage Foundation: 04_2023_IndexOfEconomicFreedom_PREFACE, pp. 1–1 (2023)
28. Li, Y., Hu, Z., Liu, J., Deng, J.: A note on regression kink model. Commun. Stat. Theory Methods **51**(23), 8246–8263 (2022). https://doi.org/10.1080/03610926.2021.1890780
29. Martin, A.D., Quinn, K.M., Park, J.H.: MCMCpack: Markov chain Monte Carlo in R. J. Stat. Softw. Softw. **42**(9), 1–21 (2011)

# Household Socio-Economic Characteristics of NEETs in Thailand

**Supanika Leurcharusmee**

**Abstract** Due to the heterogeneous socio-economic backgrounds and multifaceted causes contributing to the phenomenon of NEETs (youths not in education, employment, or training), this study classifies youths into six groups by their status including in-school, employed, NEET due to job seeking, NEET due to family responsibility, NEET due to disability, and other NEETs. Subsequently, we analyze a comprehensive list of socio-economic factors associated with different groups of youths in Thailand. The empirical analysis employs the Generalized Maximum Entropy (GME) model for the multinomial choice to analyze Thailand's socio-economic survey data. Consistent with prior research, the findings reveal that youths living in poverty or lower consumption expenditure households are more inclined to leave formal education to work or become NEETs, except in the case of NEETs due to disability. This highlights the necessity to extend support beyond education and employment opportunities to address family responsibilities, particularly for youths in households with high dependency ratios or infants. Moreover, the study underscores the diversity of NEET situations among youths from other different socio-economic backgrounds. Notably, there is evidence linking NEET status due to job-seeking with youths from higher socio-economic backgrounds, suggesting potential labor market challenges. Additionally, youths with disabilities encounter barriers not only due to their condition but also in job-seeking. While common policy measures exist, tailored interventions are crucial to facilitate their reintegration into formal education or the labor market.

**Keywords** Youth · NEET · Poverty · Socio-economic class · Generalized maximum entropy

S. Leurcharusmee (✉)
Center of Excellence in Econometrics, Faculty of Economics, Chiang Mai University, Chiang Mai, Thailand
e-mail: supanika.l@cmu.ac.th



V. Kreinovich et al. (eds.), *Applications of Optimal Transport to Economics and Related Topics*, Studies in Systems, Decision and Control 556,
https://doi.org/10.1007/978-3-031-67770-0_29

# 1 Introduction

The transition from education to employment can pose a significant difficulty for certain segments of the youth population, resulting in a phenomenon commonly referred to as ***youths not in employment, education, or training or NEETs***. The definition of NEETs varies slightly across countries. According to the International Labour Organization (ILO), NEETs are individuals aged 15–24 who meet two criteria: (1) they are not employed (either unemployed or inactive) and (2) they have not received any education or training in the four weeks prior to a survey. In 2019, the global NEET rate was 22.8%, which means that more than one in five young people were NEETs [11]. The NEET situation worsened significantly during the COVID-19 pandemic, with the global NEET rate increasing to 24.9% in 2020, and then slightly decreasing to 23.8% and 23.5% in 2021 and 2022, respectively [13].

The presence of NEETs can potentially cause adverse effects on both the economy and the well-being of individuals. Therefore, many countries have made it a policy priority to tackle youth unemployment and decrease the prevalence of NEETs [15]. However, the heterogeneity of NEETs in terms of their socio-economic backgrounds and underlying reasons for disengagement presents a significant challenge for policymakers. To address the heterogeneity within the NEETs, this study first categorizes NEETs in Thailand into distinct subgroups. Subsequently, our analysis identifies individual and household socio-economic attributes that are linked to these various NEET subgroups in Thailand.

The situation of NEETs is oftentimes associated with youth exclusion. Bell and Blanchflower [1] discovered that youth unemployment rates are commonly higher than those of adults, and during economic downturns, young people often face a disproportionately higher level of unemployment. This may be due to their lower levels of specific human capital that is relevant to employment, as well as weaker job search skills. At the national level, NEETs are more prevalent in lower-middle and low-income countries, with NEET rates of 27.4% and 26.7%, respectively. Upper-middle and high-income countries experience lower NEET rates of 18.6% and 10.5%, respectively [13]. Furthermore, studies demonstrate a correlation between NEET status and low socio-economic characteristics such as living in a deprived neighborhood and having limited education [3, 14].

Regardless of the association with the low socio-economic background, the literature also acknowledges that NEETs are a diverse group [14]. Various studies have shown that different groups of NEETs have distinct needs and require specific policies to help them re-engage in education or employment. For instance, Ralston et al. [17] reviews that the NEET situation can be voluntary or involuntary. Some may require assistance to re-enter the labor market, while others, such as those with disabilities, may require different forms of support. Given this diversity, it is crucial to disaggregate the NEET population based on their characteristics. Failure to do so would only serve to stigmatize this population with a negative label [21].

To address the diversity among NEETs, the ILO [11] categorizes NEETs into two main groups: unemployed NEETs, who are seeking or available for work, and inactive

NEETs, who are out of the labor force. Many academic works adopt the classification framework proposed by the ILO, which distinguishes NEETs as either unemployed or inactive [11]. Eurofound [5] takes this framework further by dividing NEETs into seven subgroups based on their unemployment or inactivity status, which are re-entrants, short-term unemployed, long-term unemployed, those affected by illness or disability, those with family responsibilities, discouraged workers, and other NEETs.

For Thailand, the NEET rate in Thailand was 14.9% [13]. Similar to other countries, the COVID-19 pandemic exacerbated the situation. According to a report, the NEET rate in Thailand increased by 7%. Although the number of inactive NEETs decreased by 4.6%, the rise in the NEET rate was primarily due to a 26.2% increase in youth unemployment [12]. Despite the fact that the NEET rate in Thailand is lower than the global average, it is a cause for concern in light of the country's low fertility rate and aging population. This demographic shift presents a challenge in the labor market, as the senior population is growing while the youth population is declining [20]. Therefore, it is critical to prioritize the development of human capital among young people in Thailand.

For the analysis of the heterogeneity of NEETs in Thailand, most studies rely on the labor force survey (LFS), which is the primary national household survey that reports labor market conditions in the country [11, 19]. The LFS allows researchers to classify NEETs based on their labor force status and reasons for being inactive, as shown in previous studies (see [11]). However, the LFS has some limitations in terms of the socio-economic variables it includes, such as income data only being collected from employees and not from business owners or self-employed individuals. Additionally, the LFS does not collect data on household consumption or total income, thereby constraining its potential to examine crucial socio-economic factors related to the households of NEETs, such as their poverty status and socio-economic class. Therefore, to complement the LFS data and perform a socio-economic classification of NEETs in Thailand, this study will use the Socio-economic Survey (SES), which is the national household survey that includes comprehensive socio-economic variables. Specifically, this study compiles a wide range of individual and household socio-economic characteristics to identify the factors that are most strongly linked with the NEET status in Thailand.

As this study includes a large number of socio-economic variables that could influence NEET status and classification, many of the variables are correlated and the estimation faces a severe problem of multicollinearity. With high multicollinearity, estimates of traditional econometric methods, such as least squared estimators, would have high variance and unstable [6]. An alternative method that is known to be robust to the multicollinearity problem is the maximum entropy (ME) method, introduced by Jaynes [9, 10], which is based on the principle that the model that is most consistent with the data is the one with the highest level of entropy, subject to constraints imposed by the available information. Golan, Judge and Miller [8] extended the ME method to the Generalized Maximum Entropy (GME) by adding the disturbance terms allowing the GME estimator to perform better for finite samples compared to the maximum likelihood and ME estimators [4]. In addition to the robustness against the issues of multicollinearity and small sample sizes, the GME method is semi-parametric, which means that the model requires no distributional assumption, making it even more robust against model misspecification. For the NEET classification problem, this study adopts the GME estimation method for the multinomial choice model developed by Golan, Judge and Perloff [7].

Given the limited knowledge about the diversity of NEETs, the main contribution of this study is to leverage the extensive socio-economic information of the SES data and the robustness of the GME method to classify NEETs into subgroups and identify key socio-economic factors that are linked to each NEET subgroups in Thailand.

## 2 Data

This study uses data from the 2019 Socio-economic Survey (SES) of Thailand, which is a comprehensive national household survey conducted by the National Statistical Office (NSO). The data encompasses a total of 124,874 individuals. However, this study only focuses on a sample of 11,740 young people aged 15–24 for analysis.[1]

## 3 Methodology

### 3.1 Youth Classification

Following ILO [11], youths are classified into three main groups, which are in-school, employed and NEETs. The in-school group includes all youths that are in the formal education system regardless of their employment status. The employed group includes all youths not in the formal education system and work in the labor market. Youths not in-school and not employed are considered NEETs.

As this study focuses to examine heterogeneity among NEETs, we decompose the NEET group further into four subgroups based on their unemployment status and reasons for inactivity.[2] The first subgroup of NEETs is the NEET-seeking group, which comprises those who are actively seeking employment.[3] The remaining three

[1] The 2019 SES data comprises of 11,801 individuals aged between 15 and 24. However, 14 observations were excluded because they resided in the households as housemaids and, thus, the household characteristics did not accurately reflect their socio-economic status. Additionally, 47 observations were excluded due to inconsistencies in their reported educational status. As a result, the analysis in this study was conducted on a sample of 11,740 youths.

[2] Due to the restrictions of the SES data, this study is unable to categorize NEETs into the seven categories outlined by Eurofound [5] and can only classify NEETs into four categories.

[3] As per the definition of unemployment, which encompasses both (1) individuals actively seeking employment, and (2) those not actively seeking employment but willing and able to work if an opportunity arises, the seeking group represents only a subset of the overall unemployed NEET population, and not the entire group.

subgroups include youths who are inactive due to family responsibilities (NEET-family), illness or disability (NEET-disability), and other unspecified reasons (NEET-others), respectively. That is, in this study, Thai youths are classified into six observed groups including (1) in-school, (2) employed, (3) NEET-seeking, (4) NEET-family, (5) NEET-disability, and (6) NEET-others.

### *3.2 Generalized Maximum Entropy (GME) Estimation for Youth Status*

In order to determine the individual and household socio-economic factors that are associated with the NEET status and classification, this study employs Golan, Judge and Perloff [8]'s Generalized Maximum Entropy (GME) estimation method for the multinomial choice model. For the NEET classification problem, let $y_{ij}$ be a dummy variable, which takes the value 1 if individual $i$ is in group $j$, and 0 otherwise. In this study, there are six possible groups including (1) in-school, (2) employed, (3) NEET-seeking, (4) NEET-family, (5) NEET-disability, and (6) NEET-others. The model is $y_{ij} = p_{ij} + e_{ij}$, where $p_{ij} = \Pr\{y_{ij} = 1 \mid x_i\}$ is the probability that individual $i$ is in group $j$ given a set of characteristics $x_i$, and $e_{ij}$ is the noise component. As the entropy modelling requires that $p_{ij}$ and $e_{ij}$ be in a probability form, $e_{ij}$ is transformed to $e_{ij} = \sum_h v_{ijh} w_{ijh}$, where $v_{ij} = (v_{ij1}, v_{ij2}, \ldots, v_{ijH})$ is the finite possible realization or the support of $e_{ij}$ and $w_{ij} = (w_{ij1}, w_{ij2}, \ldots, w_{ijH})$ is the corresponding unknown weights of $v_{ij}$. The GME problem is to maximize the entropy $H(p, w)$, subject to the information-moment conditions and the normalization constraints as follows.

$$\max_{p,w} H(p, w) = -\sum_{ij} p_{ij} \ln p_{ij} - \sum_{ijh} w_{ijh} \ln w_{ijh} \tag{1}$$

subject to

$$\sum_i y_{ij} x_{ik} = \sum_i p_{ij} x_{ik} + \sum_i e_{ij} x_{ik} \tag{2}$$

$$\sum_j p_{ij} = 1 \tag{3}$$

$$\sum_h w_{ijh} = 1. \tag{4}$$

As the GME problem is a constrained optimization problem, it can be solved using the Lagrange method as follows.

$$\max_{p,w,\lambda,\mu,\rho} L = \left( -\sum_{ij} p_{ij} \ln p_{ij} - \sum_{ijh} w_{ijh} \ln w_{ijh} \right) + \sum_{jk} \lambda_{jk} \left( \sum_i y_{ij} x_{ik} - \sum_i p_{ij} x_{ik} - \sum_i e_{ij} x_{ik} \right) + \sum_i \mu_i \left( 1 - \sum_j p_{ij} \right) + \sum_{ij} \rho_{ij} \left( 1 - \sum_h w_{ijh} \right). \quad (5)$$

Solving the Lagrange problem with $\hat{\beta}_{jk} = -\hat{\lambda}_{jk}$, the solutions are $\hat{p}_{ij} = \frac{\exp\left(x_i'\hat{\beta}_j\right)}{\sum_j \exp\left(x_i'\hat{\beta}_j\right)}$ and $\widehat{w}_{ijm} = \frac{\exp\left(x_i'\hat{\beta}_j v_j\right)}{\sum_j \exp\left(x_i'\hat{\beta}_j v_j\right)}$.

For the information-moment conditions, $x_{ik}$ is the standardized value of socio-economic factor $k$ of individual $i$.[4] This study includes 72 individual and household socio-economic variables, focusing on the poverty situation and socio-economic classes.

For the poverty situation of each individual, this study follows the standard poverty definition, which is an individual living in a household with consumption under the poverty line[5] [16]. In addition to poverty status, this study also includes dummy variables for consumption quintiles to examine the distribution of youths across different consumption quintiles within both the NEET and non-NEET groups. The descriptions and mean statistics of the dummy variables for poverty status and consumption quintiles are presented in Table 1.

For socio-economic classes, the SES data include a comprehensive socio-economic class variable, which is constructed based on the source of income and occupational status of the household head, who is the member with the highest income. Although the SES data offer a detailed breakdown of 35 different socio-economic classes, this study consolidated into 5 groups of 15 broader classes. The five groups depending on the household head's occupation status include (1) economically inactive, (2) farm operators, (3) fishery, forestry, hunting, agricultural services, (4) entrepreneurs for non-agricultural business, and (5) employees with the detailed class as shown in Table 2.

Other socio-economic factors included in this study comprise individual demographic including disability conditions, household size and dependency situation, education and occupation of household members, household different types of income, assets and debt, as well as regions.

---

[4] For the result comparison purpose, all $x_{ik} = \frac{X_{ik} - \bar{X}_{ik}}{SD(X_{ik})}$ are the standardized value of the corresponding explanatory variable $X_{ik}$.

[5] To calculate the poverty dummy variable, this study compares each household's per-capita consumption with the regional poverty line reported by NESDC (2019). All youths residing in a household that falls below the poverty line is considered to be in a state of poverty.

**Table 1** Descriptions and mean statistics of the poverty status and consumption quintiles

| Variables | Description | Mean | | |
|---|---|---|---|---|
| | | NEET | Non-NEET | All youths |
| *Poverty status* | | | | |
| Poverty | Poverty status | 0.206 | 0.098 | 0.111 |
| *Quintile of per-capita household consumption* | | | | |
| qconsumpt1 | Consumption quintile 1 | 0.423 | 0.262 | 0.282 |
| qconsumpt2 | Consumption quintile 2 | 0.255 | 0.248 | 0.248 |
| qconsumpt3 | Consumption quintile 3 | 0.167 | 0.210 | 0.205 |
| qconsumpt4 | Consumption quintile 4 | 0.099 | 0.172 | 0.163 |
| qconsumpt5 | Consumption quintile 5 | 0.057 | 0.108 | 0.102 |

*Source* Calculated from the 2019 SES (without population weighting)

**Table 2** Dummy variables for socio-economic class of NEETs and non-NEETs

| Variables | Description | Mean | | |
|---|---|---|---|---|
| | | NEET | Non-NEET | All youths |
| *Economically inactive* | | | | |
| socioclass1 | Receiving assistance or pensions | 0.162 | 0.158 | 0.158 |
| socioclass2 | Property income | 0.007 | 0.002 | 0.003 |
| socioclass3 | Other non-labor income | 0.008 | 0.007 | 0.007 |
| *Farm operator (Including marine culture)* | | | | |
| socioclass4 | Own the majority of the land (<20 Rai) | 0.047 | 0.067 | 0.065 |
| socioclass5 | Own the majority of the land (>=20 Rai) | 0.031 | 0.045 | 0.044 |
| socioclass6 | Rent the majority of the land (<20 Rai) | 0.011 | 0.011 | 0.011 |
| socioclass7 | Rent the majority of the land (>=20 Rai) | 0.008 | 0.013 | 0.012 |
| *Fishery, forestry, hunting, agricultural services* | | | | |
| socioclass8 | Fishery, forestry, hunting, agricultural services | 0.027 | 0.018 | 0.019 |
| *Entrepreneurs for non-agricultural business* | | | | |
| socioclass9 | Entrepreneurs with managerial skill | 0.011 | 0.015 | 0.014 |
| socioclass10 | Entrepreneurs with technical skill | 0.003 | 0.005 | 0.005 |
| socioclass11 | Other entrepreneurs | 0.154 | 0.185 | 0.181 |
| *Employees* | | | | |
| socioclass12 | Executives and managers | 0.021 | 0.020 | 0.020 |
| socioclass13 | Professionals | 0.032 | 0.035 | 0.035 |
| socioclass14 | Technicians | 0.028 | 0.024 | 0.024 |
| socioclass15 | Other employees | 0.449 | 0.395 | 0.402 |

*Source* Calculated from the 2019 SES (without population weighting)

To determine which factors are strongly associated with each of the youth statuses, this study uses the Bayes factor upper bound (BFUB), which is the upper bound of the odds in favor of the alternative hypothesis ($H_A : \beta_{jk} \neq 0$) relative to the null hypothesis ($H_0 : \beta_{jk} = 0$), instead of the traditional p-value.[6] In contrast to p-value, the BFUB provides evidence for both the null and alternative hypothesis by quantifying the strength of evidence in favor of one hypothesis relative to another. Bayes factor greater than 1 indicates that there is some evidence in favor of the alternative hypothesis over the null hypothesis. In the case of this study, the Bayes factor of $\hat{\beta}_{jk}$ being higher than 1 indicates that there is an evidence that socio-economic factor $k$ is associated with the youth status $j$. The higher Bayes factor indicates a stronger evidence in favor of the alternative hypothesis.

## 4 Results

As the objectives of this study are to (1) classify youth into NEET and Non-NEET subgroups and (2) identify key socio-economic factors that are linked to each youth status in Thailand, the results are shown in two sections. Section 4.1 illustrates youth status in Thailand and Sect. 4.2 analyzes socio-economic factors associated with different youth statuses.

### *4.1 Youth Status in Thailand*

From the 2019 SES sample, Thai youths can be classified into six observed groups including (1) in-school, (2) employed, (3) NEET-seeking, (4) NEET-family, (5) NEET-disability, and (6) NEET-others as shown in Table 1. Overall, 50.4% of all youths were in school, 37.4% were employed and 12.2% were NEETs. Among the NEETs, family responsibility is the leading reason for being NEETs, which was accounted for 5.6% of all youths. Other reasons were currently seeking for a job and inactive due to illness or disability, which were accounted for 2.5% and 1.3%, respectively. In addition, there were also 2.9% who reported to be NEETs for other unspecified reasons.

The composition of youths across education, employment and NEET differs across age groups. Specifically, for youths age 15–19 years old, 76.1% were in school, 15.6% were employed, and only 8.3% were NEETs. For youths aged 20–25 years old, only 19.3% were in school, 63.8% were employed, and 17.0% were NEETs. For both age groups, family responsibility was the main reason for being NEETs following by seeking for jobs and disability (Table 3).

[6] See Benjamin and Berger [2] for more explanation on the Bayes factor upper bound (BFUB).

**Table 3** Classification of youths in Thailand

| | Age 15–19 | | Age 20–24 | | All | |
|---|---|---|---|---|---|---|
| | Freq | % | Freq | % | Freq | % |
| In-school | 4,892 | 76.1% | 1,022 | 19.3% | 5,914 | 50.4% |
| Employed | 1,006 | 15.6% | 3,386 | 63.8% | 4,392 | 37.4% |
| NEET-Seeking | 89 | 1.4% | 203 | 3.8% | 292 | 2.5% |
| NEET-Family | 197 | 3.1% | 456 | 8.6% | 653 | 5.6% |
| NEET-Disability | 58 | 0.9% | 90 | 1.7% | 148 | 1.3% |
| NEET-Others | 190 | 3.0% | 151 | 2.8% | 341 | 2.9% |
| Total | 6,432 | 100% | 5,308 | 100% | 11,740 | 100% |

*Source* Calculated from the 2019 SES (without population weighting)

## *4.2 Socio-Economic Factors Associated with Youth Status*

For the GME results, the poverty and socio-economic class associated with different youth statuses are discussed in Sect. 4.2.1. Other socio-economic factors associated with different youth statuses are discussed in Sect. 4.2.2. It should be noted that the GME estimation uses the in-school group as the base group, as education should be considered the default for youths. Therefore, comparisons of GME coefficients shown in Table 4 should be made with the in-school group.

### 4.2.1 Poverty and Socio-Economic Class

When considering the factors with high BFUBs,[7] the poverty status is strongly associated with being employed and being NEETs due to job seeking, family responsibility and other reasons that is not disability. This suggests that youths living in poverty are more likely to leave the formal education system to either work or become NEETs.

In order to explore the correlation between youths and their economic status in a more comprehensive manner, beyond solely poverty classification, this study incorporates per-capita household consumption levels into the regression analysis. The findings illustrate that, not only are youths residing in poverty more likely to be in employment or NEETs, but those living in households with lower consumption quintiles or near-poor households also encounter similar challenges. Specifically, the empirical evidence indicates that youths residing in households belonging to quintiles 1–4 of household consumption exhibit a higher likelihood of being employed in comparison to those in quintile 5 households. Additionally, youths residing in quintiles 1 and 2 households also have a higher likelihood to be NEETs due to job seeking, family responsibility and other reasons.

[7] There is no specific cut-off point for a strong evidence. However, for a reference point, the p-value of 0.01 corresponds to a BFUB of 8.13 [2].

**Table 4** Effects of poverty and socio-economic class on youth status from the GME multinomial choice model

| | (1) Employed | (2) NEET-Seeking | (3) NEET-Family | (4) NEET-Disability | (5) NEET-Others |
|---|---|---|---|---|---|
| *Poverty status* | | | | | |
| Poverty | 0.188<br>(999.0) | 0.184<br>(3.7) | 0.183<br>(59.6) | 0.187<br>(1.1) | 0.206<br>(36.4) |
| *Quintile of per-capita household consumption (Base: consumption_q5)* | | | | | |
| qconsumpt1 | 0.795<br>(999.0) | 0.671<br>(228.1) | 0.626<br>(198.6) | 0.782<br>(1.5) | 0.652<br>(165.3) |
| qconsumpt2 | 0.624<br>(999.0) | 0.627<br>(951.2) | 0.408<br>(9.6) | 0.910<br>(2.7) | 0.529<br>(64.2) |
| qconsumpt3 | 0.402<br>(999.0) | 0.355<br>(11.3) | 0.166<br>(1.1) | 0.785<br>(2.3) | 0.311<br>(4.0) |
| qconsumpt4 | 0.202<br>(487.5) | 0.129<br>(1.1) | 0.054<br>(1.4) | 0.627<br>(1.8) | 0.177<br>(1.4) |
| *Socio-economic class (Base: socioclass15)* | | | | | |
| socioclass1 | −0.653<br>(999.0) | −0.018<br>(3.3) | −0.155<br>(1.6) | −0.328<br>(1.3) | −0.090<br>(1.0) |
| socioclass2 | 0.016<br>(1.3) | 0.115<br>(20.7) | 0.087<br>(1.6) | 0.061<br>(1.1) | 0.066<br>(1.4) |
| socioclass3 | −0.046<br>(1.2) | −0.239<br>(1.2) | 0.098<br>(1.5) | 0.076<br>(1.1) | 0.042<br>(1.2) |
| socioclass4 | −0.180<br>(999.0) | −0.104<br>(1.0) | −0.105<br>(1.6) | 0.240<br>(1.9) | −0.232<br>(6.9) |
| socioclass5 | −0.108<br>(37.6) | −0.095<br>(1.0) | 0.064<br>(1.0) | −0.260<br>(1.0) | −0.064<br>(1.0) |
| socioclass6 | −0.057<br>(1.8) | −0.382<br>(1.0) | 0.019<br>(1.5) | 0.122<br>(1.0) | −0.082<br>(1.0) |
| socioclass7 | −0.131<br>(999.0) | −0.405<br>(1.0) | 0.009<br>(2.8) | −0.240<br>(1.1) | −0.466<br>(1.0) |
| socioclass8 | −0.050<br>(1.5) | 0.041<br>(1.2) | 0.113<br>(7.9) | 0.035<br>(1.9) | −0.083<br>(1.0) |
| socioclass9 | −0.102<br>(28.7) | −0.089<br>(1.0) | −0.005<br>(6.0) | 0.022<br>(3.5) | -0.050<br>(1.1) |
| socioclass10 | −0.025<br>(1.0) | −0.006<br>(6.3) | −0.039<br>(1.1) | −0.163<br>(1.4) | 0.013<br>(2.2) |
| socioclass11 | −0.368<br>(999.0) | 0.009<br>(5.3) | −0.154<br>(2.4) | −0.102<br>(1.3) | −0.106<br>(1.1) |
| socioclass12 | −0.100<br>(20.7) | 0.002<br>(14.0) | −0.086<br>(1.0) | 0.144<br>(1.1) | 0.142<br>(20.0) |
| socioclass13 | −0.184<br>(999.0) | 0.012<br>(2.5) | −0.050<br>(1.1) | −0.179<br>(1.0) | 0.053<br>(1.0) |
| socioclass14 | −0.071<br>(4.8) | 0.008<br>(3.1) | 0.025<br>(1.5) | −0.283<br>(1.3) | 0.131<br>(8.6) |

*Source* Calculated from the 2019 SES (without population weighting)

*Note* (1) The Bayes factor upper bound (BFUB) in parenthesis (The BFUBs that exceed 999 are reported as 999). (2) The GME estimation uses the in-school group as the base group. (3) Other explanatory variables included in this model comprise individual demographic including disability conditions, household size and dependency situation, education and occupation of household members, household different types of income, assets and debt, as well as regions

Regarding the NEETs due to illness or disability group, the results show less association with poverty and household consumption levels. Despite the fact that household consumption levels typically serve as an indicator of a household's overall well-being, this may not hold true for households with members who have illnesses or disabilities, as they may incur greater healthcare expenditures. As a result, the lack of association between poverty and consumption levels in this group of NEETs may not indicate that they are more financially secure than other groups of NEETs.

For the socio-economic class analysis, this study examines whether youths residing in households belonging to various socio-economic classes exhibit varying probabilities of leaving the formal education system to either work or become NEETs. When compared to those living with a household head working as a low to medium skilled employee (socioclass15),[8] the results of the GME analysis reveal that individuals in socio-economic classes 1, 4, 5, 6, 7, 9, 11, 12, 13, and 14 are less likely to work. As no socio-economic class exhibit a higher likelihood of working than the socio-economic class 15, youths living in the households with household heads being low to medium skilled employees are most likely to leave the formal education system to work.

In order to explore the correlation between youths and their economic status in a more comprehensive manner, beyond solely poverty classification, this study incorporates per-capita household consumption levels into the regression analysis. The findings illustrate that, not only are youths residing in poverty more likely to be in employment or NEETs, but those living in households with lower consumption quintiles or near-poor households also encounter similar challenges. Specifically, the empirical evidence indicates that youths residing in households belonging to quintiles 1–4 of household consumption exhibit a higher likelihood of being employed in comparison to those in quintile 5 households. Additionally, youths residing in quintiles 1 and 2 households also have a higher likelihood to be NEETs due to job seeking, family responsibility and other reasons.

Regarding the NEETs due to illness or disability group, the results show less association with poverty and household consumption levels. Despite the fact that household consumption levels typically serve as an indicator of a household's overall well-being, this may not hold true for households with members who have illnesses or disabilities, as they may incur greater healthcare expenditures. As a result, the lack of association between poverty and consumption levels in this group of NEETs may not indicate that they are more financially secure than other groups of NEETs.

For the socio-economic class analysis, this study examines whether youths residing in households belonging to various socio-economic classes exhibit varying probabilities of leaving the formal education system to either work or become NEETs. When compared to those living with a household head working as a low to medium skilled employee (socioclass15) , the results of the GME analysis reveal that individuals in socio-economic classes 1, 4, 5, 6, 7, 9, 11, 12, 13, and 14 are less likely to work. As no socio-economic class exhibit a higher likelihood of working than

[8] socioclass15 is selected as the base group for the analysis because it is the largest group (40.2% of the sample).

the socio-economic class 15, youths living in the households with household heads being low to medium skilled employees are most likely to leave the formal education system to work.

Regarding NEET status, different socio-economic classes are associated with different types of NEETs. The socio-economic class most highly associated with being NEETs due to job-seeking is class 2, which comprises households with property income as their primary income source. This suggests that youths residing in households with substantial assets and passive income may be able to afford more time to look for jobs and, thus, more likely to be NEETs. For NEETs with family responsibilities, the class that shows the strongest association is class 8, which includes youths living in households with heads working in fishery, forestry, hunting, and agricultural services. Classes 1, 4, and 11, which include those living with heads of households receiving primary income from assistance or pensions, farm operators cultivating less than 20 Rai of their own land, and lower-skilled entrepreneurs, respectively, are less likely to be NEETs due to family responsibilities. For NEETs due to disability, the class that shows the strongest association is class 4, which comprises farm operators cultivating less than 20 Rai of their own land. Classes 12 and 14, which include youths living with employees in management positions and technicians, are more likely to be NEETs for reasons other than job seeking, family responsibility, and disability.

### 4.2.2 Other Factors Associated with Youth Status

For youths' individual characteristics associated with youth status, the results show that female are less likely to leave education to work or to be NEETs due to job seeking, disability or other reasons. However, female youths are more likely to be NEETs due to family responsibility. Age has a positive correlation of being employed and NEETs of all types. Furthermore, young people living in urban areas are less likely to be employed, but their likelihood of being NEETs does not differ from that of their rural counterparts. Notably, young people with both physical and mental disabilities, in addition to being NEETs due to disability, are also more likely to be NEETs due to job seeking.

To consider household structure factors, household size is negatively correlated to being employed and NEETs due to family responsibility. Married, divorced, and widow youths are more likely to be employed and NEETs compared to never-married youths. With regard to dependency, youths residing in households with higher ratios of infants (aged 3 and under), children (aged 4–15), and seniors (aged 65 and over) are more likely to work and be NEETs due to family responsibilities, with the coefficients being highest for those with infants.[9] Moreover, young people living in households with higher ratios of children and seniors are less likely to be NEETs due to job

[9] The dependency ratios considered in this study include the ratio of infants (aged 3 and under), children (aged 4–15), and seniors (aged 65 and over), which are calculated from the numbers of infants, children, and seniors in the households divided by household size.

seeking. Notably, there is no evidence that the presence of individuals with disabilities in the household affects the likelihood of being employed or NEETs. Furthermore, the education level of household heads is also a significant factor. Youths living with household heads who have attained at least a high school degree are less likely to be employed or NEETs.

Regarding households' financial status, youths living in households in the lower income quintiles are less likely to be employed and more likely to be NEETs due to family responsibility. Levels of current and non-current incomes does not affect the youth status. The associations with income have to be carefully interpreted due to the causality issue. In particular, while household income affects youth status, youth status also affects household income. For assets, youths in households with higher values of real estates are less likely to work and less likely to be NEETs due to family responsibility. However, youths in households with a higher value of vehicles are more likely to be NEETs due to family responsibility. The sizes of financial asset and debt (both formal and informal debts) have no association with employment or NEET status.

## 5 Conclusion

This study classifies youths in Thailand into six groups including (1) in-school, (2) employed, (3) NEET-seeking, (4) NEET-family, (5) NEET-disability, and (6) NEET-others. Taking advantage of the comprehensive list of socio-economic variables in Thailand's socio-economic survey (SES), this study analyzes a comprehensive list of individual and household socio-economic factors associated with education, employment, and NEET statuses, along with distinct NEET subgroups. To estimate the model, the study employs a multicollinear-robust method, Generalized Maximum Entropy (GME) for multinomial choice model developed by Golan, Judge and Perloff [7].

The findings indicate that youths living in poverty are more likely to leave formal education to work or become NEETs of all groups except NEETs due to disability, which shows no association with poverty and levels of per-capita household consumption. In addition, youths from households in different socio-economic classes are linked to different types of NEETs. Specifically, NEETs due to family responsibilities are linked to youths in lower socio-economic households with heads employed in fishery, forestry, hunting, and agricultural services, and disabled NEETs are associated with youths living with farm operators who cultivate less than 20 Rai of land. Job-seeking NEETs, on the other hand, are strongly linked to youths in higher socio-economic households where property income is the main source of income. Other NEETs are also associated with youths living with household heads employed in management and technical positions.

The empirical evidence that the NEET status is associated with low socio-economic characteristics in general are consistent with previous studies [3, 14].

Simultaneously, this study reaffirms the diverse socio-economic backgrounds of various NEET subgroups in Thailand. While common needs exist, tailored interventions are essential to facilitate their reintegration into formal education or the labor market.

In particular, for youths from lower-income households, the study underscores the importance of offering assistance that extends beyond education and employment opportunities. It emphasizes the significance of addressing family responsibilities, as they are linked to NEET status due to familial obligations. This need is particularly pronounced for youths residing in households characterized by a high dependency ratio or those with infants aged three years and younger. For youths with physical and mental disabilities, the findings show that they are not only more likely to be NEETs due to their disability but also due to job seeking, indicating a degree of exclusion. Among youths from a higher socio-economic background, there is some evidence linking to the NEET due to job seeking status, suggesting potential challenges in the labor market.

A main limitation of this study is the inability to capture a more comprehensive picture of NEETs due to data limitation. While the largest subgroup comprises NEETs due to family responsibility, the second-largest subgroup falls under the category of NEETs due to other reasons. According to the classification by Eurofound [5], it is likely that this other NEETs would encompass long-term unemployed individuals and discouraged workers. Additionally, it is important to note that Thailand's employment definition includes those with part-time jobs, which may not always imply a decent employment with human capital development and longer-term prospect. Given the limited evidence concerning the NEETs due to job-seeking status, it becomes evident that labor market challenges affect youths from diverse backgrounds. Consequently, there is a pressing need for more comprehensive studies that encompass NEETs with different unemployment statuses and reasons. Such research should also delve into the specific assistance and support mechanisms required to address the unique needs of these various NEET subgroups.

## References


1. Bell, D.N., Blanchflower, D.G.: Young people and the great recession. Oxford Rev. Econ. Policy **27**(2), 241–267 (2011)
2. Benjamin, D.J., Berger, J.O.: Three recommendations for improving the use of p-values. Amer. Stat. **73**(sup1), 186–191 (2019)
3. Cieslik, K., Barford, A., Vira, B.: Young people not in employment, education or training (NEET) in Sub-Saharan Africa: Sustainable development target 8.6 missed and reset. J. Youth Stud. **25**(8), 1126–1147 (2022)
4. Corral, P., Terbish, M.: Generalized maximum entropy estimation of discrete choice models. Stata J. **15**(2), 512–522 (2015)
5. Eurofound: Exploring the Diversity of NEETs. European Union, Luxembourg (2016)
6. Fraser, I.: An application of maximum entropy estimation: the demand for meat in the United Kingdom. Appl. Econ. **32**(1), 45–59 (2000)
7. Golan, A., Judge, G., Miller, D.: Maximum Entropy Econometrics: Robust Estimation with Limited Data. Wiley, New York (1996)

8. Golan, A., Judge, G., Perloff, J.M.: A maximum entropy approach to recovering information from multinomial response data. J. Amer. Stat. Assoc. **91**(434), 841–853 (1996)
9. Jaynes, E.T.: Information theory and statistical mechanics. Phys. Rev. **106**(4), 620 (1957)
10. Jaynes, E.T.: Information theory and statistical mechanics II. Phys. Rev. **108**(2), 171 (1957)
11. ILO: Global Employment Trends for Youth 2022. International Labour Organization, Geneva (2022)
12. ILO: Youth Not in Employment, Education or Training in Asia and the Pacific: Trends and Policy Considerations. International Labour Organization, Geneva (2022)
13. ILOSTAT: Indicator description: share of youth not in employment, education or training (youth NEET rate) (2023)
14. Lőrinc, M., Ryan, L., D'Angelo, A., Kaye, N.: De-individualising the 'NEET problem': an ecological systems analysis. Eur. Educ. Res. J. **19**(5), 412–427 (2020)
15. Mawn, L., Oliver, E.J., Akhter, N., Bambra, C.L., Torgerson, C., Bridle, C., Stain, H.J.: Are we failing young people not in employment, education or training (NEETs)? A systematic review and meta-analysis of re-engagement interventions. Syst. Rev. **6**, 1–17 (2017)
16. NESDC: Poverty and Inequality Report. National Economic and Social Development Council, Bangkok (2019)
17. Ralston, K., Everington, D., Feng, Z., Dibben, C.: Economic inactivity, not in employment, education or training (NEET) and scarring: the importance of NEET as a marker of long-term disadvantage. Work Employ. Soc. **36**(1), 59–79 (2022)
18. Tamesberger, D., Bacher, J.: NEET youth in Austria: a typology including socio-demography, labour market behaviour and permanence. J. Youth Stud. **17**(9), 1239–1259 (2014)
19. TDRI: Youth Employability Scoping Study. UNICEF, Bangkok (2020). https://www.unicef.org/thailand/media/4771/file/Youth%20Employability%20Scoping%20Study.pdf
20. UNICEF and College of Population Studies, Social Research Institute, Chula Unisearch: In-depth Research on Youth Not in Employment, Education or Training (NEET) in Thailand. UNICEF, Bangkok (2023). https://www.unicef.org/thailand/media/10746/file/In-depth%20research%20on%20youth%20NEET%20in%20Thailand.pdf
21. Yates, S., Payne, M.: Not so NEET? A critique of the use of 'NEET'in setting targets for interventions with young people. J. Youth Stud. **9**(3), 329–344 (2006)

# Economic Growth in Sub-Saharan Africa: An Analysis of the Technical Efficiency of Natural Resource Rent and Institutions Based on a Copula Stochastic Frontier Model

**Augustine Kwabena Sowu, Chanamart Intapan, Chukiat Chaiboonsri, and Emmanuel Mensaklo**

**Abstract** This study employs a copula-based stochastic frontier modeling (copula-SFM) framework to empirically assess resource-rich countries' efficiency in utilizing resource rent within Sub-Saharan Africa. The primary objective is to provide robust scientific evidence to aid these nations in maximizing their mineral resource utilization. Model selection criteria, specifically the Bayesian Information Criterion (BIC) and Akaike Information Criterion (AIC), identify the Normal copula-SFM as the superior choice. The findings emphasize the pivotal role of government effectiveness and population growth as strong, positive determinants of per capita income in the selected countries. In terms of average technical efficiency (TE) scores, Equatorial Guinea leads the rankings with an impressive score of 0.9907, while Tanzania trails with the lowest average TE score of 0.8177. These results underscore the potential for enhancing resource-rich countries' economic performance through effective governance and managing population growth.

**Keywords** Sub-Saharan Africa · Stochastic frontier analysis · Technical efficiency · Copula

A. K. Sowu (✉) · C. Intapan · C. Chaiboonsri · E. Mensaklo
Faculty of Economics, Chiang Mai University, Chiang Mai 50200, Thailand
e-mail: augustine_sowu@cmu.ac.th

C. Intapan
e-mail: chanamart.i@cmu.ac.th

C. Chaiboonsri
e-mail: chukiat.chai@cmu.ac.th

E. Mensaklo
e-mail: emmanuel.mensaklo@epuc.edu.gh



V. Kreinovich et al. (eds.), *Applications of Optimal Transport to Economics and Related Topics*, Studies in Systems, Decision and Control 556,
https://doi.org/10.1007/978-3-031-67770-0_30

# 1 Introduction

Africa is endowed with a diverse range of precious natural resources. The region is rich in valuable resources, ranging from precious metals and minerals to fossil fuels. Moreover, Africa possesses numerous other essential metals that serve as crucial inputs in manufacturing processes across the globe. The mining and extractive sectors significantly contribute to the economic landscape of Africa, contributing substantially to its annual exports, revenue, and GDP. In the year 2019, fossil fuels and minerals constituted over a third of the overall exports for a majority of African nations, amounting to at least 60% of countries on the continent. The presence of mineral resources in Africa has resulted in substantial revenue for government budgets, foreign currency reserves, and employment opportunities for many countries in African (Signé and Johnson [37]).

Despite the existence of substantial and expanding reserves, the successful transformation of underground resources into lasting prosperity and alleviating poverty has been limited. For instance, nations endowed with abundant resources in Sub-Saharan Africa have witnessed a deceleration in expanding their economies compared to the regional average following the decrease in commodity prices in 2014. Furthermore, these countries have struggled to make progress in alleviating poverty, exhibiting poorer performance in this area. According to estimates from the World Bank, by the year 2030, over 80% of the global population that lives in poverty will be concentrated within Sub-Saharan Africa, with resource-rich nations accounting for almost 75% of the globe's impoverished (Cust et al. [15]).

According to the World Bank (2023), as the world undergoes as the world shifts towards alternative energy sources and reduces reliance on fossil fuels, there is an increasing demand for minerals and metals as well as other rare earth elements. These resources, abundant in Africa, are crucial to producing green technologies. However, noteworthy is the fact that the existence of natural wealth alone may not ensure comprehensive growth and prosperity. Many countries in sub-Saharan Africa heavily rely on minerals, oil, and gas, with these resources accounting for a significant portion of their exports. Yet, converting this wealth into sustainable growth has proven challenging in Sub-Saharan Africa (Smith et al. [40]).

Policymakers, scholars, and governments alike have dedicated significant efforts to understanding the factors that contribute to economic growth within their respective nations. However, achieving a thorough comprehension of the drivers of the growth of the economy remains a complex and multifaceted endeavor (Kwakwa et al. [28]).

The impact of natural wealth on a country's growth is a topic of continuous discussion for a considerable time. Despite extensive research, economists and experts have yet to reach a consensus regarding the effect that natural wealth has on the progress of nations (Kwakwa et al. [28]). Advocates for natural resource abundance contend that the availability of resources inherently leads to increased wealth and purchasing power, consequently driving economic growth (Holden [22]). The viewpoint above is further supported by recent research, which suggests the crucial role played by natural resources in stimulating investment in physical infrastructure and the enhancement of human capital development. Additionally, they provide investable cash and local markets, contributing to growth (Alexeev and Conrad [2]; Brunnschweiler and Bulte [11]; North and Thomas [32]; Sinha and Sengupta [38]; Zeynalov [42]).

On the contrary, some researchers argue that nations heavily reliant on resource extraction and exports from nature tend to have lower economic growth (Badeeb et al. [7]; Haggard and Tiede [20]; Sachs and Warner [36]). In fact, resource-rich countries often encounter a paradoxical phenomenon known as the natural resource curse, where they face sluggish or negative growth (Badeeb et al. [7]).

Despite the commonly used Dutch disease perspective for elucidating the connection between growth of nations' economies and their natural wealth, the economic outcomes of nations abundant in natural resources, like Canada as well as Australia, has challenged conventional understanding. This discrepancy has prompted scholars to question the adequacy of purely economic models in elucidating the factors contributing to the contrasting achievements of different resource-rich nations. Recent research has highlighted the significant importance of institutions in influencing the ability of nations to effectively handle their resources and achieve sustainable economic development. It is asserted that institutions have a pivotal role in ensuring long-term prosperity of resource-rich nations (Brunnschweiler [10]; Bulte et al. [12]; Hartwell [21]; Isham et al. [24]).

Consequently, institutional quality has emerged as one crucial element challenging the concept of the resource curse and blurring the differentiation between wealthy and developing nations, as highlighted by (Alexeev and Conrad [2]; Bennett et al. [9]). While developed regions are generally perceived to have superior institutions compared to countries in Sub-Saharan Africa (SSA) and countries in the developing world, it is noteworthy that resource-rich nations in Sub-Saharan Africa often exhibit weaker institutional quality in contrast to nations lacking significant resources in the same region (Gueye and Lee [19]).

Natural resources' effects on the growth of an economy remains a persistent topic of ongoing debate, despite the multitude of studies conducted on this subject. The beginnings of this debate can be attributed to the mid-seventeenth century, when William Petty, the influential figure in British classical political economy, introduced the famous concept that "labour is the father of wealth, and the land is the mother of prosperity." This seminal work laid the groundwork for exploring the connection between natural wealth and economy's growth (Li et al. [29]).

However, even after centuries of research and scholarly inquiry, there remains a lack of consensus among experts concerning the degree to which natural resources affect economic growth. Literature on the topic is characterized by conflicting findings and varying perspectives. While certain research indicates natural resources can significantly favor economic growth, others contend for a detrimental association. The complexity of this debate underscores the multifaceted nature of natural resource-related dynamics and their relationship with economic growth.

This study is situated within a well-established collection of research that has extensively explored the connection between revenue from natural resources and economic growth, particularly emphasizing the significance of the quality of institutions. Over the years, numerous researchers have contributed to this field, shedding light on various aspects of the relationship. Notable among these scholars are (Brunnschweiler and Bulte [11]; Ciriacy-Wantrup [13]; Karabegović [25]; Mehlum et al. [30]).

Numerous studies analyzing the correlation between revenue from resources and a nation's growth within Sub-Saharan Africa have yielded diverse outcomes. Several of these studies have primarily focused on a link between natural wealth and economy's growth within the sub region without considering the influence of institutional quality in these countries (Ampofo et al. [4]; Asiamah et al. [6]; Kurečić and Seba [27]; Onyejiuwa [34]).

Despite the extensive research conducted on revenue generated from resource and the growth in the Sub-Saharan African nations' economy, there continues to be limited studies exploring the impact of institutions' quality on this relationship. A few studies have made the attempt to examine the interplay between institutions' quality, resource rent, and the growth of an economy within the subregion.

The research conducted by (Abdulahi et al. [1]; Raggl [35]) exclusively focused on rule of law and corruption respectively as an indicator of institutional quality, neglecting other vital dimensions including factors like political stability, effectiveness of governance, and regulatory quality. Firstly, political stability significantly influences resource management and investment decisions, as unstable environments can disrupt resource extraction and deter foreign investments, hampering economic growth potential. Secondly, the effectiveness of governance directly impacts how resource revenues are utilized, invested, and distributed among key sectors, playing a vital role in fostering sustainable economic development. Additionally, regulatory quality ensures transparent and efficient resource allocation, minimizing corruption and promoting investor confidence, which in turn, enhances economic growth prospects in countries with abundant resources (Arvanitis and Weigert [5]; Carothers [13]; Destek et al. [16]).

Numerous other scholars (Abdulahi et al. [1]; Ampofo et al. [4]; Moshiri and Hayati [31]) have conducted studies on how revenue from natural resources affects economic growth. However, it is essential to acknowledge that these studies primarily focus on establishing relationships without extensively delving into other related factors or aspects. One crucial aspect that the researchers have overlooked is measuring the efficiency level of countries that are rich in resources in the sub-region concerning their utilization of natural resources, despite facing diverse economic and political constraints.

Human welfare is significantly influenced by natural resources, providing essential raw materials like land, water, energy, food, and environmental services. It is however evident that many of these resources are limited in supply, and there is a growing awareness that human society is nearing the limits of their supply. Consequently, the promotion of resource efficiency becomes paramount in achieving sustainable development (Behrens et al. [8]; OECD [33]).

The study aims to fill a research gap by focusing on the specific institutional factors (government effectiveness, political stability, regulatory quality, and control of corruption) that impact the link between revenue from natural resource and economic growth in countries in sub-Saharan Africa. The research aims to quantitatively analyze the distinct institutional characteristics that influence the link between natural wealth and economic growth. Through rigorous empirical analysis, the study seeks to discern and measure the specific institutional qualities that mediate how natural resources influence economic growth.

In employing Copula Stochastic Frontier Model, the study aims at estimating the efficiency levels of resource utilization among the countries included in the analysis, considering their unique institutional constraints. By doing so, it contributes to existing research by providing valuable insights concerning how institutional factors shape the efficiency and performance of resource-rich countries in Sub-Saharan Africa.

Furthermore, aside exploring how the revenue from resources influences economic growth and the role of institutions' quality, this study will simultaneously investigate the potential effects of variables such as population growth, human capital formation, and physical capital formation on economic growth. By considering these additional factors, the study aims to provide a comprehensive analysis that captures the multifaceted dynamics influencing economic growth in the context of resource-rich sub-Saharan African countries. This study is structured as follows: Sect. 2 outlines the methodology and data sources, Sect. 3 presents the results and corresponding discussions, while Sect. 4 encapsulates the conclusions and policy implications.

# 2 Methodology and Data

## 2.1 Theoretical Background

### 2.1.1 The Stochastic Frontier Analysis (SFA)

Stochastic frontier analysis (SFA) is a method used for frontier estimation, which assumes a predetermined functional form for the connection between inputs and output. Aigner et al. [3] independently proposed the stochastic production function model (Coelli et al. [14]). Consider the following stochastic frontier model (SFM):

$$y_t = x_t'\beta + \varepsilon_t, \qquad (1)$$
$$\varepsilon_t = v_t - u_t$$

In the stochastic production function model, $t = 1, 2, 3, \ldots, T$, $y_t$ is the dependent variable that is the logarithm of output or input, $x_t$ is the$K \times 1$ vector that contains the explanatory variables, $\beta$ is a $K \times 1$ vector that represents unknown parameters that need to be estimated. Additionally, $u_t$ is non-negative unobservable random variable that represent technical inefficiency, while $v_t$ represents symmetric random error utilized to accommodate statistical noise.

The uniqueness of this study lies in its approach of incorporating resource rents, political factors, and various other economic variables as inputs into a production function. This approach is grounded in the framework derived from Solow's growth model of 1956 (Durlauf et al. [17]). The technical efficiency $TE_t$ can be calculated by

$$TE_t = \frac{exp\left(x_t'\beta + v_t - u_t\right)}{x_t'\beta + v_t} \tag{2}$$

The Stochastic frontier model exists a very strict assumption that the $u_t$ and $v_t$ are independent. Thus, we allow copula to make optimum use of stochastic frontier model. The copula-based stochastic frontier model sufficiently regards dependence structure between $u$ and $v$, with independence between $u$ and $v$ using product copula. First, the density function of $f(u, v)$ can be transformed into $f(u, \varepsilon)$ according to (Smith [39]).

$$f(u, v) = f(u, u + \varepsilon) = f_u(u) \times f_v(u + \varepsilon) \times c_\theta\left(F_u(u), F_v(u + \varepsilon)\right). \tag{3}$$

Then, the probability density function can be expressed as:

$$f_\theta(\varepsilon) = E_U\left[f_V(U + \varepsilon)c_\theta(F_U(U), F_V(U + \varepsilon)\right] \tag{4}$$

Where $E_U[.]$ denotes the expected value pertaining to the distribution of technical inefficiency $U$. We use log maximum likelihood to estimate the copula-based stochastic frontier model. Assume that there are time series observations of $n$ individuals or decision-making units, thus the likelihood function is given by:

$$L(\beta, \sigma_u, \sigma_v, \theta) = \prod_{t=1}^{n} f_\theta(\varepsilon_t) = \prod_{t=1}^{n} f_\theta(y_t - \beta_0 - x_t'\beta_t), \tag{5}$$

where $y_t$ is the output in logarithm form at time $t$, $x_t$ is explanatory variable vector, $\sigma_u$ and $\sigma_v$ are from marginal distributions of $U$ and $V$, respectively. Smith [39] showed that there is very few density function of $\varepsilon$ existing closed form solution. Then, an alternate approach, maximum simulated likelihood method, is used to estimate the unknown parameters in copula-based stochastic frontier model (Greene [23]; Smith [39]). If $U$ and $V$ are assumed to be half-normal and normal distribution, then the density function of $\varepsilon$ can be transformed as follows:

$$
\begin{aligned}
f_{(\varepsilon)} &= \int_0^{\infty} f_U(u) \times f_V(u+\varepsilon) x C_\theta \left(F_U(u), F_V(u+\varepsilon)\right) du \\
&= \int_0^{\infty} \frac{2\exp\left(-\frac{u^2}{2\sigma_u^2}\right)}{\sqrt{2\pi}\sigma_u} \times f_V(u+\varepsilon) \times c_\theta \left(F_U(u), F_V(u+\varepsilon)\right) du \\
&= \int_0^{\infty} \frac{2\exp\left(-(\sigma_u u_0)^2/2\sigma_u^2\right)}{\sqrt{2\pi}\sigma_u} \times f_V(\sigma_u u_0+\varepsilon) \times c_\theta \left(F_U(\sigma_u u_0), F_V(\sigma_u u_0+\varepsilon)\right) d\sigma_u u_0 \\
&= \int_0^{\infty} \frac{2\exp\left(-u_0^2/2\right)}{\sqrt{2\pi}} \times f_V(\sigma_u u_0+\varepsilon) \times c_\theta \left(F_U(\sigma_u u_0), F_V(\sigma_u u_0+\varepsilon)\right) du_0 \\
&\approx \frac{1}{R}\sum_{r=1}^{R} f_V\left(\sigma_u u_{0,r}+\varepsilon\right) \times c_\theta \left(F_U\left(\sigma_u u_{0,r}\right), F_V\left(\sigma_u u_{0,r}+\varepsilon\right)\right),
\end{aligned} \tag{6}
$$

where $u_0$ is a sequence of $R$ random draws from the standard half normal distribution. The simulated log-likelihood of the copula-based stochastic frontier approach can be expressed as follows:

$$
L_s(\beta, \sigma_u, \sigma_v, \theta) \approx
$$

$$
\sum_{t=1}^{N} \log\left\{\frac{1}{R}\sum_{r=1}^{R} f_V\left(\sigma_u u_{0,tr}+\varepsilon\right) \times c_\theta\left(F_U\left(\sigma_u u_{0,tr}\right), F_V\left(\sigma_u u_{0,tr}+\varepsilon\right)\right)\right\}, \tag{7}
$$

where $\varepsilon_t = y_t - \beta' x_t$. In addition, the relevant parameters, $\lambda, \gamma$ and $\sigma^2$, can be defined by $\lambda = \sigma_u/\sigma_v$, $\gamma = \sigma_u^2/\left(\sigma_u^2+\sigma_v^2\right)$, and $\sigma^2 = \sigma_u^2+\sigma_v^2$. Note that the standard errors of these parameters can be calculated by applying the invariance property.

### 2.1.2 Copula

Copulas are employed to express the interdependence among random variables, connecting various marginal distributions (one-dimensional distributions) to establish a multivariate joint distribution (Kreinovich and Sriboonchitta [26]). Given the collective distribution $H$, the system function $C(\cdot,\cdot)$, is represented as:

$$
H = (X_1, X_2) = H\left(F_1^{-1}(u_1), F_2^{-1}(u_2)\right) = C_\alpha\left(F_1(x_1), F_2(x_2)\right) \tag{8}
$$

In this context, $u_1 = F_1(x_1)$ and $u_2 = F_2(x_2)$, where $\alpha$ represents the parameter vector of the copula, commonly known as the dependence parameter vector. $F_1(\cdot)$ and $F_2(\cdot)$ represent uniform distributions. Bivariate copulas exhibit the following characteristics:

(1) $C(u_1, 0) = C(0, u_2)$;
(2) $C(u_1, 1) = u_1$ and $C(0, u_2) = u_2$;
(3) If u, v, w, and z belong to [0, 1], such that $u \le w$ and $v \le z$, then

$$C(w,z) - C(w,v) - C(u,z) + C(u,v) \geq 0 \tag{9}$$

The Sklar Theorem plays a crucial role in establishing a connection between copula functions and distribution functions, utilizing the three features above. This theorem states that any multivariate probability distribution function can be represented by a combination of a dependent structure and a marginal distribution, as illustrated below:

$$\begin{aligned} f(x_1, \ldots, x_n) &= \frac{\partial F(x_1, \ldots, x_n)}{\partial x_1 \ldots \partial x_n} \\ &= \frac{\partial C(u_1, \ldots, u_n)}{\partial u_1 \ldots \partial u_n} \times \Pi \frac{\partial F_i x_i}{\partial x_i} \\ &= c(u_1, \ldots, u_n) \times \Pi f_i(x_i) \end{aligned} \tag{10}$$

Hence, a bivariate density function of $x_1$ and $x_2$ can be expressed as the product of two components: the marginal densities and the copula density.

$$f(x_1, x_2) = c(u_1, u_2) \times f_1(x_1) \times f_2(x_2) \tag{11}$$

Pearson's linear correlation is commonly used to assess dependency, even though it lacks information about asymmetrical distributions. To overcome this limitation, the United States has adopted rank correlations, specifically Kendall's tau and Spearman's rho, to measure nonlinear dependence. The calculations for Kendall's tau and Spearman's rho provide the following results.

$$\begin{aligned} \tau_{x_1 x_2} &= 4 \iint_{[0,1]^2} C(u_1, u_2) - 1 = 4E[C(u_1, u_2) - 1 \\ S_{\rho(x_1 x_2)} &= 12 \iint_{[0,1]^2} u_1 u_2 d[C(u_1, u_2)] - 3 \\ &= 12 \iint_{[0,1]^2} C(u_1, u_2)\, du_1 du_2 - 3 \\ &= 12E[(U_1, U_2)] \end{aligned} \tag{12}$$

Numerous copulas can capture different dependency structures, but only a few copulas can accurately measure the full spectrum of dependent structures, spanning from lower to upper boundaries. As a result, the main properties of copula families that is used in the study are Gaussian copulas, T copula, Clayton copula, Frank copula, Gumbel copula, and Joe copula as they retain their usefulness in capturing both negative and positive dependencies (Kreinovich and Sriboonchitta [26]; Wiboonpongse et al. [41]).

The Akaike Information Criterion (AIC) and Bayesian Information Criterion (BIC) are two criteria that is used for selecting the best-fitting model among various alternatives, ensuring optimal fit for all variables (refer to Eqs. (12) and (13)). In the case of a large class of models fitted using maximum likelihood, the AIC is defined as follows:

$$AIC = -2\log L + 2k \tag{13}$$

Here, L represents the maximum likelihood function for the estimated model. The BIC value as well is defined as follows:

$$BIC = \frac{1}{n}\left(RSS + \log(n)k\hat{\sigma}^2\right) \tag{14}$$

Here, "$n$" represents the number of observations in the dataset, while "RSS" refers to the Residual Sum of Squares. The RSS measures the sum of squared differences between the observed and predicted values in the regression model. The variable "k" represents the number of parameters in the model, which corresponds to the number of variables or predictors. Additionally, $\hat{\sigma}^2$ represents the variance of the error term in the regression model. Preference is given to the models with the lowest AIC and BIC values, indicating that these models are considered the best in the context of the Copula Stochastic Frontier Model computation.

Preference is given to the models with the lowest AIC and BIC values, indicating that these models are considered the best in the context of the Copula Stochastic Frontier Model computation.

### 2.1.3 Empirical Model and Estimation Procedure

The stochastic frontier model is employed to estimate the technical efficiency analysis of the subregion's utilization of natural resource rent using the Stochastic Frontier Based Copula Approach. This approach follows the linear translog production model and utilizes equation (14) to estimate the Cobb–Douglas production function. The goal is to identify the significant independent factors that contribute to the economic growth of resource-rich countries within each country.

$$\begin{aligned}\ln(GDPC)_{it} =& \ln\beta_0 + \beta_1\ln(RR)_{it} + \beta_2(PS)_{it} + \beta_3(\mathrm{GE})_{it} + \beta_4(RQ)_{it} + \beta_5(CC)_{it}\\ &+ \beta_6(PG)_{it} + \beta_7(AE)_{it} + \beta_8\ln(MYS)_{it} + V_{it} - U_{it}\end{aligned}$$

In this study, $i$ represents the ten selected countries in sub-Saharan Africa, while $t$ denotes the year from 2002 to 2021.

### 2.1.4 Description of Variables and Data Source

I. GDP per capita (current US$) is sourced from the World Bank's World Development Indicators database from the year 2002 to 2021. The GDP per capita is the dependent variable.

II. Total natural resources rents (% of GDP) which is the independent variable in this study to investigate the impact of resource rent on the economic growth of the selected countries. Total natural resources rents are the sum of oil rents, natural gas rents, coal rents (hard and soft), mineral rents, and forest rents for the years 2002–2021.

III. Political stability measures a country's potential for political unrest and violence, including terrorism. Scores range from about −2.5 to 2.5 on a standard scale, with higher scores indicating more stability and lower scores suggesting greater instability.
IV. Government Effectiveness measures public service quality, civil service integrity, policy credibility, and implementation. Scores, on a −2.5 to 2.5 scale, indicate overall performance in these areas.
V. Regulatory Quality assesses government competence in creating policies for private sector growth. Scores, −2.5 to 2.5, reflect this ability on a standard scale.
VI. Control of Corruption measures if public power benefits private interests. Scores (−2.5 to 2.5) show corruption perception on a standard scale.
VII. Population growth rate (%) for year t is the percentage increase in midyear population from t − 1 to t, following the exponential growth model. This definition includes all residents regardless of citizenship or legal status.
VIII. Access to electricity is the portion of the population with the ability to use electricity, expressed as a percentage.
IX. Mean years of schooling is a statistical gauge of the average education duration completed by individuals aged 25 and above.

# 3 Results and Discussion

## 3.1 *Descriptive Statistics and Levin, Lin and Chu Panel Root Test on Variables*

See Table 1 .

### 3.1.1 The Stochastic Frontier Model Estimated Results

Six families of copulas were considered for estimating the stochastic frontier model. Additionally, we employed the Bayesian Information Criterion (BIC) and Akaike Information Criterion (AIC) as criteria to select the best-fitting model from various alternatives, ensuring optimal fit for all variables. The results are presented in Table 3 where it becomes evident that the Normal copula yields the lowest BIC and AIC values of −128.7470 and −132.0453, respectively. Consequently, for this study, the Normal copula SFM stands out as the most appropriate model. Furthermore, we determine the parameters for both the stochastic frontier model without a copula and the copula-based stochastic frontier model in order to identify the optimal model for our study (Table 2).

**Table 1** Descriptive statistics and Levin, Lin and Chu panel root test on variables

| | Log (GDP) | Log (Rents) | Govt | RegQ | Cor | Pop | Elec | Log (Sch) | Pol |
|---|---|---|---|---|---|---|---|---|---|
| Mean | 4150.586 | 19.733 | −0.742 | −0.67 | −0.728 | 2.799 | 50.993 | 6.362 | −0.445 |
| Median | 2606.129 | 16.571 | −0.939 | −0.752 | −1.015 | 2.771 | 47.31 | 5.953 | −0.261 |
| Maximum | 19849.72 | 56.289 | 0.62 | 0.9 | 1.245 | 4.78 | 91.806 | 11.373 | 1.111 |
| Minimum | 399.099 | 0.656 | −1.639 | −1.713 | −1.628 | 0.387 | 10.707 | 0 | −2.665 |
| Std. Dev. | 3768.237 | 15.433 | 0.611 | 0.712 | 0.718 | 0.87 | 22.074 | 2.13 | 0.949 |
| Skewness | 1.713 | 0.514 | 0.952 | 0.67 | 1.064 | −0.166 | 0.197 | 0.315 | −0.563 |
| Kurtosis | 6.598 | 2.024 | 2.714 | 2.441 | 3.21 | 3.057 | 2.03 | 2.674 | 2.559 |
| Unit root test | −2.489 | −2.027 | −2.921 | −1.442 | −8.531 | −2.659 | −5.978 | −2.277 | −10.318 |

**Table 2** Stochastic frontier without copula

| | Estimate | Std. error | z value | MBF |
|---|---|---|---|---|
| Intercept | 3.0349 | 0.2451 | 12.3804 | 0 |
| log (Rent) | 0.0049 | 0.0051 | 0.9569 | 0.6303 |
| Government effectiveness | −0.0096 | 0.0266 | −0.3603 | 0.937 |
| Regulatory quality | 0.042 | 0.0229 | 1.8398 | 0.186 |
| Corruption | 0.0141 | 0.021 | 0.6739 | 0.7982 |
| Population growth | 0.0508 | 0.0098 | 5.1696 | 0 |
| Access to electricity | −0.0027 | 0.0007 | −3.6878 | 0.0006 |
| log (means years of schooling) | −0.334 | 0.1053 | −3.1727 | 0.0065 |
| Political stability | 0.0162 | 0.0114 | 1.4206 | 0.3643 |
| $\sigma^2$ | 0.4547 | 0.3342 | 1.3608 | 0.3963 |
| $\gamma$ | 0.9966 | 0.0025 | 401.1257 | 0.000 |
| Time | 0.0156 | 0.0026 | 5.9278 | 0.000 |

Significant codes: 0 '***' 0.001 '**' 0.01 '*' 0.05 '.' 0.1 '' 1

**Table 3** Information criterion for model selection

| Copula families | BIC | AIC |
|---|---|---|
| Normal copula | −128.7470* | −132.0453* |
| t-copula | −123.4238 | −130.0453 |
| Clayton copula | −44.4577 | −47.756 |
| Frank copula | −123.4242 | −126.7225 |
| Gumbel copula | 5.2983 | 2.000 |
| Joe copula | 5.2983 | 2.000 |

*Note* * Indicates the lowest value of BIC and AIC

Where the minimum Bayes factor $MBF(y) = \exp(-t^2/2)$ as suggested by (Edwards et al. [18]).

Table 1 presents the descriptive statistics and the results of the Panel Root Test conducted on the dataset. Notably, GDP per capita exhibits the highest standard deviation, while government effectiveness displays the lowest. Additionally, the unit root test confirms that these variables are stationary. As a further step, we transformed GDP per capita, resource rent, and mean years of schooling into natural logarithms. The estimation of the stochastic frontier copula model is as follows.

The table provided above illustrates the average efficiency values for each year, along with the mean efficiency calculated across the years, for both the stochastic frontier model without a copula and the copula-based stochastic frontier model. The observations gleaned from the table unequivocally affirm that, when considered both on an annual average basis and in terms of the overall mean, the copula-based stochastic frontier model consistently outperforms its counterpart, the stochastic frontier model without a copula.

**Table 4** The estimated parameters of copula based SFM

| | Estimate | Std. error | z value | MBF |
|---|---|---|---|---|
| Intercept | 2.831 | 0.309 | 9.1629 | 0 |
| log (Rent) | 0.0814 | 0.0526 | 1.549 | 0.3013 |
| Government effectiveness | 0.6472 | 0.2735 | 2.3664 | 0.0608 |
| Regulatory quality | 0.0858 | 0.2089 | 0.4106 | 0.9191 |
| Corruption | −0.1981 | 0.1965 | −1.0079 | 0.6017 |
| Population growth | 0.1993 | 0.0819 | 2.4314 | 0.052 |
| Access to electricity | −0.0001 | 0.0026 | −0.0224 | 0.9997 |
| log (means years of schooling) | −0.6812 | 0.3909 | −1.7427 | 0.2191 |
| Political stability | −0.1613 | 0.0783 | −2.0607 | 0.1197 |
| $\sigma^2$ | 0.2619 | 0.0262 | 9.987 | 0.000 |
| $\gamma$ | 0.000 | 0.000 | 0.1917 | 0.9818 |
| Time | 0.4793 | 0.1413 | 3.3915 | 0.0032 |

Significant codes: 0 '***' 0.001 '**' 0.01 '*' 0.05 '.' 0.1 '' 1

Specifically, when examining the average efficiency figures for each individual year, it becomes evident that the copula-based stochastic frontier consistently yields higher values than the stochastic frontier model without a copula. Moreover, the copula-based stochastic frontier model demonstrates a noteworthy mean efficiency of 0.9569295, a value that stands in stark contrast to the mean efficiency of 0.4774676 obtained from the stochastic frontier model lacking a copula.

The data presented above unequivocally emphasizes the superiority of the copula-based stochastic frontier model, thereby shedding light on the significant improvement in mean efficiency when compared to its non-copula counterpart. Consequently, the copula-based stochastic frontier model is chosen as the preferred model.

The Normal copula parameter estimates of the SFM are presented in Table 4. It is evident that all independent variables, except control of corruption, access to electricity, mean years of schooling and political stability, exhibit positive coefficients. Notably, government effectiveness, population growth and political stability exert a significant influence on GDP per capita. Given that resource rent and mean years of schooling are log-transformed, a 1% increase in resource rent leads to a 0.08141% rise in GDP per capita. Conversely, a 1% increase in mean years of schooling is associated with a 0.68121% decrease in GDP per capita, with a statistical significance at the 10% level (Table 5).

Furthermore, a one-unit improvement in government effectiveness, regulatory quality and a unit increase in population contributes to a 0.6472%, 0.0858% and 0.19925% increase in GDP per capita respectively. Conversely, an increase of one unit in corruption, access to electricity, and political stability results in a decrease of 0.1981%, 0.000059234%, and 0.16129% in GDP per capita, respectively. This underscores a clear indication of inefficiency in resource utilization in Sub-Saharan Africa.

**Table 5** The average efficiency of each year and the mean efficiency

| Year | Stochastics frontier without copula | Stochastics frontier with copula |
|---|---|---|
| | Efficiency | Efficiency |
| 2002 | 0.4258 | 0.70256 |
| 2003 | 0.4313 | 0.7982 |
| 2004 | 0.4368 | 0.8674 |
| 2005 | 0.4423 | 0.9147 |
| 2006 | 0.4478 | 0.9459 |
| 2007 | 0.4533 | 0.9660 |
| 2008 | 0.4588 | 0.9787 |
| 2009 | 0.4642 | 0.9868 |
| 2010 | 0.4697 | 0.9918 |
| 2011 | 0.4751 | 0.9949 |
| 2012 | 0.4805 | 0.9968 |
| 2013 | 0.4859 | 0.9980 |
| 2014 | 0.4912 | 0.9988 |
| 2015 | 0.4966 | 0.9992 |
| 2016 | 0.5019 | 0.9995 |
| 2017 | 0.5072 | 0.9997 |
| 2018 | 0.5124 | 0.9998 |
| 2019 | 0.5177 | 0.9999 |
| 2020 | 0.5229 | 0.9999 |
| 2021 | 0.5281 | 0.9999 |
| Mean efficiency | 0.4774676 | 0.9569295 |

Our results have confirmed a significant positive dependence in the noise-inefficiency measure, with a value of 0.9418. This finding serves as compelling evidence, demonstrating that a copula based SFM should be favored over the traditional SFM.

### 3.1.2 Technical Efficiency Estimates (TE)

The table below presents the technical efficiency estimates, which gauge the ratio of observed output to the maximum feasible output. These technical efficiency values offer insights into the resource utilization efficiency of the selected countries in Sub-Saharan Africa. Values less than 1 signify that resource utilization falls short of realizing its full potential efficiency, while a value of 1 indicates that resource utilization has reached its maximum production capability. The findings reveal that Equatorial Guinea boasts the highest average technical efficiency score at 0.9907, while Tanzania records the lowest average technical efficiency score at 0.8177 (Table 6).

**Table 6** Average TE of countries in Sub-Saharan Africa

| Country | Average TE |
|---|---|
| Angola | 0.9387 |
| Botswana | 0.9786 |
| Congo Republic | 0.9205 |
| Equatorial Guinea | 0.9907 |
| Gabon | 0.9325 |
| Nigeria | 0.9255 |
| South Africa | 0.9774 |
| Sudan | 0.9565 |
| Tanzania | 0.8177 |
| Zambia | 0.8192 |

## 4 Conclusion

In this research paper, we employed a copula-based Stochastic Frontier Model (SFM) to investigate the utilization of natural resource rent in the context of GDP per capita in Sub-Saharan Africa. To select the most appropriate model, we relied on the criteria of the lowest Bayesian Information Criterion (BIC) and Akaike Information Criterion (AIC). Our study reveals that the Normal copula SFM is the most suitable model for our analysis.

Apart from control of corruption, access to electricity, mean years of schooling and political stability, our findings suggest that various factors, such as Resource Rent, Government Effectiveness, Regulatory quality and Population Growth, have been identified as having positive effects on the GDP per capita of the selected countries within the subregion.

Turning to the assessment of average Technical Efficiency (TE) scores, Equatorial Guinea leads the pack with the highest average technical efficiency score at 0.9907. It is followed closely by Botswana, South Africa, Sudan, Angola, Gabon, Nigeria, Congo Republic, and Zambia, with average technical efficiency scores of 0.9786, 0.9774, 0.9565, 0.9387, 0.9325, 0.9255, 0.9205, and 0.8192, respectively. In contrast, Tanzania records the lowest average technical efficiency score at 0.8177.

The results of our study carry significant implications as they provide practical guidelines for the efficient utilization of natural resource rent in Sub-Saharan Africa, as well as in countries endowed with substantial natural resources. Furthermore, the TE score of the top-performing country in the subregion can serve as a benchmark and a reference point for other nations seeking to replicate the successful factors behind its high efficiency. Importantly, our findings concerning the low-performing countries offer valuable insights and critical policy opportunities. We hope that policymakers will consider these findings in shaping resource rent utilization policies, aiming to enhance the efficient management and utilization of resource rent in these countries.

In future research endeavors, we envision the application of the copula-SFM model to study resource rent utilization in countries not covered in the present study, encompassing other resource-rich African nations and resource-endowed countries across the globe.

**Acknowledgements** The author is pursuing a Master's Degree in Economics at the Faculty of Economics, Chiang Mai University, under the CMU Presidential Scholarship. The author gratefully acknowledges the support of the CMU Presidential Scholarship in the publication of this work.

## References

1. Abdulahi, M.E., Shu, Y., Khan, M.A.: Resource rents, economic growth, and the role of institutional quality: a panel threshold analysis. Resour. Policy **61**, 293–303 (2019)
2. Alexeev, M., Conrad, R.: The natural resource curse and economic transition. Econ. Syst. **35**(4), 445–461 (2011)
3. Aigner, D., Lovell, C.K., Schmidt, P.: Formulation and estimation of stochastic frontier production function models. J. Econom. **6**(1), 21–37 (1977)
4. Ampofo, G.M.K., Laari, P.B., Ware, E.O., Shaw, W.: Further investigation of the total natural resource rents and economic growth nexus in resource-abundant sub-Saharan African countries. Miner. Econ. **36**(1), 97–121 (2023)
5. Arvanitis, Y., Weigert, M.: Turning resource curse into development dividends in Guinea-Bissau. Resour. Policy **53**, 226–237 (2017)
6. Asiamah, O., Agyei, S.K., Bossman, A., Agyei, E.A., Asucam, J., Arku-Asare, M.: Natural resource dependence and institutional quality: evidence from sub-Saharan Africa. Resour. Policy **79**, 102967 (2022)
7. Badeeb, R.A., Lean, H.H., Clark, J.: The evolution of the natural resource curse thesis: a critical literature survey. Resour. Policy **51**, 123–134 (2017)
8. Behrens, A., Giljum, S., Kovanda, J., Niza, S.: The material basis of the global economy: worldwide patterns of natural resource extraction and their implications for sustainable resource use policies. Ecol. Econ. **64**(2), 444–453 (2007)
9. Bennett, D.L., Faria, H.J., Gwartney, J.D., Morales, D.R.: Economic institutions and comparative economic development: a post-colonial perspective. World Dev. **96**, 503–519 (2017)
10. Brunnschweiler, C.N.: Cursing the blessings? natural resource abundance, institutions, and economic growth. World Dev. **36**(3), 399–419 (2008)
11. Brunnschweiler, C.N., Bulte, E.H.: The resource curse revisited and revised: a tale of paradoxes and red herrings. J. Environ. Econ. Manag. **55**(3), 248–264 (2008)
12. Bulte, E.H., Damania, R., Deacon, R.T.: Resource intensity, institutions, and development. World Dev. **33**(7), 1029–1044 (2005)
13. Carothers, T.: Promoting the rule of law abroad: the problem of knowledge (2003). Ciriacy-Wantrup, S.V.: Natural Resource Economics: Selected papers. Routledge (2019)
14. Coelli, T.J., Rao, D.S.P., O'Donnell, C.J., Battese, G.E.: An Introduction to Efficiency and Productivity Analysis. Springer Science & Business Media (2005)
15. Cust, J., Ballesteros, A.R., Zeufack, A.: The Dog that Didn't Bark (2022)
16. Destek, M.A., Adedoyin, F., Bekun, F.V., Aydin, S.: Converting a resource curse into a resource blessing: the function of institutional quality with different dimensions. Resour. Policy **80**, 103234 (2023)
17. Durlauf, S.N., Kourtellos, A., Minkin, A.: The local solow growth model. Eur. Econ. Rev. **45**(4–6), 928–940 (2001)
18. Edwards, W., Lindman, H., Savage, L.J.: Bayesian statistical inference for psychological research. Psychol. Rev. **70**(3), 193 (1963)

19. Gueye, C.A., Lee, M.: Natural Resource Wealth: Making Resource Windfalls Work for Sub-Saharan African Countries. International Growth Centre Blog (2015)
20. Haggard, S., Tiede, L.: The rule of law and economic growth: where are we? World Dev. **39**(5), 673–685 (2011)
21. Hartwell, C.A.: The institutional basis of efficiency in resource-rich countries. Econ. Syst. **40**(4), 519–538 (2016)
22. Holden, S.: Avoiding the resource curse the case Norway. Energy Policy **63**, 870–876 (2013)
23. Greene, W.H.: The econometric approach to efficiency analysis. In: The Measurement of Productive Efficiency and Productivity Growth, vol. 1, no. 1, pp. 92–250 (2008)
24. Isham, J., Woolcock, M., Pritchett, L., Busby, G.: The varieties of resource experience: natural resource export structures and the political economy of economic growth. World Bank Econ. Rev. **19**(2), 141–174 (2005)
25. Karabegović, A.: Institutions, Economic Growth, and the "Curse" of Natural Resources. Fraser Institute (2009)
26. Kreinovich, V., Sriboonchitta, S.: Structural Changes and their Econometric Modeling, vol. 808. Springer (2018)
27. Kurečić, P., Seba, M.: The resource curse in sub-Saharan Africa: a reality corroborated by the empirical evidence. In: 15th International Scientific Conference on Economic and Social Development-Human Resources Development Proceedings (2016)
28. Kwakwa, P.A., Adusah-Poku, F., Adjei-Mantey, K.: Towards the attainment of sustainable development goal 7: what determines clean energy accessibility in sub-Saharan Africa? Green Finance **3**(3), 268–286 (2021)
29. Li, Y., Naqvi, B., Caglar, E., Chu, C.-C.: N-11 countries: are the new victims of resource-curse? Resour. Policy **67**, 101697 (2020)
30. Mehlum, H., Moene, K., Torvik, R.: Institutions and the resource curse. Econ. J. **116**(508), 1–20 (2006)
31. Moshiri, S., Hayati, S.: Natural resources, institutions quality, and economic growth. A cross-country analysis. Iran. Econ. Rev. **21**(3), 661–693 (2017)
32. North, D.C., Thomas, R.P.: The Rise of the Western World: A New Economic History. Cambridge University Press (1973)
33. OECD: Measuring material flows and resource productivity. In: The OECD Guide, vol. 1. OECD Paris, France (2008)
34. Onyejiuwa, D.C.: Resource rents and dependence in sub-Saharan African countries economies. Open Access Libr. J. **3**(4), 1–11 (2016)
35. Raggl, A.K.: Natural Resources, Institutions, and Economic Growth: The Case of Nigeria. World Bank Policy Research Working Paper (8153) (2017)
36. Sachs, J.D., Warner, A.: Natural Resource Abundance and Economic Growth. National Bureau of Economic Research Cambridge, Mass., USA (1995)
37. Signé, L., Johnson, C.: Africa's Mining Potential: Trends. Opportunities, Challenges and Strategies (2021)
38. Sinha, A., Sengupta, T.: Impact of natural resource rents on human development: what is the role of globalization in Asia Pacific countries? Resour. Policy **63**, 101413 (2019)
39. Smith, M.D.: Stochastic frontier models with dependent error components. Economet. J. **11**(1), 172–192 (2008)
40. Smith, R., Ilas, M.A., Peszko, G.: The Changing Wealth of Nations 2021: Managing Assets for the Future-Technical Report: Valuation of Renewable Energy Resources in the Context of the Changing Wealth of Nations (2023)
41. Wiboonpongse, A., Liu, J., Sriboonchitta, S., Denoeux, T.: Modeling dependence between error components of the stochastic frontier model using copula: application to intercrop coffee production in Northern Thailand. Int. J. Approx. Reas. **65**, 34–44 (2015)
42. Zeynalov, A.: The gravity of institutions in a resource-rich country: the case of Azerbaijan. IEEP **14**, 239–261 (2017)

# Analysis of Agricultural Production in China and Measurement of Technical Efficiency Using Copula-Based Stochastic Frontier Model

**Yueyi Chen, Paravee Maneejuk, and Woraphon Yamaka**

**Abstract** The purpose of this paper is to evaluate the efficiency of agricultural production in China, analyze China's production function, and identify the primary factor affecting efficiency. Methodologically, we employ the stochastic frontier model, accounting for the interrelation between the two-sided error term (U) and the one-sided inefficiency factor (V). To address this complex relationship, we introduce a recent approach known as the Copula-based stochastic frontier model. After obtaining technical efficiency (TE) values, we classify the 31 provinces in China into three distinct groups using the K-means clustering method, each exhibiting unique characteristics. Subsequently, we conduct a LASSO regression analysis to examine the key factors influencing TE values. Our findings reveal that the most significant factor impacting TE values is the average household size, which refer to the average number of laborers per rural household in each province. Furthermore, we observe a negative correlation between the level of education among the labor force and TE values. We analyze that this negative correlation primarily results from the agricultural sector's limited appeal to individuals with higher levels of education.

**Keywords** Stochastic frontier · Frontier production function · Copula · Technical efficiency

Y. Chen · P. Maneejuk (✉) · W. Yamaka
Faculty of Economics, Chiang Mai University, Chiang Mai, Thailand
e-mail: mparavee@gmail.com

Y. Chen
e-mail: chen_yueyi@cmu.ac.th

W. Yamaka
e-mail: woraphon.yamaka@cmu.ac.th

 
V. Kreinovich et al. (eds.), *Applications of Optimal Transport to Economics and Related Topics*, Studies in Systems, Decision and Control 556,
https://doi.org/10.1007/978-3-031-67770-0_31

## 1 Introduction

China has the world's largest population, and ensuring a stable and sufficient supply of food is critical for national stability and security. Grain, including rice, wheat, and corn, is a staple in the Chinese diet. With that in mind, China ranks among the top grain-producing countries in the world. In 2020, China held the second position in global corn production, while securing the top spot for both wheat and rice production. The agricultural gross output value for China in 2020 reached $1.1 trillion, making it the world leader in this regard. Consequently, agriculture stands as one of China's largest and most significant sectors, with grain production being of paramount importance. For this reason, we have to take into account the productive efficiency of this sector. Thus, we attempt to analyze the agricultural production function of China and then examine the technical efficiencies of this region. To illustrate, these efficiencies can result in a surge in agricultural productivity without necessitating an expansion of the resource base. This implies the ability to generate higher output from the same input quantities. Similarly, technical efficiency can enhance a producer's competitiveness without requiring an increase in input factors, as highlighted by Bezat and Rembisz [4]. The concept of technical efficiency was originally introduced by Farrell [21], employing the frontier production function. This seminal work has spurred numerous subsequent developments in both estimating the frontier production function and quantifying technical efficiency. When technical efficiency is applied to agriculture, it has its own characteristics, Đokić et al. [16] showed that the average relative technical efficiency of agriculture in Western Balkans is noticeably worse than in the European Union countries, this difference is a consequence of agriculture's unfavorable resource structure, primarily due to high presence of small farms and the slow development of a non-agricultural sector that cannot accept a surplus of labor from agriculture. Refine agriculture to crop production, [23] showed that increased technical efficiency in rice production cannot be overlooked and animal production does appear to be more efficient than crop production in general. However, this paper takes into account a powerful model called stochastic frontier (SFM) which was proposed by [1]. The core concept of this model revolves around establishing a linear connection between output and input quantities, incorporating two distinct error terms: one signifying inefficiency and the other following a standard normal distribution. This model finds widespread use in evaluating the technical efficiency of production entities. Efficiency is succinctly gauged through the parameters of the frontier production function, wherein we compute the disparity between a nation's actual production output and the maximum attainable output for a given set of inputs, commonly referred to as the production frontier. The SFA model helps to explain variations in output and inefficiency among farmers. Tesema [46] uses stochastic frontier model to estimate the parameters of the production function as well as the level of efficiency. This is due to the fact that this technique takes into account inefficiency factors as well as technical errors that occur during measurement and observation.

Some papers discuss the use of copula-based stochastic frontier models in assessing technical efficiency in various sectors, particularly agriculture. This model relaxes the assumption that the two components of random error are independent, allowing for a more accurate representation of the correlation between these components, which suggested by [39, 49]. Some studies have also explored the use of more sophisticated analytical techniques, such as the K-Means algorithm and copula-based stochastic frontier model, to ensure accurate analysis results. Chaovanapoonphol et al. [11] tries to fill the gap in the literature by using the K-Means algorithm and copula-based stochastic frontier model, in which we relax the assumption that the two components of random error are independent to each other; furthermore, the correlation of the two components of random error is also represented by the estimation of copula. Moreover, Sriboonchitta [41] used the double-copula stochastic frontier model with sample selection combines the copula-based sample selection model with the copula-based stochastic frontier mode. This approach allows for a wide range of production data to be fitted without relying on restricted assumptions about the dependence between various random variables in the model. Hence, this study employs a Copula-based stochastic frontier model, which considers the influence of inputs on agricultural production output and efficiency scores. Furthermore, this model accommodates the interdependence between the two error components by utilizing the capabilities of a copula. As a result, this enhanced model exhibits greater flexibility and significantly mitigates the issue of efficiency overestimation often observed in the original SFM.

Besides obtaining the technological efficiency of various provinces in China, another main objective of this article is to analyze the factors that mainly influence technological efficiency. Rahman et al. [34] studied the main factors affecting the technological efficiency of Nigeria agriculture and found that Important factors related to technical efficiency were labor, fertilizer, age, gender, household size, marital status, other occupation and land ownership. Bojnec et al. [6] studied the main factors affecting the technological efficiency in 10 new EU member states and found that the farm fragmentation, which is defined as the share of area (in %) used by farms has a strong positive impact on technical inefficiency. Furthermore, Njeru [33] mentioned that Family labor play an important role in Kenya's agricultural production, due to agricultural production in Kenya is often conducted at the household level.

The outline of this paper is as follows. Section 2 we explain thoroughly about our proposed model and other necessary statistical properties related to our model including the basic idea of copula. Section 3 is the empirical study of agricultural production in China. Section 4 is conclusion.

# 2 Methodology

The Stochastic Frontier Model (SFM) featuring a disturbance term characterized by asymmetry and notable non-normality. A concise overview of this model will be provided in Sect. 2.1. Subsequently, Sect. 2.2 will revisit the fundamentals of copula, serving as foundational knowledge. These constitute the essential components for the development of the copula-based model, as elucidated in Sect. 2.3.

## *2.1 Stochastic Frontier Model*

Classical models of production consider ideal production as a function $h(x, \beta)$ of a vector $x$ of inputs, where $\beta$ is a vector of parameters. As real production $Y$ can only be less than the ideal one, it can be written as:

$$Y = h(x, \beta) \cdot TE \tag{1}$$

where $TE < 1$, called technical efficiency, is the ratio of actual output y to maximum feasible output $h(x, \beta)$. For instance, the Cobb–Douglas production model can be written as

$$\ln Y = x'\beta - W \tag{2}$$

where $\beta$ is a vector of coefficients and $W = -\ln(TE)$ is a non-negative error term. However, a theoretical problem with this approach is that any measurement error on $Y$ must be embedded in the disturbance $W$, making the estimation of $\beta$ very sensitive to outliers. To solve this problem, Aigner et al. proposed to add a symmetric random noise $V$ to the right-hand side of (2), resulting in the following model,

$$\ln Y = x'\beta + \varepsilon \tag{3a}$$

$$\varepsilon = V - W \tag{3b}$$

where the two error components W and V are assumed to be independent. In this model, the frontier $\exp\left(x'\beta + V\right)$ is stochastic, hence the term "stochastic frontier". The disturbances $\varepsilon$ are now assumed to arise from two sources: (1) productive inefficiency, resulting in a non-negative error term $W$, and (2) firm-specific effects $V$, which can enter the model with either sign. The technical efficiency $TE$ can then be written as

$$TE = \frac{\exp\left(x'\beta + V - W\right)}{\exp(x'\beta + V)} = \exp(-W) \tag{4}$$

The inefficiency error term $W$ is usually assumed to have a gamma, exponential, or half-normal distribution (defined as the distribution of the absolute value of a normal variable). In contrast, the symmetric error term $V$ is usually assumed to have a normal or logistic distribution.

## 2.2 Copula

A copula connects a given number of one-dimensional marginal distributions to form a joint multivariate distribution. In the following, we will limit the presentation to bivariate copula, which will be used later. Sklar's theorem states that any cumulative distribution function $F(x_1, x_2)$ of a two-dimensional random vector $(x_1, x_2)$ can be expressed as

$$F(x_1, x_2) = C(F_1(x_1), F_2(x_2)) \tag{5}$$

where $F1(\cdot)$ and $F2(\cdot)$ are the marginal cdfs of $x_1$ and $x_2$, and $C$ is a bivariate function, called a copula. If $x_1$ and $x_2$ are independent, then $C$ is the product. *A* function $C : [0, 1]^2 \to [0, 1]$ a copula if and only if it satisfies the following properties:

$C(u_1, 0) = C(0, u_2)$ for all $u_1$ and $u_2$ in [0, 1];

$C(u_1, 1) = u_1$ and $C(1, u_2) = u_2$ for all $u_1$ and $u_2$ in [0, 1];

For all $0 \le u_1 \le u_2 \le 1$ and $0 \le v_1 \le v_2 \le 1$,

$$C(u_2, v_2) - C(u_2, v_1) - C(u_2, v_1) - C(u_1, v_1) \ge 0 \tag{6}$$

The Fréchet–Hoeffding theorem states that the following bounds hold any copula $C$:

$$\max(u_1 + u_2 - 1, 0) \le C(u_1, u_2) \le \min(u_1, u_2) \tag{7}$$

for any $(u_1, u_2) \in [0, 1]^2$. The lower and upper Fréchet–Hoeffding bounds correspond to two extreme forms of dependence in which the two variables are, respectively, countermonotonic and comonotonic.

If the random vector $(x_1, x_2)$ has a joint density $f(x_1, x_2)$, it can be expressed as a function of the copula density,

$$c(u_1, u_2) = \frac{\partial^2 C(u_1, u_2)}{\partial u_1 \partial u_2} \tag{8}$$

by the following formula,

$$f(x_1, x_2) = \frac{\partial^2 F(x_1, x_2)}{\partial x_1 \partial x_2} \tag{9a}$$

$$= \frac{\partial^2 C(u_1, u_2)}{\partial u_1 \partial u_2} \frac{\partial F_1(x_1)}{\partial x_1} \frac{\partial F_2(x_2)}{\partial x_2} \tag{9b}$$

$$= c(u_1, u_2) f_1(x_1) f_2(x_2) \tag{9c}$$

where $f_1(x_1)$ and $f_2(x_2)$ are the marginal densities.

The most common measure of dependence between random variables is Pearson's correlation coefficient. However, it only measures linear dependence and is not very informative for asymmetric distributions. To measure nonlinear dependence, rank correlation coefficients such as Kendall's tau and Spearman's rho are more suitable. They can be expressed in terms of the copula as

$$\tau(x_1, x_2) = 4 \iint\limits_{[0,1]^2} C(u_1, u_2) dC(u_1, u_2) - 1 = 4E[C(U_1, U_2)] - 1 \tag{10}$$

$$\rho(x_1, x_2) = 12 \iint\limits_{[0,1]^2} C(u_1, u_2) dC(u_1, u_2) - 3 = 12E[C(U_1, U_2)] - 3 \tag{11}$$

where $(U_1, U_2)$ is a two-dimensional random vector with $C(u_1, u_2)$ as a cdf. Another important concept is that of tail dependence, which describes extreme comovements in the tail of the joint distribution of $(x_1, x_2)$. The lower and upper tail dependence coefficients are defined, respectively, as

$$\lambda_L = \lim_{u \to 0^+} \Pr\left[x_2 \le F_2^{-1}(u) | x_1 \le F_1^{-1}(u)\right] = \lim_{u \to 0^+} \frac{C(u, u)}{u} \tag{12}$$

$$\lambda_U = \lim_{u \to 1^-} \Pr\left[x_2 \le F_2^{-1}(u) | x_1 \le F_1^{-1}(u)\right] = \lim_{u \to 1^-} \frac{1 - 2u + C(u, u)}{u} \tag{13}$$

where $F_1$ and $F_2$ are the marginal cumulative distribution function of $x_1$ and $x_2$, respectively.

To fit a copula to data, one commonly uses families of copula $\{C_\theta\}$ depending on some parameter $\theta$. Copula families can be selected based on their ability to capture positive and negative dependence, as well as tail dependence. Some copulas, such as the Gaussian and Frank copulas, have the property $\lambda_U = \lambda_L = 0$, while most copulas can capture upper or lower tail dependence. For instance, Clayton copulas can measure lower tail dependence, while Gumbel and Joe copulas can measure upper tail dependence. Student-t copula reflects symmetric tail dependence. All of these copulas, except the Gaussian, T and Frank copulas, can only capture positive dependence. However, a copula may be "rotated". There are three rotated forms, with angles 90, 180 and 270°, defined as follows,

$$C_{90}(u_1, u_2) = u_2 - C(1 - u_1, u_2) \tag{14a}$$

$$C_{180}(u_1, u_2) = u_1 + u_2 - 1 + C(1 - u_1, u_2) \tag{14b}$$

$$C_{270}(u_1, u_2) = u_1 - C(u_1, 1 - u_2) \tag{14c}$$

Rotation by 90 and 270° allows for the modeling of negative dependence.

To summarize, different copulas have different characteristics, such as upper tail dependence, lower dependence, positive and negative dependences, etc. Therefore, the above copula families and relevant rotated copula can potentially capture the appropriate dependence between two random variables. Other popular copula families, such as the Farlie–Gumbel–Morgenstern (FGM) and Ali–Mikhail–Haq (AMH) copulas, although widely used in the econometrics literature, have been discarded because they cannot achieve the Fréchet bounds and they can only accommodate relatively weak dependence between the margins.

## *2.3 Copula-Based Stochastic Frontier Model*

The classical SFM makes the strong assumption that the error components $V$ and $W$ are independent. Smith proposed to relax this assumption and to model the dependence between $V$ and $W$ using a copula. The classical model is recovered as a special case corresponding to the product copula. Following Smith, the joint density $f(w, \varepsilon)$ can be obtained from $f(w, v)$ using as

$$f(w, \varepsilon) = f(w, w + \varepsilon) = f_W(w) f_V(w + \varepsilon) c_\theta(F_W(w), F_V(w + \varepsilon)) \tag{15}$$

Marginalizing out $W$ then yields

$$f_\Theta(\varepsilon) = \int_0^{+\infty} f(w, \varepsilon) dw \tag{16}$$

or, equivalently,

$$f_\Theta(\varepsilon) = E_W\left[f_V(W + \varepsilon) c_\theta(F_W(W), F_V(W + \varepsilon))\right] \tag{17}$$

where $E_W[\cdot]$ denotes the expectation with respect to the technical inefficiency $W$ and $\Theta$ represents the vector of all parameters of the marginals and the copula function. Assuming the data to consist of independent cross-sectional observations of n individuals or firms, the log-likelihood function for model (3) is given by

$$L(\beta, \sigma_w, \sigma_v, \theta) = \sum_{i=1}^{n} \log f_\Theta(\varepsilon_i) = \sum_{i=1}^{n} \log f_\Theta\left(y_i - x_i' \beta\right) \tag{18}$$

where $y_i$ is the realization of the output from individual or firm $i$, $x_i$ is explanatory variable vector for individual i, and $\sigma_w$ and $\sigma_v$ are scale parameters of the marginal distributions of $W$ and $V$, respectively. As discussed by Smith, there are very few densities function of $\varepsilon$ for which the maximum likelihood estimate has a closed-form expression. However, each term $f(\varepsilon_i)$ can be written as an expectation, and can easily be approximated using Monte Carlo simulation. The log-likelihood can then be maximized using the maximum simulated likelihood method. Assuming $W$ and $V$ to have, respectively, half-normal and normal distributions, the density function of ε can be written as

$$f(\varepsilon) = \int_0^\infty f_W(w) f_V(w+\varepsilon) c_\theta(F_W(w), F_V(w+\varepsilon))dw \tag{19a}$$

$$= \int_0^\infty \frac{2\exp\left(\frac{-w^2}{2\sigma_w^2}\right)}{\sqrt{2\pi}\sigma_w} f_V(w+\varepsilon) c_\theta(F_W(w), F_V(w+\varepsilon))dw \tag{19b}$$

$$= \int_0^\infty \frac{2\exp\left(\frac{-(\sigma_w w_0)^2}{2\sigma_w^2}\right)}{\sqrt{2\pi}\sigma_w} f_V(\sigma_w w_0+\varepsilon) c_\theta(F_W(\sigma_w w_0), F_V(\sigma_w w_0+\varepsilon))d\sigma_w w_0 \tag{19c}$$

$$= \int_0^\infty \frac{2\exp\left(\frac{-w_0^2}{2}\right)}{\sqrt{2\pi}\sigma_w} f_V(\sigma_w w_0+\varepsilon) c_\theta(F_W(\sigma_w w_0), F_V(\sigma_w w_0+\varepsilon))dw_0 \tag{19d}$$

$$\hat{f}(\varepsilon) = \frac{1}{N}\sum_{r=1}^{N} f_V\left(\sigma_w w_{0,r}+\varepsilon\right) c_\theta\left(F_W\left(\sigma_w w_{0,r}\right), F_V\left(\sigma_w w_{0,r}+\varepsilon\right)\right) \tag{20}$$

where $w_{0,r}, r = 1, \ldots, N$, is a sequence of $N$ random draws from the standard half-normal distribution. The simulated log-likelihood is then

$$\begin{aligned} &L_s(\beta, \sigma_w, \sigma_v, \theta) \\ &= \sum_{i=1}^{n} \log\left[\frac{1}{N}\sum_{r=1}^{N} f_V\left(\sigma_w w_{0,r}+\varepsilon\right) c_\theta\left(F_W\left(\sigma_w w_{0,r}\right), F_V\left(\sigma_w w_{0,r}+\varepsilon\right)\right)\right] \end{aligned} \tag{21}$$

Following Battese and Coelli, the parameters $(\sigma_w, \sigma_v)$ can be transformed to $(\lambda, \sigma)$ with $\lambda = \sigma_w/\sigma_v$ and $\sigma = \sqrt{\sigma_w^2+\sigma_v^2}$. The larger $\lambda$, the greater the inefficiency component in the model. We can also measure the global inefficiency by $\gamma = \sigma_w^2/\left(\sigma_w^2+\sigma_v^2\right)$. The values of $\lambda$ and $\gamma$ reveal whether inefficiency plays an important role in the composite error term.

In stochastic frontier analysis, the technical efficiency terms are of primary interest. They are not observed, but we can estimate their conditional expectations given $\varepsilon$,

$$TE_{\Theta} = E(\exp(-W)|\varepsilon) \tag{22a}$$

$$= \frac{1}{f_{\Theta}(\varepsilon)} \int_{0}^{+\infty} \exp(-w) f(w, \varepsilon) dw \tag{22b}$$

$$= \frac{E_W \exp(-W) f_V(W + \varepsilon) c_\theta(F_W(W), F_V(W + \varepsilon))}{E_W\left[f_V(W + \varepsilon) c_\theta(F_W(W), F_V(W + \varepsilon))\right]} \tag{22c}$$

The nominator and denominator on the right-hand side of (22c) can be approximated using Monte Carlo simulation by

$$TE_{\Theta} = \frac{A}{B} \tag{23a}$$

with

$$A = \frac{1}{N} \sum_{r=1}^{N} \exp(\sigma_w w_{0,ir}) f_V\left(\sigma_w w_{0,ir} + \varepsilon_i\right) c_\theta\left(F_W\left(\sigma_w w_{0,ir}\right), F_V\left(\sigma_w w_{0,ir} + \varepsilon_i\right)\right) \tag{23b}$$

and

$$B = \frac{1}{N} \sum_{r=1}^{N} f_V\left(\sigma_w w_{0,ir} + \varepsilon_i\right) c_\theta\left(F_W\left(\sigma_w w_{0,ir}\right), F_V\left(\sigma_w w_{0,ir} + \varepsilon_i\right)\right). \tag{23c}$$

# 3 Empirical Results

## *3.1 Dataset*

The data used in this study is China's agricultural panel data, sourced from the China Agricultural Statistical Yearbook. It encompasses grain production and five key factors influencing grain production in 31 major Chinese provinces spanning from 1999 to 2019. The descriptive statistics and correlations of variables are in the Appendix A. This paper takes into account five critical input variables in estimating China's grain production function: labor force, land area, fertilizer used, electricity used, and disaster area (Table 1).

**Table 1** Variable description for variables used in production equation

| Variable | Variable | Description | Source |
|---|---|---|---|
| Dependent variable | Grain production output (Y) | Grain production output refers to China's grain production for each year from 1999 to 2019. It includes all crops. (The unit is 10000 tons) | China Agricultural Statistical Yearbook |
| Independent variable | Labor force (L) | Labor force refers to the number of people engaged in agricultural pro-duction in China. (The unit is 10000 people) | |
| | Land area (LA) | Land area refers to the share of land area that is used to cultivate food crops in China's provinces. (The unit is 1000 hectares) | |
| | Fertilizer used (F) | Fertilizer used refers to the total amount of chemical fertilizers used by agriculture in China' provinces every year. (The unit is 10000 tons) | |
| | Electricity used (E) | Electricity used refer to the total amount of electricity used by agriculture in China's provinces every year. (The unit is billion KWH) | |
| | Power of farm machinery (P) | Power of farm machinery, which refer to the sum of the power of various power machinery used in agriculture in China's provinces. (The unit is 10000 KW) | |
| | Disaster area (D) | Disaster area refer to the area of disaster-affected crops in each province of China every year. (The unit is 1000 hectares) | |

## 3.2 Panel Data Unit Root Test

Before we use our data, we need to do panel data unit root test to check whether our variables are robustness. In this study, we use Choi test, which considered by Choi [12], as robustness test. Ex. for our variable $L_{it}$, with $i = 1,2,\ldots,N, t = 1,2,\ldots,T$ and:

$$L_{it} = \rho_i L_{i(t-1)} + u_{it} \tag{24}$$

and $u_{it}$ is integrated of order zero. The null hypothesis is:

$$H_0 : \rho_i = 1 \text{ for all } i \tag{25}$$

which implies that all the time series are unit root nonstationary. The alternative hypothesis is:

**Table 2** The result of Pm test

| | ln $L$ | ln $LA$ | ln $F$ | ln $E$ | ln $P$ | ln $D$ |
|---|---|---|---|---|---|---|
| Test statistic | 1.673 | 3.113 | 4.017 | 8.540 | 6.349 | 26.374 |
| MBF | 0.013 (Strong) | 0.000 (Decisive) | 0.000 (Decisive) | 0.000 (Decisive) | 0.026 (Strong) | 0.000 (Decisive) |

*Note* $^{***}$ indicates significance at the 1% level, $^{**}$ indicates significance at the 5% level

$$H_0 : |\rho_i| < 1 \text{ for at least one } i \text{ for finite } N \tag{26}$$

that is some time series are nonstationary, while others are not. The results are presented in Table 2.

From the results in Table 2, we observe that all variables are stationary.

## 3.3 Model Specification

To analyze the factors affecting agricultural output and measure the technical efficiency of China's grain production, we considered the following production model. The model primarily takes the form of Cobb–Douglas production function where Labor force (L), Land area (LA), fertilizer (F), Electricity used (E), Power of farm machinery (P) and Disaster area (D) are inputs.

$$Y_{it} = \alpha L_{it}^{\beta_1} LA_{it}^{\beta_2} F_{it}^{\beta_3} E_{it}^{\beta_4} P_{it}^{\beta_5} D_{it}^{\beta_6} \tag{27}$$

Transform function into a log production frontier as:

$$\begin{aligned} \ln Y_{it} = \alpha + \beta_1 \ln L_{it} + \beta_2 \ln LA_{it} + \beta_3 \ln F_{it} \\ + \beta_4 \ln E_{it} + \beta_5 \ln P_{it} + \beta_6 \ln D_{it} + U_{it} - V_{it} \end{aligned} \tag{28}$$

where $i$ represents different provinces in China ($i = \{1, 2, \ldots, 31\}$), and $t$ represents different years ($t = \{1, 2, \ldots, 21\}$).

## 3.4 Model Selection

As a sequential estimation method, the copula has to be selected before estimation. So, this part is also about choosing a copula that best fits the data. Before we start use copula, the marginal distribution about $W$ and $V$ are normal and half-normal. This study uses the Akaike Information Criterion (AIC) and the Bayesian Information Criterion (BIC) to pick the best copula among the most common used copula, i.e., Gaussian, Student's t, Clayton, Gumbel, Frank, and Joe (see Appendix B). The results are presented in Table 2.

**Table 3** AIC and BIC for Each Copula-based SFM

| Copula | AIC | BIC |
|---|---|---|
| Gaussian | −442.3532 | −397.5681 |
| Student-t | −419.8965 | −370.6329 |
| Clayton | −443.2086 | −398.4235 |
| Gumbel | −433.6795 | −388.8944 |
| Frank | −439.8509 | −395.0658 |
| Joe | **−444.3026** | **−399.5175** |

Table 3 shows the values of AIC and BIC for each Copula-based stochastic frontier model, where the minimum values are bold numbers. According to both criteria, the best model is Joe copula.

## 3.5 Estimated Results of Copula-Based SFM

This part presents the benchmark result of this paper, where we estimate the stochastic frontier model based on the copulas we chose in the previous section. The results are presented in Table 4. After obtaining the minimum Bayes factor, we observe that most parameters provide decisive evidence for $H_1$ ($H_1 : \beta \neq 0$), with only $\beta_5$ showing moderate evidence for $H_1$. The estimated parameter of the Joe copula is 2.4, providing decisive evidence for $H_1$. This result confirms the significant dependence between error components, justifying the use of the copula-based SFM.

From the above results, we can observe a negative correlation between the number of laborers engaged in grain cultivation and grain output. According to data from the National Bureau of Statistics, the proportion of agricultural workers in the total workforce decreased from 70% in 1978 to 59% in 1990 and further declined to 31.4% in 2013. As many farmers left rural areas to move to cities, they not only

**Table 4** Estimated parameters and minimum bayes factor for copula-based SFM

| Parameter | Estimated parameters | Minimum Bayes factor |
|---|---|---|
| $\alpha$ | 0.3063 | 0.0000 (Decisive) |
| $\beta_1$ | −0.0636 | 0.0017 (Decisive) |
| $\beta_2$ | 0.8882 | 0.0000 (Decisive) |
| $\beta_3$ | 0.1510 | 0.0000 (Decisive) |
| $\beta_4$ | 0.0306 | 0.0000 (Decisive) |
| $\beta_5$ | 0.0295 | 0.2300 (Moderate) |
| $\beta_6$ | −0.0287 | 0.0000 (Decisive) |
| $\sigma_V$ | 0.1304 | 0.0000 (Decisive) |
| $\sigma_U$ | 0.3840 | 0.0000 (Decisive) |
| $\theta$ | 2.4024 | 0.0000 (Decisive) |

improved their own livelihoods but also helped address the labor shortage issues faced by cities due to labor released through reforms and opening-up policies. The increasing migration of farmers from rural areas did not lead to a contraction in agricultural production. On the contrary, total grain output consistently reached new highs. For instance, China's total grain output was 30,476.50 million tons in 1978, but by the end of 2013, it had risen to 60,193.84 million tons. The main reason for this phenomenon is the advancement of agricultural technology. In 2022, China's contribution rate of agricultural technological advancement reached 62.4% (The agricultural technological advancement contribution rate refers to the proportion of agricultural technological progress to the growth rate of total agricultural output). The comprehensive mechanization rate of crop cultivation has reached 73%. The improvement in agricultural technology plays a crucial role in enhancing agricultural product yields.

We also found that the impact of grain sowing area on grain yield is the most significant among all variables (with a coefficient of 0.87). This is primarily because expanding the grain sowing area provides the conditions for concentrated and contiguous grain production, facilitating the advancement of large-scale production of grains. This, in turn, effectively enhances grain yields. Lei et al. [28] utilized a probit model to discover that China's grain production exhibits a clustering trend. This clustering effect is manifested by simultaneous increases in grain cultivation area and yield in certain grain production regions ($I_{\text{it}}$ is the spatial transfer coefficient of grain production, when $I_{\text{it}}$ >1.007, indicating the presence of clustering effect).

## 3.6 Estimate of Technical Efficiency

The efficiency range of TE value that we got in Sect. 3.5 is 0.39–0.96 (average 0.75). After we got TE value, we use K-means clustering method (KCM) to group the result and discuss the characteristics of each group.

K-Means clustering, proposed by MacQueen in 1967, is an unsupervised learning algorithm used for clustering data into a predetermined number of clusters (k clusters). It starts by defining k centroids, ideally positioned to minimize differences in clustering outcomes. Each data point is assigned to the nearest centroid, forming initial clusters. The centroids are then recalculated as cluster centers, iteratively adjusting until they no longer change locations, optimizing clustering. Finally, this algorithm aims at minimizing an objective function, in this case a squared error function. The objective function

$$W(S, C) = \sum_{k=1}^{K} \sum_{i \in S_k} \|TE_i - c_k\|^2 \tag{29}$$

where $S$ is a K-cluster partition of the entity set represented by TE value $TE_i(i \in I)$ in the M-dimensional feature space, consisting of non-empty non-overlapping clusters $S_k$, each with a centroid $c_k$ $(k = 1, 2, \ldots K)$.

Before clustering, we need to determine the number of clusters (k). To do this, we calculate silhouette scores. According to [29], the optimal k is the one that maximizes the silhouette score because it indicates well-separated and interpretable clusters while providing a quantitative and robust measure of clustering quality. For each sample in our dataset, calculate the average distance to other samples within the same cluster (intra-cluster average distance), denoted as $a$. Then, calculate the average distance to all samples in the nearest different cluster (inter-cluster average distance), denoted as $b$. For each sample, the silhouette coefficient $s$ is defined as:

$$s = \frac{b - a}{\max(a, b)} \tag{30}$$

From Table 5, we can see that in this study, the optimal k is determined to be 3.

Then we ahead to K-means clustering progress and get the clustering result (Table 6).

Within-cluster sum of squares (WCSS) is a metric used to evaluate the quality of a clustering solution, particularly in the context of K-means clustering. It measures the compactness or cohesion of the clusters. Typically, a lower WCSS indicates that the data points within clusters are closer to their respective centroids, implying better separation between clusters. From Table 5, we can observe that the WCSS for all three clusters is lower than 2, indicating the reasonableness of our clustering results.

Next, we display the TE value according to the grouping results (Fig. 1).

From this picture we can see that different groups have different characteristics, we renamed Group 3 as Low TE value group, Group 1 as High TE value group, Group 2 as Median TE value group. The TE value of each group are shows below (Table 7).

**Table 5** Silhouette score of each number of clusters

| Clusters number | k = 2 | **k = 3** | k = 4 | k = 5 | k = 6 | k = 7 | k = 8 | k = 9 |
|---|---|---|---|---|---|---|---|---|
| Silhouette score | NA | **0.47** | 0.35 | 0.28 | 0.25 | 0.22 | 0.20 | 0.19 |

**Table 6** Cluster result

| Cluster name | Cluster size | Cluster member | WCSS |
|---|---|---|---|
| 1 | 12 | Shandong, Zhejiang, Xinjiang, Sichuan, Jiangxi, Hubei, Liaoning, Jiangsu, Hunan, Shanghai, Xizang, Jilin | 0.7788 |
| 2 | 8 | Heilongjiang, Chongqing, Guangxi, Fujian, Beijing, Henan, Anhui, Guangdong | 0.6876 |
| 3 | 11 | Shanxi, Shaanxi, Gansu, Neimenggu, Qinhai, Ningxia, Hebei, Yunnan, Tianjing, Guizhou, Hainan | 1.5091 |

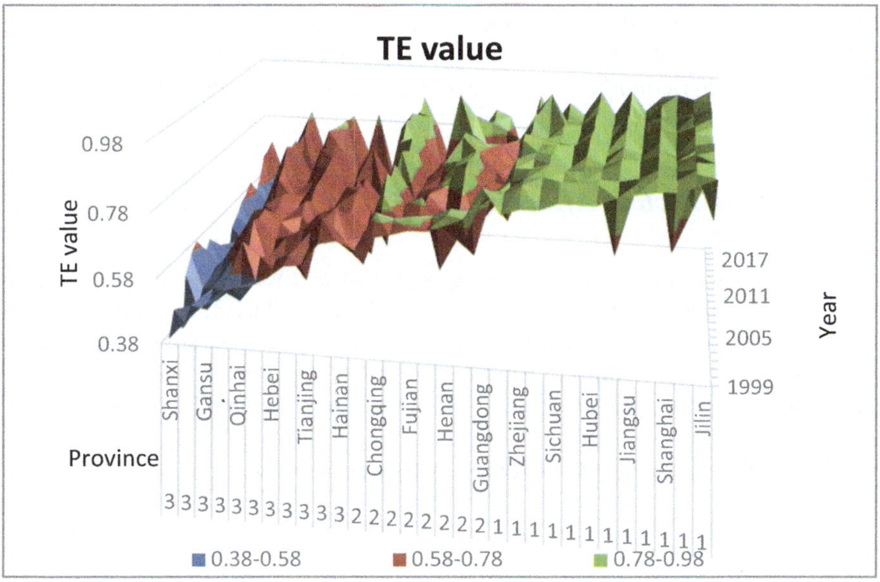


**Fig. 1** TE value of panel data

**Table 7** TE value of each group

| Group name | Minimum TE value | Maximum TE value | Average TE value |
|---|---|---|---|
| Low TE value group | 0.39 | 0.82 | 0.57 |
| High TE value group | 0.64 | 0.98 | 0.87 |
| Median TE value group | 0.55 | 0.92 | 0.75 |

**Low TE value group**: The efficiency range in this group is 0.39–0.82 (average 0.57). This may indicate that the provinces in this group experience low crop yields, inefficient water management, poor soil health, ineffective pest and disease management, inefficient use of fertilizers, pesticides, and other agricultural inputs, a lack of modern agricultural machinery, insufficient knowledge and training, and unfavorable environmental conditions.

**High TE value group**: The efficiency range in this group is 0.64–0.98 (average 0.87). The characteristics of this group are the opposite of those in the Low TE value group, suggesting that the provinces in this group likely achieve higher crop yields and employ efficient water management practices, among other factors.

**Median TE value group**: The efficiency range in this group is 0.55–0.92 (average 0.75). The characteristics of the provinces in this group fall between those of the Low TE value group and the High TE value group.

**TE value change over time**: According to the graph, the overall TE value showed an upward trend from 1999 to 2019. This indicates that China's agricultural production technology and agricultural management level are continuously improving.

## 3.7 Analyze Factors Affecting TE Value

In the last part we got the result of TE value, in this part we will analysis the factors that affect TE value, since we select nine variables, we will use LASSO regression to mitigate the problem of overfitting caused by having too many variables. The descriptive statistics and correlations of variables are in the Appendix A. The description of our variables are presented in Table 8. Then, we can write our TE equation as.

$$TE = \gamma_1 Z_1 + \gamma_2 Z_2 + \gamma_3 Z_3 + \gamma_4 Z_4 + \gamma_5 Z_5 + \gamma_6 Z_6 + \gamma_7 Z_7 + \gamma_8 Z_8 + \gamma_9 Z_9 \quad (31)$$

After using ridge regression, we get the coefficient of our variable in this model:

From the results of the estimated parameters in Table 9, we have observed that $\gamma_1$ is relatively large, indicating that average household size has a significant negative impact on technical efficiency. One crucial reason for this is that the increase in assets input by households exceeded the decrease in labor force. From 1999 to 2012, the ratio of rural household members engaged in agricultural production decreased from 35.69 to 27.85%. However, the average total value of agricultural production assets per rural household increased from 2.92 thousand yuan to 11.41 thousand yuan. As a result, agricultural technology efficiency has actually improved.

We also found that for $\gamma_6$, the coefficients are very small and the minimum bayes factor is large, those indicate that electricity used are not highly correlated with technical efficiency. It is not difficult to notice that in the results, the education level of the rural labor force is negatively correlated with the efficiency of agricultural technology. There may be several reasons for this:

1. Farmers with higher levels of education might be more inclined to explore opportunities outside of agriculture. In 2022, more than 950,000 rural students are expected to leave the countryside and move to the city by taking the college entrance examination. These students are likely to stay in the city for work after graduation and may not return to the countryside, let alone engage in farming. Additionally, farmers with higher levels of education may lack practical experience compared to those who have spent more time directly working in agriculture. This is because educated farmers might diversify their activities, investing time in non-agricultural ventures, which could result in divided attention and reduced efficiency in agricultural operations. Unlike some other large agricultural countries, China, despite having a substantial population engaged in agricultural production, has a relatively low proportion of professional farmers. According to the sample survey data from the Ministry of Agriculture of China in 2020, the number of professional farmers in the country was approximately 17 million,

**Table 8** Variable description for variables used in TE equation

| Variable | Variable | Description | Source |
|---|---|---|---|
| Dependent variable | Technical efficiency (TE) | Technical efficiency that we calculated in 3.5 | Calculated in 3.5 |
| Independent variables | Average household size ($Z_1$) | The average number of laborers per rural household in each province. (The unit is people) | National Bureau of Statistics of China |
| | Total value of agricultural production assets ($Z_2$) | Total value of agricultural production assets per rural household on average representing each province, including agricultural machinery, transportation tools, and production-related buildings. (The unit is 1000 yuan per household) | |
| | Education level of rural labor force ($Z_3$) | Education level of rural labor force refers to average years of education of the rural labor force. (The unit is year) | |
| | Water management ($Z_4$) | Water management refers to irrigated area of farmland. (The unit is 100000 hectares) | China Agricultural Statistical Yearbook |
| | Pesticides used ($Z_5$) | Pesticides used refers to total amount of pesticides used by each province. (The unit is 1000 ton) | |
| | Electricity used ($Z_6$) | Electricity used refer to the total amount of electricity used by agriculture in China's provinces every year. (The unit is billion KWH) | |
| | Power of farm machinery ($Z_7$) | Power of farm machinery refers to the sum of the power of various power machinery used in agriculture in China's provinces. (The unit is 10000 KW) | |
| | Fertilizer used ($Z_8$) | Fertilizer used refers to the total amount of chemical fertilizers used by agriculture in China' provinces every year. (The unit is 10000 tons) | |

**Table 9** Estimated parameters and minimum bayes factor for LASSO regression

| Parameter | Estimated parameters | Minimum Bayes factor |
|---|---|---|
| $\gamma_0$ | 0.7538 | 0.0000 (Decisive) |
| $\gamma_1$ | −0.3170 | 0.0000 (Decisive) |
| $\gamma_2$ | 0.0377 | 0.0000 (Decisive) |
| $\gamma_3$ | −0.2806 | 0.0000 (Decisive) |
| $\gamma_4$ | 0.0854 | 0.0000 (Decisive) |
| $\gamma_5$ | 0.0753 | 0.0000 (Decisive) |
| $\gamma_6$ | −0.0019 | 0.2638 (Moderate) |
| $\gamma_7$ | −0.0755 | 0.0000 (Decisive) |
| $\gamma_8$ | −0.0582 | 0.0016 (Decisive) |

accounting for only 9.6% of the agricultural labor force. Most other agricultural laborers work part-time during busy farming periods, and these laborers may not necessarily possess professional knowledge and experience in agricultural production. Experience often plays a crucial role in understanding the nuances of farming practices, leading to better resource management and higher technical efficiency.

2. Highly educated farmer may tend to adopt new technologies or advanced technologies that may not be suitable or compatible with their specific agriculture environment. These technologies might require more significant investment or have a steeper learning curve, leading to suboptimal resource utilization and lower technical efficiency.
3. Education in rural areas, where most agricultural activities occur, may not always be readily available or accessible. Farmers in such regions might not have access to the same level of education and training as those in urban or more developed areas. According to statistics from the Ministry of Education of China in 2013, during the pre-school education stage, the average teacher in urban areas is responsible for 16.42 students, while in rural areas, the average teacher is responsible for 40.22 students. However, at the primary school stage, the ratio changes to 18.91 in urban areas and 14.63 in rural areas. This shift indicates that the educational imbalance during the pre-school education stage has caused some rural children to receive low-quality education, limiting their ability to pursue higher education. Consequently, they may face challenges in adopting modern agricultural practices, which can impact their technical efficiency.

## 4 Conclusion

The Stochastic Frontier Model (SFM) is a linear regression model that assumes the error component as a combination of a randomly distributed error term $V$ and a non-negative inefficiency term $W$. Traditionally, these two terms are considered independent. However, we have relaxed this assumption and adopted a model that incorporates the relationship between these two error components through a copula. The approach copula based Stochastic Frontier Model employed in this paper involves using a diverse range of copula families to capture various dependence patterns between $V$ and $W$. Model selection is based on the AIC and BIC criteria. For this study, the errors' marginal distributions were assumed to follow normal and half-normal distributions. However, alternative distributions could be considered and jointly tested alongside different copula families.

This study applied this methodology to grain production data from 31 provinces of China. For these data, all investigated copula families outperformed the independence copula according to the AIC and BIC criteria, the best results being obtained for the Joe copula model. The main finding of this study is that TE values vary significantly among provinces in China, and there is a negative correlation between the educational level of the rural labor force and TE values. How to reduce the loss of agricultural labor force while ensuring the education level of the rural population, or in other words, how to make agriculture more attractive to a highly educated labor force, remains a topic for further research.

## Appendix A: Descriptive Statistics and Correlations of Variables

The descriptive statistics and correlation of variables used in production and TE equations are presented in Tables 10, 11,12 and 13.

**Table 10** Descriptive statistics for variables used in production equation

| Statistic | ln *L* | ln *LA* | ln *F* | ln *E* | ln *P* | ln *D* |
|---|---|---|---|---|---|---|
| Mean | 7.3289 | 7.6692 | 4.6302 | 4.2860 | 7.3956 | 5.4502 |
| Median | 7.5843 | 8.0469 | 4.8790 | 4.3289 | 7.5959 | 6.1247 |
| Maximum | 8.9872 | 9.5707 | 6.5738 | 7.5751 | 9.4995 | 8.3333 |
| Minimum | 4.7062 | 3.8395 | 0.9163 | −1.2040 | 4.5430 | −9.2103 |
| Std. Dev | 1.0199 | 1.2299 | 1.2111 | 1.5886 | 1.1004 | 2.5826 |
| Skewness | −0.6251 | −0.9651 | −1.0401 | −0.6516 | −0.5348 | −3.7865 |
| Kurtosis | −0.7230 | −0.1616 | 0.3563 | 1.0905 | −0.0608 | 18.1814 |

**Table 11** Descriptive statistics for variables used in TE equation

| Statistic | ln $Z_1$ | ln $Z_2$ | ln $Z_3$ | ln $Z_4$ | ln $Z_5$ | ln $Z_6$ | ln $Z_7$ | ln $Z_8$ |
|---|---|---|---|---|---|---|---|---|
| Mean | 1.31 | 1.94 | 1.97 | 2.55 | 3.31 | 4.32 | 7.42 | 4.64 |
| Median | 1.32 | 1.94 | 2.00 | 2.72 | 3.84 | 4.36 | 7.61 | 4.89 |
| Maximum | 1.75 | 3.88 | 2.27 | 4.12 | 5.16 | 7.58 | 9.50 | 6.57 |
| Minimum | 0.86 | −0.17 | 0.80 | 0.09 | −0.54 | −1.20 | 4.54 | 0.92 |
| Std. Dev | 0.14 | 0.81 | 0.15 | 1.04 | 1.39 | 1.58 | 1.10 | 1.21 |
| Skewness | −0.18 | 0.17 | −2.30 | −0.61 | −0.90 | −0.64 | −0.56 | −1.04 |
| Kurtosis | 0.58 | −0.34 | 9.32 | −0.68 | −0.23 | 1.05 | −0.03 | 0.34 |

**Table 12** Correlations of variables for variables used in production equation (1)

| Statistic | ln *L* | ln *LA* | ln *F* | ln *E* | ln *P* | ln *D* |
|---|---|---|---|---|---|---|
| ln *L* | 1 | | | | | |
| ln *LA* | 0.7587 | 1 | | | | |
| ln *F* | 0.6728 | 0.8228 | 1 | | | |
| ln *E* | 0.2908 | 0.4642 | 0.5476 | 1 | | |
| ln *P* | 0.7005 | 0.7881 | 0.8010 | 0.3631 | 1 | |
| ln *D* | 0.2863 | 0.4842 | 0.3065 | 0.1404 | 0.2795 | 1 |

**Table 13** Correlations of variables for variables used in TE equation

| Statistic | $\ln Z_1$ | $\ln Z_2$ | $\ln Z_3$ | $\ln Z_4$ | $\ln Z_5$ | $\ln Z_6$ | $\ln Z_7$ | $\ln Z_8$ |
|---|---|---|---|---|---|---|---|---|
| $\ln Z_1$ | 1 | | | | | | | |
| $\ln Z_2$ | 0.174 | 1 | | | | | | |
| $\ln Z_3$ | -0.607 | -0.083 | 1 | | | | | |
| $\ln Z_4$ | -0.026 | 0.161 | 0.277 | 1 | | | | |
| $\ln Z_5$ | -0.026 | -0.087 | 0.414 | 0.816 | 1 | | | |
| $\ln Z_6$ | -0.563 | -0.231 | 0.618 | 0.581 | 0.673 | 1 | | |
| $\ln Z_7$ | -0.025 | 0.253 | 0.283 | 0.925 | 0.802 | 0.363 | 1 | |
| $\ln Z_8$ | -0.059 | 0.033 | 0.402 | 0.836 | 0.900 | 0.548 | 0.801 | 1 |

*Note* To simplify the results, we highlight the strength of the correlation in color for the absolute values of the correlation ranging between 0 and 0.30 (green), between 0.31 and 0.80 (yellow), and between 0.81and 1.0 (red) which indicate the weak, moderate, and strong correlation, respectively. From the results in the table above, we can observe that some variables exhibit a high degree of correlation. Therefore, the panel ridge double-kink model used in this study aims to mitigate multicollinearity and enhance the robustness of the model

## Appendix B: Copula Families Used in This Study

In the following, we briefly summarize the main properties of copula families used in this study.

### B.1. Gaussian Copulas

Gaussian copulas take the form

$$C_{Ga}(u_1, u_2; \rho) = \int_{-\infty}^{\Phi^{-1}(u_1)} \int_{-\infty}^{\Phi^{-1}(u_2)} \frac{1}{2\pi\sqrt{1-r^2}} \exp\left[\frac{x_1^2 + x_2^2 - 2rx_1x_2}{2(1-r^2)}\right] dx_1 dx_2 \quad (32)$$

where $-1 < r < 1$ is Pearson's correlation coefficient and $\Phi$ is the cdf of the standard normal distribution function. This is the copula pertaining to a bivariate normal distribution with standard normal marginals and Pearson's linear correlation coefficient $r$. Parameter $\rho$ is related to Kendall's $\tau$ and Spearman's $\rho$ coefficients by the relations $\tau = (2/\pi)\sin^{-1}(r)$ and $\rho = (6/\pi)\sin^{-1}(r/2)$.

## B.2. T Copula

As Gaussian copulas, T copulas belong to the class of elliptical copula (i.e., they are the copulas of elliptically contoured distributions). However, T copulas can capture tail dependence. They are defined as

$$C(u_1, u_2|r, v) = \int_{-\infty}^{T_v^{-1}(u_1)} dx \int_{-\infty}^{T_v^{-1}(u_1)} dy \left[1 + \frac{x^2 - 2rxy + y^2}{v(1 - r^2)}\right]^{-\frac{v+2}{2}} \tag{33}$$

where $r$ is the Pearson's correlation coefficient and $T_v$ is the cdf of a Student distribution with $v$ degrees of freedom. When $v$ tends to infinity, the T copula converges to the Gaussian copula. The symmetric tail dependence coefficient can be calculated as a function of parameters $r$ and $v$ as

$$\lambda_U = \lambda_L = 2T_{v+1}\left(-\sqrt{v+1}\sqrt{\frac{1-r}{1+r}}\right). \tag{34}$$

## B.3. Clayton Copula

A copula of the form $C(u_1, u_2) = \varphi^{-1}(\varphi(u_1) + \varphi(u_2))$ is said to be Archimedean and function $\varphi$ is said to be its generator [32]. There are different Archimedean copulas based on different generators, such as the Clayton, Gumbel, Frank, Joe and AMH copulas, among others. The Clayton family can reflect lower tail dependence for $\theta > 0$. It is characterized by the following formula,

$$C_{Cl}(u_1, u_2|\theta) = \left(u_1^{\theta} + u_2^{\theta} - 1\right)^{-1/\theta} \tag{35}$$

This copula can only capture a strong lower tail and positive dependence, but it can be rotated and used in capture negative dependence or reflect strong upper tail dependence (see [10, 30]. The corresponding Kendall's tau and lower tail dependence coefficients are, respectively, $\tau_{CL} = \theta/(\theta + 2)$ and $\lambda_L = 2^{-1/\theta}$. The expression of Spearman's rho as a function of $\theta$ is more complicated.

## B.4. Frank Copula

Copulas in the Frank family are the only Archimedean copulas that attain the lower and upper Fréchet bounds, thus allowing for positive and negative dependence. The corresponding copula function is given by

$$C_{Fr}(u_1, u_2) = -\frac{1}{\theta} \ln\left(1 + \frac{(\exp(-\theta u_1) - 1)(\exp(-\theta u_2) - 1)}{\exp(-\theta) - 1}\right) \tag{36}$$

where $\theta \in (-\infty, +\infty) \setminus \{0\}$. Positive (resp., negative) values of $\theta$ correspond to positive (resp., negative) dependence. The independence copula is recovered in the limit when $\theta \to 0$. The rank correlation coefficients are given by

$$\tau = 1 - \frac{4}{\theta} + 4\frac{D_1(\theta)}{\theta} \tag{37}$$

and

$$\rho = 1 - \frac{12}{\theta}[D_1(\theta) - D_2(\theta)] \tag{38}$$

where $D_k(\theta)$ is Debye function

$$D_k(\theta) = \frac{k}{\theta^k} \int_0^\theta \frac{t^k}{\exp(t) - 1} dt, k = 1, 2. \tag{39}$$

## B.5. Gumbel Copula

The bivariate Gumbel copula is given by

$$C_{Gum}(u_1, u_2|\theta) = \exp\left[-\left((-\ln u_1)^{1/\theta} + (-\ln u_2)^{1/\theta}\right)^{\theta}\right] \tag{40}$$

where $\theta \in (1, +\infty)$. It is an asymmetric Archimedean copula that allows for strong upper tail dependence. The rotated Gumbel copula can be applied to capture negative dependence as well. The Kendall's tau of Gumbel copula is given by $\tau_{Cum} = 1 - 1/\theta$, but the Spearman's rho does not have a closed form. The upper tail dependence coefficient is $\lambda_U = 2 - 2^{1/\theta}$.

## B.6. Joe Copula

The Joe copula is defined as follows:

$$C_{joe}(u_1, u_2|\theta) = 1 - \left[(1-u_1)^\theta + (1-u_2)^\theta - (1-u_1)^\theta(1-u_2)^\theta\right]^{1/\theta} \quad (41)$$

where $\theta \geq 1$. This copula can capture upper tail dependence, as does Gumbel copula. But it can capture a stronger upper tail dependence than does Gumbel copula (see [5]); the coefficient of upper tail dependence is $\lambda u = 2 - 2^{1/\theta}$. The Kendall's tau is related to parameter $\theta$ by

$$\tau_{joe} = 1 + \frac{4}{\theta^2}\int_0^1 t\ln(t)(1-t)^{2(1-\theta)/\theta}dt \quad (42)$$

The relationship between Spearman's rho and parameter $\theta$ does not have a closed form expression. The rotated Joe copula can describe negative dependence as well.

## References


1. Aigner, D.J., Lovell, C.A.K., Schmidt, P.S.: Formulation and estimation of stochastic frontier production function models. J. Econ. **6**, 21–37 (1977)
2. Ayuttaya, K.N.: Planting season choice and technical efficiency in Thailand's arabica coffee and macadamia nut production: case study Northern of Thailand, Diploma thesis, Chiang Mai University (2011)
3. Battese, G.E., Coelli, T.J.: Frontier production functions, technical efficiency and panel data: with application to paddy farmers in India. J. Product. Anal. **3**, 153–169 (1992)
4. Bezat-Jarzębowska, A., Rembisz, W.: Efficiency-focused economic modeling of competitiveness in the agri-food sector. Procedia. Soc. Behav. Sci. **81**, 359–365 (2013)
5. Bhat, C.R., Eluru, N.: A copula-based approach to accommodate residential self-selection effects in travel behavior modeling. Transp. Res., Part B, Methodol. **43**(7), 749–765 (2009)
6. Bojnec, Š, Fertő, I., Jámbor, A., et al.: Determinants of technical efficiency in agriculture in new EU member states from Central and Eastern Europe. Acta Oeconomica **64**(2), 197–217 (2014)
7. Boyer, B.H., Gibson, M.S., Loretan, M.: Pitfalls in tests for changes in correlations, IFS Discussion Paper No 597R, Federal Reserves Board (1999)
8. Brechmann, E.C., Schepsmeier, U.: Modeling dependence with C- and D-Vine Copulas: the R package CDVine. J. Stat. Softw. **52**(3), 1–29 (2013)
9. Burns, R.: The simulated Maximum Likelihood Estimation of Stochastic Frontier Models with Correlated Error Components. The University of Sydney, Sydney (2004)
10. Cech, C.: Copula-based top-down approaches in financial risk aggregation (2006). Available from http://ssrn.com/abstract=953888 or https://doi.org/10.2139/ssrn.953888

11. Chaovanapoonphol, Y., Singvejsakul, J., Sriboonchitta, S.: Technical efficiency of rice production in the Upper North of Thailand: clustering copula-based stochastic frontier analysis. Agriculture **12**(10), 1585 (2022)
12. Choi, I.: Unit root tests for panel data. J. Int. Money Financ. **20**(2), 249–272 (2001)
13. Coelli, T.J., Rao, D.S.P., Battese, G.E.: An Introduction to Efficiency and Productivity Analysis. Kluwer, Boston (1998)
14. Coelli, T.J., Rahman, S., Thirtle, C.: Technical, allocative, cost and scale efficiencies in Bangladesh rice cultivation: a non-parametric approach. J. Agric. Econ. **53**, 607–626 (2002)
15. Cullinane, K., Song, D.W.: Estimating the relative efficiency of European container ports: a stochastic frontier analysis. Res. Transp. Econ. **16**, 85–115 (2006)
16. Đokić, D., Zekić, S., Jurjević, Ž., & Matkovski, B.: Drivers of technical efficiency in agriculture in the Western Balkans and New EU Memeber States. Custos Agronegocio Online, **16**, 2–15 (2020)
17. Dou, H.W., Ming, T.Z., Xu, J., et al.: Numerical simulation of pollutant propagation characteristics in a three-dimensional urban traffic system. China Environ. Sci. **38**(1), 51–58 (2018)
18. Greene, W.: Distinguishing between heterogeneity and inefficiency: stochastic frontier analysis of the World Health Organization's panel data on national health care systems. Health Econ. **13**, 959–980 (2004)
19. Greene, W.H.: Econometric Analysis, 6th edn. Prentice Hall, Upper Saddle River, NJ, USA (2008)
20. Greene, W.: A stochastic frontier model with correction for sample selection. J. Product. Anal. **34**(1), 15–24 (2010)
21. Farrell, M.J.: The measurement of productive efficiency. J. R. Stat. Soc: series A (General) **120**(3), 253–281 (1957)
22. Jondrow, J., Lovell, C.A.K., Materov, I.S., Schmidt, P.: On the estimation of technical inefficiency in the stochastic frontier production function model. J. Econ. **19**, 233–238 (1982)
23. Justice, G., Djokoto, F.K., Gidiglo, F.Y., Srofenyoh, K.A., Agyei-Henaku, A.-O., Afrane Arthur, A.A., Badu-Prah, C.|John Fry (Reviewing editor).: Sectoral and spatio-temporal differentiation in technical efficiency: a meta-regression. Cogent Econ. Financ. **8**, 1 (2020)
24. Kao, C., Lee, L., Pitt, M.M.: Simulated maximum likelihood estimation of the linear expenditure system with binding non-negativity constraints. Ann. Econ. Financ. **2**, 203–223 (2001)
25. Kumar, P.: Probability distributions and estimation of Ali–Mikhail–Haq copula. Appl. Math. Sci. (Ruse) **4**(14), 657–666 (2010)
26. Kumbhakar, S.C., Lovell, C.A.K.: Stochastic Frontier Analysis. Cambridge University Press, Cambridge, UK (2000)
27. Latruffe, L., Bravo-Ureta, B.E., Carpentier, A., et al.: Subsidies and technical efficiency in agriculture: Evidence from European dairy farms. Am. J Agric. Econ. **99**(3), 783–799 (2017)
28. Lei, C., Qi, Q., Jiang, L., Fu, Y., & Liang, Q.: Analysis of current and qualified major grain producing areas in China in the last 30 years. Sustainability, **14**(5), 2909 (2022)
29. Litster, A.L., Buchanan, J.W.: Radiographic and echocardiographic measurement of the heart in obese cats. Vet. Radiol. Ultrasound **41**(4), 320–325 (2000)
30. Luo, J.: Stepwise estimation of D-Vines with arbitrary specified copula pairs and EDA Tools, Diploma thesis, Technische Universität München (2010)
31. Meeusen, W., van den Broeck, J.: Efficiency estimation from Cobb-Douglas production functions with composed error. Int. Econ. Rev. **18**, 435–444 (1977)
32. Nelsen, R.B.: An Introduction to Copulas, 2nd edn. Springer, New York (2006)
33. Njeru, J.: Factors influencing technical efficiencies among selected wheat farmers in Uasin Gishu District, Kenya. In: AERC Research paper, p. 206 (2010)
34. Rahman, S., Wiboonpongse, A., Sriboonchitta, S., Chaovanapoonphol, Y.: Production efficiency of jasmine rice producers in Northern and North-eastern Thailand. J. Agric. Econ. **60**(2), 419–435 (2009)
35. Rahman, S., Wiboonpongse, A., Sriboonchitta, S., Kanmanee, K.: Total factor productivity growth and convergence in Northern Thai agriculture. Afr. J. Agric. Res. **17**(7), 2689–2700 (2012)

36. Rahman, S., Hasan, M.K.: Impact of environmental production conditions on productivity and efficiency: a case study of wheat farmers in Bangladesh. J. Environ. Manag. **88**(4), 1495–1504 (2008)
37. Rahman, S.A., Umar, H.S.: Measurement of technical efficiency and its determinants in crop production in Lafia local government area of Nasarawa State, Nigeria. Agro-Sci. **8**(2) (2009)
38. Sklar, M.: Fonctions de repartition à n dimensions et leurs marges. Publ. Inst. Stat. Univ. Paris **8**, 229–231 (1959)
39. Smith, M.D.: Stochastic frontier models with dependent error components. Econ. J. **11**, 172–192 (2008)
40. Sriboonchitta, S.: Evaluation of cost efficiency of Thai public universities. Int. J. Intell. Syst. Technol. Appl. **5**, 361–374 (2012)
41. Sriboonchitta, S., Liu, J., Wiboonpongse, A., Denoeux, T.: A double-copula stochastic frontier model with dependent error components and correction for sample selection. Int. J. Approx. Reason. **80**, 174–184 (2017)
42. Sriboonchitta, S., Nguyen, H.T., Wiboonpongse, A., Liu, J.: Modeling volatility and dependency of agricultural price and production indices of Thailand: static versus time-varying copulas. Int. J. Approx. Reason. **54**, 793–808 (2013)
43. Sriboonchitta, S., Wiboonpongse, A.: On the estimation of stochastic production frontiers with self-selectivity: jasmine and non-jasmine rice in Thailand. CMU J. Nat. Sci. **4**, 105–124 (2005)
44. Sriboonchitta, S., Wiboonpongse, A.: Technical efficiency of rural micro and community enterprises in the Upper North of Thailand. In: Asia-Pacific Productivity Conference, pp. 17–19. Seoul National University, Seoul, Korea (2006)
45. Sriboonchitta, S., Wiboonpongse, A.: The effects of production inputs, technical efficiency and other factors on jasmine and non-jasmine rice yields in Thailand, In: Asia-Pacific Productivity Conference, pp. 14–16. Brisbane, Australia (2004)
46. Tesema, T.: Determinants of production efficiency of maize-dominated farmers in western parts of Ethiopia in Gudeya Bila district: evidence under shifting cultivation area. Sci. World J. **2022**(1), 3355224 (2022)
47. Trivedi, P., Zimmer, D.: Copula modeling: an introduction for practitioners. Found. Trends Econ. **1**(1), 1–111 (2005)
48. Tiedemann, T., Latacz-Lohmann, U.: Production risk and technical efficiency in organic and conventional agriculture–the case of arable farms in Germany. J. Agric. Econ. **64**(1), 73–96 (2013)
49. Wiboonpongse, A., Liu, J., Sriboonchitta, S., et al.: Modeling dependence between error components of the stochastic frontier model using copula: application to intercrop coffee production in Northern Thailand. Int. J. Approx. Reason. **65**, 34–44 (2015)
50. Wiboonpongse, A., Sriboonchitta, S,: Agricultural production growth assessment for agroeconomic zones in Northern Thailand using statistical data. In: Asia Pacific Productivity Conference 2008. Taipei, Taiwan, Academic Sinica (2008)
51. Wiboonpongse, A., Sriboonchitta, S.: Impact of technical skill on potato production efficiency in Thailand. In: Asia Pacific Productivity Conference 2008. Taipei, Taiwan, Academic Sinica (2008)
52. Wiboonpongse, A., Sriboonchitta, S., Battese, E.G.: The effects of production input, technical inefficiency and biological risk on jasmine and non-jasmine rice yields in Thailand. Empir. Econ. Lett. **4**(3) (2005)
53. Wiboonpongse, A., Sriboonchitta, S., Rahman, S., Calkins, P.: Thanes Sriwichailumphun, Joint determination of the choice of planting season and technical efficiency of potato in Northern Thailand: a comparison of Greene's versus Heckman's sample selection approach. Afr. J. Bus. Manag. **6**(12), 4504–4513 (2012)
54. Zellner, A., Kmenta, J., Drèze, F.: Specification and estimation of Cobb-Douglas production functions. Econometrica **34**, 784–795 (1966)

# Analysis of Factors Affecting China's Demographic Fertility Transition: A Random Forest Algorithm Approach

**Xiang Qing Lu, Roengchai Tansuchat, and Jian Xu Liu**

**Abstract** The aims of this research are to analyze the four sectors of influences on China's fertility transition in terms of social development, economy, education, and policies by random forest algorithm, and combine them with the classical theory of demographic transition to compare the differences among the three stages of China's demographic transition periods, 1949–1977, 1978–2001, and 2002–2022. This paper collects annual data from different databases on 18 influential features including fertility. The data were preprocessed by data cleaning, data filling, and other methods to improve the accuracy of the model. According to the results of Gini important features ranking, the main reasons affecting the fertility transition in China have gone through three stages of evolution, from "satisfying basic survival needs" to "pursuing self-development and wealth accumulation in order to improve the standard of living" to "Self-value realization and improvement of the quality of life". Based on these findings, several insights are provided in terms of government policy to current policies on fertility and social welfare systems.

**Keywords** Chinese fertility transition · Demographic dividend · Random forest algorithm

X. Q. Lu
Faculty of Economics, Chiang Mai University, Chiang Mai 50200, Thailand
e-mail: xiangqing_lu@cmu.ac.th

R. Tansuchat (✉)
The Center of Excellence in Econometrics, Faculty of Economics, Chiang Mai University, Chiang Mai 50200, Thailand
e-mail: roengchaitan@gmail.com

J. X. Liu
School of Economics, Shandong University of Finance and Economics, Jinan 250014, China
e-mail: 20180881@sdufe.edu.cn

 
V. Kreinovich et al. (eds.), *Applications of Optimal Transport to Economics and Related Topics*, Studies in Systems, Decision and Control 556,
https://doi.org/10.1007/978-3-031-67770-0_32

# 1 Introduction

According to the World Bank analysis, China is currently one of the fastest growing and largest economies in the world. It contributed an average of 38.6% to world economic growth during the period 2013–2021, making it a major engine of global economic growth. Why is China growing so fast? There is now a widespread perspective that China's rapid economic growth over the past few decades comes from the "demographic dividend" [1–3] Bloom et al. [4] estimate the contribution of China's "demographic dividend" to GDP per capita growth between the early 1980s and early 2000s lies in the range of one-six to one-quarter, which is by no means negligible. The "demographic dividend" generally occurs during the demographic transition of a country or region from stage I to stage III, that is, the transition from high birth and high death rates to low birth and low death rates [5]. During this period, the decline in mortality and fertility underpins a rise in the proportion of the population of working age. For example, the decline in fertility leads to a significant reduction in the burden of child-rearing and a rise in the proportion of women participating in the social labor force; while those born under the high fertility of the first stage reach adulthood one after another, and the proportion of the working population continues to expand; and at the same time, even though the population of the elderly is gradually increasing, the burden of old-age rearing is not too heavy, thus contributing to the growth of growth in per capita income. The whole country's economy has shown a high savings, high investment, and high growth trend.

While this transition has been influenced by a variety of factors, such as improvements in health care, increased life expectancy, and changes in societal norms regarding family size [6]. The longevity of the transition, in turn, implies the longevity of the demographic dividend that can be capitalized on. China's demographic transition is characterized by two main features: an extremely rapid transition, which took only 30 years that makes fertility rate decreased from 5 to 2.1, compared to developed countries where the demographic transition lasted for one or two hundred years; and mandatory governmental interventions in individual childbearing (Table 1). The introduction and implementation of the family planning policy led to a significant decline in the fertility rate, which stimulated the rate of demographic transition and drove China's high economic growth. This high economic growth has thus been achieved to some extent at the expense of the duration of the demographic dividend and the potential for future development [7].

It is important to note that the "demographic dividend" itself is only a "temporary opportunity". "Temporary" means that, although declining fertility reduces the burden of child-rearing and increases the share of the working population and productivity during the dividend period, sooner or later that part of the working population will grow old and become a "debt" to society. The more this demographic dividend, the heavier the "demographic debt" that need to be repaid afterwards. The faster the demographic structure shift, the faster the population debt will come in. As the world's most populous country, China, while accelerating the enjoyment of this great and temporary "demographic dividend", is about to face the ensuing problems

**Table 1** Timeline of Government Fertility Policy in China, 1949–2023

| Year | Government Fertility Policies (family planning policy) |
|---|---|
| 1949 | Encourage and support childbearing |
| 1953 | Proper birth control |
| 1962 | Advocating family planning |
| 1964 | "One is not a lot, two is just right, three is more" |
| 1973 | "Later, longer, fewer" policy |
| 1978 | Propose "one preferably, two at most" |
| 1980 | One child policy, The most restrictive decade, 1983–1993 |
| 1982 | Family planning was set as the basic state policy |
| 1984 | Proposing to take care of rural single girls to have a second child |
| 2001 | Parents who are both only children can have a second child |
| 2002 | Advocates late marriage and late childbearing |
| 2013 | Parents can have a second child as long as one of them is an only child |
| 2015 | The two-child policy was fully lifted |
| 2021 | The implementation of a comprehensive three-child policy |

of excessively low fertility, population shrinkage and population aging as the "demographic dividend" gradually disappeared [8, 9]. The Chinese government abandoned the one child policy in 2011 and fully liberalized the two-child policy from 2015. The two-child policy has had some positive effects on the fertility which causes fertility constantly to increase until 2018, but henceforth fertility started to decline. Even though Chinese government continued issued a policy to encourage three children in 2021; however, as of the end of 2022, China's total population had declined by 850,000, the first negative population growth since 1960. The total fertility rate (TFR) is only 0.68, well below the population maintenance rate of 2.1 and, according to the projections in this paper, it will decrease to 0.319 by the end of 2023 (Fig. 1).

So, is this current low birth rate in China good or bad? [6] argue that low fertility favors economic growth in low-income countries due to low dependency ratios, and therefore maintaining a low fertility rate shares in more savings and investment, which in turn raises living standards. China's rapid growth seems to validate this. Other scholars believe that China has inevitably fallen into the "low fertility trap" [10]. That is, as fertility decreased to 1.5 children per woman, due to factors such as changing views on fertility and increasing life pressure, fertility will continue to decline, which is difficult or even impossible to reverse to some extent. The lack of a newborn working population will be accompanied by a rising proportion of elderly people which will eventually lead to a series of social problems such as economic stagnation, deflation, and pressure on social welfare. In the long run, therefore, the excessive decline in China's fertility rate is not conducive to economic development and will become one of an increasingly thorniest socioeconomic issue.

To explore this issue, it is necessary to understand the intrinsic influence mechanisms that affect fertility change in China. However, previous studies have often

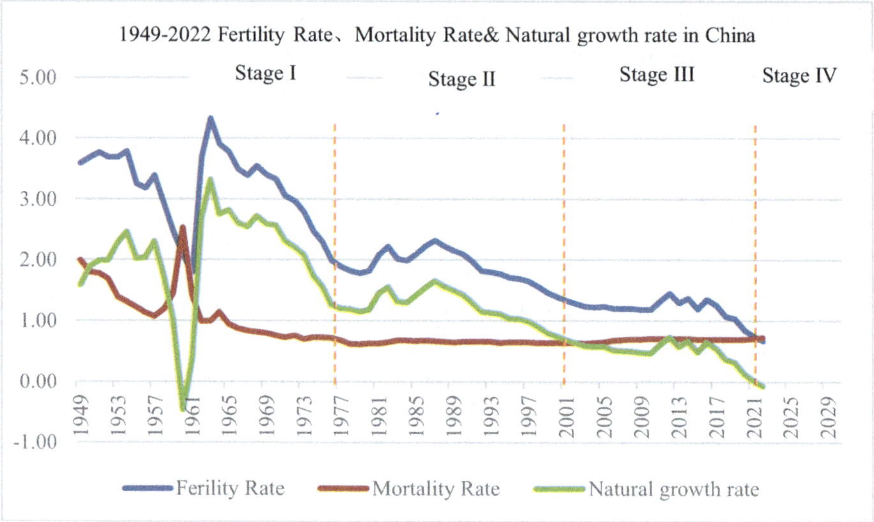


**Source:** China Statistical Yearbook & CEIC database

**Fig. 1** Population fertility rate, mortality rate and natural growth rate in China

neglected the fact that the main factors affecting fertility in different stages or phases of the demographic transition are different. Besides, most of studies use linear model or parametric models, which require high data accuracy, are sensitive to outliers, are easily affected by data noise, and are prone to problems such as overfitting, multicollinearity, and bias prediction. Random Forest is a popular machine learning algorithm that reduces the risk of overfitting of a single decision tree by combining multiple decision trees, thus improving the accuracy of the overall model. It is relatively insensitive to noise and outliers in the data, which makes it excellent at handling complex data in the real world. Random forests can also handle missing data and estimate the importance of each feature, which helps researchers identify the most critical influences and perform feature selection, which improves the interpretability of the model. Therefore, in this paper, we fill in the missing values and analyze the influencing factors of the fertility transition at three different periods of China's "demographic transition" from 1949 to 2022 based on the Random Forest algorithm. This is conducive to a more accurate understanding of the influencing mechanisms of fertility fluctuations in China at different phases of the demographic transition, to formulate corresponding fertility policies in a targeted manner.

The following is the paper's outline. We explain the support demographic transition theory of this paper and conduct some literature reviews in Sect. 2. The data and methodology are covered in Sect. 3. The findings of the empirical study are presented in Sect. 4. Section 5 concludes.

## 2 Theory and Literature Review

### 2.1 Theoretical Foundation

Demographic transition is a phenomenon and theory which refers to the historical shift from high birth rates and high death rates in societies with minimal technology, education (especially of women), and economic development, to low birth rates and low death rates in societies with advanced technology, education, and economic development, as well as the stages between these two scenarios [11]. The theory was proposed by American demographer Warren Thompson in 1929 [12]. Adolphe Landry of France made similar observations on demographic patterns and population growth potential around 1934 [13]. In the 1940 and 1950s, Frank W. Notestein developed a more formal theory of demographic transition [14]. In the 2000s, Oded Galor researched the various influence mechanisms that may induce demographic transition, texting their empirical validity and their potential role in the transition from stagnation to growth, and emphasized the importance of demographic transition for economic growth [15]. In 2011, the unified growth theory was completed, and the demographic transition became an important part of the unified growth theory [16]. By 2009, the existence of a negative correlation between fertility and industrial development had become one of the most widely accepted findings in social science [17].

It is important to note that the Classic Model of Demographic Transition divides the demographic transition into three or five stages in different papers, which are like the four stages in this paper and not fundamentally different. We separate China's demographic transition into four stages based on its own policies and age structure as shown in Figs. 1 and 2.

Taking 1978 as the boundary of China's reform and opening-up policy, the first phase (Stage I) covers the period 1949–1977. During this period the overall average fertility rate was at a relatively high level. China began to control the population in a systematic way, so the fertility rate was on a downward trend during this period in general. During the three years of the Great Famine from 1959 to 1961, the fertility rate fell sharply to 1.813% in 1961, and after 1961, when the economic situation improved, the fertility rate rebounded sharply, to a historical peak of 4.36% in 1963, and then declined gradually, to 1.89% in 1977. Therefore, we add the GF as a dummy variable to measure the impact of this period of the Great Famine on fertility between 1959 and 1961. Overall, the population structure has a more standardized shape of an expanding pyramid as Fig. 2, the census pyramid in the years 1953 and 1964.

In 2001, more than 7% of China's population aged over 65 officially entered an aging society. Taking this as a threshold, the second phase (Stage II) lasted from 1978 to 2001. During this period the fertility rate showed a fluctuating and slowly declining trend, and the mortality rate declined due to improvements in health care or other reasons, resulting in a positive rate of natural population growth, and overall slower and more stable population growth. The bottom of the demographic pyramid began to shrink slightly, and the pyramid's middle and top began to thicken, with a

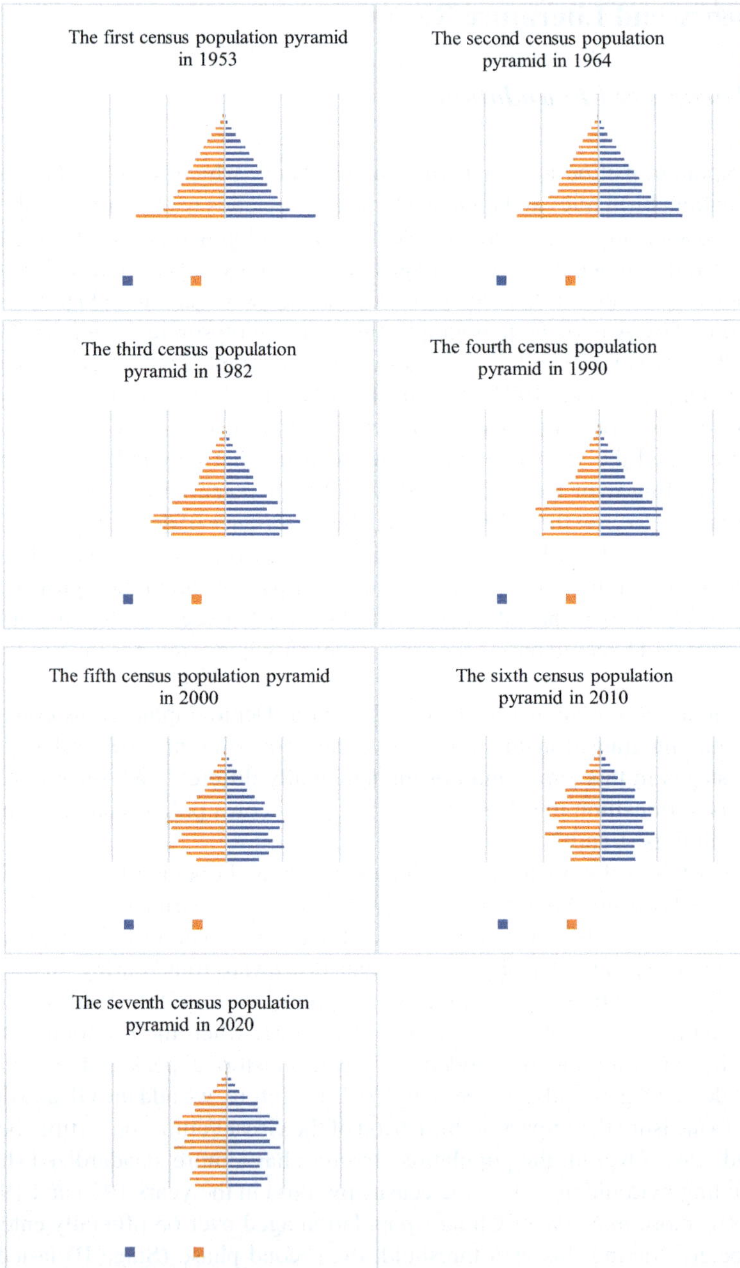


**Source:** China Statistical Yearbook

**Fig. 2** Seven national population census pyramids in China higher proportion of the population aged 20–35 years, as seen in the census pyramid among the years 1982, 1990, and 2000.

Taking 2022 as the cut-off point for the first negative population growth, the third phase (Stage III) is 2002–2022, during which, owing to changes in life attitudes and economic development, the birth and death rates have both declined, and the natural population growth rate is still positive, but the rate of growth has slowed down considerably. The population structure is further transformed into a contracting pyramid, with the bottom of the pyramid collapsing further, the middle and top thickening further, and a higher proportion of the population aged 35–59 years, as reflected in the census pyramid under the years 2000 and 2010.

The fourth phase (Stage IV) is after 2023 when China is expected to enter a super-aged society. The demographic pyramid will further shift to a rectangular or even funnel shape, which will further aggravate the problems of population shrinkage, lack of economic development potential, and social age. This paper identifies the main factors affecting fertility in China based on the first three stages so that policymakers can formulate more effective policies to cope with the challenges of the upcoming fourth stage.

## 2.2 *Literature Review*

Research on the factors influencing fertility involves the intersection of several fields such as social development, economics, education, and policies. Firstly, the influence of socio-economic development on fertility is becoming more and more prominent, especially income per capita, the economic cost of raising children, and the cost of time. Li [18] concluded that the relationship between economic factors, the degree of social security, the cost of raising children, and the fertility rate is negatively correlated. Liu [19] used Lasso regressions to study the effects of GDP, per capita wage, and other factors on the birth rates and concluded that both per capita wage and child support ratio have a positive effect on the birth rate. Sun and Wei [20] used the Barro-Becker modeling Framework and modified the model of de Silva and Tenreyro [21] by introducing endogenous parental time share. The result is that the high-income countries' fertility rate declines as the number of children is further substituted by the intimate relationship with children, and the human capital investment rises as parents balance the marginal utility of parental timeshare with the marginal utility of the child's future income.

Secondly, there is a change in the fertility rate brought about by advances in medical and health conditions, related fertility policies, and the increase in female education. Narayan and Peng [22] analyzed the determinants of fertility in China from 1957 to 2000 by using the Cointegration Tests and the ECM model and concluded that family planning policies have had a large impact on fertility in China, while long-term socio-economic development, such as further female education and mortality decline will be essential for sustaining the low fertility levels. Pezzulo et al. [23] Used the Geographical clustering method to analyze panel data in 2010–2016 through 70 countries, the results showed that areas with high-fertility clusters were mostly associated with areas of low prevalence of women with secondary or higher education, low use of contraception, and high prevalence of sexual intercourse, education, low use of contraception, and high unmet needs for family planning, although exceptions existed. Quan [24] used Ridge regression to analyze the factors influencing birth rates from 1999 to 2020, showing that the rates of female illiteracy, the elderly dependency ratio, the juvenile dependency ratio, and elderly dependency have a major impact on fertility rates in China.

In addition, there are fluctuations in fertility due to the rise in women's social status, the increase in women's labor force participation, and changes in the concept of marriage and childbearing. Chen and Deng [25] used discrete choice models such as Logit and Probit models to investigate the methodology of variables affecting fertility intentions of married, divorced, and widowed women under 52 years of age in China. Based on this information, the results show that there is a negative relationship between age, education level, and fertility intentions. Using the ARDL model, Khattak [26] argued that fertility in Pakistan during 1980–2018 is associated with the infant mortality rate, and female labor force participation rate and negatively associated with real income, unemployment rate, inflation rate, and GDP growth rate. Chen et al. [27] using factor and cluster analysis, it is argued that per capita disposable income, the number of marriages registered, and the index of practitioners are the most important indicators affecting the birth rate in China during the period 2002–2020.

However, previous studies have often neglected the fact that the main influences on fertility in different stages of the demographic transition are different. As a result, there is often a situation in which the relationship between fertility and a certain variable is ambiguous. For example, some scholars believe that per capita wages in China have a positive effect on fertility [19], while others believe that fertility declines with rising income in high-income countries [20]. This is precisely due to the fact that the data they use and the countries or regions they study are at different stages of demographic transition. Therefore, this paper will analyze the factors influencing the transition of China's birth rate based on different stages of the transition, which can complement previous studies to a certain extent.

Moreover, the models routinely used by most scholars are still linear or parametric models such as linear regression, Lasso, Ridge, ECM, ARDL, or discrete choice such as Logit and Probit models. These models assume specific functional forms and have limited ability to model nonlinear relationships. They may not be able to capture complex nonlinear effects. And being sensitive to outliers, in some cases, the models may suffer from covariance problems, leading to problems such as unstable parameter estimates or the need for a large number of samples to avoid overfitting. In this study, the random forest model with the lowest mean square error (MSE) is selected from several candidate models. The random forest model can effectively capture nonlinear relationships and is more suitable for realistic various irregular data. In particular, when analyzing population macroeconomic data, the data samples are usually annual, which makes the introduction of random forests significant. By introducing the bootstrap repeated sampling technique, random forest can effectively avoid the overfitting problem caused by insufficient sample capacity. In addition, this paper combines several imputation methods with random forest to help fix the missing values, which makes the overall model have stronger robustness, nonlinear modeling capability, ability to adapt to high-dimensional data, reduce the risk of overfitting, and thus provide more reliable prediction results.

In conclusion, this paper analyzes the factors influencing the fertility transition in three different stages of China's "demographic transition" from 1949 to 2022 by filling in the missing values and based on the random forest algorithm. On the basis of the existing literature, the analysis of the selection of features and the ranking of the importance of the features of the three stages of China's demographic transition is supplemented.

# 3 Methodology

## *3.1 Data and Variables*

The purpose of this study is to examine the multifaceted influences on fertility at different stages of the demographic transition, categorizing them into a number of dimensions, including social development, the economy, education, policy, and external shocks. In the area of social development, we consider a range of characteristic variables including the infant mortality rate (IMR), mortality rates (MR), elderly dependency ratios (EDR), child and adolescent dependency ratios (CDR), suicide rates (SR), proportion of female in the Labor force (PFLF), registered unemployment rates (RUR), the proportion of women in the childbearing age (PWCA), marriage rates (MaR), and divorce rates (DiR). In the economic domain, we looked at characteristic variables such as GDP per capita (GDPPC), GDP growth rate (GDPGR), disposable income per capita (DIPC), consumer price index (CPI), and Real estate prices growth rate of residential (REP). In the education domain, we focus on the proportion of Females with Secondary education and higher (PFSE), Ellis (1993) suggests that educated women lose their expected income due to childbearing. In addition, we consider the impact of China's family planning policies (PP), which include the one-child policy (implemented since 1980), the two-child policy (liberalized since 2015), and the three-child policy (encouraged since 2021). In addition, we consider the impact of external shocks, such as the Great Famine of 1959–1961 (GF) and the COVID-19 epidemic of 2019–2021 (COVID), on fertility. Table 2 provides the names, symbols, descriptions, and sources of the data to better understand the characteristics of the study data.

The time frame of our data is annual from the founding of China (1949) to the end of 2022. We start with data preprocessing and directly deleted variables (e.g., female labor force participation rate, consumer price index, etc.) with more than 20% frequency of missing values in the first phase of the demographic transition (1949–1977). Since the missing rate is too high and the method of filling missing values

**Table 2** Data definition

| Sectors | Variables | Description | Unit | Source |
|---|---|---|---|---|
| Social development | $FR_t$ | The total Fertility Rate in China | % | China Statistical Yearbook |
| | $MR_t$ | The total Mortality rate in China | % | China Statistical Yearbook |
| | $IMR_t$ | Infant Mortality rate | Per-Thousand Live Births | Bloomberg database |
| | $EDR_t$ | Elderly population dependency ratio, which is Ratio of people older than 64 to the working population 15–64 years | % | National Bureau of Statistics |
| | $CDR_t$ | Child and adolescent dependency ratio, which is Ratio of people younger than 15 to the working population 15–64 years | % | National Bureau of Statistics |
| | $SR_t$ | Suicide rate, measuring mental health and social stress | % | World Bank |
| | $PFLF_t$ | Female Labor Force Participation Rate, which is the Proportion of female in the Labor force | % | UN database |
| | $RUR_t$ | Registered unemployment rate | % | CEIC database |
| | $PWCA_t$ | Proportion of women in the childbearing age | % | China Statistical Yearbook |
| | $MaR_t$ | Marriage Rate | % | CEIC database |
| | $DiR_t$ | Divorce Rate | % | CEIC database |
| Economics | $GDPPC_t$ | Gross Domestic Product Per Capita | RMB | CEIC database |
| | $GDPGR_t$ | Gross Domestic Product Growth rate | % | National Bureau of Statistics |
| | $DIPC_t$ | Disposable Income per Capita | % | National Bureau of Statistics |

(continued)

**Table 2** (continued)

| Sectors | Variables | Description | Unit | Source |
|---|---|---|---|---|
| | $CPI_t$ | Chinese Consumer Price Index | % | CEIC database |
| | $REP_t$ | Real estate: residential prices growth rate | % | CEIC database |
| Education | $PFSE_t$ | Proportion of Female with Secondary education and higher | % | CEIC database |
| Policy | $PP_t$ | China's Population policies as dummy variables | | Family planning commission |
| Shocks | $GF_t$ | The great famine within 1959–1961 as dummy variable | | None |
| | $COVID_t$ | The shock of Covid-19 within 2019–2022 as dummy variable | | None |

introduces a lot of noise and retained only eight variables in the first phase, namely, the infant mortality rate (IMR), the child-adolescent dependency ratios (CDR), the mortality rate (MR), the GDP growth rate (GDPGR), the disposable income per capita (DIPC), the registered unemployment rate (RUR), great famine shocks (GF), and the family planning policy of fewer and better births (PP), which are the eight variables. Meanwhile, the frequency of missing values of IMR, GDPGR, and RUR are 17.2, 10.3, and 6.9%, respectively. we compare five methods to fill the missing values: replace by mean, replace by the median, replace by mode, replace by recursive partitioning and regression tree (RPART), and replace by multivariate imputation by chained equations (MICE). Finally, the missing values of IMR are filled by the mode method, missing values of GDPGR are filled by the median method, and missing values of RUR are filled by the MICE method as shown in Fig. 3 and Table 3. In the second phase of the demographic transition (1978–2001), we define the 10 years from 1983 to 1993 when the one-child policy was most strictly enforced as 1 in the dummy variable, and the rest of the time period is defined as 0. In the third phase of the demographic transition (2002–2022), we set up two dummy variables to denote the full liberalization of the two-child vs. three-child policy and the main shock event of the COVID-19 epidemic in 2019–2021.

## Model Selection

Based on previous research and literature review, this study carefully selected a series of well-validated machine learning algorithms in terms of model performance

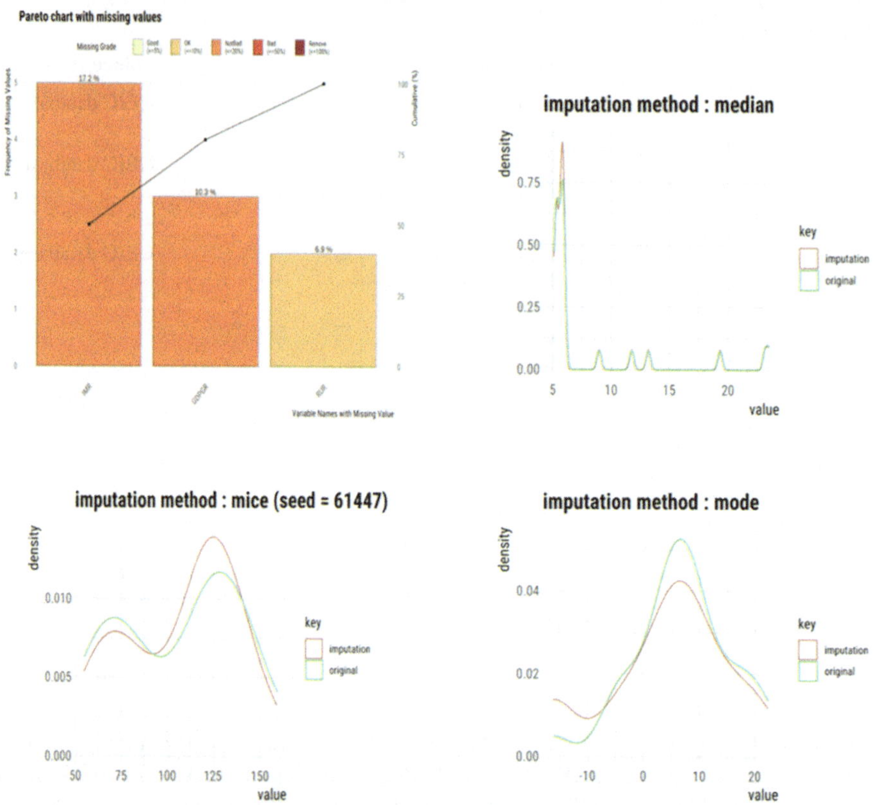


**Fig. 3** Missing values frequency and imputation shape

comparison, which in total included twelve different models. These models cover a variety of common machine learning methods such as Linear regression, Linear regression with exhaustive search, Ridge regression with minimum cross-validation error, Ridge regression based on one standard error rule, Lasso regression (regularization parameters using lambda.min and lambda.1se), Prune tree, Bagging, Random Forest, Linear Support Vector Machine (SVM), Nonlinear SVM with Radial Basis Function (RBF) kernel, and Neural Network algorithms (NN). We first carefully preprocessed the original dataset by randomly dividing it into two parts, occupying 2/3 and 1/3 proportions of the total data, respectively. The former is used as training data to ensure that the model has sufficient data information to effectively learn patterns and correlations in the data. The latter is used as test data, also called the Out-of-Bag (OOB) sample, aiming to evaluate the performance of the model on untouched data to verify its generalization ability.

In the process of model performance comparison, we adopted the mean square error (MSE) as the main evaluation metric to measure the magnitude of the model's

**Table 3** Fixing missing value within 1949–1977

| Method | | Skewness | | Kurtosis | | Results |
|---|---|---|---|---|---|---|
| | | Original | Imputation | Original | Imputation | |
| $GDPGR_t$ | Mean | −0.3908 | −0.4101 | 0.6436 | 1.0593 | Mode |
| | Median | −0.3908 | −0.4376 | 0.6436 | 1.0704 | |
| | Mode | −0.3908 | −0.5261 | 0.6436 | −0.2020 | |
| | RPART | −0.3908 | −0.4047 | 0.6436 | 1.0560 | |
| | MICE | −0.3908 | −0.4072 | 0.6436 | 1.0537 | |
| $RUR_t$ | Mean | 2.2022 | 2.2728 | 3.7651 | 4.2180 | Median |
| | Median | 2.2022 | 2.3205 | 3.7651 | 4.3473 | |
| | Mode | 2.2022 | 2.3206 | 3.7651 | 4.3494 | |
| | RPART | 2.2022 | 2.3176 | 3.7651 | 4.3475 | |
| | MICE | 2.2022 | 2.1174 | 3.7651 | 3.5653 | |
| $IMR_t$ | Mean | −0.1929 | −0.2095 | −1.4358 | −1.0278 | MICE |
| | Median | −0.1929 | −0.4250 | −1.4358 | −1.0408 | |
| | Mode | −0.1929 | −0.5017 | −1.4358 | −1.2954 | |
| | RPART | −0.1929 | −0.0352 | −1.4358 | −1.1246 | |
| | MICE | −0.1929 | −0.3675 | −1.4358 | −1.1077 | |

prediction error. The MSE is an intuitive metric that allows us to quantify the difference between the model's predictions and the actual observations. When the MSE value is small, this means that the difference between the model's predicted values and the actual observations is small, reflecting the superior performance of the model. Therefore, we chose the model with the smallest MSE as the best selection criterion to ensure sufficient scientific basis and statistical credibility in the process of model selection, as well as to better achieve the achievement of the research objectives. Table 4 shows the results of those algorithms, which ultimately show the optimal performance of the random forest algorithm.

## 3.2 Random Forest Algorithm

The Random Forest algorithm is a potent machine learning technique designed to build an integrated model comprising multiple decision trees through the application of two layers of randomness. These layers involve randomization in both data and features. In the first layer of randomness, known as Data Sampling, Random Forest randomly selects n samples from the original training dataset using bootstrap repetitive sampling, thereby creating a new set of samples for training individual decision trees. This process is iterated to generate multiple decision trees, collectively forming a "forest". In the second layer of randomness, known as Feature Extraction, each decision tree is constructed by randomly selecting a subset of features for

**Table 4** Model selection

| Method | Description | MSE |
|---|---|---|
| LS | The Least square model | 0.0466 |
| ELR | Linear regression with exhaustive search | 0.0356 |
| Ridge 1 | Ridge with minimum Cross Validation error (min) | 0.0148 |
| Ridge 2 | Ridge with one Standard error rule (1se) | 0.0149 |
| Lasso 1 | Lasso With lambda.min | 0.0392 |
| Lasso 2 | Lasso With lambda.1se | 0.0373 |
| Prunetree | Decision tree algorithm | 0.1782 |
| Bagging | Bootstrap aggregating algorithm | 0.0166 |
| RF | Random Forest algorithm | 0.0112 |
| SVM.lm | Linear support vector machine algorithm (SVM) | 0.0116 |
| SVM.nlm | Non-Linear SVM with an RBF kernel | 0.0152 |
| NN | Neural Network algorithm | 0.0387 |

*Note* MSE is a mean square error, the smaller the better $MSE_i = \frac{\sum(y_i - \hat{y}_i)^2}{n}$

training, as opposed to using all available features. This ensures that each tree is not overly reliant on specific features, contributing to increased model diversity. Once the construction of each decision tree within the Random Forest is complete, the classification or prediction results for new data points are determined through voting or averaging across these decision trees. In regression problems, the final result is the average of predicted values from all decision trees, while in classification problems, the final result is determined by a vote based on the predictions of all decision trees. This study primarily focuses on regression problems. The detailed workflow of the Random Forest algorithm is illustrated in Fig. 4 below.

Based on the data that have been pre-processed, different numbers of feature variables were selected at different stages of demographic transition. Specifically, we selected eight key feature variables at the first stage of the demographic transition, 17 feature variables at the second stage, and 18 feature variables at the third stage. These feature variables represent the factors that we believe may have an impact on fertility. One of the reasons for the popularity of random forests is the availability of variable importance measures (VIMs), and there are two widely used VIMs called impurity importance and permutation importance [28]. Impurity Importance is directly derived from tree-based models, such as decision trees and random forests. Therefore, when utilizing these types of models, Impurity Importance is a natural and intuitive choice. On the other hand, Permutation Importance is more applicable to non-tree models, such as linear models and support vector machines.

Calle and Urrea [29] showed that rankings based on the impurity VIM can be more robust to perturbations of the data compared with those obtained with the permutation importance. Besides, since the Gini impurity index is commonly used as the splitting criterion in classification trees, the corresponding impurity importance

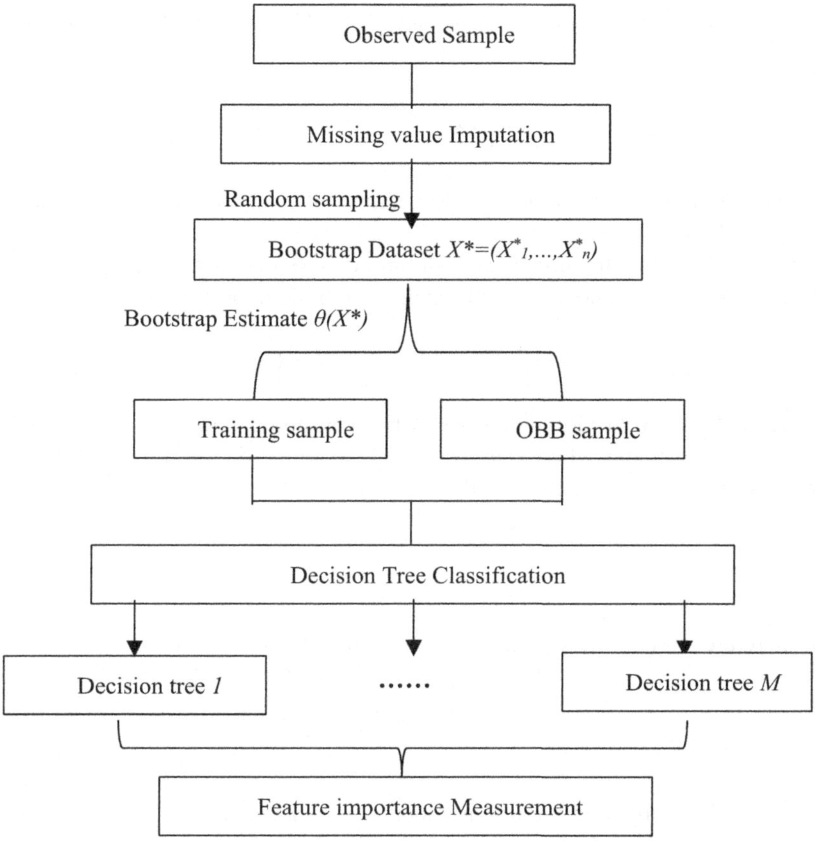


**Fig. 4** The flow of the random forest algorithm

is often called the Gini importance score. To identify the key features influencing fertility at different stages, we utilized the Gini impurity index of the Random Forest Algorithm to assess the importance of the characteristic variables in this study.

The Gini impurity index reflects the probability that a randomly selected sample in the sample set will be misclassified. In other words, the smaller the Gini impurity index, the lower the probability that the selected sample in the set is misclassified, indicating a higher purity of the set, i.e., the more significant the effect of the feature variable on the target variable. On the contrary, if the Gini impurity index is larger, it indicates the more impure the set is, i.e., the weaker the influence of the characteristic variables on the target variable. By using the Random Forest algorithm and the Gini impurity index, we are able to identify the characteristic variables that have a significant impact on fertility in different stages of the demographic transition, which contributes to a deeper understanding of the drivers of fertility.

Assuming that there are $m$ features $X_1, X_2, X_3, \ldots, X_m$, the Gini impurity index or Gini importance score for each feature $X_j$, i.e., the average change in node splitting impurity of the *jth* feature across all decision trees of a random forest as below [30]:

$$GI_m = \sum_{k=1}^{|k|}\sum_{k\prime\neq k} P_{mk}P_{mk\prime} = 1 - \sum_{k=1}^{|k|} P_{mk}^2 \tag{1}$$

where $GI_m$ is Gini impurity index; $k$ is the number of categories. $P_{mk}$ is the proportion of category $k$ in node $m$.

The importance of feature $X_j$ at node $m$, i.e., the amount of change in the Gini impurity index before and after the branching of node m as shown:

$$VIM_{jm}^{(Gini)} = GI_m - GI_l - GI_r \tag{2}$$

Here $GI_l$ and $GI_r$ are the Gini indices of the two new nodes before and after branching, respectively. The nodes where feature $X_j$ appears in the decision tree are the set $M$, then at the first $i$ the importance of the tree is like:

$$VIM_{ij}^{(Gini)} = \sum_{m\in M} VIM_{jm}^{(Gini)} \tag{3}$$

If there are a total of $n$ trees in a random forest, then we have:

$$VIM_{j}^{(Gini)} = \sum_{i=1}^{n} VIM_{ij}^{(Gini)} \tag{4}$$

At last, all the obtained importance scores were normalized by:

$$VIM_{j}^{(Gini)\prime} = VIM_{j}^{(Gini)}\left(\sum\nolimits_{i=1}^{m} VIM_{i}^{(Gini)}\right)^{-1} \tag{5}$$

# 4 Empirical Results and Analysis

## *4.1 Random Forest Regression*

In the previous Random Forest data preprocess, we randomly divided the observations into a training set and a test set and applied random forests to the training set for 5–6 different values of the number of splitting variables *m*. Figure 5 shows the number of decision trees, it can be seen that, in China's demographic transition *Phase 1*, when the number of decision trees exceeds 200, the error of the model is basically unchanged; similar in *Phase 2*, when the number of decision trees exceeds 200, the error of the model basically remains stable; in *Phase 3*, when the number of trees exceeds 350, the error of the model is basically become horizontal. With the increase in the number of trees, the error is showing a downward trend. When the number of decision trees is between 400 and 500 the error fluctuation rate is less than 0.1%, and therefore we are using 500 decision trees in this case can ensure the stability and reliability of the results.

Table 5 presents the Random Forest Regression among with the results, where *% Var explained* is expressed as the overall explanation of the variance of the response variable by the predictor variables, in this paper, *Phase 1* predictor variables explained about 85.46% of the total variance, i.e., $R^2 = 0.8546$ for this regression; *Phase 2* predictor variables explained about 90.84% of the total variance, i.e., $R^2 = 0.9084$ for this regression; *Phase 3* predictor variables explained about 89.78% of the total variance, i.e., $R^2 = 0.8978$ for this regression. All are sizable numbers, suggesting that fertility is strongly influenced with these features.

## *4.2 Feature Importance Ranking of Influencing Factors*

Each decision tree in the integrated learning process of the random forest is assigned a Gini importance score for each feature. The final feature importance score is the average or weighted mean of the scores from all the decision trees. In this paper Gini importance score (based on the reduction of Gini impurity) is mainly used to evaluate the importance of features as we mentioned on 3.3. These metrics take into account how the features reduce the impurity of the nodes during the decision tree splitting process, and thus measure the contribution of the features to the classification task [31]. IncNodePurity in Fig. 6 refers to increase in node purity, node purity is actually the reduction of RSS (residual sum of squares), that is, the data or class in the node are the same. Breiman [28] pointed that the impurity importancc is also known as the mean decrease of impurity (MDI), the permutation importance as mean decrease of accuracy (MDA). Hence Mean Decrease Gini impurity index which is a measure of the degree of mixing of different categories of samples in a node. The larger the value of IncNodePurity, the more important the feature is.

### 4.2.1 Phase 1

In *Phase 1* (1949–1977), the important features affecting fertility in China were, in decreasing order of importance, as follows: GDP per capita (GDPPC), infant mortality rate (IMR), child and adolescent dependency ratio (CDR), mortality rate (MR), GDP growth rate (GDPGR), disposable income per capita (DIPC), elderly dependency ratio (EDR), registered unemployment rate (RUR), great famine (GF), and family planning policy (PP). The importance ratings of these features are all greater than 0.2 except for the family planning policy, reflecting the extent of their impact on fertility.

At first, China's relatively low economic level and low per capita GDP in the early 20th Century resulted in many families relying on labor to sustain the family. As a

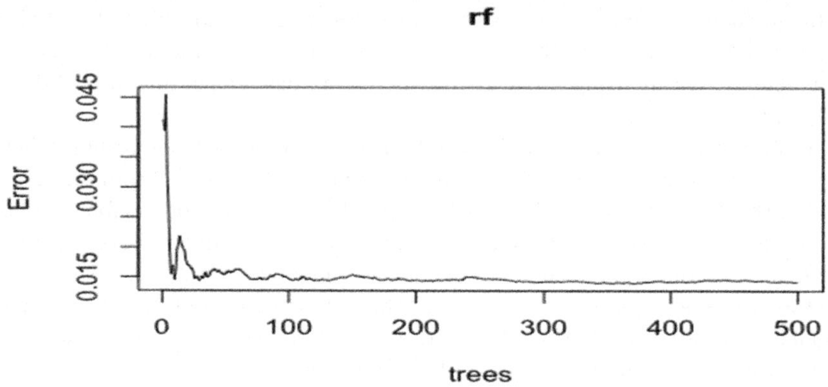


**Phase 1 (1949-1977)**

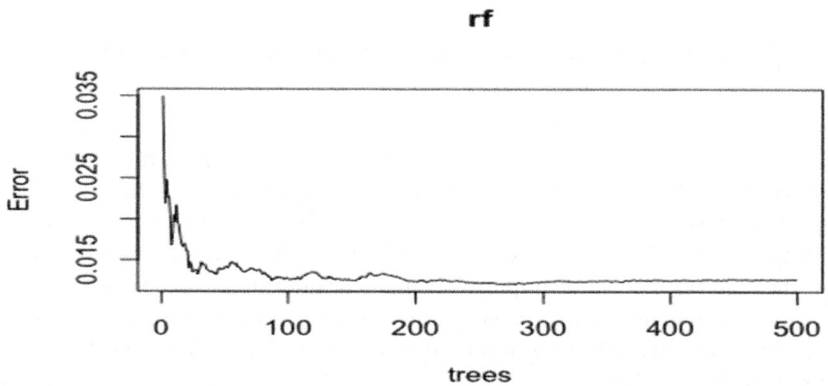


**Phase 2 (1978-2001)**

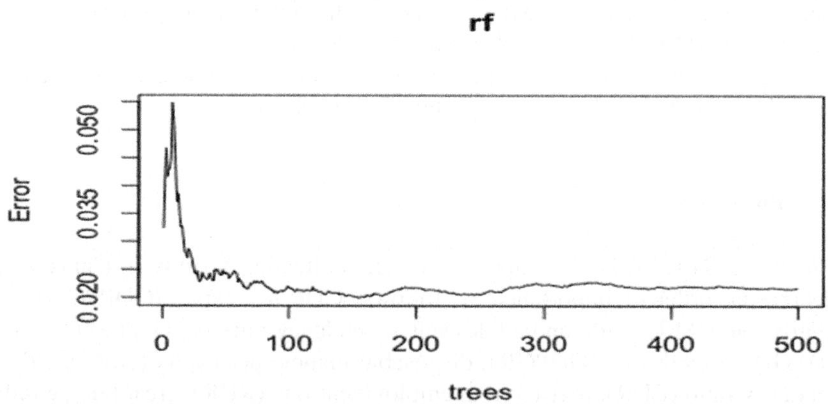


**Phase 3 (2002-2022)**

**Fig. 5** MSE of the RF algorithm for three demographic transition phases

**Table 5** Random forest regression

| No. of tree | | No. of variables tried at each split | Mean of squared residuals | %Var explained |
|---|---|---|---|---|
| Phase 1 | 500 | 6 | 0.01238884 | 85.46 |
| Phase 2 | 500 | 5 | 0.01273147 | 90.84 |
| Phase 3 | 500 | 6 | 0.02205127 | 89.78 |

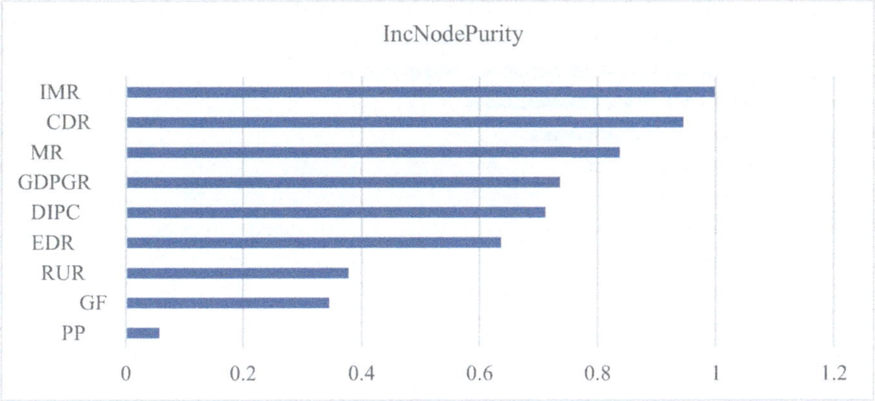


**Phase 1 (1949-1977)**

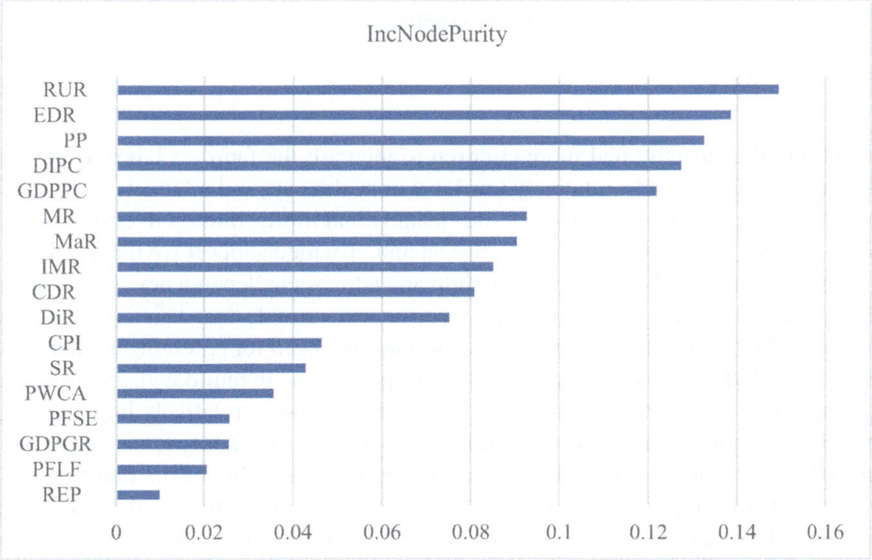


**Phase 2 (1978-2001)**

**Fig. 6** Feature importance ranking of influencing factors among three Phases

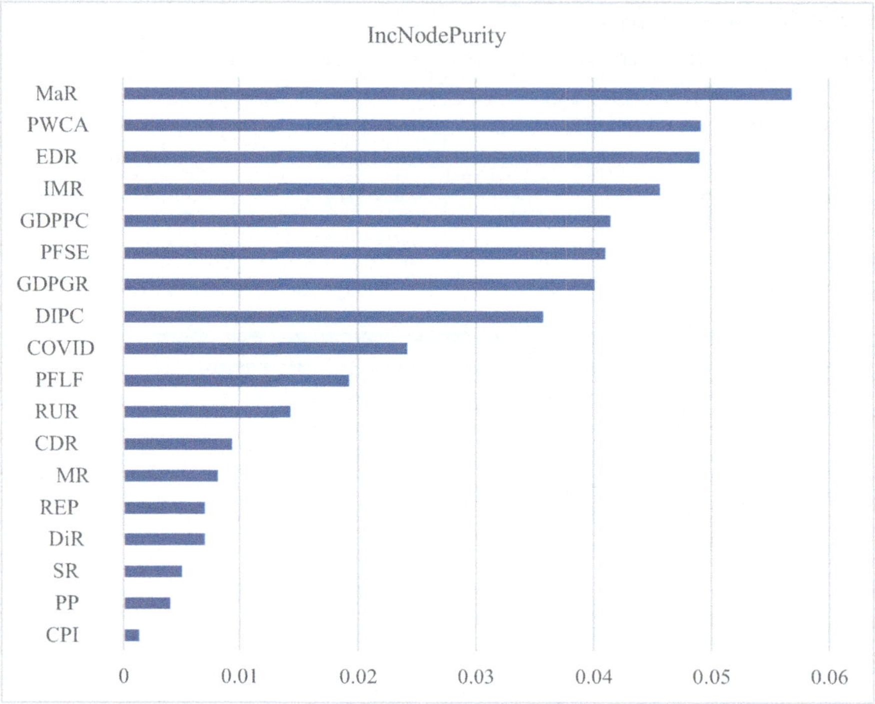


**Phase 3 (2002-2022)**

**Fig. 6** (continued)

result, families usually had more children to increase the family's labor resources. At that time, China had a relatively low level of health care and relatively high infant mortality rates. This meant that families usually had more children to ensure that at least some of them survived into adulthood and provided support for the family.

Secondly, relatively high levels of child and adolescent dependency (CDR) mean that there are more young people to care for and provide for. This is one of the reasons why families have more children to meet the need for parenting adolescents. Such high CDRs are usually associated with a population pyramid with an expansive pattern, i.e., with high birth and death rates. In this case, a relatively large number of children being born compensates for the possible loss of family labor due to high mortality rates, thus ensuring the economic and social stability of the family.

Overall, the imperfections of the social welfare system, the need for families for the elderly, and the high level of CDRs provide incentives for families to have more children. This incentive is particularly evident in the context of an expansive population pyramid, where families are more inclined to have more children in order to sustain the family's livelihood and social support system. As a result, the EDRs rank relatively low in terms of their impact on fertility during this period.

Although China-proposed "One is not a lot, two is just right, three is more" in 1964 for planned population control, there was a lack of explicit government interventions to limit births, and thus the policy impacts had a relatively small degree of influence on the fertility rate at that time. It is noteworthy that the Great Famine of 1959–1961 has a degree of impact score on fertility of 0.345, which is much higher than that of the family planning policy, which is 0. 058. This suggests that the Great Famine had a significant impact on fertility because the extreme food shortages that occurred in this period led to a significant decline in the fertility rate. This result is consistent with the trend in fertility fluctuations shown in Fig. 1.

### 4.2.2 Phase 2

In the Second Phase (1978–2001), the important features affecting China's fertility rate, in decreasing order of importance, are as follows: the registered unemployment rate (RUR), the elderly dependency ratio (EDR), family planning policy (PP), per capita disposable income (DIPC), per capita GDP (GDPPC), the mortality rate (MR), marriage rate (MaR), the infant mortality rate (IMR), the child-adolescent dependency ratio (CDR), the divorce rate (DiR), the consumer price index (CPI), the suicide rate (SR), the proportion of women in childbearing age (PWCA), the proportion of Female with Secondary education and higher (PFSE), the GDP growth rate (GDPGR), the proportion of female in the Labor force (PFLF), and real estate prices growth rate of residential (REP).

In the early years of China's reform and opening, reforms were mainly concentrated in rural areas and some cities. By the early 1980s, the scope of reform was expanded to include more areas, including state-owned enterprises. During this period, some SOEs began to face declining efficiency and economic downturns, which led to a wave of layoffs and layoffs. In the late 1980s, the Chinese government embarked on a reform of SOEs, which involved large-scale enterprise restructuring, privatization, and layoffs. Many uncompetitive SOEs were closed or downsized, resulting in millions of workers being laid off. At the same time, large-scale rural–urban labor migration increased competition for jobs in cities, leaving some urban residents unemployed. This trend also led some people to change from stable employment to self-employment, choosing to be laid off or leave their old jobs. As a result, unemployment became the highest-rated impact feature in this phase.

The population age pyramid showed a tendency to contract at the bottom and thicken in the middle and top during this period, especially evident in the 2000s, due to the decline in fertility and the rise in the share of the labor force and the old-age population. This has also led to a rise in the EDRs, which has become an important cause of fertility decline. From 1983 to 1993, China implemented a stricter one-child policy, a family planning policy that included measures such as the implantation of an intrauterine device in one child and the ligation of a second child, and contraceptive and birth control measures were considered to be the most stringent. This policy had a significant impact on fertility decline, ranking third.

Rapid socio-economic development, especially the increase in disposable income per capita (DIPC) and GDP per capita (GDPPC) ranked high in its impact on fertility. This suggests that families are more likely to delay or reduce their decisions to have children when they are more economically stable and affluent.

It is important to note that the Chinese government only began to adjust its policies in the real estate market after 1998 to encourage the development of the real estate industry. Therefore, the impact of house prices on fertility during the period 1978–2001 is scored low. This means that real estate prices had a relatively small impact on fertility during this phase.

### 4.2.3 Phase 3

During *Phase 3* (2002–2022), the important features affecting China's fertility rate, in decreasing order of importance, are as follows: marriage rate (MaR), the proportion of women of childbearing age (PWCA), the elderly dependency ratio (EDR), the infant mortality rate (IMR), GDP per capita (GDPPC), and proportion of Female with Secondary education and higher (PFSE), GDP growth rate (GDPGR), disposable income per capita (DIPC), COVID-19, the proportion of female in the Labor force (PFLF), registered unemployment rate (RUR), child adolescent dependency ratio (CDR), the mortality rate (MR), Real estate prices growth rate of residential (REP), the divorce rate (DiR), suicide rate (SR), family planning policy (PP), and consumer price index (CPI).

During this period, people's views on marriage changed as society developed and values evolved. More and more people pay more attention to personal freedom and independence and no longer regard marriage as an inevitable life choice. This led to a decline in the marriage rate as some people chose not to get married or delayed marriage. This period has seen an increase in the number of women choosing not to marry or not to have children, an increase in cohabitation outside of marriage, and an increase in births to single mothers. The increase in female education and the rise in the number of professional careers has led many women to choose to delay childbearing in order to pursue education and careers, which is associated with an increase in individualism and autonomy, resulting in a decrease in the number of women of childbearing age in a given age bracket. MaR and the PWCA thus become the top two most important influencing features.

Both fertility and mortality continue to decline during this stage, and the demographic pyramid takes on a contractionary character. Hence the sandwich generation appears, that middle-aged adults must feed both their children and their aging parents, and are living under greater pressure, which reduces fertility to a certain extent. Secondly, the increase in the proportion of the elderly population may have led to more medical resources and attention being focused on the elderly, while the medical care of newborns may have been neglected to a certain extent. The development policy of "the rich first pushing those being rich later" has actually led to a widening gap between the rich and the poor. China's medical resources are unevenly distributed among different regions. Environmental factors such as pollution, declining air quality, and food safety issues during industrialization, as well as some infectious diseases or unhealthy social and lifestyle patterns such as smoking and alcohol abuse, may have a negative impact on the health of newborns. Thus, the rise in IMR has brought about a decline in fertility.

In addition, China's rapid economic growth, which has brought about a rise in per capita GDP and per capita disposable income, has effectively improved the status of women and the proportion of women with higher education, and women with higher education are usually more inclined to delay childbearing. Besides, the degree of impact of COVID-19 from 2019 to 2021 on fertility rate scores 0.024, which is due to the scarcity of healthcare resources and regulations at that time that had an impact on fertility rates.

It can be seen that compared to *Phase 2*, the decline in unemployment makes it no longer the main influential feature affecting fertility, and the decline in fertility brings about a decline in the CDR, but the cost of adolescent upbringing is still an issue that deserves to be continued to be explored in the future. The CPI, as a proxy variable for inflation, suggests that despite the rise in prices and the cost of living over the course of the 21st century, it does not usually directly affect fertility rates from the graph. Of the second and third phases, EDRs are among the top three features affecting fertility and are indicative of the increasing pressures of China's aging society. At last, policy influences such as the long-standing one-child policy may still have an impact on the fertility decisions of women of childbearing age. Despite the release of the comprehensive two-child and three-child policies in 2015 and 2021, we can see that the score is only 0.0039, which may be due to the lag effect of the policies, but for now, the impact of the policy is minimal.

## 4.3 RF Partial Dependency Diagram

Partial Dependency Plots, or PDPs for short, show the functional relationship between the influence of one or two feature variables on the model's predictions: approximately linear, monotonic, or more complex. When the PDP curve of a feature is almost horizontal or jittering irregularly, this feature may be a useless feature. When the PDP curve of a feature is very steep, it means that the contribution of this feature is relatively large. The resultant graph can be seen in Fig. 7. We select the GDPPC, IMR in *Phase 1*, RUR, EDR in *Phase 2*, and MaR, PWCA, PFSE, DIPC in *Phase 3*. All of these features are variables that contribute more to the results.

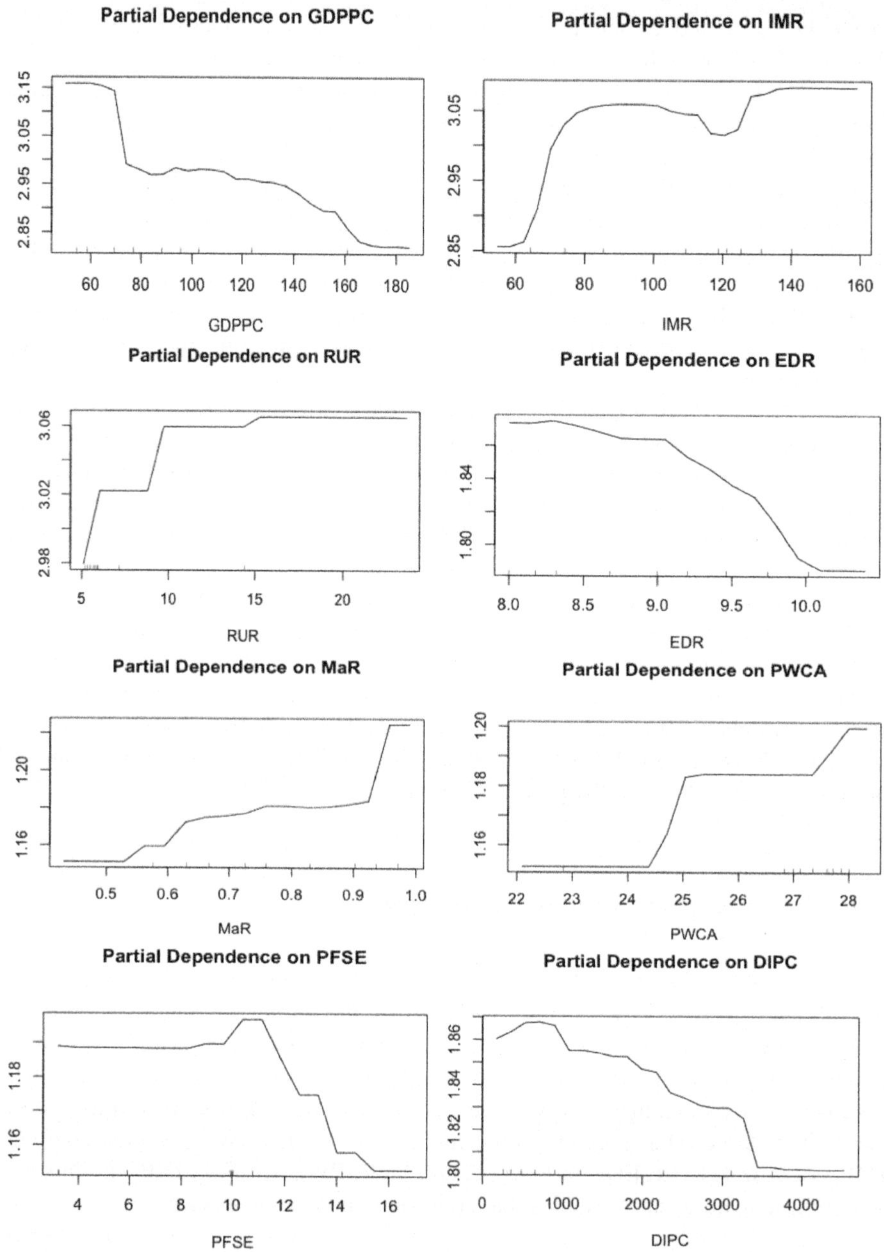


**Fig. 7** Partial dependency diagram for selected features

## 5 Conclusion

This study applies the random forest algorithm to analyze the four sectors of influences on China's fertility transition in terms of social development, economy, education, and policies, and combine them with the classical theory of demographic transition to compare the differences among the three stages of China's demographic transition, 1949–1977, 1978–2001, and 2002–2022. This paper collects annual data from different databases on 18 influential features including fertility. The data were preprocessed by data cleaning, data filling, and other methods to improve the accuracy of the model. According to the results of Gini important features ranking, it can be seen that in the first stage or *Phase 1*, GDPPC, IMR, CDR, and MR are the core features affecting fertility, which we categorize as the fertility response caused by physiological survival needs. In the second stage or *Phase 2*, the core features are RUR, EDR, PP, and DIPC, which we categorize as the fertility response to the pursuit of economic development and accumulation of wealth. And *Phase 3*, is mainly affected by MaR, PWCA, EDR, PFSE, and other features, which we categorize as fertility responses in pursuit of a person's self-value satisfaction and quality of life.

In general, we can see that the main reasons affecting the fertility transition in China have gone through three stages of evolution, from "satisfying basic survival needs" to "pursuing self-development and wealth accumulation in order to improve the standard of living" to "Self-value realization and improvement of the quality of life". This is similar to Maslow's Hierarchy of Needs Theory, and we can also build an interesting pyramid to describe our conclusions as follows in Fig. 8. That is, as the demographic transition progresses from the first to the third phase, fertility is bound to show a downward trend. This demographic evolution will also affect the intrinsic dynamics of fertility decisions.

It's important to be aware of several limitations of this study. In fact, the figure does not show a fourth phase, or Stage IV, although our research suggests that the stage of China's fertility transition is entering the fourth stage and is irreversible. Besides, in

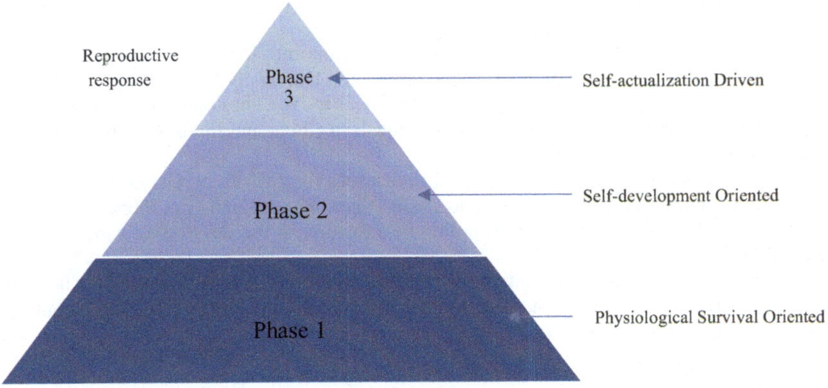


**Fig. 8** Drivers of different phases of fertility transition

this study, the rate of economic growth (GDPGR) is the fifth most important feature of fertility in the first phase and is positively dependent. In the second phase, it decreases to the third last and is negatively dependent. In the third phase, it rises to seventh place and is still negatively dependent. It is difficult to speculate on the degree of partial dependence of fertility on economic growth in *Phase 4*. Since fertility intention, preference preparation, and fertility decision-making include not only social development, economic status, education, employment opportunities, and policies, but also social views and household planning, this paper does not quantify the characteristics of social views and does not analyze the micro-data of residents' family planning. Also, there are differences in fertility rates across Chinese provinces, including gender preferences and differences in fertility costs. We suggest that future research could use China's inter-provincial panel data to analyze the impact of regional differences on fertility, among other things.

Finally, since this paper has found that childbearing decisions are currently driven by "Self-value realization and improvement of the quality of life", then there is a needed suggestion for the coming fourth stage. Firstly, to encourage society and the Government to provide more family support measures, including parental leave, childcare services, and family-rearing subsidies, in order to alleviate the financial and time pressures of family childbearing. Secondly, improve working conditions, including the provision of flexible working hours, childcare services, and career advancement opportunities, so as to help parents better balance their work and family responsibilities. Thirdly, formulating social policies to support childbearing, including lowering the cost of childbearing and providing support for children's education and health care.

## References


1. Asian Development Bank: Asian development outlook 2011 update: Preparing for demographic transition. Asian Development Bank, Philippines (2011)
2. Bloom, D.E., Williamson, J.G.: Demographic transitions and economic miracles in emerging Asia. World Bank Econ. Rev. **12**(3), 419–455 (1998)
3. Coale, A.J., Hoover, E.M.: Population Growth and Economic Development in Low-Income Countries: A Case Study of India's Prospects. Princeton University Press, Princeton (1958)
4. Bloom, D.E., Canning, D., Hu, L., Liu, Y., Mahal, A., Yip, W.: The contribution of population health and demographic change to economic growth in China and India. J. Comp. Econ. **38**, 17–33 (2010)
5. Cai, F., Wang, D.: Demographic transition: Implications for growth. In: Garnaut, R., Song, L. (eds.) The China Boom and its Discontents. Asia-Pacific Press, Canberra (2005)
6. Bongaarts, J., Sinding, S.: Population policy in transition in the developing world. Science **333**(6042), 574–576 (2011)
7. Wang, F., Mason, A.: Demographic dividend and prospects for economic development in China. In: UN Expert Group Meeting on Social and Economic Implications of Changing Population Age Structures. Mexico City (2007)
8. Wei, Z., Hao, R.: Demographic structure and economic growth: evidence from China. J. Comp. Econ. **38**, 472–491 (2010)
9. Golley, J., Zheng, W.E.I.: Population dynamics and economic growth in China. China Econ. Rev. **35**, 15–32 (2015)

10. Lutz, W., Skirbekk, V., Testa, M.R.: The low-fertility trap hypothesis: forces that may lead to further postponement and fewer births in Europe. Vienna Yearb. Popul. Res. 167–192 (2006)
11. Models of Demographic Transition [Biz/ed Virtual Developing Country]. web.csulb.edu. Accessed 16 June 2021
12. Thompson, W.: Encyclopedia of Population, vol. 2, pp. 939–940. Macmillan Reference (2003). ISBN 978-0-02-865677-9
13. Landry, A.: Adolphe landry on the demographic transition revolution. Popul. Dev. Rev. **13**(4), 731–740 (1987). https://doi.org/10.2307/1973031.JSTOR1973031
14. Woods, R.: The Demography of Victorian England and Wales, p. 18. Cambridge University Press (2000). ISBN 978-0-521-78254-8
15. Oded, G.: The demographic transition and the emergence of sustained economic growth. J. Eur. Econ. Assoc. **3**(2), 494–504 (2005)
16. Galor, O.: Unified Growth Theory. Princeton University Press, Princeton (2011). ISBN 9781400838868
17. Myrskylä, M., Kohler, H.P., Billari, F.C.: Advances in development reverse fertility declines. Nature **460**(7256), 741–743 (2009)
18. Li, J.: Factors influencing China's birth rate and its development trend. Dev. Stud. 71–78 (In Chinese) (2014)
19. Liu, L.: Analysis of the factors influencing the birth rate based on the Lasso regression method. Mudanjiang Normal College J. (Natural Science Edition) 1–3 (In Chinese) (2018)
20. Sun, T., Wei, S.: Longer parental time and lower fertility rate. J. Econ. Ageing. **22**, 100382 (2022)
21. De Silva, T., Tenreyro, S.: The fall in global fertility: a quantitative model. Am. J. Econ.: Macroeconomics **12**(3), 77–109 (2020)
22. Narayan, P.K., Peng, X.: An econometric analysis of the determinants of fertility for China, 1952–2000. J. Chin. Econ. Bus. Stud. **4**(2), 165–183 (2006)
23. Pezzulo, C., Nilsen, K., Carioli, A., Tejedor-Garavito, N., Hanspal, S.E., Hilber, T., ... Tatem, A.J.: Geographical distribution of fertility rates in 70 low-income, lower-middle-income, and upper-middle-income countries, 2010–16: a subnational analysis of cross-sectional surveys. Lancet. Glob. Health. **9**(6), e802–e812 (2021)
24. Quan, Y.: A study on the influencing factors of fertility rate in China. Highlights Sci. Eng. Technol. **49**, 142–147 (2023). https://doi.org/10.54097/hset.v49i.8494
25. Chen, Z., Deng, C.: Analysis of factors influencing women's fertility intentions in China. China Popul. Sci. (In Chinese) (2007)
26. Khattak, S.W:. Determinants of fertility intentions in Pakistan. Putaj Human. Soc. Sci. **26**(2), (2019)
27. Chen, R., Xu, P., Song, P., Wang, M., He, J.: China has faster pace than Japan in population aging in next 25 years. Biosci. Trends **13**(4), 287–291 (2019)
28. Calle, M.L., Urrea, V.: Stability of random forest importance measures. Brief. Bioinform. **12**(1), 86–89 (2011)
29. Breiman, L.: Classification and Regression Trees. Routledge (2017)
30. Ishwaran, H.: The effect of splitting on random forests. Mach. Learn. **99**, 75–118 (2015)
31. Wright, M.N., Dankowski, T., Ziegler, A.: Unbiased split variable selection for random survival forests using maximally selected rank statistics. Stat. Med. **36**(8), 1272–1284 (2017)
32. Choi, K.H., Shin, S.: Population aging, economic growth, and the social transmission of human capital: an analysis with an overlapping generations model. Econ. Model. **50**, 138–147 (2015)
33. Yingchun, J., Zhenzhen, Z.: China's low fertility rate from the perspective of gender and development. Soc. Sci. China **41**(4), 169–184 (2020)
34. Arnold, F., Zhaoxiang, L.: Sex Preference, Fertility, and Family Planning in China. In: The Population of Modern China, pp. 491–523. Springer, US, Boston, MA (1992)
35. Lesthaeghe, R.: The second demographic transition in Western countries: an interpretation. In: Mason, K.O., Jensen, A.M., (eds.) Gender and Family Change in Industrial Countries, pp. 17–62. Clarendon Press, Oxford (1995)

36. Caldwell, J.C., Schindlmayr, T.: Explanations of the fertility crisis in modern societies: a search for commonalities. Popul. Stud. **57**(3), 241–263 (2003)
37. Menze, B.H., Kelm, B.M., Masuch, R., Himmelreich, U., Bachert, P., Petrich, W., Hamprecht, F.A.: A comparison of random forest and its Gini importance with standard chemometric methods for the feature selection and classification of spectral data. BMC Bioinf. **10**, 1–16 (2009)
38. Speiser, J.L., Miller, M.E., Tooze, J., Ip, E.: A comparison of random forest variable selection methods for classification prediction modeling. Expert Syst. Appl. **134**, 93–101 (2019)
39. Nembrini, S., König, I.R., Wright, M.N.: The revival of the Gini importance? Bioinformatics **34**(21), 3711–3718 (2018)
40. Wang, C., Kan, A.K., Zeng, Y., Li, G., Wang, M., Ci, R.: Population distribution pattern and influencing factors in Tibet based on random forest model. Acta Geogr. Sin. **74**(4), 664–680 (2019)
41. Wang, L., Yang, Y., Xu, L., Ji, T.: Application of random forest algorithm in the detection of foreign objects in wine. Appl. Math. Nonlinear Sci.
42. Breiman, L.: Random forests. Mach. Learn. 45, 5–32 (2001)
43. Bailey, A.J., Ellis, M.: Going home: the migration of Puerto Rican-born women from the United States to Puerto Rico. Prof. Geogr. **45**(2), 148–158 (1993)

# Quantile Prediction in the Capital Asset Pricing Model Using Histogram-Valued Data

**Wilawan Srichaikul, Somsak Chanaim, Worrawat Saijai, and Woraphon Yamaka**

**Abstract** The purpose of this paper is to introduce an approach to prediction quantile on Capital Asset Pricing Model (CAPM) using histogram-valued data. This study applies this purpose to predict the 5-min returns of two prominent stocks, namely Apple (AAPL) and Microsoft (MSFT), within the S&P500 market covering from January 1, 2019, to June 30, 2023. Moreover, we obtain data for the S&P500 index and US government bonds, representing the market return and risk-free rates, respectively. In this study we found that all quantiles of AAPL stock returns exhibit beta values greater than one, indicating that investing in AAPL will have a higher return than market return, while all quantiles of MSFT stock returns exhibit beta values less than one, indicating that investing in MSFT will have a less return than market return. Additionally, the results of the R-squared value of AAPL confirm the dynamic relationship between the asset and the market return at various levels of quantiles. On the other hand, the R-squared value for MSFT exhibits peaks in both the lower and upper quantile ranges of the U-shape curve which provides insights into the dynamics of MSFT stock returns in relation to the market returns at various quantile levels.

**Keywords** CAPM · Histogram-Valued Data · Quantile

---

W. Srichaikul · W. Saijai · W. Yamaka
Faculty of Economics, Chiang Mai University, Chiang Mai 50200, Thailand
e-mail: srichaikul.w@gmail.com

W. Saijai
e-mail: worrawat.s@cmu.ac.th

W. Yamaka
e-mail: woraphon.econ@gmail.com

S. Chanaim (✉)
International College of Digital Innovation, Chiang Mai University, Chiang Mai 50200, Thailand
e-mail: somsak.chanaim@cmu.ac.th



V. Kreinovich et al. (eds.), *Applications of Optimal Transport to Economics and Related Topics*, Studies in Systems, Decision and Control 556,
https://doi.org/10.1007/978-3-031-67770-0_33

## 1 Introduction

Uncertainty plays a crucial role in the world of financial markets, exerting a profound influence on investment decisions. In this regard, the ability to navigate murky waters and make informed choices is essential. Often, we are faced with only approximations of asset returns, albeit with the best available estimates. Economists and investors often equate this uncertainty with risk—the realm of outcomes governed by probabilities [12, 13].

The Capital Asset Pricing Model (CAPM), introduced by Sharpe [11] and Lintner [6], offers a framework to gauge risk in financial markets and assets. At its core, CAPM is a tool for evaluating investment portfolios. Its fundamental premise is that investors should be compensated for both the time value of money and the inherent risks associated with their investments. The risk-free rate embodies the time value of money, remunerating investors for committing their funds over time, as opposed to a risk-free asset. The residual components of the model delineate risk, quantifying the compensation an investor demands for embracing greater risk when investing in risky assets.

The assessment of beta risk, a crucial element of CAPM, is typically conducted through linear regression models [13]. Beta, a measure of an asset's volatility relative to the market, is not directly observable. It's usually estimated using the standard linear regression model, as expressed in Eq. (1):

$$r_{l,t,f} = \beta_{0,l} + \beta_{1,l} \times r_{m,t,f} + \varepsilon_{l,t}, \quad t = 1, 2, \ldots, T \tag{1}$$

Here, $r_{l,t,f}$ represents the return on asset $l$ minus with the risk free $r_{f,t}$ at time $t$. $r_{l,t}$ is the return on asset $l$ at time $t$, where $r_{l,t,f} = r_{l,t} - r_{f,t}$. $r_{m,t}$ is the market return at time $t$, $r_{m,t,f} = r_{m,t} - r_{f,t}$. $\varepsilon_{i,t}$ is the error term assumed to follow an independent and identically distributed normal distribution with a mean of zero and variance $\sigma^2$. $\beta_{1,l}$ characterizes an asset's riskiness—a value less than 1 signifies lower volatility compared to the market, while a value greater than 1 indicates higher volatility.

There has been numerous studies applying CAPM to dissect financial markets. Autchariyapanitkul et al. [1] employed quantile regression within an asymmetric Laplace distribution framework to predict stock returns and estimate beta coefficients within the classical CAPM. Phadkantha et al. [9] scrutinized the risk, rate of return, and dependencies of REITs in Asia using CAPM. Hundal et al. [5] delved into the relationship between return and risk in the Finnish stock market using CAPM. Munir et al. [7] explored the impact of stock returns utilizing CAPM. Rakpho et al. [10] assessed the efficacy of MS-CAPM in various ASEAN stock market specifications, using the models to investigate the relationship between expected return and risk in ASEAN-6 financial markets. Phadkantha et al. [9] introduced a Bayesian approach to quantile regression for intervalvalued data, applying it to CAPM to measure beta risk in Thai stock returns. Additionally, Zarifhonarvar [14] examined the correlation between return rate and systematic risk to elucidate CAPM in the Tehran Stock Exchange.

However, in financial markets, asset prices (such as stocks and bonds) are often viewed as high-frequency data, often analyzed tick by tick during intraday trading. Nevertheless, many academics advocate for weekly or monthly analysis, rather than relying solely on daily closing prices. The rationale behind this is that tick-by-tick data generates a substantial volume, making it challenging to discern signal from noise. On the other hand, focusing solely on closing prices may result in the loss of valuable intraday insights.

To address this challenge, our study proposes an alternative approach for data collection, creating daily histogram-valued time series (HTS) from intraday CAPM data. In contrast to traditional datasets composed of single-valued observations, such as cross-sectional or time series data, histogram-valued time series offering a richer and more sophisticated data representation. For example, in a time series context, the high/low price interval of a specific stock on a given day can serve as the datum at time $t$, thus forming a time series of intervals indexed by $t$. The collection of histograms indexed by $t$ constitutes a histogram time series. For instance, one might consider constructing a daily histogram of intraday prices or returns. Histogram data analysis, a promising technique, can effectively manage these extensive datasets [3].

The primary innovation of this paper lies in its capacity to promptly analyze specific datasets within predefined quantiles when utilizing histogram data. For example, it is recommended to leverage histogram data containing low quantile values of intraday returns, such as the 5-minute returns on a given day, to serve as a representative value for that day's return. This approach becomes especially valuable when considering the development of the Capital Asset Pricing Model (CAPM) in situations where market or stock prices experience substantial declines. Conversely, when the market maintains a steady course, it's essential to consider that histogram data tends to cluster around the median value. In contrast to the traditional CAPM, which relies solely on closing price data, histogram data should exhibit high quantile values during periods of upward price volatility [1]. In this case, it becomes evident that employing histogram-valued data allows for the direct estimation of quartile values from the desired quantile.

The subsequent sections of this paper are organized as follows: Sect. 2 provides a concise overview of the methodologies employed in our study, while Sect. 3 offers a detailed description of the data. The empirical results and conclusions of our study are presented in Sects. 4 and 5, respectively.

## 2 Methodology

### *2.1 Histogram-Valued Time Series*

As previously noted, our examination involves high-frequency data, observed at the tick-bytick granularity. While this kind of data offers a wealth of insights, handling its vast quantity presents a formidable task. To tackle this challenge, we convert the data into a histogram format and identify appropriate values within the histogram's spectrum. This approach is in accordance with the definition of histogram-valued variables, as articulated by Dias and Brito [2]:

**Definition 2.1** $H$ is a histogram-valued variable when to each date $t$ corresponds a empirical distribution $Y(t)$, that can be represented by a histogram:

$$H_{Y(t)} = \left\{\left[\underline{I}_{Y(t)_1}, \bar{I}_{Y(t)_1}\right), p_{t1}; \left[\underline{I}_{Y(t)_2}, \bar{I}_{Y(t)_2}\right), p_{t2}; \ldots; \left[\underline{I}_{Y(t)n_t}, \bar{I}_{Y(t)n_t}\right], p_{tn_t}\right\}. \quad (2)$$

where $\underline{I}_{Y(t)_i}$ and $\bar{I}_{Y(t)_i}$ represent the lower and upper bound of the interval $i$; $p_{ti}$ is the frequency associated to the subinterval $\left[\underline{I}_{Y(t)_i}, \bar{I}_{Y(t)_i}\right)$ with $i \in \{1, 2, \ldots, n_t\}$, $n_t$ is the number of subintervals for the $t^{\text{th}}$ unit, $t = 1, \ldots, T$ $\sum_{i=1}^{n_t} p_{it} = 1$, $\underline{I}_{Y(t)_i} \le \bar{I}_{Y(t)_i}$ and $\bar{I}_{Y(t)_i} \le \underline{I}_{Y(t)_{i+1}}$.

Alternatively, $Y(t)$ can be represented by the inverse of the cumulative empirical distribution function, also called quantile function $\Psi^{-1}_{Y(t)}(\alpha)$, $\alpha \in (0, 1)$ :

$$\Psi^{-1}_{Y(t)}(\alpha) = \begin{cases} \underline{I}_{Y(t)_1} + \dfrac{\alpha}{w_{t1}} \Delta I_{Y(t)_1} & \text{if } 0 \le \alpha < w_{t1} \\ \underline{I}_{Y(t)_2} + \dfrac{\alpha - w_{t1}}{w_{t2} - w_{t1}} \Delta I_{Y(t)_2} & \text{if } w_{t1} \le \alpha < w_{t2} \\ \vdots & \\ \underline{I}_{Y(t)_{n_t}} + \dfrac{\alpha - w_{tn_t-1}}{1 - w_{tn_t-1}} \Delta I_{Y(t)_{n_t}} & \text{if } w_{tn_t-1} \le \alpha \le 1 \end{cases} \quad (3)$$

where $w_{tl} = \begin{cases} 0 & \text{if } l = 0 \\ \sum_{h=1}^{l} p_{th} & \text{if } l = 1, \ldots, n_t \end{cases}$ and $\Delta I_{Y(t)_i} = \bar{I}_{Y(t)_i} - \underline{I}_{Y(t)_i}$ with $i \in \{1, \ldots, n_t\}$; $n_t$ is the number of subintervals in $Y(t)$.

For the time series of the quantile histogram value data at level $\alpha$ is $\{\Psi^{-1}_{Y(1)}(\alpha), \Psi^{-1}_{Y(2)}(\alpha), \ldots, \Psi^{-1}_{Y(t)}(\alpha)\}$

Considering the sub time series $y_{t-1}$ at date $t-1$ and the the sub time series $y_t$ at date $t$, we can define the set of all possible return by

$$r_t(i, j) = \log\left(\frac{y_t(i)}{y_{t-1}(j)}\right), i = 1, 2, \cdots, K, j = 1, 2, \cdots, K \quad (4)$$

where $y_{t-1}(j)$ is the the value at time $j$ from date $t-1$ and $y_t(i)$ is the the value at time $i$ from date $t$, $i, j = 1, 2, \cdots, K$ , so that the number of all possible return $r_t$ at date $t$ is $K \times K$ denote by $R(t)$ or $R(t) = \{r_t(1, 1), r_t(1, 2), \cdots, r_t(K, K)\}$, and by definition 2.1 we can transform $R(t), t = 2, 3, \cdots, T$ to histogram value data and have $H_{R(t)}$ and $\Psi^{-1}_{R(T)}(\alpha)$.

So that the set of histogram value time series data is $\{H_{R(2)}, H_{R(3)}, \cdots, H_{R(T)}\}$ and we choose the level $\alpha$ to select the set of quantile histogram value time series data $\{\Psi^{-1}_{R(2)}(\alpha), \Psi^{-1}_{R(3)}(\alpha), \cdots, \Psi^{-1}_{R(T)}(\alpha)\}$. In this paper we select $\alpha = 0.1,\ 0.5$ and 0.9 to represent the quantile of histogram value time series data.

## 2.2 CAPM Model

Sharpe [11] is credited with introducing the original Capital Asset Pricing Model (CAPM), which was developed to examine the relationship between the expected return of an asset and its level of risk. The core idea at the heart of CAPM revolves around evaluating the balance between compensation for risk and potential return, relative to the market's overall performance. This balance is assessed by equating the expected return of the asset to the riskfree rate plus a multiplication of the market risk premium and the asset's systematic risk. The model equation is expressed as (1). To estimate the CAPM model with histogram value time series, we derive equation as follows:

$$\Psi^{-1}_{R(l,t,f)}(\alpha) = \beta_{0,l} + \beta_{1,l} \times \Psi^{-1}_{R(m,t,f)}(\alpha) + \varepsilon_{l,t}, \tag{5}$$

where $\Psi^{-1}_{R(l,t,f)}(\alpha) = \Psi^{-1}_{R(l,t)}(\alpha) - r_{f,t}$ and $\Psi^{-1}_{R(m,t,f)}(\alpha) = \Psi^{-1}_{R(m,t)}(\alpha) - r_{f,t}$, $\beta_{0,l}$ and $\beta_{1,l}$ are a constant and explained parameters for asset $l$. $\Psi^{-1}_{R(l,t)}(\alpha)$ is the quantile at level $\alpha$ of histogram return on the asset $l$ at time $t$. $\Psi^{-1}_{R(m,t)}(\alpha)$ is the quantile at level $\alpha$ of histogram return on the market index at time $t$. $r_{f,t}$ is the risk-free asset return at time $t$. The error term $\varepsilon_{l,t}$ is assumed to be i.i.d. $N\left(0, \sigma^2\right)$. The beta coefficient signifies the connection between the risk associated with a specific asset and the overall risk of the entire market. In another context, it also acts as an indicator of the systematic risk inherent to that asset.

# 3 DATA

In our research, we analyze the 5-minute returns of two prominent stocks, namely Apple (AAPL) and Microsoft (MSFT), within the S&P500 market. The dataset covers the time span from January 1, 2019, to June 30, 2023, and is collected from the Bloomberg database. Additionally, we obtain data for the s&P500 index and US government bonds, representing the market return and risk-free rates, respectively.

The selection of Apple and Microsoft as the focus of our study is motivated by their well-established status as major US stocks with substantial trading volumes. Subsequently, we transform these time series variables into log-return rates for further analysis. You can find a statistical summary of these two stocks in Table 1.

Our study employs the minimum Bayes factor (MBF) as a diagnostic tool to assess the statistical significance of our results. Notably, all data series at quantiles 10, 50, and 90 are stationary, as indicated by the MBF values. It's worth noting that the

**Table 1** Statistical summary for histogram-values at quantile 0.1, 0.5, and 0.9

| | Mean | Median | Min. | Max. | Std. Dev. | Skew. | Kurt. | JB | Unit root test (ADF) |
|---|---|---|---|---|---|---|---|---|---|
| At quantile 0.1 | | | | | | | | | |
| AAPL | −0.0114 | −0.0072 | −0.1280 | 0.0620 | 0.0185 | −1.4474 | 5.3638 | 1760.81 (0.0000) | 0.0000 (0.0000) |
| S&P500 | −0.0074 | −0.0038 | −0.1095 | 0.0566 | 0.0126 | −2.1956 | 10.558 | 6193.13 (0.0000) | 0.0000 (0.0000) |
| MSFT | −0.0107 | −0.0071 | −0.0997 | 0.0517 | 0.016 | −1.2319 | 3.4447 | 850.47 (0.0000) | 0.0000 (0.0000) |
| At quantile 0.5 | | | | | | | | | |
| AAPL | 0.0002 | 0.0006 | −0.0787 | 0.0902 | 0.0161 | −0.0279 | 3.2837 | 512.17 (0.0000) | 0.0000 (0.0000) |
| S&P500 | −0.0003 | 0.0004 | −0.0793 | 0.0801 | 0.0105 | −0.5279 | 8.4084 | 3403.53 (0.0000) | 0.0000 (0.0000) |
| MSFT | −0.0001 | 0.0003 | −0.0667 | 0.0704 | 0.0143 | −0.1278 | 2.342 | 263.89 (0.0000) | 0.0000 (0.0000) |
| At quantile 0.9 | | | | | | | | | |
| AAPL | 0.0120 | 0.0100 | −0.042 | 0.1304 | 0.017 | 1.2935 | 4.7695 | 1395.67 (0.0000) | 0.0000 (0.0000) |
| S&P500 | 0.0071 | 0.0051 | −0.0399 | 0.1002 | 0.0108 | 1.8353 | 12.0035 | 7463.3 (0.0000) | 0.0000 (0.0000) |
| MSFT | 0.0107 | 0.0079 | −0.0401 | 0.0900 | 0.0153 | 1.2005 | 3.2720 | 781.01 (0.0000) | 0.0000 (0.0000) |
| Closed price return | | | | | | | | | |
| BOND | 0.0001 | 0.0000 | 0.0000 | 0.0001 | 0.0000 | 0.3916 | −0.9891 | 74.821 (0.0000) | 0.0000 (0.0000) |

Note: ( ) is Minimum Bayes factor (MBF) (see, [4])

skewness of our returns exhibits both negative and positive values, implying that the distribution of our returns may not follow a normal distribution. This observation is further corroborated by the Jarque-Bera (JB) test, which confirms that the data doesn't have a normal distribution property, a finding consistent with the results obtained from the minimum Bayesian coefficient (MBF) analysis.

**Table 2** The parameter estimation for $\Psi^{-1}_{R(\text{AAPL},t,f)}(\alpha)$ at $\alpha = 0.1, 0.5$ and $0.9$

| Parameter | $\alpha = 0.1$ | $\alpha = 0.5$ | $\alpha = 0.9$ |
|---|---|---|---|
| $\beta_{0,\text{AAPL}}$ | −0.0028 | 0.0005 | 0.0041 |
| | [0.0004] | [0.0003] | [0.0004] |
| | (0.0000) | (0.2902) | (0.0000) |
| $\beta_{1,\text{AAPL}}$ | 1.1593 | 1.1460 | 1.1184 |
| | [0.0268] | [0.0306] | [0.0332] |
| | (0.0000) | (0.0000) | (0.0000) |
| $R^2$ | 0.6235 | 0.5534 | 0.5018 |

Note: [ ] is standard errors and ( ) is Minimum Bayes factor (MBF) (see, [4])

**Table 3** The parameter estimation for $\Psi^{-1}_{R(\text{MSFT},t,f)}(\alpha)$ at $\alpha = 0.1, 0.5$ and $0.9$

| Parameter | $\alpha = 0.1$ | $\alpha = 0.5$ | $\alpha = 0.9$ |
|---|---|---|---|
| $\beta_{0,\text{MSFT}}$ | −0.0056 | 0.0000 | 0.0064 |
| | [0.0005] | [0.0003] | [0.0005] |
| | (0.0000) | (0.9928) | (0.0000) |
| $\beta_{1,\text{MSFT}}$ | 0.6929 | 0.5917 | 0.6159 |
| | [0.0318] | [0.0367] | [0.0381] |
| | (0.0000) | (0.0000) | (0.0000) |
| $R^2$ | 0.2964 | 0.1873 | 0.1878 |

**Note:** [ ] is standard errors and ( ) is Minimum Bayes factor (MBF) (see, [4])

# 4 Results

## 4.1 Empirical Estimation Results

In our research, we conducted quantile predictions using the Capital Asset Pricing Model (CAPM) with histogram-valued data for the returns of AAPL and MSFT stocks within the context of the S&P500 market. The estimated parameter values are presented in Tables 2 and 3.

Table 2 displays the outcomes of parameter estimation for CAPM when utilizing histogram values at different quantile levels (0.1, 0.5, and 0.9). It is noteworthy that, based on the results in Table 2, all quantiles of AAPL stock returns exhibit beta values greater than 1 $\left(\beta_{1,\text{AAPL}} > 1\right)$, indicating that these assets are more volatile compared to the overall market. From Table 3, the results of parameter estimation for MSFT within the CAPM framework, using histogram values at different quantile levels (0.1, 0.5, and 0.9), are presented. It's noteworthy that in all quantiles, $\beta_{1,\text{MSFT}} < 1$, indicating that the asset exhibits lower volatility compared to the market.

Figure 1 depict scatter plots illustrating histogram-valued data at quantiles 0.1, 0.5, and 0.9. These plots represent the relationship between the daily response of the

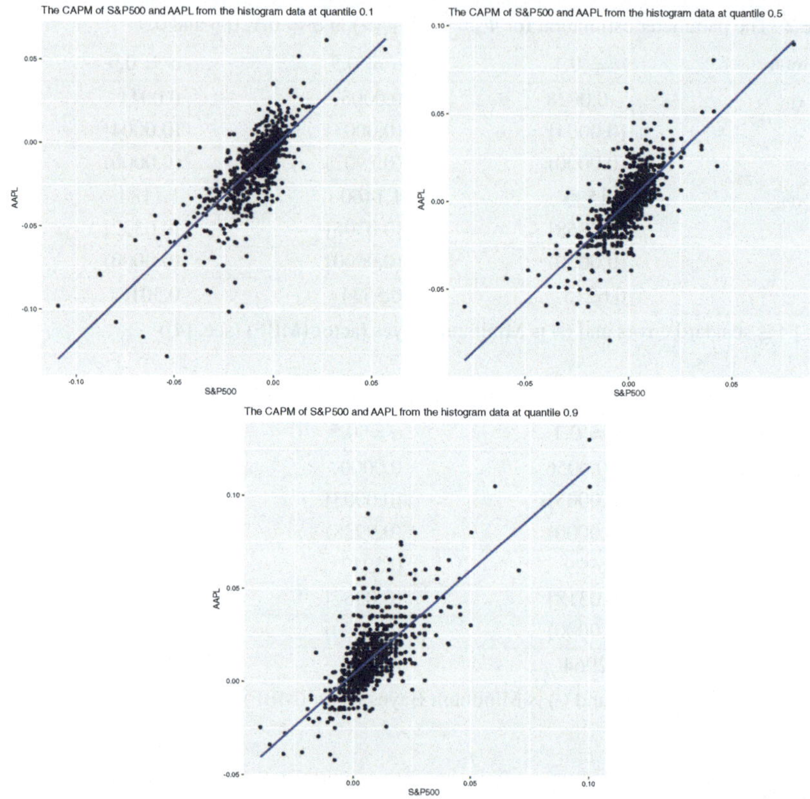


**Fig. 1** The CAPM of S&P500 and AAPL from the histogram data at quantile 0.1, 0.5, and 0.9

S&P500 and the daily response of AAPL after subtracting the risk-free asset return $r_{f,t}$. The figures provide a visual representation of how these variables interact at different quantile levels.

Figure 2 present scatter plots that visualize histogram-valued data at quantiles 0.1, 0.5, and 0.9. These plots illustrate the relationship between the daily response of the S&P500 and the daily response of MSFT after subtracting the risk-free asset return $r_{f,t}$.

Figure 3 displays parameter estimations encompassing the entire range of quantiles. Based on the information presented in these graphs, we can summarize the results as follows: In graph (a), it is observable that the value exhibits a gradual increase as the quantile level rises. However, the difference of $\beta_{0,\text{AAPL}}$ between quantile zero and quantile 1.00 is relatively small in magnitude. In graph (b), when considering quantile levels below 0.50, it's evident that the $\beta_{1,\text{AAPL}}$ value hovers around approximately 1.150. However, once the quantile value reaches 0.5, $\beta_{1,\text{AAPL}}$ starts to rise and gradually decreases afterward. Nevertheless, it remains greater

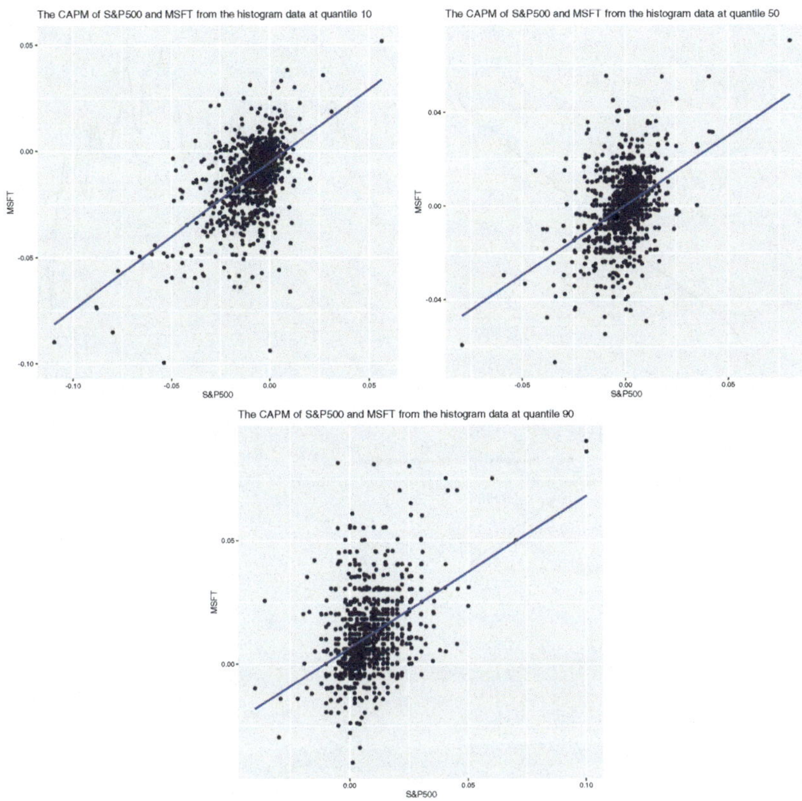


**Fig. 2** The CAPM of S&P500 and MSFT from the histogram data at quantile 0.1, 0.5, and 0.9

than 1. This implies that investing in AAPL is expected to yield higher returns than the market returns across all quantiles. In graph (c), The R-squared value for AAPL exhibits the U-shape curve between the lowest and the highest quantiles, and the bottom of the U-sharp (low parameter values) are between quantile 0.5 and 0.75. The results confirm that the estimated parameter values vary significantly across different quantiles. Each quantile offers a distinct value for the beta coefficients $(\beta_{0,\mathrm{AAPL}}$ and $\beta_{1,\mathrm{AAPL}})$, indicating a dynamic relationship between the asset and the market return at various levels of quantiles.

Figure 4 presents parameter estimations spanning the entire range of quantiles. Based on the information provided by these graphs, we can summarize the results as follows: In graph (a), The value of the parameter $\beta_{0,\mathrm{MSFT}}$ increases consistently with the rise in quantile levels. In graph (b), In the quantile range between 0.1 and 0.5, the parameter $\beta_{1,\mathrm{MSFT}}$ gradually decreases and stabilizes around 0.6. Beyond a quantile of 0.9 the parameter value starts to increase. It's notable that the value of this parameter for every quantile is less than 1, implying that the return of MSFT stock

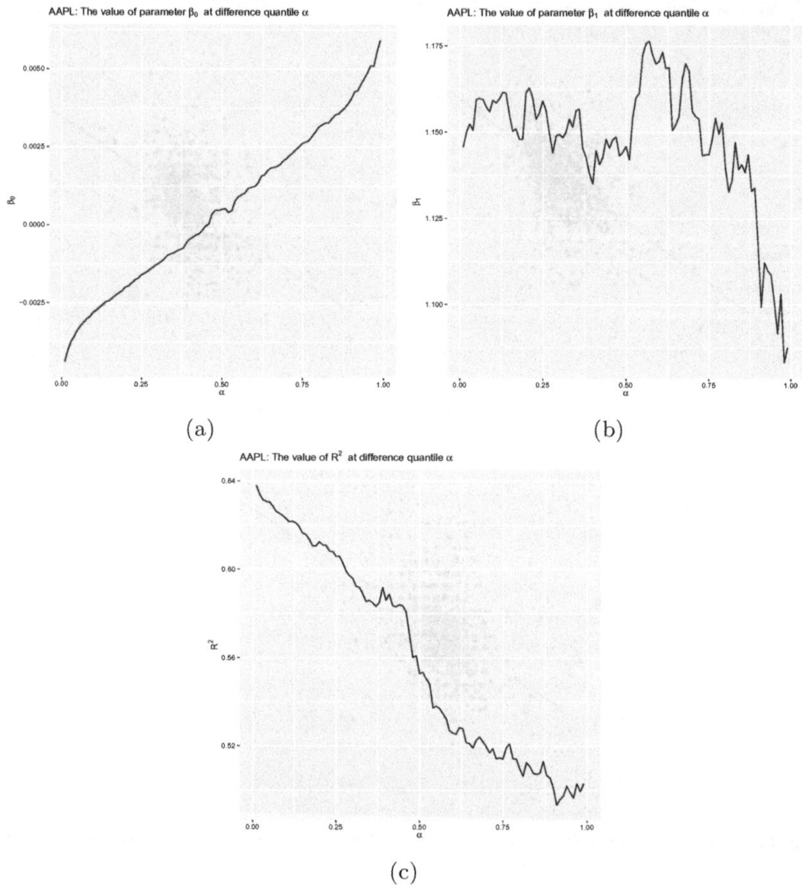


**Fig. 3** Marginal parameter plots of AAPL stock at difference quantile ($\alpha$). **a** $\beta_{0,\text{AAPL}}$, **b** $\beta_{1,\text{AAPL}}$, **c** $R^2$ of AAPL

is expected to be lower than that of the market for each quantile. In graph (c), The R-squared value for MSFT exhibits peaks in both the lower and upper quantile ranges of the U-shape curve. In essence, these findings highlight how parameter values vary across different quantiles and provide insights into the dynamics of MSFT stock returns in relation to the market returns at various quantile levels.

In the next step of our analysis, we will utilize all the estimated parameters obtained thus far. This step will involve measuring volatility, where the estimated parameters will play a crucial role in assessing the volatility of the data or assets under consideration.

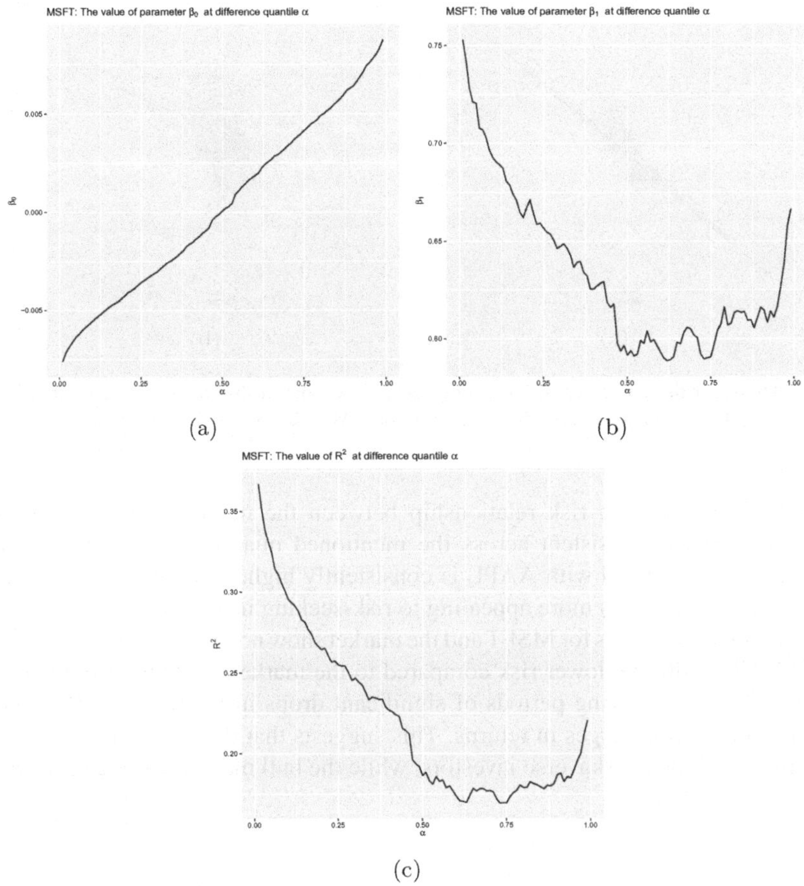


**Fig. 4** Marginal parameter plots of MSFT stock at difference quantile ($\alpha$). **a** $\beta_{0,\text{MSFT}}$, **b** $\beta_{1,\text{MSFT}}$, **c** $R^2$ of MSFT

## *4.2 Measures the Volatility of Stock*

The slope of these lines, which signifies the value of $\beta_1$, serves as a metric for risk. It delineates the trade-off between risk and return in our analysis. A high beta suggests greater risk, which may align with the preferences of aggressive investors seeking potentially higher returns. Conversely, a low beta indicates lower risk, which may be more suitable for defensive investors looking for stability and lower volatility in their investments.

In Fig. 5, we incorporate the parameters acquired from Tables 2 and 3. Specifically, the orange line corresponds to quantile $\alpha = 0.1$, the green line represents $\alpha = 0.5$, the blue line depicts $\alpha = 0.9$.

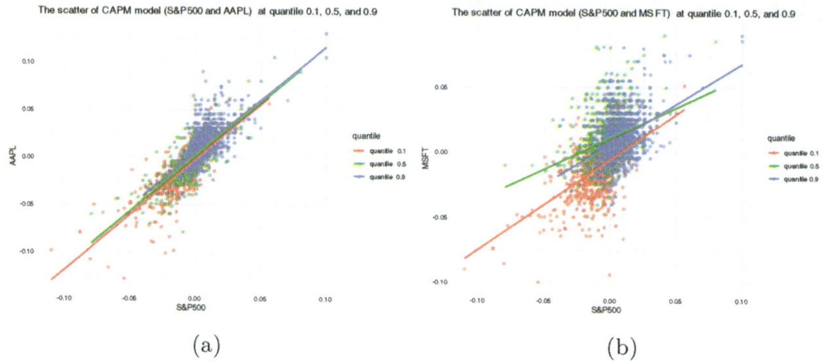


**Fig. 5** The securities characteristic line representing the relationship between the returns of two markets at different quantiles ($\alpha$). **a** Security line for AAPL. **b** Security line for MSFT

We observe that the risk relationship between the market and AAPL returns remains relatively consistent across the mentioned quantiles (Fig. 5a). Additionally, the risk associated with AAPL is consistently higher than that of the market, indicating that AAPL is more appealing to risk-seeking investors.

In contrast, the results for MSFT and the market show notable differences (Fig. 5b), with MSFT exhibiting lower risk compared to the market. Furthermore, the risk is significantly lower during periods of significant drops in returns ($\alpha = 0.1$), but it rises notably during surges in returns. This suggests that the bear market of MSFT is more appealing to riskaverse investors, while the bull market is more attractive to risk-seeking investors.

## 5 Conclusion

In this paper, we present an approach for predicting quantiles within the Capital Asset Pricing Model (CAPM) framework using histogram-valued data for two stocks within the S&P500 market. The dataset spans from January 1, 2019, to June 30, 2023. Our analysis encompasses high, median, and low quantiles. Our findings include parameter estimations for CAPM based on histogram-valued data at quantiles 0.1, 0.5, and 0.9. Remarkably, it's worth noting that the beta values for AAPL stock returns across all quantiles consistently surpass unity, and these values appear to be fluctuating rather than stabilizing at a specific value or within a defined range, indicating that these assets are more volatile than the overall US market. In simpler terms, investing in AAPL is expected to yield higher returns than the market across all quantiles. Additionally, we observe that market return effectively explains the returns of AAPL stocks, particularly at lower quantile levels. Conversely, our analysis reveals a different pattern for MSFT. In the case of MSFT, market returns tend to be higher than the stock returns, suggesting a distinct relationship between the two assets.

However, it's crucial to emphasize that the explained parameter and R-squared values do not demonstrate stability; instead, they exhibit a U-shaped pattern. Significantly, it's noteworthy that the quantile R-squared values exhibit a consistent pattern in both AAPL and MSFT cases. They demonstrate high R-squared values at lower quantiles, with the R-squared value gradually decreasing as the quantile number increases. This pattern indicates a stronger impact between the returns during periods when the markets experiences significant drops in returns. However, it's important to observe that as the quantile increases, the impact of the relationship appears to diminish.

In this study, we have discovered that opting for histogram data empowers investors to utilize data that is better suited to the prevailing market conditions when selecting the CAPM model. This approach is more advantageous than relying on a single data point, such as closing prices or uniformly spaced interval data, and subsequently applying a central method to represent the data in consideration of varying market conditions.

In future research, we can apply the CAPM base histogram-valued data with other model such as Fama and French model, Markov-switching model, and GARCH model.

**Acknowledgements** This research work was partially supported by Chiang Mai University and Center of Excellence in Econometrics, Faculty of Economics, Chiang Mai University.

## References

1. Autchariyapanitkul, K., Chanaim, S., Sriboonchitta, S.: Quantile regression under asymmetric Laplace distribution in capital asset pricing model. Econ. Risk 219–231 (2015)
2. Dias, S., Brito, P.: Distribution and symmetric distribution regression model for histogram-valued variables (2013). arXiv:1303.6199
3. Gonzalez-Rivera, G., Arroyo, J.: Time series modeling of histogram-valued data: The daily histogram time series of S&P500 intradaily returns. Int. J. Forecast. **28**(1), 20–33 (2012)
4. Held, L., Ott, M.: On p-values and Bayes factors. Ann. Rev. Stat. Appl. **5**, 393–419 (2018)
5. Hundal, S., Eskola, A., Tuan, D.: Risk-return relationship in the Finnish stock market in the light of Capital Asset Pricing Model (CAPM). J. Trans. Manag. **24**(4), 305–322 (2019)
6. Lintner, J.: Security prices, risk, and maximal gains from diversification. J. Finan. **20**(4), 587–615 (1965)
7. Munir, S., Sajjad, M., Humayon, A., Chani, M.: Determination of stock returns using five-factor CAPM: evidence from PSX. Int. Trans. J. Eng. Manag. Appl. Sci. Technol. **11**(8), 1–10 (2020)
8. Phadkantha, R., Yamaka, W., Sriboonchitta, S.: A Bayesian approach to quantile regression for interval-valued data: application to CAPM. In: International Conference of the Thailand Econometrics Society, pp. 313–324. Springer International Publishing, Cham (2022)
9. Phadkantha, R., Yamaka, W., Tansuchat, R.: Analysis of risk, rate of return and dependency of REITs in ASIA with capital asset pricing model. In: Predictive Econometrics and Big Data TES2018, pp. 536–548. Springer International Publishing (2018)
10. Rakpho, P., Yamaka, W., Sriboonchitta, S.: Testing CAPM using markov switching models: application to ASEAN-6 stock markets. In: International Conference of the Thailand Econometrics Society, pp. 301–311. Springer International Publishing, Cham (2022)
11. Sharpe, W.F.: Capital asset prices: a theory of market equilibrium under conditions of risk. J. Finan. **19**(3), 425–442 (1964)

12. Wang, J., Zhou, M., Guo, X., Qi, L., Wang, X.: A Markov regime switching model for asset pricing and ambiguity measurement of stock market. Neurocomputing **435**, 283–294 (2021)
13. Yamaka, W., Phadkantha, R.: A convex combination approach for Markov switching CAPM of interval data. Soft Comput. **25**(12), 7839–7851 (2021)
14. Zarifhonarvar, A.: The Capital Asset Pricing Model: A New Empirical Investigation. ZBW - Leibniz Information Centre for Economics, Kiel, Hamburg (2023)

# Spatial Regression Analysis of FDI and Economic Development: The World Perspective

**Rossarin Osathanunkul, Jirapa Inthisang Trochim, and Woraphon Yamaka**

**Abstract** Foreign direct investments (FDIs) play a vital role in promoting an open and effective international economic system and are essential for economic development. However, the benefits of FDI are not equally distributed across countries, sectors, and locations. Theoretical approaches highlight the significance of complex integration strategies of FDI inflows and the interdependence between locations. Yet, little attention has been paid to incorporating the potential cross-country dependencies into the empirical analysis of the determinants of foreign direct investment. Therefore, this paper aims to ex-amine the relationship between foreign direct investment and economic development, utilizing a spatial panel data technique. The empirical analysis uses panel data for 166 countries from 2001 to 2017. Our findings suggest a weak spatial dependence on foreign direct investment and economic development at the international level.

**Keywords** Foreign direct investments · Economic growth · International · Spatial models

## 1 Introduction

The relationship between foreign direct investment (FDI) and economic growth has been the subject of intensive empirical scrutiny over the past few decades. The theoretical underpinnings of this research rest largely on the neoclassical and endogenous growth models that link FDI to economic growth [21]. Scholars have widely examined the potential benefits of FDI to host countries, which include creating job opportunities, heightening competitiveness, and exploiting advanced technology and

R. Osathanunkul · J. I. Trochim
Faculty of Economics, Chiang Mai University, Chiang Mai 50200, Thailand

W. Yamaka (✉)
Center of Excellence in Econometrics, Faculty of Economics, Chiang Mai University, Chiang Mai 50200, Thailand
e-mail: woraphon.econ@gmail.com

 
V. Kreinovich et al. (eds.), *Applications of Optimal Transport to Economics and Related Topics*, Studies in Systems, Decision and Control 556,
https://doi.org/10.1007/978-3-031-67770-0_34

other home countries' intangible assets. Additionally, FDI may serve as a vital source of external financing, thus potentially enhancing the economic condition of the host nation [18].

The empirical literature employing country-specific data has overwhelmingly produced positive findings for a favorable relationship between FDI inflows and economic growth in the host country [1]. For example, Asheghian [2] and Salehizadeh [19] report a statistically significant positive association between FDI and economic growth in the United States. Kornecki and Raghavan [14] document that Central and Eastern European (CEE) countries perceive FDI as an indispensable instrument for economic modernization and development. Moreover, utilizing panel data analysis, there was evidence revealing bidirectional causality and long-run relationships between inward FDI and economic growth in China ([11, 20]) and Malaysia [10]. Almfraji et al. [1] also discovered that FDI inflows and economic growth in Qatar manifest a mutually reinforcing association over the long term, leveraging time-series data from 1990 to 2010 and deploying VAR impulse responses and the Granger causality test. Furthermore, Simionescu (2016) established a bidirectional relationship between FDI inflows and economic growth using Bayesian regression models. Nonetheless, some empirical re-search reports mixed findings on the relationship between FDI and economic growth. Reference [5], for instance, reported minimal or no evidence of FDI contributing to economic growth in India. A few studies suggested that economic growth may attract more FDI inflows ([4, 6, 12]). Conversely, Xu [22] disclosed that FDI in some countries may exert a negative influence on GDP growth.

The varying results in studies examining the relationship between foreign direct investment (FDI) and economic growth might be attributable to differences in country contexts and time periods. The exclusion of spatial impacts in the FDI-growth nexus is another possible factor. The proliferation of capital flows and the emergence of novel capital transfer mechanisms may engender spatial autocorrelation, which could pose challenges to statistical analyses. Disregarding spatial effects may also result in omitted variable bias [3]. As FDI spillovers are likely to possess spatial dimensions, the significance of these spillovers is anticipated to diminish with distance. Consequently, a spatial approach is imperative to appraise the implications of FDI in the global economy.

In a recent study, Mahran [16] further highlights that spatial dependence may occur when data is gathered from neighboring countries, resulting in variations in the dependent variable across different neighboring countries. The determination of a country's gross domestic product (GDP) is not solely based on its own GDP per capita, but is also affected by variables present in neighboring countries, especially those in close proximity. Additionally, the spatial influence of other locations is believed to decrease with distance, with the effect of a distant country considered less significant than that of a nearby one. Thus, spatial dependence may weaken or generate biased estimators of the ordinary least squares (OLS) method [15].

To address the aforementioned concerns, we have employed a spatial panel data model to examine the impact of FDI on the global economy. Furthermore, we have classified all countries into three distinct groups according to the World Bank classification, namely low-income, middle-income, and high-income, and the world to better comprehend the effects of FDI across different economic categories. This study is the first of its kind to investigate the spatial influence of FDI on GDP growth at the international level.

The present study proceeds as follows. Section 2 introduces the spatial panel data model. Section 3 presents the data used in the analysis, and Sect. 4 provides the estimated results. Finally, Sect. 4.4 concludes the paper and provides policy implications based on the research findings.

## 2 Methodology

### 2.1 Spatial Panel Data Models

In this study, we explore the spatial impacts of foreign direct investment on economic development. Therefore, the spatial panel data model is considered, and the model can be written in the matrix form as:

$$y = \lambda(I_T \otimes W_N)\mathbf{y} + X'\beta + \varepsilon, \tag{1}$$

where $y$ and $X$ are the vector of the dependent variable and the matrix in-dependent variables, respectively. $\otimes$ is the Kronecker product. $\mathbf{y}$ is a vector of autoregressive spatial lagged dependent variable. $W_N$ is known as an $N \times N$ dimensional weight matrix; and the parameters $\lambda$ and $\beta$ correspond to the coefficient of the spatially lagged dependent variable and independent variables, respectively. $I_T$ is an identity matrix of dimension $T$. We note that if we do not consider an additional $\lambda(I_T \otimes W_N)\mathbf{y}$ term, Eq. (1) may face the omitted variables bias and thereby leading to a bias of $\beta$ and all statistical inferences are invalid.

In the panel data model, the disturbance vector is the sum of two terms, which are

$$\varepsilon = v + u, \tag{2}$$

where $v$ is a vector of time invariant individuals specific effects and $u$ is the spatially autocorrelated innovations which are usually assumed as a random variable of normal distribution, $u \sim N(0, \sigma_u^2)$. We can define $u$ as

$$u = (I_N \otimes \rho W_N)^{-1} v, \tag{3}$$

where $-1 < \rho < 1$. Thus, both $\lambda$and $\rho$ reflect the strength of the dependent variable's spatial autocorrelation. Each parameter indicates the responsiveness of the dependent variable to a change in the independent variables; $\lambda$ is the coefficient showing the spillover effects of the independent variables on the dependent variable. The key point of the model is the weight matrix $W_N$ which is used to present the contiguity of the international countries. In the traditional approach of defining weight parameters, the element on the $i$th row and $j$th column equals 1 if the $i$th country and $j$th country have a mutual border, otherwise, it equals 0. As a routine, $W_N$ is always row-normalized in spatial econometric analysis, and the elements in normalized $W_N$ are between 0 and 1 [7]. However, in this study, the construction of the weight matrix of the conventional approach may not be reliable as it is evident that in many situations, spatial influence may exhibit aspects of *both* distance and boundary relations [8]. Getis [8] suggested that the best weighting scheme for capturing spatial autocorrelation effects can be given by the combination of *power-distance* and *boundary-shares*. Thus, we capture the spatial autocorrelation and spillover effects in our panel model using the weight matrix which is computed by

$$w(d_{k,l}) = \frac{\min(d_{k,l})}{d_{k,l}}, \tag{4}$$

where $d_{k,l}$ is the distance between country $k$ and $l$, measured between capital cities; $\min(d_{k,l})$ is the minimum distance in our sample. Under the above rule, a non-zero entry in the $k$th column of row $l$ indicates that the $k$th observation will be used to adjust the prediction of the $l$th observation, $(l \neq k)$. Hence, $W_N$is a square matrix

$$W_N = \begin{bmatrix} 0 & w(d_{12}) & \cdots & w(d_{1n}) \\ w(d_{21}) & 0 & & w(d_{2n}) \\ \vdots & & 0 & \vdots \\ w(d_{n1}) & w(d_{n2}) & \cdots & 0 \end{bmatrix}., \tag{5}$$

where $w(d_{kl}) = w(d_{kl})$. A row-standardized weighting matrix is now applied. $W_N$ is normalized so that each row sums to unity.

In the estimation stage, we employ the maximum likelihood estimation method for both the random and the fixed effects models. The likelihood function, which is conditional on the weight matrix, is maximized with respect to all unknown parameters in Eq. 1. To assess the model's performance, we utilize the Akaike Information Criterion (AIC), where the model with the lowest AIC value is favored. It is worth noting that the likelihood for these models can be maximized by utilizing a general optimization algorithm on a concentrated version of the likelihood function.

## 3 Data and Variables

This paper aims to analyze the impact of foreign investment on international economic development using a balanced panel dataset of 166 countries worldwide from 2001 to 2017. The dataset includes variables such as real economic development (proxied by GDP per capita), country population (proxied by POP), net inflows of

**Table 1** Descriptive statistics by economic development levels

| | GDP per capita (US$) (GDP) | lnGDP | TRADE (%) | FDI (%) | Population(person) | lnPOP |
|---|---|---|---|---|---|---|
| World (N = 166) | | | | | | |
| Mean | 11673.88 | 9.3651 | 89.1437 | 5.7858 | 37981700 | 15.7151 |
| Median | 26412.62 | 10.7453 | 80.3201 | 3.0977 | 8009142 | 15.8961 |
| Maximum | 114234.24 | 11.6401 | 442.6200 | 165.2798 | 1390000000 | 21.0500 |
| Minimum | 172.9679 | 5.1531 | 1.5480 | −58.3229 | 19404 | 9.8732 |
| HIGH-INCOME (N = 57) | | | | | | |
| Mean | 28732.85 | 10.265 | 112.4698 | 8.3851 | 20309903 | 15.2540 |
| Median | 38993.39 | 10.5712 | 93.3591 | 3.4460 | 5313399 | 15.4857 |
| Maximum | 114234.24 | 11.6401 | 442.6200 | 165.2798 | 326000000 | 19.6016 |
| Minimum | 13191.9440 | 9.4873 | 79.5500 | −58.3229 | 19404 | 9.8732 |
| MIDDLE-INCOME (N = 82) | | | | | | |
| Mean | 5767.47 | 8.6599 | 84.2786 | 4.7140 | 56793674 | 15.8167 |
| Median | 3505.34 | 8.1620 | 82.2469 | 3.2792 | 7403584 | 15.8175 |
| Maximum | 12001.50 | 9.3927 | 210.3738 | 55.0759 | 1390000000 | 21.0500 |
| Minimum | 1024.58 | 6.9320 | 50.5481 | −37.1548 | 69670 | 11.1515 |
| LOW-INCOME (N = 27) | | | | | | |
| Mean | 422.58 | 6.0463 | 61.6668 | 4.4295 | 14101751 | 16.1093 |
| Median | 410.58 | 6.0175 | 56.2583 | 2.1513 | 11556763 | 16.2628 |
| Maximum | 908.88 | 6.8122 | 111.3541 | 103.3374 | 57310019 | 17.8640 |
| Minimum | 172.96 | 5.1531 | 1.5480 | −4.8523 | 555888 | 13.2283 |

foreign direct investment as a percentage of GDP (proxied by FDI), and trade (measured as the sum of exports and imports of goods and services as a share of GDP) (proxied by TRADE). All data were obtained from the World Bank database website. Table 1 provides descriptive statistics for the variables used in the analysis. Some variables were transformed using natural logarithm, while others were processed using percentage transformation to mitigate potential estimation bias that may arise from the wide range of data values.

Table 1 summarizes country-group statistics for world, high-income group, middle income group, and low-income group classified according to the World Bank. Economies are divided according to 2017 GDP per capita, calculated using the World Bank Atlas method. Low-income economies are defined as those with a GDP per capita, calculated using the World Bank Atlas method, of $995 or less in 2017; middle-income economies are those with a GDP per capita between $996 and $12,055; and high-income economies are those with a GDP per capita of $12,056 or more. This Table also suggests that there is a large dif-ference among the country's net FDI inflows as there is a wide range between maximum and minimum values from the world perspective. As expected, FDI inflows are indeed higher for high-income small economies in which the maxi-mum FDI inflow is 165.279% of GDP (Malta). Interestingly, despite their high GDP per capita, some high-income countries have a negative FDI net inflow. It is obvious that the characteristics of the FDI net inflows are different across coun-tries within the group. Likewise, the middle-income and the low-income groups also show a similar pattern. However, we can observe a difference in size of FDI net inflows from the quite different values of mean percentage of FDI net inflows. For the trade indicator, as expected, the high-income group possess the highest mean value of percentage of trade, whereas the low-income group has the lowest level. Lastly, our last variable, population of each group shows a similar size.

Table 2 provides the results derived from panel unit root test for each variable in each income group. The results reject the null hypothesis of unit root in the level, as most MBF values are approaching to zero, indicating substantial evi-dence supporting a stationary variable. Thus, we conclude that all variables are I(0). We next consider the spatial panel data models for group $g$ as

$$\ln GDP^g = \beta_0^g + \lambda^g (I_T \otimes W_N) \ln GDP^g + \beta_1^g \ln POP + \beta_2^g TRADE + \beta_3^g FDI + \varepsilon^g \qquad (6)$$

# 4 Empirical Results

## 4.1 Model Selection

Before interpreting the spatial panel models, it is important to assess the performance of the models to ensure reliable results. In this section, three types of estimations—Pooled data, Fixed effects, and Random effects—are applied to fit the models based on panel data. Table 3 shows the AIC values for each specification. It is observed that the Fixed effects model has the smallest AIC for all groups, which is reasonable since the panel data considers both cross-sectional and time-series factors. Therefore, the Fixed effects spatial panel model will be used to interpret the empirical results.

## 4.2 Estimated Parameter Results

Table 4 presents the results of the spatial panel regression analysis. The Table is divided into four panels based on the World Bank classification, including the world, low-income, middle-income, and high-income groups, and each panel contains three columns corresponding to three model specifications: pooled, fixed effects, and random effects. As the fixed effects model provides the best estimation performance according to Table 3, we interpret the results based on this model specification. The coefficient estimates from the fixed effects spatial model suggest a negative relationship between foreign direct investment (FDI) net inflows and GDP per capita

**Table 2** Panel Unit root test

| Variable | World | | |
|---|---|---|---|
| | None | Intercept | Intercept & trend |
| lnGDP | 21.8480 | −15.5711 | −4.2500 |
| | (1.0000) | (0.0000) | (0.0000) |
| lnPOP | 2.2282 | −18.3658 | −22.0145 |
| | (0.9871) | (0.0000) | (0.0000) |
| TRADE | 0.9136 | -−3.1694 | −2.1383 |
| | (0.8195) | (0.0008) | (0.0162) |
| **FDI** | −9.9417 | −15.3971 | −12.4225 |
| | (0.0000) | (0.0000) | (0.0000) |
| Variable | High income | | |
| | None | Intercept | Intercept & trend |
| lnGDP | 11.4150 | −6.0712 | −5.2028 |
| | (1.0000) | (0.0000) | (0.0000) |
| lnPOP | 0.0274 | −7.2765 | −6.4202 |
| | (0.5109) | (0.0000) | (0.0000) |
| TRADE | 3.0013 | −2.8527 | −2.1901 |
| | (0.9987) | (0.0022) | (0.0143) |
| **FDI** | −6.5287 | −6.7871 | −5.8250 |
| | (0.0000) | (0.0000) | (0.0000) |
| Variable | Middle income | | |
| | None | Intercept | Intercept & trend |
| lnGDP | 1.0472 | −4.4382 | −4.9585 |
| | (0.8525) | (0.0000) | (0.0000) |
| lnPOP | 0.2550 | −18.1079 | −13.5954 |
| | (0.6007) | (0.0000) | (0.0000) |
| TRADE | 0.6237 | −0.7127 | −0.1825 |
| | (0.7336) | (0.2380) | (0.4276) |
| **FDI** | −4.7304 | −4.2323 | −1.77297 |
| | (0.0000) | (0.0000) | (0.0381) |
| Variable | Low-income | | |
| | None | Intercept | Intercept & trend |
| lnGDP | 17.2841 | −12.2732 | −1.0670 |
| | (1.0000) | (0.0000) | (0.1430) |
| lnPOP | 2.5595 | −8.1698 | −15.6913 |
| | (0.9948) | (0.0000) | (0.0000) |
| TRADE | −1.9724 | −0.9026 | −0.2269 |
| | (0.0243) | (0.1833) | (0.4102) |
| **FDI** | −6.1811 | −13.5585 | −11.540 |
| | (0.0000) | (0.0000) | (0.0000) |

*Note* The values in parentheses represent the MBF (Maximum Bayes Factor), which is used to calibrate the p-values. The coefficients in bold indicate the coefficients used for interpretation (Maneejuk and Yamaka 2021)

**Table 3** Model selection (AIC) for each group

| | World | Low income | Middle income | High income |
|---|---|---|---|---|
| Pooled OLS | 35451.82 | 4342.009 | 15380.29 | 10029.7 |
| Fixed effects | **15875.82** | **3476.78** | **8276.509** | **3506.165** |
| Random effect | 37173.79 | 4142.109 | 16292.06 | 10835.16 |

for all groups but the low-income group. A potential explanation for this adverse effect is that substantial capital inflows or outflows can result in short-term volatility, which the home country can quickly withdraw, causing significant harm to the host country's economy, especially through the exodus of private debt. In the case of high-income nations, their robust domestic financial systems may render them less dependent on foreign direct investment (FDI) for their income, as they possess strong domestic financial resources.

Additionally, longrun FDI inflows may also negatively impact local firms by providing a more competitive market. In particular, a more developed economic system may attract FDI inflows under better economic conditions, but large withdrawals in foreign investment may damage financial institutions [9]. Therefore, it is crucial to control FDI flows, especially in high-income and middle-income countries. Alternatively, in less advanced economies, the potential gain from FDI inflow may outweigh the potential loss in financial instability, as a result, attracting FDI inflow would be their better strategy.

Furthermore, the control variables of population and trade show a positive effect on the economy. Larger population and trade size lead to higher consumption of products and thus contribute to economic development. Overall, these findings provide important insights for policymakers to consider when formulating FDI policies and strategies.

From a spatial analysis point of view, we can find the negative rho as spatial error autocorrelation reflects a pattern of intra -country competition within the same in-come group. The result of the coefficient $\lambda$ of the spatially-lagged dependent variable $(I_T \otimes W_N) \ln GDP$ is positive for all groups indicating a positive economic relationship among the world. However, the positive magnitude of low-income group is the smallest compared to those of the others. This indicates that there is a low economic relationship among low-income countries. In addition, we can observe that the MBF values of these lambdas are zero, indicating that there is strong evidence of spatial spillover effect for the dependent variables.

## 4.3 Spatial Spillover Effects Estimation and Analysis

It is important to note that the interpretation of the estimated coefficients in the spatial panel regression models should be done with caution, as the effects of ex-planatory variables are doubled due to the presence of dependent variables on both sides of

**Table 4** Estimation result for the panel from the world

| | World (Panel A) | | | Low-income (Panel D) | | |
|---|---|---|---|---|---|---|
| | Pooled | Fixed effects | Random effect | Pooled | Fixed effects | Random effect |
| $\beta_0$ | 2.4180 | | −5.7023 | 3.8368 | | 3.8368 |
| | (0.0000) | | (0.0000) | (0.1910) | | (0.1910) |
| lnPOP | 0.1405 | **0.6032** | 0.4959 | 0.1965 | **3.7181** | 0.1965 |
| | (0.0000) | **(0.0000)** | (0.0000) | (0.3749) | **(0.3998)** | (0.3749) |
| TRADE | 0.0013 | **0.0007** | −0.0007 | 0.0016 | **0.0047** | 0.0016 |
| | (0.2882) | **(0.0386** | (0.0621) | (1.0000) | **(0.9910)** | (1.0000) |
| **FDI** | −0.0018 | **−0.0003** | −0.0003 | 0.0097 | **0.0062** | 0.0097 |
| | (1.0000) | **(1.0000)** | (1.0000) | (0.9992) | **(1.0000)** | (0.9992) |
| $\lambda$ | 0.5994 | **0.8063** | 0.8183 | 0.4020 | **0.1900** | 0.4020 |
| | (0.0000) | **(0.0000)** | (0.0000) | (0.0592) | **(1.0000)** | (0.0590) |
| $\rho$ | −0.6505 | **−0.4171** | −0.4537 | −0.2504 | **−0.0608** | −0.2504 |
| $\sigma_v^2$ | (0.0000) | **(0.0000)** | (0.0000) | (0.9047) | **(1.0000)**<br>82.7008<br>(0.0000) | (0.9043)<br>0.0001<br>(1.0000) |
| | **High-income (Panel B)** | | | Middle-income (Panel C) | | |
| | Pooled | Fixed effects | Random effect | Pooled | Fixed effects | Random effect |
| $\beta_0$ | 3.4984 | | −1.7214 | 1.6445 | | −8.1198 |
| | (0.0000) | | (0.0500) | (0.0040) | | (0.0000) |
| **lnPOP** | 0.1473 | **0.3097** | 0.2837 | 0.2071 | **1.2894** | 0.6554 |
| | (0.0000) | **(0.0000)** | (0.0000) | (0.0000) | **(0.0000)** | (0.0000) |
| **TRADE** | −0.0006 | **0.0002** | 0.0003 | −0.0001 | **0.0021** | 0.00145 |
| | (1.0000) | **(1.0000)** | (0.9844) | (1.0000) | **(0.0279)** | (0.2838) |
| **FDI** | −0.0014 | **−0.0003** | −0.0003 | 0.0326 | **−0.0047** | −0.0039 |
| | 1.0000 | **0.7517** | 0.7984 | 0.0382 | **0.5269** | 0.7335 |
| $\lambda$ | 0.4927 | **0.7699** | 0.7707 | 0.5442 | **0.7376** | 0.7846 |
| | (0.0000) | **(0.0000)** | (0.0000) | (0.0000) | **(0.0000)** | (0.0000) |
| $\rho$ | −0.3047 | **−0.3005** | −0.3184 | −0.7264 | **−0.6291** | −0.6797 |
| $\sigma_v^2$ | 0.0331 | **0.0016** | 0.0007<br>84.0595<br>(0.0000) | 0.0000 | **0.0000** | 0.0000<br>37.3845<br>(0.0000) |

*Notes* The values in parentheses represent the MBF (Maximum Bayes Factor), which is used to calibrate the p-values. The coefficients in bold indicate the coefficients used for interpretation (Maneejuk and Yamaka 2021)

**Table 5** Test results of spatial spillover effect (Time-fixed effects)

| Variable | Total effects | Direct effects | Indirect effects |
|---|---|---|---|
| World | | | |
| lnPOP | 3.1113 | 0.8237 | 2.2876 |
| TRADE | 0.0036 | 0.0010 | 0.0026 |
| FDI | −0.0015 | −0.0004 | −0.0011 |
| High-income | | | |
| lnPOP | 1.3458 | 0.4094 | 0.9364 |
| TRADE | 0.0010 | 0.0003 | 0.0007 |
| FDI | −0.0015 | −0.0004 | −0.0011 |
| Middle-income | | | |
| lnPOP | 4.9655 | 1.6370 | 3.3285 |
| TRADE | 0.0086 | 0.0028 | 0.0058 |
| FDI | −0.0178 | −0.0059 | −0.0119 |
| Low-income | | | |
| lnPOP | 4.5904 | 3.7527 | 0.83777 |
| TRADE | 0.0058 | 0.0048 | 0.0010 |
| FDI | 0.0075 | 0.0062 | 0.0013 |

the equation [15]. Therefore, in this section, we use the partial differential method to estimate the direct, indirect, and total effects of FDI net inflows, population, and trade on economic development. The results of these calculations are presented in Table 5, and they allow for a more meaningful interpretation of the impact of independent variables on the dependent variable.

Table 5 presents the direct and spillover effects estimated based on the results of Table 4. The findings reveal notable differences between the direct and indirect ef-fects of each explanatory variable. Direct effects refer to changes in the dependent variable (lnGDP) caused by explanatory variables within a given country, while indirect effects capture changes in the dependent variable in neighboring countries due to changes in the explanatory variable of the province under consideration. The results indicate that the lnPOP and TRADE variables have positive estimation for direct effect and indirect effect coefficients, suggesting that these factors positively impact the world's economic development and neighboring countries, with no negative spatial spillover effect. Specifically, the direct effect coefficient of lnPOP is estimated to be 0.8237, 0.4094, 1.6370, and 3.7527 for the world group, high-income group, mid-dle-income group, and low-income group, respectively. This indicates that the lnPOP factor contributes the highest direct effect on low-income countries, with a 1.

## 4.4 Conclusion

This study focuses on investigating the international interdependencies and spillover effects of foreign direct investment (FDI) net inflows on economic development. Unlike the conventional econometric approaches, we employ spatial econometric techniques to explore the relationship between FDI and economic development on a global scale. Our paper hypothesizes that FDI inflows lead to spillovers of technological and managerial knowledge to local firms, which promote the economic development of the recipient country. Moreover, we assume that these spillover effects have a spatial dimension, which means that the relevance of spillovers should decline with distance. To test our hypothesis, we utilize a spatial panel model that allows us to examine the spatial interdependence of FDI and economic development. Additionally, we compare the results of different spatial panel estimators, namely Pooled OLS, Fixed Effects, and Random Effects. Ultimately, we find that the parameters estimated by the spatial fixed effects panel model are the most reliable as evidenced by the lowest Akaike Information Criterion (AIC).

Based on our analysis using the spatial fixed effects panel model, we find that the parameters estimated are more reliable than the pooled and random effects estimations, as evidenced by the lowest AIC value. Thus, our interpretation of the results is based on this fixed effects panel model. Our findings suggest that the adjustment of the structure of FDI net inflows is not an effective factor for boosting economic development in each group. Rather, we find that the population factor plays a crucial role in economic development for all groups. Specifically, increasing the number of people can lead to an increase in the labor supply, which can be a driving force for economic growth.

Based on these findings, we recommend that policymakers in low, middle, and high-income countries focus on policies that increase their population size, such as encouraging immigration and promoting family planning. Additionally, policies that support the growth of local domestic firms, such as providing access to financing and improving the business environment, may be more effective in promoting economic development than policies focusing solely on attracting FDI.

However, this study has several limitations that should be noted. First, our analysis only focuses on the relationship between FDI net inflows and economic development and does not consider other factors that may affect economic growth, such as political stability or natural resources. Second, the spatial econometric techniques used in this study have certain assumptions and limitations, which may affect the robustness of our findings.

In future research, it may be beneficial to incorporate additional variables that may influence economic development, such as political stability or natural resource endowments. Additionally, alternative spatial econometric techniques could be used to test the robustness of our findings.

**Acknowledgements** We are grateful for financial support from Centre of Excellence in Econometrics, Faculty of Economics, Chiang Mai University.

## References

1. Almfraji, M.A., Almsafir, M.K., Yao, L.: Economic growth and foreign direct investment inflows: the case of Qatar. Procedia Soc. Behav. Sci. **109**, 1040–1045 (2014)
2. Asheghian, P.: Determinants of economic growth in the United States: the role of foreign direct investment. Int. Trade J. **18**(1), 63–83 (2004)
3. Baltagi, B.H., Egger, P., Pfaffermayr, M.: Estimating models of complex FDI: Are there third-country effects? J. Econ. **140**(1), 260–281 (2007)
4. Carkovic, M., Levine, R.: Does foreign direct investment accelerate economic growth. Does Foreign Direct Investment Promote Development **195**, 220 (2005)
5. Chakraborty, C., Nunnenkamp, P.: Economic reforms, FDI, and economic growth in India: a sector level analysis. World Dev. **36**(7), 1192–1212 (2008)
6. Choi, C.: Foreign direct investment and income convergence. Appl. Econ. **36**(10), 1045–1049 (2004)
7. Elhorst, J.P.: Spatial Econometrics: From Cross-sectional Data to Spatial Panels, vol. 479. Springer, Heidelberg (2014)
8. Getis, A.: Reflections on spatial autocorrelation. Reg. Sci. Urban Econ. **37**(4), 491–496 (2007)
9. Desbordes, R., Wei, S.J.: The Effects Of Financial Development On Foreign Direct Investment. The World Bank (2014)
10. Fadhil, M.A., Yao, L., Ismeal, W.: Causality relation between FDI inflows and economic growth in Qatar. Int. J. Bus. Manag. **7**(14), 40–45 (2012)
11. Fang, Q., Liu, Y.: Empirical analysis: business cycles and inward FDI in China. Am. J. Appl. Sci. **4**(8), 800–804 (2007)
12. Kherfi, S., Soliman, M.: FDI and economic growth in CEE and MENA countries: a tale of two regions. Int. Bus. Econ. Res. J. (IBER), **4**(12) (2005)
13. Kopczewska, K., Kudła, J., Walczyk, K.: Strategy of spatial panel estimation: spatial spillovers between taxation and economic growth. Appl. Spat. Anal. Policy **10**(1), 77–102 (2017)
14. Kornecki, L., Raghavan, V.: Inward FDI stock and growth in Central and Eastern Europe. Int. Trade J. **25**(5), 539–557 (2011)
15. Lesage, J.P., Pace, R.K.: Pitfalls in higher order model extensions of basic spatial regression methodology. Rev. Reg. Stud. **41**(1), 13–26 (2011)
16. Mahran, H.A.: The impact of governance on economic growth: spatial econometric approach. Rev. Econ. Political Sci. **8**(1), 37–53 (2023)
17. Maneejuk, P., Yamaka, W.: Significance test for linear regression: how to test without P-values? J. Appl. Stat. **48**(5), 827–845 (2021)
18. Maneejuk, P., Yamaka, W.: The role of economic contagion in the inward investment of emerging economies: the dynamic conditional copula approach. Mathematics **9**(20), 2540 (2021)
19. Salehizadeh, M.: Foreign direct investment inflows and the US economy: An empirical analysis. Econ. Issues **10**(2), 29–50 (2005)
20. Sharahili, Y., Liu, Y.: Empirical analysis II: Business cycles and inward FDI in China. Am. J. Appl. Sci. **5**(10), 1409–1414 (2008)
21. Shaw, G.K.: Policy implications of endogenous growth theory. Econ. J. **102**(412), 611–621 (1992)
22. Xu, B.: Multinational enterprises, technology diffusion, and host country productivity growth. J. Dev. Econ. **62**(2), 477–493 (2000)

# Enhancing Time Series Forecasting in Foreign Exchange Markets with a Hybrid Model Based on Histogram-Valued Data

**Wilawan Srichaikul, Somsak Chanaim, Worrawat Saijai, and Woraphon Yamaka**

**Abstract** This study explores exchange rate forecasting using a hybrid model with equal weight, alongside traditional models like ARIMA(p, d, q), ETS(A, N, N), TBATS, and the NNAR(p, k) model. We evaluate and compare their performance using RMSE, MAE, and MAPE measures on data spanning from January 2018 to July 2023. Our findings highlight the hybrid model with equal weight as an effective choice for USD/JPY and USD/CAD exchange rate prediction, outperforming other models in terms of lower RMSE, MAE, and MAPE values. In contrast, the ARIMA(p, d, q) model excels in forecasting EUR/USD and USD/CHF, aligning closely with true values. The hybrid model, while not always the best, consistently offers competitive performance, providing a versatile tool for Forex rate prediction. Future research can extend this hybrid model to forecast other financial instruments, including stock indices, digital assets, or commodity markets, offering valuable insights for investors and policymakers.

**Keywords** Hybrid model · Histogram · Time series forecasting · Forex · Currency

## 1 Introduction

The foreign exchange market, often referred to as the Forex market, holds a prominent position in the world of finance due to its unparalleled global liquidity. In this market, exchange rates, represented as base and quote currency pairs, signify the number of

W. Srichaikul · W. Saijai · W. Yamaka
Faculty of Economics, Chiang Mai University, Chiang Mai 50200, Thailand
e-mail: srichaikul.w@gmail.com

W. Saijai
e-mail: worrawat.s@cmu.ac.th

W. Yamaka
e-mail: woraphon.econ@gmail.com

S. Chanaim (✉)
International College of Digital Innovation, Chiang Mai University, Chiang Mai 50200, Thailand
e-mail: somsak.chanaim@cmu.ac.th

 
V. Kreinovich et al. (eds.), *Applications of Optimal Transport to Economics and Related Topics*, Studies in Systems, Decision and Control 556,
https://doi.org/10.1007/978-3-031-67770-0_35

units of quote currency that can be obtained in exchange for a single unit of the base currency [4].

Since the collapse of the Bretton Woods system in the early 1970s, the foreign exchange market has been a focal point of both academic research and practical analysis. There are multiple reasons for the paramount importance of the Forex market, with one of the most critical being its role in determining the values of international investments. Consequently, Forex serves as the cornerstone of international investments and global trade. Moreover, the fluctuations in currency values exert a significant influence on the competitiveness of multinational corporations and nations by affecting the value of imported and exported goods and services. Therefore, the volatility of Forex rates is a subject of considerable concern for both scholars and practitioners. Forecasting Forex volatility has emerged as a vital financial issue due to its far-reaching implications [7]. Accurate predictions of forex rates play a pivotal role in the effective execution of monetary policies and the development of successful trading and hedging strategies within the foreign exchange market. Hedging is a risk management technique employed to mitigate the impact of adverse currency movements. Regardless of their specific objectives, individuals involved in forex trading recognize that precise forecasting of forex rate trends is a fundamental aspect of any robust forex trading strategy [10].

As mentioned earlier, the forecasting of exchange rates is of paramount importance. While numerous factors contribute to the fluctuation of exchange rates, encompassing political, economic, interest rate, and economic indicators such as GDP growth, employment data, inflation rates, and trade balances, other factors such as market sentiment, news, events, and speculation also exert substantial influence. Market sentiment, shaped by speculators' perceptions and sentiments toward a currency, can trigger short-term fluctuations in exchange rates. Traders and investors frequently engage in speculative activities within the forex market, making buy or sell decisions based on their expectations of future price movements. Market liquidity, particularly in emerging market currencies, can lead to more significant price swings when it is low. The advent of algorithmic trading and high-frequency trading has introduced additional complexity to forex markets, resulting in rapid price movements. Furthermore, investor trading psychology plays a crucial role, and the combined impact of these diverse factors makes prediction challenging.

Indeed, in this complex environment, the selection of appropriate forecasting techniques is essential when aiming to predict exchange rates. Over the past decades, numerous econometric models and machine learning approaches have been employed to analyze exchange rates in the field of foreign exchange. These models include traditional methods like Autoregressive (AR), Moving Average (MA), ARIMA, as well as modern machine learning models. Here are some examples of studies using various techniques: Pong et al. [14] employed models such as ARMA, ARFIMA, GARCH, and option implied volatilities to investigate the volatility of exchange rates for currencies like the pound, mark, and yen against the dollar. Pradhan and Kumar [15] used Artificial Neural Networks (ANN) for forecasting foreign exchange rates, including the US dollar, British pound, euro, and Japanese yen. Ngan [12] applied the ARIMA model to forecast the foreign exchange rate in Vietnam, specifically the VND/USD exchange rate. Rout et al. [16] utilized the ARMA model in combination with differential evolution-based training to predict currency exchange rates. Nanayakkara et al. [11] used the GARCH and ANN models to estimate the daily rate of the US dollar against the Sri Lankan rupee (USD/LKR). Tlegenova [19] employed the ARIMA approach to predict exchange rates involving the tenge (KZT) against the US dollar (USD), euro (EUR), and Singapore dollar (SGD). Shen et al. [18] explored exchange rate prediction using deep learning techniques and the conjugate gradient method. Parot et al. [13] combined an Artificial Neural Network (ANN) with Vector Auto Regressive (VAR), Vector Error Corrective model (VECM), and post-processing to enhance the prediction accuracy of EUR/USD exchange rate returns. Liao et al. [8] studied the forecasting of USD to RMB exchange rates using Recurrent Neural Network (RNN) and Support Vector Machine (SVM). Maneejuk and Srichaikul [9] evaluated the predictability of foreign exchange markets with ARIMA and deep learning techniques, including artificial neural networks (ANN), recurrent neural networks (RNN), long short-term memories (LSTM), and support vector machines (SVM) using daily data. Erem [3] conducted research using daily data of Swiss Franc (CHF), British Pound Sterling (GBP), and United States Dollar (USD) per Euro (EUR) to forecast daily, weekly, and monthly exchange rates. These studies illustrate the diverse range of methodologies employed in the field of exchange rate forecasting, reflecting the ongoing efforts to improve predictive accuracy in this important financial domain.

Building upon the existing literature, it's noteworthy that none of the models explored hybrid approaches or the combination of multiple models to enhance prediction accuracy based on histogram-valued data. Given the persistent challenges associated with exchange rate forecasting, this paper endeavors to investigate and compare the predictive performance of several hybrid models, including auto-regressive integrated moving average (ARIMA), ETS(A, N, N) model, TBATS model, and NNAR(p, k) model. Our evaluation will consider various criteria, such as Root Mean Square Error (RMSE), Mean Absolute Error (MAE), and Mean Absolute Percentage Error (MAPE). By incorporating these hybrid models and assessing them against these metrics, we aim to contribute to the advancement of exchange rate forecasting techniques in the context of histogram-valued data, recognizing the inherent complexity of this field.

Based on our literature review, it became evident that prior research predominantly relied on closing prices for their analyses. However, closing prices, being a single point in time, may not consistently capture the nuances of daily market fluctuations. Therefore, opting for a histogram approach enables us to select data that reflects a wider spectrum of market scenarios. In this study, we specifically chose data at the 50th quartile, as it affords us the ability to examine market conditions under more typical circumstances.

The primary objective of this study was to employ a time series forecasting hybrid model to predict foreign currency market movements. The hybrid model approach, which involves amalgamating forecasts generated by multiple models, is known to enhance the accuracy of predictions. This practice has become increasingly prevalent as it offers improved forecasting performance. The application of hybrid models holds significant potential for aiding policymakers in strategic planning and decision-making, as well as assisting investors in devising effective financial strategies. In this study, we implement the hybrid model methodology using histogram-valued data for four currency pairs: Euro (EUR/USD), Japanese yen (USD/JPY), Swiss franc (USD/CHF), and Canadian dollar (USD/CAD). By doing so, we aim to provide valuable insights into the forecasting of these currency markets.

The remaining sections of this study are organized in the following format. In Sect. 2, we present a brief outline of the methodologies used in the study, whilst Sect. 3 presents estimation procedures. In the last two sections, Sects. 4 and 5, empirical results and conclusions of our study are presented respectively.

## 2 Methodology

### 2.1 Histogram-Valued Time Series

As previously mentioned, our analysis involves the examination of high-frequency data at the tick-by-tick level. While this type of data offers a rich source of information, its sheer volume can present significant challenges. To overcome this challenge, we employ a data transformation technique, converting the data into a histogram format. This transformation allows us to identify appropriate values within the range of the histogram. As described by Dias and Brito [2], histogram-valued variables can be defined as follows:

**Definition 2.1** $H$ is a histogram-valued variable when to each date $t$ corresponds a empirical distribution $Y(t)$, that can be represented by a histogram:

$$H_{Y(t)} = \left\{\left[\underline{I}_{Y(t)_1}, \bar{I}_{Y(t)_1}\right), p_{t1}; \left[\underline{I}_{Y(t)_2}, \bar{I}_{Y(t)_2}\right), p_{t2}; \ldots; \left[\underline{I}_{Y(t)n_t}, \bar{I}_{Y(t)n_t}\right], p_{tn_t}\right\}. \quad (1)$$

where $\underline{I}_{Y(t)_i}$ and $\bar{I}_{Y(t)_i}$ represent the lower and upper bound of the interval $i$; $p_{ti}$ is the frequency associated to the subinterval $\left[\underline{I}_{Y(t)_i}, \bar{I}_{Y(t)_i}\right)$ with $i \in \{1, 2, \ldots, n_t\}$, $n_t$ is the number of subintervals for the $t$th unit, $t = 1, \ldots, T$ $\sum_{i=1}^{n_t} p_{it} = 1$, $\underline{I}_{Y(t)_i} \leq \bar{I}_{Y(t)_i}$ and $\bar{I}_{Y(t)_i} \leq \underline{I}_{Y(t)_{i+1}}$.

Alternatively, $Y(t)$ can be represented by the inverse of the cumulative empirical distribution function, also called quantile function $\Psi^{-1}_{Y(t)}(\alpha)$, $\alpha \in (0, 1)$:

$$\Psi^{-1}_{Y(t)}(\alpha) = \begin{cases} \underline{I}_{Y(t)_1} + \dfrac{\alpha}{w_{t1}} \Delta I_{Y(t)_1} & \text{if } 0 \leq \alpha < w_{t1} \\ \underline{I}_{Y(t)_2} + \dfrac{\alpha - w_{t1}}{w_{t2} - w_{t1}} \Delta I_{Y(t)_2} & \text{if } w_{t1} \leq \alpha < w_{t2} \\ \vdots & \\ \underline{I}_{Y(t)_{n_t}} + \dfrac{\alpha - w_{tn_t-1}}{1 - w_{tn_t-1}} \Delta I_{Y(t)_{n_t}} & \text{if } w_{tn_t-1} \leq \alpha \leq 1 \end{cases} \quad (2)$$

where $w_{tl} = \begin{cases} 0 & \text{if } l = 0 \\ \sum_{h=1}^{l} p_{th} & \text{if } l = 1, \ldots, n_t \end{cases}$ and $\Delta I_{Y(t)_i} = \bar{I}_{Y(t)_i} - \underline{I}_{Y(t)_i}$ with $i \in \{1, \ldots, n_t\}$; $n_t$ is the number of subintervals in $Y(t)$.

For the time series of the quantile histogram value data at level $\alpha$ is $\{\Psi^{-1}_{Y(1)}(\alpha), \Psi^{-1}_{Y(2)}(\alpha), \ldots, \Psi^{-1}_{Y(t)}(\alpha)\}$. In this paper we select $\alpha = 0.5$ represent the median of histogram value time series data.

## 2.2 ARIMA(p, d, q) Model

The stochastic process $\Psi^{-1}_{Y(t)}(\alpha)$ can be expressed in the following form.

$$\phi(L)\Delta^d \Psi^{-1}_{Y(t)}(\alpha) = \mu + \theta(L)\varepsilon_t, \forall t \geq 0 \tag{3}$$

where $\varepsilon_t$ is an error term with a mean of zero and a variance of $\sigma^2$, independently and identically distributed (iid) for all $t$. $\phi(L) = 1 - \phi_1 L - \phi_2 L^2 - \cdots - \phi_p L^p$ and $\theta(L) = 1 + \theta_1 L + \theta_2 L^2 + \cdots + \theta_q L^q$, where $\Delta$ is the lag operator. If $d = 0$, the data is stationary. If not, we need to transform the data by taking the difference $d$ times to satisfy the stationarity assumption.

## 2.3 ETS(A, N, N) Model

This method was introduced by Hyndman et al. [6]. The ETS model consists of three components: Error, Trend, and Seasonality, and it defines how these components interact with each other. In the case of ETS(A, N, N), it represents an additive error with no trend and nonseasonal components. The ETS(A, N, N) can be write in the following equations

$$\Psi^{-1}_{Y(t)}(\alpha) = l_{t-1} + \varepsilon_t, \tag{4}$$

$$l_t = l_{t-1} + \alpha\varepsilon_t, \tag{5}$$

where $l_t = \frac{1}{t}\sum_{i=1}^{t} \Psi^{-1}_{Y(i)}(\alpha)$, $\alpha$ is the smoothing parameter and $\alpha \geq 0$. We need to estimate $l_{t-1}, \alpha$ by $\hat{l}_{t-1}$, $\hat{\alpha}$ and we can get $\hat{\Psi}^{-1}_{Y(t)}(\alpha) = \hat{l}_{t-1}$ because $E(\varepsilon_t) = 0$, $\varepsilon_t \sim N(0, \sigma^2)$. So that

$$\Psi^{-1}_{Y(t)}(\alpha) = \hat{\Psi}^{-1}_{Y(t)}(\alpha) + \varepsilon_t$$

From (5) we have $\hat{\Psi}^{-1}_{Y(t)}(\alpha) = \hat{\Psi}^{-1}_{Y(t-1)}(\alpha) + \hat{\alpha}\varepsilon_{t-1}$. Then

$$\Psi^{-1}_{Y(t)}(\alpha) = \hat{\Psi}^{-1}_{Y(t-1)}(\alpha) + \hat{\alpha}\varepsilon_{t-1} + \varepsilon_t \tag{6}$$

the stationary assumption is not required in this model.

## 2.4 TBATS Model

The TBATS (Trigonometric seasonality, Box-Cox transformation, ARMA errors, Trend, and Seasonal) model is an advanced time series forecasting model developed to address various aspects of time series data, including seasonality, trends, and nonlinearity.

The formula for the TBATS model can be intricate because it is designed to capture various components of time series data. The general representation of the TBATS model is as follows:

$$\begin{aligned}\Psi^{-1}_{Y(t)}(\alpha) =& (\text{Box-Cox Transformation}) \times (\text{Trend Component}) \\ & \times (\text{Seasonal Component}) \times (\text{ARMA Errors})\end{aligned} \tag{7}$$

The TBATS (Trigonometric seasonality, Box-Cox transformation, ARMA errors, Trend, and Seasonal) model combines several key components to provide a comprehensive representation of time series data. These components include: Box-Cox Transformation: This component is used to stabilize the variance of the original time series data. It involves applying a power transformation to the data, helping to make it more suitable for modeling. Trend Component: The trend component captures the long-term trends or patterns present in the time series. It can be modeled using various techniques such as exponential smoothing or autoregressive components. Seasonal Component: Accounting for periodic patterns or seasonality in the data, the seasonal component in the TBATS model is capable of capturing multiple seasonal patterns if they exist in the dataset. ARMA Errors: The ARMA (Autoregressive Moving Average) component represents the residual errors remaining in the time series after accounting for the trend, seasonality, and Box-Cox transformation. It captures any remaining autocorrelation or temporal dependencies in the data.

The TBATS model leverages these components to provide a comprehensive understanding of the time series data, making it useful for both point forecasting and prediction intervals. Importantly, it does not require the assumption of stationarity, which is a significant advantage. For detailed mathematical expressions and further insights, refer to the work by De Livera et al. [1].

## 2.5 NNAR(p, k) Model

The NNAR(p, k) model incorporates autoregressive terms and potentially nonlinear transformations to make predictions. The general representation of the NNAR(p, k) model is as follows:

$$\Psi_{Y(t)}^{-1}(\alpha) = f\left(\Psi_{Y(t-1)}^{-1}(\alpha), \Psi_{Y(t-2)}^{-1}(\alpha), \ldots, \Psi_{Y(t-p)}^{-1}(\alpha), \varepsilon_t\right) = \hat{\Psi}_{Y(t)}^{-1}(\alpha) + \varepsilon_t, \quad (8)$$

where $\hat{\Psi}_{Y(t)}^{-1}(\alpha)$ is the prediction value of quantile histogram at level $\alpha$ at time t. $\Psi_{Y(t-1)}^{-1}(\alpha), \Psi_{Y(t-2)}^{-1}(\alpha), \ldots, \Psi_{Y(t-p)}^{-1}(\alpha)$ are lagged values of the quantile histogram time series at level = $\alpha$, with "p" being the order of the autoregressive component. These lagged values are used to capture the temporal dependencies in the data. $\varepsilon_t$ represents the error term, which accounts for the residuals or unexplained variation in the model. $f(\cdot)$ is a nonlinear function that encapsulates the relationships and dependencies among the variables. This function is crucial for modeling complex patterns and nonlinearities in the time series. Commonly, neural networks are employed as a technique to estimate the NNAR(p, k) model, allowing it to capture intricate relationships and patterns in the data. Neural networks are particularly well-suited for handling nonlinearities and complex dependencies in time series forecasting.

## 2.6 Hybrid Model with Equal Weight

Hybrid models are employed in time series forecasting by combining multiple models to enhance the accuracy and reliability of predictions. These hybrid models can utilize equal weighted or optimal-weighted approaches to assign significance to each individual model's output. The choice of weights can be based on the performance of the models using validation data or optimization techniques like minimizing MSE, Root Mean RMSE, MAE, among others.

In our research, we employ an ensemble model for time series forecasting represented by Eq. (9):

$$\Psi_{Y(t)}^{-1}(\alpha) = \sum_{i=1}^{m} w_i \hat{\Psi}_{Y(t),i}^{-1}(\alpha) + \varepsilon_t. \quad (9)$$

Where $w_i = \frac{1}{m}$ represents the equal weight assigned to each model, with $m$ denoting the total number of models, $\hat{\Psi}_{Y(t),i}^{-1}(\alpha)$ corresponds to the quantile price of histogram prediction at time $t$ at level $\alpha$ generated by the $i$th model, where $i$ ranges from 1 to $m$, and $\varepsilon_t$ signifies the error value derived from the ensemble model. For this particular study, we select four models and denote their respective predictions as follows: $\hat{\Psi}_{Y(t),1}^{-1}(\alpha)$ originates from the ARIMA model, $\hat{\Psi}_{Y(t),2}^{-1}(\alpha)$ is derived from the ETS model, $\hat{\Psi}_{Y(t),3}^{-1}(\alpha)$ is based on the TBATS model, and $\hat{\Psi}_{Y(t),4}^{-1}(\alpha)$ is the outcome

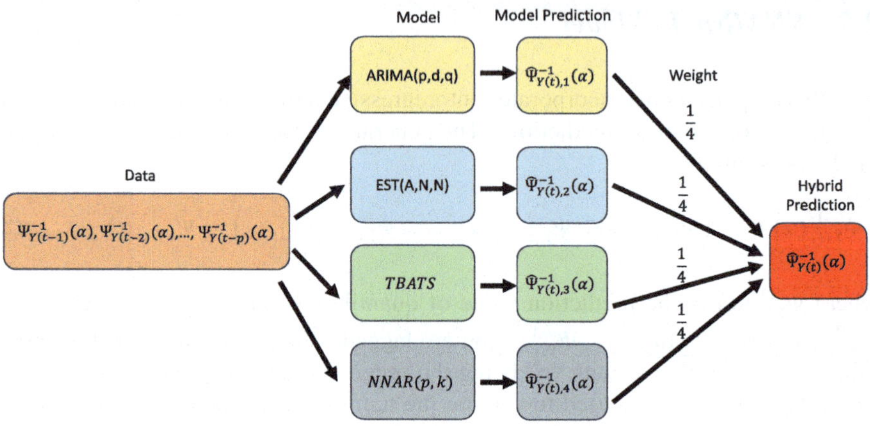


**Fig. 1** The flowchart of the hybrid model with equal weight model

of the NNAR model. The process for ensemble modelling is defined as shown in Fig. 1.

To implement both single models and hybrid methods, the "forecastHybrid" package developed by Shaub and Ellis [17] in 2020 is utilized.

## *2.7 Forecasting Comparison Criteria*

In our study, the forecasting models employ the following evaluation metrics: RMSE, MAE, and MAPE. These metrics can be calculated as follows:

a. Root Mean Squared Error (RMSE),

$$\mathrm{RMSE} = \sqrt{\frac{1}{N}\sum_{t=1}^{N}\left(\Psi_{Y(t)}^{-1}(\alpha) - \hat{\Psi}_{Y(t)}^{-1}(\alpha)\right)^2} \tag{10}$$

b. Mean absolute error (MAE)

$$\mathrm{MAE} = \frac{1}{N}\sum_{t=1}^{N}\left|\Psi_{Y(t)}^{-1}(\alpha) - \hat{\Psi}_{Y(t)}^{-1}(\alpha)\right| \tag{11}$$

c. Mean absolute percentage error (MAPE)

$$\mathrm{MAPE} = \frac{1}{N}\sum_{t=1}^{N}\left|\frac{\Psi_{Y(t)}^{-1}(\alpha) - \hat{\Psi}_{Y(t)}^{-1}(\alpha)}{\Psi_{Y(t)}^{-1}(\alpha)}\right|, \tag{12}$$

where $\Psi_{Y(t)}^{-1}(\alpha)$ and $\hat{\Psi}_{Y(t)}^{-1}(\alpha)$ are the true and predicted value of quantile histogram value at level $\alpha$ at time $t$, respectively. $N$ is the total number of forecasting data points.

## 3 DATA

This paper's primary focus is on assessing the forecasting the quantile histogram price for four currency pairs: Euro (EUR/USD), Japanese yen (USD/JPY), Swiss franc (USD/CHF), and Canadian dollar (USD/CAD) at level 0.5. These currency pairs are chosen as they represent some of the most actively traded pairs in the Foreign Exchange (FOREX) market. The data utilized in this study are sourced from the Bloomberg database, encompassing 5-minute intervals from January 2, 2018, to July 31, 2023. A comprehensive description of the data is provided in Table 1.

To ensure data stationarity, we employ the Augmented Dickey-Fuller test (ADF). Instead of using p-values, we utilize the Minimum Bayes factor (MBF) to assess stationarity (For more details about Minimum Bayes factor (MBF), see Held and Ott, 2018). The MBF values, ranging from 0.4582 to 0.9723, indicate that all three time series sets are non-stationary. However, we choose not to transform the data into log-return rates because the ETS(A, N, N), TBATS, and NNAR(p, k) models do not require stationarity assumptions, except for ARIMA (p, d, q) models where we apply differencing to achieve stationarity.

Table 1 displays quantile histogram-valued data at level 0.5 or the median. The skewness of the data exhibits both negative and positive values, suggesting that the returns may not adhere to a normal distribution. This observation is corroborated by the Jarque-Bera (JB) test, confirming that the data does not follow a normal distribution.

**Table 1** Descriptive statistics

| | $\Psi_{Y_{EUR/USD}(t)}^{-1}(0.5)$ | $\Psi_{Y_{USD/JPY}(t)}^{-1}(0.5)$ | $\Psi_{Y_{USD/CHF}(t)}^{-1}(0.5)$ | $\Psi_{Y_{USD/CAD}(t)}^{-1}(0.5)$ |
|---|---|---|---|---|
| Mean | 1.1302 | 115.7680 | 0.9508 | 1.3078 |
| Median | 1.1300 | 110.3200 | 0.9523 | 1.3130 |
| Maximum | 1.2501 | 150.2580 | 1.0204 | 1.4503 |
| Minimum | 0.9568 | 102.5380 | 0.8578 | 1.2042 |
| Std. Dev. | 0.0593 | 12.0292 | 0.0374 | 0.0424 |
| Skewness | −0.3402 | 1.2434 | −0.1165 | 0.0719 |
| Kurtosis | 2.8465 | 3.0936 | 1.7822 | 2.9080 |
| Jarque-Bera | 35.3406 | 449.7800 | 111.6460 | 2.1185 |
| Unit root test | 0.4501 (0.8265) | 0.9466 (0.4582) | 0.4339 (0.8072) | 0.7549 (0.9723) |

*Note* ( ) is Minimum Bayes factor (MBF) (see, [5])

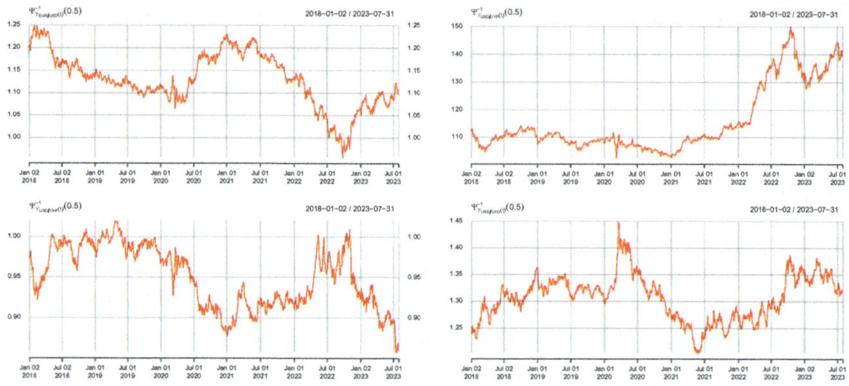

**Fig. 2** The quantile histogram prices for EUR/USD, USD/JPY, USD/CHF, and USD/CAD at level 0.5

In Fig. 2, we present quantile histogram-valued data of prices $\Psi^{-1}_{Y(t)}(\alpha)$ for EUR/USD, USD/JPY, USD/CHF, and USD/CAD at level 0.5. Additionally, Fig. 3 displays density plots of histogram-valued data for all currency pairs.

The subsequent process comprises two main components. The first part involves model training, utilizing data from January 2, 2018, to December 31, 2022. The second part entails model testing, using data from January 1, 2023, to July 31, 2023. Each model trained in the first part is then employed to predict the market for the four currency pairs in the second part, aiming to avoid overfitting issues.

# 4 Results

## *4.1 Empirical Estimation Results*

In this section, we evaluate the overall performance of the forecasting models using three criteria: RMSE, MAE and MAPE. The dataset for each currency pair is divided into two parts: a training set and a testing set. We report the RMSE, MAE, and MAPE for both the training and testing data in Tables 2, 3, 4 and 5, which correspond to the individual currency pairs EUR/USD, USD/JPY, USD/CHF, and USD/CAD.

To conduct this evaluation, we follow a two-step process. First, we use the training set to train five different models: ARIMA(p, d, q), ETS(A, N, N), TBATS, NNAR(p, k), and a hybrid model with equal weights. After the training phase, we assess the performance of each model using the testing data. If the best-performing model on the training data matches the best-performing model on the testing data, we conclude that the hybrid model with equal weights is suitable for forecasting in this currency market. However, if the best models differ between the training and testing datasets,

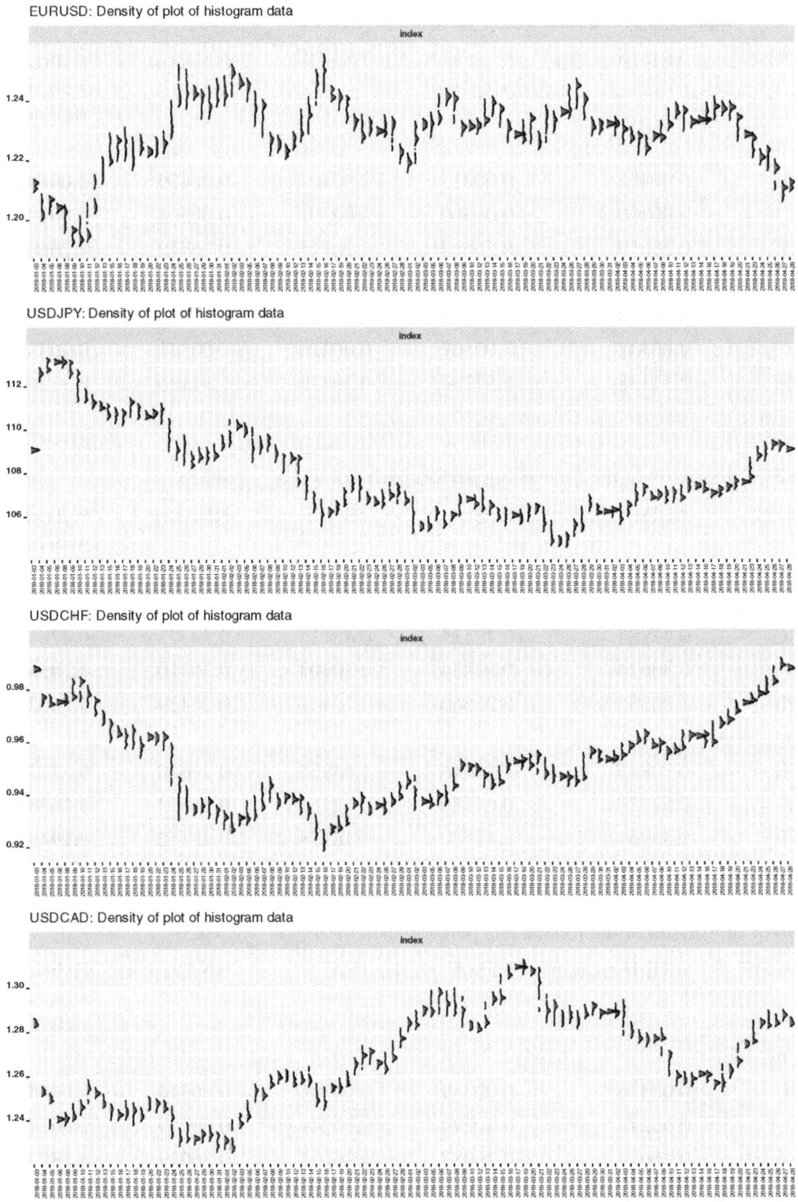


**Fig. 3** Example of density plot of histogram-value data for all currency pairs from 3 Jan 2018 to 28 Apr 2018

**Table 2** Comparison of the forecasting performance for $\Psi^{-1}_{Y_{\text{EUR/USD}}(t)}(0.5)$

| EUR/USD | ARIMA(p, d, q) | ETS(A, N, N) | TBATS | NNAR(p, k) | Hybrid model with equal weight |
|---|---|---|---|---|---|
| Training | | | | | |
| RMSE | 0.00408 | 0.00413 | 0.00409 | 0.00409 | **0.00407** |
| MAE | 0.00300 | 0.00300 | 0.00301 | 0.00301 | **0.00299** |
| MAPE | 0.26659 | 0.26687 | 0.26713 | 0.26712 | **0.26603** |
| Testing | | | | | |
| RMSE | **0.00445** | 0.01089 | 0.01168 | 0.00448 | 0.00447 |
| MAE | **0.00338** | 0.00409 | 0.00426 | 0.00340 | 0.00339 |
| MAPE | **0.31158** | 0.37797 | 0.39420 | 0.31426 | 0.31266 |

**Table 3** Comparison of the forecasting performance for $\Psi^{-1}_{Y_{\text{USD/JPY}}(t)}(0.5)$

| USD/JPY | ARIMA(p, d, q) | ETS(A, N, N) | TBATS | NNAR(p, k) | Hybrid model with equal weight |
|---|---|---|---|---|---|
| Training | | | | | |
| RMSE | 0.47884 | 0.48435 | 0.48113 | 0.48773 | **0.47883** |
| MAE | 0.30624 | 0.30854 | 0.30703 | 0.30710 | **0.30564** |
| MAPE | 0.26385 | 0.26560 | 0.26441 | 0.26439 | **0.26320** |
| Testing | | | | | |
| RMSE | 0.70865 | 1.51646 | 1.63250 | 0.70640 | **0.35733** |
| MAE | 0.52737 | 0.64180 | 0.64761 | 0.51633 | **0.26380** |
| MAPE | 0.38964 | 0.47700 | 0.48178 | 0.38152 | **0.19490** |

**Table 4** Comparison of the forecasting performance for $\Psi^{-1}_{Y_{\text{USD/CHF}}(t)}(0.5)$

| USD/CHF | ARIMA(p, d, q) | ETS(A, N, N) | TBATS | NNAR(p, k) | Hybrid model with equal weight |
|---|---|---|---|---|---|
| Training | | | | | |
| RMSE | 0.00346 | 0.00349 | 0.00346 | 0.00346 | **0.00345** |
| MAE | 0.00248 | 0.00249 | 0.00249 | 0.00249 | **0.00247** |
| MAPE | 0.26013 | 0.26087 | 0.26055 | 0.26029 | **0.25937** |
| Testing | | | | | |
| RMSE | **0.00413** | 0.00561 | 0.00521 | 0.00424 | 0.00439 |
| MAE | **0.00314** | 0.00341 | 0.00340 | 0.00323 | 0.00334 |
| MAPE | **0.34632** | 0.37563 | 0.37443 | 0.35609 | 0.36794 |

**Table 5** Comparison of the forecasting performance for $\Psi^{-1}_{Y_{\text{USD/CAD}}(t)}(0.5)$

| USD/CAD | ARIMA(p, d, q) | ETS(A, N, N) | TBATS | NNAR(p, k) | Hybrid model with equal weight |
|---|---|---|---|---|---|
| Training | | | | | |
| RMSE | 0.00474 | 0.00478 | 0.00471 | 0.00474 | **0.00473** |
| MAE | 0.00343 | 0.00344 | 0.00342 | 0.00343 | **0.00341** |
| MAPE | 0.26201 | 0.26278 | 0.26170 | 0.26213 | **0.26131** |
| Testing | | | | | |
| RMSE | 0.00456 | 0.00884 | 0.00910 | 0.00457 | **0.00408** |
| MAE | 0.00336 | 0.00393 | 0.00448 | 0.00336 | **0.00304** |
| MAPE | 0.24994 | 0.29191 | 0.33260 | 0.25001 | **0.22626** |

we select the best-performing model based on the testing data for predicting the currency market.

The forecasting accuracy results for the three criteria are presented in Tables 2, 3, 4 and 5, providing insights into the performance of each model for each currency pair.

In Table 2, we present an evaluation of model prediction performance for EUR/USD. When trained on the available data, the hybrid model with equal weight exhibited the lowest RMSE, MAE, and MASE values, indicating its superior performance during training. However, when applied to the testing dataset, we observed that the ARIMA(p, d, q) model outperformed the other models, demonstrating the lowest RMSE, MAE, and MASE values. This suggests that the ARIMA model is more accurate than the hybrid model with equal weight when making predictions on unseen data. Nevertheless, it's noteworthy that the testing data results for the hybrid model with equal weight were not the poorest among the models tested. This implies that the hybrid model maintains competitive performance in comparison to other individual models.

Table 3 displays the model prediction performance for USD/JPY. In the training data, the hybrid model with equal weight demonstrates the lowest values across all criteria, suggesting its superior predictive capability. Importantly, when tested on unseen data, the hybrid model with equal weight continues to exhibit the lowest values for all evaluation criteria. This indicates that the hybrid model with equal weight outperforms ARIMA(p, d, q), ETS(A, N, N), TBATS, and NNAR(p, k) in predicting USD/JPY exchange rates.

Table 4 showcases the model prediction performance for USD/CHF. In the training data, the hybrid model with equal weight demonstrates the lowest values for all evaluation criteria, suggesting its strong predictive performance. However, when tested on new data, ARIMA(p, d, q) outperforms all other models, exhibiting the lowest values for all criteria. This indicates that the ARIMA model is more effective than the hybrid model with equal weight in predicting USD/CHF exchange rates.

Nevertheless, it's worth noting that the performance of the hybrid model with equal weight in the testing data is still competitive when compared to the other models (ETS(A, N, N), TBATS, and NNAR(p, k)).

Table 5 displays the model prediction performance for USD/CAD. When using the training data, the hybrid model with equal weight achieves the best values for all evaluation criteria, affirming its superior predictive capabilities. Remarkably, during testing on new data, the hybrid model with equal weight continues to outperform all other models, maintaining the best values for all criteria. This reinforces the conclusion that the hybrid model with equal weight is superior to ARIMA(p, d, q), ETS(A, N, N), TBATS, and NNAR(p, k) for predicting USD/CAD exchange rates.

Furthermore, Figs. 4, 5, 6 and 7 present the training data results for ARIMA(p, d, q), ETS(A, N, N), TBATS, NNAR(p, k), and the hybrid model with equal weight across EUR/USD, USD/JPY, USD/CHF, and USD/CAD. Additionally, Figs. 8, 9, 10 and 11 showcase the testing data outcomes for these models in the same currency pairs. In these figures, the red line represents the predicted values, while the blue line represents the true values.

## 5 Conclusion

This research underscores the effectiveness of the hybrid model with equal weight in predicting foreign exchange markets, including EUR/USD, USD/JPY, USD/CHF, and USD/CAD, using Histogram-valued data spanning from January 2, 2018, to

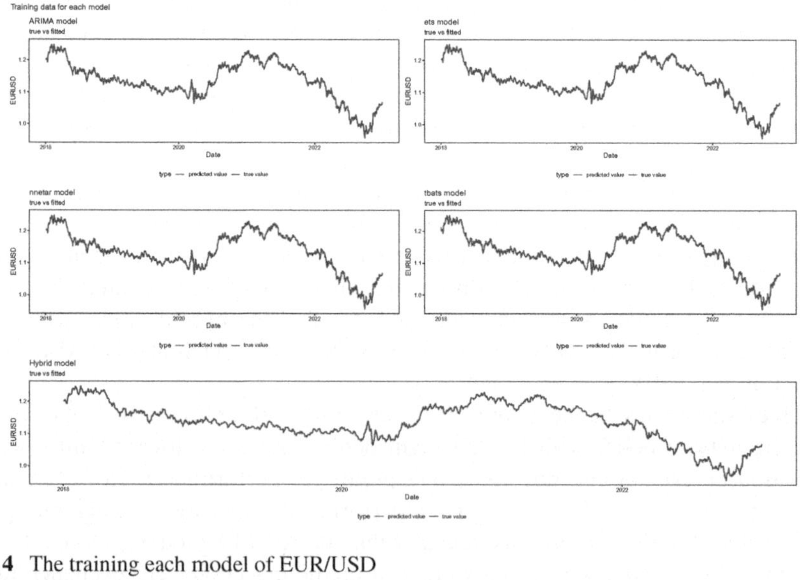


**Fig. 4** The training each model of EUR/USD

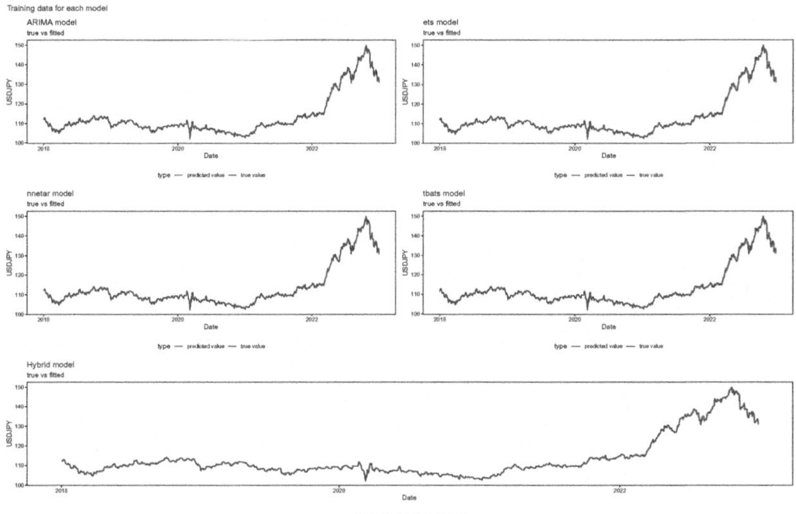


Fig. 5 The training each model of USD/JPY

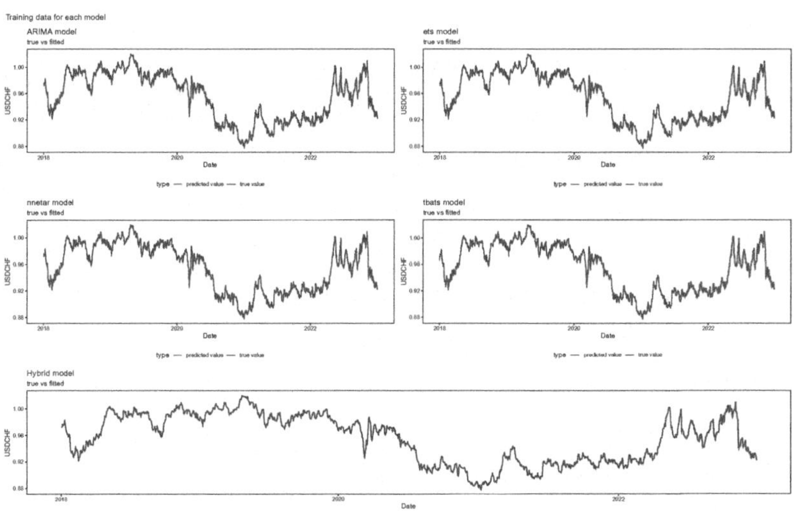


Fig. 6 The training each model of USD/CHF

July 31, 2023. We conducted a comprehensive comparison of the forecasting performance between the hybrid model with equal weight and several other models, namely ARIMA(p, d, q), ETS(A, N, N), TBATS, and NNAR(p, k).

For each currency pair, our analysis involved two phases. First, we utilized training data from January 2, 2018, to December 31, 2022. Subsequently, we assessed the

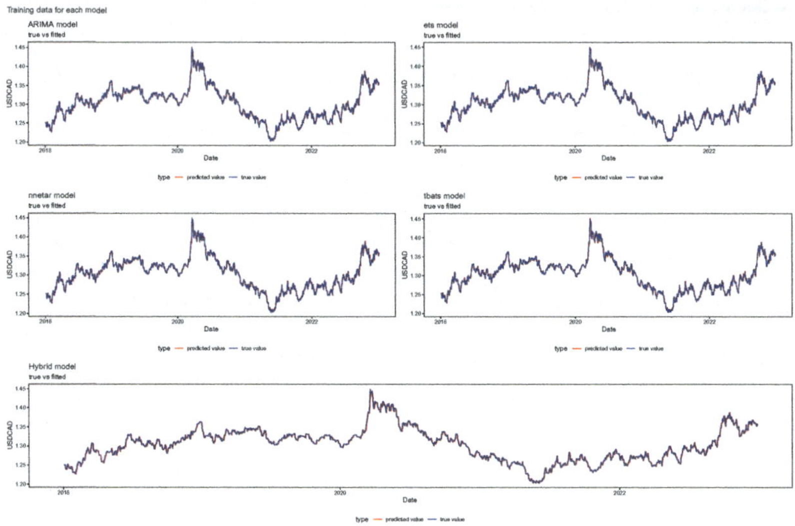


**Fig. 7** The training each model of USD/CAD

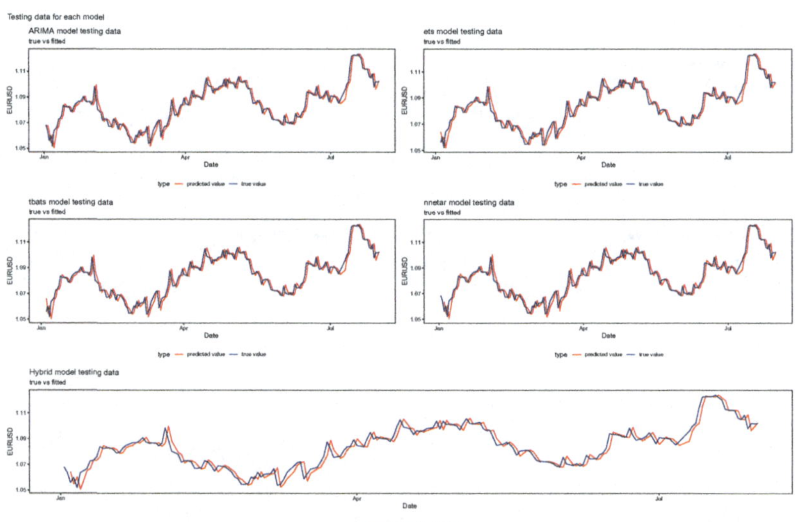


**Fig. 8** The testing each model of EUR/USD

models' performance using testing data from January 1, 2023, to July 31, 2023. Our evaluation was based on key metrics, including RMSE, MAE, and MAPE.

The study's findings underscore the effectiveness of the hybrid model with equal weight, particularly in predicting USD/JPY and USD/CAD exchange rates, where it consistently outperforms the other models by exhibiting significantly lower RMSE,

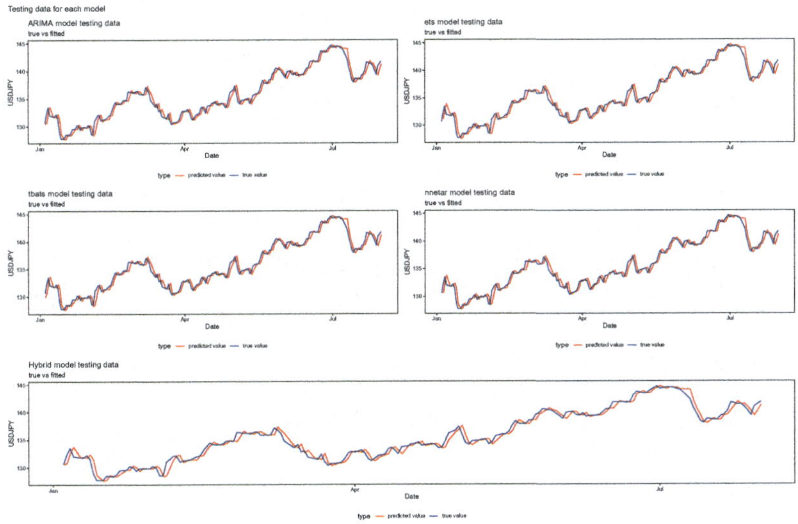


**Fig. 9** The testing each model of USD/JPY

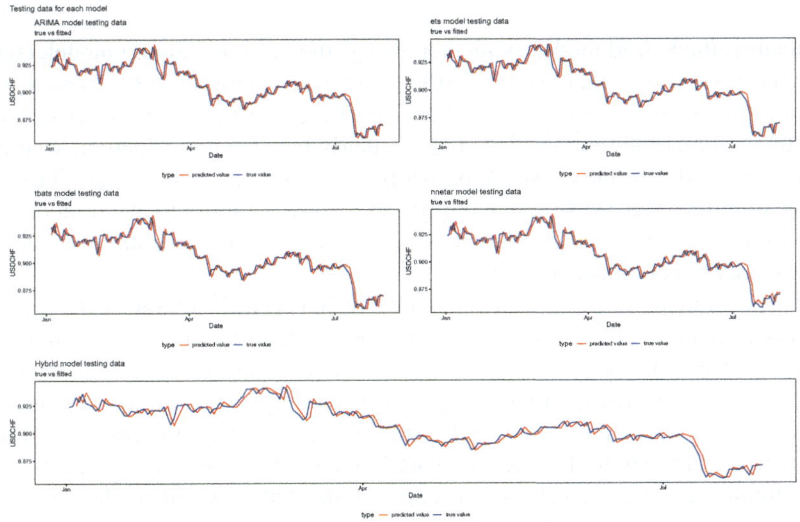


**Fig. 10** The testing each model of USD/CHF

MAE, and MAPE values. Conversely, the ARIMA(p, d, q) model demonstrates admirable performance in forecasting EUR/USD and USD/CHF, especially when evaluated with testing data. Furthermore, the graphical representation of predicted values closely matches the true values, providing additional evidence of the models' efficacy.

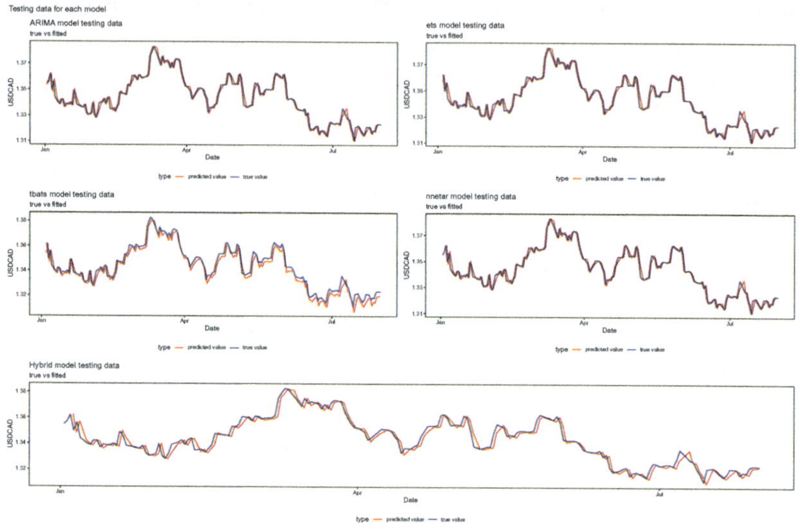


**Fig. 11** The testing each model of USD/CAD

Notably, the hybrid model, while not always the best-performing model, consistently proves to be a reliable and robust option for forecasting. It offers an average level of effectiveness, making it a valuable tool for researchers and investors when selecting a forecasting model. This is particularly beneficial in situations where it's challenging to determine the single best-fitting model for a given dataset. The hybrid model simplifies the decision-making process, allowing users to avoid models that may not be well-suited to the data at hand and facilitating more informed choices when model selection is uncertain.

Due to limitations in our understanding of applying histogram data to neural networks, we are leaving these forecasting models for further study. It's important to note that our primary focus is on recommending the use of histogram data in time series models and demonstrating how we combine forecasting models into a hybrid model for predicting time series data. Our contribution showcases the superior performance of the hybrid model compared to individual forecasting models.

In future research endeavors, the application of the hybrid model with equal weight in conjunction with additional regression models holds promise for predicting a diverse range of financial instruments. This could encompass stock market indices, digital assets, or commodity markets. Such an approach may yield valuable insights and improved forecasting accuracy across various financial domains. The flexibility of the hybrid model, combined with its consistent performance, makes it a versatile tool for tackling forecasting challenges in different sectors of the financial markets. As financial data continues to evolve and become more complex, innovative modeling techniques like the hybrid model can play a pivotal role in enhancing predictive capabilities and informing investment decisions.

**Acknowledgements** This research work was partially supported by Chiang Mai University and Center of Excellence in Econometrics, Faculty of Economics, Chiang Mai University.

## References

1. De Livera, A.M., Hyndman, R.J., Snyder, R.D.: Forecasting time series with complex seasonal patterns using exponential smoothing. J. Am. Stat. Assoc. **106**(496), 1513–1527 (2011)
2. Dias, S., Brito, P.: Distribution and symmetric distribution regression model for histogram-valued variables (2013). arXiv:1303.6199
3. Erem, E.: Forecasting exchange rates using artificial neural networks. J. Financ. Econ. **11**(1), 58–67 (2023)
4. Galeshchuk, S., Mukherjee, S.: Deep networks for predicting direction of change in foreign exchange rates. Intell. Syst. Account., Financ. Manag. **24**(4), 100–110 (2017)
5. Held, L., Ott, M.: On p-values and Bayes factors. Annu. Rev. Stat. Its Appl. **5**, 393–419 (2018)
6. Hyndman, R.J., Khandakar, Y.: Automatic time series forecasting: the forecast package for R. J. Stat. Softw. **27**, 1–22 (2008)
7. Jung, G., Choi, S.Y.: Forecasting foreign exchange volatility using deep learning autoencoder-LSTM techniques. Complexity **2021**, 1–16 (2021)
8. Liao, R., Boonyakunakorn, P., Harnpornchai, N., Sriboonchitta, S.: Forecasting the exchange rate for USD to RMB using RNN and SVM. J. Phys.: Conf. Ser. **1616**(1), 012050 (2020). IOP Publishing
9. Maneejuk, P., Srichaikul, W.: Forecasting foreign exchange markets: further evidence using machine learning models. Soft Comput. **25**(12), 7887–7898 (2021)
10. Menkhoff, L., Taylor, M.P.: The obstinate passion of foreign exchange professionals: technical analysis. J. Econ. Lit. **45**(4), 936–972 (2007)
11. Nanayakkara, K.A.D.S.A., Chandrasekara, N.V., Jayasundara, D.D.M.: Forecasting exchange rates using time series and neural network approaches. European Int. J. Sci. Technol. **3**(2), 65–73 (2014)
12. Ngan, T.M.U.: Forecasting foreign exchange rate by using ARIMA model: a case of VND/USD exchange rate. Methodology **2014**, 2015 (2013)
13. Parot, A., Michell, K., Kristjanpoller, W.D.: Using Artificial Neural Networks to forecast Exchange Rate, including VAR-VECM residual analysis and prediction linear combination. Intell. Syst. Account., Financ. Manag. **26**(1), 3–15 (2019)
14. Pong, S., Shackleton, M.B., Taylor, S.J., Xu, X.: Forecasting currency volatility: a comparison of implied volatilities and AR (FI) MA models. J. Bank. Financ. **28**(10), 2541–2563 (2004)
15. Pradhan, R.P., Kumar, R.: Forecasting exchange rate in India: an application of artificial neural network model. J. Math. Res. **2**(4), 111 (2010)
16. Rout, M., Majhi, B., Majhi, R., Panda, G.: Forecasting of currency exchange rates using an adaptive ARMA model with differential evolution based training. J. King Saud Univ.-Comput. Inf. Sci. **26**(1), 7–18 (2014)
17. Shaub, D., Ellis, P., Shaub, M.D.: Package 'forecastHybrid' (2020). https://cranr-project.org/web/packages/forecastHybrid/forecastHybrid.pdf
18. Shen, F., Chao, J., Zhao, J.: Forecasting exchange rate using deep belief networks and conjugate gradient method. Neurocomputing **167**, 243–253 (2015)
19. Tlegenova, D.: Forecasting exchange rates using time series analysis: the sample of the currency of Kazakhstan (2015). arXiv:1508.07534

# An Analysis of the Dynamic Impact of Oil Price Fluctuations on China's Economy

**Ruiting Xu and Paravee Maneejuk**

**Abstract** This study aims to investigate the mechanisms governing international oil price fluctuations and their dynamic repercussions on China's macroeconomy. To achieve this, we employ the BEKK-GARCH model to examine whether fluctuations in international oil prices exert an influence on China's oil prices. Furthermore, we utilize the nonlinear Markov Regime Vector Autoregression (MS-VAR) model to analyze the evolving effects of international oil price fluctuations on China's macroeconomy under varying regimes. Our findings reveal a significant two-way volatility spillover effect between international and domestic crude oil markets. Moreover, the impact of international crude oil price movements on China's macroeconomy is tempered by China's unique economic conditions. It's noteworthy that the Chinese economy tends to exhibit stability after achieving equilibrium rather than undergoing rapid shifts.

**Keywords** BEKK-GARCH · China's economy · MS-VAR · Oil price · Spillover effect

## 1 Introduction

Since the twentieth century, the international oil market has weathered three significant oil crises, with its volatility leaving profound imprints on the global economic landscape [16–18]. In recent years, the amplitude of fluctuations in international oil prices has exceeded 50%, leading to adverse effects on macroeconomic stability. China's growing oil consumption is heavily dependent on imports [12]. Between 2008 and 2022, China's foreign dependence on oil surged from 47.8 to 73.5%. This heightened reliance underscores the formidable energy security challenges faced by China. As the domestic economy continues its robust expansion and oil demand

R. Xu · P. Maneejuk (✉)
Faculty of Economics, Chiang Mai University, Chiang Mai 50200, Thailand
e-mail: mparavee@gmail.com

 
V. Kreinovich et al. (eds.), *Applications of Optimal Transport to Economics and Related Topics*, Studies in Systems, Decision and Control 556,
https://doi.org/10.1007/978-3-031-67770-0_36

grows unabated, maintaining a stable energy supply becomes pivotal for sustaining a thriving economy.

This study tries to understand the fluctuation of international oil price from two aspects, namely, the fluctuation mechanism and the transmission mechanism [20–22]. In view of the fact that domestic oil price is the most direct object for the impact and transmission of international oil price fluctuations, a lot of quantitative analysis has been done in the existing research, and the research content is about the relationship between domestic oil price and international oil price [2, 3]. Li et al. [13] uses Granger causality test when studying the interaction between foreign crude oil prices and Chinese crude oil prices, and the results show that the two are mutually Granger causality. In addition, there are also scholars using multivariate GARCH model for analysis. This model has great advantages in analyzing the agglomeration of financial assets over time and capturing the effectiveness of risk cross-transfer between different financial assets. Some scholars have used the multivariate GARCH model to conduct relevant empirical studies on price spillovers between different crude oil markets [1, 28, 29]. The research results are roughly divided into two categories. Chan and Woo [4] and Zhang and Wei [27] found that the spillover effect is two-way; that is, there is a long-term equilibrium and guidance relationship between domestic and foreign crude oil prices and domestic crude oil prices. Hou et al. [11] holds the opposite view, that is, the volatility spillover effect between the domestic market and the international market is one-way. The reason for the difference between the two conclusions may be that the multivariate autoregressive regression model was adopted in the research process, or the interval selected by the research samples was inconsistent. Starting from the dimension of volatility spillover effect, this study establishes a BEKK-GARCH model and uses the latest data to explore the oil price volatility and risk information transfer effect between domestic and foreign markets, enriching previous research conclusions.

When examining the relationship between national economic conditions and oil prices, many scholars have primarily relied on multivariate GARCH models and linear VAR models. However, there is room for expanding the perspective on explaining the relationship between economic variables. Because the economy is in a rapidly changing situation, the traditional linear model can't perfectly show this economic situation [7, 8, 28]. Therefore, if a linear model is adopted at this time, that is, an economic state is used to study the problem, the results will change, and the results obtained will not be accurate and persuasive. Therefore, this study utilizes the Markov Regime Vector Autoregression (MS-VAR) model, which categorizes international oil price volatility into "high price" and "low price" intervals. It investigates the impact of international oil prices on China's macroeconomic indicators during these two volatility intervals. The key findings suggest that rising international oil prices lead to reduced resource input, resulting in a decline in real output levels. For example, Lin et al., [15] applied the MS-VAR model, considering parameter characteristics that change with the state of the district system. Their analysis explored the impact of the Consumer Price Index (CPI) and oil price shocks on the domestic consumer market under different district systems. The study confirmed that CPI data exhibited high and low volatility during different periods, leading to policy recommendations based on regime-specific correlations and probability distributions.

However, there's a gap in research when it comes to understanding how price risks spread between crude oil markets due to their changing correlations. These market correlations are influenced by various factors and economic policies, making it essential to quantitatively analyze how they evolve. This analysis can help predict the direction of the domestic crude oil market, providing valuable insights for decision-making in China's oil sector.

This study aims to investigate whether changes in international oil prices have a significant impact on Chinese oil prices and, more importantly, how this affects China's overall economy. To achieve this, we use two methods: the BEKK-GARCH model and the MS-VAR model. These approaches allow us to examine how international oil price fluctuations affect China's economy under different conditions. However, it's important to note that there are still gaps in our understanding of this topic.

The structure of this paper is as follows: Sect. 2 provides a detailed description of the data and the variables used in the study. Section 3 outlines the methodology employed for analysis. Section 4 presents the study's findings and results. Finally, Sect. 5 offers concluding remarks.

## 2 Data Description and Variables

To examine the volatility spillover effect between domestic and international oil price markets, we utilized a time series dataset spanning from January 2007 to December 2022 to construct the BEKK-GARCH (1, 1) model. International oil prices were represented by WTI crude oil prices, sourced from the EIA website, while domestic crude oil prices in China were represented by China Daqing crude oil prices, sourced from the Wind database.

To investigate the impact of international oil prices on China's macroeconomy, we considered four key macroeconomic indicators: Daqing oil price (*DQ*), industrial added value (*IND*), the consumer price index (*CPI*), and money supply (*MS*). The MS-VAR model was applied using quarterly data ranging from 2000Q2 to 2022Q4. Table 1 provides an overview of the dataset and its description.

Both the BEKK-GARCH model and the MS-VAR model require the time series data to be stationary. To assess stationarity, we employed the Augmented Dickey-Fuller test on the selected time series. In the empirical analysis phase, we employed a seasonal smoothing technique to mitigate the influence of seasonal variation factors within the time series data. Subsequently, we calculated the natural logarithm of the smoothed data, resulting in $\ln WTI$, $\ln DQ$, $\ln IND$, $\ln CPI$, and $\ln MS$, respectively.

This transformation was performed to alleviate volatility, which aids in mitigating heteroscedasticity. Since these variables do not exhibit regressiveness in different dimensions and economic contexts, we further processed them by taking the first-order logarithmic difference. The resulting variables, namely $d\ln WTI$, $d\ln DQ$,

**Table 1** Variables selection and data source

| Variable | Description | Source | Symbol |
|---|---|---|---|
| International oil price | *WTI* stands for West Texas Intermediate (Crude Oil), it is used as the variable of international oil price | U.S. EIA website | *WTI* |
| Domestic oil price | Daqing crude oil price represents the whole crude oil price level in China, it is used as the variable of domestic oil price | Wind database | *DQ* |
| Industrial added value | The industrial added value is used to represent the actual output growth and economic growth of China | National Bureau of Statistics of China website | *IND* |
| Consumer price index | It can measures the country s inflation | National Bureau of Statistics of China website | *CPI* |
| Broad Money supply | It can explain the changes of China's monetary policy to some extent | National Bureau of Statistics of China website | *MS* |

dln*IND*, dln*CPI*, and dln*MS*, represent the rate of change in international oil prices, domestic oil prices, industrial added value, the consumer price index, and money supply, respectively.

# 3 Research Methodology

## *3.1 BEKK-GARCH(1,1) Model*

The BEKK-GARCH model is an extension of the Generalized Autoregressive Conditional Heteroskedasticity (GARCH) model, designed to handle multivariate time series data [25]. In this study, we employ a binary BEKK-GARCH(1, 1) model to examine whether fluctuations in international crude oil market prices have spillover effects on China's domestic crude oil market prices. The model comprises conditional variance and covariance matrix can be shown as follow:

$$H_t = cc' + \alpha\varepsilon_{t-1}\varepsilon'_{t-1}\alpha' + \beta H_{t-1}\beta' \tag{1}$$

where, $H_t = \begin{pmatrix} \sigma_{11,t} & \sigma_{12,t} \\ \sigma_{21,t} & \sigma_{22,t} \end{pmatrix}$ is $n \times n$ variance and covariance matrix at time t. Equation (1) has a quadratic form, thus $H_t$ is restricted to be positive. $c = \begin{pmatrix} c_{11} & 0 \\ c_{21} & c_{22} \end{pmatrix}$ is constant term of intercept coefficient, and CC′ is a positive definite matrix. $\varepsilon_{t-1}$

is the disturbance vector $n \times 1$. $\alpha = \begin{pmatrix} a_{11} & \alpha_{12} \\ \alpha_{21} & \alpha_{22} \end{pmatrix}$ stands for volatility aggregation, $\beta = \begin{pmatrix} \beta_{11} & \beta_{12} \\ \beta_{21} & \beta_{22} \end{pmatrix}$ stands for volatility persistence.

Decompose the conditional variance equation of BEKK-GARCH(1,1):

$$\begin{aligned}\sigma_{11,t} = {} & c_{11}^2 + c_{12}^2 + a_{11}^2\varepsilon_{1,t-1}^2 + 2\alpha_{11}\alpha_{21}\varepsilon_{1,t-1}\varepsilon_{2,t-1} + a_{21}^2\varepsilon_{2,t-1}^2 + \beta_{11}^2\sigma_{11,t-1} \\ & + 2\beta_{11}\beta_{21}\sigma_{12,t-1} + \beta_{21}^2\sigma_{22,t-1},\end{aligned} \tag{2}$$

$$\begin{aligned}\sigma_{12,t} = {} & c_{12}c_{22} + a_{11}a_{12}\varepsilon_{1,t-1}^2 + (a_{11}a_{22} + a_{12}a_{21})\varepsilon_{1,t-1}\varepsilon_{2,t-1} + a_{21}a_{22}\varepsilon_{2,t-1}^2 \\ & + \beta_{11}\beta_{12}\sigma_{11,t-1} + (\beta_{11}\beta_{22} + \beta_{12}\beta_{21})\sigma_{12,t-1} + \beta_{21}\beta_{22}\sigma_{22,t-1},\end{aligned} \tag{3}$$

$$\begin{aligned}\sigma_{22,t} = {} & c_{22}^2 + a_{12}^2\varepsilon_{1,t-1}^2 + 2a_{12}a_{22}\varepsilon_{1,t-1}\varepsilon_{2,t-1} + a_{22}^2\varepsilon_{2,t-1}^2 + \beta_{12}^2\sigma_{11,t-1} \\ & + 2\beta_{12}\beta_{22}\sigma_{12,t-1} + \beta_{22}^2\sigma_{22,t-1},\end{aligned} \tag{4}$$

According to Eqs. (2) to (4), the parameter $\sigma_{11,t}$ and $\sigma_{22,t}$ represent the conditional variance of the domestic crude oil market and international crude oil market return, respectively. $\sigma_{12,t}$ and $\sigma_{21,t}$ represent the conditional covariance of the domestic and international crude oil market return.

The coefficient matrix of ARCH effect term is $\alpha$, capture the effects of own past shocks on their own conditional variance, while the coefficient matrix of GARCH effect term is $\beta$, measure the effect of past fluctuations in the market on its own conditional variance.

This study aims to study the volatility spillover between domestic crude oil market and international crude oil market. Therefore, the coefficient $a_{12}$, $a_{21}$, $\beta_{12}$, $\beta_{21}$ are our interest. $a_{12}$, and $a_{21}$, reflect the overall impact spillover effect of international oil price on domestic oil price return and vice versa. Similarly $\beta_{12}$, and $\beta_{21}$ are volatility spill-over. To examine the volatility spillover effect, the Wald test forms are as follows:

$$\mathrm{H}_0 : a_{12} = \beta_{12} = \beta_{21} = \beta_{21} = 0,\ \mathrm{H}_1 : a_{12}, \alpha_{21}, \beta_{12}, \beta_{21} \neq 0 \tag{5}$$

If $a_{12}$ and $\beta_{12}$ are not zero at a given significance level, the international crude oil market has a volatility spillover effect on the domestic crude oil market. Similarly, according to whether the coefficients $a_{21}$ and $\beta_{21}$ are significantly different from zero, we can also judge whether the domestic crude oil market has a volatility spillover effect on the international crude oil market.

### 3.2 MS-VAR model

Given the binary nature of fluctuations in international oil prices, domestic oil prices, industrial added value, consumer price index, and money supply, coupled with the observed trends in international oil price dynamics over the past two decades and the intricacies of China's macroeconomic landscape, this study opts for the Markov Regime Vector Autoregression (MS-VAR) model for empirical analysis. Utilizing quarterly data spanning from 2000Q2 to 2022Q4, our research employs the MS-VAR model to identify distinct economic states and explore the nonlinear, dynamic effects of international oil price fluctuations on the macroeconomy under various economic conditions. The MS-VAR model amalgamates the Vector Autoregression (VAR) model within the time series framework while positing that the model's parameters take on functional forms that adapt to shifts in the economic environment and regional transitions within the economic system. At its core, the MS-VAR model operates on the premise that when a variable undergoes a shift in its behavior pattern after a certain period, it signifies that these variables exhibit a transfer or transition behavior between distinct Markov Regimes. In this model, the parameters depend on different regimes, that is, the state variable $S_t \in \{1, 2\}$

The MS-VAR() model can be expressed as:

$$Y_t = \sum_{i=1}^{l} A(i, S_t) * (Y_{t-i}) + \varepsilon_t \tag{6}$$

where, $Y_t$ is a $K \times 1$ vector representing the endogenous variables at time $tl$, is the order of the MS-VAR model, representing the number of lags included in the model; $A(i, S_t)$ is a $k \times k \times l$ matrix of coefficients corresponding to lag $i$ and $S_t$ is the regime or state variable at time $t$, determining the current regime; $\varepsilon_t$ is a vector representing the error term or innovations at time $t$, assumed to be normally distributed with mean zero and covariance matrix $\sum(S_{(t)})$; $S_{(t)}$ is the regime or state variable at time $t$, taking on different discrete values that represent the different economic regimes. The regime switching process is typically assumed to follow a first-order Markov chain, and the transition probabilities between different regimes are governed by a transition matrix $\pi$, where, $\pi(i, j)$ represents the probability of transitioning from regime $i$ to regime $j$, [18, 21]).

The model involves estimating the time-varying coefficients $A(i, S_t)$ and the transition probabilities $\pi$. The parameters of the model are typically estimated using maximum likelihood estimation or Bayesian methods, which involve optimizing the likelihood function based on the observed data.

The MS-VAR model offers the flexibility to adapt to evolving dynamics within a system over time, enabling distinct parameter estimates for each regime. This adaptability proves particularly valuable when modeling economic systems that demonstrate unique behaviors under specific regimes. It allows us to gain valuable insights into the nonlinear and heterogeneous responses of the system to various shocks under different economic conditions.

The data we used to construct MS-VAR model is proceed by taking the first-order logarithmic difference. And the dynamic impact of international oil price fluctuations ($\mathrm{dln}WTI_t$) on domestic oil price ($\mathrm{d}\ln DQ_t$), industrial added value ($\mathrm{d}\ln IND_t$), consumer price index ($\mathrm{d}\ln CPI_t$), and money supply ($\mathrm{d}\ln MS_t$) can be represented as follows

$$\begin{aligned}\mathrm{d}\ln IND_t = \beta_2 + \sum_{i=1}^{l}(A_{2i}(s_t) * \mathrm{dln}WTI_{t-i}) + \sum_{i=1}^{l}(B_{2i}(s_t) * \mathrm{dln}IND_{t-i}) \\ + \sum_{i=1}^{l}(C_{2i}(s_t) * \mathrm{dln}CPI_{t-i}) + \sum_{i=1}^{l}(D_{2i}(s_t) * \mathrm{d}\ln MS_{t-i}) + \varepsilon_2(t),\end{aligned} \tag{7}$$

$$\begin{aligned}\mathrm{d}\ln IND_t = \beta_2 + \sum_{i=1}^{l}(A_{2i}(s_t) * \mathrm{dln}WTI_{t-i}) + \sum_{i=1}^{l}(B_{2i}(s_t) * \mathrm{dln}IND_{t-i}) \\ + \sum_{i=1}^{l}(C_{2i}(s_t) * \mathrm{d}\ln CPI_{t-i}) + \sum_{i=1}^{l}(D_{2i}(s_t) * \mathrm{dln}MS_{t-i}) + \varepsilon_2(t),\end{aligned} \tag{8}$$

$$\begin{aligned}\mathrm{dln}CPI_t = \beta_3 + \sum_{i=1}^{l}(A_{3i}(s_t) * \mathrm{dln}WTI_{t-i}) + \sum_{i=1}^{l}(B_{3i}(s_t) * \mathrm{dln}IND_{t-i}) \\ + \sum_{i=1}^{l}(C_{3i}(s_t) * \mathrm{dln}CPI_{t-i}) + \sum_{i=1}^{l}(D_{3i}(s_t) * \mathrm{dln}MS_{t-i}) + \varepsilon_3(t),\end{aligned} \tag{9}$$

$$\begin{aligned}\mathrm{dln}MS_t = \beta_4 + \sum_{i=1}^{l}(A_{4i}(s_t) * \mathrm{dln}WTI_{t-i}) + \sum_{i=1}^{l}(B_{4i}(s_t) * \mathrm{dln}IND_{t-i}) \\ + \sum_{i=1}^{l}(C_{4i}(s_t) * \mathrm{dln}CPI_{t-i}) + \sum_{i=1}^{l}(D_{4i}(s_t) * \mathrm{dln}MS_{t-i}) + \varepsilon_4(t),\end{aligned} \tag{10}$$

where $t$ represents the time point, $i$ represents the lag order, $A, B, C, D$ are the coefficient matrices, $\beta$ is the intercept term, $\varepsilon$ is the error term.

In this model, we consider the dynamic impact of international oil price fluctuations on the domestic oil price, industrial added value, consumer price index, and money supply. The coefficient matrices $A, B, C, D$ capture the magnitude and direction of the dynamic effects of each variable on the others.

# 4 Empirical Results

## 4.1 The Spillover Effect of Oil Price Fluctuation Between Domestic and International Oil Market

In this section, we use Daqing crude oil price (*DQ*) to represent the domestic crude oil market price and *WTI* crude oil price (*WTI*) to represent the international crude oil market price. Our objective here is to explore the presence of volatility spillover effects between the domestic and international crude oil markets.

It can be seen from the results in Table 2 that there is decisive strong evidence shows that both $a_{11}$ and $a_{22}$ are significantly different from zero, indicating that both the *DQ* and the *WTI* have significant ARCH volatility spillover effects on themselves, and the price changes are significantly influenced by their own previous fluctuations. Through ARCH coefficient matrix, it can be seen that *WTI* is more sustained by its own impact than *DQ* ($a_{22} > a_{11}$); There also have very strong evidence to support that $\beta_{11}$ and $\beta_{22}$ are significantly different from zero, indicating that *DQ* and *WTI* have GARCH volatility spillover effects on themselves. $a_{12}^2$ and $a_{21}^2$ respectively represent the volatility spillover of *WTI* and *DQ* from the impact of new information of the other side, and the impact of one unit from *DQ* price will increase the volatility of *WTI* price by 0.5621 units ($a_{11}^2 = 0.5621$). The impact of one unit from *WTI* price will increase the volatility of *DQ* price by 0.0399 units ($a_{21}^2 = 0.0399$). It shows that the volatility spillo-ver effect of *WTI* on *DQ* is stronger than that of *DQ* on *WTI*. However, there is no evidence to support that $b_{12}$ and $b_{21}$ are significantly different from 0, indicating that there is no ARCH volatility spillover effect between *DQ* and *WTI*, but there is a significant GARCH volatility spillover effect.

**Table 2** BEKK-GARCH model estimation results

| Variables | Coeff | t-Statistic |
|---|---|---|
| $C_{11}$ | 0.0371 | 2.1379*(0.1018) |
| $C_{21}$ | 0.0607 | 5.7552***(0.0000) |
| $C_{22}$ | 0.0091 | 0.2091(0.9874) |
| $a_{11}$ | 0.7497 | 4.5204***(0.0000) |
| $a_{12}$ | −0.1133 | −0.8482(0.6979) |
| $a_{21}$ | 0.1997 | 1.0142(0.5979) |
| $a_{22}$ | 1.0502 | 6.0692***(0.0000) |
| $\beta_{11}$ | 1.1125 | 4.2519***(0.0001) |
| $\beta_{12}$ | 0.8967 | 4.7695***(0.0000) |
| $\beta_{21}$ | −1.5941 | −6.6596***(0.0000) |
| $\beta_{22}$ | −1.3196 | −7.3161***(0.000) |

*Note* 1 and 2 denote domestic and internatinal oil maket returns,respectively. The MBF in parentheses. (), "*","**" strong, very strong, decisive strong evidence supports $H_1$, respectively

**Table 3** Wald test results

| H0 | Chi-Squared | MBF |
|---|---|---|
| $a_{12} = \alpha_{21} = \beta_{12} = \beta_{21} = 0$ | 69.5902*** | |

*Note* "***" indicates rejection of the null hypothesis with decisive evidence

In order to further investigate the direction of volatility spillover effect between domestic and international crude oil markets, we follow the hypothesis in Eq. (6), and Wald test in Table 3. shows that the hypotheses reject the null hypothesis at the 1% significance level, indicating a significant two-way volatility spillover effect between *DQ* and *WTI* markets. The test outcomes for the hypotheses suggest that it's not just *WTI* prices influencing *DQ* prices; conversely, *DQ* prices also influence *WTI* prices. Although we cannot discern from these results the precise mechanisms at play, such as supply, demand, financial markets, or their interactions across different countries, it is relatively certain that international oil price fluctuations impact China's domestic oil prices.

## *4.2 The Impact of International Oil Price Fluctuation on China's Macro Economy*

In accordance with statistical theory, the MS-VAR model can be configured as various Markov Regime Vector Autoregression models, denoted as MSM-VAR, MSI- VAR, MSA-VAR, and MSH-VAR. Given that both oil price trends and numerous macroeconomic variables typically exhibit characteristics of both high and low volatility, low and high correlation, we have opted for a two-regime model with a lag order of 1 based on information criteria such as AIC(Akaike information criterion) and SC(Schwarz Criterion) (Table 4). After comprehensive consideration of various models, this study has established the MSIH(2)-VAR(1) model (Table 5). The model's performance was assessed and compared using maximum likelihood values and information screening criteria. The results are presented in the Table 6.

The regression analysis examining changes in international oil prices reveals several significant relationships. Notably, there is decisive and strong evidence supporting these findings: a 1% increase in international oil prices during a lag period is strongly associated with a substantial 36.96% increase in the current *DQ* oil price, while also leading to a significant 9.21% decrease in the current money supply. Additionally, a 1% increase in international oil prices during a lag period corresponds to a notable 2.98% increase in its own current oil prices and a 0.31% increase in the current *CPI*. Furthermore, a 1% increase in international oil prices during a lag period is strongly linked to an 8.29% increase in industrial production. This implies that rising international oil prices positively impact the growth rate of industrial added value (*IND*) and the consumer price index (*CPI*). Interestingly, the analysis reveals a negative correlation between the growth rate of international

**Table 4** Lag order selection

| Lag | LogL | LR | FPE | AIC | SC | HQ |
|---|---|---|---|---|---|---|
| 1 | 929.2825 | 85.45138* | 2.67e-16* | −21.66946* | −20.79518 | −21.31822* |
| 2 | 944.3187 | 26.08701 | 3.42E-16 | −21.42937 | −19.82652 | −20.78543 |
| 3 | 958.7504 | 23.29925 | 4.49E-16 | −21.17471 | −18.84329 | −20.23808 |
| 4 | 970.7824 | 17.97554 | 6.33E-16 | −20.86223 | −17.80225 | −19.6329 |
| 5 | 987.2827 | 22.66307 | 8.19E-16 | −20.65741 | −16.86887 | −19.13539 |
| 6 | 1000.525 | 16.59275 | 1.18E-15 | −20.3741 | −15.85698 | −18.55937 |
| 7 | 1023.983 | 26.56636 | 1.38E-15 | −20.33693 | −15.09125 | −18.22951 |
| 8 | 1052.388 | 28.74821 | 1.50E-15 | −20.419 | −14.44475 | −18.01888 |

**Table 5** Statistical values of each MS-VAR model

| | VAR(1) | MSI(2)-VAR(1) | MSIH(2)-VAR(1) | MSIA(2)-VAR(1) |
|---|---|---|---|---|
| Loglikelihood | 1007.885 | 1022.2174 | 1052.1817 | 1050.1131 |
| AIC | −21.7308 | −21.5604 | −21.8485 | −21.6692 |
| SC | −20.8975 | −20.1161 | −19.9875 | −17.1954 |

*Note* 2 represents two regimes; 1 represents a lag of one order

oil prices and the growth rate of the money supply. This unexpected result may be attributed to the time lag in economic cycle fluctuations and the implementation of macroeconomic monetary policies.

The regression analysis concerning changes in domestic oil prices (*DQ*) reveals significant relationships, supported by decisive and very strong evidence: For every 1% increase in domestic oil prices during a lag period, there is decisive evidence to suggest that the current money supply will increase by 7.46%. Moreover, there is very strong evidence to indicate that a 1% increase in domestic oil prices during a lag period corresponds to a significant 8.98% decrease in the current industrial added value. This aligns with the widely acknowledged concept that the monetary transmission channel, primarily operating through interest rate adjustments, plays a crucial role. Fluctuations in international oil prices prompt adjustments in a country's monetary policy, subsequently influencing its domestic economic output and price levels.

From the regression equation of changes in industrial production value *IND*, there is decisive strong evidence support that every 1% rise in industrial added value lagging a period, the current consumer price *CPI* will rise 5.39%. From the regression equation of changes in *CPI*, there is decisive strong evidence that the consumer price with a lag of one period has a positive effect on the industrial added value and consumer price of the current period. From the regression equation of the change of money supply *MS*, there is a decisive strong evidence support that the money supply lagging a period has a positive promoting effect on the current money supply.

**Table 6** Parameter estimation of MSIH(2)-VAR(1) model

| | dln*WTIt* | dln*DQt* | dln*INDt* | dln*CPIt* | dln*MS*t |
|---|---|---|---|---|---|
| Const (St,1) | −0.1012* | −0.1022* | −0.0060 | 0.0026 | 0.01945*** |
| | (−1.7878) | (−1.6558) | (−0.7918) | (1.3405) | (4.3979) |
| | [0.2023] | [0.2539] | [0.7309] | [0.4072] | [0.0000] |
| Const (St,2) | 0.027545 | 0.057132* | 0.01706* | 0.005409*** | 0.016682*** |
| | (0.7218) | (1.5498) | (1.6961) | (2.9724) | (5.6803) |
| | [0.7707] | [0.3009] | [0.2373] | [0.0121] | [0.0000] |
| dln*WTI*t−1 | 0.029823** | 0.369571*** | 0.082959* | 0.003067** | −0.092142*** |
| | (0.105) | (1.2821) | (1.9397) | (0.0828) | (−3.3782) |
| | [0.0945] | [0.0396] | [0.1524] | [0.9294] | [0.0033] |
| dln*DQ*t−1 | −0.035841 | −0.377291 | −0.089783** | 0.000668 | 0.074623*** |
| | (−0.1296) | (−1.3361) | (−2.1855) | (0.0838) | (3.0382) |
| | [0.9916] | [0.4096] | [0.0918] | [0.9965] | [0.0099] |
| dlnINDt−1 | −0.324714 | −0.255942 | −0.250912** | 0.053924*** | 0.039533 |
| | (−0.6511) | (−0.5252) | (−2.2286) | (2.6972) | (1.1761) |
| | [0.8090] | [0.8712] | [0.0835] | [0.0263] | [0.5008] |
| dlnCPIt−1 | 1.524279 | 0.380175 | 1.120071*** | 0.246632*** | −0.110536 |
| | (0.7043) | (0.1830) | (2.6039) | (3.0022) | (−0.7658) |
| | [0.7803] | [0.9834] | [0.0337] | [0.0110] | [0.7459] |
| dlnMSt−1 | 1.101029 | 0.797769 | 0.347869* | −0.032061 | 0.475356*** |
| | (1.0550) | (0.7534) | (2.0312) | (−0.8984) | (5.5723) |
| | [0.5732] | [0.7529] | [0.1271] | [0.6679] | [0.0000] |
| Variance (St, 1) | 0.1909 | 0.2140 | 0.0189 | 0.0036 | 0.0158 |
| Variance (St, 2) | 0.1095 | 0.0965 | 0.0265 | 0.0062 | 0.0067 |

*Note* t-Statistic in parentheses (); MBF in parentheses []; strong, very strong, decisive strong evidence supports H1, respectively

Analysis of the regression equation concerning changes in industrial production value (*IND*) reveals a compelling and decisive link: every 1% rise in industrial added value lagging a period corresponds to a significant 5.39% increase in the current consumer price index (*CPI*). Similarly, the regression equation for changes in *CPI* provides decisive and robust evidence that a 1% increase in consumer prices with a lag of one period has a positive impact on both the industrial added value and consumer prices of the current period. Additionally, the analysis of the change in money supply (*MS*) demonstrates a decisive and strong association: a 1% increase in money supply lagging a period has a positive and promoting effect on the current money supply. These findings illustrate the intricate interplay between industrial production, consumer prices, and money supply, highlighting the dynamic nature of these economic variables and their impacts on one another.

The result in Table 7 reveals noteworthy insights into the economic system's transition probabilities between regime 1 and regime 2. Specifically, regime 1 exhibits a relatively high conversion probability of 0.7191, while regime 2 boasts an even higher probability of 0.7895. A comparative analysis of these probabilities highlights that the likelihood of regime 2 prevailing surpasses that of regime 1. This suggests that there is a considerable probability that China's economy will persist in its current regime, with a relatively low probability of transitioning to an alternative regime. This insight underscores the robustness of China's economic system, indicating that once a stable feedback mechanism between the growth rate of international oil prices and the domestic macroeconomy is established, it tends to endure, regardless of whether the economic environment experiences low or high volatility. Moreover, it shows that it takes time for economic activity, which is affected by various factors, to change, require time to manifest changes, and the establishment of a second equilibrium necessitates a period of stability. Adjustments to the internal mechanisms also demand time for accumulation.

Table 7 also summarizes the sample summary and duration of states under different regimes. It can be seen that under regime 1, the number of sample observations falling into the regime is roughly 38, which is about three quarters of that under regime 2. Similarly, the unconditional probability under regime 1 is 0.4284, which is significantly lower than that under regime 2. This confirms the longer duration of low volatility mentioned above. In terms of duration, in regime 1, the average duration was 3.56 quarters, indicating a duration of less than one year. The duration under regime 2 is 4.75, indicating that the duration of low volatility sample is significantly longer than that under regime 1, which confirms that the self-persistence of low volatility is stronger than that of high volatility.

Figure 1 depicts the volatility patterns of *DQ*, *IND*, *CPI*, *MS*, and *WTI* within the MSIH(2)-VAR(1) framework. The analysis spans from the third quarter of 2000 to the fourth quarter of 2022. Notably, the figure reveals a striking resemblance in the trends of international and domestic oil prices, accompanied by noticeable internal correlations and resonance between the two. This trend is particularly pronounced during key time periods. In contrast, the volatility of *IND*, *CPI*, and *MS* appears relatively subdued, making it challenging to discern a distinct relationship with *WTI*. Nonetheless, there is a prevailing positive correlation among these variables in the broader context.

Figures 2 and 3 illustrate the smoothed probability distributions characterizing China's economic fluctuations during the sample period from the third quarter of 2000 to the fourth quarter of 2022 for regime 1 and regime 2, respectively. The horizontal axis denotes time, while the vertical axis represents the probability of

**Table 7** Probabilistic transformation matrix and regime characteristics

| | Regime 1 | Regime 2 | Nobs | Prob | Duration |
|---|---|---|---|---|---|
| Regime 1 | 0.7191 | 0.2809 | 38.2 | 0.4284 | 3.56 |
| Regime 2 | 0.2105 | 0.7895 | 51.8 | 0.5716 | 4.75 |

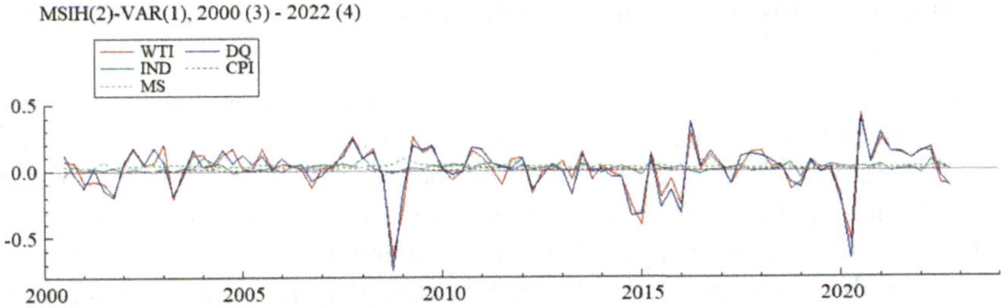


**Fig. 1** Volatility of sample data in the model

economic variables conforming to a specific regime system. To discern which regime system the data points from various samples belong to, one can rely on the height of the straight line in the smoothed probability graph. If the height of the straight line exceeds 0.5, it signifies that the economic state at that time aligns with the corresponding regime system [22]. Analyzing the smoothed probability distribution graphs, we observe that China's economy exhibited periods of high volatility within the regime system in 2005, the end of 2008 to 2009, 2015, and 2016. In contrast, during other years, China's economy experienced lower volatility, which broadly corresponds to the overall trajectory of China's economic development.

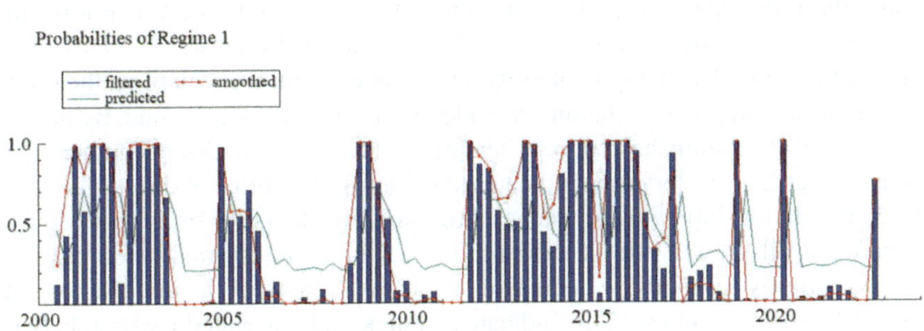


**Fig. 2** Smooth probability distribution of sample data in regime 1

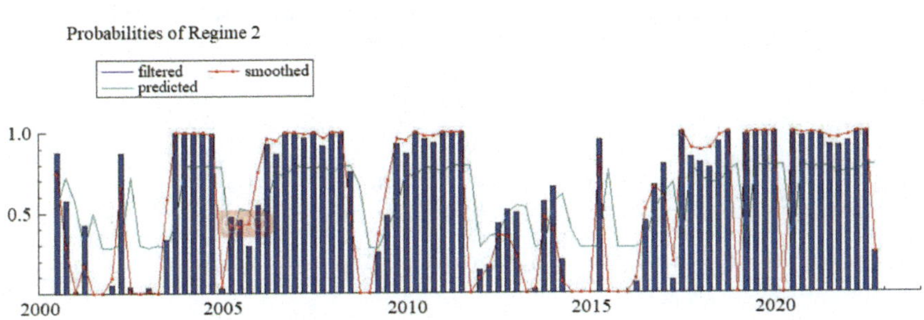


**Fig. 3** Smooth probability distribution of sample data in regime 2

## 4.3 Impulse response function analysis

Impulse response analysis is a valuable method for directly uncovering the short-term dynamic interactions among multiple variables within an economic system. In the context of studying the dynamic relationship between fluctuations in international crude oil prices and China's macroeconomic variables under distinct regimes, we employ impulse response analysis. Institution-dependent impulse response analysis typically involves utilizing models like the Markov Regime Vector Autoregression model. These models enable the identification of different economic regimes and the estimation of individual impulse response functions for each regime. This approach is instrumental in capturing the nonlinearity and heterogeneity of how shocks affect the economy, providing policymakers and researchers with more refined and precise insights.

Through the impulse response function, we can assess the dynamic impact relationships among various macro variables in the economic system. This allows us to delve deeper into the interaction between fluctuations in international oil prices and China's macroeconomic variables. Moreover, it facilitates a comparative analysis of the dynamic relationships between international oil price fluctuations and the fluctuations of variables such as *DQ*, *IND*, *CPI*, and *MS* under different regime systems.

Figure 4 illustrates the orthogonalized impulse response function, taking into account the immediate effect without cumulative considerations. When a positive shock of one standard deviation in the change rate of domestic oil prices occurs during both regime 1 with high volatility and regime 2 with low volatility, the growth rate of domestic oil prices in the subsequent period experiences a decline. By the third period, volatility diminishes entirely, leading to the complete disappearance of the impulse effect. This observation indicates that domestic oil prices exhibit a feedback-correcting relationship with themselves. Additionally, it's noteworthy that when the international oil growth rate is influenced by changes in the rate of domestic oil price, it promptly responds by decreasing its growth rate until the impulse response vanishes by the third phase. This indicates a quick and correlated reaction between international oil prices and the rate of change of domestic oil prices.

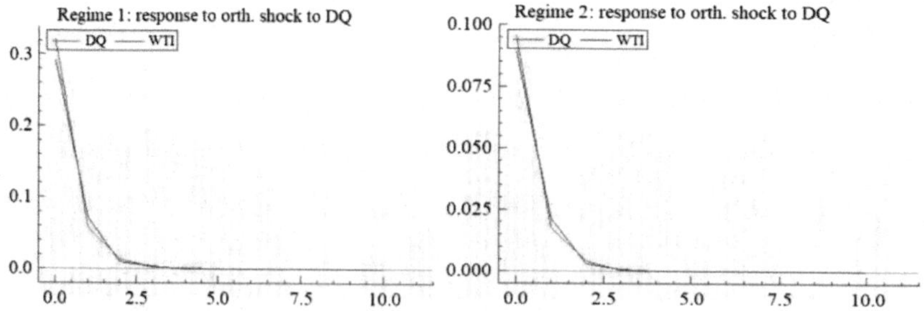


**Fig. 4** Impulse response diagram of *DQ* under 2 regimes

Figure 5 depicts the response to a positive shock of one standard deviation in the current growth rate of *IND* during the high volatility period. In this scenario, the subsequent growth rate of *IND* experiences a decline and gradually diminishes in the second phase. This pattern suggests that *IND* exhibits a feedback-correcting relationship with itself. Moreover, when the growth rate of *WTI* oil prices is affected by the current growth rate of *IND*, it responds swiftly by decreasing its growth rate until the third phase, at which point the impulse effect entirely dissipates. During the low volatility period, a shock to the current growth rate of *IND* results in a rapid decline in the growth rate of *WTI* oil prices in the first phase, reaching negative values. In the second phase, it begins to rise and then fluctuates steadily until the pulse effect vanishes entirely in the third phase.

Figure 6 shows that in the period of high volatility, when China's *CPI* growth rate is shocked by an external positive standard deviation, the growth rate of *CPI* and oil price will decline rapidly, and the degree of oil price change is greater than the degree of *CPI* change. After the third period, the *CPI* fluctuation declines to zero. After the fifth phase, the pulse effect of *WTI* fluctuation disappeared. Under the period of low volatility, under the external positive shocks, the growth rate of *WTI* rapidly declines after rising in the next period, until the shock decays to 0 after the seventh period. The *CPI* growth rate will gradually decline from the first phase to the fourth period, until the pulse effect completely disappears in the fifth phase.

Figure 7 reveals intriguing insights. During high volatility periods, a positive shock in the current period's *MS* leads to an initial decline in the growth rate of *MS*

**Fig. 5** Impulse response diagram of IND under 2 regimes

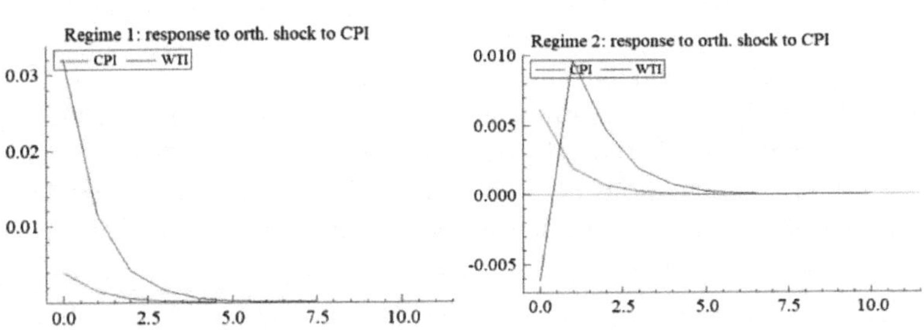


**Fig. 6** Impulse response diagram of CPI under 2 regimes

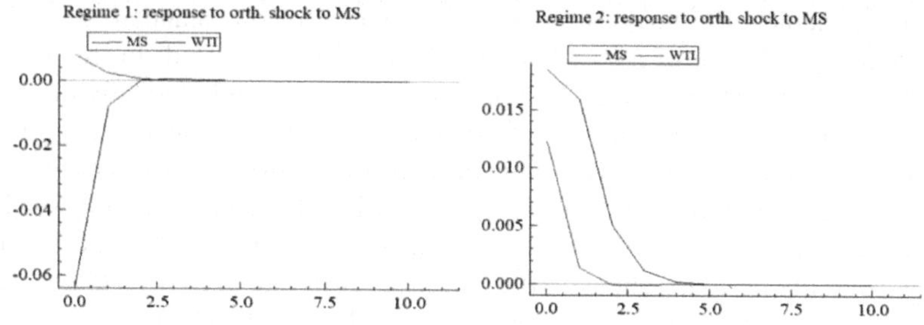


**Fig. 7** Impulse response diagram of MS under 2 regimes

for the first two periods, followed by a return to equilibrium. Simultaneously, the *WTI* growth rate experiences an increase, with the impulse response diminishing in the second phase. In contrast, during low volatility periods, a positive shock in the current period's *MS* triggers a rapid decline in the growth rates of both *MS* and *WTI* in the subsequent period. Notably, the magnitude of change in oil prices exceeds that of *MS*. After the initial phase, *MS* continues to decline until it reaches zero in the second phase, while *WTI* decays to zero after the fifth phase, ultimately extinguishing the impulse response.

This impulse response chart underscores the intricate relationship between international oil price fluctuations and China's macroeconomy. In high volatility regimes, a positive impact on the change rates of *DQ*, *IND*, and *CPI* in the current period results in a rapid decline in *WTI*'s growth rate. Interestingly, the degree of *WTI*'s decline surpasses the magnitude of change in *DQ*, *IND*, and *CPI*. Conversely, during low volatility regimes, a positive shock to the change rate of *DQ*, *IND*, and MS in the current period leads to a sharp initial decline in *WTI*'s growth rate, followed by a gradual decay to zero, ultimately extinguishing the impulse effect.

## 5 Conclusion

This paper has conducted a comprehensive analysis, beginning with data processing and utilizing the BEKK-GARCH(1,1) and MS-VAR models. It has partitioned the economic system and international oil prices into two distinct regimes, one characterized by high volatility and the other by low volatility. The findings reveal a significant two-way volatility spillover effect between international and domestic crude oil markets, with international crude oil prices exerting a stronger influence on domestic prices. These price fluctuations not only impact China's economic growth but also exhibit a notable lag effect. The study further demonstrates that increasing international crude oil prices are closely associated with rising domestic crude oil prices, positively affecting industrial added value and the consumer price index (*CPI*), albeit inversely correlated with the money supply's growth rate. This inverse relationship is attributed to the lag in economic cycle fluctuations and macroeconomic policy implementation.

Additionally, the study underscores the regime-specific nature of these relationships, with high and low volatility regimes exhibiting different correlations. Importantly, China's own economic conditions moderate the impact of international oil price fluctuations, highlighting the country's propensity for stability once equilibrium is reached. Lastly, it notes that high volatility regimes are relatively short-lived, while periods of low volatility are marked by greater economic stability. These findings offer valuable insights into the intricate interplay between international oil prices and China's macroeconomic variables, providing a nuanced understanding of these dynamics across various economic conditions and regimes.

Based on the research findings, several key recommendations can be derived:

(1) Improve China's Oil Pricing Mechanism and Enhance Domestic Oil Capacity.

    Given the increasing impact of international oil price fluctuations on China's macroeconomy, it is crucial to review and adjust the current international oil pricing mechanism. Greater emphasis should be placed on transitioning towards a more market-oriented approach, where factors such as supply, demand, and global market dynamics play a significant role in determining prices. Moreover, policy support and incentives should be provided to encourage energy enterprises to engage in crude oil exploration, development, and production. Initiatives such as tax incentives and reduced administrative approvals can stimulate crude oil production capacity, reducing dependence on international markets.

(2) Increase China's Crude Oil Reserves and Enhance Emergency Preparedness.

    Expanding crude oil reserves is essential to ensure the stability of the country's energy supply and reduce reliance on imported oil. This strategic move can mitigate energy-related risks, particularly during periods of sharp international oil price fluctuations. By releasing reserves during such times, China can stabilize domestic market prices and shield the economy from adverse shocks. However, the government should carefully weigh the costs and risks associated with increasing reserves, considering factors like storage expenses, resource utilization efficiency, and potential market impacts to ensure the policy's effectiveness and economic rationale.

(3) Adjust the Energy Structure and Promote Sustainable Development.

    In the midst of accelerated industrialization, China should prioritize investments in research and development to advance crude oil exploration, extraction, and refining technologies, optimizing energy utilization in the process. To reduce vulnerability to international oil price fluctuations, it is imperative to boost the share of renewable energy sources like solar, wind, hydro, and biomass in the overall energy supply. Policy incentives, technological innovation, and substantial investments should be deployed to scale up the use of renewable energy. Concurrently, China should work towards diminishing its reliance on non-renewable energy sources, specifically oil and coal, as part of a broader strategy to safeguard against the negative impacts of international oil price volatility.

**Acknowledgements** The author extends heartfelt gratitude to the Center of Excellence in Econometrics, Faculty of Economics at Chiang Mai University for their invaluable financial support.

## References

1. Alenoghena, R.O.: Oil price shocks and macroeconomic performance of the Nigerian economy: a structural VAR approach. Facta Univ. Ser.: Econ. Org. **1**, 299–316 (2021)
2. Bildirici, M.: Economic growth and electricity consumption in Africa: MS-VAR and MS-Granger Causality analysis. Soc. Sci. Res. Netw.Netw. (2012). https://doi.org/10.2139/ssrn.2129017
3. Bildirici, M., Gokmenoglu, S. M.: Environmental pollution, hydropower energy consumption and economic growth: Evidence from G7 countries. Renew. Sustain. Energy Rev. **75**, 68–85 (2017). https://doi.org/10.1016/j.rser.2016.10.052
4. Chan, H.L., Woo, K.Y.: An investigation into the dynamic relationship between international and China's crude oil prices. Appl. Econ. **48**(24), 2215–2224 (2016)
5. Chen, X.: Analysis on causes, effects and countermeasures of international crude oil price fluctuations [J]. Price Monthly **08**, 46–49 (2016)
6. Cheng, W.L.: Quantitative analysis of factors affecting international oil price [J]. International Petroleum Economics,(08):40–43+71 (2005)
7. Ehrmann, M., Ellison, M., Valla, N.: Regime-dependent impulse response functions in a Markov-switching vector autoregression model. Econ. Lett. **78**(3), 295–299 (2003). https://doi.org/10.1016/s0165-1765(02)00256-2
8. Gong, X., Guan, K., Chen, L., Liu, T., Zheng, S.X.: What drives oil prices?—A Markov switching VAR approach. Resour. Policy. Policy **74**, 102316 (2021). https://doi.org/10.1016/j.resourpol.2021.102316
9. Hameed, Z., Shafi, K., Nadeem, A.: Volatility spillover effect between oil prices and foreign exchange markets. Energ. Strat. Rev. Strat. Rev. **38**, 100712 (2021). https://doi.org/10.1016/j.esr.2021.100712
10. Herrera, A.M., Karaki, M.B., Rangaraju, S.K.: Oil price shocks and US economic activity. Energy Policy **129**, 89–99 (2019)
11. Hou, Y., Wang, L., Guo, B.: As Tar. Price Risk transmission between crude oil futures markets: an analysis based on BEKK-GARCH Model [J]. Price Theory Pract. **10**, 93–97 (2021)
12. Huang, Y., Li, F.: Research on the time-varying spillover effect of international curde oil price on China's exchange rate. Energy Rep. **8**, 138–148 (2022). https://doi.org/10.1016/j.egyr.2022.03.067
13. Li, J., Tang, L., Li, L.: The Co-movements between crude oil price and internet concerns: causality analysis in the frequency domain. J. Syst. Sci. Inf. **8**(3), 224–239 (2020)
14. Li, L., Yin, L., Zhou, Y.: Exogenous shocks and the spillover effects between uncertainty and oil price. Energy Econ. **54**, 224–234 (2016). https://doi.org/10.1016/j.en-eco.2015.11.017
15. Lin, L., Zhou, Z., Jiang, Y., Ou, Y.: Risk spillovers and hedge strategies between global crude oil markets and stock markets: Do regime switching processes combining long memory and asymmetry matter?. N. Am. J. Econ. Finance. **57**, 101398 (2021)
16. Naccache, T.: Oil price cycles and wavelets. Energy Econ. **33**(2), 338–352 (2011)
17. Nasir, M.A., Al-Emadi, A.A., Shahbaz, M., Hammoudeh, S.: Importance of oil shocks and the GCC macroeconomy: a structural VAR analysis. Res. Policy. Policy **61**, 166–179 (2019)

18. Nimmonrat, N., Pastpipatkul, P., Yamaka, W., Maneejuk, P.: Price transmission mechanism for natural gas in Thailand. In: Econometrics for Financial Applications, pp. 684–697. Springer International Publishing (2018)
19. Nyangarika, A.: Influence Oil Price Towards Macroeconomic Indicators in Russia (2019)
20. Ruyuehui: Domestic and foreign crude oil futures prices and spot prices of crude oil in our country correlation research [D]. Shanghai International Studies University (2021)
21. Pastpipatkul, P., Yamaka, W., Sriboonchitta, S.: Effect of quantitative easing on ASEAN-5 financial markets. Causal Inf. Economet. 525–543 (2016)
22. Pastpipatkul, P., Yamaka, W., Wiboonpongse, A., Sriboonchitta, S.: Spillovers of quantitative easing on financial markets of Thailand, Indonesia, and the Philippines. In: Integrated Uncertainty in Knowledge Modelling and Decision Making: 4th International Symposium, IUKM 2015, Nha Trang, Vietnam, 15–17 Oct. 2015, Proceedings 4, pp. 374–388. Springer International Publishing
23. Wei, Y., Qin, S., Li, X., Zhu, S., Wei, G.: Oil price fluctuation, stock market and macroeconomic fundamentals: evidence from China before and after the financial crisis. Financ. Res. Lett. Res. Lett. **30**, 23–29 (2019)
24. Yildirim, Z., Arifli, A.: Oil price shocks, exchange rate and macroeconomic fluctuations in a small oil-exporting economy. Energy **219**, 119527 (2021)
25. Yu, L., Zha, R., Stafylas, D., He, K., Liu, J.: Dependences and volatility spillovers between the oil and stock markets: new evidence from the copula and VAR-BEKK-GARCH models. Int. Rev. Financ. Anal. Financ. Anal. **68**, 101280 (2020). https://doi.org/10.1016/j.irfa.2018.11.007
26. Zhang, Y., Fan, Y., Tsai, H., Wei, Y.: Spillover effect of US dollar exchange rate on oil prices. J. Policy Model. **30**(6), 973–991 (2008). https://doi.org/10.1016/j.jpolmod.2008.02.002
27. Zhang, Y.J., Wei, Y.M.: The crude oil market and the gold market: Evidence for cointegration, causality and price discovery. Resour. Policy **35**(3), 168–177 (2010)
28. Zheng, C.: An innovative MS-VAR model with integrated financial knowledge for measuring the impact of stock market bubbles on financial security. J. Innov. Knowl. Innov. Knowl. **7**(3), 100207 (2022). https://doi.org/10.1016/j.jik.2022.100207
29. Zhipeng, Y., Shenghong, L.: Hedge ratio on Markov regime-switching diagonal Bekk-Garch model. Financ. Res. Lett. Res. Lett. **24**, 49–55 (2018). https://doi.org/10.1016/j.frl.2017.06.015

# Deep Reinforcement Learning for Automated of Asian Stocks Trading

**Todsapon Panya and Manad Khamkong**

**Abstract** In a complex and changeable stock market, algorithmic stock trading has firmly established itself as a fundamental aspect of the present-day financial market, where most transactions are now fully automated. Additionally, Deep Reinforcement Learning (DRL) agents, renowned for their exceptional performance in intricate games such as chess and Go, are increasingly impacting the stock market. In this paper, we examine the potential of deep reinforcement learning to optimize the portfolio returns of 15 Asian stocks. We model stock trading as a Markov decision process problem because of its stochastic and interactive nature. Furthermore, we train a deep reinforcement learning agent using three actor-critic-based algorithms: proximal policy optimization (PPO), advantage actor-critic (A2C), and deep deterministic policy gradient (DDPG). We tested the algorithm on Asian stocks to see how well it performed pre-COVID, during COVID, and post-COVID. The trading agent's performance using various reinforcement learning algorithms is assessed and compared to the traditional min-variance portfolio allocation strategy. The proposed three individual algorithms are above the minimum variance in risk-adjusted return as evaluated by the Sharpe ratio.

## 1 Introduction

An effective automated stock trading strategy is significant for investment firms and hedge funds. This strategy enhances capital allocation and optimizes investment performance, aiming for higher expected returns. Achieving maximum returns involves assessing potential returns and associated risks. Nonetheless, analysts need help in comprehensively evaluating all pertinent variables within the intricate and ever-changing stock market environment [19, 37].

Due to this rationale, multiple endeavors have been undertaken to construct an automated stock trading system capable of securing a positive profit. Among these

T. Panya · M. Khamkong (✉)
Department of Statistics, Faculty of Science, Chiang Mai University, Chiang Mai 50200, Thailand
e-mail: Manad.k@cmu.ac.th

 
V. Kreinovich et al. (eds.), *Applications of Optimal Transport to Economics and Related Topics*, Studies in Systems, Decision and Control 556,
https://doi.org/10.1007/978-3-031-67770-0_37

endeavors, reinforcement learning emerges as one of the prevailing and essential concepts driving these efforts.

Deep reinforcement learning (DRL), which effectively manages the interplay between exploration (venturing into unexplored territory) and exploitation (leveraging existing knowledge), holds significant potential for automating trading within quantitative finance. DRL algorithms exhibit robust capabilities in resolving complex issues associated with dynamic decision-making through interactions within unfamiliar environments and offer two significant advantages of portfolio scalability and market model independence.

Deep reinforcement learning (DRL) involves employing a reward system to enhance forthcoming rewards. In numerous challenging games, DRL systems have demonstrated the capacity to outperform human players. To illustrate, DeepMind's AlphaGo, powered by a DRL algorithm, achieved victory over the Go world champion, Lee Sedol, in March 2016. Notably, DRL does not demand extensive labeled training datasets, which is advantageous given the escalating challenges associated with labeling massive datasets due to the exponential growth of data. This alleviates significant time and labor constraints.

The exploration-exploitation technique within deep reinforcement learning (DRL) achieves equilibrium by managing the experimentation of approaches and exploiting existing knowledge. This distinct characteristic sets it apart from other learning algorithms. Additionally, the agent is motivated to delve into uncharted territories that human specialists have yet to explore during the exploration phase. When simulating the stock trading process, a Markov Decision Process (MDP) can serve as the foundational framework for DRL.

As highlighted in the works of [2, 24], the actor-critic approach within deep reinforcement learning (DRL) has recently found application in the field of finance. Reference [24] elaborate that the actor-critic method has been effectively employed in contexts like video games such as [34], demonstrating its capability to comprehend and adapt to intricate and extensive environments. Consequently, the actor-critic technique has emerged as a fitting choice for addressing stock trading scenarios involving substantial portfolios. Moreover, the application of Deep Reinforcement Learning (DRL) methods to automate the trading of individual stocks [1, 9, 14, 23].

Anticipated is a scenario where the COVID-19 pandemic leads to substantial and erratic variations in returns and volatility within financial assessments, driven by the structural modifications in data caused by the pandemic [26]. These fluctuations could present unique challenges for investors and analysts, making it crucial to adopt adaptive strategies to navigate the stock market. As traditional models might struggle to capture this unprecedented volatility, a comprehensive understanding of the underlying factors and a flexible approach to risk management will be pivotal in achieving long-term financial stability. Furthermore, the structural shift may introduce a vulnerability to conventional trading models, which typically operate under the assumption of either minimal change or a static trend in volatility. These factors underscore our rationale for opting to employ the DRL algorithm.

This study examines three Deep Reinforcement Learning (DRL) algorithms (A2C, DDPG, and PPO). These algorithms are designed to autonomously develop effective stock trading strategies by maximizing investment returns. This exploration focuses on enhancing the return of the Asian stock portfolio through optimization. The study involves a thorough comparison and evaluation of their respective performance metrics.

The rest of this document is structured as follows: Sect. 2 presents an overview of prior research. Section 3 furnishes essential context, concisely explaining reinforcement learning and automated stock trading agents. Moving on to Sect. 4, we elucidate our findings from empirical analysis. Finally, in Sect. 5, we wrap up this paper and propose potential avenues for future research.

## 2 Literature Review

In recent times, deep reinforcement learning has been applied to financial markets. These applications involve dealing with either discrete or continuous state and action spaces. To tackle this, three main learning strategies are used: the critic-only approach, the actor-only approach, and the actor-critic approach [13]. Learning models with continuous action space provide nuanced and precise control compared to discrete action space.

The critic-only approach trains a value function to guide decisions in reinforcement learning by estimating future rewards from states or actions. These aid learning in complex environments, particularly with continuous or high-dimensional action spaces. It complements various reinforcement learning methods applied in finance, robotics, and games [6, 10, 17].

In summary, the critic-only approach emphasizes value function approximation. At the same time, deep Q-learning is a specific algorithm that leverages deep neural networks to approximate the Q function, enabling effective decision-making in complex and high-dimensional environments. These concepts represent essential building blocks in reinforcement learning, contributing to advancements in autonomous agents and intelligent systems. However, the primary drawback of relying solely on the critic-oriented methodology is its applicability being restricted to scenarios where state and action spaces are both discrete and finite. This constraint becomes impractical when dealing with an extensive array of stocks within a portfolio, given that stock prices inherently exist within a continuous spectrum.

The actor-only approach utilizing reinforcement learning involves training an algorithmic trading agent to make optimal decisions directly without explicitly estimating the value of different states or actions [11, 18, 29].

This method focuses on learning a policy that directly maps market conditions to trading actions, such as buying, selling, or holding assets. The agent explores the market environment, learns from rewards gained or losses incurred, and adjusts its policy to maximize cumulative profits. In portfolio management, the actor-only approach allows the agent to dynamically allocate assets based on learned policies, adjusting the portfolio composition to changing market trends or risk profiles. This can lead to more responsive and efficient management of investment portfolios.

Rather than utilizing a neural network to acquire knowledge about the Q-value, the neural network is employed to grasp the policy. The policy denotes a distribution of probabilities which essentially represents a strategy for a specific state, indicating the probability of selecting a permissible action. To counteract the challenges posed by high-dimensional spaces and enhance the effectiveness of trading, recurrent reinforcement learning is implemented as outlined in [29]. This approach that relies solely on the actor can adeptly navigate environments characterized by continuous action spaces. On the other hand, the optimization methods (typically called policy gradient methods) suffer from high variance in the gradient estimates, leading to slow learning.

The actor-critic approach, a hybrid of the actor-only and critic-only methods, has gained traction in finance [3, 20, 24, 38]. It combines direct action policy learning with value estimation. This dynamic interaction enables the agent to refine trading strategies, optimize portfolio allocations, and manage risks effectively.

The actor-network determines actions, while the critic network evaluates their value, fostering adaptive decision-making in complex and volatile financial markets. This approach is particularly beneficial for continuous or high-dimensional action spaces, enhancing the precision of trading decisions and portfolio management.

The actor-critic approach in adapting to complex environments and its success in playing challenging video games [34]. It proposes that this approach holds promise when applied to trading with an extensive stock portfolio.

# 3 Methods

## 3.1 *Markowitz Mean-Variance Portfolio Theory*

### 3.1.1 The Minimum Risk Mean-Variance Portfolio

The inflection point on the hyperbola (efficient frontier) marks the location of the minimum variance portfolio. Solving a fundamental quadratic optimization problem using standard Lagrange multiplier methods yields a single, distinct solution for this portfolio allocation [27]. The optimization problem can be expressed as follows:

$$\begin{array}{ll} \min_{\omega} & \frac{1}{2}\omega^{\mathrm{T}}\Omega\omega \\ subject\ to & \omega^{\mathrm{T}}r = E\left(r_p\right) \\ and & \omega^{\mathrm{T}}1 = 1. \end{array}$$

The vector $\omega$ represents the individual investments (weights of the assets) with the constraint that the entire available capital is invested, denoted by the equation $\omega^{\mathrm{T}}1 = 1$. The condition expresses the desired minimum target return $E\left(r_p\right)$ is expressed by the condition $\omega^{\mathrm{T}}r = E\left(r_p\right)$, where the vector r estimates the expected mean of the assets $(E\left(r_i\right))$.

## 3.2 Markov Decision Process for Stock Trading

Using a Markov Decision Process (MDP), we represent the stock trading procedure and establish our trading objective to maximize the expected return. Furthermore, to account for the inherent unpredictability of the dynamic stock market, we apply a Markov Decision Process (MDP) as follows [16]:

- State $s = [p, h, b]$: a vector that includes stock prices $p \in R_+^D$, the stock shares $h \in Z_+^D$, and the remaining balance $b \in R_+$, where $D$ denotes the number of stocks and $Z_+$ denotes non-negative integers.
- Action a: a vector of actions over $D$ stocks. The allowed actions on each stock include selling, buying, or holding, resulting in decreasing, increasing, and no change of the stock shares $h$, respectively.
- Reward $r\left(s, a, s'\right)$: the direct reward of taking action a at state $s$ and arriving at the new state $s$.
- Policy $\pi(s)$: the trading strategy at the state $s$, which is the probability distribution of actions at the state $s$.
- Q-value $Q_\pi(s, a)$: the expected reward of taking action a at state $s$ following policy $\pi(s)$.
  The state transition of a stock trading process is shown in Fig. 2. At each state, one of three possible actions is taken on stock $d(d = 1, \ldots, D)$ in the portfolio.
- Holding, $\mathbf{h}_{t+1}[d] = \mathbf{h_t}[d]$.
- Buying $\mathbf{k}[d]$ shares results in.

At the time $t$ an action is taken and the stock prices update at $t + 1$. Accordingly, the portfolio values may change from "portfolio value 0" to "portfolio value 1", "portfolio value 2", or "portfolio value 3", respectively, as illustrated in Fig. 1. Note that the portfolio value is $\mathbf{p}^{\mathrm{T}}\,\mathbf{h} + b$.

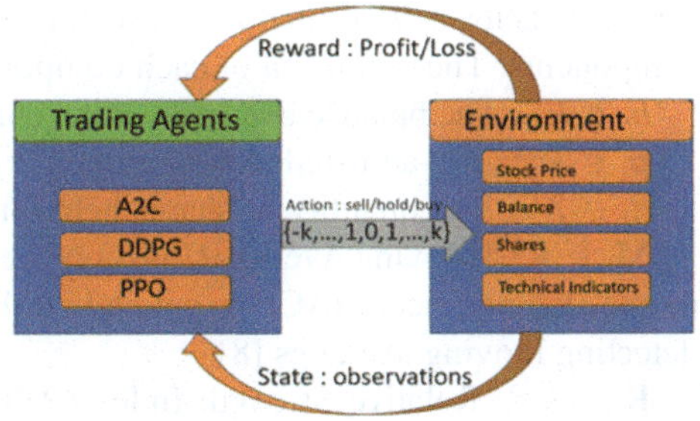


**Fig. 1** Overview of reinforcement learning-based stock trading strategy

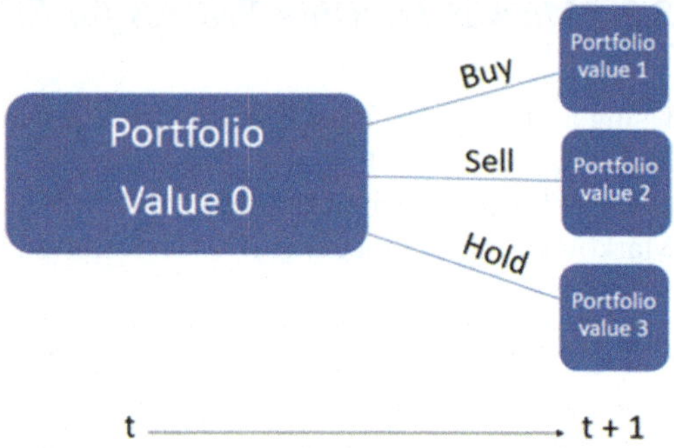


**Fig. 2** A starting portfolio value with three actions results in three possible portfolios

## 3.3 Stock Market Environment

Before instructing a deep reinforcement trading agent, we meticulously construct an environment that emulates real-world trading conditions. This environment facilitates the agent's engagement and learning process. In the realm of actual trading, myriad factors are demanding consideration. For instance, historical stock prices, existing shareholdings, technical indicators, and more necessitate attention. Our trading agent is tasked with acquiring this pertinent information from the environment and executing actions outlined in the preceding section. To facilitate this endeavor, we utilize OpenAI Gym for environment implementation and agent training [4, 30, 31].

We employ a continuous action space to simulate the trading dynamics of numerous stocks. Our framework accommodates a portfolio comprising a total of fifteen distinct stocks.

We employ a continuous action space to simulate the trading dynamics of numerous stocks. Our framework accommodates a portfolio comprising a total of fifteen distinct stocks.

(1) State Space: We represent the state space of the multi-stock trading environment using a 91-dimensional vector composed of seven distinct information components. The definition of each component is as follows:

$b_t \in R_+$: The balance available at the current time step $t$.

$\mathbf{p}_t \in R_+^{15}$: The adjusted closing price for each stock.

$h_t \in Z_+^{15}$: The quantity of shares held for each specific stock.

$\mathbf{M}_t \in R^{15}$: Moving Average Convergence Divergence (MACD) is computed based on the closing price. MACD is a widely utilized momentum indicator renowned for detecting moving averages [8].

$\mathbf{R}_t \in R_+^{15}$: Relative Strength Index (RSI) is derived from the closing price and serves as a metric for assessing the magnitude of recent price fluctuations. When the price hovers near the support line, it signifies an oversold condition in the stock, prompting a potential buying action. Conversely, when the price approaches the resistance level, it indicates an overbought state, prompting a possible selling action [8].

$C_t \in R_+^{15}$: Commodity Channel Index (CCI) is computed utilizing the high, low, and closing prices. CCI involves contrasting the present price with the average price within a specified time frame, indicating potential buying or selling actions [25].

$\mathbf{X}_t \in R^{15}$: Average Directional Index (ADX) is determined using the high, low, and closing prices. ADX aids in identifying the strength of a trend by quantifying the extent of price movement [15].

(2) Action Space: For a single stock, the action space is defined as $\{-k, \ldots, -1, 0, 1, \ldots, k\}$, where $k$ and $-k$ presents the number of shares we can buy and sell, and $k \leq h_{\max}$ while $h_{\max}$ is a predefined parameter that sets as the maximum amount of shares for each buying action. Therefore, the size of the entire action space is $(2k + 1)^{15}$. The action space is then normalized to $[-1, 1]$. Because the reinforcement learning algorithms A2C and PPO establish the policy directly using a Gaussian distribution, it is necessary to ensure that this distribution is both normalized and symmetrical [31].

## 3.4 Agents for Automated Stock Trading

### 3.4.1 A2C

Advantage Actor-Critic (A2C) algorithm is a popular reinforcement learning method that combines Actor-Critic architecture elements with an advantage function for bootstrapping.

The Actor-Critic framework has two main components: the actor and the critic. The actor is responsible for selecting actions based on the current policy, while the critic evaluates the value or expected return of a state or state-action pair. By incorporating both actor and critic, A2C aims to learn a more stable and efficient policy [28].

The advantage function in A2C estimates how good a particular action is compared to the average action for a given state. It quantifies the advantage of taking a specific action rather than following the average policy in that state. The advantage function is crucial because it helps reduce the variance when updating the policy, which improves the learning process's stability. The advantage function can be mathematically defined as follows:

$$A(s_t, a_t) = Q(s_t, a_t) - V(s_t)$$

Where:

$Q(s_t, a_t)$ is the value after executing decision action at (buy, sell, hold) in state $s_t$ or the asset return value.

$V(s_t)$ is the state value function.

### 3.4.2 DDPG

DDPG (Deep Deterministic Policy Gradients) is an off-policy actor-critic algorithm designed for environments with continuous action spaces. The primary distinction between DDPG and A2C lies in their approach to policy representation [22].

In DDPG, the objective is to learn a deterministic policy, where, for a given state, the agent consistently selects the same action deterministically. In other words, the deterministic policy directly maps each state to a particular action without any probability distribution. This means that if the agent encounters the same state multiple times, it will always produce the same action in response.

On the other hand, in A2C (Advantage Actor-Critic) and similar stochastic policybased methods, the policy is represented as a probability distribution over actions given a specific state. This stochastic policy maps each state to a set of action probabilities, allowing for randomization in action selection. Consequently, if the agent encounters the same state multiple times, it may choose different actions based on the probabilities associated with those actions. DDPG assumes $a_t = \pi_\theta(s_t)$ with deterministic policy. DDPG updates $\theta \longleftarrow \theta + \alpha \nabla_\theta J(\theta)$, and its objective function is as follows:

$$J(\theta) = E_{\tau \sim p_\theta(\tau)}\left[\sum_{t=0}^{\infty} \gamma^t r(s_t, a_t)\right]$$

and the objective function gradient is

$$\nabla_\theta J(\theta) = \sum_{t=0}^{T} E\left[\nabla_\theta \pi_\theta(s_t) \nabla_{s_t} Q^{\pi_\theta}(s_t, a_t)\right]$$

Where:

$\theta$ is the set of parameters of the actor network.

J($\theta$) represents the expected return (or some variant of the return) based on the critic's Q-value estimates.

$\alpha$ is the learning rate, controlling the step size of the updates.

### 3.4.3 PPO

We explore and use PPO as a component in the ensemble method. PPO [22] is introduced to control the policy gradient update and ensure that the new policy will be similar to the previous one. PPO tries to simplify the objective of Trust Region Policy Optimization (TRPO) by introducing a clipping term to the objective function [32].

Suppose we represent the probability ratio between the old and new policies as:

$$r_t(\theta) = \frac{\pi_\theta(a_t \mid s_t)}{\pi_{\theta_{old}}(a_t \mid s_t)}$$

The clipped surrogate objective function of PPO is:

$$J^{\text{CLIP}}(\theta) = E_t\left[\min\left(r_t(\theta)\hat{A}(s_t, a_t), clip(r_t(\theta), 1-\varepsilon, 1+\varepsilon)\hat{A}(s_t, a_t)\right)\right]$$

where $r_t(\theta)\hat{A}(s_t, a_t)$ is the standard policy gradient objective, and $\hat{A}(s_t, a_t)$ is the estimated advantage function. The function $clip(r_t(\theta), 1-\varepsilon, 1+\varepsilon)$ clips the ratio $r_t(\theta)$ to be within $[1-\varepsilon, 1+\varepsilon]$. The objective function of PPO takes the minimum of the clipped and standard objective. PPO discourages significant policy change moves outside of the clipped interval. Therefore, PPO improves the policy network's training stability by restricting the policy update at each training step. We select PPO for stock trading because it is stable, fast, and more straightforward to implement and tune.

## 4 Empirical Results

### *4.1 Stock Data Preprocessing*

The research utilized daily data from 15 prominent stock markets in Asia throughout the study period, including the Stock Exchange of Thailand (SET), Singapore's Straits Times (ST), the Philippine Stock Exchange Index (PSEI), Vietnam's Ho Chi Minh City Stock Exchange (HCMC), Malaysia's FTSE Bursa Malaysia Index Series (FTSE), the Jakarta Stock Exchange Composite (JKT), India's Bombay Stock Exchange (BSE), Pakistan's Karachi Stock Exchange (KSE), the China AMC CSI 300 Index ETF (CSI), Japan's Nikkei Stock Average (NIKKIE), Korea Stock Exchange's KOSPI Index, Taiwan Stock Exchange (TWSE), Turkey's Borsa İstanbul Stock Exchange (BIST), Kazakhstan Stock Exchange (KASE), and the Russian Trading System (RTS).

The study employed training data covering the period from 2014 to 2022, which was segregated into three distinct sets: "Before COVID-19" (spanning from August 25, 2014, to February 28, 2019), "During COVID-19" (encompassing the time from August 25, 2014, to February 28, 2020), and "After COVID-19" (comprising the period from August 25, 2014, to February 28, 2022). then, the test data set was also categorized into three sets: "Before COVID-19" (ranging from March 01, 2019, to August 30, 2019), "During COVID-19" (covering the interval from March 01, 2020, to August 30, 2020), and "After COVID-19" (extending from March 01, 2022, to August 30, 2022). Furthermore, the algorithms are optimized using the A2C PPO DDPG agent, and their effectiveness is assessed using various performance metrics such as Sharpe ratio, learning rate, and number of episodes.

The ultimate goal of the study is to determine if these deep learning algorithms can be considered viable tools for trading, providing an advantage by leveraging accurate trading data and enabling adaptability to market changes. Additionally, in cases where no information is available due to holidays or other reasons, the stock index price is calculated based on the previous day's trading price [35], with the stock index prices sourced from the Bloomberg database.

## *4.2 Performance Comparisons*

Evaluation metrics express the performance of each agent in each stock market, and we include cumulative return, annual return, annual volatility, Sharpe ratio, and max drawdown.

4.2.1 The change in the entire investment price over a particular period is called cumulative return, also known as total return. Use the following formula:

$$Cumulative\ Return = \frac{(Final\ Price\ of\ Stock - Initial\ Price\ of\ Stock)}{Initial\ Price\ of\ Stock} \times 100$$

4.2.2 The annual return is computed as a percentage of the initial investment and is the return on investment made during the year. In other words, the annual return equals the cumulative yearly return. Utilizing the following formula:

$$Anunual\ Return = \frac{(Eng\ of\ Year\ Stock\ Price - Beginning\ of\ Year\ Stock\ Price)}{Beginning\ of\ Year\ Stock\ Price} \times 100$$

4.2.3 Annual Volatility refers to the yearly volatility metric, reflecting the price swings an asset or market index may experience over a single calendar year. It is an essential tool for investors, as it allows them to evaluate an investment's potential risks and rewards over a longer-term horizon. (add) The formula for calculating the Annual Volatility is as follows:

$$Annual\ Volatility = \sqrt{\sum (P_{av} - P_i)^2}$$

Where:
$P_i$ = the portfolio's daily price % change on the $i$th day
$P_{av}$ = the portfolio's average price % change overall trading days.

4.2.4 The Sharpe ratio is a mathematical metric that represents the average return earned above the risk-free rate for each unit of volatility. It provides valuable insights into the relationship between excess returns during a specific time frame and the potential increase in volatility and risk associated with an investment strategy. The formula for calculating the Sharpe ratio is as follows:

$$sharpe\ Ratio = \frac{R_p - R_f}{\sigma_p}$$

Where:
$R_p$ = the return of portfolio (annual return)
$R_f$ = the risk-free rate
$\sigma_p$ = standard deviation of the portfolio value.

4.2.5 The max drawdown measures the potential downside risk during a specific timeframe. It represents the most significant observed loss from the highest point (peak Value) to the lowest point (trough Value) in the portfolio before a new peak is achieved. The formula for calculating the maximum drawdown is as follows:

$$Max\ Drawdown = \frac{(P - T)}{P}$$

Where:
P = Peak value of the portfolio (highest point before a new peak)
T = Trough value of the portfolio (lowest point after the peak).

In summary, as shown in Table 1, DDPG had the highest cumulative and annual returns, the highest Sharpe ratio, and a relatively moderate level of volatility, suggesting it outperformed the other strategies during this period. On the other hand, min-variance had the worst performance in terms of both cumulative and annual returns, and it exhibited the highest volatility and the largest maximum drawdown, indicating significant losses and risk. When evaluating these portfolio strategies, investors should consider their risk tolerance and investment goals.

Table 2 and Fig. 3 illustrate that the A2C, PPO, and DDPG agents exhibit superior risk-adjusted performance. Moreover, Fig. 3 demonstrates that the DDPG approach

**Table 1** Performance metrics comparison on Asian stocks from 01, March 2019, to 30, September 2019 (before COVID-19)

| | A2C | DDPG | PPO | Min-variance |
|---|---|---|---|---|
| Cumulative returns (%) | −0.70 | 3.00 | −2.86 | −7.88 |
| Annual return (%) | −1.45 | 6.35 | −5.86 | −15.71 |
| Annual volatility (%) | 5.35 | 6.83 | 4.86 | 7.73 |
| Sharpe ratio | −0.25 | 0.94 | −1.22 | −2.17 |
| Max drawdown (%) | −3.51 | −4.38 | −5.65 | −12.11 |

**Table 2** Performance metrics comparison on Asian stocks from 01, March 2020, to 30, September 2020 (during COVID-19)

| | A2C | DDPG | PPO | Min-variance |
|---|---|---|---|---|
| Cumulative returns (%) | 20.10 | 27.47 | 21.69 | 23.77 |
| Annual return (%) | 46.92 | 66.47 | 51.02 | 56.49 |
| Annual volatility (%) | 9.50 | 14.21 | 8.91 | 21.96 |
| Sharpe ratio | 4.10 | 3.66 | 4.67 | 2.15 |
| Max drawdown (%) | −2.55 | −3.64 | −3.35 | −12.41 |

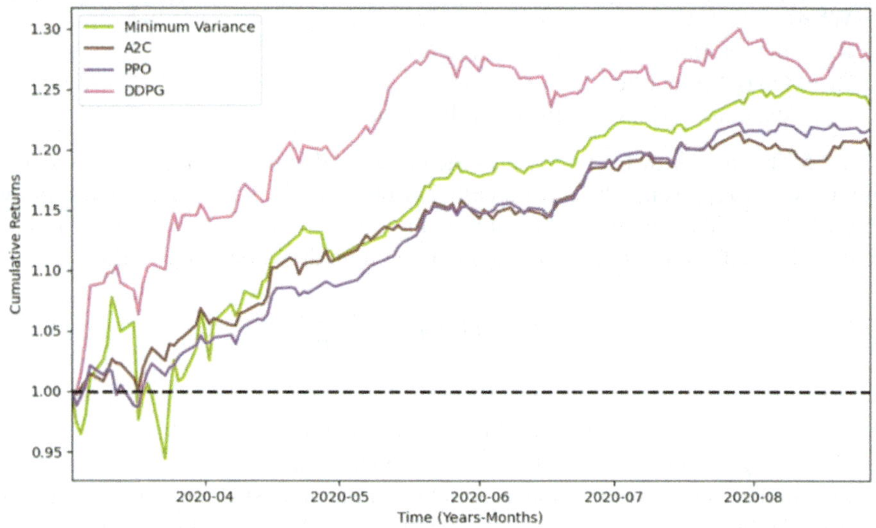


**Fig. 3** Cumulative return curves of A2C, DDPG, and PPO algorithm strategy, the min-variance portfolio allocation strategy, and the Asian Stocks

significantly surpasses Asian Stocks and the performance of the minimum variance portfolio allocation techniques.

Regarding cumulative returns, DDPG has the highest percentage gain over the entire period, followed by min-variance, PPO, and A2C in order of decreasing cumulative return.

DDPG has the highest average annual returns percentage gain, followed by min-variance, PPO, and A2C. Moreover, DDPG demonstrates superior performance compared to A2C in cumulative and annual returns.

While min-variance showcases the highest annual volatility at 21.96%, implying substantial fluctuations, this outcome seems contrary to the conventional notion that minimizing variance should lead to lower volatility. In contrast, the potential perception of A2C and PPO demonstrate relatively smoother performance, suggesting a more consistent and less volatile pattern regarding return variability.

PPO has a higher Sharpe Ratio of 4.67, indicating a more favorable risk-to-return tradeoff. This suggests that PPO generates a more significant return per unit of risk than minvariance.

A2C has a relatively small Max Drawdown of −2.55%, indicating a lower peak-totrough loss than min-variance, indicating that A2C experienced less severe losses during the crisis. Therefore, A2C shows better downside protection with a smaller Max Drawdown than min-variance, which experienced a more substantial peak-to-trough loss.

**Table 3** Performance metrics comparison on Asian stocks from 01, March 2022, to 30, September 2022 (after covid-19)

| | A2C | DDPG | PPO | Min-variance |
|---|---|---|---|---|
| Cumulative returns (%) | −11.42 | 5.71 | −3.82 | −2.70 |
| Annual return (%) | −22.64 | 12.48 | −7.92 | −5.63 |
| Annual volatility (%) | 14.56 | 20.43 | 17.27 | 18.79 |
| Sharpe ratio | −1.69 | 0.68 | −0.39 | −0.21 |
| Max drawdown (%) | −16.60 | −10.37 | −11.67 | −19.00 |

To summarize, DDPG performs well regarding cumulative and annual returns but has slightly higher volatility and a lower Sharpe ratio than PPO and A2C. Additionally, PPO has the highest Sharpe ratio and performs well regarding annual return and drawdown. Furthermore, A2C has the lowest volatility and negligible drawdown, making it a relatively stable option. However, min-variance appears to have the highest volatility and most significant drawdown among the strategies mentioned. Your choice among these strategies might depend on your risk tolerance, return objectives, and investment horizon. This suggests that PPO was more resilient during the crisis and could be a better risk management choice.

From Table 3, DDPG was the only strategy with positive cumulative and annual returns, though it still had significant annualized volatility. In contrast, A2C had the worst performance, with the most significant negative cumulative and annual returns and a negative Sharpe ratio. On the other hand, PPO and min-variance also had negative returns but performed better than A2C in terms of cumulative and annual returns.

In conclusion, Before COVID-19, DDPG had the highest annual return and Sharpe ratio, indicating better risk-adjusted performance. However, min-variance had the highest volatility and the worst performance overall. During COVID-19, all strategies performed significantly better, with higher returns and Sharpe ratios, despite min-variance higher volatility. After COVID-19, A2C had the highest negative returns and the lowest Sharpe ratio, while DDPG still managed positive returns and a positive Sharpe ratio, showcasing its relative resilience. It is important to note that various factors can influence investment performance, including market conditions, investment strategy, and the specific assets or stocks being traded. Past performance does not guarantee future results, and investors should consider various factors when making investment decisions.

## 5 Conclusion

In this study, we have investigated the viability of employing actor-critic algorithms, namely Proximal Policy Optimization (PPO), Advantage Actor-Critic (A2C), and Deep Deterministic Policy Gradient (DDPG) agents, for acquiring expertise in stock trading strategies. This adaptation is aimed at accommodating varying market conditions. The study's findings reveal the effective performance of three Reinforcement Learning (RL) algorithms throughout various testing periods: pre-COVID, during COVID-19, and post-COVID-19. Specifically, DDPG demonstrated the highest Sharpe ratio before the pandemic, while PPO exhibited a Sharpe ratio aligned with Islamic stock market activities during the COVID-19 period. Moreover, DDPG continued to display a favorable Sharpe ratio in the post-COVID-19 period. This aligns with prior research by [12, 14], indicating RL's potential in navigating abnormal market conditions, such as the COVID-19 pandemic, and suggests its effective application in managing abnormal events in financial markets. However, more research is needed to determine how well reinforcement learning algorithms would perform in real-world stock markets [9]. In future research, it would be intriguing to investigate two areas: firstly, delving into more intricate models, and secondly, addressing the challenges posed by vast datasets, such as those related to Asian stocks [5, 33]. Furthermore, we can explore the incorporation of additional elements into the analysis of the state space. This could encompass examining natural language processing applied to financial market news [21], amalgamating fundamental and technical analysis indicators [36], incorporating ESG (Environmental, Social, and Governance) features [7], and refining the design of more efficient reward functions.

## References


1. Ataiefard, F.: Adversarial robustness testing of deep reinforcement learning based automated trading software (2022)
2. Bekiros, S.D.: Fuzzy adaptive decision-making for boundedly rational traders in speculative stock markets. Eur. J. Oper. Res. **202**(1), 285–293 (2010)
3. Bekiros, S.D.: Heterogeneous trading strategies with adaptive fuzzy actor-critic reinforcement learning: a behavioral approach. J. Econ. Dyn. Control **34**(6), 1153–1170 (2010)
4. Brockman, G., Cheung, V., Pettersson, L., Schneider, J., Schulman, J., Tang, J., Zaremba, W.: Openai gym (2016). arXiv:1606.01540
5. Burda, Y., Edwards, H., Pathak, D., Storkey, A., Darrell, T., Efros, A.A.: Large-scale study of curiosity-driven learning (2018). arXiv:1808.04355
6. Chen, L., Gao, Q.: Application of deep reinforcement learning on automated stock trading. In: 2019 IEEE 10th International Conference on Software Engineering and Service Science (ICSESS), pp. 29–33. IEEE (2019)
7. Chen, Q., Liu, X.Y.: Quantifying ESG alpha using scholar big data: an automated machine learning approach. In: Proceedings of the First ACM International conference on AI in Finance, pp. 1–8 (2020)
8. Chong, T.T.L., Ng, W.K., Liew, V.K.S.: Revisiting the performance of MACD and RSI oscillators. J. Risk Finan. Manag. **7**(1), 1–12 (2014)
9. Danish, M.: Optimization of stock trading strategy with reinforcement learning (Doctoral dissertation) (2023)
10. Dang, Q.V.: Reinforcement learning in stock trading. In: International Conference on Computer Science, Applied Mathematics and Applications, pp. 311–322. Springer International Publishing, Cham (2019)
11. Deng, Y., Bao, F., Kong, Y., Ren, Z., Dai, Q.: Deep direct reinforcement learning for financial signal representation and trading. IEEE Trans. Neural Netw. Learn. Syst. **28**(3), 653–664 (2016)

12. Faturohman, T., Nugraha, T.: Islamic stock portfolio optimization using deep reinforcement learning. J. Islam. Monetary Econ. Fin. **8**(2), 181–200 (2022)
13. Fischer, T.G.: Reinforcement learning in financial markets-a survey (No. 12/2018). FAU Discussion Papers in Economics (2018)
14. Ge, J., Qin, Y., Li, Y., Huang, Y., Hu, H.: Single stock trading with deep reinforcement learning: a comparative study. In: 2022 14th International Conference on Machine Learning and Computing (ICMLC), pp. 34–43 (2022)
15. Gurrib, I.: Performance of the Average Directional Index as a market timing tool for the most actively traded USD based currency pairs. Banks Bank Syst. **13**(3), 58–70 (2018)
16. Ilmanen, A.: Expected Returns: An Investor's Guide to Harvesting Market Rewards, vol. 535. Wiley (2011)
17. Jeong, G., Kim, H.Y.: Improving financial trading decisions using deep Q-learning: predicting the number of shares, action strategies, and transfer learning. Expert Syst. Appl. **117**, 125–138 (2019)
18. Jiang, Z., Liang, J.: Cryptocurrency portfolio management with deep reinforcement learning. In: 2017 Intelligent Systems Conference (IntelliSys), pp. 905–913. IEEE (2017)
19. Kim, Y., Ahn, W., Oh, K.J., Enke, D.: An intelligent hybrid trading system for discovering trading rules for the futures market using rough sets and genetic algorithms. Appl. Soft Comput. **55**, 127–140 (2017)
20. Li, J., Rao, R., Shi, J.: Learning to trade with deep actor critic methods. In: 2018 11th International Symposium on Computational Intelligence and Design (ISCID), vol. 2, pp. 66–71. IEEE (2018)
21. Li, X., Li, Y., Yang, H., Yang, L., Liu, X.Y.: DP-LSTM: Differential privacy-inspired LSTM for stock prediction using financial news (2019). arXiv:1912.10806
22. Lillicrap, T.P., Hunt, J.J., Pritzel, A., Heess, N.M.O., Erez, T., Tassa, Y., ... Wierstra, D.P.: U.S. Patent No. 10,776,692. U.S. Patent and Trademark Office, Washington, DC (2020)
23. Liu, X.Y., Xia, Z., Rui, J., Gao, J., Yang, H., Zhu, M., Guo, J.: FinRL-Meta: Market environments and benchmarks for data-driven financial reinforcement learning. Adv. Neural Inf. Proc. Syst. **35**, 1835–1849 (2022)
24. Liu, X.Y., Xiong, Z., Zhong, S., Yang, H., Walid, A.: Practical deep reinforcement learning approach for stock trading (2018). arXiv:1811.07522
25. Maitah, M., Prochazka, P., Cermak, M., Šrédl, K.: Commodity channel index: evaluation of trading rule of agricultural commodities. Int. J. Econ. Finan. Issues **6**(1), 176–178 (2016)
26. Malini, H.: Behaviour of stock returns during Covid-19 pandemic: evidence from six selected stock market in the world. Jurnal Ekonomi Indones. **9**(3), 247–263 (2020)
27. Markowitz, H.M.: Portfolio Selection. Yale University Press (1968)
28. Mnih, V., Badia, A.P., Mirza, M., Graves, A., Lillicrap, T., Harley, T., ... Kavukcuoglu, K.: Asynchronous methods for deep reinforcement learning. In: International Conference on Machine Learning, pp. 1928–1937. PMLR (2016)
29. Moody, J., Saffell, M.: Learning to trade via direct reinforcement. IEEE Trans. Neural Netw. **12**(4), 875–889 (2001)
30. Prafulla, D., Christopher, H., Oleg, K., Alex, N., Matthias, P., Alec, R., ... Peter, Z.: Openai baselines. GitHub (2017). https://github.com/openai/baselines
31. Raffin, A., Hill, A., Gleave, A., Kanervisto, A., Ernestus, M., Dormann, N.: Stable-baselines3: reliable reinforcement learning implementations. J. Mach. Learn. Res. **22**(1), 12348–12355 (2021)
32. Schulman, J., Levine, S., Abbeel, P., Jordan, M., Moritz, P.: Trust region policy optimization. In: International Conference on Machine Learning, pp. 1889–1897. PMLR (2015)
33. Wang, L., Zhang, W., He, X., Zha, H.: Supervised reinforcement learning with recurrent neural network for dynamic treatment recommendation. In: Proceedings of the 24th ACM SIGKDD International Conference on Knowledge Discovery & Data Mining, pp. 2447–2456 (2018)
34. Wu, Y., Tian, Y.: Training agent for first-person shooter game with actor-critic curriculum learning. In: International Conference on Learning Representations (2016)

35. Yamaka, W., Maneejuk, P.: Analyzing the causality and dependence between gold shocks and Asian emerging stock markets: a smooth transition copula approach. Mathematics **8**(1), 120 (2020)
36. Yang, H., Liu, X.Y., Wu, Q.: A practical machine learning approach for dynamic stock recommendation. In: 2018 17th IEEE International Conference on Trust, Security and Privacy in Computing and Communications/12th IEEE International Conference on Big Data Science and Engineering (TrustCom/BigDataSE), pp. 1693–1697. IEEE (2018)
37. Zhang, Y., Yang, X.: Online portfolio selection strategy based on combining experts' advice. Comput. Econ. **50**, 141–159 (2017)
38. Zhang, Z., Zohren, S., Stephen, R.: Deep reinforcement learning for trading. J. Finan. Data Sci. (2020)

# Soft Skills and Work Efficiency Across Different Generations: Evidence from Thailand

**Pradthana Jaipong, Rossarin Osathanunkul, Nootchanat Pirabun, Suchanya Waewa, Pimpilai Utiem, Prasoet Chaiwong, Prassanee Sinpimolboon, Todsapol Srinuch, Kampol Woradit, and Nuttee Suree**

**Abstract** This study aims to assess the influence of soft skill learning programs on worker efficiency in Thailand. We examine several soft skills essential for enhancing workplace capabilities, including Planning and Ways of Working, Communication, Critical Thinking, and Mental Flexibility. Employing a Structural Equation Model and utilizing data from 220 training workers, we investigate the relationship between these soft skills and job efficiency. Our findings provide evidence that for every one-point increase in soft skills performance score, there is a corresponding 0.24%

P. Jaipong · R. Osathanunkul · S. Waewa · P. Utiem · P. Chaiwong · P. Sinpimolboon · T. Srinuch · K. Woradit · N. Suree
CMU School of Lifelong Education, Chiang Mai University, Chiang Mai 50200, Thailand
e-mail: pradthana.j@cmu.ac.th

R. Osathanunkul
e-mail: rossarin.o@cmu.ac.th

S. Waewa
e-mail: suchanya.wae@cmu.ac.th

P. Utiem
e-mail: pimpilai.u@cmu.ac.th

P. Chaiwong
e-mail: prasoet.c@cmu.ac.th

P. Sinpimolboon
e-mail: prassanee.sin@cmu.ac.th

T. Srinuch
e-mail: todsapol.s@cmu.ac.th

K. Woradit
e-mail: kampol.w@cmu.ac.th

N. Suree
e-mail: nuttee.suree@cmu.ac.th

N. Pirabun (✉)
Office of Research Administration, Chiang Mai University, Chiang Mai 50200, Thailand
e-mail: nootchanat.pi@cmu.ac.th

 
V. Kreinovich et al. (eds.), *Applications of Optimal Transport to Economics and Related Topics*, Studies in Systems, Decision and Control 556,
https://doi.org/10.1007/978-3-031-67770-0_38

increase in job efficiency. Notably, Mental Flexibility has a more pronounced impact on worker learning performance, leading to higher working efficiency, while Critical Thinking exhibits weaker support for improved learning performance. Further-more, our study underscores the variations in the impact of learning skills across different worker generations, such as Generation X, Millennials, and Generation Z, highlighting their unique contributions to learning performance.

**Keywords** Efficiency · SEM · Soft skills · Thailand · Worker

## 1 Introduction

In the business world today, there has been a notable shift towards prioritizing worker knowledge as a means to gain a competitive edge [9]. This emphasis on worker knowledge has underscored the growing importance of both hard and soft skills within the workforce [3]. A global employer survey found that fifty percent of respondents identified a skills gap, particularly in the realm of soft skills [11]. The primary challenge in the business environment is addressing the deficiency in soft skills among certain workers, necessitating targeted interventions. Furthermore, a specific challenge prevalent across numerous industries is the absence of well-defined strategies for providing soft skills training to workers.

In the recent 21st-century globalization, the concept of labor competition has come to the forefront. As industries expand across geographical borders and markets become more interconnected, traditional workforce paradigms are rapidly evolving. It is clear that the pervasive influence of digital and AI technologies is driving a transformation in the world of work. This transformation demands that today's workforce acquire new skills and cultivate a continuous readiness for adaptation as emerging career paths take shape. Simultaneously, the challenges posed by the COVID-19 crisis have accelerated and intensified this transformation [8]. A competitive edge is no longer solely determined by technical expertise but by the harmonious integration of knowledge, skills, and particularly, soft skills [20]. These soft skills encompass a spectrum of interpersonal, communication, adaptability, and problem-solving abilities that empower individuals to excel in an increasingly diverse and collaborative global marketplace.

While soft skills have gained significant recognition for their profound influence on both personal and professional success, hard skills-defined as technical expertise or specialized knowledge acquired through education and experience-remain important [19]. Soft skills encompass a range of competencies that are not tied to specific tasks but are essential for effective interaction and relationships within an organization. These skills are essential in any role as they involve how individuals relate to and collaborate with others within the organizational context. In contrast, hard skills refer to specific technical or task-oriented capabilities required to perform a particular job [3]. Similarly, Symonds et al. [21] stated that both hard and soft skills are essential for labor markets and employers in the twenty-first century. In a rapidly changing and technology-driven world, it is becoming increasingly evident that merely developing either hard or soft skills is barely adequate for thriving. Consequently, soft skills have emerged as the foremost competencies in today's global job market, particularly within the realm of entrepreneurship, in an era characterized by rapid technological advancement [13]. Heckman and Kautz [10] said that soft skills are character traits, objectives, and preferences valued not only in the workplace but also in academic settings and other contexts. "Soft skills should be taken into account in public policies on the development and investment in training". Soft skills are a predictor of success in life. Furthermore, it has been established that soft skills are essential in a variety of fields and industries, even in those that are highly technical [4, 10].

Effective soft skills training strategies play a pivotal role in successfully managing the diverse workforce, including both seasoned professionals from various generations and newcomers entering the workforce for the first time. The modern workplace consists of four distinct generational cohorts: Baby Boomers (born 1946–1964), Generation X (born 1965–1979), Millennials (born 1980–1994), and Generation Z (born 1995–2010) [17]. Each of these generational groups possesses unique personality traits, values, and priorities. However, it's worth noting that the challenge of harmoniously integrating these generations in the workplace is identified by Bursch and Kelly [2]. For instance, Baby Boomers may perceive younger generations (Generation X and Millennials) as lacking self-discipline and displaying arrogance. Conversely, Generation X might resist change and appear less creative when compared to the Baby Boomer generation. Additionally, Generation X may give the impression of having subpar problem-solving skills and slower response times. In light of these differences, the development of soft skills is essential for fostering better communication, collaboration, and understanding among the various generations in the workforce. It can help bridge gaps, dispel stereotypes, and promote a more inclusive and harmonious work environment.

Prior research on soft skills and generational differences yielded a range of findings. Magano et al. [15] explored the traits and competencies of Generation Z in the context of project management. They found that Generation Z individuals had a strong awareness of traits such as optimism, resilience, responsibility, and a results-oriented nature, but they often lacked recognition of traits like individualism and interpersonal and social skills. The significance of soft skills for Generation Z was underscored due to the significant challenges posed by the contemporary era of disruption. Cuic Tankovic et al. [5] further elaborated on this notion, asserting that Generation Z individuals needed to possess robust soft skills to effectively navigate the complexities of their rapidly evolving environment. In alignment with this perspective, Tulgan [22] observed a growing demand for soft skills from Generation Z across various sectors and vocations. These findings were complemented by Ismail et al. [12] who reinforced the importance of soft skills, especially in leadership contexts, highlighting their vital role within the Generation Z. Moore and Frazier [18] argued that both Generation Z and Millennial cohorts had to equip themselves with Future Skill Generation competencies to effectively address the demands of the Society 5.0 era. These competencies included Complex problem-solving, Critical Thinking,

Creativity, People Management, Coordinating With Others, Emotional Intelligence, Judgment and Decision-Making Skills, Service Orientation, Negotiation, and Cognitive Flexibility. Bussin [1] concurred with this perspective, referencing the World Economic Forum's recommended skill set for millennials, which closely aligned with the aforementioned skills. It is noteworthy that these competencies are inherently human, resistant to automation, and irreplaceable by machines. Their development required ongoing and persistent efforts to remain competitive on a global scale as we transitioned from the Industry 4.0 era toward the envisioned Society 5.0 paradigm. In the recent study of Dondi et al. [8] they collected data from 18,000 participants to identify a set of skills essential for professional success. They identified 56 foundational skills associated with higher employment likelihood, income, and job satisfaction. These skills were categorized into four major groups: "cognitive," "digital," "interpersonal," and "self-leadership" (See Fig. 1).

In this study, we focus on the cognitive dimension of soft skills, recognizing it as the essential foundation for skill development. Within this cognitive domain, we explore a set of skills and abilities that encompass critical thinking, effective problem-solving, and the valuable quality of mental flexibility. This cognitive realm further branches into four distinct subcategories: critical thinking, planning and effective ways of working, mental flexibility, and communication. The significance of cognitive skill proficiency cannot be emphasized enough, as it forms the basis for individuals to tackle complex challenges. It empowers them to systematically break down

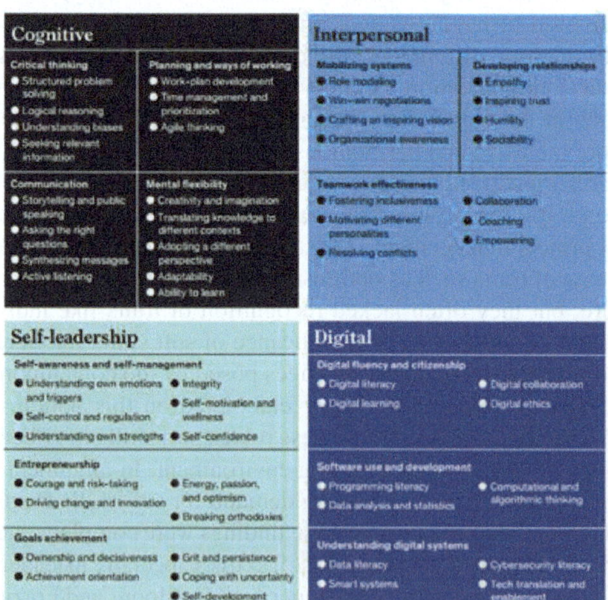


**Fig. 1** 56 DELTAS across 13 skill groups and four categories [8]

intricate problems into simpler components, identify root causes, and develop innovative and practical solutions. Our study strongly supports the idea that higher levels of cognitive skill proficiency are associated with improved performance among soft-skill learners. This performance enhancement, in turn, leads to higher income levels, highlighting the crucial role of cognitive skills in shaping both individual success and organizational progress. Our findings underscore the importance of investing in the development of cognitive skills, emphasizing their essential impact on individual and collective achievements.

The study was conducted in Thailand between 2021 and 2022. Thailand is selected as the research location due to the fact that its working-age population comprises the vast majority of the country's population, accounting for approximately 66.5%. This age group possesses significant potential for productivity, making it a crucial driving force behind the nation's economic system. Recognizing this importance, it is imperative to prioritize the development of the working-age population to ensure high quality. This entails enhancing their potential skills and competencies to align with individual abilities and the evolving demands of the labor market. However, existing literature primarily emphasizes the significance of soft skills for individual workers, often overlooking the relative importance of the cognitive dimension of soft skills across different generations, specifically Generation X, Millennials, and Generation Z. Consequently, this paper seeks to address the following research question: In Thailand, which cognitive skills hold the greatest significance for each of these generations?

To address the research question posed above, we utilize a Structural Equation Model (SEM) to explore the influence of soft skills on worker learning performance. This performance is measured through post-test scores obtained after completing four courses, encompassing planning and ways of working, communication, critical thinking, and mental flexibility. Subsequently, we extend our examination to deter-mine whether achieving higher test scores translates into enhanced working efficiency.

The structure of this paper is as follows: Sect. 2 provides details on the data source, while Sect. 3 outlines the methodology employed. In Sect. 4, we delve into the examination of the impact of soft skill courses on worker learning performance as well as worker efficiency. Finally, Sect. 5 draws conclusions based on our findings.

## 2 Data

The dataset consists of 220 observations, each corresponding to a participant active-ly engaged in the "Skills for Life and Work (Skills4Life)" program, which Chiang Mai University's School of Lifelong Education offers. This program encompasses four core courses: critical thinking, planning and effective ways of working, mental flexibility, and communication. Within these four courses, there are a total of 16 sub-courses, presenting a diverse array of skill development opportunities for participants. It's imperative to underline that all individuals represented in this dataset are members of the workforce who have enrolled in at least one of the four primary courses offered by the Skills4Life program. Furthermore, we have included a category for students, specifically those in the final year of their education who will soon transition into the workforce. This inclusion ensures coverage of the Generation Z demographic.

Table 1 provides a comprehensive overview of all variables included in the questionnaire, while Table 2 offers a summary of the dataset's key statistical characteristics.

## 3 Methodology

Our analysis employs a Simultaneous Equation Model (SEM) to assess the combined impact of soft skills on worker learning performance and subsequently the influence of worker performance on efficiency, as illustrated in Fig. 2. In the first stage, we examine how various characteristics influence the performance of Soft Skill Learners, which is represented as an individual's average post-test score (Eq. 1). In the second stage, we delve into investigating the effects of Soft Skill Learners' performance on job efficiency, which is quantified as the percentage of knowledge acquired through learning (Eq. 2).

$$learnersPerf = \alpha_0 + \beta_{1i} gender_i + \beta_{2j} gen_j + \beta_{3k} status_k + \beta_{4l} subcategory_l + \beta_{5m} gen\_sub_m + \beta_6 numcourse + \mu \quad (1)$$

$$efficiency = \alpha_1 + \beta_{1i} learnersPerf_i + \nu \quad (2)$$

To evaluate the impact of soft skills on efficiency, we consider a range of factors. These include demographic variables such as gender, age group (referred to as Generation), and Respondent's status. Additionally, we examine course-related variables, including the number of courses taken, the status of the type of cognitive category, and the status of the generation that falls under each cognitive category.

## 4 Results

In this section, we present the estimated results of the Simultaneous Equation Model (SEM), as summarized in Table 3. These results shed light on the influence of demographic variables and course-related factors, including generation, gender, the number of courses taken, the status of the type of cognitive category, and the generation that engages in different soft-skill categories.

Our findings reveal several significant insights. Initially, all the variables related to generation did not demonstrate statistical significance. This suggests that the specific generation a participant belongs to did not exert a statistically significant impact

**Table 1** Summary of variables

| | Variable | Explanations | | | Freq | % |
|---|---|---|---|---|---|---|
| Dependent variable | $efficiency$ | Percentage of efficiency gained from soft skill learning | | | | |
| | $learnersPerf$ | Soft skill Learners' performance as a person's average post-test score (point) | | | | |
| Independence variable | $gender_i$ | Gender of respondents | $i = 0$ | LGBTQ+ | 15 | 6.82 |
| | | | $i = 1$ | Female | 137 | 62.27 |
| | | | $i = 2$ | Male | 68 | 30.91 |
| | $gen_j$ | Generation of respondents | $j = 0$ | Gen X (Born: 1965-1979) | 58 | 26.36 |
| | | | $j = 1$ | Millennials (Born: 1980–1994) | 106 | 48.18 |
| | | | $j = 2$ | Gen Z (Born: 1995–2010) | 56 | 25.45 |
| | $status_k$ | Respondent's status at Chiang Mai University (CMU) | $k = 0$ | Student | 37 | 16.82 |
| | | | $k = 1$ | Outsider | 98 | 44.55 |
| | | | $k = 2$ | Worker | 72 | 32.73 |
| | | | $k = 3$ | Alumni | 13 | 5.91 |
| | $subcategory_l$ | Soft skills (Skills4Life) courses | $l = 0$ | Planning and Ways of Working | 118 | 53.64 |
| | | | $l = 1$ | Communication | 129 | 58.64 |
| | | | $l = 2$ | Critical Thinking | 119 | 54.09 |
| | | | $l = 3$ | Mental Flexibility | 122 | 55.45 |

(continued)

**Table 1** (continued)

| | Variable | Explanations | | | Freq | % |
|---|---|---|---|---|---|---|
| | $gen\_sub_m$ | $gen_j \times subcategory_l$ | $m = 0$ | Millennials who enrolled in the Communication | 55 | 25.00 |
| | | | $m = 1$ | Millennials who enrolled in Critical Thinking | 53 | 24.09 |
| | | | $m = 2$ | Millennials who enrolled in Mental Flexibility | 56 | 25.45 |
| | | | $m = 3$ | Millennials who enrolled in Planning and Ways of Working | 64 | 29.09 |
| | | | $m = 4$ | Gen X who enrolled in the Communication | 39 | 17.73 |
| | | | $m = 5$ | Gen X who enrolled in Critical Thinking | 37 | 16.82 |
| | | | $m = 6$ | Gen X who enrolled in Mental Flexibility | 34 | 15.45 |
| | | | $m = 7$ | Gen X who enrolled in Planning and Ways of Working | 38 | 17.27 |
| | | | $m = 8$ | Gen Z who enrolled in Communication | 32 | 14.55 |
| | | | $m = 9$ | Gen Z who enrolled in Critical Thinking | 27 | 12.27 |
| | | | $m = 10$ | Gen Z who enrolled in Mental Flexibility | 28 | 12.73 |
| | | | $m = 11$ | Gen Z who enrolled in Planning and Ways of Working | 14 | 6.36 |
| | $numcourse$ | The number of soft skills courses enrolled in. | | | | |

*Note* dummy variable = 0 is baseline comparison

**Table 2** Summary statistic

| | Variable | Obs | Mean | Std. | Min | Max | | Variable | Obs | Mean | Std. | Min | Max |
|---|---|---|---|---|---|---|---|---|---|---|---|---|---|
| 1 | $efficiency$ | 220 | 75.5175 | 15.9376 | 25.6923 | 100 | 13 | $gen\_sub_1$ | 220 | 0.2409 | 0.4286 | 0 | 1 |
| 2 | $learnersPerf$ | 220 | 89.8591 | 9.695 | 70 | 100 | 14 | $gen\_sub_2$ | 220 | 0.2545 | 0.4366 | 0 | 1 |
| 3 | $gender_1$ | 220 | 0.6227 | 0.4858 | 0 | 1 | 15 | $gen\_sub_3$ | 220 | 0.2909 | 0.4552 | 0 | 1 |
| 4 | $gender_2$ | 220 | 0.3091 | 0.4632 | 0 | 1 | 16 | $gen\_sub_4$ | 220 | 0.1773 | 0.3828 | 0 | 1 |
| 5 | $gen_1$ | 220 | 0.4818 | 0.5008 | 0 | 1 | 17 | $gen\_sub_5$ | 220 | 0.1682 | 0.3749 | 0 | 1 |
| 6 | $gen_2$ | 220 | 0.2545 | 0.4366 | 0 | 1 | 18 | $gen\_sub_6$ | 220 | 0.1545 | 0.3623 | 0 | 1 |
| 7 | $status_1$ | 220 | 0.4455 | 0.4981 | 0 | 1 | 19 | $gen\_sub_7$ | 220 | 0.1727 | 0.3789 | 0 | 1 |
| 8 | $status_2$ | 220 | 0.3273 | 0.4703 | 0 | 1 | 20 | $gen\_sub_8$ | 220 | 0.1455 | 0.3534 | 0 | 1 |
| 9 | $status_3$ | 220 | 0.0591 | 0.2363 | 0 | 1 | 21 | $gen\_sub_9$ | 220 | 0.1227 | 0.3289 | 0 | 1 |
| 10 | $subcategory_1$ | 220 | 0.5727 | 0.4958 | 0 | 1 | 22 | $gen\_sub_{10}$ | 220 | 0.1273 | 0.334 | 0 | 1 |
| 11 | $subcategory_2$ | 220 | 0.5318 | 0.5001 | 0 | 1 | 23 | $gen\_sub_{11}$ | 220 | 0.0636 | 0.2447 | 0 | 1 |
| 12 | $subcategory_3$ | 220 | 0.5364 | 0.4998 | 0 | 1 | 24 | $numcourse$ | 220 | 5.6364 | 4.5304 | 1 | 16 |

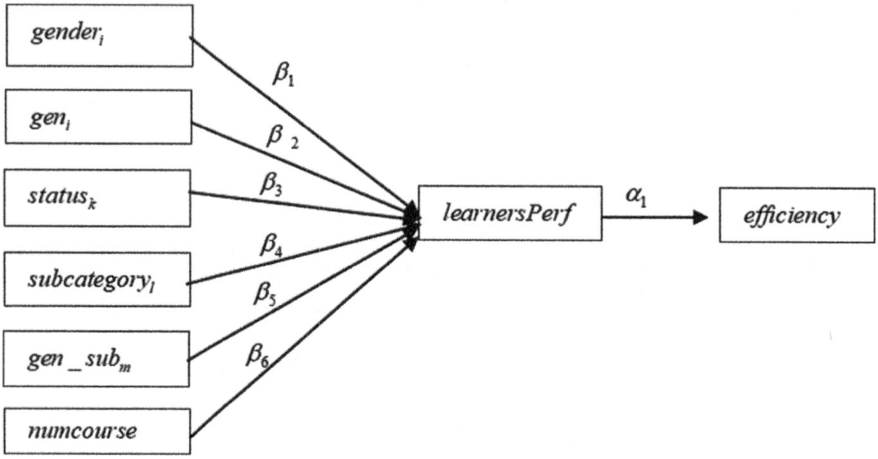


**Fig. 2** The methodology framework

on their soft skills learning performance. This observation could imply a relatively consistent pattern of soft skills development across generations in our study. In contrast, gender emerged as a notable and statistically significant factor. The coefficients exclusively for males displayed statistical significance, providing robust evidence that soft skill Learners' outcomes vary significantly between LGBTQ+ individuals and males. Typically, for the reference category (LGBTQ+), the coefficient is set to zero as a point of reference. The coefficient for males, which stands at -4.5175 points, suggests that, on average, male individuals scored 4.5175 points lower in performance compared to LGBTQ+ individuals. This implies that, within the scope of our study, being male is associated with a modest decrease in soft skill Learners' performance when compared with LGBTQ+ individuals.

Turning to the status of the respondent, we identified noteworthy patterns. The coefficients for workers and alumni of CMU (Chiang Mai University) were also found to be statistically significant, indicating that these roles significantly influence the performance of soft-skilled learners. The coefficient for "Worker" (3.9116) suggests that CMU workers, on average, reported a 3.9116-point higher performance score compared to CMU students. Similarly, alumni displayed a significantly higher performance score, approximately 7.8930 points higher than students. This implies that assuming a role other than that of a student is associated with higher soft skills learning performance.

In terms of Soft skills courses, the coefficients for the type of cognitive category exhibited statistical significance, with the exception of critical thinking. Communication emerged as a pivotal factor, supported by strong evidence indicating that individuals with strong communication skills outperformed their peers by 7.1158 points on the soft skill Learners' performance scale, when all other factors remained constant. Furthermore, Mental Flexibility displayed compelling evidence, signifying that nearly all cognitive categories exerted a significant impact on problem-

**Table 3** Estimation results

| | Coef. | Std. Err. | z-stat | MBF | |
|---|---|---|---|---|---|
| *learnersPerf* **equation** | | | | | |
| $generation_1$ (Millennials) | −2.0508 | 3.2714 | −0.63 | 0.82 | |
| $generation_2$ (Gen Z) | 3.1959 | 3.7668 | 0.85 | 0.6968 | |
| $gender_1$ (Female) | −2.6437 | 2.3524 | −1.12 | 0.5341 | |
| $gender_2$ (Male) | −4.5175 | 2.4698 | −1.83 | 0.1874 | * |
| $status_1$ (Outsider) | 1.009 | 1.9829 | 0.51 | 0.8781 | |
| $status_2$ (Worker) | 3.9116 | 2.1737 | 1.8 | 0.1979 | * |
| $status_3$ (Alumni) | 7.893 | 2.9257 | 2.7 | 0.0261 | * |
| $subcategory_1$ (Communication) | 7.1158 | 1.2296 | 5.79 | 0 | *** |
| $subcategory_2$ (Critical Thinking) | 0.7557 | 1.218 | 0.62 | 0.8251 | |
| $subcategory_3$ (Mental Flexibility) | 2.2962 | 1.2241 | 1.88 | 0.1708 | * |
| $gen\_sub_1$ (Millennials x Critical Thinking) | 0.0311 | 1.8283 | 0.02 | 0.9998 | |
| $gen\_sub_2$ (Millennials x Mental Flexibility) | 5.903 | 1.8331 | 3.22 | 0.0056 | ** |
| $gen\_sub_3$ (Millennials x Planning and Ways of Working) | 3.2942 | 1.7304 | 1.9 | 0.1645 | * |
| $gen\_sub_4$ (Gen X x Communication) | 4.9306 | 2.5578 | 1.93 | 0.1553 | * |
| $gen\_sub_5$ (Gen X x Critical Thinking) | −3.045 | 2.5605 | −1.19 | 0.4926 | |
| $gen\_sub_6$ (Gen X x Mental Flexibility) | 2.512 | 2.4702 | 1.02 | 0.5944 | |
| $gen\_sub_7$ (Gen X x Planning and Ways of Working) | 4.6603 | 2.6184 | 1.78 | 0.2051 | * |
| $gen\_sub_8$ (Gen Z x Communication) | 8.3317 | 2.4697 | 3.37 | 0.0034 | ** |
| $gen\_sub_9$ (Gen Z x Critical Thinking) | −1.09 | 2.4236 | −0.45 | 0.9037 | |
| $gen\_sub_{10}$ (Gen Z x Mental Flexibility) | −4.0194 | 2.2888 | −1.76 | 0.2125 | * |
| $gen\_sub_{11}$ (Gen Z x Planning and Ways of Working) | 2.2159 | 2.6456 | 0.84 | 0.7027 | |
| *numcourse* | 0.2539 | 0.1861 | 1.36 | 0.3966 | |
| intercept | 84.718 | 3.8002 | 22.29 | 0 | *** |
| *efficiency* **equation** | | | | | |
| *learnersPerf* | 0.1667 | 0.1103 | 1.51 | 0.3198 | * |
| intercept | 60.534 | 9.9651 | 6.07 | 0 | *** |

*Note* (1) We evaluate the significant factors and examine the influence of variables identified by the Minimum Bayes Factor (MBF). MBF indicates the likelihood of adopting the null hypothesis. (2) *, **, *** denote moderate, strong, and decisive evidence supporting the alternative hypotheses of [16] solving abilities. The coefficient for "Mental Flexibility" suggests that individuals with strong mental flexibility performed approximately 2.2962 points better than those with strong critical thinking skills.

In summary, while specific generations did not significantly influence soft skills learning performance, our results highlight the significant roles played by gender, roles (Worker and Alumni), and the type of cognitive category in shaping soft skills learning outcomes. These findings provide valuable insights into the multifaceted factors that influence the development of soft skills among our study participants.

When we consider the interaction term between generation and course in our model, it uncovers intriguing coefficients that shed light on the relationship between course selection and academic performance across different generations. Here's a break-down of the key findings: Millennials who enrolled in Mental Flexibility and Planning and Ways of Working demonstrated significant performance advantages, with coefficients of 5.9030 (decisive evidence) and 3.2942 (strong evidence) respectively. This suggests that Millennials excelled academically when they chose these courses compared to their peers who opted for Communication. Interestingly, "Gen X individuals who enrolled in Communication and Planning and Ways of Working" displayed higher academic performance scores compared to Millennials who chose Communication. This implies that, for Gen X, specific course choices led to better academic outcomes compared to Millennials in similar courses. Notably, Gen Z participants who enrolled in Mental Flexibility exhibited a lower academic performance score compared to Millennials who chose Communication. This finding underscores the nuanced impact of course selection on different generations, emphasizing the need to tailor educational programs to accommodate the unique needs and strengths of each generation and their cognitive category enrollments. Such customization can effectively enhance soft skill performance across the board. In the final part of our estimation, we shift our focus to examining the impact of Soft Skill Learners' performance on efficiency, which is measured as the percentage of knowledge gained from learning. On average, our findings reveal that for every one-point increase in soft skills, there is a corresponding 0.2439% increase in efficiency. This result is supported by strong evidence, highlighting the significant influence of soft skills on enhancing efficiency in the learning process.

## 5 Conclusion

This study aimed to investigate the impact of soft skills on the job efficiency of Skill for Life learners in Thailand using survey data from 220 respondents collected from Chiang Mai University's School of Lifelong Education. The study employs Structural Equation Modeling (SEM) to examine the influence of soft skills on worker learning performance and subsequently assess whether learning performance leads to higher worker efficiency.

Our results provide decisive evidence supporting the notion that Millennials who enrolled in the 'Mental Flexibility' course exhibit superior learning performance.

This particular skill is crucial for them due to the need to adapt and embrace innovative problem-solving approaches, making cognitive flexibility a vital asset for this generation. Additionally, our study highlights 'Planning and Ways of Working' as an effective course for Millennials, aligning with Twenge & Campbell [24] assertion that Millennials often prioritize clear structures and guidelines in their learning and work environments.

Additionally, our observations reveal that Gen X workers who enrolled in 'Communication' and 'Planning and Ways of Working' courses tend to excel in these classes compared to other courses. Gen X typically places a strong emphasis on teamwork and collaboration, and effective communication plays a pivotal role in their ability to work cohesively in teams and achieve common objectives [14].

On the other hand, it's noteworthy that Gen Z participants who enrolled in 'Communication' and 'Mental Flexibility' courses displayed lower performance scores. This phenomenon can be attributed to Gen Z's proficiency in digital communication tools. However, this proficiency sometimes results in an excessive reliance on text-based communication methods, such as texting and messaging apps. Consequently, they may face challenges in developing robust face-to-face communication skills or navigating complex offline interpersonal dynamics [23].

In summary, these results emphasize the connections between generation and soft skills, particularly cognitive skills, and their influence on worker performance and efficiency. As advocated by Disler et al. [7] tailoring educational approach-es to align with the unique characteristics and preferences of different generational cohorts can significantly enhance learning outcomes. This highlights the paramount importance of flexibility in curriculum design and delivery to meet the diverse needs of learners across generations.

**Acknowledgements** The paper is supported by School of Lifelong Education, Chiang Mai University and Center of Excellence in Econometrics, Faculty of Economics, Chiang Mai University.

## References

1. Bussin, M.: Remuneration and talent management. eBook Partnership (2014)
2. Bursch, D., Kelly, K.: Managing the multigenerational workplace. Tersedia secara online di (2014). http://www.kenanflagler.unc.edu/ [diakses di Surabaya, In-donesia: 23 Oktober 2017]
3. Cimatti, B.: Definition, development, assessment of soft skills and their role for the quality of organizations and enterprises. Int. J. for Quality Res. **10**(1), 97 (2016)
4. Cinque, M.: "Lost in translation". Soft skills development in European countries. Tun. J. Higher Edu. **3**(2), 389–427 (2016)
5. Cuic Tankovic, A., Kapeš, J., Benazić, D.: Communication skills in generation Z as future tourism employees. Commun. Res. Pract. **8**(1), 86–102 (2022)
6. Deloitte: The 2021 Deloitte Global Resilience Report (2021). https://www2.deloitte.com/global/en/pages/about-deloitte/articles/global-resilience.html
7. Disler, R.T., Spiliopoulos, N., Inglis, S.C., Currow, D.C., Davidson, P.M.: Cognitive screening in chronic obstructive pulmonary disease: patient's perspectives. Disabil. Rehabil. **42**(9), 1233–1239 (2020)

8. Dondi, M., Klier, J., Panier, F., Schubert, J.: Defining the Skills Citizens Will Need in the Future World of Work. McKinsey & Company, 25 (2021)
9. Halawi, L.A., McCarthy, R.V., Aronson, J.E.: Knowledge management and the competitive strategy of the firm. Learn. Organ. **13**(4), 384–397 (2006)
10. Heckman, J.J., Kautz, T.: Hard evidence on soft skills. Labour Econ. **19**(4), 451–464 (2012)
11. Hurrell, S.A.: Rethinking the soft skills deficit blame game: employers, skills withdrawal and the reporting of soft skills gaps. Human Relat. **69**(3), 605–628 (2016)
12. Ismail, D.H., Nugroho, J., Rohayati, T.: Literature Review: Soft Skill Needed by Gen Z in the Era RI 4.0 and Society 5.0. Majalah Ilmiah Bijak **20**(1), 119–131 (2023)
13. Jain, V.: What are soft skills (2009). Retrieved from http://schoolofeducators.com/2009/02/Importance-of-soft-skills-development-in-education/
14. Kupperschmidt, B.R.: Multigeneration employees: strategies for effective management. Health News **19**(1), 65–76 (2000)
15. Magano, J., Silva, C., Figueiredo, C., Vitória, A., Nogueira, T., Pimenta Dinis, M.A.: Generation Z: fitting project management soft skills competen-cies-A mixed-method approach. Educ. Sci. **10**(7), 187 (2020)
16. Maneejuk, P., Yamaka, W.: Significance test for linear regression: how to test without P-values? J. Appl. Stat. **48**(5), 827–845 (2021)
17. McCrindle, M., Wolfinger, E.: Generations defined. Ethos **18**(1), 8–13 (2010)
18. Moore, K., Frazier, R.S.: Engineering education for generation Z. Am. J. Eng. Educ. (AJEE) **8**(2), 111–126 (2017)
19. Robles, M.M.: Executive perceptions of the top 10 soft skills needed in to-day's workplace. Bus. Commun. Q. **75**(4), 453–465 (2012)
20. Stock, W.A., Hansen, W.L.: Ph. D. program learning and job demands: How close is the match?. Am. Econ. Rev. **94**(2), 266–271 (2004)
21. Symonds, W.C., Schwartz, R., Ferguson, R.F.: Pathways to prosperity: Meeting the challenge of preparing young Americans for the 21st century (2011)
22. Tulgan, B.: Meet Generation z: the second generation within the giant " Millennial" cohort. Rainmaker Thinking **125**(1), 1–13 (2013)
23. Twenge, J.M.: iGen: Why today's super-connected kids are growing up less rebellious, more tolerant, less happy-and completely unprepared for adult-hood. Atria Books (2017)
24. Twenge, J.M., Campbell, W.K.: Generational differences in psychological traits and their impact on the workplace. J. Manag. Psychol. **23**(8), 862–877 (2008)

# The Impact of Green Finance Investments on Carbon Emissions Reduction: A Finding of High-Performance Stocks in the S&P Global Clean Energy Index Using Machine Learning with Bayesian Additive Regression Trees

**Terdthiti Chitkasame and Pathairat Pastpipatkul**

**Abstract** This study aims to verify the effectiveness of carbon emissions reduction through investments in Green finance, focusing on the S&P Global Clean Energy Index as a representative of the global Green finance sector. Specifically, the research identifies constituent stocks within this index that hold the potential to serve as prototypes for carbon emissions reduction initiatives. Utilizing monthly data for 99 stocks, the S&P Global Clean Energy Index, and industrial carbon emissions, this study employs the BART machine model, a novel approach combining Bayesian additive regression trees (BART) and machine learning. The findings of the study reveal a significant relationship between the log prices of AVANGRID Inc.'s stocks and carbon emissions. Notably, as the log prices of these stocks increase, carbon emissions demonstrate a corresponding increase. This effect is particularly pronounced when their price ranges from 44.70 to 46.99 US dollars. This suggests a substantial impact of AVANGRID Inc.'s stock prices on the rise in carbon emissions. Furthermore, the company's significant investments in cutting-edge technology to boost grid capacity in New York for renewable energy adoption in 2023 may result in short-term cost increases, potentially impacting profits, and investor outlook. However, these initiatives are expected to yield long-term environmental benefits and sustainable returns, positively influencing both the company's stock price and shareholder value in the future.

**Keywords** Green finance · BART machine · Variable selection · The S&P global clean energy index

T. Chitkasame · P. Pastpipatkul (✉)
Faculty of Economics, Chiang Mai University, Chiang Mai, Thailand
e-mail: Ppthairatp@hotmail.com

T. Chitkasame
e-mail: terdthiti_c@cmu.ac.th

 
V. Kreinovich et al. (eds.), *Applications of Optimal Transport to Economics and Related Topics*, Studies in Systems, Decision and Control 556,
https://doi.org/10.1007/978-3-031-67770-0_39

## 1 Introduction

Green finance is a crucial tool in addressing the issue of carbon emissions and promoting environmental improvement. Green finance plays a pivotal role in addressing the issue of carbon emissions and promoting environmental improvement targeting private companies [26]. By providing financing for investments that provide environmental benefits, green finance support projects and initiatives that contribute to the reduction of carbon emissions [4]. This can include investments in renewable energy, energy efficiency measures, sustainable infrastructure development, and other initiatives aimed at mitigating climate change, reducing carbon emission, and promoting sustainable development [21].

The emergence of green finance for addressing climate change and promoting sustainable development has garnered significant attention in recent years. Governments and society have recognized the importance of green finance in tackling carbon emissions and achieving environmental improvements. It is widely believed that the development of green finance can effectively reduce carbon emissions and promote environmental improvement on a larger scale [25]. Green finance has the potential to redirect capital towards environmentally friendly projects and initiatives, ultimately contributing to the reduction of carbon emissions [20]. Exploring the Potential of Green Finance in Carbon Emission Reduction Given the increasing intensity of carbon dioxide emissions globally, there is a pressing need for effective environmental governance and green development. Green finance policies have emerged as a potent tool to support these goals, with an emphasis on sustainable economic growth [27].

In recent times, a substantial body of research has surfaced, underscoring the increasing attention directed towards the interaction between $CO_2$ emission allowances and clean energy assets as a Green finance. This phenomenon brings forth advantages not only for shareholders and the companies issuing these assets but also for the nations where these entities operate. As an illustrative instance, Dutta et al. [5] utilize the bivariate VAR-GARCH model as a means to investigate the connections between returns and volatility in the context of $CO_2$ emissions and clean energy prices. Their findings reveal a noteworthy positive influence of fluctuations in EUA prices on the returns of clean energy stocks [11].

From the literature review in the previous section, the initial challenge is understanding what green finance entails. Green finance covers a wide range of financial products and services designed to support environmentally sustainable projects and initiatives. This includes green bonds, green loans, sustainable investment funds, or Company doing on clean energy [14]. Secondly, the challenge at hand pertains to the adequacy of available data for the utilization of conventional model estimations. This arises due to the multifaceted nature of green development, covering guidelines to foster green finance. One notable milestone in this context is the advent of the Paris Agreement in 2015. Therefore, this study was decided to use the S&P Global Clean Energy Index to represent Green finance because it is composed of firms producing clean energy, which is an important area of research in the context of environmental

[6]. This index covers a group of 99 global stocks that are verified to be companies (stocks) producing and generating clean energy. However, previous research has analyzed the overall index group that may affect carbon emissions. There has small amount of research looked at operations at the individual company level.

Ensemble-of-trees methods have gained significant popularity for predictive modeling in both regression and classification tasks. Established algorithms like random forests [3] and stochastic gradient boosting [7] have become widely adopted. Furthermore, there have been notable advancements in ensemble methods, including dynamic trees [23] and Bayesian additive regression trees. These novel approaches distinguish themselves by relying on a fundamental Bayesian probability model rather than a standalone algorithm.

The limitation of the traditional BART model is the absence of a prediction function. To utilize the model for making predictions on new or unseen data, one must supply the test data as an input during the model's training phase. Consequently, generating forecasts for future data points necessitates refitting the entire model. Given that the runtime for this process can be non-trivial, making predictions becomes a laborious undertaking [19]. Then, Urbanek [24] introduced an innovative approach to enhance the conventional BART model by incorporating machine learning techniques, resulting in what is now referred to as "Machine Learning with Bayesian Additive Regression Trees" or BART Machine. Notably, in terms of runtime performance, our algorithm exhibits significantly enhanced speed and parallelization capabilities, enabling computations to be conducted across multiple processor cores as needed. This parallelization extends beyond just model construction; it encompasses other functionalities such as prediction, variable selection, and notably, the computation of partial effects, which provides insights into the average impact on dependent variables [2].

Hence, the principal objective of this research is to assess the efficacy of carbon emissions reduction strategies through investments in Green finance, with a specific emphasis on the S&P Global Clean Energy Index. Furthermore, this study conducts a detailed evaluation of the 99 constituent companies (stocks) featured within this index. This leads to conclusions about which companies' activity patterns (stock prices) are significantly more effective in reducing carbon emissions. Additionally, our selection of the time frame for analysis, spanning from the year 2020 to the present, aligns with the establishment of the European Environment Agency and the inception of the European Green Deal. These developments mark the initiation of robust efforts to enforce guidelines outlined in the Paris Agreement, underscoring the relevance of this timeframe to our research. Finally, this study employs the BART machine model, a novel approach combining Bayesian additive regression trees (BART) and machine learning.

The following sections outline the research methods used in this work. The second section discusses the methodology utilized in this investigation. Section 3 contains the data sources and an explanation of the data. Section 4 discusses the empirical findings. Finally, Sect. 5 summarizes some key findings and conclusions.

# 2 Methodology

## *2.1 Bayesian Additive Regression Trees (BART)*

BART represents a Bayesian method for estimating nonparametric functions using regression trees. The predictor space is divided recursively into hyper-rectangles, serving the purpose of approximating an unknown function ($f$). The dimension of predictor space corresponds to the number of variables, which we represent as "$p$". Tree-based regression models possess the capacity to adaptively accommodate interactions and nonlinear patterns. Models constructed by combining multiple regression trees exhibit a heightened capability compared to individual trees in capturing interactions, non-linear relationships, and additive effects within a function. ($f$) [16].

BART can be viewed as an ensemble of trees, utilizing a unique estimation method based on a complete Bayesian probability model. In essence, the BART model can be represented as follows:

$$Y = f(X) + \varepsilon \approx T_1^{\mathcal{M}}(X) + T_2^{\mathcal{M}}(X) + \cdots + T_m^{\mathcal{M}}(X) + \varepsilon, \varepsilon \sim \mathcal{N}_n\left(0, \sigma^2 I_n\right) \quad (1)$$

where $Y$ is the $n \times 1$ vector of responses, $X$ is the $n \times p$ design matrix (the predictors column-joined), $\varepsilon$ is the $n \times 1$ vector of noise. Here we have m distinct regression trees, each composed of a tree structure, denoted by $\mathcal{T}$, and the parameters at the terminal nodes (also called leaves), denoted by $\mathcal{M}$. The two together, denoted as $\mathcal{T}^{\mathcal{M}}$ represents an entire tree with both its structure and set of leaf parameters [4].

Additionally, the ensemble's number of trees, denoted as $\mathcal{M}$, is a predetermined hyperparameter. While BART typically aims for superior predictive performance by allowing increased flexibility with a larger, Chipman et al. [4] advocate for a smaller $\mathcal{M}$ (defaulted to $\mathcal{M} = 20$) when emphasizing variable selection. With fewer trees in the ensemble, predictor variables vie for inclusion as they have fewer chances to appear in the model. Yet, if m is excessively small, the Gibbs sampler in BART may frequently get stuck in local modes, leading to potential instability in estimation results [4]. Additionally, a very small $\mathcal{M}$ limits the model's flexibility to capture diverse complex functions. Conversely, an overly large m introduces the risk of insignificant variables entering the model without significantly affecting the overall fit, posing challenges for effective variable selection.

## 2.2 The Priors of BART Machine

According to BART model, the prior consists of three components as: (1) the tree structure of itself, (2) the leaf parameters conditioned on the tree structure, and (3) the error variance denoted as $\sigma^2$, which remains independent of both the tree structure and leaf parameters.

$$\mathbb{P}\left(\mathcal{T}_1^{\mathcal{M}}, \ldots, \mathcal{T}_m^{\mathcal{M}}, \sigma^2\right) = \left[\prod_t \mathbb{P}(\mathcal{T}_t^{\mathcal{M}})\right]\mathbb{P}\left(\sigma^2\right) = \left[\prod_t \mathbb{P}(\mathcal{M}_t|\mathcal{T}_t)\mathbb{P}(\mathcal{T}_t)\right]\mathbb{P}\left(\sigma^2\right)$$
$$= \left[\prod_t \prod_\ell \mathbb{P}(\mu_{t,\ell}|\mathcal{T}_t)\mathbb{P}(\mathcal{T}_t)\right]\mathbb{P}\left(\sigma^2\right) \quad (2)$$

Firstly, the $\mathbb{P}(\mathcal{T}_t)$ is a critical element of the prior impacting the node placements within the tree. Nodes situated at depth "$d$" possess a nonterminal status, determined by the probability $\alpha(1+d)^{-\beta}$. It's important to note that $\alpha$ falls within the range of (0, 1), while $\beta$ spans from $[0, \infty]$. In this context, depth refers to the node's distance from the root of the tree. For instance, the root itself holds a depth value of 0, and its immediate child node would be assigned a depth of 1, and so on. This particular form of the prior is capable of promoting the creation of shallow tree structures, effectively curbing the complexity typically associated with individual trees. As a standard practice, $\alpha$ and $\beta$ are assigned default hyperparameter values of $\alpha = 0.95$, and $\beta = 2$.

The prior distribution ($\mathbb{P}(\mathcal{T}^{\mathcal{M}})$) specifies the probability of any sum-of-trees model, which is composed of $m$ trees ($\mathcal{T} = (\mathcal{T}_1, \ldots, \mathcal{T}_m)$) and their corresponding leaf parameters $\mathcal{M} = (\mathcal{M}_1, \ldots, \mathcal{M}_m)$. The prior distribution $\mathbb{P}(\mathcal{T}^{\mathcal{M}})$ is factorized into $\mathbb{P}(\mathcal{T})\mathbb{P}(\mathcal{M}|\mathcal{T})$, where $\mathbb{P}(\mathcal{T})$ is the product of independent priors for each tree and $\mathbb{P}(\mathcal{M}|\mathcal{T})$ is the product of independent priors for each leaf parameter given the tree structure. The priors for the tree structure ($\mathbb{P}(\mathcal{T}_t)$) depend on two hyperparameters: $\alpha$ and $\beta$, which control the depth and size of the trees. The priors for the leaf parameters $\mathbb{P}(\mathcal{M}_t|\mathcal{T}_t)$ are normal distributions with mean ($\mu$) is zero and variance ($\sigma^2$) [4].

## 2.3 Posterior Distribution of BART Machine

A Gibbs sampler introduced by Geman and Geman [8]. It is utilized for generating samples from the posterior distribution denoted as $\mathbb{P}\left(\mathcal{T}_1^{\mathcal{M}}, \ldots, \mathcal{T}_m^{\mathcal{M}}, \sigma^2|y\right)$. The aspect of the Gibbs sampler is applied to BART involves a technique referred to as "Bayesian backfitting," which was proposed by Hastie and Tibshirani [11]. In this process, when fitting the $j$th tree, the remaining $m-1$ trees are held constant iteratively. This is achieved by considering solely the residual response that has not yet been modeled.

$$R_{-j} = y - \sum_{t \neq j} \mathcal{T}^{\mathcal{M}}(X) \tag{3}$$

The Gibbs sampler,

$$\begin{aligned} 1 &: \mathcal{T}_1 \big| R_{-1}, \sigma^2 \\ 2 &: \mathcal{M}_1 \big| \mathcal{T}_1, R_{-1}, \sigma^2 \\ 3 &: \mathcal{T}_2 \big| R_{-2}, \sigma^2 \\ 4 &: \mathcal{M}_2 \big| \mathcal{T}_2, R_{-2}, \sigma^2 \\ &\vdots \\ 2m-1 &: \mathcal{T}_m \big| R_{-m}, \sigma^2 \\ 2m &: \mathcal{M}_m \big| \mathcal{T}_m, R_{-m}, \sigma^2 \\ 2m+1 &: \mathcal{T}_m | \mathcal{T}_1, \mathcal{M}_1, \ldots, \mathcal{T}_m, \mathcal{M}_m, \varepsilon \end{aligned} \tag{4}$$

The process begins with proposing a modification to the structure of the first tree, denoted as $\mathcal{T}$. These proposed changes are either accepted or rejected through a Metropolis–Hastings step, as outlined by Hastings [13]. It's important to note that sampling from the posterior distribution of the tree structure is independent of the leaf parameters, as these parameters can be mathematically integrated out of the computation. Once the tree structure is determined, samples are drawn from the posterior distribution of the $b$ leaf parameters, represented as $\mathcal{M}_1 := \{\mu_1, \ldots, \mu_b\}$. This iterative procedure is repeated for each tree, utilizing the updated set of partial residuals ($R_{-j}$). Finally, considering the revised tree structures and leaf parameters, a draw from the posterior distribution of $\sigma^2$ is made based on the residuals of the full model as $\varepsilon := y - \sum_{t=1}^{m} \mathcal{T}^{\mathcal{M}}(X)$.

In a terminal node, the leaf parameter's posterior distribution in $\mathcal{M}$ is normally distributed, with its mean being a weighted combination of the likelihood and prior parameters, thanks to the normal distribution's conjugate property. Similarly, due to the conjugacy between normal and inverse gamma distributions, the posterior distribution of $\sigma^2$ is also inverse gamma [8].

### *2.4 Variable Selection in BART Machine*

The variable selection methods which form the BART machine introduced in Bleich et al. [1, 2] work best when dealing with datasets where the number of influential covariates affecting the response is considerably smaller than the total number of covariates.

Variable inclusion proportions have been applied for variable selection. This technique represents the proportion of times each predictor is chosen as a splitting rule divided by the total number of splitting rules in the model. The variable selection process is as follows:

1. Calculate the model's variable inclusion proportions
2. Randomly permute the response vector to break the relationship between covariates and the response. Reconstruct the model and compute the "null" variable inclusion proportions. Repeat this process multiple times to create a null permutation distribution.
3. Depending on the desired level of selection strictness, three selection rules can be applied:
   (a) Local Threshold: Include a predictor $x_k$ if its variable inclusion proportion exceeds the $1 - \alpha$ quantile of its own null distribution.
   (b) Global Max Threshold: Include a predictor $x_k$ if its variable inclusion proportion exceeds the $1 - \alpha$ quantile of the distribution of the maximum null variable inclusion proportions from each permutation of the response.
   (c) Global SE Threshold: Select $x_k$ if its variable inclusion proportion surpasses a threshold determined from its own null distribution mean and standard deviation, with a shared global multiplier applicable to all predictors.

The Local procedure (Local Threshold) is the least strict in selection, while the Global Max procedure is the most stringent. The Global SE procedure is a compromise but shares similarities with the Global Max approach. Bleich et al. [2] demonstrate that the best procedure depends on the underlying sparsity of the problem, which is often unknown.

## 2.5 *Partial Dependence*

A data analyst might to explore how the stock ($x_k$) influences the response on average while considering carbon emission. This can be investigated using Friedman [7] Partial Dependence Function (PDP).

$$f_k(x_k) = \mathrm{E}_{x_{-k}}\left[f(x_k, x_{-k})\right] = \int f(x_k, x_{-k}) d\mathrm{P}(x_{-k}) \tag{5}$$

Here, $x_{-k}$ represents all variables except $x_k$. . The Partial Dependence Function (PDP) of $x_k$ calculates the average value of $f$ when $x_k$ is held constant while $x_{-k}$ varies across its marginal distribution ($d\mathrm{P}(x_{-k})$). Since we don't have knowledge of the true model f or the predictor distribution ($d\mathrm{P}(x_{-k})$), we estimate Eq. 5 by performing the following computation:

$$\hat{f}(x_k) = \frac{1}{n}\sum_{i=1}^{n}\hat{f}\left(x_k, x_{-k,i}\right) \tag{6}$$

Here, $n$ represents data's number of observations (stock), and $\hat{f}$ represents the BART Machine model. BART offers an estimated posterior distribution, enabling us to create credible bands for the PDP function. In Eq. 6, $\hat{f}$ can be substituted with a function that computes the $q$th quantile of the post-burned-in MCMC iterations for $\hat{y}$.

## 3 Data

In this study the methodology examined the effects of green stock price on the carbon emission reduction by investigating the impact of significant stock price on carbon emissions only in industrial sectors. This study utilized monthly data from a chosen set of stocks constituting the S&P Global Clean Energy Index. This covers 99 stocks originating from various countries. All of which met specific criteria related to their utilization of renewable or clean energy sources within their operations and products.

The listing in Table 1 presents solely the only 15 constituent stocks because of the page limitation. However, it's essential to note that our study collected a total of 99 stocks. It is impractical to display the entire list of these stocks here, but interested readers can access the comprehensive list at in appendix part or www.spglobal.com.

**Table 1** The list of 15 constituents of the S&P global clean energy index

| Constituents | Ticker | Sectors |
|---|---|---|
| First Solar Inc | FSLR | Information Technology |
| Consolidated Edison Inc | ED | Utilities |
| Enphase Energy Inc | ENPH | Information Technology |
| SolarEdge Technologies Inc | SEDG | Information Technology |
| Vestas Wind Systems AS | VWS | Industrials |
| Iberdrola SA | IBE | Utilities |
| China Yangtze Power Co A | 600,900 | Utilities |
| Energias de Portugal SA | EDP | Utilities |
| Orsted | ORSTED | Utilities |
| Chubu Electric Power Co | 9502 | Utilities |
| Avangrid Inc | AGR | Utilities |
| NEL | NEL | Industrials |
| GoldWind Science & Technology co | 2208 | Industrials |
| Power cell Sweden | PCELL | Industrials |
| Zhejiang Akcome New Energy Tech | 002,610 | Information Technology |

All stock price data and S&P Global Clean Energy Index used start from January 2022 to May, 2023 total 42 observation received from the Thomson Reuters database. While the carbon dioxide emission (only focus on global industrial sectors; CARBON) has been collected from EIA (U.S. Energy Information Administration).

Each data has been transformed into a logarithm for achieve data normalization. The summary statistics for 15 constituent stocks, including the S&P Global Clean Energy Index (INDEX) and industrial carbon emissions (CARBON), are presented in Table 2. All variables exhibit positive average rates. However, when subjected to the Jarque–Bera normality test, it is evident that none of the variables follow a normal distribution, as indicated by the Jarque–Bera test probabilities, which are in close proximity to zero.

**Table 2** The data descriptive

| | Mean | Maximum | Minimum | Std. Dev | Skewness | Kurtosis | JB | Prob |
|---|---|---|---|---|---|---|---|---|
| FSLR | 4.501 | 5.363 | 3.585 | 0.432 | 0.095 | 2.596 | 0.340 | 0.000 |
| ED | 4.416 | 4.611 | 4.184 | 0.121 | 0.125 | 1.695 | 3.016 | 0.001 |
| ENPH | 4.960 | 5.770 | 3.466 | 0.616 | −1.013 | 3.040 | 7.017 | 0.003 |
| SEDG | 2.267 | 2.761 | 1.693 | 0.284 | −0.216 | 2.101 | 1.699 | 0.000 |
| VWS | 5.493 | 5.886 | 4.402 | 0.349 | −1.645 | 4.843 | 24.296 | 0.000 |
| IBE | 2.258 | 2.440 | 2.016 | 0.092 | −0.492 | 3.354 | 1.868 | 0.003 |
| 600,900 | 6.684 | 7.126 | 6.359 | 0.198 | 0.165 | 2.270 | 1.098 | 0.008 |
| EDP | 1.508 | 1.649 | 1.242 | 0.079 | −0.924 | 4.829 | 11.547 | 0.003 |
| ORSTED | 3.017 | 3.173 | 2.835 | 0.096 | −0.275 | 2.149 | 1.755 | 0.006 |
| 9502 | 7.193 | 7.420 | 7.044 | 0.080 | 0.453 | 3.404 | 1.682 | 0.000 |
| AGR | 3.840 | 4.003 | 3.616 | 0.095 | −0.379 | 2.332 | 1.745 | 0.000 |
| NEL | 2.773 | 3.426 | 2.176 | 0.278 | 0.297 | 3.029 | 0.604 | 0.000 |
| 2208 | 2.530 | 2.919 | 2.249 | 0.175 | 0.471 | 2.404 | 2.124 | 0.000 |
| PCELL | 5.224 | 5.935 | 4.370 | 0.345 | −0.335 | 2.899 | 0.784 | 0.000 |
| 2610 | 2.678 | 3.727 | 2.044 | 0.461 | 0.411 | 2.033 | 2.751 | 0.000 |
| INDEX | 7.116 | 7.551 | 6.407 | 0.256 | −1.081 | 3.746 | 8.939 | 0.001 |
| CARBON | 4.727 | 4.834 | 4.538 | 0.062 | −1.115 | 4.504 | 12.365 | 0.002 |

**Table 3** The model selection

| | Ridge regression | LASSO regression | BART machine |
|---|---|---|---|
| MSE | 2.998 | 2.445 | 1.997 |

# 4 Result

## *4.1 The Model Selection*

The assessment of estimator performance involves the utilization of the mean squared error (MSE), which quantifies the average of squared errors. This represents the average squared disparity between predicted values and the ground truth. Consequently, this study employed the MSE as a model evaluation measure for regression models and the lower value indicates a better fit model.

Table 3 presents the Bayesian Additive Regression Trees (BART) machine model as the most appropriate model in relation to prediction accuracy and sparsity, as evidenced by its highest MSE value. Conversely, the ridge method performs the least optimistically indicated by its lowest MSE value. Our study not only demonstrates the superior fit of the BART Model, but it aligns with findings from Bleich et al. [1], who extensively tested various machine learning models and concluded that the BART Model is more suitable compared to others. Notably, the BART machine model stands out for its ability to conduct variable selection and find how the stocks have influenced the response on average carbon emission. Consequently, the subsequent interpretation of the study's findings will focus on the Bayesian Additive Regression Trees (BART) machine model.

## *4.2 The Variable Selection Results*

Firstly, this study presents a novel approach for variable selection utilizing the nonparametric Bayesian model, Bayesian Additive Regression Trees (BART). The methodology is grounded in the Tree-Based method, with a particular focus on Tree-based variable selection. This our variable selection technique employs the intricacies of the decision tree structure, as elucidated briefly. The process initiates with all observations in a single root node. The root node's decision rule is determined, comprising a splitting variable ($x_k$) and a split point ($c$) where the Mean Square Error (MSE) has been represented as a split point by BART. Subsequently, the observations in the root node are divided into two groups based on the condition $x_k \geq c$ (MSE) or $x_k < c$ (MSE), forming a right daughter node and a left daughter node. Further binary splits can be chosen within each of these nodes.

This evaluation is conducted by analyzing the splitting rules in the m trees over the post-burn-in MCMC iterations, commonly referred to as "inclusion proportions" [4].

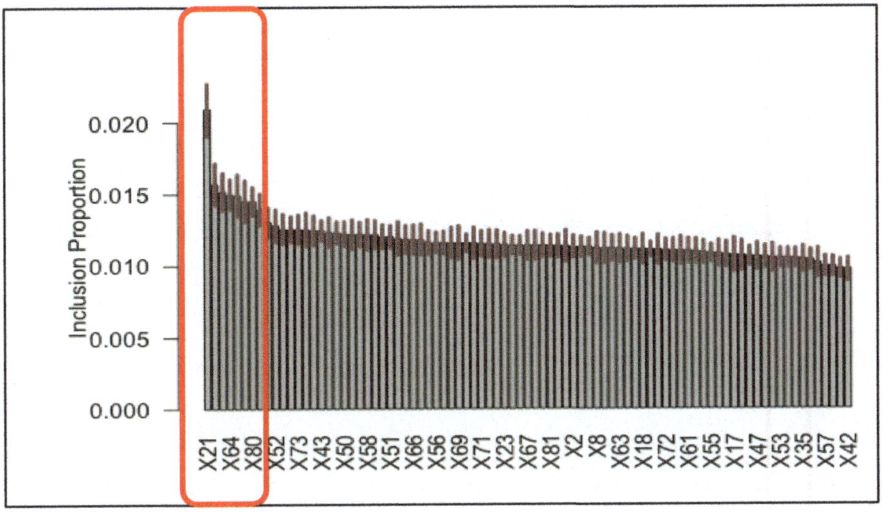


**Fig. 1** The variable importance by inclusion proportions

The inclusion proportion associated with a particular predictor denotes the frequency with which that variable is selected as a splitting rule out of all the splitting rules generated from the posterior draws of the sum-of-trees model. Figure 1 provides a visual representation of these inclusion proportions for all variables. The initial findings from this research indicate that variables X21, X64, and X80 exhibit noteworthy inclusion proportions at a significance level more than 99%. These variables are likely to have an impact on carbon emissions within the company.

After we demonstrate the principled variable selection procedure introduced in the previous part. We have employed in detail following the Bleich et al. [2] for selecting variables based on the three thresholds as 1. Local threshold 2. Global threshold and 3. Global SE Threshold. It also displays the plot in Fig. 2.

To determine an appropriate selection threshold, a permutation-based approach is employed to establish a null distribution for variable inclusion proportions ($P = (p_1, \ldots, p_k)$). This involves creating ($p$) permutations of the response vector $\left(y_1^*, y_2^*, \ldots, y_p^*\right)$, and for each permuted response vector ($y_p^*$), a BART model is run using $y_p^*$ as the response and the original predictor variables $(x_1, \ldots, x_k)$. This permutation strategy preserves potential dependencies among predictors while eliminating any correlation between predictors and the response. The variable inclusion proportions estimated from each BART run with permuted responses ($p_{k,p}^*$ for predictor ($x_k$) from the $p$th permuted response, and $p_P^*$ for the vector of all variable inclusion proportions) are retained. The variable inclusion proportions across all P permutations $\left(p_1^*, \ldots, p_P^*\right)$ form the null distribution for variable inclusion proportions ($p$) from the real (unpermuted) response $y$. The challenge lies in selecting an appropriate threshold for predictor ($x_k$) based on the permutation null distribution $\left(p_1^*, \ldots, p_P^*\right)$.

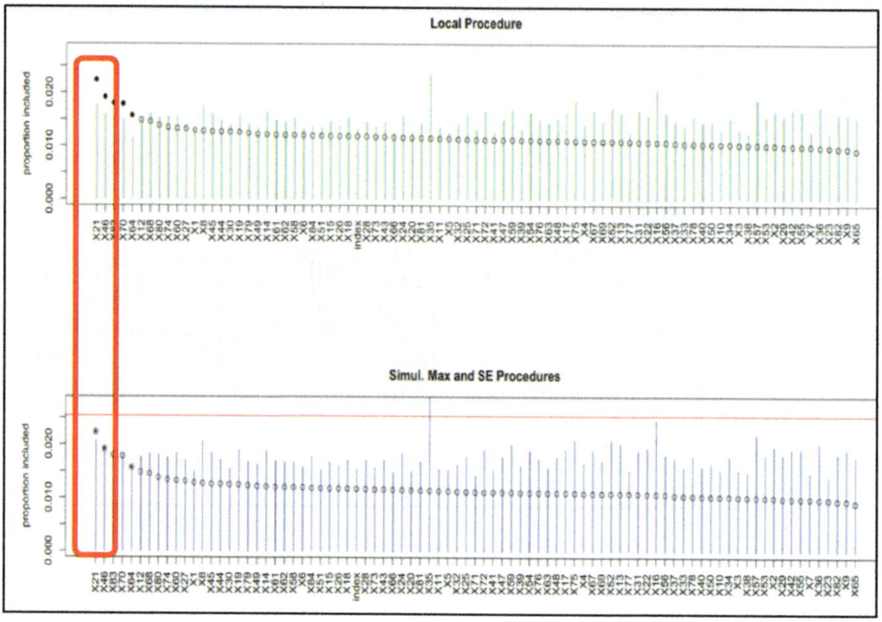


**Fig. 2** Visualization of the three variable selection procedures

Three threshold strategies will be considered, differing in the stringency of their resulting variable selection procedures [1].

The visualization of the three distinct variable selection procedures has been presented in Fig. 2, with a significance level ($\alpha$) set at 0.05. The upper plot depicts the "Local thresholds" procedure, where the green lines represent threshold levels derived from permutation distributions. To be selected, a variable must surpass these thresholds. The plotted points denote the variable inclusion proportions based on the observed data. If the observed value exceeds the threshold (indicated by the green bar), the variable is included and is depicted as a solid dot; conversely, if it does not exceed the threshold, it is not included and is represented as an open dot.

Additionally, it's important to highlight that the overall average inclusion proportion $\overline{p}_{...}$ is 0.025, equivalent to $^1/_k$. This suggests that, on average across datasets and BART runs, the variable inclusion proportions are accurate. Furthermore, the small standard deviation across predictors indicates that $\overline{p}_{k}...$ are approximately $^1/_k$ as well. This observation implies that there is no systematic bias favoring different covariates once the effects of overfitting by dataset and remaining in local modes by run are averaged out.

While the lower plot provides an overview of both the "Global SE" and "Global Max" thresholds. The red line serves as the cutoff for the "Global Max" threshold, and variables that surpass this threshold are represented as solid dots. In contrast, the blue lines represent the thresholds for the "Global SE" procedure. Variables that

**Table 4** The variable selection results

| Symbol | Company name | Ticker |
|---|---|---|
| 1. Local Thresholds | | |
| X21 | Avangrid Inc. | AGR |
| X46 | Nel | NEL |
| X64 | Goldwind Science &Technolog | 2208 |
| X70 | Powercell Sweden | PCELL |
| X83 | Zhejiang Windey | 002,610 |
| 2. Global Max and Global SE Thresholds | | |
| X21 | Avangrid Inc. | AGR |

exceed the "Global SE" cutoff but not the "Global Max" threshold are denoted as asterisks, while open dots signify variables that do not exceed either threshold.

Subsequently, following the identification of significant inclusion proportions among all relevant variables, these outcomes are considered to explore significant associations with carbon emissions using the concept of variable selection. The findings of this research can be categorized into two key criteria. Firstly, under the "Local Thresholds" criterion, it was discerned that five variables representing stock prices, namely X21 (AVANGRID Inc.), X46 (NEL), X64 (GOLDWIND SCIENCE&TECHNOLOG), X70 (POWERCELL SWEDEN), and X83 (ZHEJIANG WINDEY), exert a significant influence on industrial carbon emissions. Secondly, when considering the "Global Max and Global SE Thresholds," it was observed that only the price of X21 (AVANGRID Inc.) has a notable impact on industrial-level carbon emissions. Obviously, only 5% of stocks have a significant effect on carbon emissions from the S&P Global Clean Energy Index. This result has shown in Table 4.

Bleich et al. [2] demonstrate that the Local procedure (Local Threshold) represents the least stringent approach to variable selection, whereas both the Global Max and Global SE threshold procedures emerge as the most stringent alternatives. Their study also recommends careful consideration of the Global Max and Global SE threshold procedures when conducting variable selection. Hence, this conclusion highlights the significant influences of AVANGRID Inc.'s stock price on carbon emissions. However, it's important to note that the direction of this relationship remains indeterminate based solely on our findings. Consequently, further examination of the relationship model will be discussed in the subsequent section.

## 4.3 The Partial Dependence Result

Since BART (machine) model offers an estimated posterior distribution, it becomes possible to construct credible intervals for the Partial Dependence Function (PDP). As outlined in Eq. 5 in the methodology section, the $\hat{f}$ can be substituted with a

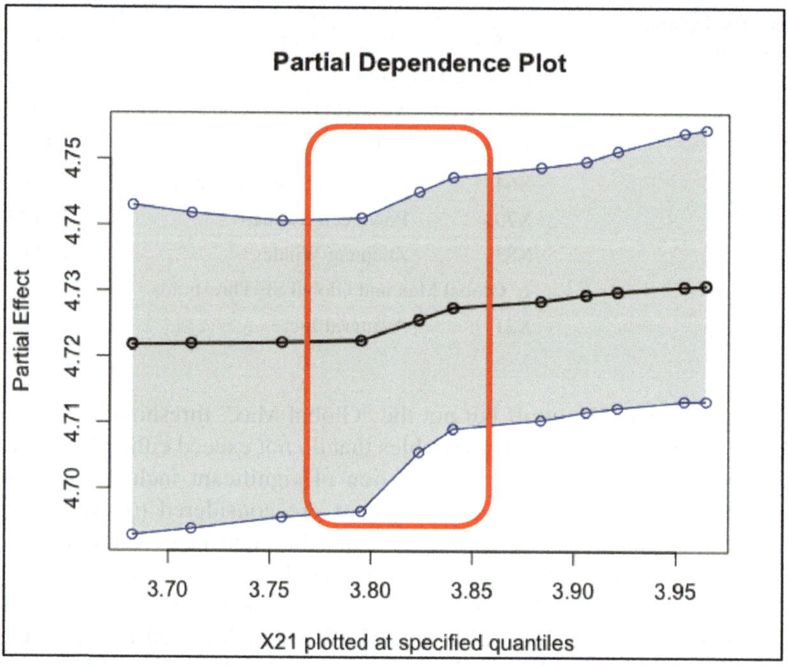


**Fig. 3** Partial dependence plot

function that computes the q-th quantile of the post-burned-in MCMC iterations for $\hat{y}$, which represents carbon emissions. It is shown in Fig. 3.

Figure 3 illustrates the relationship between the logarithmic price range of AVANGRID Inc.'s stocks (X-Axis) and log-transformed Carbon Emissions (Y-Axis). Obviously, it's evident that the study's findings do not change over time; instead, they portray the influence of varying levels of the independent variable (log price) on different levels of the dependent variable (log Carbon Emissions).

The study's findings, as illustrated in Fig. 3, indicate that as the log prices of AVANGRID Inc.'s stocks increase, carbon emissions also demonstrate a corresponding increase. This effect is particularly pronounced during the period when the log price changes from 3.80 to 3.85, which translates to a price range of 44.70 to 46.99 US dollars. These results lead to the conclusion that AVANGRID Inc.'s stock prices have a significant impact on the increase in carbon emissions.

## *4.4 Discussion*

Generally, company's stock price and its impact on carbon emissions are not directly related. A company's stock price reflects the perceived value of the company by investors, and this value is based on a variety of financial and market factors.

Nonetheless, our research has unveiled a finding regarding the substantial association between stock prices and carbon emissions. Our results also consistent recent research conducted by Ma et al. [17] and Lu et al. [14]. They both have also indicated that the stock market, particularly the S&P500, exhibits responsiveness to climate change. Additionally, Wegener et al. [25] conducted a comprehensive investigation into the influence of corporate disclosure of carbon-related data and carbon emission reduction efforts within Canadian firms. Their findings suggest that voluntary carbon information disclosure and active carbon emission reduction initiatives can yield positive effects on stock market valuation.

Intriguingly, our study has been surprising outcomes as the increase in AVANGRID Inc.'s stock prices (an aspect of considerable interest to investor) has been correlated with a concurrent rise in carbon emissions. The initial step in comprehending these unexpected effects involves considering the temporal scope of the data. It was observed that this study's dataset spans from January 2020 to May 2023. During this period, a decline in stock prices is evident, commencing from the outset of 2020 and continuing through to the present. This decline can be attributed to the repercussions of the COVID-19 outbreak, coupled with the company's strategic shift towards sustainability, a development announced in 2021. Additionally, the company's plans for substantial investments in the expansion of pioneering technology, aimed at augmenting grid capacity in New York to expedite the adoption of renewable energy in 2023, contribute to this context. Consequently, in the short term, these initiatives may incur increased costs, potentially reducing profits and leading to a less favorable short-term outlook for investors. However, the implementation of these activities is anticipated to yield positive long-term environmental outcomes. Moreover, it holds the potential to generate sustainable returns, benefitting both the company's stock price and its shareholders over the long run [6].

Finally, the findings of this study hold valuable insights for other companies, especially those listed on the S&P Global Clean Energy Index. AVANGRID Inc. can serve as a role model for implementing activities aimed at substantial carbon emissions reduction. While these actions may have short-term implications for stock prices during their execution, they are deemed essential for contributing to the development of a sustainable economic system over the long term. This demonstrates a commitment to environmental responsibility and aligning business practices with broader sustainability goals.

## 5 The Conclusion

This study aims to verify the effectiveness of carbon emissions reduction through investments in Green finance. Green finance comprises a diverse range of variables and indexes. This research has selection the S&P Global Clean Energy Index. This index is chosen due to its comprehensive representation of global companies involved in the Green finance sector. Furthermore, the study directs its attention towards the constituent stocks that contribute to the computation of the S&P Global Clean Energy

Index. The aim is to identify those specific companies (stocks) that exhibit significant potential to serve as prototype examples in the execution of initiatives geared towards carbon emissions reduction. Then, this covers monthly 99 stock price data, S&P Global Clean Energy Index and industrial carbon emissions (CARBON) have been focusing on this study estimated by BART machine model. The BART machine model is the novel model that can provide a fast, easy-to-use by combining the benefit implementation of Bayesian additive regression trees (BART) and machine learning.

The findings of this research can be highlight that when the price of AVANGRID Inc. changes from 44.70 to 46.99 US dollars will increase the industrial carbon. These results lead to the conclusion that AVANGRID Inc.'s stock prices have a significant impact on the increase in carbon emissions. Therefore, AVANGRID Inc. emerges as an exemplar for implementing initiatives geared toward significant carbon emissions reduction. The AVANGRID Inc will be an exemplar for implementing initiatives geared toward significant carbon emissions reduction to other company.

## Appendix

| Ticker | Company name | Sector | Ticker | Company name | Sector |
|---|---|---|---|---|---|
| FSLR | FIRST SOLAR INC | Information Technology | NDX1 | NORDEX | Industrials |
| ED | CONSOLIDATED EDISON INC | Utilities | 600,905 | CHINA THREE GORGES RENEWABLES (GROUP) | Utilities |
| ENPH | ENPHASE ENERGY INC | Information Technology | ERG | ERG | Utilities |
| VWS | VESTAS WIND SYSTEMS | Industrials | 300,274 | SUNGROW POWER SUPPLY LTD A | Industrials |
| SEDG | SOLAREDGE TECHNOLOGIES INC | Information Technology | MBTN | MEYER BURGER TECHNOLOGY AG | Information Technology |
| IBE | IBERDROLA SA | Utilities | CPFE3 | CPFL ENERGIA SA | Utilities |
| 600,900 | CHINA YANGTZE POWER LTD A | Utilities | NOVA | SUNNOVA ENERGY INTERNATIONAL INC | Utilities |
| EDP | EDP ENERGIAS DE PORTUGAL SA | Utilities | 112,610 | CS WIND CORP | Industrials |
| 9502 | CHUBU ELECTRIC POWER INC | Utilities | NHPC | NHPC LTD | Utilities |

(continued)

(continued)

| Ticker | Company name | Sector | Ticker | Company name | Sector |
|---|---|---|---|---|---|
| ORSTED | ORSTED | Utilities | BEPC | BROOKFIELD RENEWABLE SUBORDINATE V | Utilities |
| PLUG | PLUG POWER INC | Industrials | S92 | SMA SOLAR TECHNOLOGY AG | Information Technology |
| ORA | ORMAT TECH INC | Utilities | SLR | SOLARIA ENERGIA Y MEDIO AMBIENTE S | Utilities |
| CMIG4 | CIA ENERGETICA DE MINAS GERAIS PRE | Utilities | BLX | BORALEX INC CLASS A | Utilities |
| SHLS | SHOALS TECHNOLOGIES GROUP INC CLAS | Industrials | 2129 | TCL ZHONGHUAN RENEWABLE ENERGY TEC | Information Technology |
| ARRY | ARRAY TECHNOLOGIES INC | Industrials | CSIQ | CANADIAN SOLAR INC | Industrials |
| RUN | SUNRUN INC | Industrials | ECV | ENCAVIS AG | Utilities |
| EA.R | ENERGY ABSOLUTE NON-VOTING DR PCL | Utilities | VBK | VERBIO VEREINIGTE BIOENERGIE AG | Energy |
| NPI | NORTHLAND POWER INC | Utilities | 6865 | FLAT GLASS GROUP LTD H | Information Technology |
| ENGI11 | ENERGISA UNITS SA | Utilities | 2459 | JA SOLAR TECHNOLOGY LTD A | Information Technology |
| EDPR | EDP RENOVAVEIS SA | Utilities | ENLT | ENLIGHT RENEWABLE ENERGY LTD | Utilities |
| BKW | BKW N AG | Utilities | AURE3 | AUREN ENERGIA SA | Utilities |
| 9830 | HANWHA SOLUTIONS CORP | Materials | NEOEN | NEOEN SA | Utilities |
| VER | VERBUND AG | Utilities | 586 | CHINA CONCH VENTURE HOLDINGS LTD | Industrials |
| GPRE | GREEN PLAINS INC | Energy | 688,599 | TRINA SOLAR LTD A | Information Technology |
| AGR | AVANGRID INC | Utilities | INE | INNERGEX RENEWABLE ENERGY INC | Utilities |

(continued)

(continued)

| Ticker | Company name | Sector | Ticker | Company name | Sector |
|---|---|---|---|---|---|
| 601,012 | LONGI GREEN ENERGY TECHNOLOGY LTD | Information Technology | 336,260 | DOOSAN FUEL CELL LTD | Industrials |
| ANE | CORPORACION ACCIONA ENERGIAS RENOV | Utilities | NEL | NEL | Industrials |
| CWEN | CLEARWAY ENERGY INC CLASS C | Utilities | BLDP | BALLARD POWER SYSTEMS INC | Industrials |
| 9958 | CENTURY IRON AND STEEL INDUSTRIAL | Materials | 300,118 | RISEN ENERGY LTD A | Information Technology |
| 3576 | UNITED RENEWABLE ENERGY LTD | Information Technology | 688,390 | JIANGSU GOODWE POWER SUPPLY TECHNO | Industrials |
| SPWR | SUNPOWER CORP | Industrials | 601,016 | CECEP WIND-POWER CORP A | Utilities |
| MEL | MERIDIAN ENERGY LTD | Utilities | AMPS | ALTUS POWER INC CLASS A | Utilities |
| 1798 | CHINA DATANG CORPORATION RENEWABLE | Utilities | 2056 | HENGDIAN GROUP DMEGC MAGNETICS LTD | Information Technology |
| 600,674 | SICHUAN CHUANTOU ENERGY LTD A | Utilities | 2487 | DAJIN HEAVY INDUSTRY LTD A | Industrials |
| NEOE3 | NEOENERGIA SA | Utilities | TPIC | TPI COMPOSITES INC | Industrials |
| 600,025 | HUANENG LANCANG RIVER HYDROPOWER I | Utilities | 600,116 | CHONGQING THREE GORGES WATER CONSE | Utilities |
| REX | REX AMERICAN RESOURCES CORP | Energy | 2610 | ZHEJIANG AKCOME NEW ENERGY TECHNOL | Information Technology |
| AESB3 | AES BRASIL SA | Utilities | 300,827 | SINENG ELECTRIC LTD A | Industrials |
| RNW | RENEW ENERGY GLOBAL PLC CLASS A | Utilities | 322,000 | HD HYUNDAI ENERGY SOLUTIONS LTD | Information Technology |
| ENELAM | ENEL AMERICAS SA | Utilities | 930 | COFCO BIOTECHNOLOGY LTD A | Materials |

(continued)

(continued)

| Ticker | Company name | Sector | Ticker | Company name | Sector |
|---|---|---|---|---|---|
| 6443 | TSEC CORP | Information Technology | 601,827 | CHONGQING SANFENG ENVIRONMENT GROU | Utilities |
| ENRG | ENERGIX RENEWABLE ENERGIES LTD | Utilities | 300,772 | ZHEJIANG WINDEY LTD A | Industrials |
| CEN | CONTACT ENERGY LTD | Utilities | 300,443 | JINLEI TECHNOLOGY LTD A | Industrials |
| FCEL | FUELCELL ENERGY INC | Industrials | | | |
| 2208 | GOLDWIND SCIENCE & TECHNOLOGY CO L | Industrials | | | |
| 601,615 | MING YANG SMART ENERGY GROUP LTD A | Industrials | | | |
| 9519 | RENOVA INC | Utilities | | | |
| GEVO | GEVO INC | Energy | | | |
| GUNKUL.R | GUNKUL ENGINEERING NON-VOTING DR P | Utilities | | | |
| SCATC | SCATEC SOLAR | Utilities | | | |
| 300,763 | GINLONG TECHNOLOGIES LTD A | Industrials | | | |
| GVOLT | GREENVOLT ENERGIAS RENOVAVEIS SA | Utilities | | | |
| PCELL | POWERCELL SWEDEN | Industrials | | | |
| 3868 | XINYI ENERGY HOLDINGS LTD | Utilities | | | |
| 591 | CECEP SOLAR ENERGY LTD A | Utilities | | | |
| 2531 | TITAN WIND ENERGY (SUZHOU) LTD A | Industrials | | | |

## References


1. Bleich, J., Kapelner, A., Jensen, S., George, E.: Variable selection inference for bayesian additive regression trees. ArXiv e-prints (2013)
2. Bleich, J., Kapelner, A., Jensen, S., George, E.: Variable selection for BART: an application to gene regulation.". Ann. Appl. Stat. **8**(3), 1750–1781 (2014)
3. Breiman, L.: Statistical Modeling: The Two Cultures." Statistical Science, 16(3), 199–231 (2001)
4. Chen, X., Chen, Z.: Can Green Finance Development Reduce Carbon Emissions? Empirical Evidence from 30 Chinese Provinces (2021)
5. Chipman, H., George, E., McCulloch, R.: BART: Bayesian additive regressive trees. Ann. Appl.; Chipman, H., McCulloch, R.: BayesTree: Bayesian methods for tree based models. Statistics **4**(1), 266–298 (2010). Rpackage version 0.3–1.1. http://CRAN.R-project.org/package=BayesTree
6. Dutta, A., Bouri, E., Noor, M.H.: Return and volatility linkages between $CO_2$ emission and clean energy stock prices. Energy **164**, 803–810 (2018)
7. Ferreira, P., Loures, L.C.: An econophysics study of the S&P global clean energy index. Sustainability **12**(2), 662 (2020)
8. Friedman, J.: Greedy function approximation: a gradient boosting machine. Ann. Stat. **29**(5), 1189–1232 (2002)
9. Gelman, A.: Parameterization and Bayesian modeling. J. Am. Stat. Assoc. **99**(466), 537–545 (2004)
10. Geman, S., Geman, D.: Stochastic relaxation, gibbs distributions, and the Bayesian restoration of images. IEEE Trans. Pattern Anal. Mach. Intell. **6**, 721–741 (1984)
11. Gibson, R., Krueger, P.: The sustainability footprint of institutional investors. Paper 571 (2018)
12. Hammoudeh, S., Ajmi, A.N., Mokni, K.: Relationship between green bonds and financial and environmental variables: a novel time-varying causality. Energy Econ. **92**, 104941 (2020)
13. Hastie, T., Tibshirani, R.: Bayesian backfitting. Stat. Sci. **15**(3), 196–213 (2000)
14. Hastings, W.: Monte Carlo sampling methods using Markov Chains and their applications. Biometrika **57**(1), 97–109 (1970)
15. Lu, W., Zhu, N., Zhang, J.: The Impact of Carbon Disclosure on Financial Performance Under Low Carbon Constraints (2021)
16. Kanamura, T.: A model of price correlations between clean energy indices and energy commodities. J. Sustain. Financ. Invest. **12**(2), 319–359 (2022); Gelman, A., Carlin, J., Stern, H., Rubin, D.: Bayesian Data Analysis, 2nd ed. Chapman & Hall/CRC (2004)
17. Kapelner, A., Bleich, J.: Bartmachine: Machine learning with bayesian additive regression trees (2013). arXiv:1312.2171
18. Ma, G., Liang, M., Sun, W.: Effect Analysis of Carbon Information on Enterprise Value Based on Big Data (2022)
19. Martin, V.: Green finance: regulation and instruments. J. Central Bank. Theory Practice **12**(2), 185–209 (2023)
20. Pratola, M., Chipman, H., Higdon, D., McCulloch, R., Rust, W.: Parallel Bayesian Additive Regression Trees. Technical report, University of Chicago (2013)
21. Risal, N., Joshi, S.: Measuring Green Banking Practices on Bank's Environmental Performance: Empirical Evidence from Kathmandu Valley (2018)
22. Surya, B., Muhibuddin, A., Suriani, S., Rasyidi, ES., Baharuddin, B., Fitriyah, AT., Abubakar, H.: Economic Evaluation, Use of Renewable Energy, and Sustainable Urban Development Mamminasata Metropolitan, Indonesia (2021)
23. Taddy, M., Gramacy, R., Polson, N.: Dynamic trees for learning and design. J. Am. Stat. Assoc. **106**(493), 109–123 (2011)
24. Urbanek, S.: rJava: Low-level R to Java interface. R package version 0.9–6 (2013). http://CRAN.R-project.org/package=rJava
25. Wegener, M., Amin, G.R.: Minimizing greenhouse gas emissions using inverse DEA with an application in oil and gas. Expert Syst. Appl. **122**, 369–375 (2019)

26. Zhang, J., Ke, H.: The Moderating Effect and Threshold Effect of Green Finance on Carbon Intensity: From the Perspective of Capital Accumulation (2022)
27. Zhang, R., Xu, H., Zhang, Z., Lyu, Y., Lu, Y., Duan, H.: Whether Green Finance Improves Green Innovation of Listed Companies—Evidence from China (2022)

# Analysis of Time—Varying Coefficients and Forecasting Effects Between Greenhouse Gas Emissions and Its Determinants in Thailand

**Pathairat Pastpipatkul and Panicha Subsai**

**Abstract** This paper uses a dynamic linear model (DLM) to find the relationship between greenhouse gas emissions and its determinants in Thailand. From the Bayesian model averaging (BMA) analysis was found that only forest, GDP, and industrial are sufficient for further investigation and can explain best the GHG emission. To explain the correlation, the time-varying coefficients of forest and GDP strongly explain and change in the same direction with GHG emission. In contrast, intercept term and industrial variables significantly explain and change in the opposite direction with GHG emission. There are some significant events that affect time-varying coefficients such as climate policy and the economic crisis in 2009. Moreover, The DLM used to be a forecasting tool for the next 10 years. The time-varying coefficients from intercept, GDP, and industrial can have a potential effect on GHG emissions. Forest will potentially explain GHG emissions in a better direction. On the contrary, the coefficients from the intercept term are in opposite directions.

**Keywords** Greenhouse gas · Forest · Economic growth · Bayesian dynamic linear model · Bayesian · Bayesian model averaging · Time-varying coefficients

## 1 Introduction

Climate change and global warming are currently a challenging responsibility across the world. The United Nations has established Sustainable Development Goals (SDGs) to promote the significance of sustainable development and green development is urgent as environmental problems. Greenhouse gases (GHG) store significant heat in the Earth's atmosphere, which is part of the greenhouse gas effect. Human activities are the most important actions that cause greenhouse gas effects [13]. Since the Industrial Revolution, human activities such as the combustion of oil, coal, and

P. Pastpipatkul · P. Subsai (✉)
Center of Excellence in Econometrics, Faculty of Economics, Chiang Mai University, Chiang Mai 52000, Thailand
e-mail: panicha.s.y@gmail.com



V. Kreinovich et al. (eds.), *Applications of Optimal Transport to Economics and Related Topics*, Studies in Systems, Decision and Control 556,
https://doi.org/10.1007/978-3-031-67770-0_40

gas as well as economic activity have rapidly accelerated the release of GHG into the atmosphere and increased the severity of climate change [5].

Thailand continuously requires energy to drive its economy, as a developing country, is one of the fastest-growing energy-intensive economies in Southeast Asia [28]. Such adverse effects major economic sectors, including agriculture, tourism, and industry. Thailand consumes energy and releases a large amount of GHG emissions, especially in the energy sector [17]. The industrial sector was the second largest energy-consuming sector. In the last decade, the industrial sector experienced a major revolution due to rapid economic development [6].

United Nations Framework Convention on Climate Change (UNFCC) mentioned in 2021 that Thailand only contributes less than one percent of the global greenhouse gas emissions, it has set up the goal to become a low-carbon society and efforts have been made to shift its energy policy towards green energy over the past decade. The renewable energy has been expanded to achieve the goal of sustainable green growth. Many studies analyzed renewable energy would substantially reduce GHG emissions [24]. The process of urbanization with migration from rural areas to cities provides chances to access for public utilities. As a result, urbanization raises energy demand, resulting in carbon emissions [35]. Reference [29] mentioned urbanization is one factor contributing to Thailand's greenhouse gas emissions. Currently, technological innovations are the most important factor in combating climate change [16]. Technological advances have decreased the need for fossil fuels and, as a result, GHG emissions [31].

Moreover, Thailand is one of the top ten destination countries globally considering the number of international tourist arrivals ([30]). Additionally, The usage of electricity in businesses intensifies the emissions as well as increases the use of home electricity [18]. Tourism is associated with an increase in energy consumption and the UNWTO reports that the tourism industry is accountable for about 5% of global emissions [14]. Many Thai businesses rely on the agricultural sector as a supplier of some raw materials [25]. The increasing demand for housing, agriculture, transportation, and other infrastructure facilities is putting a strain on forest land [21]). Hence, the depletion of forested land has been exacerbated by urbanization, industrialization, settlements, mining, and agriculture. Changes in land use, such as rapid urbanization and forest depletion, can result in large amounts of climate change. Forests, on the other hand, play an important role in the climatic system when trees absorb carbon dioxide from the atmosphere and store it in their biomass. This contributes to a slower rate of climate change ([20]). Agriculture, Forestry, and Other Land Use (AFOLU) have a linkage with global climate change as a major contributor [14]. In the case of economic performance, Gross Domestic Product (GDP) is one of the indicators of a country's overall performance [8] and one of the main signs used to evaluate the situation of any nation's economic system. With the above support, these variables would affect the nation's growth and applied study on the impact of reducing GHG emissions.

When there are many potential explanatory variables, it becomes crucial to determine which variables should be included in the model and how important they are for econometric analysis. Bayesian Model Averaging (BMA) figures the problem out by using models' estimations [34]. The BMA model applies Bayes' theorem to weights for this averaging stem. This study also applied Dynamic linear models (DLM) which general regression models have coefficients that can vary over time [15, 26].

Although many previous studies have investigated the economy-energy-environment nexus in different countries, there is a few of empirical research in the context of Thailand. Currently, Thailand has limited studies on climate policies. For example, Pagnarith et al. use the Long-range Energy Alternatives Planning (LEAP) system studied the effects of RE utilization and CO2 mitigation in the power sectors and Winyuchakrit et al. used the Extended SnapShot Tool (ExSS), a macro-economic model, to estimate future energy demand and Carbon dioxide emissions. Moreover, there is a lack of econometric studies investigating the determinants of GHG emissions and their impacts on environmental quality. To fill a gap in the literature, this study used the Dynamic Linear Model (DLM) to examine the changing effects in the usage of renewable energy, urbanization, industrialization, tourism, agricultural productivity, forest area, economic growth, and technological innovations on GHG emissions in Thailand.

The rest of this paper is organized as follows. Section 2 provides data description in detail and presents the methodology. Results from model estimations and forecasts are shown and discussed in Sect. 3. In Sect. 4, conclusions are presented and lastly, references are shown.

## 2 Data and Methodology

### *2.1 Data Description*

This article investigates the dynamic effects of renewable energy, urbanization, industrialization, tourism, agriculture, forest, economic growth, and technological innovations on Thailand's GHG emissions. In Table 1, this study employed World Development Indicators (WDI) data in the Data project (1991–2020). Using the natural logarithm, which ensures the data normality, all important data were integrated into the model.

### *2.2 Bayesian Inference*

There is a method Bayes theorem that improves the probabilities for a hypothesis when there is more available data, it is called Bayesian inference. In the analysis, Bayesian updates the distribution of which this distribution is called the prior distribution. Based on the observed data, the updated distribution of is obtained from the Bayes' Rule and the posterior distribution is found using Bayes' formula. The

**Table 1** Specification of the variables

| Variables (Logarithmic forms) | Descriptions | Units |
|---|---|---|
| LGHG | GHG emissions | Kt of $CO_2$ equivalent |
| LRenewable | Renewable energy use | Percent of total final energy use |
| LUrbanization | Urbanization | Number of urban population |
| LIndustrial | Industrialization | Industry value added (% of GDP) |
| LTourism | International tourism | Number of tourist arrivals |
| LAgriculture | Agricultural productivity | Agricultural value added (% of GDP) |
| LForest | Forest area | Square kilometers (sq. km) |
| LGDP | Gross Domestic Product | Constant Thai baht |
| LTechnology | Technological innovations | Number of patent applications |

posterior distribution of parameters can be computed, and the Monte Carlo method is commonly used to draw the sample from the posterior distribution of interest ([33]). The most commonly used technique is to implement a Gibbs sampler, and the Gibbs simulation model frequently simplifies implementation and accelerates the mixing of the resulting Markov chain.

## 2.3 Bayesian Model Averaging

In standard statistical procedures, model uncertainty is frequently ignored. In general, data analysts choose a model from a set of models and proceed with that model as if it had generated the data [11]. To deal with the model uncertainty problem in regression, this study used Bayesian estimation, specifically the Bayesian Model Averaging (BMA) approach, which has become a popular alternative to model selection [34]. A linear model structure has the following general form:

$$y = \alpha_\gamma + \beta_\gamma X + \varepsilon, \varepsilon \sim N(0, \sigma^2 I), \quad (1)$$

where $y$ is the dependent variable, $\alpha_\gamma$is a constant, $\beta_\gamma$is the coefficients, and $\varepsilon$ is an error term with variance $\sigma^2$. BMA estimates models for all possible X variable combinations and computes a weighted average for each. If X has K possible variables, the estimations will have $2^K$ variable combinations and $2^K$ models. The model weights for this averaging are derived from the posterior model probabilities derived from Bayes' theorem:

$$p(M_\gamma|y,X) = \frac{p(y|M_\gamma,X)p(M_\gamma)}{p(y|X)} = \frac{p(y|M_\gamma,X)p(M_\gamma)}{\sum_{S=1}^{2K} p(y|M_s,X)p(M_s)} \tag{2}$$

$$p(\theta|y,X) = \sum_{\gamma=1}^{2^K} p(\theta|M_\gamma,y,X)\frac{p(M_\gamma|X,y)p(M_\gamma)}{\sum_{S=1}^{2K} p(M_S|y,X)p(M_S)} \tag{3}$$

As a result in model (2), $p(y|X)$denotes the integrated likelihood, $2^K$is the number of estimated models from K potential variables, and the posterior model probability (PMP; $p(M_\gamma|y,X)$) is the marginal likelihood of the model $p(y|M_\gamma,X)$multiplied by$p(M_\gamma)$. The condition then leads to PMP, and model (3) illustrates the model weighted posterior distribution for any statistic $\theta$.

## 2.4 Zellner's g Prior

The Bayesian regression model employs a specific prior structure known as "Zellner's g prior" [34]. The prior coefficients have the specified mean and variance and are distributed normally. In this case, the prior mean is zero, and its variance ($\sigma^2$) is expressed in model (4) and the regression coefficient prior; $\beta_\gamma$ is in model (5).:

$$Zellner's\ g \sim \sigma^2\left(\frac{1}{g}X'X\right)^{-1}. \tag{4}$$

$$\beta_\gamma|g \sim N\left(0,\sigma^2\left(\frac{1}{g}X'X\right)^{-1}\right) \tag{5}$$

where g is the hyperparameter indicating how certain the coefficients are zero. The value of g also indicates the prior coefficient variances, for example, a small g indicates that there are few prior coefficient variances and that the result is quite certain. A large g, on the other hand, denotes high uncertainty. Prior uncertainty is reflected in the posterior distribution of coefficients.

## 2.5 Dynamic Linear Models

The dynamic linear model (DLM) is a linear regression model in which the coefficients can change over time. The linear regression model is the most used tool for relating the dependent variable y to the explanatory variables x. Consider the below model:

$$y_t = F_t'\theta_t + v_t,\quad v_t \sim N(0,V_t), \tag{6}$$

$$\theta_t = G_t\theta_{t-1} + w_t, w_t \sim N(0, W_t); t = 1, 2, ..., T, \tag{7}$$

$$y_t = \alpha_t + x_t\beta_t + v_t, v_t \sim N(0, V_t), \tag{8}$$

$$\alpha_t = \alpha_{t-1} + w_{\alpha,t}, w_{\alpha,t} \sim N(0, w_{\alpha,t}), \tag{9}$$

$$\beta_t = \beta_{t-1} + w_{\beta,t}, w_{\beta,t} \sim N(0, w_{\beta,t}). \tag{10}$$

where $y_t$ denotes the observations through time that are conditionally independent give $\theta_t$, $V_t$and $\theta_t$denotes the vector of parameters at time t. $F_t^{'}$and $G_t$are the system matrices, $V_t$and $W_t$are the covariance matrices, and $v_t$and $w_t$ are mutually independent errors.$F_t$, $V_t$, $G_t$,or $W_t$change with time in a time varying. A dynamic linear regression can be represented mathematically in model (8) to model (10). The Kalman filter method was used to estimate the parameters ([26]). In this study, we consider 5,000 rounds of sampler, with 2,000 representing the burn-in sample.

## 2.6 Filtering, Smoothing, and Forecasting

Assuming that a DLM is completely specified and that there are no unknown parameters in its definition, the filtering and smoothing algorithms are used to obtain the conditional distributions' means and variances. The conditional distribution of $\theta_t$ given $y_{1:t}$ is the filtering distribution. The smoothing distribution at time s, on the other hand, is the conditional distribution of $\theta_t$ given $y_{1:s}$ and $s \geq t$. All of these distributions are assumed to be Gaussian and completely determined by mean and variance.

The forecast distribution denotes a conditional distribution of future states, given the data up to time t by using a prior with mean, and variance ($C_0$). Several criteria assess forecasting accuracy including the mean absolute deviation (MAD), the mean square error (MSE), and the mean absolute percentage error (MAPE), defined respectively by the following model (11)–(13).

$$MAD = \frac{1}{n}\sum\nolimits_{t=1}^{n} |e_t| \tag{11}$$

$$MSE = \frac{1}{n}\sum\nolimits_{t=1}^{n} |e_t^2| \tag{12}$$

$$MAPE = \frac{1}{n}\sum\nolimits_{t=1}^{n} \frac{|e_t|}{y_t} \tag{13}$$

An additional statistic to assess the forecasting performance which is Theil's U (Theil; 1966), shown in model (14). Theil's U compares the MSE of the model with the MSE of the trivial "no-change" model that predicts the next observation to be the same as the current one.

**Table 2** Descriptive statistics of dependent variable and independent variables (1991–2020)

| Variable | Mean | Sd | Maximum | Minimum | Skewness | Kurtosis |
|---|---|---|---|---|---|---|
| LGHG | 0.61 | 6.13 | 9.87 | −15.84 | −0.39 | −0.35 |
| LRenewable | 0.43 | 5.85 | 12.45 | −13.3 | −0.01 | −0.37 |
| LUrbanization | 1.43 | 31.05 | 137.31 | −54.39 | 2.38 | 9.11 |
| LIndustrial | 3.54 | 1.77 | 6.37 | −0.17 | −0.55 | −0.48 |
| LTourism | 0.07 | 6.52 | 12.3 | −14.05 | −0.29 | −0.63 |
| LAgriculture | −0.69 | 6.94 | 15.74 | −12.71 | 0.51 | −0.26 |
| LForest | −0.41 | 8.3 | 15.78 | −19.51 | −0.04 | −0.37 |
| LGDP | −0.28 | 8.37 | 13.42 | −18.97 | −0.05 | −0.92 |
| LTechnology | −0.32 | 7.49 | 25.31 | −17.62 | 0.65 | 3.09 |

$$U = \sqrt{\frac{\sum_{t=2}^{n} (y_t - f_t)^2}{\sum_{t=2}^{n} (y_t - y_{t-1})^2}} \tag{14}$$

A value of U less than one means that the entertained model produces better forecasts, on average, than the no-change model. In Bayesian analysis of Dynamic Linear Models, all the parameters, and the functions for smoothing and forecasting are needed to perform a Bayesian analysis. The Markov chain Monte Carlo (MCMC) method is used to generate a sample from the posterior distribution [?].

# 3 Results

## *3.1 Descriptive Statistics*

Table 2 shows GHG emissions, renewable energy, urbanization, industrialization, tourism, agriculture, forest, economic growth, and technological innovations. At average value, there are four variables that have positive signs while those of the other four—agriculture, forest, GDP, and technology—have negative signs. The most extreme difference between maximum and minimum values can be observed in urbanization. Most variables have negatively skewed distribution in data sets. The excess kurtosis statistic, equal to 3.09, and 9.11 for technology and urbanization, respectively, indicates the Leptokurtic characteristics of the distribution.

**Table 3** Standardized coefficients estimation

| Variable | PIP | Post mean | Post SD | Cond.Pos. sign |
|---|---|---|---|---|
| LForest | 0.97 | 0.52 | 0.19 | 1 |
| LGDP | 0.52 | 0.17 | 0.21 | 1 |
| LIndustrial | 0.38 | −0.1 | 0.18 | 0.01 |
| LAgriculture | 0.28 | 0.04 | 0.12 | 1 |
| LTech | 0.25 | −0.03 | 0.1 | 0 |
| LUrban | 0.25 | −0.03 | 0.1 | 0.02 |
| LRenew | 0.23 | −0.03 | 0.1 | 0.04 |
| LTourism | 0.22 | 0 | 0.09 | 0.53 |
| (Intercept) | 1 | 0.04 | – | – |

## 3.2 *Bayesian Model Averaging*

The Bayesian Model Averaging method helps statistical analysts solve the problem of the small observations in the model estimation [34]. This approach determines the independent variables that have the potential to describe the dependent variable and select appropriate model. The results showing the determinants of GHG emission are shown in Table 3.

Table 3 shows the effect of each explanatory variable on GHG. Start with $PIP$ shows independent variables' posterior probabilities; for example, LForest that can explain the dependent variable, nearly 97%. The next column shows all models' averaged coefficients with the relatively larger value corresponding to the more significant explanatory variable. In this study, LForest exhibits a large coefficient value of 0.52. In terms of the column Cond.Pos.Sign, the posterior probability coefficient is strongly positive when the coefficient is close to one, (LForest, LGDP, and LAgriculture). On the other hand, if the coefficient Cond.Pos.Sign close to zero which is inferred a negative sign.

In this paper, a total of 8 independent variables are considered. From the Bayesian Model Averaging analysis, it was found that three independent variables are appropriate to describe the dependent variables for having $PIP$ higher than 0.35 according to Table 3. Therefore, based on the 8 datasets of independent variables, there are

**Table 4** Estimation of forecasting performance

| Variable | MAD | MAPE | U |
|---|---|---|---|
| Intercept | 0.0974 | −0.3824 | 1.1587 |
| LForest | 0.3376 | 0.0398 | 1.2669 |
| LGDP | 0.0275 | 0.0649 | 0.8932 |
| LIndustrial | 0.0403 | −0.2368 | 1.5325 |

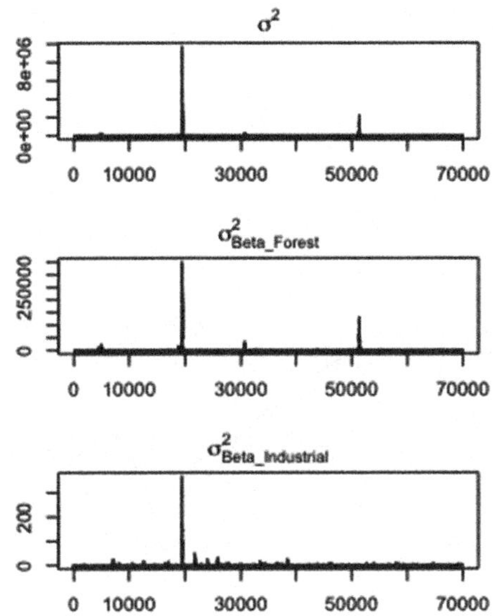


**Fig. 1** Trace plots of MCMC output

LForest, LGDP and LIndustrial are sufficient to estimate the best model of GHG emission.

## 3.3 *Bayesian Estimation*

In this study, the DLM using the Bayesian method was estimated to establish the relationship of GHG with the independent variables determined as suffice by the estimated results from the statistical test by the BMA method in the previous subsection. The estimated results from the Bayesians analysis of DLM—also known as the state space approach are presented below.

Figure 1 shows the variance of the equation, intercept term, forest, GDP, and Industrial from 1991 to 2020. The five series are shown in Fig. 1.

## 3.4 *Parameters' Estimation*

In Fig. 2, the line graphs present time-varying coefficients estimated by Bayesian DLM of the intercept term, forest, GDP, and industrial. The graph represents the time interval and the value of the time-varying coefficient on the x-axis and the y-axis, respectively.

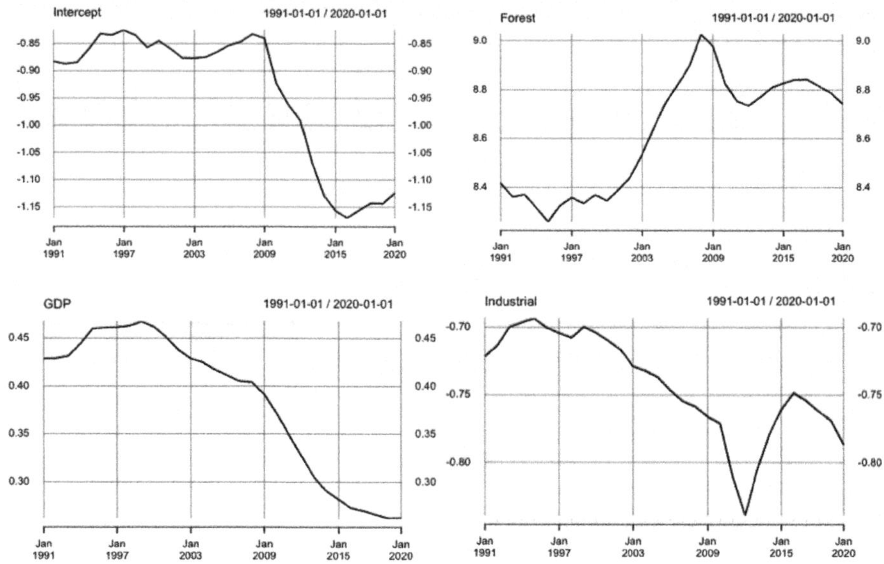


**Fig. 2** Estimated time-varying coefficients by Bayesian DLM

The first graph shows the downward trend of the intercept term with negative time-varying coefficients. There was a fluctuation around -0.85 in intercept term during 1991. In 1994, Thailand was ratified to become a party to the Convention on Climate Change and was effective for Thailand, as a result the coefficients decreased in 1995. Moreover, the intercept term remained constant at approximately −0.90 until 2009. Since 2007, Thailand has incorporated climate change into its national economic and social development plans (Ministry of Energy, [1]). The rate of the intercept term then plummeted to −1.15 in 2015. The second graph reveals positive time-varying coefficients of forest rocketed to 9.0 in 2009. After that, the time-varying coefficients dropped and stayed at approximately 8.8%. The third graph went constantly down from 0.45 to 0.30. The last figure, time-varying coefficients of industrial was a downward trend to −0.75 in 2009 before reaching a trough at −0.80 in 2012. These time-varying coefficients increased rapidly and were almost steady at about −0.75 from the year 2016 to 2020.

To forecast time-varying coefficient for 10 years of independent variables, this study used the state space method. There are the criteria commonly used for model selection are the mean absolute deviation (MAD), the mean square error (MSE), the mean absolute percentage error (MAPE), and Theil's U (U) as shown in Table 4.

Table 4 presents the statistics for the model selection about forecasting performance that predicts the next observation. In the model selection was performed upon the expected value of the mean absolute deviation (MAD), the mean absolute percentage error (MAPE) and the Theil's U (U) of each variable that the best performance due to the minimization of the values [27].

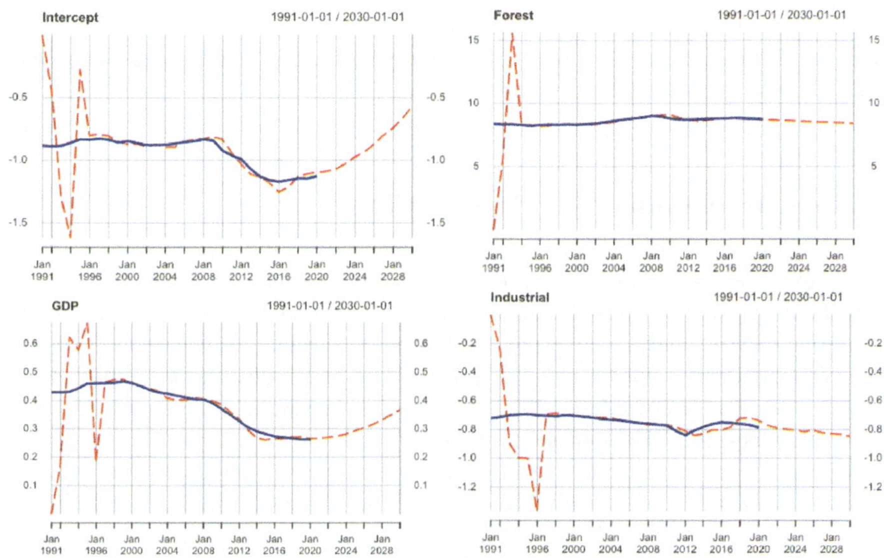


**Fig. 3** forecasting time-varying coefficients. Notes: Blue solid line shows time-varying coefficient from the DLM model. Red dashed line shows the 10 years from the forecasting time-varying coefficient

Figure 3, considering time-varying coefficients of intercept term, forest, GDP, and industrial. The time series shows a fluctuation of linear in each period. In first graph, time-varying coefficients of the intercept term move around −0.50 to −1.0. Then, the second graph illustrates time-varying coefficients of forest moved approximately at 8. The third graph shows time-varying coefficients of GDP varied steadily between 0.2 and 0.4 which means the graph moves close to zero and has fewer explanations for GHG emissions. In contrast, coefficients of industrial moved steadily at −0.8. It is the consequence that coefficients from industrial varied close to −1 and have more explanation for GHG emissions. Moreover, the high negative value of the time-varying coefficient is decreasing which means the industrial variable explains the GHG emissions more than ever. There are some significant events in Thailand such as the impact of the recession in 2009 on industrial production and overall energy use ([3]), and gas reduction plans according to the Paris Agreement to the UNFCCC ([1]). In 2009, greenhouse gas emissions declined, because of the impact of the recession on industrial production and overall energy use. So, the GHG emissions have dropped even more sharply.

The red dashed line in Fig. 3, the linear time-varying coefficients are forecasted for the future (10 years). The time-varying coefficients can tell the sign of changes in the predictor variable. The time-varying coefficients from intercept term, forest, GDP, and industrial were −1, 8, 0.3, and −0.8 respectively. It has the potential to be more important to GHG emissions. Forest has strongly positive coefficients; it can significantly affect GHG emissions. Furthermore, the coefficient of industrial values that move close to zero indicates that in the future, industrial will potentially explain

GHG emissions better. On the contrary, the coefficient from GDP is 0.3 forecasted to affect GHG in a better direction. Surprisingly, this tool predicts the result of GHG and GDP will grow in an upward direction.

## 4 Conclusions and Policy Implications

This paper uses a dynamic linear model (DLM) to find the relationship of between greenhouse gas emissions and its determinants in Thailand. From the Bayesian model averaging (BMA) analysis covering 8 independent variables, it was found that only three independent variables namely forest, GDP, and industrial are sufficient for further investigation considering their $PIP$ having the values higher than 0.35 which can explain best the GHG emission.

To explain the correlation, the DLM found the time-varying coefficient between the dependent and independent variables. The time-varying coefficients of forest and GDP strongly explains and changes in the same direction with GHG emission. In contrast, intercept term and industrial variable significantly explains and change in the opposite direction with GHG emission. There are some significant events that affect to time-varying coefficients such as climate policy and the economic crisis in 2009.

Moreover, The DLM used to be a forecasting tool for next 10 years. The time-varying coefficients from intercept, GDP, and industrial can have a potential strong significant affect to GHG emission. Especially forest will potentially explain GHG emissions in a better direction. On the contrary, the coefficients from the intercept term are in opposite directions. Results presented in this paper have implications for strategic policy, particularly for future National Plans. For the suggestions, the related agencies or government should consider encouraging economic activities.

## References

1. Alternative Energy Development Plan: AEDP. Department of Renewable Energy Development and Energy Efficiency, Thailand (2015)
2. Ates, S.A.: Energy efficiency and $CO_2$ mitigation potential of the Turkish iron and steel industry using the LEAP (long-range energy alternatives planning) system. Energy **90**, 417–428 (2015)
3. Broder, J.M.: Emissions Fell in 2009, Showing Impact of Recession. New York Times (2011). Accessed from 16 Feb 2011. https://www.nytimes.com/2011/02/17/science/earth/17emit.html
4. Carter, C.K., Kohn, R.: On Gibbs sampling for state space models. Biometrika **81**(3), 541–553 (1994)
5. Cramer, W., Yohe, G.W., Auffhammer, M., HuggelC., Molau U., Dias M.d.S., Leemans R.: Detection and attribution of observed impacts, Climate Climate Change 2014: Impacts, Adaptation, and Vulnerability. Cambridge University Press, United Kingdom (2014)
6. Department of Alternative Energy Development and Efficiency (DEDE): Energy Balance of Thailand 2017. DEDE, Bangkok (2018)
7. Durbin, J., Koopman, S.J.: Time Series Analysis by State Space Methods, vol. 38. OUP Oxford (2012)

8. Feng, L., Zhang, J.: Application of artificial neural networks in tendency forecasting of economic growth. Econ. Model. **40**, 76–80 (2014)
9. Gasimli, H., Gamage, S., Rajapakshe, S.: Energy, Trade, Urbanization and Environmental Degradation Nexus in Sri Lanka: Bounds Testing Approach. Energies, 12(9), 1655. MDPI AG (2019)
10. Harvey, A.C.: Forecasting, structural time series models and the Kalman filter (1990)
11. Hoeting, J.A., Madigan, D., Raftery, A.E., Volinsky, C.T.: Bayesian model averaging: a tutorial (with comments by M. Clyde, David Draper and EI George, and a rejoinder by the authors. Stat. Sci. **14**(4), 382–417 (1999)
12. Hong, J.H., Kim, J., Son, W., Shin, H., Kim, N., Lee, W.K., Kim, J.: Long-term energy strategy scenarios for South Korea: transition to a sustainable energy system. Energy Policy **127**, 425–437 (2019)
13. Intergovernmental Panel on Climate Change (IPCC): Climate change 2013: the physical science basis: Working Group I contribution to the Fifth assessment report of the Intergovernmental Panel on Climate Change. Cambridge University Press, United Kingdom (2014)
14. IPCC, in: Contribution of Working Groups I, II and III to the Fifth Assessment Re- port of the Intergovernmental Panel on Climate Change, IPCC, Geneva, Switzerland, p. 151 (2014)
15. Laine, M.: Introduction to Dynamic Linear Models for Time Series Analysis. Geodetic Time Series Analysis in Earth Sciences, pp. 139–156 (2020)
16. Lee, C.C., Yuan, Z., Lee, C.C., Chang, Y.F.: The impact of renewable energy technology innovation on energy poverty: Does climate risk matter? Energy Econ. **116**, 106427 (2022)
17. Office of Natural Resources and Environmental Policy and Planning (ONEP): Thailand's thirdnational communication. ONEP, Bangkok (2018)
18. Raihan, A., Muhtasim, D., Farhana, S., Hasan, A., Faruk, O., Paul, A.: Toward environmental sustainability: Nexus between tourism, economic growth, energy use and carbon emissions in Singapore. Global Sustainability Research (2022)
19. Raihan, A., Muhtasim, D.A., Farhana, S., Hasan, M.A.U., Pavel, M.I., Faruk, O., Rahman, M., Mahmood, A.: Nexus between economic growth, energy use, urbanization, agricultural productivity, and carbon dioxide emissions: New insights from Bangladesh. Energy Nexus. 8 (2018)
20. Raihan, A., Begum, R., Nizam, M., Mastura, S., Abdullah, S., Sharifah.: A review of emission reduction potential and cost savings through forest carbon sequestration. Asian J. Water, Environ. Pollut. **16**, 1–7 (2019)
21. Raihan, Asif, Tuspekova, Almagul: Nexus between energy use, industrialization, forest area, and carbon dioxide emissions: New insights from Russia. J. Environ. Sci. Econ. **1**, 1–11 (2022)
22. Pagnarith, K., Limmeechokchai, B.: Renewable energy utilization and $CO_2$ mitigation in the power sector: A case study in selected GMS countries. Songklanakarin J. Sci. Technol. **33**(3) 2011
23. Pan, L.J., Xie, Y.B., Li, W.: An analysis of emission reduction of chief air pollutants and greenhouse gases in Beijing based on the LEAP model. Proc. Environ. Sci. **18**, 347–352 (2013)
24. Park, S., Lee, S., Jeong, S.J., Song, H.-J., Park, J.-W.: Assessment of CO2 emissions and its reduction potential in the Korean petroleum refining industry using energy-environment models. Energy **35**(6), 2419–2429 (2010)
25. Pastpipatkul, P., Piboonrungroj, P., Subsai, P.: An Analysis of the Effects of Tourism Demand, Yield Curve, and Stock Returns on Economic Growth of Thailand: A Comparison Between the Bayesian DCC-GARCH and Bayesian Change-Point Methods. In: Sriboonchitta, S., Kreinovich, V., Yamaka, W. (eds.) Credible Asset Allocation, Optimal Transport Methods, and Related Topics. TES 2022. Studies in Systems, Decision and Control, vol. 429. Springer, Cham (2022)
26. Petris, G., Petrone, S., Campagnoli, P.: Dynamic Linear Models with R. Springer Science & Business Media (2009)
27. Petris, G.: An R package for dynamic linear models. J. Stat. Softw. **36**(12), 1–16 (2010)
28. Shrestha R.M., Malla S., Liyanage M.H.: Scenario-based analyses of energy system development and its environmental implications in Thailand. Energy Policy **35**(6) (2007)

29. United Nations Framework Convention on Climate Change: Mid-century, Long-term Low Greenhouse Gas Emission Development Strategy, Thailand Submitted under the Paris Agreement. Climate Change Management and Coordination Division Office of Natural Resources and Environmental Policy and Planning, Bangkok , Thailand (2021)
30. UNWTO: Tourism highlights 2019 edition. United Nations World Tourism Organisation (2019). Retrieve from https://www.e-unwto.org/doi/pdf/10.18111/9789284421152
31. Voumik, L.C., Islam, M.J., Raihan, A.: Electricity production sources and $CO_2$ emission in OECD countries: static and dynamic panel analysis. Global Sustain. Res. **1**(2), 12–21 (2022)
32. Winyuchakrit, P., Limmeechokchai, B., Matsuoka, Y., Gomi, K., Kainuma, M., Fujino, J., Suda, M.: Thailand's low-carbon scenario 2030: analyses of demand side $CO_2$ mitigation options. Energy Sustain. Dev. **15**(4), 460–466 (2011)
33. West, M., Harrison, J.: Bayesian Forecasting and Dynamic Models, 2nd edn. first edition: 1989 edn, Springer, New York (1997)
34. Zeugner, S., Feldkircher, M.: Bayesian model averaging employing fixed and flexible priors: The BMS package for R. J. Stat. Softw. **68**, 1–37 (2015)
35. Zheng, W., Walsh, P.P.: Economic growth, urbanization and energy consumption - A provincial level analysis of China. Energy Econ. Elsevier **80**(C), 153–162 (2019)

# Relationships Among Financial and Commodity Markets on Economic Growth: New Evidence from Bayesian Estimation of Seemingly Unrelated Time Series Equations

**Pathairat Pastpipatkul, Songsak Sriboonchitta, and Panicha Subsai**

**Abstract** This paper uses Bayesian estimation of the Seemingly Unrelated Time Series Equations (BSUTSE) model to present the relationships of explanatory variables on economic growth. From the Bayesian model average (BMA) analysis, the total considered independent variables is eleven. It was found that there are three variables that can explain the best model of yield curves and real GDP that consists of gold price, and stock from industrial, resource, financials, property and construction, and technology. The BSUTSE presents time-varying coefficients estimation on each economic growth variable. For the yield curve, most variables can explain economic growth more than ever. For real GDP, the return from property construction, industry, and resources significantly explains economic growth and changes in the opposite direction with real GDP. Gold price and the financial industry can explain changes in the same direction with real GDP. Only coefficients from technology parameters have more explain real GDP in the same direction. The BSUTSE is the model that allows coefficients to vary in time in an equation system. The advantage of The BSUTSE model, it can estimate time-varying coefficients.

**Keywords** Economic growth · Crude oil' price · Gold price · Inflation · Stock return · Bayesian dynamic linear model · Bayesian · Bayesian model average · Time varying coefficients

## 1 Introduction

The effects on economic growth are reflected in the coincident economic index (CEI) and leading economic index (LEI), which provide valuable growth forecasting. Growth in Gross Domestic Product (GDP) is commonly used. Many researchers have documented growth in real economic activity by focusing on the yield curve [13].

---

P. Pastpipatkul
Center of Excellence in Econometrics, Chiang Mai 52000, Thailand

S. Sriboonchitta · P. Subsai (✉)
Faculty of Economics, Chiang Mai University, Chiang Mai 52000, Thailand
e-mail: panicha.s.y@gmail.com



V. Kreinovich et al. (eds.), *Applications of Optimal Transport to Economics and Related Topics*, Studies in Systems, Decision and Control 556,
https://doi.org/10.1007/978-3-031-67770-0_41

Many studies have shown that the slope of the yield curve, or the difference in yields between long-term and short-term interest rates, is a good predictor of economic activity. According to [11], the yield spread between the ten-year treasury bond rate and the three-month treasury bill rate is a good predictor of future output growth. The yield curve is one of the LEIs that has valuable growth forecasting to implication capital indicator. GDP is one of the most important indicators of a country's overall performance ([12, 15, 27]), and one of the primary indicators used to assess the state of any country's economic system. Economic growth is influenced by a number of interconnected factors, including economic, political, and social developments, and there is a complex relationship between these factors. This paper broadens the definition of economic growth to include real GDP and the yield curve.

Volatility in one market may have an impact on the price index of another as well as financial and commodity markets which are some of the most difficult issues for investors. The volatility of oil and gold affects stock prices, has implications in the US capital market, and can be seen indirectly in inflation and unemployment [10]. Consumption, industrial production, and investment in both markets are all directly affected. The purpose of this paper is to investigate the relationship between crude oil price, gold price, inflation, and stock return of eight industry groups in Thailand's stock market and economic growth.

As part of the commodity sector, oil has the largest market share and accounts for roughly 40% of global energy consumption [25]. As a result, oil prices are an important factor in the production of goods and services in many sectors and have long been regarded as a leading indicator of stock price movements in the global economy. Furthermore, it has an impact on production costs as well as the prices of goods and services. According to [17]), the oil price serves as a barometer for economic prospects. Reference [2] mentions that changes in oil prices have an impact on macroeconomic events, investment costs, firm production structures and unemployment, consumer situation, monetary policies, interest rates, and inflation. Recent research has focused on the negative effects of oil prices on economic activity (e.g., [22, 30]). However, oil has a positive impact on economic activity ([21]) and [18] used a multivariate VAR model to investigate the relationship between oil price, real stock price, real economic activity, and interest rates in Greece. The results show that changes in the oil price significantly explain changes in the stock price return.

The stock market is influenced by macroeconomic factors such as the gold price, crude oil price, gold and oil price volatility, inflation rate, and exchange rate. The Thai financial market has grown significantly in the last decade by providing investors with the opportunity to earn high returns ([19]), indicating that the Thai financial market remains appealing to many investors. The Stock Exchange of Thailand (SET), which acts as a financial intermediary by channeling funds and returns, is the focus of this study. Precious metals, particularly gold, are traded in financial markets as well. Given the peaks and troughs in commodity prices, both individual and institutional investors are concerned about the relationship between commodities and stock prices ([1]). The gold price is influenced by macroeconomic factors such as the price of crude oil, the inflation rate, GDP, stock prices, the consumer price index, and the money supply (see [4, 28]). Reference [8] investigated the relationships between GDP, inflation, and gold price volatility and concluded that GDP and inflation have a significant impact on gold price volatility. Reference [3] used the VAR-GARCH model to investigate the effect of gold price volatility on Chinese stock market returns and their findings demonstrated a significant impact of gold price volatility on Chinese stock market returns.

This study used Bayesian Model Averaging (BMA) to determine which explanatory variables should be included in the model and how important they are. BMA estimates models for all possible explanatory variable combinations and computes a weighted average over all of them ([32]). The model weights for this averaging are derived from Bayes' theorem posterior model probabilities. This study also used Bayesian estimation of the Seemingly Unrelated Time Series Equations model (BSUTSE), which can be thought of as a general regression model with time-varying coefficients ([16]). The use of a dynamic model in time series analysis is well documented in statistical literature, but different terminology and notation may be used. Furthermore, by focusing on Bayesian analysis, this study solves the consistency of the estimator. The Bayesian estimation of Seemingly Unrelated Time Series Equations (BSUTSE) model starts with a univariate formulation and then abruptly converts the scalar variables to vectors. Reference [13] refers to them as structural time series, [9] use the state space approach, and [20] use the acronym SUTSE model.

Recent studies have found correlations on economic growth, especially in the context of Thailand. It provides attention to the economic indicators, namely real GDP and the yield curve. This paper has first objective to find the potential explanatory variables from many explanatory variables (crude oil, gold price, inflation, and stock from eight industries) that have the potential to describe both dependent variables by using BMA. After, getting the best models, this paper applies the BSUTSE model to estimate time-varying coefficients from prior sampling. The BSUTSE model contributes to time series data analysis that usually has only one coefficient to explain variables in all periods. Hence, the second objective is focused on BSUTSE to estimate time-varying coefficients. The BSUTSE model estimates time-varying coefficients in contrast, the SUR model has non-varying coefficients to explain variables in all periods. The last part compares the forecasting consequences from BSUTSE and Seemingly unrelated regression (SUR) by using time-varying coefficients to forecast each dependent variable.

The rest of this paper is organized as follows. Section 2 provides data description in detail and presents the methodology. Results from model estimations and forecasts are shown and discussed in Sect. 3. In Sect. 4, conclusions are presented and lastly, references are shown in the last part.

# 2 Data and Methodology

## 2.1 Data Description

This paper examines Thailand's economic growth, crude oil price, gold price, stock return, and inflation using quarterly data from 2013 to 2022. In the event of a stock return, the Stock Exchange of Thailand (SET) has eight industry groups, namely agro and food industry, consumer products, financials, industrials, property and construction, resources, services, and technology. The yield curve is defined as the 10-year government bond yield minus the 3-month treasury yield, with 2013 as the base year. The yield curve is calculated using the following formula:

$$y_t^n = \frac{(y_i + y_{t=1}^e + \ldots\ldots + y_{t+n-1}^e)}{n} = L_{t'}^n \quad (1)$$

where $y_t^n$ refers to the yield on a bond at maturity $n$, at time $t$ period $t$ in country $i$, which represents the expected yield at time period $t + j$ based on information available at time t. At time $t$, $L_t^n$ represents the liquidity premium for an $n$-period bond. All variables are measured in percentages and the data is retrieved from CEIC, Thomson reuters Eikon, and investing.com, in that order.

In this study can be divided the methodologies into four main parts. First, this study selects the BMA method, which is a method that determines the independent variables that have the potential to describe the dependent variables. After, determining the appropriate models, the second part uses the BSUTSE model to estimate time-varying coefficients from prior sampling. Next, this study uses time-varying coefficients for forecasting in each dependent variable and compares the forecasting performance between BSTUSE and SUR.

## 2.2 Bayesian Inference

Bayesian inference is a method that uses the Bayes theorem to improve the probabilities of a hypothesis as more data becomes available. Bayesian updates are especially useful in dynamic sequence analysis, where there are some initial guesses about the distribution of $\theta_t$. This is referred to as the prior distribution. The Bayes' Rule is used to obtain the updated distribution of $\theta_t$ (based on the observed data) after observing some data. The Bayes' formula is commonly used to calculate the posterior distribution. The posterior distribution of parameters can be computed, and the Monte Carlo method is commonly used to draw the sample from the posterior distribution of interest ([29]). The most commonly used technique is to implement a Gibbs sampler, and the Gibbs simulation model frequently simplifies implementation and accelerates the mixing of the resulting Markov chain.

## 2.3 Bayesian Model Averaging

In the first part, this study selects the BMA method, which is a method that determines the independent variables that have the potential to describe the dependent variables. According to a large set of variables based on the theory as well as on previous empirical analysis, the Bayesian Model Averaging (BMA) method applies to select the independent variables that have the potential to describe the dependent variables.

In general, data analysts select a model from a certain category of models and continue with the chosen model as if it had generated the data ([14]). For the reason of data limitation, this study used Bayesian estimation and specifically the BMA approach that has become a popular alternative to model selection ([32]) to handle the model uncertainty problem in regression.

The general form of a linear model structure can be expressed as follows:

$$y = \alpha_{\gamma} + \beta_{\gamma} X + \varepsilon,\ \varepsilon \sim N(0, \sigma^2 I), \tag{2}$$

where y is the dependent variable, $\alpha_{\gamma}$ is a constant, $\beta_{\gamma}$is the coefficients, and $\varepsilon$ is an error term with variance $\sigma^2$. BMA estimates models for all possible combinations of $X$variables and builds a weighted average for all.

If $X$contains $K$ possible variables, it means that the estimations have $2^K$variable combinations and $2^K$models. The model weights for this averaging come from posterior model probabilities that emerge from Bayes' theorem:

$$p(M_{\gamma}|y, X) = \frac{p(y|M_{\gamma}, X)p(M_{\gamma})}{p(y|X)} = \frac{p(y|m_{\gamma}, X)p(M_{\gamma})}{\sum_{s=1}^{2^K} p(y|M_s, X)p(M_s)} \tag{3}$$

Therefore, $p(y|X)$ denotes the integrated likelihood, $2^K$ is the number of estimated models from $K$ potential variables, $p(M_{\gamma}|y, X)$or the posterior model probability (PMP) is the marginal likelihood of the mode$p(y|M_{\gamma}, X)$ times a prior model probability$p(M_{\gamma})$. Then, the condition leads to PMP and Eq. (3) shows the model weighted posterior distribution for any statistic $\theta$.

$$p(\theta|y, X) = \sum_{\gamma=1}^{2^K} p(\theta|M_{\gamma}, y, X)\frac{p(M_{\gamma}|X, y)p(M_{\gamma})}{\sum_{s=1}^{2^K} p(M_s|y, X)p(M_s)} \tag{4}$$

Bayesian regression model has a specific prior structure to use that is called "Zellner's g prior" ([32]). The prior coefficients are in the normal distribution with the specified mean and variance. Here, the prior mean is zero, and its variance ($\sigma^2$) is formulated in Eq. (4):

$$Zel\ln er's\ g \sim \sigma^2(\frac{1}{g}X'X)^{-1} \tag{5}$$

The important prior is the one of regression coefficient;$\beta_\gamma$.

$$\beta_\gamma | g \sim N\left(0, \sigma^2\left(\frac{1}{g}X'X\right)^{-1}\right) \tag{6}$$

where $g$ is the hyperparameter that reflects how certain the coefficients are zero. A small g means few prior coefficient variances and implies quite certain. In contrast, a large g means very uncertain. The posterior distribution of coefficients reflects prior uncertainty.

## 2.4 Dynamic Linear Models

In the second part uses the BSUTSE model to estimate time-varying coefficients from prior sampling. Dynamic linear model (DLM) is the usual linear regression model, where the coefficients are allowed to change over time. The linear regression model is the popular tool for relating the dependent variable $y$ to explanatory variables $x$.

Consider a univariate model

$$y_t = F_t^{'}\theta_t + v_t, v_t \sim N(0, V_t) \tag{7}$$

$$\theta_t = G_t\theta_{t-1} + w_t, w_t \sim N(0, W_t) \tag{8}$$

$$t = 1, 2, ..., T$$

Here, $y_t$is the observations through time, conditionally independent given $\theta_t$ and $V_t$, $\theta_t$ is the vector of parameters at time t. The system matrices are $F_t^{'}$ and $G_t$, the covariance matrices are $V_t$and $W_t$, and the errors $v_t$ and $w_t$ are mutually independent. In a time-varying $F_t, V_t, G_t,$ *or* $W_t$ changes with time. A dynamic linear regression can be modeled as

$$y_t = \alpha_t + \chi_t\beta_t + v_t, \ v_t \sim N(0, V_t), \tag{9}$$

$$\alpha_t = \alpha_{t-1} + w_{\alpha,t'} \ \ w_{\alpha,t} \sim N(0, w_{\alpha,t}), \tag{10}$$

$$\beta_t = \beta_{t-1} + w_{\beta,t}, \ \ w_{\beta,t} \sim N(0, w_{\beta,t}). \tag{11}$$

For the estimation of the parameters, the Kalman filter method was used ([20]). In this study, we consider 50,000 rounds of sampler with 20,000 being the sample of burn-in.

## 2.5 Comparing and Forecasting

After, determining the best dependent variable from BMA method, the rest of this study estimates time-varying coefficients with Bayesian estimation of the Seemingly Unrelated Time Series Equations model (BSUTSE) for forecasting. The BSUTSE is the model that allows coefficients to vary in time in an equation system. This study also compares the forecasting performance with SUR which is the basic structural model. The advantage of The BSUTSE model, it can estimate time-varying coefficients in contrast, the SUR model has non-varying coefficients to explain variables in all periods.

### 2.5.1 Seemingly Unrelated Regressions Model

A seemingly unrelated regression (SUR) model is a generalization of a linear regression model that consists of several regression equations ([31]). Each equation has its own dependent variable and may have a different set of external explanatory variables. Suppose the model consists of $j = l...m$ linear regression equations for $i = l...N$ individuals. The jth equation for individual $i$ is

$$y_{ij} = X_{ij}^{'}\beta_j + \varepsilon_{ij}. \tag{12}$$

The model for the $j$th equation with all observations stacked, can be written as

$$y_j = X_i^{'}\beta_j + \varepsilon_{ij}. \tag{13}$$

The $m$ equations are stacked into an SUR model:

$$\begin{bmatrix} y_1 \\ y_2 \\ \vdots \\ y_m \end{bmatrix} = \begin{bmatrix} X_1 & 0 \cdots & 0 \\ . & \vdots \ddots & \vdots \\ 0 & 0 \cdots & X_m \end{bmatrix} \begin{bmatrix} \beta_1 \\ \beta_2 \\ \vdots \\ \beta_m \end{bmatrix} + \begin{bmatrix} \varepsilon_1 \\ \varepsilon_2 \\ \vdots \\ \varepsilon_m \end{bmatrix} \tag{14}$$

The error terms are assumed to have zero mean and to be independent across individuals and homoscedastic. For a given individual, the errors are correlated across equations,

$$E(\varepsilon_{ij}\varepsilon_{ij}{'}|X) = \sigma_{jj}{'} and\ \sigma_{jj}^{'} \neq 0\ where\ j \neq j. \tag{15}$$

### 2.5.2 Seemingly Unrelated Time Series Equations Model

Seemingly Unrelated Time Series Equations (SUTSE) is a multivariate model used to describe univariate series and comes from the structural time series literature, that uses a state-space formulation. SUTSE is a class of models that specify the dependence structure among the state vectors $\theta_{1,t}$, ..., $\theta_{m,t}$ $\theta_{1,t}$, ..., $\theta$ as follows. The equation (7) corresponds to the qualitative assumption that all series follow the same type of dynamics, and that the components of the state vectors have similar interpretations across the different DLMs. Clearly, the individual DLMs can be combined to give a DLM for the multivariate observations. This study modified SUTSE from [20] to get time-varying coefficients.

$$y_{ti} = F\theta_t^{(i)} + v_{ti} \tag{16}$$

$$\theta_t^{(i)} = G\theta_{t-1}^{(i)} + w_t^{(i)} \tag{17}$$

Assuming in Eq. (16), $y_t$ denote the multivariate observation at time $t$, and suppose that the $i$th component of $y_t$ follows the DLM with $\theta_t^{(i)} = (\theta_t^{(i)} = \theta_{t1}^{(i)}, ...\theta_{tp}^{(i)})'$ for $i = l, ..., m$. Then, a model of SUTSE for $y_t$ is

$$y_t = (F \otimes I_m)\theta_t + v_t, \quad v_t \sim N(0, V) \tag{18}$$

$$\theta_t = (G \otimes I_m)\theta_{t-1} + w_t, w_t \sim N(0, W) \tag{19}$$

When the have diagonal variances, it can assume for $W$ a block-diagonal structure with $p$ blocks of $m$size. The forecasting made at time $t$ of $\theta_{t+k}^{(i)}$ $or$ $y_{t+k,i}$ are based only on the distribution $\theta_t^{(i)}$. This paper uses the probability law of the process specified by assigning transition densities which are the continuous states of the probabilities to describes a curve under area.

### 2.5.3 The Forecasting Performance

The functions for forecasting are needed to perform a Bayesian analysis. A common approach is to use the Markov chain Monte Carlo (MCMC) method to generate a sample from the posterior distribution ([20]). One way of obtaining a sample from the joint posterior distribution of parameters and unobservable states is to run a Gibbs sampler. There are several criteria to assess forecasting accuracy including the mean absolute deviation (MAD), the mean square error (MSE), and the mean absolute percentage error (MAPE), defined respectively by the following formulae (20)–(22).

$$MAD = \frac{1}{n}\sum\nolimits_{t=1}^{n} |e_t| \tag{20}$$

$$MSE = \frac{1}{n}\sum\nolimits_{t=1}^{n} |e_t^2| \tag{21}$$

$$MAPE = \frac{1}{n}\sum\nolimits_{t=1}^{n} \frac{|e_t|}{y_t} \tag{22}$$

$$U = \sqrt{\frac{\sum_{t=2}^{n} (y_t - f_t)^2}{\sum_{t=2}^{n} (y_t - y_{t-1})^2}} \tag{23}$$

Reference [20] refer to an additional statistic to assess the forecasting performance which is Theil's U ([26]), shown in Eq. (23). The MSE of the model with the MSE of the trivial "no-change" model is compared with Theil's U that predicts the next observation to be the same as the current one. A value of U less than one means that the entertained model produces better forecasts, on average, than the no-change model.

## 3 Results

This section shows results from the study in the Bayesian Model Average method for determining the independent variables. After that, in next process is the Bayesian analysis of SUTSE (BSUTSE) to interpret the effect of potential variables on the yield curve and real GDP. Then, the comparison between the forecasting performance of BSUTSE and SUR for each dependent variable is shown. Lastly, this paper also compares the graph obtained from BSUTSE and SUR to capture the real situation.

### *3.1 Descriptive Summary*

In Table 1, the description shows that, as time passed from 2013 to 2022, the growth in real GDP, yield curve, crude oil price, gold price, US inflation, and return of eight industry groups in the stock market. On average, there are four industry groups have positive signs, while the other four (consumer finance, industry, and property and construction) have negative signs. The highest and lowest values were observed from crude oil prices. Some variables showed positive skewness. The excess kurtosis statistic, which is equal to 9.11, 3.09, and 12.14 for the crude oil price, resource, and property and construction respectively indicate, a high-fat tail and more Standard Deviation than the normal distribution, the Leptokurtic characteristics of the distribution.

### *3.2 Bayesian Model Average*

This study selects the Bayesian Model Average method, which is a method that determine the independent variables that have the potential to describe the dependent variables and it's used to solve the problem of the small observations in the model estimation for getting the appropriate models ([32]). The result of the variables'

**Table 1** Statistics of crude oil, gold price, inflation, economic growth, and Thai stock market

| Variable | Mean | Sd | Maximum | Minimum | Skewness | kurtosis |
|---|---|---|---|---|---|---|
| Real GDP | 0.61 | 6.13 | 9.87 | −15.84 | −0.39 | −0.35 |
| Yield | 1.17 | 0.5 | 0.5 | 0.07 | 0.06 | −0.22 |
| Gold price | 0.43 | 5.85 | 12.45 | −13.3 | −0.01 | −0.37 |
| Crude oil | 1.43 | 31.05 | 137.31 | −54.39 | 2.38 | 9.11 |
| US inflation | 3.54 | 1.77 | 6.37 | −0.17 | −0.55 | −0.48 |
| Stock market | | | | | | |
| Agriculture | 0.07 | 6.52 | 12.3 | −14.05 | −0.29 | −0.63 |
| Consumer | −0.69 | 6.94 | 15.74 | −12.71 | 0.51 | −0.26 |
| Finance | −0.41 | 8.3 | 15.78 | −19.51 | −0.04 | −0.37 |
| Industry | −0.28 | 8.37 | 13.42 | −18.97 | −0.05 | −0.92 |
| Construction | −0.32 | 7.49 | 25.31 | −17.62 | 0.65 | 3.09 |
| Resource | 2.1 | 13.18 | 63.59 | −16.56 | 3.01 | 12.14 |
| Service | 1.48 | 7.15 | 20.8 | −17.97 | −0.18 | 1.5 |
| Technology | 0.26 | 7.69 | 21.58 | −14.17 | 0.51 | 0.14 |

*Source* Own calculation

selection by using the Bayesian Model Average method. The results corresponded to yield curve and real GDP are shown in Table 2 .

Table 2 shows the corresponding statistics about the yield curve and real GDP alternately. Starting with real GDP, the data is given inside the first column $PIP_{rgdp}$ which represents posterior probabilities, e.g., posterior probabilities of financial stock return can interpret dependent variable nearly 99%.

The next column, $Post\ Mean_{rgdp}$ displays the averaged coefficients of all models. There is the financial stock return that shows a relatively large coefficient appears to be the most important. In addition, the coefficients' posterior standard deviations ($Post\ SD_{rgdp}$) reflect further evidence: all variable is certainly positive. The column $Cond.Pos.Sign_{rgdp}$ is inferred model's direction by considering the value of coefficients. If the value is close to one, the model will infer a positive direction, for example, return from consumer products and finance. On the other hand, there are some variables that have coefficients close to zero which is inferred a negative sign.

In the case of the yield curve, the data start to be presented in the second column $PIP_{yield}$, approximately 93% of posterior model response on models that include gold price. The next column, $PostMean_{yield}$, property and construction are inferred to be the most important to the model. In contrast, the gold price is nearly zero that means insignificantly to this model. In addition, the coefficients' posterior standard deviations ($PostSD_{yield}$) reflect further evidence: all variable is certainly positive. The column $Coud.Pos.Sign_{yield}$ can be inferred model' direction, there are gold price, return from consumer products which have positive coefficient values. On the other hand, their coefficients of return from agriculture, technology, and service are zero, their signs are negative.

**Table 2** Standardized coefficients of all variables

| Variable | $PIP_{rgdp}$ | $PIP_{yield}$ | $PostMean_{rgdp}$ | $PostMean_{yield}$ | $PostSD_{rgdp}$ | $PostSD_{yield}$ | $CondPosSign_{rgdp}$ | $CondPosSign_{yield}$ |
|---|---|---|---|---|---|---|---|---|
| Gold price | 0.17 | 0.93 | 0 | −0.41 | 0.06 | 0.18 | 0.68 | 1 |
| Crude oil | 0.18 | 0.16 | −0.01 | 0 | 0.07 | 0.06 | 0.16 | 0.61 |
| US inflation | 0.17 | 0.48 | −0.01 | −0.12 | 0.06 | 0.16 | 0 | 0 |
| Agriculture | 0.25 | 0.17 | −0.05 | 0 | 0.14 | 0.09 | 0 | 0.02 |
| Consumer | 0.21 | 0.26 | 0.03 | −0.05 | 0.09 | 0.12 | 0.98 | 1 |
| Finance | 0.99 | 0.17 | 0.87 | 0 | 0.23 | 0.09 | 1 | 0.13 |
| Industry | 0.19 | 0.22 | −0.02 | 0.05 | 0.09 | 0.16 | 0.01 | 0.91 |
| Construction | 0.81 | 0.33 | −0.52 | 0.08 | 0.33 | 0.16 | 0 | 0 |
| Resource | 0.25 | 0.28 | −0.05 | −0.07 | 0.13 | 0.16 | 0 | 0 |
| Service | 0.24 | 0.17 | −0.05 | −0.01 | 0.16 | 0.1 | 0.07 | 0.07 |
| Technology | 0.59 | 0.17 | −0.19 | −0.01 | 0.2 | 0.07 | 0 | 0.05 |
| (Intercept) | 1 | 1 | 0.12 | 2.37 | – | – | – | – |

*Source* Own calculation
*Note* $PIP$, Post Mean, Post SD, Cond.Pos.Sign corresponds to real GDP variable, respectively

This paper totally considered eleven independent variables. From the Bayesian Model Average analysis, it was found that three independent variables had been appropriate to describe the dependent variables by considering from $PIP_{rgdp}$. In this dataset, there are variables from eight industry groups namely financials, property and construction, and technology explain the best model of real GDP. Also, it was found that three independent variables had been appropriate to describe the dependent variables by considering from $PIP_{yield}$. In this dataset, there are variables from gold price, US inflation, and recourse that can explain the best model of the yield curve.

### 3.3 Bayesian Estimation of Parameters

In Fig. 1, the line graphs represent coefficient estimation by Bayesian analysis of SUTSE on each economic growth variable that shows coefficients of intercept term, gold price, US inflation, and return from finance, property construction, resources, and technology from 2013 to 2022. Here, the x-axis represents the time interval, and the y-axis represents the value of the coefficient. The results are presented in two parts which are vertically divided into columns. The column shows variables that are related to the yield curve and real GDP respectively. Starting with the yield curve, the graph showed about positive and negative coefficients.

These independent variables (except technology) varied in negative which means these independent variables move in the opposite direction with the yield curve and have more explanation with the yield curve. The various negative factors such as political instability in Yingluck Shinawatra's government ([7]), Fed's raised interest rates, Greece's debt crisis ([5]), the US anti-trade policy, the Political uncertainty in the Euro (BOT, 2017) and COVID-19 pandemic.

The second economic growth variable, real GDP, the graph showed positive and negative coefficients. Return from property construction and technology was negative hence, these variables strongly significantly explain economic growth and changes in the opposite direction with real GDP.

Gold price and financial industry were positive, both variables can explain and change in the same direction with real GDP. Only coefficients from the US inflation parameter varied around 0 and had more explained real GDP in the same direction. The affecting factors such as global crisis ([5]) and the COVID-19 pandemic. In the next section compare the performance of the BSUTSE model and SUR for forecasting each dependent variable.

Table 3 presents the statistics for the model comparison of performance between Bayesian analysis of SUTSE and SUR. In the model, selection was performed upon the criteria in the order selection process commonly used namely the mean absolute deviation (MAD), the mean square error (MSE), the mean absolute percentage error (MAPE), and Theil's U (U).

This section compares the values of MAD, MSE, and U from the Bayesian analysis of SUTSE and SUR. Both estimations seem to have in common that they work with systems. In contrast, SUR shows a constant coefficient however, the BSUTSE

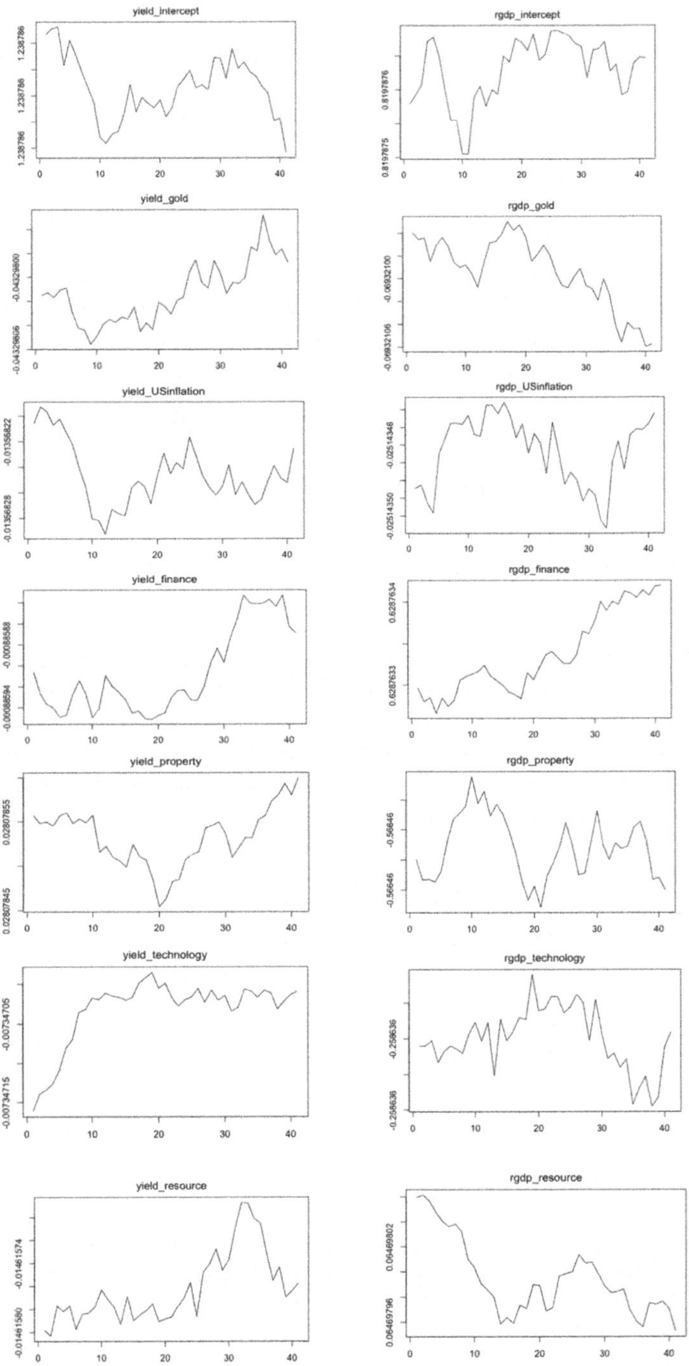


**Fig. 1** Estimation of the variable by Bayesian analysis of SUTSE. Note: The first vertical column shows variables that affect to yield curve. The second vertical column shows variables affecting real GDP

**Table 3** Estimation of forecasting performance

| Variable | SUTSE MAD | SUR MAD | SUTSE MAPE | SUR MAPE | SUTSE U | SUR U |
|---|---|---|---|---|---|---|
| Yield | 0.33105 | 0.33156 | 0.54743 | 0.54819 | 1.19416 | 1.19414 |
| Real GDP | 4.02645 | 4.02572 | −0.00444 | −0.00321 | 0.56984 | 0.56985 |

model presents time-varying coefficients. It is customary to minimize functions in the optimization by default it seeks a minimum negative likelihood of a specified DLM for a given data set ([20]). In this study, results from the BSUTSE model are better to interpret economic growth than SUR estimation.

## 3.4 *Forecasting in Dependent Variable*

This part compares the consequences from the BSUTSE model and SUR model. The difference between the two estimations is the obtained coefficient from both estimations. SUR shows constant coefficients and the BSUTSE model presents time-varying coefficients.

Figure 2 compares the forecasting results from the BSUTSE model (the blue lines), the actual values of economic growth (the black lines), and the value from the estimation of SUR (the red lines). Based on the consequence, the SUR model (the red lines) is as efficient as the BSUTSE model (the blue lines) in capturing real situations (the black lines). Also, the results from Table 3 indicated that results from the BSUTSE model are better to interpret economic growth. This study agrees with [20] that SUTSE can be applied to more general DLMs than the SUR or basic structural model and the SUTSE seems natural to allow the coefficients to vary in time. In brief, the BSUTSE model might be more effective in capturing real situations. However, in shock period has severe changes that strongly affect to coefficients.

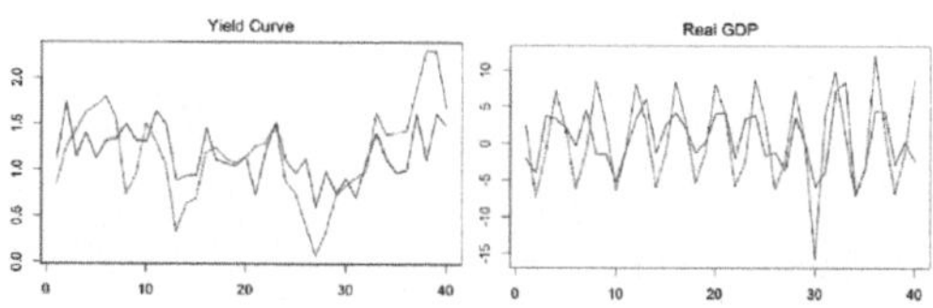


**Fig. 2** Forecasting in time-varying coefficient. Note: Black line shows the actual values of the yield curve and real GDP, respectively. Blue line shows the forecasting values of the yield curve and real GDP that come from the BSUTSE model. Red line shows the values of the yield curve and real GDP from SUR estimation

## 4 Conclusions and Policy Implications

This paper uses Bayesian analysis of the SUTSE model to present the relationship between financial and commodity markets on economic growth (real GDP and yield curve). The Bayesian model average (BMA) analysis considered eleven independent variables and it was found that there are three independent variables from gold price, and eight industry groups namely industrial, and resource can explain the best model of yield curve, by considering from $PIP_{yield}$. It was found that three independent variables from eight industry groups namely financials, property and construction, and technology explain the best model of real GDP by considering from $PIP_{rgdp}$.

The BSUTSE model presents coefficient estimation on each economic growth variable that shows coefficients of intercept term, gold price, and return from five industries including finance, property construction, technological, industrial, and resource industries. For the yield curve, there are positive and negative coefficients. These independent variables (except property construction) move close to zero and have more explanation with the yield curve. The negative value of the coefficient is decreasing which means variables can explain the economic growth more than ever. In the second economic growth variable, real GDP, return from three industries including property construction, industry, and resource was negative. These three variables strongly explain economic growth and changes in the opposite direction with real GDP. Gold price and financial industry were positive which means both variables can explain and change in the same direction with real GDP. Only coefficients from technology varied close to 0 and had more explained real GDP in the same direction. The affecting factors such as politics ([7]), the global crisis ([5]) and the COVID-19 pandemic.

In this article, it can be seen that the same dependent variable might affect the independent variables in different directions even though they measure at the same time. The BSUTSE model might be more effective in capturing real situations. Also, in the part comparing the performance of forecasting each dependent variable, this method is better for interpreting economic growth. This study agrees with [20] that the SUTSE can be applied to more general DLMs than the basic structural model and the SUTSE allows the coefficients to vary in time. However, in shock period has severe changes that strongly affect to coefficients. However, when the BSUTSE was used, in shock period severely changed that strong effect on coefficients. For recommendation, the BSUTSE model can be one of the interesting tools for studying in time-varying coefficient field.

## References


1. Akkoc, U., Civcir, I.: Dynamic linkages between strategic commodities and stock market in Turkey: evidence from SVAR-DCC-GARCH model. Res Policy **62** (2019)
2. Arouri, M.: Dose crude oil move stock markets in Europe? Econ. Model. **28**, 1716–1725 (2011)

3. Arouri, M.E., Lahiani, A., Nguyen, D.K.: World gold prices and stock returns in China: insights for hedging and diversification strategies. Econ. Model. **44**, 273–282 (2015)
4. Bapna, I., Sood, V., Totala, N.K., Saluja, H.S.: Dynamics of macroeconomic variables affecting price innovation in gold: a relationship analysis. Pac. Bus. Rev. Int. **5**, 1–9 (2012)
5. Bank of Thailand: Annual Report Year 2014. Bank of Thailand, Thailand (2015)
6. Bank of Thailand: Annual Report Year 2016. Bank of Thailand, Thailand (2017)
7. BBC News Services: Thailand crisis: Protesters launch Bangkok 'shutdown' (2014). Retrieve from https://www.bbc.com/news/world-asia-25708092
8. Cai, J., Cheung, Y.L., Wong, M.C.: What moves the gold market? J. Futures Mark. **21**, 257–278 (2001)
9. Durbin, T., Koopman, S.: Time Series Analysis by State Space Methods, 2nd edn. Oxford Statistical Science Series. Oxford University Press (2012)
10. Ebrahim, Z., Inderwildi, O.R., King, D.A.: Macroeconomic impacts of oil price volatility: mitigation and resilience. Front. Energy **8**, 9–24 (2014)
11. Estrella, A., Hardouvelis, G.A.: The term structure as a predictor of real economic activity. J. Finan. **46**(2), 555–576 (1991)
12. Feng, Lihua, Zhang, Jianzhen: Application of artificial neural networks in tendency forecasting of economic growth. Econ. Model. **40**, 76–80 (2014)
13. Harvey, A.C.: Forecasting. Cambridge University Press, Structural Time Series Models and the Kalman Filter (1991)
14. Hoeting, J.A., Madigan, D., Raftery, A.E., Volinsky, C.T.: Bayesian model averaging: a tutorial. Stat. Sci. **14**(4), 382–417 (1999)
15. Kordanuli, Bojana, Barjaktarović, Lidija, Jeremić, Ljiljana, Alizamir, Meysam: Appraisal of artificial neural network for forecasting of economic parameters, p. 465. Statistical Mechanics and its Applications, Physica A (2016)
16. Laine, M.: Introduction to Dynamic Linear Models for Time Series Analysis. Geodetic Time Series Analysis in Earth Sciences (2019)
17. Lang, K., Auer, B.R.: The economic and financial properties of crude oil: a review. N.Am. J. Econ. Finan. (2019)
18. Papapetrou, E.: Oil price shocks, stock market, economic activity and employment in Greece. Energy Econ. **23**, 511–532 (2001)
19. Pastpipatkul, P., Yamaka, W., Sriboonchitta, S.: Co-Movement and Dependency Between New York Stock Exchange, London Stock Exchange, Tokyo Stock Exchange, Oil Price, and Gold Price. IUKM (2015)
20. Petris, G., Petrone, S.,and Campagnoli, P. (2007). Dynamic linear models. In: Dynamic Linear Models with R. Use R. Springer, New York, NY
21. Raza, S.A., Shahbaz, M., Amir-ud-Din, R., Sbia, R., Shah, N.: Testing for wavelet based time-frequency relationship between oil prices and US economic activity. Energy, Elsevier **154**(C), 571–580 (2018)
22. Sadorsky, P.: Oil price shocks and stock market activity. Energy Econ. **21**(5), 449–69 (1999)
23. Su, C-W., Huang, S-W., Qin, M., Umar, M.: Does crude oil price stimulate economic policy uncertainty in BRICS? Pac.-Basin Finan. J. Elsevier **66**(C) (2021)
24. Su, C.W., Qin, M., Tao, R., Umar, M.: Financial implications of fourth industrial revolution: can bitcoin improve prospects of energy investment? Technol. Forecasting Soc. Change **158**, 120178 (2020)
25. Su, C.-W., Sun, T., Ahmad, S., Mirza, N.: Does institutional quality and remittances inflow crowd-in private investment to avoid Dutch Disease? A case for emerging seven (E7) economies. Res. Policy **72**, 102111 (2021)
26. Thiel, H.: Applied Economic Forecasting, p. 26. North Holland, Amsterdam (1966)
27. Tumer, Abdullah, Akkuş, Aytekin: Forecasting Gross Domestic Product per Capita Using Artificial Neural Networks with Non-Economical Parameters, p. 512. Statistical Mechanics and its Applications, Physica A (2018)
28. Wang, Y.: An Empirical study in the relationship between crude oil and gold futures. Res. Policy **35**, 168–177 (2013)

29. West, M., Harrison, J.: Bayesian Forecasting and Dynamic Models, 2nd edn. first edition: 1989 edn, Springer, New York (1997)
30. Eyden, V.R., Difeto, M., Gupta, R., Wohar, M.E.: Oil price volatility and economic growth: evidence from advanced economies using more than a century's data. Appl. Energy Elsevier **233**(612–621), 2 (2019)
31. Zellner, A.: An efficient method of estimating seemingly unrelated regressions and tests for aggregation bias. J. Am. Stat. Assoc. **57**(298), 348–368 (1962)
32. Zeugner, S., Feldkircher, M.: Bayesian model averaging employing fixed and flexible priors: the BMS package for R. J. Stat. Softw. **68**(4) (2015)

# A Quantile Regression for Computer Crimes Act in Thailand

**Suphanit Chansong, Rungrapee Phadkantha, and Puntoon Wiranya**

**Abstract** In this study of 411 university students in the Chiang Mai region, the focus was on assessing their awareness and understanding of computer related offenses under the Computer Crime Act of 2017. Utilizing the Quantile regression model, the research identified influential factors. Age was found to positively correlate with increased awareness and comprehension of computer related offenses, attributed to life experiences and extended exposure to digital technologies. Additionally, higher GPAs were associated with greater awareness, suggesting academic institutions' role in shaping digital ethics understanding. In addition, the greater the experience seen in the EXPERIAN variable, the greater the level of awareness and understanding of computer crimes. This study offers valuable insights into the nuanced dynamics of digital ethics awareness and understanding, with age, academic performance, and internet experience as pivotal factors.

**Keywords** Computer crimes · Quantile regression · Online crime · Cyber crime

## 1 Introduction

Cybercrime is a rapidly growing and significant issue; cybercrime is a new type of crime with the entry of computers are used as tools and another where they are the direct targets. Examples of cybercrimes in the former category include online

S. Chansong
Faculty of Management Sciences, Department of Communication Arts,
Chiang Mai Rajabhat University, Chiang Mai, Thailand
e-mail: suphanit_cha@cmru.ac.th

R. Phadkantha
Faculty of Economics, Center of Excellence in Econometrics, Chiang Mai University,
Chiang Mai, Thailand
e-mail: rungrapee.ph@gmail.com

P. Wiranya (✉)
Office of Research Administration, Chiang Mai University, Chiang Mai 50200, Thailand
e-mail: wiranya.pun@gmail.com

 
V. Kreinovich et al. (eds.), *Applications of Optimal Transport to Economics and Related Topics*, Studies in Systems, Decision and Control 556,
https://doi.org/10.1007/978-3-031-67770-0_42

harassment, threats via the internet or mobile phones, and copyright infringement. Therefore, there is a necessity and rationale for legislating to control online media, as well as actions that cause harm to data, computers, or computer systems related to the security and public safety of the country or its economic stability or public infrastructure.

In Thailand, there are the Computer Crime Act of 2007 and the Computer Crime Act (Amendment) of 2017, which deal with offenses related to computers found that the problem that occurs mostly is the problem of data dissemination that spreads widely on the Internet, which arises from the lack of measures in Thailand to control the dissemination of information on the Internet for commercial or profit seeking purposes, although criminal laws have established penalties. However, the law still does not cover the behaviors that occur in the present. Furthermore, the Computer Related Offenses Act, 2nd edition B.E. 2017, which is a specific law used to enforce crimes committed on the Internet or computer related offenses, does not have any provisions for disseminating computer data that spreads widely for commercial or profit seeking purposes, which should be considered an offense under the Computer Related Offenses Act. Therefore, the researchers suggest finding a suitable legal solution to the current situation by proposing additional amendments to the law that are appropriate and cover the current behaviors.

Therefore, awareness and understanding of the Computer Related Offenses Act, version 2 of 2017, is necessary for the general public. This is because the use of the internet can be just as dangerous to users as other technologies and media, but the danger from the internet can spread quickly and violently. This is because the internet is like a large library without central control and content organization. Therefore, while the internet provides high quality and fast information, there is also a chance that incorrect or inaccurate information can be provided. Because it is a system that is uncontrolled and unfiltered, the internet is a source that makes it easier for children and youth to access inappropriate media content. Content that is violent or contagious can have an impact on the attitudes and behaviors of users [2]. Besides attitude, several other factors seemed to significantly impact computer usage. These factors include the level of computer proficiency, the extent of computer access, and the quantity of computer related courses completed [1]. For the example case of committing an offense, the court verdict No. 6681/2562 regarding the conversation through Facebook Messenger application, which is a chat program through the internet system that is a closed system and has three members. External parties cannot access or read the conversation messages. The conversation only takes place among the people in the group and is not considered to be spoken or disseminated to a third party, according to the Commercial and Civil Code, Sect. 423 [10].

From the study on the perception and understanding of computer related offenses according to the Computer Related Offenses Act of 2017, by the general public in Nonthaburi and Bangkok [9], it was found that the majority of the sample group, which consisted mostly of students, used the internet an average of 2–3 h per session for online chatting, conversation, information searching, and that the majority of the sample group had a low level of awareness and understanding regarding computer related offenses under the Computer Related Offenses Act of 2017. Additionally, a study on computer and internet usage behaviors that are at risk of computer related offenses under the Computer Related Offenses Act of 2007 [7] among students of Songkhla Rajabhat University found that the majority of the sample group had only moderate knowledge about this law.

Chiang Mai is a large city that is important in determining its developmental position. Additionally, it serves as an educational center and a hub of the region, hosting numerous educational institutions at the preparatory and tertiary levels that contribute to developing education and human resources [8]. The researcher chose this sample because the group consists of tertiary learners, representing a new generation of youth immersed in the digital age. They extensively utilize social media for various advantages compared to previous generations [12]. Understanding this Act is crucial. Subsequently, the researcher opted for a stratified sample based on seven university groups, adjusting the proportions to match the student population. The selection of group members was then done randomly within each group.

Therefore, the objective of this study is to investigate the level of awareness and understanding of computer related offenses according to the Computer Crime Act of 2017, and to compare the level of awareness and understanding of computer related offenses according to the Computer Crime Act of 2017, as well as to study the relationship between perception and understanding, personal factors of university students, and the level of awareness regarding computer related offenses according to the Computer Crime Act of 2017, using the logit-lasso method.

Quantile regression is a statistical technique designed to assess the association between independent variables and specific quantiles of the dependent variable. Koenker and Bassett's seminal work in 1978 [4] introduced Quantile Regression (QR) as an effective alternative to traditional mean regression methods. In contrast to conventional regression, which primarily focuses on estimating the conditional mean, quantile regression empowers us to estimate conditional quantiles, thereby offering a more comprehensive insight into the relationship being analyzed. In quantile regression, distinct parameters are computed for various quantiles, allowing us to capture valuable information about the entire conditional distribution of the dependent variable. This approach proves particularly valuable when dealing with asymmetric distributions or when our primary interest lies in specific quantiles. It enables us to discern how the relationship between variables evolves across different segments of the distribution. The estimation of quantile regression models involves minimizing a loss function, which quantifies the disparities between observed and predicted quantiles. Diverse techniques, such as linear programming or optimization algorithms, can be employed for this estimation process. Consequently, QR has achieved widespread acclaim across various domains, as exemplified by its incorporation in studies conducted by [3, 11, 15–17].

Therefore, in this study, we will use the quantile regression model to explain the level of awareness and understanding of computer related offenses according to the Computer Crime Act of Thailand. The rest of the paper is organized as follows: in Sect. 2 we discuss methodology. In Sect. 3, we outline the data while we offer results of this study in Sect. 5.

## 2 Quantile Regression

To explain the basic concept of quantile regression, consider the following model:

$$y_t = x'_{i,t}\beta_i^{\tau} + \varepsilon_t \; ; \; i = 1, \ldots, k; \; j = 1, \ldots, n \tag{1}$$

where $x'_{i,t}$ is $n \times k$ independent variables, $\beta_i^{\tau}$ is $1 \times k$ vector of coefficients at given $\tau$. Note that unlike the traditional quantile regression, we consider the quantile $\tau$ as an unknown parameter and estimate it jointly with other parameters. $\varepsilon_t$ is the error which does not assume any distributions. Thus, $\tau$th$(0 < \tau < 1)$ conditional quantile of $y_t$ is given $x'_{i,t}$ is simply

$$Q_y(\tau|x) = x'_{i,t}\beta_i^{\tau} \tag{2}$$

In this study, we aim to relax the assumption in the conventional quantile regression by regarding $\tau$ as the unknow parameter. In the traditional estimation of quantile regression model, the focus is on using ordinary least squares (OLS) with a general technique for estimating families of conditional quantile functions [5]. The $\tau$ specific coefficient vector $\beta^{\tau}$ can be estimated by minimizing the loss function:

$$\beta^{\tau} = \underset{\beta(\tau)}{\arg\min} \sum_{j=1}^{n} \rho^{\tau}(y_t - x_{i,t}\beta_i^{\tau}), \tag{3}$$

where $\rho^{\tau}(L) = L(\tau - I(L < 0))$ is called the check function. $L$ is a loss function which is corresponding to $y_t - x_{i,t}\beta_i^{\tau}$ and $I(.)$ is the indicator function. In addition, the quantile regression model can be estimated by maximizing the likelihood based on the asymmetric Laplace density (ALD):

$$L(\beta^{\tau}, \sigma|y) = \frac{\tau(1-\tau)}{\sigma} \exp\left(-\sum_{t=1}^{T} \rho^{\tau} \frac{(y_t - x_{i,t}\beta_i^{\tau})}{\sigma}\right), \tag{4}$$

where $\sigma$ is a nuisance parameter. Note that, the maximization of the likelihood in Eq. (4) with respect to the parameter $\beta^{\tau}$ is equivalent to the minimization of the objective function in Eq. (3) [14].

## 3 Data

The study collected data from a survey questionnaire, which included responses from a sample size of 411 individuals. Here is an overview of the key demographic and academic characteristics of the respondents based on the data presented in Table 1:

**Table 1** Data statistics

| General data | Quantity | Percentage |
|---|---|---|
| *1. Gender* | | |
| Men | 167 | 40.63 |
| Women | 211 | 51.34 |
| LGBT | 33 | 8.03 |
| *2. Age* | | |
| 18 years old | 37 | 9 |
| 19 years old | 97 | 23.6 |
| 20 years old | 105 | 25.55 |
| 21 years old | 73 | 17.76 |
| 22 years old | 41 | 9.98 |
| Over 22 years old | 58 | 14.11 |
| *3. Current academic year* | | |
| First year/ Freshman | 91 | 22.14 |
| Second year/Sophomore | 141 | 34.31 |
| Third year/Junior | 100 | 24.33 |
| Fourth year/Senior | 66 | 16.06 |
| Over fifth year/ Super senior | 13 | 3.16 |
| *4. Faculty* | | |
| Faculty of management science | 249 | 60.58 |
| Faculty of communication arts | 101 | 24.57 |
| Faculty of humanities | 31 | 7.54 |
| Faculty of mass communication | 3 | 0.73 |
| Faculty of social sciences | 3 | 0.73 |
| Faculty of liberal arts | 24 | 5.84 |
| *5. Grade point average: GPA* | | |
| GPA 1.51–2.00 | 15 | 3.65 |
| GPA 2.01–2.50 | 80 | 19.46 |
| GPA 2.51–3.00 | 142 | 34.55 |
| GPA 3.01–3.50 | – | – |
| GPA 3.51–4.00 | 174 | 42.34 |
| *6. Experian in using the internet* | | |
| less than 1 year | 5 | 5.11 |
| 2–3 years | 21 | 7.06 |
| 4–5 years | 29 | 1.22 |
| More than 5 years | 356 | 86.62 |
| *7. Time (number of hours spent using the internet per time)* | | |
| 1–2 h | 32 | 7.79 |
| 3–4 h | 62 | 15.09 |
| 5–6 h | 95 | 23.11 |
| More than 6 h | 222 | 54.01 |

**Gender Distribution**—Females accounted for the majority of respondents at 51.34%, approximately 211 individuals. Males made up about 40.63% of the sample, which corresponds to roughly 167 individuals. A smaller portion, approximately 8.03%, or around 33 individuals, identified themselves as gender diverse or gender non-conforming.

**Age**—The most common age group among respondents was 20 years old, comprising 25.55% of the sample, or approximately 105 individuals. The second largest age group was 19 years old, representing 23.60% of respondents, or roughly 97 individuals. The age group with the fewest respondents was 18 years old, accounting for 9.00%, or approximately 37 individuals.

**Academic Year**—The distribution of respondents across academic years revealed that 34.31% were in their second year of study, equivalent to around 141 individuals. The second highest percentage, 24.33%, or approximately 100 individuals, were in their third year. Only a small fraction, about 3.16%, or around 13 individuals, were studying in their fifth year or higher.

**Faculty Distribution**—The majority of respondents were enrolled in the Faculty of Management Sciences, accounting for 60.58%. The Faculty of Communication Arts had the second highest representation, at 24.57%. The Faculties of Social Sciences and Mass Communication had the lowest representation, each accounting for 0.73%.

**Grade Point Average (GPA)**—The majority of respondents fell within the GPA range of 3.51–4.00, comprising 42.34% of the sample. The next most common GPA range was 2.51–3.00, with 34.55% of respondents. The lowest percentage, 3.65%, corresponded to the GPA range of 1.51–2.00.

**Internet Usage Experience**—A significant majority of students reported having 5 years or more of internet usage experience, accounting for 86.62% of respondents. Additionally, most of the students reported spending an average of 6 h or more per internet session, which represented 54.01% of the sample.

These demographic and academic characteristics provide essential context for understanding the respondent population and the factors that might influence their awareness and understanding of computer related offenses under the Computer Crime Act of 2017.

## 4 Empirical Results

Table 3 presents the results of quantile regression analysis conducted across various quantiles, ranging from 0.1 to 0.8, to assess the level of awareness and understanding of crimes related to the Computer Crimes Act B.E. 2017 among university students in the Chiang Mai area. This study aimed to uncover the factors influencing students' perception and comprehension of computer related offenses under the Computer Related Offenses Act. Instead of depending on p-values to determine the importance of the findings, this study uses the Minimal Bayes factor (MBF) [6, 13]. The magnitude of MBF (Minimum Bayes Factor) values must be considered while interpreting

**Table 2** Evidence categories for Minimum Bayes Factor (MBF)

| Bayes factor | Strength of evidence |
|---|---|
| 0.33 < MBF < 1 | "Weak evidence" |
| 0.1 < MBF < 0.33 | "Moderate evidence" |
| 0.033 < MBF < 0.1 | "Substantial evidence" |
| 0.01 < MBF < 0.033 | "Strong evidence" |
| 0.003 < MBF < 0.01 | "Very strong evidence" |
| MBF < 0.003 | "Decisive evidence" |

them, as different ranges represent varying levels of evidence. This is explained below (Table 2):

**At the 0.1 Quantile Level** (Lowest Awareness and Understanding)—Increasing the AGE of the respondents by 1 year resulted in a statistically significant increase of 2.5028 points in their scores of awareness and understanding of computer related actions. Similarly, increasing GPA by 1 level led to a statistically significant increase of 3.3564 points in awareness and understanding scores. However, an increase in the STUDYYEAR level did not positively affect the level of awareness and understanding of computer related actions.

**At the 0.2 and 0.3 Quantile Levels** (Slightly Higher Awareness and Understanding)—AGE and GPA continued to have statistically significant positive impacts. At the 0.2 quantile, the impact of AGE was 1.9047 points, and the impact of GPA was 1.3491 points. Meanwhile, at the 0.3 quantile, the impact of AGE was 1.6882 points, and the impact of GPA was 0.6493 points.

**At the 0.4–0.6 Quantile Levels** (Moderate Awareness and Understanding)—In this range, only the factor of AGE was found to have a statistically significant impact. When the AGE of the respondents increased by 1 year, it had varying impacts: 1.2986 points at the 0.4 quantile, 1.0468 points at the 0.5 quantile, and 0.6282 points at the 0.6 quantile.

**At the 0.8 Quantile Level** (Highest Awareness and Understanding)—The significant factor here is EXPERIENCE, implying that if respondents have a longer history of using the internet, it leads to a statistically significant increase of 1.9407 points in their awareness and understanding scores of computer related actions.

Therefore, the Quantile model analysis conducted on the sample group of university students in the Chiang Mai area provides a detailed understanding of the factors influencing the level of awareness and understanding of computer related wrongful actions under the Computer Crimes Act B.E. 2017. It highlights those various factors such as age, GPA, study year, and internet experience have differential impacts across different quantile levels. This analysis allows us to pinpoint which factors are most influential at each level of awareness and understanding and quantify the magnitude of their impact, providing valuable insights for further research and policy considerations. Furthermore, there is evidence indicating that the confidence interval for the estimated coefficient from the quantile regression model is within the 95% confidence range. This can be seen in Fig. 1, where the orange line represents the confidence interval, and the blue line represents the estimated coefficient value.

**Table 3** Results of quantile regression for quantiles ranging from 0.1 to 0.8

| Variables | Coefficient | Std. error | T-Statistic | MBF |
|---|---|---|---|---|
| $\tau = 0.1$ | | | | |
| C | 34.2114 | 8.7381 | 3.9152 | 0.0005 |
| TIME | 0.3218 | 0.9767 | 0.3295 | 0.9472 |
| STUDYYEAR | −2.2134 | 1.0419 | −2.1244 | 0.1047 |
| GPA | 2.5028 | 0.6125 | 4.0865 | 0.0002 |
| GENDER | −0.6515 | 0.7459 | −0.8734 | 0.6829 |
| INTERNET EXPERIAN | 0.0167 | 2.5052 | 0.0067 | 1.0000 |
| AGE | 3.3564 | 0.7470 | 4.4929 | 0.0000 |
| $\tau = 0.2$ | | | | |
| C | 44.4452 | 4.1134 | 10.8049 | 0.0000 |
| TIME | 0.3968 | 0.7175 | 0.5531 | 0.8582 |
| STUDYYEAR | −0.8333 | 0.9443 | −0.8824 | 0.6775 |
| GPA | 1.3491 | 0.7323 | 1.8424 | 0.1832 |
| GENDER | 0.3571 | 0.6335 | 0.5638 | 0.8531 |
| INTERNET EXPERIAN | 0.0794 | 1.3105 | 0.0606 | 0.9982 |
| AGE | 1.9047 | 0.6665 | 2.8577 | 0.0169 |
| $\tau = 0.3$ | | | | |
| C | 47.2734 | 3.3601 | 14.0691 | 0.0000 |
| TIME | 0.7792 | 0.5870 | 1.3274 | 0.4144 |
| STUDYYEAR | −0.9090 | 0.5604 | −1.6221 | 0.2683 |
| GPA | 0.6493 | 0.3690 | 1.7599 | 0.2125 |
| GENDER | −0.3896 | 0.6648 | −0.5860 | 0.8422 |
| INTERNET EXPERIAN | 0.7792 | 0.8356 | 0.9325 | 0.6474 |
| AGE | 1.6882 | 0.3794 | 4.4497 | 0.0001 |
| $\tau = 0.4$ | | | | |
| C | 52.4679 | 3.3947 | 15.4559 | 0.0000 |
| TIME | 0.2597 | 0.4933 | 0.5265 | 0.8706 |
| STUDYYEAR | −0.5195 | 0.5300 | −0.9801 | 0.6186 |
| GPA | 0.5195 | 0.3485 | 1.4905 | 0.3293 |
| GENDER | 0.0000 | 0.6315 | 0.0000 | 1.0000 |
| INTERNET EXPERIAN | 0.5195 | 0.7289 | 0.7127 | 0.7757 |
| AGE | 1.2986 | 0.3863 | 3.3614 | 0.0035 |

(continued)

**Table 3** (continued)

| Variables | Coefficient | Std. error | T-Statistic | MBF |
|---|---|---|---|---|
| $\tau = 0.5$ | | | | |
| C | 55.0493 | 3.4967 | 15.7433 | 0.0000 |
| TIME | −0.0218 | 0.4917 | −0.0444 | 0.9990 |
| STUDYYEAR | −0.7851 | 0.5139 | −1.5277 | 0.3113 |
| GPA | 0.4362 | 0.3554 | 1.2272 | 0.4709 |
| GENDER | 0.3053 | 0.5931 | 0.5148 | 0.8759 |
| INTERNET EXPERIAN | 0.9596 | 0.7542 | 1.2723 | 0.4451 |
| AGE | 1.0468 | 0.3942 | 2.6556 | 0.0294 |
| $\tau = 0.6$ | | | | |
| C | 56.7940 | 2.8057 | 20.2421 | 0.0000 |
| TIME | −0.0299 | 0.4920 | −0.0608 | 0.9982 |
| STUDYYEAR | −0.1496 | 0.4513 | −0.3314 | 0.9466 |
| GPA | 0.3740 | 0.3384 | 1.1051 | 0.5430 |
| GENDER | 0.3291 | 0.5947 | 0.5533 | 0.8581 |
| INTERNET EXPERIAN | 0.9922 | 0.6224 | 1.5943 | 0.2806 |
| AGE | 0.6282 | 0.3663 | 1.7152 | 0.2297 |
| $\tau = 0.7$ | | | | |
| C | 61.9047 | 2.6533 | 23.3315 | 0.0000 |
| TIME | 0.0000 | 0.5000 | 0.0000 | 1.0000 |
| STUDYYEAR | −1.00E-15 | 0.4000 | 0.0000 | 1.0000 |
| GPA | −3.00E-15 | 0.3000 | 0.0000 | 1.0000 |
| GENDER | 3.00E-15 | 0.5000 | 0.0000 | 1.0000 |
| INTERNET EXPERIAN | 1.0000 | 0.7000 | 1.0000 | 0.6065 |
| AGE | −9.00E-17 | 0.3000 | 0.0000 | 1.0000 |
| $\tau = 0.8$ | | | | |
| C | 60.9523 | 2.7535 | 22.1364 | 0.0000 |
| TIME | −1.90E-17 | 0.5600 | 0.0000 | 1.0000 |
| STUDYYEAR | 8.80E-16 | 0.4300 | 0.0000 | 1.0000 |
| GPA | 1.20E-15 | 0.3900 | 0.0000 | 1.0000 |
| GENDER | −9.70E-16 | 0.6800 | 0.0000 | 1.0000 |
| INTERNET EXPERIAN | 1.9047 | 0.6760 | 2.8173 | 0.0189 |
| AGE | −2.30E-15 | 0.4000 | 0.0000 | 1.0000 |

*Note* MBF represents the minimum bayes factor

# 5 Conclusion

The Quantile model used to analyze the level of awareness and understanding of computer related wrongful actions in the year 2017 B.E. yielded valuable insights by

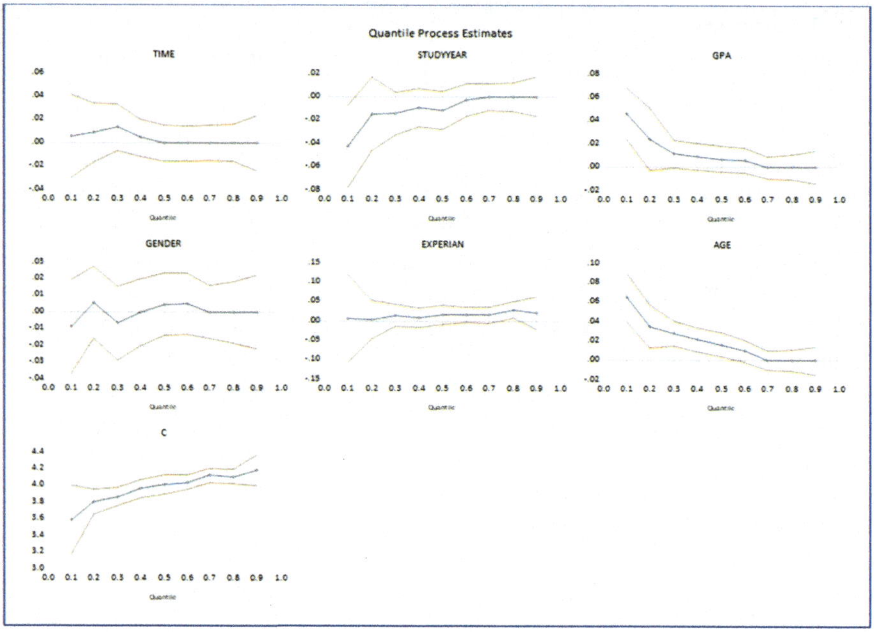


**Fig. 1** Quantile regression results for university students in the area of Chiang Mai

identifying key factors that influenced awareness and understanding across different quantiles. Here's an expanded explanation of the findings:

**Low-Level Awareness and Understanding (Quantile 0.1–0.3)**—In this quantile, which represents individuals with lower levels of awareness and understanding regarding computer related wrongful actions, two significant factors emerged as having a statistically significant positive impact: AGE and GPA. AGE: This finding suggests that older individuals tend to have a better grasp of computer related wrongful actions compared to younger individuals within this quantile. It may be due to accumulated life experiences and exposure to computer technologies over time. GPA (Grade Point Average): The positive impact of GPA indicates that individuals with higher academic achievements are more likely to exhibit increased awareness and understanding of computer related wrongful actions. This could be because they may have received education or training that emphasizes digital ethics or security.

**Medium-Level Awareness and Understanding (Quantile 0.4–0.6)**—In the middle quantile, where individuals have a moderate level of awareness and understanding, the factor that continued to have a statistically significant positive impact was Age. AGE: This finding suggests that, even in the medium level quantile, age plays a role in influencing awareness and understanding. Older individuals in this group still tend to have better comprehension of computer related wrongful actions.

**High-Level Awareness and Understanding (Quantile 0.8)**—For individuals with a high level of awareness and understanding of computer related wrongful actions, a different factor came to the forefront: EXPERIAN. EXPERIAN: The presence of EXPERIAN as a statistically significant positive factor in this quantile indicates that individuals with extensive experience in using the internet for more than five years tend to have a deeper understanding of computer related wrongful actions. This suggests that hands on experience and exposure to online activities may enhance one's awareness and comprehension of digital ethical issues.

In conclusion, the Quantile model analysis provides valuable insights into the factors that significantly influence awareness and understanding of computer related wrongful actions across different quantiles. It not only identifies these influential factors but also quantifies the magnitude of their impact at each level, helping us gain a clearer understanding of the dynamics involved in digital ethics awareness and understanding.

**Acknowledgements** The authors thank the "Faculty of Management Science Research Fund, Chiang Mai Rajabhat University" for its financial support. This research work was partially supported by "Chiang Mai University."

## References

1. Al-Khaldi, M.A., Al-Jabri, I.M.: The relationship of attitudes to computer utilization: new evidence from a developing nation. Comput. Hum. Behav. **14**(1), 23–42 (1998)
2. Chansaeng, S., Rongjig, A., Srisodsaluk, P., Kahanak, S., Laisuwannachart, P.: Factors related to risky sexual behaviors among junior high school students in Mueang District, Suphan Buri Province, from internet use. Nurs. J. Minist. Public Health **30**(2), 121–135 (2020)
3. He, X., Zhu, L.X.: A lack-of-fit test for quantile regression. J. Am. Stat. Assoc. **98**(464), 1013–1022 (2003)
4. Koenker, R., Bassett Jr, G.: Regression quantiles. Econ.: J. Econ. Soc. 33–50 (1978)
5. Koenker, R.: Quantile regression, vol. 38. Cambridge University Press (2005)
6. Maneejuk, P., Yamaka, W.: Significance test for linear regression: how to test without P-values? J. Appl. Stat. **48**(5), 827–845 (2021)
7. Namsai, A., Chantawi, P.: A Study of knowledge level and computer and internet usage behavior at risk of committing offenses under the computer crime act of 2007, among Students of Songkhla Rajabhat University. In: Proceedings of the 6th National Academic Conference, pp. 1479–1489 (2015)
8. Office of Provincial Commercial Affairs Chiang Mai: Summary of Chiang Mai Province (2022). https://chiangmai.moc.go.th/th/content/page/index/id/1055
9. Pariyawisit, C., Rewadee, S.: Perception and Understanding of the Public Regarding Offenses under the Computer Crime Act of 2017, in Nonthaburi Province and Bangkok. Rajabhat University, Nonthaburi, Thailand (2016)
10. Pengniti, P.: Explanation of the Civil and Commercial Code on Torts, the 1996 Liability of Officials Act, and Other Related Laws, 10th ed., revised. The Thai Bar Under The Royal Patronage, Bangkok (2017)
11. Phadkantha, R., Yamaka, W., Sriboonchitta, S.: Analysis of herding behavior using Bayesian quantile regression. In: Beyond Traditional Probabilistic Methods in Economics, vol. 2, pp. 795–805. Springer International Publishing (2019)

12. Sarawanawong, J., Fyeted, K., Ngernpoolsap, D., Chantrdee, W.: Online social media usage behaviors of undergraduate student, Kasetsart University. J. Libr. Inf. Sci. Srinakharinwirot Univ. **10**(2), 16–31 (2017)
13. Sellke, T., Bayarri, M.J., Berger, J.O.: Calibration of p values for testing precise null hypotheses. Am. Stat. **55**(1), 62–71 (2001)
14. Snchez, B.L., Lachos, H.V., Labra, V.F.: Likelihood based inference for quantile regression using the asymmetric Laplace distribution. J. Stat. Comput. Simul. **81**, 1565–1578 (2013)
15. Wang, H.J., Zhu, Z., Zhou, J.: Quantile regression in partially linear varying coefficient models. Ann. Stat. 3841–3866 (2009)
16. Wei, Y., Carroll, R.J.: Quantile regression with measurement error. J. Am. Stat. Assoc. **104**(487), 1129–1143 (2009)
17. Yamaka, W., Autchariyapanitkul, K., Meneejuk, P., Sriboonchitta, S.: Capital asset pricing model through quantile regression: an entropy approach. Thai J. Math. 53–65 (2017)

# How to Deal with Tail Dependence of Mixture Copula with an Extreme Weight

**Sundusit Saekow, Phisanu Chiawkhun, Woraphon Yamaka, Nawapon Nakharutai, and Parkpoom Phetpradap**

**Abstract** Tail dependence is the tool to measure dependence in the extremes of a bivariate distribution. Copula is the function used to describe the dependence between the variables. Therefore, copula is used in order to determine the tail dependence. In this work, we focus on dealing with the case of an extreme weight of mixture copula. We provide the transformed Bayesian model averaging to estimate the tail dependence comparing with the conventional ways to estimate the mixture copula by simulation study.

## 1 Introduction

There are couple ways to measure the risk in finance such as Pearson's product moment correlation coefficient and Spearman's rank order correlation coefficient [1]. However, they are controlled by small movements around the mean and fail to describe dependence between extreme events which has become a major issue in financial risk management. Facing asset selection and allocation with respect to portfolio management extreme events are primarily represented by jump risks and default risks. Extreme events appear in the tails of return distributions of assets and because

S. Saekow (✉) · P. Chiawkhun · N. Nakharutai
Faculty of Science, Department of Statistics, Chiang Mai University, Chiang Mai, Thailand
e-mail: sundusit_sae@cmu.ac.th

P. Chiawkhun
e-mail: phisanu.c@cmu.ac.th

N. Nakharutai
e-mail: nawapon.nakharutai@cmu.ac.th

W. Yamaka
Faculty of Economics, Chiang Mai University, Chiang Mai, Thailand
e-mail: woraphon.yamaka@cmu.ac.th

P. Phetpradap
Faculty of Science, Department of Mathematics, Chiang Mai University, Chiang Mai, Thailand
e-mail: parkpoom.phetpradap@cmu.ac.th



V. Kreinovich et al. (eds.), *Applications of Optimal Transport to Economics and Related Topics*, Studies in Systems, Decision and Control 556,
https://doi.org/10.1007/978-3-031-67770-0_43

of the fact that asset return distributions are heavy tailed, extraordinary downside losses are more likely to happen than expected assuming Gaussian distributions [2]. Therefore, we focus on the concept of tail dependence.

The coefficients of tail dependence are measures of pairwise dependence that depend only on the copula of a pair of random variables $X_1$ and $X_2$ with continuous marginal distribution functions [1]. The motivation for looking at these coefficients is that they provide measures of extremal dependence or, in other words, measures of the strength of dependence in the tails of a bivariate distribution.

To know the tail dependence, we need to calculated the tail dependence coefficient (TDC). It is shown that the tail dependence coefficient is related to the copula function. Copula functions are used to describe the dependence between random variables in general since it is the function that represents the joint distribution function of its marginal distribution functions [1]. In fact, the tail dependence can be shown as the limit of copula function. That means, in order to determine the tail dependence, the estimation of the copula function must be considered. Usually, there are various copulas considered in tail dependence, for example, Frank copula, Clayton copula and Gumbel copula [3]. We can choose the best copula as the true copula or we can combine the selected copulas using mixture copula [4] and Bayesian model averaging [5].

Our study is aimed at estimating tail dependence within mixture copulas, a challenging endeavor conventionally involving the simultaneous optimization of mixed copula weights and parameters. This becomes particularly demanding due to the absence of a closed-form solution for the weight parameter. It's important to highlight that this weight parameter is crucial for calculating the expected tail dependence of multiple copula functions within mixture copula models.

In our research, we propose an alternative approach. We advocate the use of Bayesian Model Averaging, a versatile statistical technique, to precisely determine the optimal weight for mixture copulas. This novel method simplifies the estimation process by zeroing in on the weight of the mixture copula. This simplification makes it easier to gain insight into tail dependence within mixture copula models, promising to provide a more accessible and intuitive understanding of this complex statistical problem.

This study is organized as follow, Sect. 2 is copula review, Sect. 3 is tail dependence, Sect. 4 is Bayesian model averaging, Sect. 5 is simulation study and Sect. 6 is conclusion and future work.

## 2 Copula Review

A copula function is the joint distribution of the uniform distributions. Copulas are used to described the dependence between random variables. It shows the connection between each marginal distribution which shows as the joint distribution.

**Definition 1** ([6]) A two-dimensional copula $C$ is the function from $I^2$ to $I$ with the following properties;

1. For every $v, z \in I$,
$$C(v, 0) = 0 = C(0, z)$$
and
$$C(v, 1) = v \text{ and } C(1, z) = z.$$
2. For every $v_1, v_2, z_1, z_2 \in I$ such that $v_1 \leq v_2$ and $z_1 \leq z_2$,
$$C(v_2, z_2) - C(v_2, z_1) - C(v_1, z_2) + C(v_1, z_1) \geq 0.$$

**Theorem 1** (Sklar's Theorem [7]) *Let $F_1(x)$, $F_2(y)$ be marginal distribution functions. Then, for every $(x, y) \in \mathbb{R}^2$:*

1. *If $\mathcal{C}$ is any subcopula whose domain contains Range $F_1 \times$ Range $F_2$,*
$$\mathcal{C}(F_1(x), F_2(y))$$
*is a joint distribution function with margins $F_1(x)$ and $F_2(y)$.*
2. *Conversely, if $F(x, y)$ is a joint distribution function with margins $F_1(x)$ and $F_2(y)$, there exists a unique subcopula $\mathcal{C}$, with domain Range $F_1 \times$ Range $F_2$, such that*
$$F(x, y) = \mathcal{C}(F_1(x), F_2(y)).$$
*If $F_1(x)$ and $F_2(y)$ are continuous, the subcopula is a copula. Otherwise, there exists a copula $C$ such that*
$$C(v, z) = \mathcal{C}(v, z)$$
*for every $(v, z) \in$ Range $F_1 \times$ Range $F_2$.*

Sklar's theorem shows the connection between two or more joint distributions by a copula function. It also shows that we can form them into the joint distribution function using a copula function. This result in the various application of copula function.

The class of copulas that we are interested in this work is Archimedean copula which is defined as follows.

**Definition 2** ([6]) Let $\phi$ be a continuous strictly decreasing function from $I$ to $[0, \infty]$ such that $\phi(1) = 0$ and let $\phi^{[-1]}$ denotes the pseudo-inverse of $\phi$. That is $\phi^{[-1]}(t) = \phi^{-1}(t)$ for $t \in [0, \phi(0)]$, and $\phi^{[-1]} = 0$ for $t \geq \phi(0)$. Then $C(v, z) = \phi^{[-1]}(\phi(v) + \phi(z))$ is a copula and called **Archimedean copula**.

Clayton copula and Gumbel copula are **Archimedean copula** generated by these generators.

1. Clayton copula is generated by $\phi(t) = (t^{-\theta} - 1)/\theta, \theta \in [-1, \infty)$, i.e.,

$$C_\theta(v, z) = \max\left[(v^{-\theta} + z^{-\theta} - 1)^{-\frac{1}{\theta}}, 0\right]. \quad (1)$$

2. Gumbel-Hougaard copula is generated by $\phi(t) = (-\ln t)^\theta, \theta \in [1, \infty)$, i.e.,

$$C_\theta(v, z) = \exp\left(-\left((-\ln v)^\theta + (-\ln z)^\theta\right)^{\frac{1}{\theta}}\right). \quad (2)$$

## 3 Tail Dependence

There are a number of ways to measure dependence among risks [1], notably Kendall's tau or Spearman's rho, which focus on global dependence. The concept of tail dependence is a sufficient and more specialized measure of dependence in the tails, which makes it a key concept in many financial and insurance applications.

**Definition 3** ([1]) Let $X_1$ and $X_2$ be random variables with distribution functions $F_1$ and $F_2$, respectively. The coefficient of upper tail dependence ($\lambda_U$) and the coefficient of lower tail dependence ($\lambda_L$) are

$$\lambda_U = \lim_{v\to 1^-} \mathbb{P}(Y > G^{-1}(v)|X > F^{-1}(v)),$$

$$\lambda_L = \lim_{v\to 0^+} \mathbb{P}(Y \le G^{-1}(v)|X \le F^{-1}(v)),$$

provided the limit $\lambda_U(\lambda_L) \in [0, 1]$ exists. If $\lambda_U(\lambda_L) \in (0, 1]$, then $X_1$ and $X_2$ are said to show upper(lower) tail dependence. It $\lambda_U(\lambda_L) = 0$, they are asymptotically independent in the upper(lower) tail.

From Definition 3, the upper(lower) tail dependence coefficient (TDC) can be derived to the form of the copula function.

$$\begin{aligned}\lambda_U &= \lim_{v\to 1^-} \mathbb{P}(Y > G^{-1}(v)|X > F^{-1}(v)) \\ &= \lim_{v\to 1^-} \frac{\mathbb{P}(Y > G^{-1}(v), X > F^{-1}(v))}{\mathbb{P}(X > F^{-1}(v))} \\ &= \lim_{v\to 1^-} \frac{1 - 2v + C(v, v)}{1 - v}\end{aligned} \quad (3)$$

$$\begin{aligned}\lambda_L &= \lim_{v\to 0^+} \mathbb{P}(Y \le G^{-1}(v)|X \le F^{-1}(v)) \\ &= \lim_{v\to 0^+} \frac{\mathbb{P}(Y \le G^{-1}(v), X \le F^{-1}(v))}{\mathbb{P}(X \le F^{-1}(v))} \\ &= \lim_{v\to 0^+} \frac{C(v,v)}{v}\end{aligned} \tag{4}$$

If we have a copula, we can calculate the TDC of it by the copula function and Eqs. 3 and 4.

1. Clayton Copula: $C_\theta(v,z) = \max\left[(v^{-\theta}+z^{-\theta}-1)^{-\frac{1}{\theta}}, 0\right], \theta \in [-1,\infty) - \{0\}$

$$\lambda_U = 0 \text{ and } \lambda_L = 2^{-\frac{1}{\theta}}.$$

2. Gumbel Copula: $C_\theta(v,z) = \exp\left(-\left((-\ln v)^\theta + (-\ln z)^\theta\right)^{\frac{1}{\theta}}\right), \theta \in [1,\infty)$

$$\lambda_U = 2 - 2^{\frac{1}{\theta}} \text{ and } \lambda_L = 0.$$

When there are multiple copulas that all show similar goodness of fit, usually only a single best copula is selected and treated as the true copula. This approach ignores model uncertainty resulting in the under- or over-estimated. An alternative way is to combine the selected copulas.

**Lemma 1** ([6]) *Let $C_1(v,z), C_2(v,z), \ldots, C_n(v,z)$ be 2-dimensional copula and $w_1, w_2, \ldots, w_n \in [0,1]$. Then*

$$C_{mix}(v,z) = w_1C_1(v,z) + w_2C_2(v,z) + \cdots + w_nC_n(v,z)$$

*is also a copula.*

The coefficient of upper and lower tail dependence can also be written as follows (Fig. 1);

$$\lambda_{U_{C_{mix}}} = \sum_{i=1}^{n} w_i \lambda_{U_{C_i}} \tag{5}$$

$$\lambda_{L_{C_{mix}}} = \sum_{i=1}^{n} w_i \lambda_{L_{C_i}} \tag{6}$$

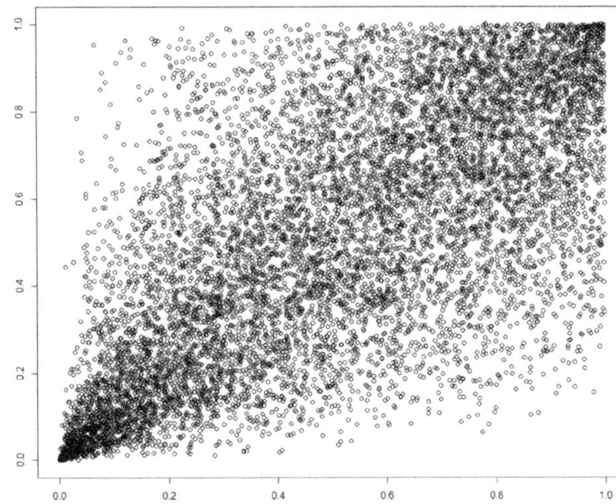


**Fig. 1** Mixture of survival clayton copula and gumbel copula with parameter (2, 2) and weight (0.5, 0.5)

## 4 Bayesian Model Averaging

The goal of using Bayesian model averaging (BMA) is to improve the estimation of TDC from the individual copula including the finite mixture copula model. This approach is based on the Bayesian model averaging [5].

Suppose that $\Delta$ is the quantity of interest, such as the TDC of the copula, and $C_1, \ldots, C_n$ are selected copulas being considered. The posterior distribution of $\Delta$ given data $\mathscr{D}$ is

$$p(\Delta|\mathscr{D}) = \sum_{i=1}^{n} p(\Delta|C_i, \mathscr{D})p(C_i|\mathscr{D}). \tag{7}$$

This is an average of the posterior distributions $p(\Delta|C_i, \mathscr{D})$ under each of the models $C_i$ considered, weighted by their posterior model probabilities $p(C_i|\mathscr{D})$.

In the context of treating TDC as the quantity of interest within Bayesian model averaging, this method directly blends the calculated $\hat{\lambda}$ values from multiple fitted copula models to produce a single unified estimate of the TDC. Suppose that $\hat{\lambda}_i$ is the TDC estimated from a fitted copula $C_i$ to the data. The BMA weighted estimate of TDC, $\hat{\lambda}_{BMA}$, across multiple fitted copulas $C_i$, for $i = 1, \ldots, n$ can be calculated directly as followed;

$$\hat{\lambda}_{BMA} = \sum_{i=1}^{n} W_{C_k}\hat{\lambda}_{C_i} \tag{8}$$

where $W_{C_k}$ can be estimated by $\dfrac{\exp(-\frac{1}{2}BIC_k)}{\sum_{k'=1}^{K}\exp(-\frac{1}{2}BIC_{k'})}$.

In this work, we also provide the transformed weight of the BMA which can be calculated using $BIC_k/\log N$ instead of $BIC_k$ where $N$ is the number of records in dataset $\mathscr{D}$ which we call BMAlog as follow;

$$\hat{\lambda}_{BMAlog} = \sum_{i=1}^{n} W_{Clog_k}\hat{\lambda}_{C_i} \tag{9}$$

where $W_{Clog_k}$ can be estimated by $\dfrac{\exp(-\frac{1}{2}\frac{BIC_k}{\log(N)})}{\sum_{k'=1}^{K}\exp(-\frac{1}{2}\frac{BIC_{k'}}{\log(N)})}$.

## 5 Simulation Study

To examine the mixture copula fitting method and Bayesian model averaging method, we aim for the simulation study using R version 4.3.2. In this case, we focus on the mixture copula from Survival Clayton copula and Gumbel Copula. Then we follow these steps;

1. Generate the datasets 1000 pairs of the mixture copula from Survival Clayton copula and Gumbel copula where parameters $(\theta_1, \theta_2)$ are (2, 2), (2, 4), (2, 6), (4, 2), (4, 4), (4, 6), (6, 2), (6, 4) and (6, 6) and weights are (0.1, 0.9), (0.3, 0.7), (0.5, 0.5), (0.7, 0.3) and (0.9, 0.1).
2. Estimate the datasets using mixture copula approach.
   a. Estimate the parameter of datasets. Obtain the $(\hat{\theta}_1, \hat{\theta}_2)$ and $(\hat{W}_1, \hat{W}_2)$ for Survival Clayton copula and Gumbel copula, respectively.
   b. Calculated the TDC of the mixture copula from (a) using Eqs. (5) and (6).
3. Estimate the datasets using BMA.
   a. Estimate the parameter of datasets. Obtain $(\hat{\theta}_1, \hat{\theta}_2)$ using maximum likelihood estimator.
   b. Use the BIC from estimating the parameter to obtain the weight of Bayesian model averaging approach which include $(W_{C_1}, W_{C_2})$ and $(W_{Clog_1}, W_{Clog_2})$.
   c. Calculated the TDC from weight and parameter from (a) and (b) using Eqs. (5) and (6).
4. Repeat step 1.–3. for 100 times.
5. Determine the bias of each estimator and compare them.

The result of the simulation study are as follow (Tables 1, 2, 3, 4 and 5);

From the simulation study, we can see that the bias of BMAlog is lower than the bias of BMA in all of the case but it still slightly higher than the bias of mixture copula in most of the case. Moreover, the BMAlog seems to perform better when the weight of the mixture copula is extreme, i.e., weights are close to the maximum and minimum weight possible. In fact, the only case that bias of BMAlog is lower

**Table 1** The weight is (0.1, 0.9)

| $\theta_1$ | $\theta_2$ | Bias of mixture copula | Bias of BMA | Bias of BMAlog |
|---|---|---|---|---|
| 2 | 2 | 0.017307 | 0.014977 | 0.013347 |
| 2 | 4 | 0.006117 | 0.010321 | 0.010288 |
| 2 | 6 | 0.004540 | 0.008466 | 0.008463 |
| 4 | 2 | 0.016541 | 0.012264 | 0.011056 |
| 4 | 4 | 0.005675 | 0.009957 | 0.009929 |
| 4 | 6 | 0.003624 | 0.008419 | 0.008418 |
| 6 | 2 | 0.012083 | 0.012177 | 0.011158 |
| 6 | 4 | 0.005676 | 0.007396 | 0.007353 |
| 6 | 6 | 0.100863 | 0.011676 | 0.007046 |

**Table 2** The weight is (0.3, 0.7)

| $\theta_1$ | $\theta_2$ | Bias of mixture copula | Bias of BMA | Bias of BMAlog |
|---|---|---|---|---|
| 2 | 2 | 0.015217 | 0.029935 | 0.018953 |
| 2 | 4 | 0.008358 | 0.031389 | 0.029352 |
| 2 | 6 | 0.006138 | 0.029023 | 0.028307 |
| 4 | 2 | 0.011331 | 0.027342 | 0.020722 |
| 4 | 4 | 0.005703 | 0.027965 | 0.025404 |
| 4 | 6 | 0.005308 | 0.027154 | 0.025177 |
| 6 | 2 | 0.011527 | 0.026528 | 0.021978 |
| 6 | 4 | 0.006373 | 0.022424 | 0.021747 |
| 6 | 6 | 0.004486 | 0.022787 | 0.021261 |

**Table 3** The weight is (0.5, 0.5)

| $\theta_1$ | $\theta_2$ | Bias of mixture copula | Bias of BMA | Bias of BMAlog |
|---|---|---|---|---|
| 2 | 2 | 0.015394 | 0.046798 | 0.024275 |
| 2 | 4 | 0.010552 | 0.046158 | 0.033039 |
| 2 | 6 | 0.007988 | 0.049454 | 0.038558 |
| 4 | 2 | 0.012178 | 0.045811 | 0.027730 |
| 4 | 4 | 0.006117 | 0.038871 | 0.025126 |
| 4 | 6 | 0.004535 | 0.032409 | 0.021475 |
| 6 | 2 | 0.007860 | 0.039979 | 0.030830 |
| 6 | 4 | 0.006169 | 0.034916 | 0.023955 |
| 6 | 6 | 0.004063 | 0.028339 | 0.018553 |

**Table 4** The weight is (0.7, 0.3)

| $\theta_1$ | $\theta_2$ | Bias of mixture copula | Bias of BMA | Bias of BMAlog |
|---|---|---|---|---|
| 2 | 2 | 0.011962 | 0.036356 | 0.018627 |
| 2 | 4 | 0.009892 | 0.034203 | 0.020088 |
| 2 | 6 | 0.008900 | 0.033763 | 0.020735 |
| 4 | 2 | 0.012058 | 0.041930 | 0.027950 |
| 4 | 4 | 0.005465 | 0.023642 | 0.020350 |
| 4 | 6 | 0.005232 | 0.015904 | 0.013671 |
| 6 | 2 | 0.010739 | 0.042413 | 0.028125 |
| 6 | 4 | 0.004957 | 0.024813 | 0.020803 |
| 6 | 6 | 0.003958 | 0.015630 | 0.014882 |

**Table 5** The weight is (0.9, 0.1)

| $\theta_1$ | $\theta_2$ | Bias of mixture copula | Bias of BMA | Bias of BMAlog |
|---|---|---|---|---|
| 2 | 2 | 0.011189 | 0.016549 | 0.012458 |
| 2 | 4 | 0.010445 | 0.014129 | 0.010926 |
| 2 | 6 | 0.011963 | 0.015586 | 0.012528 |
| 4 | 2 | 0.008090 | 0.014806 | 0.014001 |
| 4 | 4 | 0.005694 | 0.009822 | 0.009669 |
| 4 | 6 | 0.005086 | 0.007700 | 0.007549 |
| 6 | 2 | 0.005474 | 0.014604 | 0.014166 |
| 6 | 4 | 0.004730 | 0.008527 | 0.008466 |
| 6 | 6 | 0.003883 | 0.006378 | 0.006369 |

than bias of mixture copula is parameter (2, 2) with weight (0.1, 0.9), parameter (4, 2) with weight (0.1, 0.9) and parameter (6, 6) with weight (0.1, 0.9). The difference between bias on the extreme weight, which is (0.1, 0.9) and (0.9, 0.1), is also lower than other case so we will focus more on that case.

We decided to do the simulation study again. This time, we focus only on the extreme weight of the parameter (2, 2) for 100 times. We are interested in the number of times that the bias of the BMAlog is lower than bias of the mixture copula and bias of the BMA. The result are as follows (Figs. 2 and 3);

The result shows that the BMAlog is slightly better than BMA since more than 60% of the case has the lower bias. It also shows that BMAlog can compete with the mixture copula in the extreme weight.

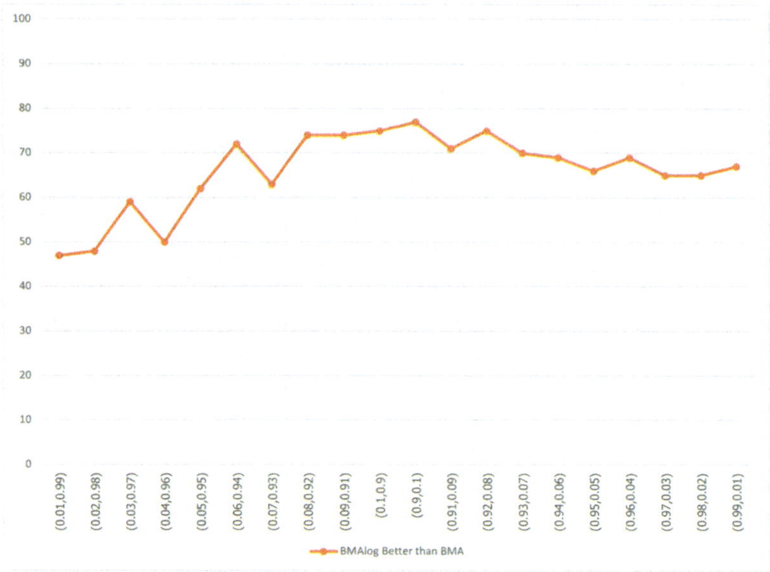


**Fig. 2** Percentage of BMAlog better than BMA

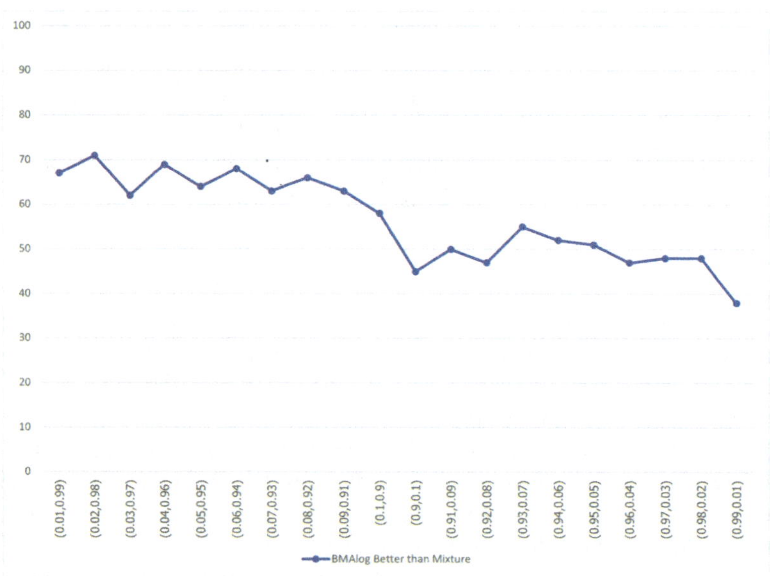


**Fig. 3** Percentage of BMAlog better than Mixture copula

## 6 Conclusion and Future Work

This study focus on estimating the weight of mixture copula for tail dependence. The conventional way to estimate the mixture copula needs to optimize weight and parameters at the same time which sometimes is difficult to obtain. Bayesian model averaging can estimate the weight of the mixture copula independently from the parameters. We also propose transformed Bayesian model averaging to compete with both the conventional way and original Bayesian model averaging.

The simulation study contain the result of estimating the tail dependence of mixture copula from 3 methods, the conventional way, Bayesian model averaging(BMA) and transformed Bayesian model averaging (BMAlog). It shows that estimating the tail dependence using BMAlog is better than using BMA overall since the mean of bias is lower. It also shows that the BMAlog performs better when the weight of the mixture copula is extreme. Moreover, it can compete with the conventional way to estimate the mixture copula. When the weight is extreme, it appears that BMAlog has lower bias than the conventional ways more than 40% of the times. However, there's improvement needed for BMAlog in order to have more consistency result.

**Acknowledgements** This work was supported by Development and Promotion of Science and Technology Talents Project (DPST).

## References

1. Maxneil, A.J., Frey, R., Embrechts, P.: Quantitative Risk Management—Concept. Techniques and Tools, Princeton University (2005)
2. Schmuki, S.N.: Tail dependence: Implementation, analysis, and study of the most recent concepts, Master's Thesis, Eidgenössische Technische Hochschule Zürich (2008)
3. Choroś, B., Ibragimov, R., Permiakov, E.: Copula estimation, copula theory and its applications. Lecture Notes Statist. **198**, 77–91 (2010)
4. Czado, C., Kastenmeier, R., Brechmann, E.C., Min, A.: A mixed copula model for insurance claims and claim sizes. Scand. Actuar. J. **2012**(4), 278–305 (2012)
5. Hoeting, et al.: Bayesan model averaging; a tutorial. Stat. Sci. **14**(4), 382–417 (1999)
6. Nelson, R.B.: An Introduction to Copulas. Springer, Berlin (2006)
7. Sklar, A.: Fonctions de Répartition án dimensions et Leurs Marges. Publications de l'Institut Statistique de l'Université de Paris **8**, 229–231 (1959)
8. Hofert, M., Kojadinovic, I., Maechler, M., Yan, J.: Copula: multivariate dependence with copulas. R package version 1.1-2 (2023). https://CRAN.R-project.org/package=copula
9. Nagler, T., Schepsmeier, U., Stoeber, U., Brechmann, E. Graeler, B., Erhardt, T.: VineCopula: statistical inference of vine copulas. R package version 2.4.5 (2023). https://CRAN.R-project.org/package=VineCopula

# An Analysis of Volatility Spillover Effect Between Energy and Agricultural Markets

**Pachraporn Arkornsakul, Tanapol Rattanasamakarn, and Konnika Palason**

**Abstract** In this investigation spanning the period from 2019 to 2023, we delve into the dynamic relationship between the energy and agriculture markets, placing a specific emphasis on the disruptive energy price crisis emanating from the Ukraine war. The article conducts a thorough analysis of volatility spillovers between these two crucial sectors, elucidating directional volatility transmission and pinpointing key sectors that serve as primary contributors to volatility shocks. Notably, the Total Spillover Index (TSI), a pivotal metric in this context, consistently records at approximately 13% throughout the study period, indicating noteworthy spillover effects and a robust inter-sector connectedness. This study not only brings to light intricate patterns of spillover effects but also provides valuable insights into their implications for financial stability.

**Keywords** Energy market · Agricultural markets · Volatility spillover effect

## 1 Introduction

The global landscape has undergone a seismic transformation in the aftermath of the COVID-19 pandemic, ushering in profound consequences for nations' economies worldwide. This study delves into the intricate relationship between two pivotal sectors, energy and agriculture, against the backdrop of the transformative period from 2019 to 2023. Our specific focus centers on the disruptive energy price crisis stemming from the Ukraine war, exploring long-term price dynamics, price volatility, and the transmission of market risk. As the pandemic unfolded, countries, including economic powerhouses like the United States, the United Kingdom, and the European Union, faced a pressing challenge-the relentless surge in food and energy inflation.

P. Arkornsakul (✉) · T. Rattanasamakarn · K. Palason
Faculty of Management Sciences, Chiangmai Rajabhat University, 202, Changpuek Rd., Muang, Chiang Mai 50300, Thailand
e-mail: pacharaporn_ark@g.cmru.ac.th



V. Kreinovich et al. (eds.), *Applications of Optimal Transport to Economics and Related Topics*, Studies in Systems, Decision and Control 556,
https://doi.org/10.1007/978-3-031-67770-0_44

This surge, largely attributed to the disruption of global supply chains that had previously operated with relative stability, underscores the complexity of the issues at hand.

The energy market, in particular, experienced a significant upheaval due to Western nations' imposition of trade sanctions on Russian energy inputs in response to Russia's incursion into Ukraine. Simultaneously, Ukraine's crucial agricultural shipments faced severe disruptions, courtesy of the Russian military's control over key transport routes. Noteworthy is the official nomenclature employed by Russian President Vladimir Putin, referring to the 2022 conflict with Ukraine as a "special military operation," which holds substantive economic implications. Wartime production cuts ensued, translating into higher sales prices for both food and energy, significantly contributing to the prevailing local inflation experienced by numerous nations.

The tandem surge in energy and food inflation raises profound questions about the resilience of economies in the face of such challenges. Russia's decision to curtail energy exports in retaliation to Western trade sanctions, despite the latter's objective of diminishing Russia's export earnings, has disproportionately impacted several European countries, exacerbating their economic woes. For instance, the United Kingdom witnessed a dramatic transformation of its energy inflation rate, skyrocketing from single-digit figures in September 2021 to an astonishing 57.6% year-on-year in July 2022. Similarly, Belgium, as a Eurozone member, grappled with energy inflation reaching a staggering 69.2% year-on-year in October 2022. While food inflation has not reached the same alarming heights as energy, it remains a persistent concern, and the elasticity of food prices warrants continued scrutiny.

Looking ahead, the World Bank's latest Commodity Market Outlook report projects historically elevated levels of price shocks in global commodity markets until the culmination of 2024. Even regions not directly embroiled in trade sanctions, such as many Asian countries, are experiencing the combined impact of reduced energy exports and food shortages, resulting in upward pressure on import prices and local prices within the region.

As we peer into the future, a sobering reality emerges-the world's reliance on Russia, a dominant player in global agricultural, fertilizer, and energy exports, coupled with the diminishing agricultural production in Ukraine, portends a prolonged period of food and energy shortages. The ongoing conflict's indeterminate nature implies that the world may grapple with subsequent waves of food and energy price hikes.

Crucially, both food and energy serve as inelastic inputs for output production, underscoring their indispensability for individuals' sustenance, making them impervious to price fluctuations. Furthermore, transportation, a critical component of the supply chain, heavily relies on oil. Given these interdependencies, it becomes evident that food and energy prices share a significant correlation, consistently moving in tandem.

Empirical evidence, as demonstrated by Parker, suggests that energy inflation is the primary driver of overall price increases, followed by food, housing, and other commodities [1]. The ramifications of rising energy prices reverberate across various sectors of the economy, impacting not only transportation costs but also broader economic aspects, particularly natural gas. Natural gas, aside from its role in heating, plays a pivotal role in fertilizer production. In light of the tightening of global fertilizer exports following the Russian-Ukrainian conflict, elevated gas prices are likely to perpetuate rising fertilizer costs, further exacerbating food inflation.

Consequently, energy prices assume a central role in elucidating the upward trajectory of agricultural and food price volatility. This suggests that energy price shocks are poised to trigger co-inflation in food, including agricultural inputs. When compounded by prolonged periods of adverse weather conditions, including hot spells and droughts, particularly in Europe, the resulting decline in agricultural production could accentuate food shortages in the months ahead.

The link between energy and food inflation is most pronounced in the second-round effect. This phenomenon occurs because energy price shocks tend to propagate to non-energy consumer prices, including food, as consumers adjust their inflation expectations in response to the initial energy inflation [3]. While energy price increases may exhibit a quicker moderation compared to food price increases, the former remains susceptible to sudden and potentially swifter surges, potentially exerting a more prolonged impact on inflation in the years to come. Although oil prices experienced a retreat in the latter part of 2022, the broader global price landscape, as indicated by inflation data, reveals a trajectory diverging from historical norms. Subsequent cost escalations in production and the cost of living, if not adequately counterbalanced by wage increases, could disrupt the economic stability of both developed and emerging markets. This underscores the significance of our study in examining the repercussions of inflation shocks on the future economy, a critical concern for contemporary policymaking. Furthermore, the interdependence of global commodity and energy prices, especially during periods of crisis such as the impending recession, as posited by Adekoya et al., accentuates the risk of a global recession stemming from persistently high inflation [3]. Against this backdrop, our study calls for a comprehensive exploration of the dual price relationship between food and energy and its consequential impact on economic development. This will be accomplished through the employment of a time-varying parametric vector autoregressive (TVP-VAR) model, capable of capturing the dynamic nature of the underlying economic structure with flexibility and robustness.

The energy-agriculture interdependence has remained an enigmatic puzzle for decades, punctuated by market uncertainties, oil crises, and geopolitical upheavals that frequently reverberate across markets, transcending national borders [4]. The pioneering work of Diebold and Yilmaz, hereafter referred to as DY, marked a significant milestone in this context, introducing a novel framework for measuring market return and volatility spillovers [5]. Grounded in the Vector Autoregressive (VAR) model, the DY framework employs Cholesky factor identification and a 200-week rolling window to visualize spillover effects. Notably, DY's groundbreaking research delved into the interdependence of global stock markets during the period from 1992 to 2007. Their findings shed light on the pronounced spillover effects between Western countries, including the United States, the United Kingdom, and Germany, and Asian countries such as Japan, Korea, and Taiwan. Importantly, their work illustrated that while an economic crisis may engender volatility spillovers, the same cannot be said for return spillovers.

Nevertheless, current research has yet to thoroughly investigate the intricate dynamics between the energy market and the agriculture market from the perspective of market trading-a matter of utmost significance for trading participants, particularly those engaged in the oil market, and researchers in this field. It is crucial to recognize that the nominal values of both the energy and agriculture markets wield substantial influence over final price outcomes. Therefore, given the pivotal role of the agriculture market as the primary invoicing currency in the energy market, this study squarely focuses on the trading landscape. Its primary objective is to reveal and elucidate the interactive relationship, commonly termed the spillover effect, between the energy market and the agriculture market.

## 2 Literature Review

The spillover effect between the energy market and the agriculture market has been a subject of growing interest in recent years. This literature review provides an overview of key studies that have explored the dynamic relationship between these two markets, focusing on mean spillover, volatility spillover, and risk spillover effects.

Understanding sectoral volatility transmission is of paramount importance as heightened market volatility can serve as a crucial precursor to financial crises. Moreover, it reflects increased interconnectedness among different assets. Studies by Laborda and Olmo underscore this significance, emphasizing that surges in market volatility can act as early warning signs for financial instability [6].

Research by Zhang et al. confirms the steady strengthening of correlations between global financial markets, driven by economic integration and financial globalization [7]. Recent events, particularly the outbreak of the COVID-19 pandemic, have had a profound impact on market volatility and volatility spillovers. Hasan et al. argue that the stock market displays higher sensitivity to pandemic outbreaks than the real economy [8]. This aligns with findings from Bissoondoyal-Bheenick et al., Aslam et al., and Akinlaso et al., all of whom report significant increases in stock return and volatility spillovers during the rapid spread of the COVID-19 pandemic [9–11].

During the pandemic, twelve European stock markets experienced persistently high directional volatility spillovers [10]. A noteworthy occurrence was the COVID-19 outbreak in Tunisia, which saw conventional stock markets transmitting negative volatility spillovers to Islamic stock markets [11]. This prompted scholars to draw comparisons between the volatility spillovers during the COVID-19 pandemic and those witnessed during the global financial crisis (GFC). Choi's research findings demonstrated that the total volatility spillovers among South Korea, Japan, China, and the US stock markets were more pronounced during the pandemic than during the GFC [20]. Of these markets, the South Korean stock market emerged as the primary transmitter of volatility. Gunay and Can's research corroborated these findings, further underscoring that the COVID-19 pandemic triggered more extensive volatility spillovers than the GFC, with advanced economies particularly affected in comparison to emerging markets [13].

Expanding beyond the realm of financial assets, Ghorbel and Jeribi observed a surge in volatility spillover across various fronts during the COVID-19 pandemic, including US market indexes, cryptocurrencies, and oil prices [14]. In China, Wen et al. reported an amplification of volatility spillovers from the stock market to the commodity market [15]. A comprehensive examination by Samitas et al. encompassed multiple asset classes, such as fine wine, equities, bonds, crude oil, commodities, gold, copper, shipping, and real estate, all of which documented peak total volatility spillovers during the pandemic [16]. Farid et al. highlighted significant bidirectional volatility spillovers among precious metals in the US during the COVID-19 period [17]. Furthermore, Elgammal et al. identified substantial bidirectional volatility spillovers between global stock markets and both energy and gold markets. Notably, cross-volatility spillovers were also observed, with the energy market influencing gold markets [18].

Additionally, numerous scholars have undertaken in-depth examinations of sectoral volatility within the stock market and its consequential spillover effects. For instance, Su and Liu conducted an analysis of sectoral volatility interconnections within China's stock market, revealing a notably high degree of connectivity [19]. Key risk transmitters in this context included the Consumer Discretionary, Industrials, and Materials sectors. Similarly, Choi's research in the US context indicated that the onset of the COVID-19 pandemic resulted in heightened volatility spillovers across 11 sectors, with the Energy, Consumer Discretionary, and Consumer Staples sectors taking on pivotal roles as primary transmitters of volatility shocks [20].

In conclusion, this body of literature revealed the critical role of understanding and analyzing volatility spillovers in financial and commodity markets, particularly during significant events such as the Energy price crisis from the Ukraine war. These findings shed light on the complex interconnectedness of markets and sectors, offering valuable insights for risk management and investment strategies in an increasingly globalized and interdependent financial landscape.

## 3 Methodology

In this empirical study, we explore the intricate relationship between the energy market and the agriculture market, examining it through three distinct dimensions: long-term price dynamics, price volatility, and the transmission of market risk. The primary objective of this research is to gain a comprehensive understanding of the interplay between these markets and to ascertain the precise extent to which fluctuations in the agriculture market impact changes in the energy market. This investigation aims to enhance our ability to forecast agricultural prices within the context of broader macroeconomic conditions.

### 3.1 The ARMA-GARCH Approach to Estimating Volatility

We employ the ARMA-GARCH model to estimate the market volatility and to determine the volatility patterns in energy market and agriculture market. We note that the GARCH model will likely capture both volatility clustering and heteroskedasticity of the variables. The GARCH mode is generally considered more advantageous than the ARCH model in estimating the residuals with unstable magnitude obtained from the AR or ARMA model. We also employ the ARMA model to analyze the market volatility. The market return for each sector is estimated as follows.

$$X_{it} = \ln\left(\frac{D_{it}}{D_{i,t-1}}\right), \tag{1}$$

The volatility of each sectoral index's return is estimated using the autoregressive moving average-generalized autoregressive conditional heteroskedasticity (ARMA-GARCH) model. The ARMA process is used to model the conditional mean of the time series, while the GARCH process

$$\begin{aligned} y_t &= \propto_0 + \sum_{i=1}^{r} \theta_i y_{t-i} + \varepsilon_t + \sum_{j=1}^{s} \varphi_t \varepsilon_{t-j} \\ \varepsilon_t &= \sigma_t u_t \text{ with } \mu_t \sim WN(0,1) \\ \sigma_t^2 &= \gamma_0 + \sum_{i=1}^{q} \alpha_i \varepsilon_{t-i}^2 + \sum_{j=1}^{p} \beta_j \sigma_{t-j}^2 \end{aligned} \tag{2}$$

The study's primary objective is to investigate volatility transmission across commodity market. So, the next step is to adopt the network analysis approach proposed by to explore the spillover effects. This approach enables a deeper examination of the association structure. Additionally, the transmission structure's direction and node weight can also be identified simultaneously. The rich information and simplicity of the interpretation provided by this method make it a perfect fit for the study's objectives.

The spillover index based on a forecast error variance decomposition in a VAR model. The method proceeds as follows. First, the VAR model of order p is fitted to the time series of volatility obtained through the ARMA-GARCH process. Second, using the data until time t, the forecast of the volatility series for h periods ahead is estimated, and the error variance decomposition of each forecast is obtained, corresponding to the shocks from the same or other network components at time t. Last, based on the obtained forecast error variance decomposition, the volatility spillover index of each time series and the total spillover index are calculated.

## 3.2 Volatility Spillover Effect Test Between Energy Market and Agriculture Market

The volatility spillover effect signifies that price volatility in distinct markets can mutually influence each other. Specifically, the magnitude of price volatility in one market may not only be influenced by its own past volatility but also by the price volatility in foreign markets. In simpler terms, volatility information can be transmitted among different markets. In the context of the two markets examined in this paper, the rapidly changing nature of market transactions has raised significant concerns. Whether fluctuations in Crude oil and Natural Gas prices will impact the international Agriculture market is the central focus of our investigation into volatility spillover effects.

To identify models with the lowest AIC values for modeling the energy and agriculture markets, we have singled out specific models. In this context, drawing inspiration from the work of Lin and Tamvakis and Hammoudeh et al. regarding the interactions among various oil markets, we have constructed a model to test the presence of the volatility spillover effect [21, 22].

## 3.3 Mean Spillover Effect is Based on a VAR Model

In terms of economics, mean spillover effect between two markets refers to the fact that the price in one market is affected not only by its own previous price movement but also by previous prices in other markets. In the long run, the test of mean spillover effect plays an important role in assessing whether the impact of change in Energy market and Agriculture market is significant, and whether the change in Energy market can help to forecast future trends of the latter.

Mean spillover effect is based on a VAR model, that is, the standard linear Granger causality test. Therefore, according to the least values of AIC and SC in a VAR model, the optimal lagged order can be set, and then the Granger causality test is carried out to detect the mean spillover effect between energy market and agriculture market. Specifically, let $Xt$ denotes energy market at time $t$ and $Yt$ be the agriculture market at time $t$, then a hypothesis $si = 0 (j = 1, 2, \ldots, m)$ is tested in a bi-variable regression equation. If the hypothesis is rejected, then we may say that the change of oil price is according to Granger, caused by the change of energy market. In the same way, we may also test whether energy market change Granger causes the change of agriculture market. In both equations, the parameter $m$ represents the largest lagged order

$$
\begin{aligned}
Y_t &= k_0 + \sum_{i=1}^{m} k_i Y_{t-i} + \sum_{j=1}^{m} s_j X_{t-j} + \varepsilon_t \\
X_t &= k_0 + \sum_{i=1}^{m} k_i X_{t-i} + \sum_{j=1}^{m} s_j Y_{t-j} + \varepsilon_t
\end{aligned}
\tag{3}
$$

# 4 Empirical Results

## 4.1 Summary Statistics of Data

The data used in this study consist of 2 parts: energy market commodity prices, which consist of WTI crude oil prices and natural gas prices, and the second part is commodity prices. In the agricultural market (Agriculture Market Commodity), which will consist of wheat prices and corn prices, all 2 products are products in the agricultural food category with volume and the highest turnover The data used in the study will be daily trading data. It was in the period from 01/01/2010 to 31/07/2023, with access to data from Refinitiv Eikon—Thomson Reuters (Figs. 1 and 2).

Let COt, WHt, WTIt and NGt represent Corn, Wheat, Crude Oil WTI and Natural Gas on date t, respectively, whose statistical characteristics are shown in Table 1. From the summary statistics, several traits can be identified. Firstly, neither of the price series follows a normal distribution. Secondly, there can be seen significant autocorrelation and heteroskedasticity in both price series; hence, volatility clustering does exist in both series. Additionally, from the results of an ADF test, it shows that at both price series are non-stationary but their first-order difference series are stationary, so both are I(1) series with decisive evidence.

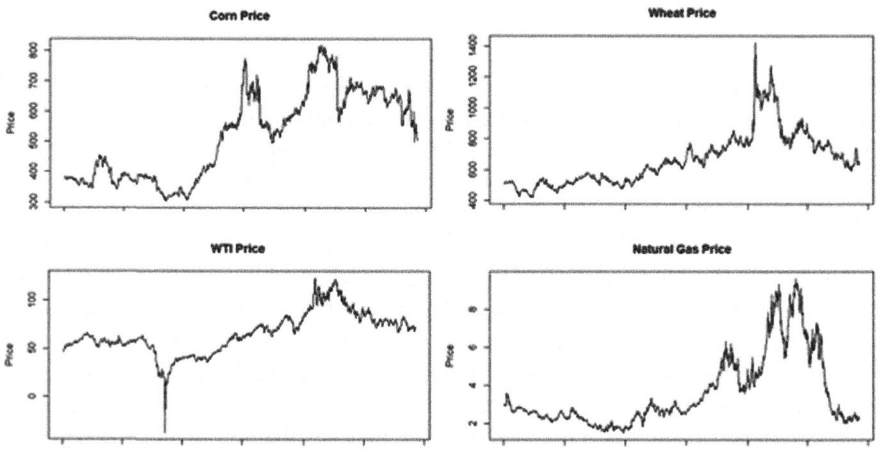


**Fig. 1** Price of corn, wheat, crude oil WTI and natural gas

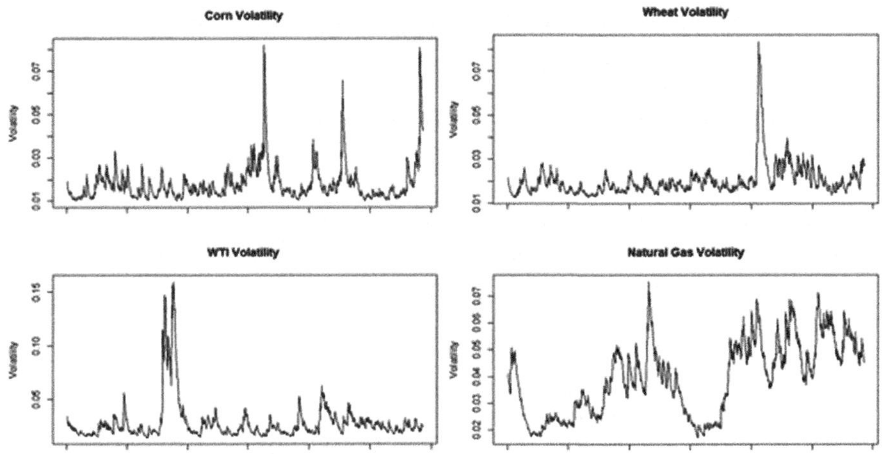


**Fig. 2** Volatility of corn, wheat, crude oil WTI and natural gas

**Table 1** Summary statistics

| Statistics | $CO_t$ | $WH_t$ | $WTI_t$ | $NG_t$ |
|---|---|---|---|---|
| Mean | 0.9214 | 0.8521 | 2.4712 | 1.8771 |
| Std. Dev. | 3.1458 | 4.5741 | 8.3745 | 7.8413 |
| Skewness | 0.9712 | 1.0024 | 0.9744 | 0.8742 |
| Kurtosis | 2.1145 | 3.1124 | 3.2746 | 3.0124 |
| JB statistic | 214.3314 | 304.2214 | 274.3287 | 211.8742 |
| ADF statistics | | | | |
| Level | −2.1021 | −3.2143 | −6.3441 | −2.7451 |
| 1st | −37.8812* | −34.1124* | −35.7421* | −35.4415* |

*Note* JBstatistic is used for normal distribution test. ADF statistic is introduced to test the stationary properties of price series, and during the test, 1st means the first-order difference. The symbol "*" is used to denote decisive evidence when the Minimum Bayes Factor (MBF) falls below the threshold of 0.001 [23]

## 4.2 *Cross-Correlation of the Energy Market and Agriculture Market*

Although the time series data for Corn, Wheat, Crude Oil WTI, and Natural Gas all display nonstationary characteristics, they do exhibit a cointegrating relationship. This cointegration satisfies the essential prerequisite for constructing a Vector Auto-Regression (VAR) model. To reinforce the appropriateness of utilizing this model, a cross-correlation test is conducted, with a specific focus on examining the relationship between the energy market and the agriculture market. In Table 2, the cross-correlation coefficients are displayed and calculated using a lagged period of 2. It's essential to highlight that all these coefficients are not only substantial but also

**Table 2** Cross-correlation coefficients between the energy market and agriculture market

| | $CO_{t-1}$ | $CO_{t-2}$ | $WH_{t-1}$ | $WH_{t-2}$ | $WTI_{t-1}$ | $WTI_{t-2}$ | $NG_{t-1}$ | $NG_{t-2}$ |
|---|---|---|---|---|---|---|---|---|
| $CO_{t-1}$ | 0.0046 | 0.0295 | 0.1560 | 0.8759 | 0.0046 | 0.0295 | 0.1560 | 0.8759 |
| $CO_{t-2}$ | −0.0218 | 0.0263 | −0.8260 | 0.4087 | −0.0218 | 0.0263 | −0.8260 | 0.4087 |
| $WH_{t-1}$ | −0.0086 | 0.0164 | −0.5210 | 0.6022 | −0.0086 | 0.0164 | −0.5210 | 0.6022 |
| $WH_{t-2}$ | 0.0138 | 0.0138 | 1.0050 | 0.3153 | 0.0138 | 0.0138 | 1.0050 | 0.3153 |
| $WTI_{t-1}$ | −0.0260 | 0.0294 | −0.8870 | 0.3754 | −0.0260 | 0.0294 | −0.8870 | 0.3754 |
| $WTI_{t-2}$ | −0.0222 | 0.0263 | −0.8440 | 0.3986 | −0.0222 | 0.0263 | −0.8440 | 0.3986 |
| $NG_{t-1}$ | −0.0206 | 0.0163 | −1.2610 | 0.2076 | −0.0206 | 0.0163 | −1.2610 | 0.2076 |
| $NG_{t-2}$ | 0.0008 | 0.0138 | 0.0590 | 0.9526 | 0.0008 | 0.0138 | 0.0590 | 0.9526 |

**Table 3** Volatility spillover effect

| | $CO_t$ | $WH_t$ | $WTI_t$ | $NG_t$ | From other |
|---|---|---|---|---|---|
| $CO_t$ | 99.54 | 0.13 | 0.24 | 0.09 | 0.46 |
| $WH_t$ | 5.96 | 93.23 | 0.15 | 0.65 | 6.77 |
| $WTI_t$ | 0.01 | 0.32 | 99.67 | 0.01 | 0.33 |
| $NG_t$ | 0.49 | 0.19 | 0.67 | 98.66 | 1.34 |
| From other | 6.46 | 0.63 | 1.06 | 0.75 | 2.23 |

statistically significant. This finding underscores the presence of significant lead-lag relationships between the two markets. Consequently, it is imperative to proceed with the construction of a VAR model to delve deeper into and analyze these intricate dynamics.

### 4.3 Volatility Spillover Effect

In this section, we delve into the analysis of volatility spillover effects between the energy market and the agriculture market over the entire period spanning from 2019 to 2023. Our approach employs the vector autoregression (VAR) methodology, and we rigorously assess the stationarity of our dataset to ensure the suitability of the VAR model. To accomplish this, we utilize the augmented Dickey-Fuller (ADF) test and confirm that all series utilized in the model exhibit stationarity. If desired, detailed results of these tests are available upon request. Table 3 provides a comprehensive representation of the dynamic interconnection and spillover network between the energy market and the agriculture market. Within the "To Others" directional row, each cell quantifies the cumulative volatility transmitted to individual sectors. Conversely, in the "From Others" column, each cell signifies the cumulative volatility received from other sectors. The "NET" directional connectedness row quantifies

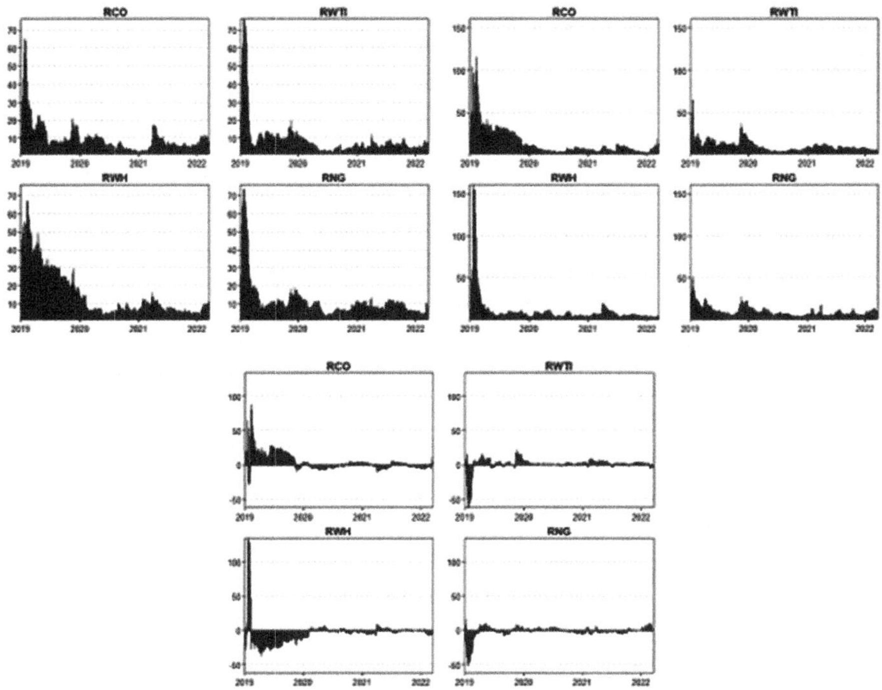


**Fig. 3** From/To/Net spillover between the energy market and agriculture market

the difference between the transmitted volatility and the received volatility for each sector.

Figure 3illustrates the specific changes in spillover effects over the entire period of each of the energy market and agriculture market. In addition, we use the vertical red line to indicate the change in spillover when the Energy price crisis from the Ukraine war case was to examine the spillover effects' differences between pre-pandemic and during the pandemic. Our results from Fig. 5 indicate that risk absorbers during the entire period (Figs. 4 and 6).

The Total spillover index (TSI), sometimes referred to as the total connectedness index (TCI), can be found in the lower right-hand corner, highlighted in shading. The TSI serves as an indicator of the overall average spillover effects encompassing both the energy market and agriculture market. Over the period spanning from 2019 to 2023, the calculated TSI for the energy market and agriculture market stands at approximately 13%. This TSI outcome signifies that the spillover effects among these sectors are considerable, underlining the robust inter-sector connectedness between them.

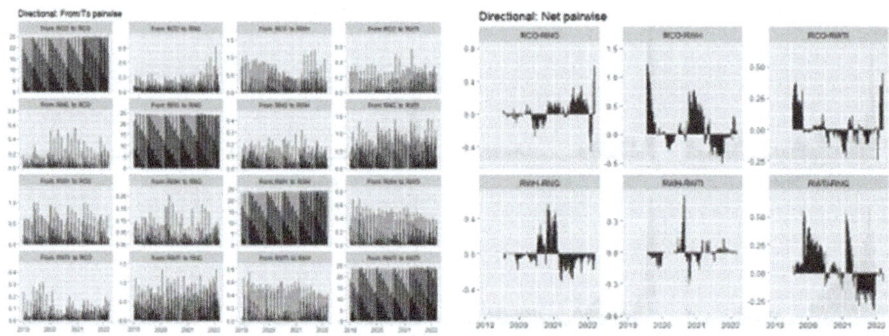


**Fig. 4** Pairwise From/To/Net spillover between the energy market and agriculture market

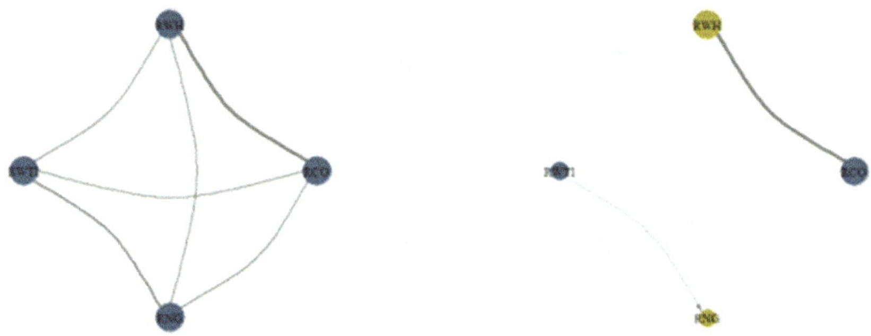


**Fig. 5** Volatility spillover between the energy market and agriculture market

## 5 Conclusion

This article delves into the dynamic relationship between two pivotal sectors: energy and agriculture markets. Spanning the period from 2019 to 2023, with a particular focus on the disruptive COVID-19 pandemic, the study uncovers the intricate patterns of spillover effects between these sectors and explores their implications for financial stability.

The research underscores the paramount importance of understanding sectoral volatility transmission, identifying it as a critical precursor to financial crises and a reflection of the growing interconnectedness among various assets. The strengthening correlations among global financial markets, driven by economic integration and financial globalization, come to the forefront. Notably, the stock market's heightened sensitivity to pandemic outbreaks emerges as a consistent theme, backed by multiple studies. The COVID-19 pandemic emerged as a significant disruptor, profoundly impacting market volatility and volatility spillovers. Surpassing the levels witnessed during the global financial crisis (GFC), volatility spillovers extend their reach from advanced to emerging economies and permeate into the realm

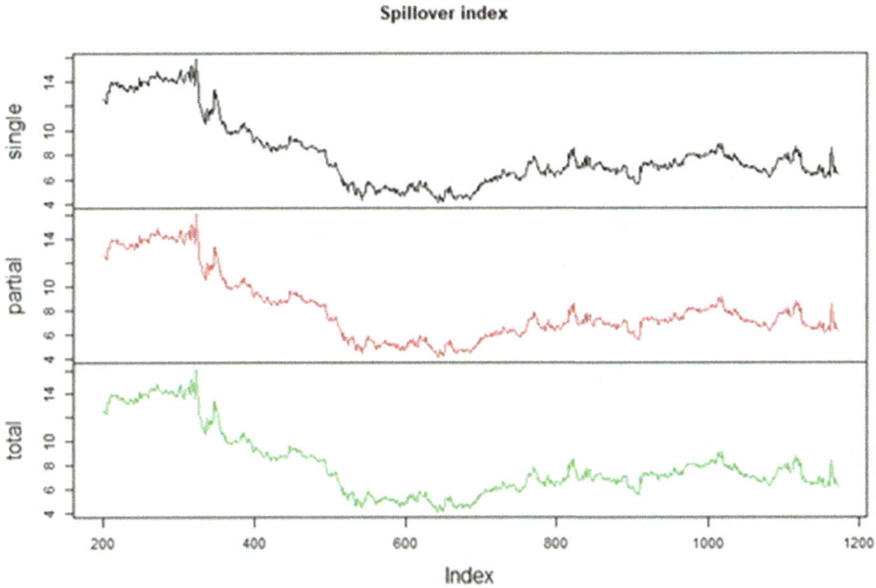


**Fig. 6** Total spillover index (TSI) between Energy and Agriculture markets

of commodities. Employing vector autoregression (VAR) methodology and rigorous stationarity tests, the research validates the suitability of the model. The results affirm that all series used in the model exhibit stationarity.

Importantly, this study thoroughly explores into the volatility spillovers between the energy and agriculture markets. It reveals the directional transmission of volatility and identifies specific sectors that play pivotal roles in transmitting volatility shocks. These findings provide a multifaceted perspective on the interconnected nature of these sectors. One crucial metric, the Total Spillover Index (TSI), registers at approximately 13% over the study period. This figure signifies substantial spillover effects and robust inter-sector interconnectedness. To summarize, this article offers a comprehensive exploration of how volatility interacts between the energy and agriculture markets. The insights gleaned from this research carry significant implications for risk management and investment strategies in an increasingly globalized and interconnected financial landscape.

## References

1. Parker, M.: Global inflation: the role of food, housing and energy prices (2017)
2. Battistini, N., Grapow, H., Hahn, E., Soudan, M.: Wage share dynamics and second-round effects on inflation after energy price surges in the 1970s and today. Economic Bulletin Boxes, vol. 5 (2022)

3. Adekoya, O.B., Oliyide, J.A., Akinseye, A.B., Ogunbowale, G.O.: Oil and multinational technology stocks: predicting fear with fear at the first and higher order moments. Financ. Res. Lett. **46**, 102210 (2022)
4. Adekoya, O.B., Oliyide, J.A., Akinseye, A.B., Ogunbowale, G.O.: Oil and multinational technology stocks: predicting fear with fear at the first and higher order moments. Financ. Res. Lett. **46**, 102210 (2022)
5. Diebold, F.X., Yilmaz, K.: Measuring financial asset return and volatility spillovers, with application to global equity markets. Econ. J. **119**(534), 158–171 (2009)
6. Laborda, R., Olmo, J.: Volatility spillover between economic sectors in financial crisis prediction: evidence spanning the great financial crisis and Covid-19 pandemic. Res. Int. Bus. Financ. **57**, 101402 (2021)
7. Zhang, D., Hu, M., Ji, Q.: Financial markets under the global pandemic of COVID-19. Financ. Res. Lett. **36**, 101528 (2020)
8. Hasan, M.B., Mahi, M., Sarker, T., Amin, M.R.: Spillovers of the COVID-19 pandemic: impact on global economic activity, the stock market, and the energy sector. J. Risk Financ. Manag. **14**(5), 200 (2021)
9. Bissoondoyal-Bheenick, E., Do, H., Hu, X., Zhong, A.: Learning from SARS: return and volatility connectedness in COVID-19. Financ. Res. Lett. **41**, 101796 (2021)
10. Aslam, F., Ferreira, P., Mughal, K.S., Bashir, B.: Intraday volatility spillovers among European financial markets during COVID-19. Int. J. Financ. Stud. **9**(1), 5 (2021)
11. Akinlaso, M.I., Robbana, A., Mohamed, N.: Volatility transmission within financial markets during the COVID-19 pandemic: are faith-based investors well off in Tunisia? J. Islam. Account. Bus. Res. **13**(1), 98–113 (2022)
12. Choi, S.Y.: Volatility spillovers among Northeast Asia and the US: evidence from the global financial crisis and the COVID-19 pandemic. Econ. Anal. Policy **73**, 179–193 (2022)
13. Gunay, S., Can, G.: The source of financial contagion and spillovers: an evaluation of the Covid-19 pandemic and the global financial crisis. PLoS ONE **17**(1), e0261835 (2022)
14. Ghorbel, A., Jeribi, A.: Volatility spillovers and contagion between energy sector and financial assets during COVID-19 crisis period. Eurasian Econ. Rev. **11**, 449–467 (2021)
15. Wen, F., Cao, J., Liu, Z., Wang, X.: Dynamic volatility spillovers and investment strategies between the Chinese stock market and commodity markets. Int. Rev. Financ. Anal. **76**, 101772 (2021)
16. Samitas, A., Papathanasiou, S., Koutsokostas, D., Kampouris, E.: Volatility spillovers between fine wine and major global markets during COVID-19: a portfolio hedging strategy for investors. Int. Rev. Econ. Financ. **78**, 629–642 (2022)
17. Farid, S., Kayani, G.M., Naeem, M.A., Shahzad, S.J.H.: Intraday volatility transmission among precious metals, energy and stocks during the COVID-19 pandemic. Resour. Policy **72**, 102101 (2021)
18. Elgammal, M.M., Ahmed, W.M., Alshami, A.: Price and volatility spillovers between global equity, gold, and energy markets prior to and during the COVID-19 pandemic. Resour. Policy **74**, 102334 (2021)
19. Su, X., Liu, Z.: Sector volatility spillover and economic policy uncertainty: evidence from China's stock market. Mathematics **9**(12), 1411 (2021)
20. Choi, S.Y.: Dynamic volatility spillovers between industries in the US stock market: Evidence from the COVID-19 pandemic and Black Monday. N. Am. J. Econ. Financ. **59**, 101614 (2022)
21. Lin, S.X., Tamvakis, M.N.: Spillover effects in energy futures markets. Energy Econ. **23**(1), 43–56 (2001)
22. Hammoudeh, S., Li, H., Jeon, B.: Causality and volatility spillovers among petroleum prices of WTI, gasoline and heating oil in different locations. N. Am. J. Econ. Financ. **14**(1), 89–114 (2003)
23. Maneejuk, P., Yamaka, W.: Significance test for linear regression: how to test without P-values? J. Appl. Stat. **48**(5), 827–845 (2021)

# Online Media Use and Risk of Cyberbullying Among Undergraduate Students of Chiang Mai Rajabhat University: Structural Equation Model

**Supatatt Dangkrueng, Rungrapee Phadkantha, and Wiranya Puntoon**

**Abstract** This study aims to investigate the relationship between online media usage and the risk of being subjected to internet harassment among undergraduate students at Chiang Mai Rajabhat University. The sample consists of 417 undergraduate students from Chiang Mai Rajabhat University, selected using accidental sampling. Data was collected through questionnaires. The independent variables involved in the research are age, gender, faculty of the respondents, time periods for using online media, the number of hours spent online per day, the number of devices assessing online media, and the number of activities engaged in on online media. Structural Equation Modeling (SEM) was employed to examine the associations among these variables. The results showed that the statistically significant factors that influence the risk of online cyberbully are different in each variable. Specifically, in the gender variable provided strong evidence, and very strong evidence was observed for the number of devices used to access online media. These were identified as significant factors affecting the risk of internet harassment among students.

**Keywords** Cyberbullying · Structural equation model · Online media

## 1 Introduction

Communication technology has rapidly changed and transformed people's way of living since the internet was invented. People spend time daily using online media in all aspects of their lives, such as for information, education, and entertainment.

---

S. Dangkrueng
Faculty of Management Sciences, Chiang Mai Rajabhat University, Chiang Mai, Thailand
e-mail: supatatt.da@cmru.ac.th

R. Phadkantha
Faculty of Economics, Center of Excellence in Econometrics, Chiang Mai University, Chiang Mai, Thailand

W. Puntoon (✉)
Office of Research Administration, Chiang Mai University, Chiang Mai 50200, Thailand
e-mail: wiranya.pun@gmail.com

 
V. Kreinovich et al. (eds.), *Applications of Optimal Transport to Economics and Related Topics*, Studies in Systems, Decision and Control 556,
https://doi.org/10.1007/978-3-031-67770-0_45

In the case of Thailand, a recent survey in Thailand shows that online population is increasing from 4.8 million users in 2002 to 52.16 million users in 2021 [12]. Online media becomes part of our everyday life and time spent on internet of Thai people reaches above global average [18]. Although the internet has numerous benefits, such as being a vast source of information, a communication tool, and a facilitator of convenience in various aspects, it also comes with its own risks. These risks include an increased potential for online harassment, potential issues with mental health due to excessive online activities, and the potential to foster conflicts within society, among others [9]. Cyberbullying, with its diverse manifestations, is a compelling subject of study, particularly online cyberbullying, given its significant impact on individuals of all ages. Cyberbullying can be categorized into types, including Verbal Bullying, Social Bullying, Physical Bullying, and Exclusion [7].

As an emergence of internet penetration, global study of new media effects has been conducted to examine and explore negative consequences to individuals and society. The effects of media use especially in terms of new media, or digital media was affecting behaviors and emotions. Reference [8] explained a term of cyberbullying as misuse of technology to harass others and it is a new form of bullying in 21st century. Some cyberbully victims have been driven to suicide, and affected to their feelings such as anger, sadness, frustration, and fear. Cyberbullying is a new phenomenon of traditional bullying which has been a severe problem over the last decade, it accounts for about one-four to one-third of all bullying [16]. Cyberbullying has emerged through electronic media, such as mobile phone to the internet. Cyberbullying refers to intentional behavior that hurts and harms another person [13].

There have been numerous previous studies examining online sexual harassment, such as those investigating the effects of online games on online sexual harassment. These studies indicate that sexual content in video games can have a greater influence on online sexual harassment directed at women compared to men [3, 6], there is a study on cyberbullying behaviors and the factors that influence cyberbullying. It has been found that cyberbullying can have severe consequences, including mental health issues, anxiety, depression, and even suicidal thoughts, with youth being the primary target of cyberbullying victimization (see, e.g., [1, 10, 15, 17, 19]). Although there are many studies that indicate the widespread use of the internet contributes to online harassment, from another perspective, there are studies that suggest that empathy towards others can help reduce online harassment [4].

Therefore, this study aims to investigate the online media use and the risk of cyberbullying among undergraduate students at Chiang Mai Rajabhat University. It employs a structural equation model (SEM) to identify factors influencing the risk of experiencing cyberbullying. The significance of studying cyberbullying lies in examining and using the findings as guidelines, contributing to the prevention or reduction of cyberbullying risk for individuals and society in the current digital era.

## 2 Structural Equation Model (SEM)

In this study, we utilize Structural Equation Modeling (SEM) to assess the consequences of cyberbullying. SEM enables the simultaneous analysis of complex relationships among multiple variables, integrating confirmatory factor analysis and path analysis. It is particularly advantageous for latent variables, assisting in estimating unobservable constructs [2]. The data for this research were obtained via a survey that explored diverse factors such as gender, age, faculty, time periods for using online media, number of hours spent online per day, number of devices accessing online media, and number of activities engaged in on online media, as follows:

$$\begin{aligned} Cyberbully = {} & \alpha_1 + \beta_{1i} Gender_i + \beta_{2i} Age_i + \beta_{3i} Faculty_i + \beta_{4i} Time_i \\ & + \beta_{5i} Hours_i + \beta_{6i} Device_i + \beta_{7i} Activities_i + \varepsilon_1 \end{aligned} \quad (1)$$

Where $Cyberbully$ is a latent variable, $Gender$, $Age$, $Faculty$, $Time$, $Hours$, $Device$, $Activities$ are observed variable, $\beta_1$, $\beta_2$, $beta_3$, $\beta_4$, $\beta_5$, $\beta_6$, and $\beta_7$ are coefficients indicating the strength of the relationships between the risk of online cyberbullying and the observed variables. $\varepsilon_1$ is the error term, capturing unobserved factors affecting the risk of online cyberbullying.

In Additional, the survey sought to gather insights into respondents' encounters with online harassment, incorporating nine questions (Eq. 2), which encompassed:

- Have you received messages from acquaintances that were "mocking" and made you feel uncomfortable? ($y_1$)
- Have you received messages from acquaintances that were "harassing" and made you feel uncomfortable? ($y_2$)
- Has someone posted messages on the online media you use that were "mocking" and made you feel uncomfortable? ($y_3$)
- Has someone posted messages on the online media you use that were "harassing" and made you feel uncomfortable? ($y_4$)
- Have you received messages from strangers that were "harassing" and made you feel uncomfortable? ($y_5$)
- Have there been posts about you on online media by others that made you feel uncomfortable? ($y_6$)
- Have there been posted "pictures" of you that were embarrassing, and you don't want anyone to see? ($y_7$)
- Have there been posted "messages" about you that were embarrassing, and you don't want anyone to know? ($y_8$)
- Do you feel that someone has mocked you or made you a target of jokes on online media? ($y_9$).

We assess using a Likert scale ranging from 1, indicating strongly disagree, to 5, indicating strongly agree, as follow:

$$Cyberbully = \alpha + \beta y_i + \varepsilon_2 \quad (2)$$

## 3 Data

In this study, data were collected using a questionnaire, which includes responses from a sample size of 417 individuals. The focus was particularly on the accidental sampling process. The dependent variable includes being a risk of online cyber-bullying (students' level of cyber victimization), while the independent variables encompass the age, gender, faculty of the respondents, time periods for using online media, number of hours spent online per day, number of devices used to access online media, and the number of activities engaged in on online media, as shown in Table 1.

The relationships of factors influencing cyberbullying are depicted in Fig. 1, with one endogenous variable measured from 9 observed variables and 18 observed exogenous variables.

## 4 Results

The results of the structural equation modeling (SEM) estimation are presented in Table 2. All variables were assessed using the Maximum Likelihood Estimation method with STATA. This research utilizes the Minimal Bayes factor (MBF) to assess the significance of the results instead of relying on p-values [11, 14]. Interpretation of MBF (Minimum Bayes Factor) values is contingent upon their magnitude, with distinct ranges indicating various degrees of evidence, as outlined below:

$0.33 < \text{MBF} < 1$: "weak evidence"
$0.1 < \text{MBF} < 0.33$: "moderate evidence"
$0.033 < \text{MBF} < 0.1$: "substantial evidence"
$0.01 < \text{MBF} < 0.033$: "strong evidence"
$0.003 < \text{MBF} < 0.01$: "very strong evidence"
$\text{MBF} < 0.003$: "decisive evidence".

These intervals suggest that as the MBF value decreases, the strength of evidence supporting the alternative hypothesis (i.e., rejecting the null hypothesis) increases. Therefore, smaller MBF values indicate stronger evidence in favor of the alternative hypothesis.

Table 2 presents the empirical results from Eq. 1 for the risk of cyberbullying among undergraduate students at Chiang Mai Rajabhat University. The observed variables exhibit varying significant effects on the risk of cyberbullying. According to the MBF, the empirical result provides very strong evidence of gender (Gender) it was observed that males ($x_1$) and females ($x_2$) had a direct negative impact on the risk of cyber-bullying compared to LGBTQ+ ($x_3$). This suggests that both males and females were less likely to experience cyberbullying compared to LGBTQ+ individuals.

When considering the variable of age, the results indicate moderate evidence, with the coefficient at -0.061. This means that an increase in age by 1 year reduces the risk

**Table 1** Variables influencing online cyberbullying ($n = 417$)

| Characteristics | $n$(%) |
|---|---|
| Gender | |
| Male ($x_1$) | 140 (33.57) |
| Female ($x_2$) | 247 (59.23) |
| LGBTQ+ ($x_3$) (Base variable) | 30 (7.19) |
| Age (Average) ($x_4$) | 20.91 |
| Faculty | |
| Faculty of Education ($x_5$) | 130 (31.18) |
| Faculty of Humanities and Social Sciences ($x_6$) | 87 (20.86) |
| Faculty of Management Science ($x_7$) | 137 (32.85) |
| Faculty of Science and Technology ($x_8$) | 41 (9.83) |
| International College ($x_9$) | 12 (2.88) |
| Faculty of Agricultural Technology ($x_{10}$) (Base variable) | 10 (2.40) |
| Time periods for using online media | |
| 10 PM–6 AM ($x_{11}$) | 253 (60.67) |
| 6 AM–2 PM ($x_{12}$) | 290 (69.54) |
| 2PM–10 PM ($x_{13}$) | 359 (86.09) |
| Number of hours spent online per day | |
| Less than 4 h per day ($x_{14}$) (Base variable) | 16 (3.84) |
| 4–7 h per day ($x_{15}$) | 109 (26.14) |
| 7–10 h per day ($x_{16}$) | 160 (38.37) |
| More than 10 h per day ($x_{17}$) | 132 (31.65) |
| Number of devices accessing online media ($x_{18}$) | |
| 1 device | 10 (2.40) |
| 2 devices | 102 (24.46) |
| 3 devices | 198 (47.48) |
| 4 devices | 79 (18.94) |
| More than 4 devices | 28 (6.71) |
| Number of activities engaged in on online media ($x_{19}$) | |
| 1 activity | 25 (6.00) |
| 2 activities | 38 (9.11) |
| 3 activities | 47 (11.27) |
| 4 activities | 58 (13.91) |
| 5 activities | 95 (22.78) |
| 6 activities | 63 (15.11) |
| 7 activities | 45 (10.79) |
| 8 activities | 22 (5.28) |
| 9 activities | 14 (3.36) |
| 10 activities | 10 (2.40) |

*Note* dummy variable = 0 is baseline comparison

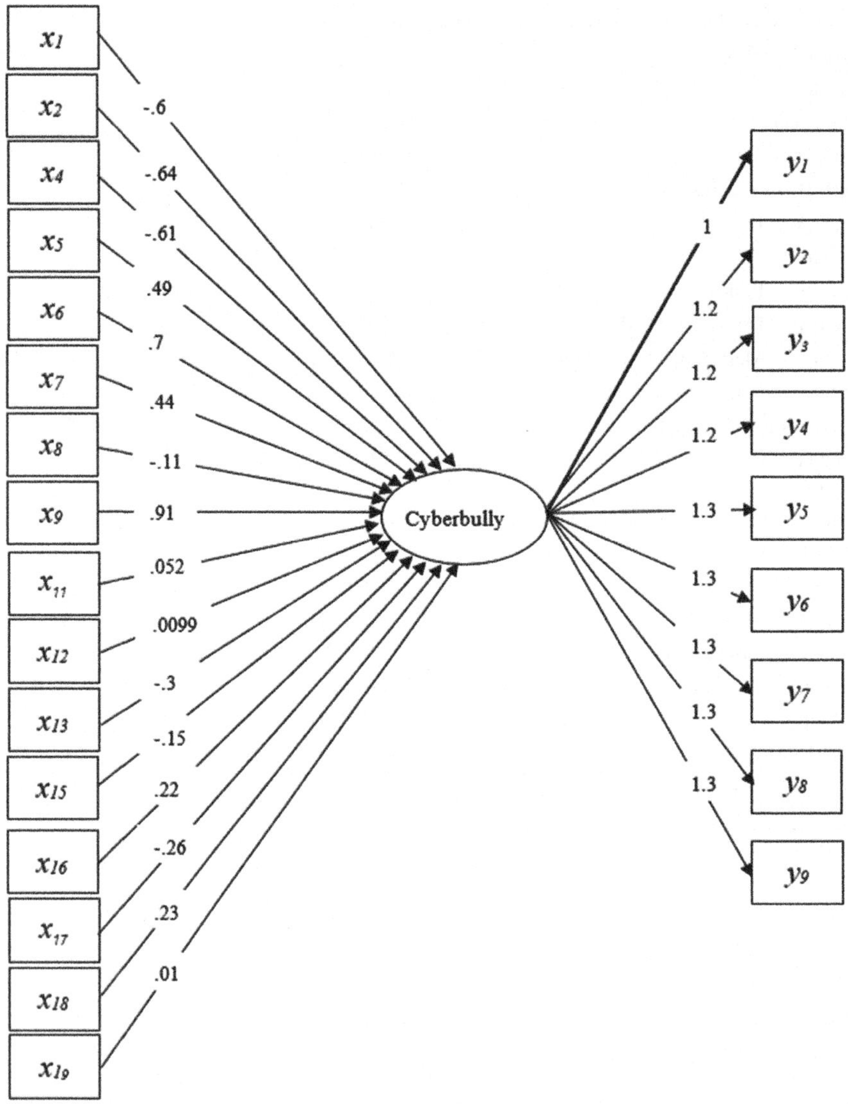


**Fig. 1** Factors affecting cyberbullying

of being an online cyberbully by 0.061 %. Regarding the variable of faculty, using the Faculty of Agricultural Technology ($x_{10}$) as the base variable, the results provide weak evidence for the Faculty of Education ($x_5$) and Faculty of Management Science ($x_7$), while they provide moderate evidence that the Faculty of Humanities and Social Sciences ($x_6$) and International College ($x_9$) have positive effects on the risk

**Table 2** The result of the parameter estimation of the observed variables is available

| | Coef. | Std. Err. | Z-value | MBF |
|---|---|---|---|---|
| Structural | | | | |
| $x_1$ | −0.604 | 0.211 | −2.870 | 0.011 |
| $x_2$ | −0.645 | 0.204 | −3.160 | 0.005 |
| *Age* | | | | |
| $x_4$ | −0.061 | 0.339 | 1.800 | 0.186 |
| *Faculty* | | | | |
| $x_5$ | 0.490 | 0.339 | 1.450 | 0.371 |
| $x_6$ | 0.696 | 0.344 | 2.020 | 0.114 |
| $x_7$ | 0.443 | 0.336 | 1.320 | 0.458 |
| $x_8$ | −0.110 | 0.362 | −0.300 | 1.000 |
| $x_9$ | 0.907 | 0.542 | 1.670 | 0.243 |
| *Time* | | | | |
| $x_1 1$ | 0.052 | 0.110 | 0.480 | 1.000 |
| $x_1 2$ | 0.010 | 0.114 | 0.090 | 1.000 |
| $x_1 3$ | −0.300 | 0.154 | −1.940 | 0.138 |
| *Hours* | | | | |
| $x_1 5$ | 0.155 | 0.279 | −0.550 | 0.990 |
| $x_1 6$ | 0.219 | 0.279 | 0.790 | 0.872 |
| $x_1 7$ | 0.262 | 0.279 | −0.940 | 0.756 |
| *Device* | | | | |
| $x_1 8$ | 0.227 | 0.651 | 3.490 | 0.000 |
| *Activities* | | | | |
| $x_1 9$ | 0.010 | 0.267 | 0.370 | 1.000 |

*Note* MBF represents the Minimum Bayes Factor. If MBF has a value between 1 and 1/3, it is considered weak evidence for the alternative hypothesis, and < 1/300 is considered decisive evidence [11, 14]

of cyberbullying. This implies that Faculty of Education ($x_5$), Faculty of Management Science ($x_7$), Social Sciences ($x_6$), and International College ($x_9$) have a higher risk of being an online cyberbully compared to students from the Faculty of Agricultural Technology ($x_{10}$) with 0.490, 0.443, 0.696, and 0.907 points, respectively. Conversely, we find weak evidence that the Faculty of Science and Technology ($x_8$) has a negative effect on the risk of cyberbullying. This implies that the Faculty of Science and Technology ($x_8$) has a lower risk of being an online cyberbully compared to students from the Faculty of Agricultural Technology ($x_{10}$) with 0.110 points.

In terms of time periods for using online media, there was a positive effect on the risk of online cyberbullying. The empirical results provided weak evidence for using online media during the time periods 10 PM–6 AM ($x_{11}$) (Coef. 0.052) and 6 AM–2 PM ($x_{12}$) (Coef. 0.010). Additionally, using online media during the time periods 2 PM–10 PM ($x_{13}$) (Coef. −0.300) provided moderate evidence regarding the risk of online cyberbullying, indicating that using online media during the time periods 2 PM–10 PM ($x_{13}$) has a lower risk of online cyberbullying compared to other time periods.

For the variable "hours," the results indicate weak evidence for the number of hours spent online per day affecting the risk of cyberbullying. The coefficients for using online media for 4-7 h per day ($x_{15}$) (0.155), using online media for 7–10 h per day ($x_{16}$) (0.219), and using online media for more than 10 h per day ($x_{17}$) (0.262) suggest that using online media for 4–7 h per day ($x_{15}$), using online media for 7-10 h ($x_{16}$), and using online media for more than 10 h per day ($x_{17}$) have a higher risk of being an online cyberbully compared to students using online media for less than 4 h per day ($x_{14}$) (base variable) with an average of 0.155, 0.219, and 0.262 points, respectively.

Concerning the devices variable ($x_{18}$), we observe that the number of devices accessing online media has a positive effect on the risk of cyberbullying, providing decisive evidence (Coef. 0.227). This suggests that students who used a greater number of devices to access online media had a higher risk of experiencing cyberbullying with 0.227 points. On the other hand, there is weak evidence for the activities variable ($x_{19}$) (Coef. 0.010) having a positive effect on the risk of cyberbullying. It indicates that students with a high number of activities engaged in online media have a higher risk of online cyberbullying with 0.010 points.

## 5 Discussion and Conclusion

This study reveals the factors influencing the risk of cyberbullying among university students in Chiang Mai Rajabhat University, Thailand, using Structural Equation Model (SEM). Data was collected through questionnaires from a Accidental sampling of 417 participants, comprising 140 males (33.57%), 247 females (59.23%), and 30 LGBTQ+ individuals (7.19%). The average age of the respondents was approximately 21 years old. There are a total of 9 endogenous observed variables used to measure the risk of becoming victims of cyberbullying, assessed on a Likert scale, ranging from 1, indicating strongly disagree, to 5, indicating strongly agree. Additionally, there are 18 exogenous observed variables, including age, gender, faculty affiliation, online media usage duration, daily online hours, number of devices for online access, and online activities.

The results from the structural analysis model constructed in this study demonstrate the statistically significant factors that influence the risk of online cyberbully are different in each variable. Our findings strong evidence the idea that individuals with diverse gender identities are more vulnerable to online cyberbullying. Specifically, LGBTQ+ individuals face a heightened risk of victimization compared to their male and female counterparts [21]. Furthermore, our results present very strong evidence that individuals with a higher frequency of online media activities are more susceptible to cyberbullying victimization compared to those who engage less frequently in online media activities [5].

However, it's important to acknowledge potential limitations in this study, such as the age range of the sampled participants (The age range is from 20 to 27 years old.) and the composition of the population sample group. This age range may not provide a comprehensive understanding of the relationships influencing online harassment. Expanding the age range of survey participants could lead to a more comprehensive exploration of these relationships. Furthermore, increasing the sample size may also help identify additional factors that have a greater influence on the risk of cyberbullying.

**Acknowledgements** The authors thank the "Faculty of Management Sciences, Chiang Mai Rajabhat University" for its financial support. This research work was partially supported by "Chiang Mai University".

## References

1. Barlett, C., Coyne, S.M.: A meta-analysis of sex differences in cyberbulling behavior: the moderating role of age. Aggress. Behav. **40**(5), 474–488 (2014)
2. Bollen, K.A.: Structural Equation with Latent Variables. Department of Sociology the University of North Carolina, Chapel Hill North Carolina (1989)
3. Burnay, J., Bushman, B.J., Larøi, F.: Effects of sexualized video games on online sexual harassment. Aggress. Behav. **45**(2), 214–223 (2019)
4. Choi, J., Kruis, N., Lee, J.: Empathy, self-control, and online harassment: a partial test of Agnew's social concern theory. Comput. Hum. Behav. **136**, 107402 (2022)
5. Choi, K.S., Cho, S., Lee, J.R.: Impacts of online risky behaviors and cybersecurity management on cyberbullying and traditional bullying vic-timization among Korean youth: application of cyber-routine activities theory with latent class analysis. Comput. Hum. Behav. **100**, 1–10 (2019)
6. Dill, K.E., Brown, B.P., Collins, M.A.: Effects of exposure to sex-stereo-typed video game characters on tolerance of sexual harassment. J. Exp. Soc. Psychol. **44**(5), 1402–1408 (2008)
7. Hinduja, S., Patchin, J.W.: Bullying Beyond the Schoolyard: Preventing and Responding to Cyberbullying. Corwin press
8. Kraft, E.: Cyberbullying: a worldwide trend of misusing technology to harass others. In: WIT Transactions on Information and Communication Technologies, vol. 36, pp. 155–166 (2006)
9. Guerra, C., Pinto-Cortez, C., Toro, E., Efthymiadou, E., Quayle, E.: Online sexual harassment and depression in Chilean adolescents: variations based on gender and age of the offenders. Child Abuse Neglect **120**, 105219 (2021)
10. Kowalski, R.M., Giumetti, G.W., Schroeder, A.N., Lattanner, M.R.: Bullying in the digital age: a critical review and meta-analysis of cyberbullying research among youth. Psychol. Bull. **140**(4), 1073 (2014)
11. Maneejuk, P., Yamaka, W.: Significance test for linear regression: how to test without P-values? J. Appl. Stat. **48**(5), 827–845 (2021)
12. Office of National Broadcasting and Telecommunications Commission (NBTC), Internet Users. (https://ttid.nbtc.go.th/internet_users) (Date accessed: 1st Sept. 2023)
13. Smith, P.K.: Cyberbullying and cyber aggression. Handbook of School Violence and School Safety. International Research and Practice, vol. 2, pp. 93–103 (2012)
14. Sellke, T., Bayarri, M.J., Berger, J.O.: Calibration of p values for testing precise null hypotheses. Am. Statist. **55**(1), 61–71 (2001)
15. Sittichai, R., Tudkuea, T.: Cyberbullying behavior among youth in the three southern border provinces, Thailand. Acad. Serv. J. **28**(1), 86–99 (2017)

16. Sittichai, R., Smith, P.K.: Bullying in south-east Asian countries: a review. Aggress. Violent Behav. **23**, 22–35 (2015)
17. Smith, P.K., Mahdavi, J., Carvalho, M., Fisher, S., Russell, S.: Cyberbullying: its nature and impact in secondary school pupils. J. Child Psychol. Psychiatry **49**(4), 376–385 (2008)
18. Thailand Internet User Behavior (2022). https://www.etda.or.th/th/Useful-Resource/documents-for-download.aspx. Date accessed: 1st Sept. 2023
19. Vandebosch, H., Van Cleemput, K.: Cyberbullying among youngsters: pro-files of bullies and victims. New Media Soc. **11**(8), 1349–1371 (2009)
20. Wick, S.E., Nagoshi, C., Basham, R., Jordan, C., Kim, Y.K., Nguyen, A.P., Lehmann, P.: Patterns of cyber harassment and perpetration among college stu-dents in the United States: a test of routine activities theory. Int. J. Cyber Criminol. **11**(1), 24–38 (2017)
21. Yokotani, K., Takano, M.: Social contagion of cyberbullying via online perpetrator and victim networks. Comput. hum. Behav. **119**, 106719 (2021)

# The Factors Influencing Consumers' Decisions to Purchase OTOP Products Through E-Marketplaces in Chiang Mai

**Walaiphon Suphan and Nootchanat Pirabun**

**Abstract** This study delves into the factors that influence consumers' decisions to purchase OTOP products through electronic commerce marketplaces (E-Marketplace) in Chiang Mai Province. It explores four key sets of factors: demographics, Consumer buying behavior through e-marketplaces, the online marketing mix (6Ps), and 5A marketing strategies that impact consumer behavior. The online marketing mix (6Ps) comprises Product, Price, Place, Promotion, Personalization, and Privacy, while the 5A marketing strategies include Awareness, Appeal, Ask, Act, and Advocate. The binary logistic regression model is used to analyze these factors' influence on consumer decisions regarding OTOP product purchases through electronic marketplaces. Our results, show the importance of personalization in marketing strategies and the role of frequency in purchases emerged as key influencers on consumer behavior. These insights can guide OTOP businesses in adapting their marketing strategies and product offerings to cater to changing consumer demands in an increasingly digital landscape.

**Keywords** One Tambon One Product (OTOP) · Binary logistic regression model · E-Marketplace · 6Ps · 5As

W. Suphan
Digital Technology in Business Program, Faculty of Management Sciences, Chiangmai Rajabhat University, Chiang Mai 50300, Thailand
e-mail: walaiphon_sup@cmru.ac.th

N. Pirabun (✉)
Office of Research Administration, Chiang Mai University, Chiang Mai 50200, Thailand
e-mail: nootchanat.pi@cmu.ac.th



V. Kreinovich et al. (eds.), *Applications of Optimal Transport to Economics and Related Topics*, Studies in Systems, Decision and Control 556,
https://doi.org/10.1007/978-3-031-67770-0_46

# 1 Introduction

The One Tambon[1] One Product (OTOP) initiative in Thailand focuses on locally produced goods that reflect the unique cultural heritage of various regions. These products have gained popularity and export potential. The Thai government has supported OTOP through investment promotion initiatives. OTOP products can be grouped into five categories: (1) Food (2) Beverages (3) Textiles (4) Souvenirs/Handicrafts and (5) Non-food and medicinal herbal items. Chiang Mai Province consistently ranks among the top five provinces in Thailand in terms of OTOP product sales. This is attributable to the province's distinctive products that are deeply rooted in Lanna cultural traditions, underpinned by a wealth of accumulated wisdom over time. However, in 2022, OTOP sales faced challenges due to the COVID-19 pandemic, with total revenue declining by 12.34% compared to 2021, dropping from 278 to 244 billion Baht. This decline was largely a result of reduced consumer spending and economic uncertainty caused by the pandemic (see Fig. 1).

In the past, the Community Development Department used to organize events where they showcased One Tambon One Product (OTOP) items. They did this a few times a year, both in big cities and local areas. However, they had to cancel the event because of COVID-19. It was risky to have lots of people gathering, as it could make the virus spread more. Nowadays, more and more people are buying things online because of COVID-19. This means that companies selling things on the internet, like e-commerce websites, are getting more customers. Furthermore, this transition has triggered a surge in logistics activities, primarily associated with the delivery of products resulting from online transactions. This surge has amplified the demand for delivery services, including courier services, facilitating the transportation of goods from sellers to buyers [1, 7]. Because of these changes, it's really important for OTOP

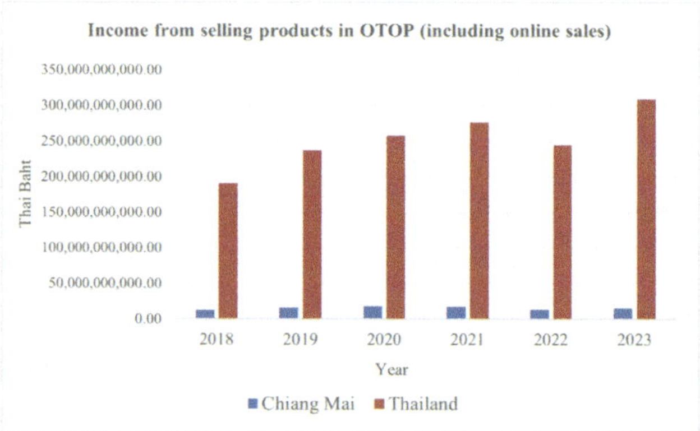


**Fig. 1** Income from selling products in OTOP (including online sales)

[1] Tambon is a term used in Thailand to refer to a subdistrict or commune.

business owners to start selling their products online too. The government is helping them with this by giving them support to sell things on the internet.

As a result, it is crucial for service providers in the field of electronic marketplace trading (E-Marketplace) to engage in developmental activities if businesses are to remain flexible in responding to technological advancements and changing consumer behaviors while meeting the most pressing consumer demands. These initiatives require an extensive investigation of the factors that influence customers' choices to make online purchases through electronic commerce marketplaces (E-Marketplace).

The Marketing Mix, often referred to as the 4Ps, is a foundational concept in marketing. It encompasses the four key elements that a company can control to influence its target market's purchasing decisions. These four elements are Product, Price, Place, and Promotion [5]. However, in recent years, the Marketing Mix has evolved to include additional Ps, making it the 6Ps. Here's a breakdown of the Marketing Mix 6Ps: (1) Product: The product is what the business is selling, whether it's a physical product, a service, or an idea. (2) Price: The price is what customers pay for the product. (3) Place: The place where customers can buy the product. (4) Promotion: Promotion is how the business communicates with customers about its product. (5) Personalization: Personalization is a modern marketing concept that involves tailoring products, services, and marketing messages to individual customer preferences and behavior. It leverages data and technology to create more relevant and engaging customer experiences. And (6) Privacy: Privacy is a critical consideration in modern marketing, especially in the context of data collection and personalization. It pertains to respecting and safeguarding customers' personal information and complying with privacy regulations to maintain trust. These 6Ps collectively shape a company's marketing strategy and help it meet customer needs and achieve its business objectives. Companies may also consider additional Ps in specific industries or situations, such as "Physical Evidence" in service marketing, which relates to the tangible cues that customers use to evaluate service quality.

Marketing 4.0: moving from Traditional to Digital by Kotler et al. [4] is a book that explores modern marketing strategies in the digital age and introduces the concept of Marketing 4.0, which is an evolution of traditional marketing principles. This book essentially advocates for a shift in marketing strategies from the traditional 4Ps (Product, Price, Place, Promotion) to a more customer-centric approach that considers the digital landscape. The book aligns with the 5A marketing strategies by emphasizing customer-centricity and the customer journey. A summary of the book's key ideas in relation to the 5A marketing strategies: (1) Awareness: Investigate how consumers become aware of products available in the electronic commerce marketplace, including the role of advertising and word-of-mouth. (2) Appeal: Explore strategies used to attract consumers' attention and interest in products, such as the use of eye-catching visuals or unique selling propositions. (3) Ask: Analyze how consumers seek information and engage with sellers or online platforms to gather details about products and make informed decisions. (4) Act: Study the factors that prompt consumers to take action, such as adding products to their cart, proceeding to checkout, and completing the purchase. And (5) Advocate: Assess the extent to which satisfied customers become advocates for products by leaving positive reviews, recommending them to others, or sharing their experiences on social media.

This study aims to investigate the factors influencing co0nsumers' decisions to purchase OTOP (One Tambon One Product) products through the electronic commerce marketplace in Chiang Mai Province. To achieve this objective, we plan to explore two key sets of factors: the online marketing mix (6Ps) and 5A marketing strategies that impact consumers' decision-making behavior in this context. To address the research objective posed above, we utilize a binary logistic regression to explore the Factors influencing the decision to purchase OTOP products through the electronic commerce marketplace in Chiang Mai Province.

The paper's structure is as follows: Sect. 2 presents information about the data source, followed by Sect. 3, which outlines the methodology used. In Sect. 4, we conduct an indepth analysis of Factors influencing the decision to purchase OTOP products through the electronic commerce marketplace. Lastly, Sect. 5 draws conclusions based on our research findings.

## 2 Data

From a study conducted with a random sample of 400 consumers residing in Chiang Mai province, using a survey as the data collection tool, our information is divided into four main sections: Part 1: Respondent demographics from the survey. Part 2: Consumer buying behavior through E-Marketplaces. Part 3: The marketing mix factors (6Ps) affecting purchasing decisions on E-Marketplaces. Part 4: Marketing strategies (5As) influencing purchase decisions. In Table 1, provides an overview of the data and an explanation of the data, along with the data frequency from the survey.

## 3 Methodology

In analyzing the factors affecting the decision to purchase OTOP products through E-Marketplaces, we employ a logistic model or binary logistic regression model, as proposed by Berkson [2]. The application of the logistic model is widespread in various research areas, such as the study of behaviors and decision-making. Binary logistic regression is preferred because it offers a straightforward and easy-to-understand method for analyzing and interpreting results.

The logistic regression model is used to establish relationships between one dependent variable and multiple independent variables. In our case, the dependent variable represents the decision to purchase OTOP products through E-Marketplaces, it can take on two values: 1 and 0. The logistic equation is shown in Eq. 1

$$y_{1a} = \alpha_0 + \alpha_{1i}x_{1i} + \alpha_{2j}x_{2j} + \alpha_{3k}x_{3k} + \alpha_{4l}x_{4l} + \alpha_{5m}x_{5m} + \alpha_{6n}x_{6n} + \alpha_{7p}x_{7p} + \alpha_{8q}x_{8q} + \alpha_{9r}x_{9r} + \alpha 6Ps + \alpha 5As + \mu \tag{1}$$

**Table 1** Summary of variables

| Independence variable | | | | | |
|---|---|---|---|---|---|
| Part 1: Respondent demographics from the survey | | | | | |
| Variable | Explanations | | | Freq | % |
| $x_{1i}$ | Gender of respondents | $i = 0$ | Male | 146 | 36.5 |
| | | $i = 1$ | Female | 254 | 63.5 |
| $x_{2j}$ | Age of respondents | $j = 0$ | Less than or equal to 20 years old | 63 | 15.75 |
| | | $j = 1$ | 21 to 30 years old | 151 | 37.75 |
| | | $j = 2$ | 31 to 40 years old | 128 | 32 |
| | | $j = 3$ | 41 to 50 years old | 38 | 9.5 |
| | | $j = 4$ | 51 to 60 years old | 20 | 5 |
| $x_{3k}$ | Education of respondents | $k = 0$ | Below a bachelor's degree | 172 | 43 |
| | | $k = 1$ | Bachelor's degree | 180 | 45 |
| | | $k = 2$ | Master's degree | 46 | 11.5 |
| | | $k = 3$ | Doctorate degree | 2 | 0.5 |
| $x_{4l}$ | Occupation of respondents | $l = 0$ | Student | 148 | 37 |
| | | $l = 1$ | Government employee | 81 | 20.25 |
| | | $l = 2$ | Public/private sector employee | 59 | 14.75 |
| | | $l = 3$ | Self-employed | 13 | 3.25 |
| | | $l = 4$ | Homemakers | 88 | 22 |
| | | $l = 5$ | Unemployed | 11 | 2.75 |
| $x_{5m}$ | Average monthly income | $m = 0$ | Not exceeding 10,000 Baht | 136 | 34 |
| | | $m = 1$ | 10,001–20,000 Baht | 126 | 31.5 |
| | | $m = 2$ | 20,001–30,000 Baht | 114 | 28.5 |
| | | $m = 3$ | 30,001–40,000 Baht | 21 | 5.25 |
| | | $m = 4$ | 40,001–50,000 Baht | 3 | 0.75 |

(continued)

**Table 1** (continued)

| | | | | | |
|---|---|---|---|---|---|
| Part 2: Consumer buying behavior through e-Marketplaces | | | | | |
| $x_{6n}$ | Channel for purchasing products | $n = 0$ | Website | 22 | 5.5 |
| | | $n = 1$ | Line MyShop | 26 | 6.5 |
| | | $n = 2$ | Facebook | 20 | 5 |
| | | $n = 3$ | Instagram | 78 | 19.5 |
| | | $n = 4$ | Twitter | 35 | 8.75 |
| | | $n = 5$ | e-Marketplace | 127 | 31.75 |
| | | $n = 6$ | Tiktok | 49 | 12.25 |
| | | $n = 7$ | General market/shops | 38 | 9.5 |
| | | $n = 8$ | Other | 5 | 1.25 |
| $x_{7p}$ | e-Marketplaces platforms | $p = 0$ | Shopee | 274 | 68.5 |
| | | $p = 1$ | Lazada | 254 | 63.5 |
| | | $p = 2$ | JD Centra | 168 | 42 |
| | | $p = 3$ | eBay | 48 | 12 |
| | | $p = 4$ | Alibaba | 49 | 12.25 |
| | | $p = 5$ | Other | 11 | 2.75 |
| $x_{8q}$ | Spending per transaction on e-marketplaces on average during the last three years | $q = 0$ | 1–500 Baht | 180 | 45 |
| | | $q = 1$ | 501–1,000 Baht | 105 | 26.25 |
| | | $q = 2$ | 1,001–1,500 Baht | 50 | 12.5 |
| | | $q = 3$ | 1,501–2,000 Baht | 34 | 8.5 |
| | | $q = 4$ | More than 4,000 Baht | 31 | 7.75 |
| $x_{9r}$ | Monthly frequency of using e-marketplaces to purchase products | $q = 0$ | Occasionally (e.g., once every 2-3 months or less frequently) | 122 | 30.5 |
| | | $q = 1$ | 1–3 times per month | 150 | 37.5 |
| | | $q = 2$ | 4–6 times per month | 78 | 19.5 |
| | | $q = 3$ | 7–9 times per month | 34 | 8.5 |
| | | $q = 4$ | More than 10 times per month | 16 | 4 |
| Part 3: Marketing mix factors (6Ps) | | | | | |
| | | | | Mean | Meaning |
| *Product* | | The level of opinion is represented on a scale from 1 to 5: | | 4.41 | Very high |
| *Price* | | 1: Very low 4: High | | 4.403 | Very high |
| *Place* | | 2: Low 5: Very high | | 4.435 | Very high |
| *Promotion* | | 3: Moderate | | 4.432 | Very high |
| *Personalization* | | | | 4.416 | Very high |
| *Privacy* | | | | 4.398 | Very high |

(continued)

**Table 1** (continued)

| | | | | | |
|---|---|---|---|---|---|
| Part 4: Marketing strategies (5As) | | | | | |
| *Aware* | | The level of opinion is represented on a scale from 1 to 5: | | 4.051 | High |
| *Appeal* | | 1: Very low 4: High | | 4.203 | High |
| *Ask* | | 2: Low 5: Very high | | 4.264 | Very high |
| *Act* | | 3: Moderate | | 4.358 | Very high |
| *Advocate* | | | | 4.244 | Very high |
| Dependence variable | | | | | |
| $y_{1a}$ | Have you ever purchased OTOP products on e-Marketplaces? | $a = 0$ | No | | |
| | | $a = 1$ | Yes | | |
| $y_{2b}$ | Do you expect to purchase OTOP products on e-marketplaces in the future? | $b = 0$ | No | | |
| | | $b = 1$ | Yes | | |

*Note* dummy variable = 0 is baseline comparison

In analyzing the factors influencing the purchase of OTOP products through electronic marketplaces, both in the past and in the future, we consider a total of four categories of factors. These include demographic variables such as gender, age, income, etc. Additionally, we examine consumer behavior, encompassing marketing mix factors (6Ps) and marketing strategies (5As), shown in Fig. 2.

## 4 Results

The results of our study are presented in this section. We have utilized the logistic regression model to estimate the factors influencing the decision to purchase OTOP products through electronic marketplaces, as outlined in Eq. 1 and shown in Fig. 2.

Our findings reveal important insights in the demographic data that influence the decision to choose OTOP products through E-Marketplaces. Both genders showed no statistically significant evidence. However, only individuals in the age group of 41-50 years displayed substantial evidence. This means that individuals aged between 41 and 50 are more likely to choose OTOP products through electronic marketplaces compared to younger individuals (below 60 years old). The marginal effect is calculated to be 0.272, indicating that if a purchaser of OTOP products falls within the 41–50 age group, it increases the likelihood of selecting OTOP products by 27.20%. Regarding the educational variables, individuals who have completed their undergraduate and master's degrees tend to show a higher inclination towards choosing

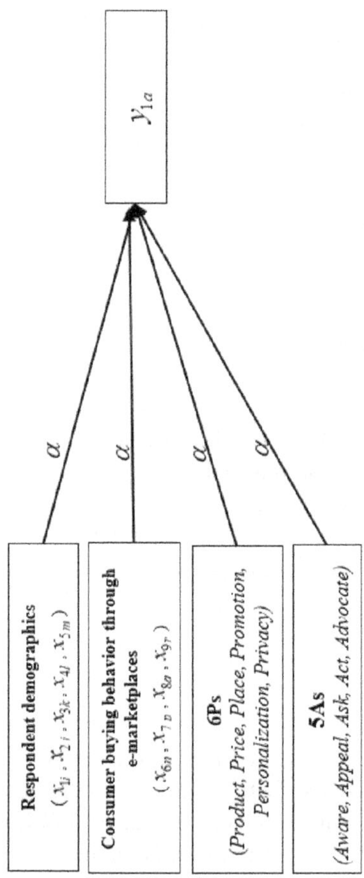


**Fig. 2** The conceptual framework

OTOP products through e-marketplaces compared to those with lower levels of education. This implies that if a buyer has completed a bachelor's degree, it increases the likelihood of selecting OTOP products by 17.5% (substantial evidence), and if they have completed a master's degree, it increases the likelihood by 28.9% (strong evidence). In the occupation segment, individuals employed in private companies and those who are unemployed are more likely to purchase OTOP products through e-marketplaces compared to students. Lastly, in terms of income, individuals with incomes between 30,000 to 40,000 Baht tend to make more purchases compared to those with incomes below 10,000 Baht.

In terms of consumer purchasing behavior, individuals who buy products through Line MyShop, Twitter, and regular markets/shops tend to purchase OTOP products through e-marketplaces less frequently than those who use websites as their channel. This observation is supported by moderate evidence, with probabilities of −0.315, −0.347, and −0.385%, respectively. This means that consumers who purchase products through Line MyShop are 0.315% less likely to buy OTOP products compared to those using websites, and the same trend applies to the other mentioned channels. Similarly, consumers who purchase through e-Marketplace and TikTok have a lower likelihood of buying OTOP products compared to those using websites. This is supported by strong evidence, with a probability of 38.5%. In other words, consumers who purchase products through e-Marketplace are 38.5% less likely to buy OTOP products compared to those using websites. However, the average spending per purchase does not significantly impact the purchasing decision. On the other hand, the frequency of purchasing OTOP products each month reveals that consumers who buy 1–3 times per month and 4–6 times per month are more likely to make purchases compared to those who buy once every 2–3 months. This observation is supported by strong and substantial evidence, respectively (Table 2).

In terms of marketing mix factors (6Ps), consumers who place significant importance on the Personalization factor are more likely to purchase OTOP products. This indicates that tailoring marketing efforts and product offerings to specific customer preferences and needs is an effective strategy in driving OTOP product sales. The marginal effect is calculated to be 0.364, indicating that consumers with a higher emphasis on Personalization are 36.4% more likely to choose to purchase OTOP products. The data suggests that Personalization is a significant driver of consumer decisions to purchase OTOP products. To capitalize on this trend, businesses should invest in strategies that customize their marketing messages and product offerings to align with the preferences and needs of various customer segments. This approach can enhance the appeal of OTOP products and increase their sales among a broader range of consumers. Lastly, within the Marketing strategies (5As) factors, consumers who consider the ASK factor to be highly important are more likely to purchase OTOP products. There is a 19.6% increase in the likelihood of choosing to purchase OTOP products (moderate evidence) for consumers who prioritize the ASK factor. Conversely, consumers who place low importance on the Advocate factor tend to purchase OTOP products more than those who consider it highly important. This means that OTOP businesses operating through e-marketplaces should prioritize the 'ASK' aspect, such as Interactive Website, Create an interactive website or online platform

**Table 2** Estimation results

| | Marginal effect | Coef. | Std. Err. | z-stat | MBF | |
|---|---|---|---|---|---|---|
| Constant | | 1.336 | 1.790 | 0.750 | 0.755 | |
| $x_{21}$ (21–30 years) | −0.029 | −0.137 | 0.444 | −0.310 | 0.953 | |
| $x_{22}$ (31–30 years) | 0.037 | 0.178 | 0.611 | 0.290 | 0.959 | |
| $x_{23}$ (41–50 years) | 0.272 | 2.008 | 0.930 | 2.160 | 0.097 | * |
| $x_{24}$ (51–60 years) | 0.215 | 1.455 | 1.154 | 1.260 | 0.452 | |
| $x_{31}$ (Bachelor) | 0.175 | 0.862 | 0.337 | 2.560 | 0.038 | * |
| $x_{32}$ (Master) | 0.289 | 2.148 | 0.704 | 3.050 | 0.010 | ** |
| $x_{33}$ (Doctorate) | 0.082 | 0.438 | 1.719 | 0.250 | 0.969 | |
| $x_{41}$ (Gov employee) | 0.099 | 0.513 | 0.587 | 0.870 | 0.685 | |
| $x_{42}$ (Private employee) | 0.241 | 1.520 | 0.711 | 2.140 | 0.101 | . |
| $x_{43}$ (Self-employed) | 0.085 | 0.452 | 0.901 | 0.500 | 0.882 | |
| $x_{44}$ (Homemakers) | 0.084 | 0.425 | 0.597 | 0.710 | 0.777 | |
| $x_{45}$ (Unemployed) | 0.281 | 2.707 | 1.216 | 2.230 | 0.083 | * |
| $x_{51}$ (10k–20k baht) | 0.085 | 0.426 | 0.438 | 0.970 | 0.625 | |
| $x_{52}$ (20k–30k baht) | 0.062 | 0.308 | 0.566 | 0.540 | 0.864 | |
| $x_{53}$ (30k–40k baht) | 0.225 | 1.573 | 1.030 | 1.530 | 0.310 | . |
| $x_{54}$ (40k–50k baht) | 0.068 | 0.356 | 1.812 | 0.200 | 0.980 | |
| $x_{61}$ (Line MyShop) | −0.315 | −1.333 | 0.886 | −1.500 | 0.325 | . |
| $x_{62}$ (Facebook) | −0.138 | −0.607 | 1.002 | −0.610 | 0.830 | |
| $x_{63}$ (Instagram) | −0.247 | −1.083 | 0.758 | −1.430 | 0.360 | |
| $x_{64}$ (Twitter) | −0.347 | −1.473 | 0.82 | −1.800 | 0.198 | . |
| $x_{65}$ (e-Marketplace) | −0.358 | −1.622 | 0.751 | −2.160 | 0.097 | * |
| $x_{66}$ (Tiktok) | −0.413 | −1.773 | 0.756 | −2.340 | 0.065 | * |
| $x_{67}$ (market/shops) | −0.385 | −1.642 | 0.839 | −1.960 | 0.146 | . |
| $x_{68}$ (Other) | −0.256 | −1.08 | 1.314 | −0.820 | 0.714 | |
| $x_{71}$ (Lazada) | 0.024 | 0.113 | 0.301 | 0.380 | 0.930 | |
| $x_{72}$ (JD Centra) | −0.178 | −0.840 | 0.328 | −2.570 | 0.037 | * |
| $x_{73}$ (eBay) | −0.012 | −0.059 | 0.505 | −0.120 | 0.993 | |
| $x_{74}$ (Alibaba) | −0.122 | −0.548 | 0.501 | −1.090 | 0.552 | |
| $x_{75}$ (Other) | −0.658 | −3.600 | 1.174 | −3.070 | 0.009 | *** |
| $x_{82}$ (501−1,000 Baht) | 0.060 | 0.299 | 0.397 | 0.750 | 0.755 | |
| $x_{83}$ (1,001–1,500 Baht) | 0.050 | 0.253 | 0.552 | 0.460 | 0.900 | |
| $x_{84}$ (1,501–2,000 Baht) | −0.069 | −0.314 | 0.592 | −0.530 | 0.869 | |
| $x_{85}$ ($\geq$ 4,000 Baht) | −0.118 | −0.525 | 0.687 | −0.770 | 0.743 | |

(continued)

**Table 2** (continued)

| | | | | | | |
|---|---|---|---|---|---|---|
| $x_{92}$ (1–3 times per month) | 0.196 | 1.002 | 0.373 | 2.680 | 0.028 | ** |
| $x_{93}$ (4–6 times per month) | 0.204 | 1.172 | 0.514 | 2.280 | 0.074 | * |
| $x_{94}$ (7–9 times per month) | 0.228 | 1.531 | 0.746 | 2.050 | 0.122 | |
| *Product* | −0.236 | −1.138 | 0.700 | −1.63 | 0.265 | . |
| *Price* | 0.027 | 0.128 | 0.638 | 0.200 | 0.980 | |
| *Place* | 0.147 | 0.707 | 0.677 | 1.040 | 0.582 | |
| *Promotion* | −0.139 | −0.672 | 0.654 | −1.030 | 0.588 | |
| *Personalization* | 0.364 | 1.755 | 0.643 | 2.730 | 0.024 | ** |
| *Privacy* | −0.092 | −0.444 | 0.592 | −0.750 | 0.755 | |
| *Aware* | −0.060 | −0.287 | 0.319 | −0.900 | 0.667 | |
| *Appeal* | 0.026 | 0.125 | 0.452 | 0.280 | 0.962 | |
| *Ask* | 0.196 | 0.944 | 0.560 | 1.690 | 0.240 | . |
| *Act* | −0.117 | −0.565 | 0.565 | −1.000 | 0.607 | |
| *Advocate* | −0.175 | −0.844 | 0.499 | −1.690 | 0.240 | . |

*Note* *, **, *** denote moderate, strong and decisive evidence [6]

where customers can ask questions, seek additional information, and engage with the brand. Implement live chat or chatbot functionalities to provide immediate assistance. Additionally, Customer Reviews, as encourage customers to leave reviews and ratings for OTOP products. Positive reviews and testimonials can answer potential customers' questions and build trust.

## 5 Conclusion

The study aims to investigate the factors influencing consumers' decisions to purchase OTOP products through electronic marketplaces in Chiang Mai province. Specifically, it focuses on examining the online marketing mix factors (6Ps) and the 5A marketing strategies that impact consumers' purchasing behavior of OTOP products through electronic commerce. The collected data from a random sample of 400 consumers residing in Chiang Mai province were analyzed using logistic regression models. The study intends to provide recommendations for OTOP businesses on how to effectively market their products through electronic marketplaces and enhance their competitiveness.

Our study has found substantial evidence that the age group of 41–50 years shows a significant inclination towards choosing OTOP products through electronic marketplaces. Marketing and product presentation efforts should focus on this age group due to their financial stability and strong purchasing power. However, it's important to note that expanding the consumer base to other age groups should still be considered as part of the marketing strategy. Consumers with higher levels of education, such as those with undergraduate and master's degrees, are more likely to choose OTOP products through e-marketplaces. Policymakers can encourage initiatives that enhance the education and digital literacy of the population, as this could lead to increased online shopping and, consequently, the purchase of OTOP products. This finding is consistent with previous studies [3] that socioeconomic characteristics of individuals have little significance in explaining the behavior of experienced e-shoppers, as previous internet use and self-efficacy in online shopping are more influential factors.

In marketing, marketing mix factors (6Ps), we suggest that sellers make personalized marketing messages and product recommendations to align with individual customer preferences. Use data-driven insights to understand customer needs and tailor offerings accordingly. And, Marketing strategies (5As), sellers could Implement interactive features on websites, such as live chat for customer support and the ability for customers to leave reviews and ratings. This can enhance customer engagement and build trust. Additionally, Minimize Emphasis on Advocacy, given that the "Advocate" factor has a negative influence on consumer decisions, businesses should not overly focus on this aspect in their marketing strategies.

**Acknowledgements** The paper is supported by Digital Technology in Business Program, Faculty of Management Sciences Chiangmai Rajabhat University. This research work was partially supported by Chiang Mai University.

## References

1. Akareem, H.S., Ferdous, A.S., Todd, M.: Impact of patient portal behav-ioral engagement on subsistence consumers' wellbeing. Int. J. Res. Mark. **38**(2), 501–517 (2021)
2. Berkson, J.: Application of the logistic function to bio-assay. J. Am. Stat. Assoc. **39**(227), 357–365 (1944)
3. Hernández, B., Jiménez, J., José Martín, M.: Age, gender and income: do they really moderate online shopping behaviour? Online Inf. Rev. **35**(1), 113–133 (2011)
4. Kotler, P., Kartajaya, H., Setiawan, I.: Marketing 4.0: moving from Traditional to Digital. Wiley (2016)
5. Kotler, P., Armstrong, G.: Pearson education limited. Principles of Marketing 17th Global Edition (2018)
6. Maneejuk, P., Yamaka, W.: Significance test for linear regression: how to test without P-values? J. Appl. Stat. **48**(5), 827–845 (2021)
7. Tran, L.T.T.: Managing the effectiveness of e-commerce platforms in a pandemic. J. Retail. Consum. Serv. **58**, 102287 (2021)

# Bayesian Information Criteria in Learning Outcome: Evidence from Vietnam

**Pham Xuan Giang and Bui Huy Khoi**

**Abstract** The learning outcome of each student depends on a lot of factors: learning spirit, level of interest in learning, training, and taking part in activities to create the best environment for developing skills, improving their academic achievement, and the influence of the learning environment, work, teamwork, etc. The paper uses Bayesian Information Criteria in Learning Outcome: Evidence from Vietnam. The learning outcome is used to discover a research model comprising 06 factors: Facilities (FAC), Lecturer (LEC), University (UNI), Learning Motivation (MOT), Learning Method (LM), and Friend (FR). We collected a sample of 98 responses in the analysis. The results show that Lecturer (LEC), University (UNI), and Friend (FR) influence learning outcome (LO). From the research results, the author has given some implications to help managers improve learning outcomes. Previous studies revealed that using linear regression. This study uses the optimal choice of Bayesian Information Criteria.

**Keywords** Lecturer · University · Friend · Learning outcome · Bayesian

## 1 Introduction

The learning outcome of each student depends a lot on factors, learning spirit, level of interest in learning, training, and taking part in activities to create the best environment for developing skills, improving their academic achievement, and the influence of the learning environment, work, teamwork, etc. [22]. The paper aims to understand and evaluate the influence on the learning outcomes of students. This study understands and evaluates the influence on the learning outcomes of students. Through the research results, analyze and process the data got during the research process to give lecturers and students a more intuitive view of the factors affecting the learning outcomes and factors affecting learning outcomes. From there, we take criterion to

P. X. Giang · B. H. Khoi (✉)
Industrial University of Ho Chi Minh City, Ho Chi Minh City, Vietnam
e-mail: buihuykhoi@iuh.edu.vn



V. Kreinovich et al. (eds.), *Applications of Optimal Transport to Economics and Related Topics*, Studies in Systems, Decision and Control 556,
https://doi.org/10.1007/978-3-031-67770-0_47

improve student outcomes to solve the urgent problem of finding out the factors that affect the students' learning outcomes to affect the learning results students and propose criterion to help improve the learning outcomes of students during their study and participation in activities to help students to improve the quality of learning and achieve better results. Helping the university to have orientations and policies in management and improve student quality. Through the research results, analyze and process the data got during the research process to give lecturers and students a more intuitive view of the factors affecting the learning outcomes. From there, take measures to improve student outcomes. The paper uses Bayesian Information Criteria in Learning Outcome: Evidence from Vietnam to solve the urgent problem of finding out the factors that affect the students' learning outcomes and propose some measures to help improve the learning outcomes of students in Vietnam during their study and participation in activities to help students improve the quality of learning and achieve better results and the university to have orientations and policies in management and improve student quality.

## 2 Literature Review

### *2.1 Learning Outcome (LO)*

Learning outcome (LO) is a term that researchers have given many definitions [5]. It is defined as the overall assessment of the students themselves about the knowledge and skills they have gained in studying subjects at university [7]. According to Winarso [21], intermediate performance is evidence of students' success in terms of knowledge, skills, and attitudes, which have been outlined in educational goals.

### *2.2 Facilities (FAC)*

University facilities and technology are systems of various physical and technical means used to serve the comprehensive education and training of students. Facilities include lecture halls, laboratories, computer rooms, and equipment [2]. Learning materials include books, journals, ebooks, open materials, specialized documents, information, and communication technology that meet training implementation and learning environment. The study of [18] has also shown that facilities are one factor affecting the learning outcomes of students.

*Hypothesis (H1): Facilities have a positive impact on student learning outcomes.*

## 2.3 University (UNI)

The impression of a brand—that of the product or the organization—reflects the customer's perception of that brand [1, 3]. As a business, a university is an organization that educates and trains human resources for the country. A university with an excellent reputation, good training quality, an increasingly modern learning environment, etc. will make a good impression on students and help students get high academic achievements. A training program is a content, organizational structure, functions, tasks, and academic-related activities of a training unit being deployed to train a discipline in a certain level of study. [13] showed that the characteristics of the university have an impact on students' learning outcomes. Therefore, hypothesis H2 is put forward:

*H2: The University has a positive impact on student learning outcomes.*

## 2.4 Lecturer (LEC)

Lecturers play an important role in student learning [9]. They are guides, directions, and suggestions for learning methods so that students can absorb knowledge and learn more easily. Therefore, the capacity of highly qualified lecturers, the ability to communicate, and the method of subject organization will affect the learning process of each student. Research by [12] also shows that the teacher's factor will affect the learning outcomes of students.

*H3: Lecturer has a positive impact on student learning outcomes.*

## 2.5 Learning Motivation (MOT)

Students' learning motivation is the desire to get something [4], be it a university scholarship or to prove themselves, the motivation, and the goal that each student aspires to. In the study of Ulyani and Qohar (2021) on factors affecting learning outcomes, learning motivation is one of the influential factors.

*H4: Motivation has a positive impact on student learning outcomes.*

## 2.6 Learning Method (LM)

Each student needs to make a study plan for themselves [14], take time to study on their own, and then organize interactive activities, and exchange with friends, and teachers to deepen their understanding of the problem and knowledge in the best way. In this study, the learning methods mentioned are self-study and teamwork-based learning. According to the research of [15], the factor of learning method influences the learning outcomes of students.

*H5: Learning methods have a positive impact on student learning outcomes.*

## 2.7 Friend (FR)

The more peers a student has to motivate their learning and who behave positively, the more it can affect improving learning outcomes [8]. There is a saying "Learning from a teacher is not good at studying friends". Friends are people of the same age and it is easier to exchange access to knowledge than with lecturers. Friends will influence each other through learning each other's knowledge and skills, sharing and learning from each other. According to a study by [8], the friend factor has a positive effect on students' learning outcomes.

*H6: Friends have a positive influence on student learning outcomes.*

# 3 Method

## 3.1 Sample

According to [19] for the best regression analysis, it is necessary to ensure the sample size is $N \geq 8m + 50$. The results of gender statistics show that out of 98 survey subjects, there are 34 respondents of the male gender, accounting for 34%. Gender Female has 64 respondents, accounting for 65.3%. This shows that there is a difference in the ratio of males and females in this survey. According to data, the number of second-year students responding to the survey accounted for the highest rate, with 68.4%, and the lowest rate of fourth-year students, accounting for only 3.1%. It can be seen that second-year students are easier to survey and receive more support. This is the same age group as the survey authors. Table 1 describes the statistics of the sample characteristics.

From the statistical results, it shows that the business administration sector accounts for the highest number and percentage of the industries because the research team mainly belongs to this industry another reason is according to secondary data collected from the business administration sector has the largest number of scholarship winners, so it is necessary to find out which factors positively affect the learning outcomes of students of the Faculty of Business Administration. Besides, the two groups of Mechanical Engineering and Fashion Garments have the smallest number and percentage (only 1 and 2%).

**Table 1** Statistics of sample

| Characteristics | | Amount | Percent (%) |
|---|---|---|---|
| Sex and School Year | Male | 34 | 34.7 |
| | Female | 64 | 65.3 |
| | First | 15 | 15.3 |
| | Second | 67 | 68.4 |
| | Third | 13 | 13.3 |
| | Fourth | 3 | 3.1 |
| Faculty | Business Administration | 44 | 44.9 |
| | Auditing and Accounting | 6 | 6.1 |
| | Tourism Business | 9 | 9.2 |
| | Mechanical Engineering | 1 | 1.0 |
| | Foreign Language | 5 | 5.1 |
| | Fashion Garment | 2 | 2.0 |
| | IT | 3 | 3.1 |
| | Food | 6 | 6.1 |
| | Construct | 6 | 6.1 |
| | Banking and Finance | 8 | 8.2 |
| | Other | 8 | 8.2 |

## *3.2 Reliability Test*

Cronbach's Alpha coefficient is a statistical test used to check the closeness with which the items in the scale correlate with each other. With this method, the analyst can eliminate inappropriate variables and limit garbage variables in the research process. In, 5 independent variables measured by 22 observed variables and 1 dependent variable measured by 3 observed variables will be put into a preliminary test by the criteria in testing the reliability of the scale by Cronbach's Alpha [6], the author gives the following rules for evaluating the reliability of the scale:

- <0.6: Unsuitable factor scale
- 0.6–07: Acceptance scale for new studies
- 0.7–0.8: Acceptable
- 0.8–0.95: The scale is good
- >0.95: Acceptable but not good.

When preliminarily evaluating the scale according to the results of Cronbach's Alpha coefficient $\geq$0.6 and the total variable correlation coefficient $\geq$0.3, the new scale is assessed as contributing value. Other researchers have previously agreed that when Cronbach's Alpha is between 0.8 and 1, the scale is said to be a good measure. Many other researchers consider that a Cronbach's Alpha coefficient of 0.6 or higher can be used when measurement concepts are new to survey participants.

### 3.3 Bayesian Information Criteria

The best model for R software was chosen to use BIC (Bayesian Information Criteria). In the theoretical environment, BIC has been used to choose models. BIC can be used as a regression model to estimate one or more dependent variables from one or more independent variables [17]. The BIC is an important and useful metric for determining a full and straightforward model. A model with a lower BIC is chosen based on the BIC information standard. When the minimal BIC value is reached, the best model will end [10, 16, 17].

The posterior probability $P(\beta_j \neq 0|D)$ given by variable $X_j$ with $(j = 1, 2, \ldots, p)$ indicates the possibility that the independent variable affects the occurrence of the event (or a non-zero effect).

$$P(\beta_j \neq 0|D) = \sum_{M_k \in A} P(M_k|D) * I_k(\beta_j \neq 0) \tag{3}$$

where A is a set of models selected in Occam's Window described in Eq. 3 and $I_k(\beta_j \neq 0)$ is 1 when $\beta_j$ in the model $M_k$ and 0 if otherwise. The term $P(M_k|D) * I_k(\beta_j \neq 0)$ in the above equation means the posterior probability of the model $M_k$ not included $X_j = 0$. The rules for explaining this posterior probability are as follows [16]: Less than 50%: evidence against impact; Between 50 and 75%: weak evidence for impact; Between 75 and 95%: positive evidence; Between 95 and 99%: strong evidence; From 99%: very strong evidence [11].

## 4 Results

### 4.1 Mean

Factors and items are in Table 2. The mean of items is from 3.02 to 4.37, being good with research data in Table 2. Facilities (4.37) are appreciated the best.

### 4.2 Bayesian Model Selection

Every stage of the search for the best model is detailed in the R report. BIC selects the best 5 models in Table 3.

There are six independent and one dependent variable. Lecturer (LEC) influences Learning outcome (LO) with a Probability is 95.3% and University (UNI) and Friend (FR) influence Learning outcome (LO) with an average probability is 55.2 and 51.1%. Facilities (FAC), Learning Method (LM), and Motivation (MOT) influence Learning outcome (LO) with low probability is 20.7, 13.7, and 7.6%.

**Table 2** Mean

| Factor | Item | Mean |
|---|---|---|
| LO | Last semester's results | 3.02 |
| LEC | Lecturers help you absorb knowledge easily and achieve good learning results | 3.13 |
| MOT | University supports scholarships to make you more interested in improving your academic results | 3.48 |
| FAC | Modern facilities will help students learn excitedly and achieve high learning results | 4.37 |
| UNI | University training helps you learn better | 3.46 |
| | The learning environment makes your study results better | |
| LM | Students who spend more time on self-study will achieve better learning results | 4.1 |
| | Teamwork will achieve high academic results | |
| FR | Students learn knowledge and skills from friends to help them achieve good academic results | 3.9 |
| | Students want to do teamwork | |

**Table 3** Bayesian Model Selection

| LO | Probability (%) | SD | Model 1 | Model 2 | Model 3 | Model 4 | Model 5 |
|---|---|---|---|---|---|---|---|
| Intercept | 100.0 | 1.21818 | −0.1680 | 1.0185 | 0.8566 | 2.1220 | −0.9408 |
| LEC | 95.3 | 0.11333 | 0.3104 | 0.2676 | 0.3305 | 0.2868 | 0.2984 |
| MOT | 7.6 | 0.02722 | | | | | |
| FAC | 20.7 | 0.08989 | | | | | 0.1763 |
| UNI | 55.2 | 0.20084 | 0.3254 | 0.3364 | | | 0.3243 |
| LM | 13.7 | 0.08037 | | | | | |
| FR | 51.1 | 0.17115 | 0.2775 | | 0.2872 | | 0.2889 |

## 4.3 Model Evaluation

According to the results from Table 4, BIC shows model 1 is the optimal selection because BIC (−4.8437) is minimum. Lecturer (LEC), University (UNI), and Friend (FR) impact learning outcome (LO) is 17.3% in Table 4. BIC finds model 1 is the optimal choice and three variables have a probability of 15.6%. The regression equation below is statistically significant, according to the analyses above.

$$LO = -0.1680 + 0.3104LEC + 0.3254UNI + 0.2775FR$$

**Table 4** Model test

| Model | nVar | $R^2$ | BIC | Post prob |
|---|---|---|---|---|
| Model 1 | 3 | 0.173 | −4.8437 | 0.156 |
| Model 2 | 2 | 0.133 | −4.7767 | 0.151 |
| Model 3 | 2 | 0.131 | −4.6220 | 0.140 |
| Model 4 | 1 | 0.088 | −4.4595 | 0.129 |
| Model 5 | 4 | 0.192 | −2.5834 | 0.050 |

$BIC = -2 * LL + \log(N) * k$

## *4.4 Discussion*

The analysis results of the Bayesian Model Selection have identified the factors affecting the students' Learning outcome (LO) at the University of Industry in Ho Chi Minh City, which are University (UNI), Lecturer (LEC), and Friend (FR). Combining the results from the model and the current learning situation of students at the University of Industry, as well as based on the beta coefficient of each independent variable, the group makes the following recommendations:

University (0.3254): First, it is necessary to adjust training programs accordingly by adding more specialized knowledge for students, and enhancing practical and practical activities associated with the needs of society to create a learning environment for students. Second, the training program is always updated and renewed so that students can study effectively and grasp the trend. Third, it always creates a modern and friendly learning environment for students and creates excitement for students.

Lecturer (0.3104): They need to provide students with full official learning resources, as well as subject-related resources. This is a condition for students to easily absorb all the knowledge of the subject, besides listening in class, students can study and learn more at home. They need to pay more attention to their students and create conditions for students to understand the content of the course by doing more practical exercises, group exercises, and specific and vivid examples to avoid situations. Students only focus on grades, not on knowledge and practical application.

Friend (0.2775): Students should build many good relationships in school and society. This will help students in study and work. When they do not understand the lesson, they will be helped by their friends, learning knowledge from their friends that they do not have. Students need to be proactive, flexible, and sociable with people around them to learn from each other, and support and help each other in learning and applying knowledge in society. Having friends, students will be compensated and learn from each other knowledge and skills in life through which they can cultivate themselves and develop more, such as learning from each other how to think critically, and communication skills.

## 5 Conclusions

This study was conducted to determine the factors that affect the learning outcomes of students by the optimal choice of Bayesian Model Selection. Students of all majors have quite good academic performance, with about 84.6% of students achieving good or higher performance. Second- and third-year students do better because they have more appropriate learning methods and are better adapted to credit-based learning. There are no students with poor academic performance. University factors such as Lecturer (LEC), University (UNI), and Friend (FR) influence the social performance of students of public universities. And the scholarship factor has an unknown influence on the financial performance of students at IUH.

**Limitations**

We found some limitations that we have not exploited in the research topic, which are that they could not calculate the student's financial performance, have not exploited as well as mentioned the influencing students' social relations, specifically online learning (because of the Covid epidemic, sometimes students have to study online), the problem of working part-time.

## References

1. Aaker, D.A.: Building Strong Brands, pp. 598–614. The Free Press, New York (1996)
2. Asselineau, M.: Educational Facilities' Performance and Lecture Halls. CRC Press, Building Acoustics (2015)
3. Balmer, J.M., Podnar, K.: Corporate brand orientation: Identity, internal images, and corporate identification matters. J. Bus. Res. **134**, 729–737 (2021)
4. Berkling, K.: Connecting peer reviews with students' motivation-onboarding, motivation and blended learning. In: International Conference on Computer Supported Education, pp. 24–33. SciTePress (2015)
5. Given, C.: Reframing Student Learning Outcome Assessment: A Faculty Leadership Initiative Focused on Student Success in Community College Gateway Courses. Northeastern University (2020). Accessed https://repository.library.northeastern.edu/files/neu:bz60cz80h/fulltext.pdf.
6. Hair, J.F.: Multivariate data analysis: an overview. Int. Encycl. Stat. Sci. 904–907 (2011)
7. Hanson, S.: Legal Method and Reasoning. Routledge (2012)
8. Hutanto, M.I.: The influence of discipline, learning facilities, and friends class to economics learning outcomes. Tarbiyah: Jurnal Ilmiah Kependidikan **10**, 10–17 (2021)
9. Islam, M.K., Sarker, M.F.H., Islam, M.S.: Promoting student-centred blended learning in higher education: a model. E-Learn. Dig. Media 20427530211027721 (2021)
10. Kaplan, D.: On the quantification of model uncertainty: a bayesian perspective. Psychometrika **86**, 215–238 (2021)
11. Khoi, B.H., Thach, N.N.: Bayesian Consideration for Analyzing Employee's Motivation: Evidence from Vietnam. Springer, Optimal Transport Statistics for Economics and Related Topics (2023)
12. Lim, D.H., Morris, M.L.: Learner and instructional factors influencing learning outcomes within a blended learning environment. J. Educ. Technol. Soc. **12**, 282–293 (2009)
13. Moely, B.E., Ilustre, V.: The impact of service-learning course characteristics on university students' learning outcomes. Mich. J. Community Serv. Learn. **21**, 5–16 (2014)

14. Nortcliffe, A.: Can students assess themselves and their peers?: a five year study. Stud. Engage. Exp. J. 1 (2012)
15. Prasetya, T.A., Harjanto, C.T.: Improving learning activities and learning outcomes using the discovery learning method. VANOS J. Mech. Eng. Educ. **5** (2020)
16. Raftery, A.E.: Bayesian model selection in social research. Soc. Methodol. 111–163 (1995)
17. Raftery, A.E., Madigan, D., Hoeting, J.A.: Bayesian model averaging for linear regression models. J. Am. Stat. Assoc. **92**, 179–191 (1997)
18. Roberts, L.W., Edgerton, J.D., Peter, T.: The importance of place: facility conditions and learning outcomes. Educ. Can. **48**, 48–51 (2008)
19. Tabachnick, B., Fidell, L.: Using Multivariate Statistics, 4th edn., pp. 139–179. HarperCollins, New York (2001)
20. Ulyani, O., Qohar, A.: Development of manipulative media to improve students' motivation and learning outcomes on the trigonometry topic. In: AIP Conference Proceedings, p. 040035. AIP Publishing LLC (2021)
21. Winarso, W.: Authentic assessment for academic performance; study on the attitudes, skills, and knowledge of grade 8 mathematics students. Malikussaleh J. Math. Learn. (MJML) 1 (2018)
22. Wulandari, A., Usman, O.: The Influence of Learning Motivation, Learning Discipline, and Learning Environment to the Student Achievement Learning State University of Jakarta (2021). Learning Discipline, and Learning Environment to the Student Achievement Learning State University of Jakarta accessed at https://repository.unja.ac.id/28193/5/BAB%20I%20nadyaa.pdf

The manufacturer's authorised representative in the EU is Springer Nature Customer Service Centre GmbH, Europaplatz 3, 69115 Heidelberg, Germany. If you have any concerns regarding our products, please contact ProductSafety@springernature.com

Printed and bound by CPI Group (UK) Ltd, Croydon, CR0 4YY
28/09/2026
02224056-0018